Phytoantioxidants and Nanotherapeutics

Phytoantioxidants and Nanotherapeutics

Edited by

Mithun Rudrapal
Pune, India

WILEY

This edition first published 2022

Registered Office
John Wiley & Sons, Inc., 111 River Street, Hoboken, NJ 07030, USA

Editorial Office
John Wiley & Sons, Inc., 111 River Street, Hoboken, NJ 07030, USA

For details of our global editorial offices, customer services, and more information about Wiley products visit us at www.wiley.com.

Wiley also publishes its books in a variety of electronic formats and by print-on-demand. Some content that appears in standard print versions of this book may not be available in other formats.

Library of Congress Cataloging-in-Publication Data
Name: Rudrapal, Mithun, editor.
Title: Phytoantioxidants and nanotherapeutics / edited by Mithun Rudrapal.
Description: Hoboken, NJ : Wiley, 2022. | Includes bibliographical references and index.
Identifiers: LCCN 2022011366 (print) | LCCN 2022011367 (ebook) | ISBN 9781119811770 (cloth) | ISBN 9781119811800 (Adobe PDF) | ISBN 9781119811831 (epub)
Subjects: MESH: Antioxidants–therapeutic use | Phytochemicals–therapeutic use | Theranostic Nanomedicine–methods
Classification: LCC QK898.A57 (print) | LCC QK898.A57 (ebook) | NLM QV 325 | DDC 613.2/86–dc23/eng/20220613
LC record available at https://lccn.loc.gov/2022011366
LC ebook record available at https://lccn.loc.gov/2022011367

Cover Image: © PhotobyTawat/Shutterstock
Cover Design by Wiley

Set in 9.5/12.5pt STIXTwoText by Straive, Pondicherry, India

Contents

Preface *vii*
List of Contributors *ix*

1 **Natural Antioxidants in Oxidative Stress-Induced Diseases** *1*
Himadri Kalita and Prashanta Kumar Deb

2 **Phytomedicines as Sources of Natural Antioxidants** *31*
Syed Mohammad Abdullah and Papiya Mitra Mazumder

3 **Herbs, Spices, and Dietary Constituents as Sources of Phytoantioxidants** *55*
N. Shaista Jabeen, V. Kiruthiga, A. Vinodhini, and Mithun Rudrapal

4 **Phytoantioxidants and Their Role in Cellular Oxidative Stress** *77*
Swarnali Bhattacharjee, Prashanta Kumar Deb, and Rajlakshmi Devi

5 **Bioactive Flavonoids as Phytoantioxidants** *99*
Joyani Das

6 **Nanoparticulate Delivery Systems for Phytoconstituents** *121*
Khomendra Kumar Sarwa, Dipali Patel, Dhaneshwar Uraon, Manmohan Singh Jangde, Vijendra Kumar Suryawanshi, and Swarnlata Saraf

7 **Nanodelivery of Herbal and Phyto-Antioxidants** *139*
Gülsen Kaya, Merve Keskin, Saban Keskin, and Mithun Rudrapal

8 **Nanodelivery of Antioxidant Herbal Extracts, Spices, and Dietary Constituents** *145*
Shiv Kumar Prajapati, Dolly Jain, Sabnam Parveen, Siddhartha Maji, and Prashanta Kumar Deb

9 **Nanophytomedicine in Disease and Therapy** *173*
Sonal Upadhyay, Ravi Bhushan, and Mithun Rudrapal

10 **Biochemical and Therapeutic Targets for Nanophytomedicines** *203*
Prithviraj Chakraborty, Debarupa Dutta Chakraborty, and Mainak Chakraborty

11 Green Approaches for Synthesis of Nanophytopharmaceuticals/Nanophytodelivery Systems *231*
Gülsen Kaya, Merve Keskin, and Mithun Rudrapal

12 Characterization of Nanophytopharmaceuticals *239*
Ashwini Kumar Mishra and Pravat Kumar Sahoo

13 Toxicity of Nanostructures and Nanodrugs *267*
Tapash Chakraborty and Anupam Sarma

14 Nanotherapeutics of Phytoantioxidants for Microbial Infections *289*
Ravi Bhushan, Sonal Upadhyay, and Mithun Rudrapal

15 Nanotherapeutics of Phytoantioxidants for Viral Infections *317*
Rakesh Sagar, Sharad P. Pandey, and Abhishek K. Sah

16 Nanotherapeutics of Phytoantioxidants for Parasitic Diseases and Neglected Tropical Diseases *351*
James H. Zothantluanga, Zonunmawii, Punamjyoti Das, Himangshu Sarma, and Abd. Kakhar Umar

17 Nanotherapeutics of Phytoantioxidants for Inflammatory Disorders *377*
Santwana Palai

18 Nanotherapeutics of Phytoantioxidants for Cardiovascular Diseases *405*
Himangshu Sarma, Pallabi Kashyap, James H. Zothantluanga, and Rajlakshmi Devi

19 Nanotherapeutics of Phytoantioxidants for Diabetes Mellitus *433*
Bui Thanh Tung and Tong Thi Thu Hoa

20 Nanotherapeutics of Phytoantioxidants for CNS Disorders *449*
Arpita Paul, Nilayan Guha, and Md Kamaruz Zaman

21 Nanotherapeutics of Phytoantioxidants for Aging and Neurological Disorders *477*
Ganesh B. Shevalkar, Mahendra K. Prajapati, Bhushan A. Bhairav, and Chandrakantsing V. Pardeshi

22 Nanotherapeutics of Phytoantioxidants in Cancer *495*
Ashish Shah, Vaishali Patel, and Ghanshyam Parmar

23 Challenges and Regulatory Issues of Nanophytotherapeutics *521*
Smriti Sharma and Vinayak Bhatia

Index *535*

Preface

Phytomedicines, herbal drugs, or dietary components have been widely used all over the world since ancient times and have been recognized by physicians, herbal practitioners, and patients for their better therapeutic effect, as they have fewer adverse effects as compared with modern medicines. Nanodelivery of phyto or herbal constituents has a potential future for enhancing the therapeutic utility and overcoming the problems associated by phytocomponents. In this context, nanodelivery systems or nanocarriers help to treat complex diseases such cancer, viral infections, neurological diseases, central nervous system disorders, and diabetes, just to name a few. Recently, there have been enormous developments in the field of nano-based drug delivery systems to provide therapeutic agents or natural-based active components to its target tissue-specific location for the treatment of various diseases. There are a large number of herbal drug delivery systems successfully employed in recent times. There are still certain challenges that need to be addressed, and therefore, advanced technologies like nanotechnology need to be applied for the successful delivery of phyto and herbal components in the therapeutic management of diseases.

The book entitled *Phytoantioxidants and Nanotherapeutics* presents a comprehensive review on the medicinal importance of phytomedicine-derived antioxidants with their development into useful nanodrug delivery systems as nanotherapeutics for potential therapeutic applications as nanobiomedicine in a wide range of oxidative-stress-induced chronic human diseases such as cancer, inflammatory disorders, viral infections, bacterial infections, parasitic infections, central nervous system disorders, cardiovascular disorders, diabetes, and neurological diseases. The chapters embedded in the book contain relevant information with latest updates on advanced topics related to the phytomedicine-derived antioxidants and nanodrug delivery systems. The chapters are presented in a clear and lucid manner in order to aid flow, continuity, and technical presentation. The chapters encompass phytomedicine-derived bioactive compounds as antioxidants (phytoantioxidants, plant polyphenols, and flavonoids) and their role in oxidative stress-induced human diseases. Herbs, plant extracts, spices, and dietary constituents as sources of natural antioxidants and their potential medicinal benefits are also covered. Nanostructured or nanoparticulate systems used for the delivery of phytomedicine (including herbal extracts, spices, and dietary constituents) and phytoconstituents are summarized in this book. Biochemical and therapeutic targets of nanodrugs along with the toxicity of nanostructures in the biological matrix are detailed herein. The latest developments on nanotherapeutics of phytoantioxidants for the treatment of certain chronic human diseases such as microbial infections, parasitic diseases, inflammatory disorders, cardiovascular disorders, diabetes, central nervous system disorders, neurological disorders, and cancer are reviewed systematically and comprehensively. Finally, challenges and regulatory issues of nanophytotherapeutics are also highlighted. Some of the key features of this book include:

- Presents comprehensive information on the potential role of bioactive phytochemicals, particularly polyphenolic compounds and flavonoids in oxidative-stress-induced human diseases.
- Illustrates nanostructured or nanoparticulate systems used for the delivery of phytomedicine and/or bioactive phytoconstituents
- Highlights biochemical and therapeutic targets of nanodrugs with the toxicity of nanostructures in the biological matrix
- Delineates latest developments on nanotherapeutics of phytoantioxidants for the treatment of certain chronic human diseases

This book would be a useful resource for academic and research studies for the development of nanodrug delivery systems using phytomedicine-derived or herbal phytoantioxidants. It is, therefore, strongly believed that through this book many stakeholders working in pharmaceutical and biomedical sectors will be benefited. This book will be particularly useful to drug developers, drug manufacturers, pharmaceutical scientists (R&D), discovery scientists, biomedical scientists, healthcare professionals, biotechnologists, chemists, biochemists, phytochemists, pharmacologists, toxicologists, researchers, academicians, and students. The exceptional design, well-defined structure, and the rich content of the proposed book is timely and the need of the hour, and at present, there is no such book available in the market. Additionally, scientists involved in drug delivery research, nanoscience and nanotechnology research, formulation development research, nanobiomedicine research, nanomaterials research, nanodrug formulation research, nanodrug delivery research, and biomedical research are also expected to be the wider audience, users, or readers of this book, regardless if whether they are engaged in basic (chemistry, nanomaterials, nanoscience) or applied (such as pharmaceutical, nanomedicine, or biomedical) research.

List of Contributors

Syed Mohammad Abdullah
Department of Pharmaceutical Sciences and Technology, Birla Institute of Technology, Mesra, Ranchi, Jharkhand, India

Bhushan A. Bhairav
Department of Pharmaceutical Quality Assurance, R. C. Patel Institute of Pharmaceutical Education and Research, Shirpur, Maharashtra, India

Vinayak Bhatia
Department of Glaucoma, ICARE Eye Hospital and Postgraduate Institute, Noida, Uttar Pradesh, India

Swarnali Bhattacharjee
Life Sciences Division, Institute of Advanced Study in Science and Technology, Guwahati, Assam, India

Ravi Bhushan
Centre for Genetic Disorder, Institute of Science, Banaras Hindu University, Varanasi, Uttar Pradesh, India

Debarupa Dutta Chakraborty
Department of Pharmaceutical Chemistry, Mata Gujri College of Pharmacy, Mata Gujri University, Kishanganj, Bihar, India

Mainak Chakraborty
Departments of Pharmaceutical Technology and Pharmacology, NSHM Knowledge Campus, Kolkata, West Bengal, India

Prithviraj Chakraborty
Department of Pharmaceutics, Mata Gujri College of Pharmacy, Mata Gujri University, Kishanganj, Bihar, India

Tapash Chakraborty
Department of Pharmaceutics, Girijananda Chowdhury Institute of Pharmaceutical Science (GIPS), Guwahati, Assam, India

Joyani Das
Sarojini Naidu Vanita Pharmacy Maha Vidyalaya, Secunderabad, Telangana, India

Punamjyoti Das
Department of Pharmaceutical Sciences, Faculty of Science and Engineering, Dibrugarh University, Dibrugarh, Assam, India

Prashanta Kumar Deb
Department of Pharmaceutical Chemistry, School of Pharmaceutical Sciences, Shoolini University, Solan, Himachal Pradesh, India
and
Life Sciences Division, Institute of Advanced Study in Science and Technology, Guwahati, Assam, India

Rajlakshmi Devi
Life Sciences Division, Institute of Advanced Study in Science and Technology, Guwahati, Assam, India

Nilayan Guha
Department of Pharmaceutical Sciences, Faculty of Science and Engineering, Dibrugarh University, Dibrugarh, Assam, India

Tong Thi Thu Hoa
VNU University of Medicine and Pharmacy, Vietnam National University Hanoi, Cau Giay, Hanoi, Vietnam

N. Shaista Jabeen
Department of Zoology, Dhnabagyam Krishnaswamy Mudaliar College for Women (Autonomous), Vellore, Tamil Nadu, India

Dolly Jain
Oriental College of Pharmacy and Research, Oriental University, Indore, Madhya Pradesh, India
and
Adina College of Pharmacy, Sagar, Madhya Pradesh, India

Manmohan Singh Jangde
Shri Balaji College of Pharmaceutical Sciences, Sakti, Chhattisgarh, India

Himadri Kalita
Department of Zoology, Life Sciences Division, Assam Don Bosco University, Guwahati, Assam, India

Pallabi Kashyap
Department of Pharmaceutical Chemistry, Girijananda Chowdhury Institute of Pharmaceutical Sciences, Guwahati, Assam, India

Gülsen Kaya
Scientific and Technological Research Center, Inonu University, Malatya, Turkey

Merve Keskin
Vocational School of Health Services, Bilecik Seyh Edebali University, Bilecik, Turkey

Saban Keskin
Vocational School of Health Services, Bilecik Seyh Edebali University, Bilecik, Turkey

V. Kiruthiga
Department of Zoology, Dhnabagyam Krishnaswamy Mudaliar College for Women (Autonomous), Vellore, Tamil Nadu, India

Ashwini Kumar Mishra
Department of Pharmaceutics, Delhi Institute of Pharmaceutical Sciences and Research University (DIPSAR), Delhi Pharmaceutical Sciences and Research University, New Delhi, Delhi, India

Siddhartha Maji
Institute of Pharmacy, Ram-Eesh Institute of Vocational and Technical Education, Greater Noida, Uttar Pradesh, India

Papiya Mitra Mazumder
Department of Pharmaceutical Sciences and Technology, Birla Institute of Technology, Mesra, Ranchi, Jharkhand, India

Santwana Palai
Department of Veterinary Pharmacology & Toxicology, College of Veterinary Science & Animal Husbandry, Orissa University of Agriculture & Technology, Bhubaneswar, Odisha, India

Sharad P. Pandey
Department of Pharmacy, Shri Govindram Seksariya Institute of Technology and Science (SGSITS), Indore, Madhya Pradesh, India

Chandrakantsing V. Pardeshi
Department of Pharmaceutics, R. C. Patel Institute of Pharmaceutical Education and Research, Shirpur, Maharashtra, India

Vaishali Patel
Department of Pharmaceutics, Laxminarayan Dev College of Pharmacy, Bharuch, Gujarat, India

Ghanshyam Parmar
Department of Pharmacy, Sumandeep Vidyapeeth, Vadodara, Gujarat, India

Sabnam Parveen
Birbhum Pharmacy School, Birbhum, West Bengal, India

Dipali Patel
Department of Pharmaceutics, Shri Rawatpura Sarkar University, Raipur, Chhattisgarh, India

Arpita Paul
Department of Pharmaceutical Sciences, Faculty of Science and Engineering, Dibrugarh University, Dibrugarh, Assam, India

Mahendra K. Prajapati
Department of Pharmaceutics, School of Pharmacy and Technology Management, SVKM's NMIMS, Shirpur, Maharashtra, India

Shiv Kumar Prajapati
Institute of Pharmacy, Ram-Eesh Institute of Vocational and Technical Education, Greater Noida, Uttar Pradesh, India

Mithun Rudrapal
Department of Pharmaceutical Chemistry, Rasiklal M. Dhariwal Institute of Pharmaceutical Education and Research, Pune, Maharashtra, India

Rakesh Sagar
Department of Pharmacy, Shri Govindram Seksariya Institute of Technology and Science (SGSITS), Indore, Madhya Pradesh, India

Abhishek K. Sah
Department of Pharmacy, Shri Govindram Seksariya Institute of Technology and Science (SGSITS), Indore, Madhya Pradesh, India

Pravat Kumar Sahoo
Department of Pharmaceutics, Delhi Institute of Pharmaceutical Sciences and Research (DIPSAR), Delhi Pharmaceutical Sciences and Research University, New Delhi, Delhi, India

Swarnlata Saraf
University Institute of Pharmacy, Pt. Ravishankar Shukla University, Raipur, Chhattisgarh, India

Anupam Sarma
Department of Pharmaceutics, Girijananda Chowdhury Institute of Pharmaceutical Science (GIPS), Guwahati, Assam, India

Himangshu Sarma
Life Sciences Division, Institute of Advanced Study in Science and Technology, Guwahati, Assam, India

Khomendra Kumar Sarwa
Department of Pharmacy, Government Girls Polytechnic, Raipur, Chhattisgarh, India

Ashish Shah
Department of Pharmacy, Sumandeep Vidyapeeth, Vadodara, Gujarat, India

Smriti Sharma
Department of Chemistry, Miranda House, University of Delhi, Delhi, India

Ganesh B. Shevalkar
Department of Pharmaceutics, R. C. Patel Institute of Pharmaceutical Education and Research, Shirpur, Maharashtra, India

Vijendra Kumar Suryawanshi
Department of Pharmaceutics, M J College, Bhilai, Chhattisgarh, India

Bui Thanh Tung
VNU University of Medicine and Pharmacy, Vietnam National University Hanoi, Cau Giay, Hanoi, Vietnam

Abd. Kakhar Umar
Department of Pharmaceutics and Pharmaceutical Technology, Faculty of Pharmacy, Universitas Padjadjaran, Jatinangor, Indonesia

Sonal Upadhyay
Centre for Genetic Disorder, Institute of Science, Banaras Hindu University, Varanasi, Uttar Pradesh, India

Dhaneshwar Uraon
Department of Pharmacy, Government Girls Polytechnic, Raipur, Chhattisgarh, India

A. Vinodhini
Department of Zoology, Dhnabagyam Krishnaswamy Mudaliar College for Women (Autonomous), Vellore, Tamil Nadu, India

Md Kamaruz Zaman
Department of Pharmaceutical Sciences, Faculty of Science and Engineering, Dibrugarh University, Dibrugarh, Assam, India

Zonunmawii
Department of Pharmaceutical Sciences, Faculty of Science and Engineering, Dibrugarh University, Dibrugarh, Assam, India

James H. Zothantluanga
Department of Pharmaceutical Sciences, Faculty of Science and Engineering, Dibrugarh University, Dibrugarh, Assam, India

1

Natural Antioxidants in Oxidative Stress-Induced Diseases

Himadri Kalita[1] and Prashanta Kumar Deb[2]

[1] *Department of Zoology, Life Sciences Division, Assam Don Bosco University, Guwahati, Assam, India*
[2] *Department of Pharmaceutical Chemistry, School of Pharmaceutical Sciences, Shoolini University, Solan, Himachal Pradesh, India*

> *Ayurvedic herbalism is not of the east or the west, of ancient or modern time. It is a science of living that encompasses the whole of life, and which relates the life of the individual to that of the universe. A knowledge that belongs to all living beings.*
>
> —Dr. David Frawley and Dr. Vasant Lad, *The Yoga of Herbs*.

1.1 Natural Antioxidants: Game Changers in Human Disease Management

1.1.1 Types and Sources of Natural Antioxidants

Antioxidants are the compounds acting as a defense system for our body. They neutralize the reactive oxygen species (ROS) generated during various metabolic processes inside our body and during disease conditions. There are various sources of antioxidants such as endogenous antioxidants, which are generated inside our body and exogenous antioxidants from various food sources (Table 1.1) [1]. In recent times, people are more toward natural antioxidant compounds rather than synthetic ones. Polyphenols are a group of compounds with high antioxidant activity. Among the sources of natural antioxidants, the most important are those coming from routinely consuming vegetables and fruits. The classification of natural antioxidants is depicted in Figure 1.1.

1.1.2 Chemical Structures of the Natural Antioxidants

1.1.2.1 Phenolic Compounds and Flavonoids

Phytochemicals, especially phenolics in fruits and vegetables, are the major bioactive compounds known for health benefits [2]. According to Bravo, the presence of polyphenols in fruits and vegetables is greatly influenced by genetic factors, environmental conditions, and degree of ripeness, for instance [3].

Many natural antioxidants exhibit a wide range of biological effects including antibacterial, antiviral, anti-inflammatory, antiallergenic, antithrombotic, and vasodilatory actions [2]. Medicinal

Phytoantioxidants and Nanotherapeutics, First Edition. Edited by Mithun Rudrapal.

Table 1.1 Some fruits and vegetables with their antioxidant compounds.

No	Fruit/vegetable	Part used	Antioxidant
1.	Banana	Unripe (green) fruit and peel	Phenols and flavonoids
2.	Mango	Peel, kernel	Gallic acid, ellagic acid, gallates, gallotannins, condensed tannins
3.	Watermelon	Peel, rinds	Citrulline, lycopene, flavonoids, and phenols
4.	Cucumber	Peel	Flavonoids and phenols
5.	Potato	Peel	Chlorogenic acid, caffeic acid, ferulic acid, and phenols
6.	Coffee	Coffee ground and residue	Polyphenols, tannins, and gallic acids
7.	Apple	Peel	Epicatechin, catechins, anthocyanins, quercetin glycosides, chlorogenic acid, hydroxycinnamates, phloretin glycosides, and procyanidins
8.	Grapes	Skin and seeds	Coumaric acid, caffeic acid, ferulic acid, chlorogenic acid, cinnamic acid, neochlorogenic acid, *p*-hydroxybenzoic acid, protocatechuic acid, vanillic acid, gallic acid, proanthocyanidins, quercetin 3-*o*-glucuronide, quercetin, and resveratrol
9.	Guava	Skin and seeds	Catechin, cyanidin 3-glucoside, galangin, gallic acid, homogentisic acid, and kaempferol
10.	Pomegranate	Peel and pericarp	Gallic acid, cyanidin-3,5-diglucoside, cyanidin-3-diglucoside, and delphinidin-3,5-diglucoside
11.	Carrot	Peel	Phenols, β-carotene
12.	Cucumber	Peel	Phenols, flavonoids, pheophytin, phellandrene, caryophyllene
13.	Potato	Peel	Gallic acid, caffeic acid, vanillic acid, chlorogenic acid, ferulic acid, and phenols
14.	Tomato	Skin and pomace	Carotenoids

Source: Anwar et al. [1]/IntechOpen/licensed under CC BY 3.0.

plants basically contain various antioxidants, e.g. vitamins, carotenoids, and phenolic compounds such as catechin, epicatechin, lignin and tannins, and anthocyanins [4].

In the group of polyphenolic compounds, flavonoids have been extensively studied and include catechins (Figure 1.2), proanthocyanins (Figure 1.3), anthocyanidins, flavones, flavonols, and their glycosides. Studies on the structure–activity relationship have afforded consistent evidence revealing the specific role of structural components and requirements for scavenger radicals, chelating action, and oxidizing activity of flavonoid compounds. In fact, the in vitro antioxidant activity of flavonoids and their metabolites depends on the arrangement of functional groups in the nuclear structure [5]. Most of the beneficial effects of flavonoids on human health are attributed to their antioxidant and chelating properties [5] and also to antimutagenic and antitumoral effects [6]. Flavonoids inhibit a variety of enzyme systems. Among them, there are several oxygenases such as prostaglandin synthase, the key enzyme in eicosanoids biosynthesis. Further, flavonoids also act by inhibiting the hyaluronidase activity helping maintain proteoglycans of connective tissue and preventing the spread of bacterial or tumor metastases [7]. By hindering the oxidation reactions, in

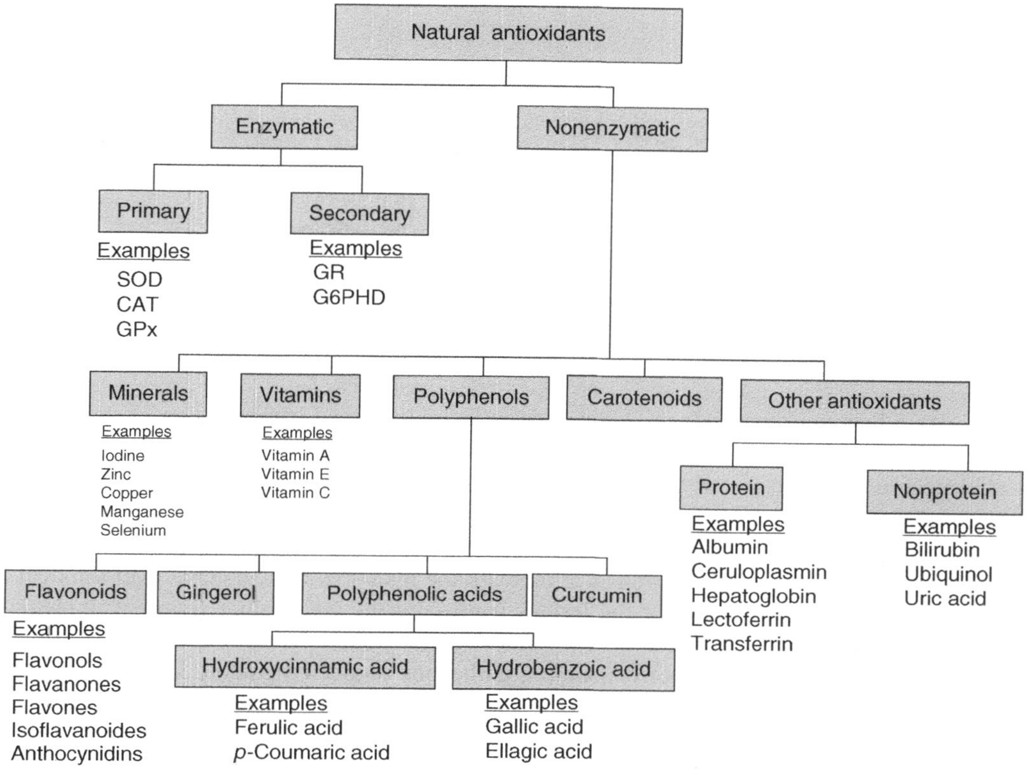

Figure 1.1 Classification of natural antioxidants. *Source:* Anwar et al. [1]/IntechOpen/licensed under CC BY 3.0.

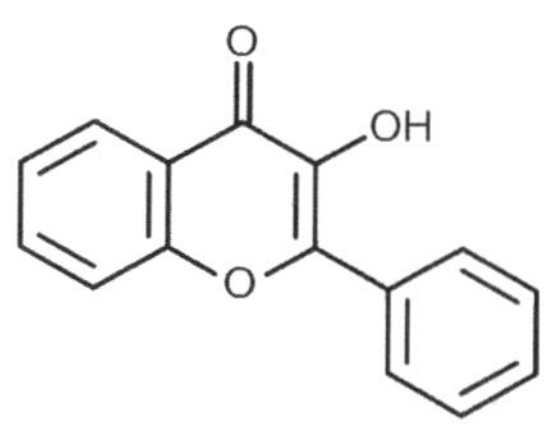

Figure 1.2 Catechin molecule.

which flavonoids are preferentially oxidized, they preserve the body's natural antioxidants such as ascorbic acid [8].

The potential of plant biomasses as a source of bioactive compounds also refers to the by-products and/or residues of a given production system. For instance, fruit residues (i.e. pomace) are inexpensive, are easily available, and contain bioactive molecules. Consequently, over the past years, research focus has shifted to such residues as a possible source of antioxidant compounds. Shui and Leong reported that antioxidant compounds, such as (−)-epicatechin and proanthocyanidins, which existed as dimers through pentamers in star fruit (*Averrhoa carambola* L.) residues, delay oxidative rancidity of soybean oil to a greater extent than butylated hydroxytoluene (BHT) [9].

Oxygen free radical processes are involved in both physiological and pathological conditions, which skin tissue repair caused mainly by trauma and burns [10]. The role of antioxidants in the removal of inflammation products is already known and these compounds are also beneficial in wound healing for other reasons. Antioxidants work against the excess of proteases and ROS, protecting protease inhibitors from oxidative damage. In addition, antioxidants can prevent destruction of fibroblasts and other cells caused by ROS over a generation, and therefore may be important in the successful treatment of lesions [11]. Another category of antioxidant leucocyanidin

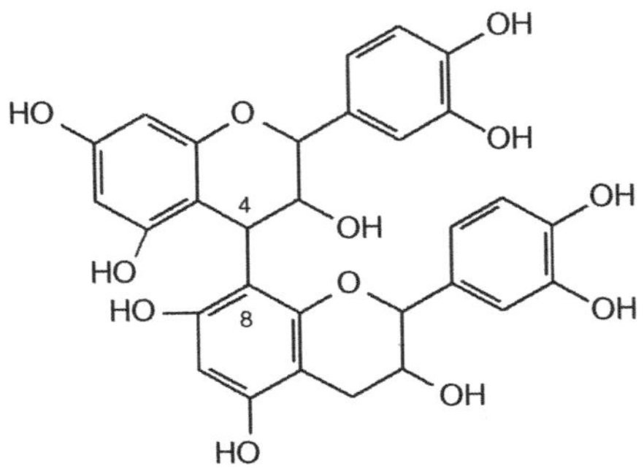

Figure 1.3 Proanthocyanin molecule.

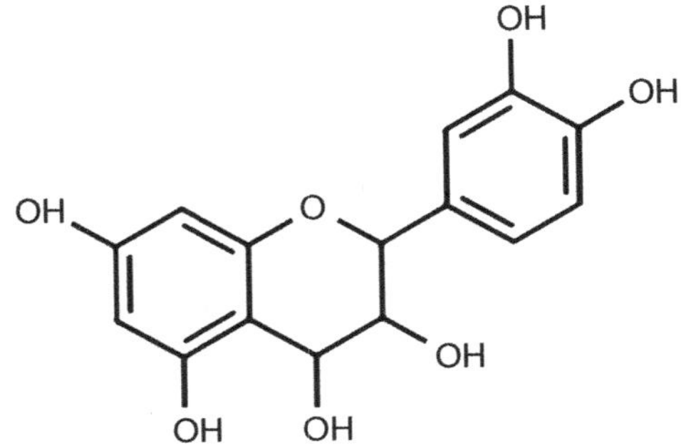

Figure 1.4 Leucocyanidin molecule.

(Figure 1.4), a flavonoid that induces cell proliferation by increased incorporation of thymidine into cellular DNA [12], accelerating the healing of skin wounds [13].

Gallocatechin (GE) consists of the largest groups of naturally occurring phenols with antioxidant potential that is widely distributed in leaves, seeds, bark, and flowers of plants [5]. GE is a good candidate for wound healing and its treatment was able to decrease the epithelization period, healing the lesions in nine days, as well as to increase the hydroxyproline content over the treatment period. Histological analysis of the lesions confirmed the GE healing potential, showing fibroblast proliferation and induction of re-epithelialization process. On the one hand, ROS are necessary for effective defense against invading pathogens and cell signaling and even in the absence of infection, low levels of ROS are necessary for cell signaling, especially angiogenesis. Consequently, a closer relationship between production and detoxification of ROS is crucial for the normal repair process of an injury.

1.1.2.2 Carotenoids

Carotenoids are also natural antioxidants and they contribute to the stability of foods. Such pigments are not evenly distributed in the food itself as various investigators have found that carotenoids are usually more concentrated in the peel than in the pulp of fruits and vegetables [12]. Three plantain varieties and two dessert banana varieties (Cavendish and Yangambi-5) were investigated by [14] for pro-vitamin A carotenoid content. Banana peel has a substantially higher carotenoid. Banana peel (*Musa* spp., cv. Prata Anã) might be considered a source of carotenes as *trans*-β-carotene, *trans*-α-carotene, and *cis*-β-carotene, which are the major carotenoids from this raw material. Banana peel is a kind of important raw material that might be more exploited regarding its carotenoid concentration as well as in the pulp.

1.1.3 Beneficial Role of Natural Antioxidants in the Management of Human Disease (Illustration and Diagram)

Free radicals are highly reactive species having unpaired electrons in their outermost shell. They react rapidly with the membranes eventually causing cellular damage and finally death. To protect the body from these radicals, the living system should generate or intake various antioxidants as nutraceuticals. Free radicals are generated in the body due to the exposure to various chemicals found in polluted air, water, etc. Most of the diseases including diabetes, hyperlipidemia, Parkinson's, and Alzheimer's are the results of the action of these free radicals. They interact with the lipid component of the cell membrane (lipid peroxidation) along with the RNA and DNA [15].

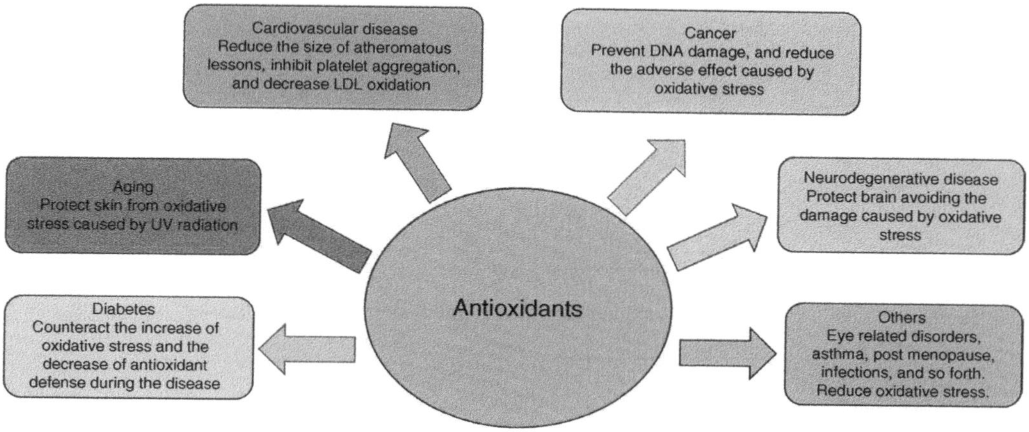

Figure 1.5 Influence of antioxidants on human health. Antioxidants can influence many aspects of human health such as diabetes, aging, cardiovascular and NDs, cancer, and other illnesses. Antioxidants produce several beneficial effects, promoting a healthy status, and reducing the oxidative stress caused by ROS. *Source:* Based on Refs. [16, 17].

An effective antioxidant complex has different types of radical catching antioxidant sites that seek and destroy free radicals at many cellular sites. There are single specific antioxidants, for example, vitamin E, specific for the protection of an outer fatty layer of cells. A number of scientific studies are going about addressing the varied health benefits of antioxidant supplementation in processes like stress, aging, pathogen infestation, apoptosis, and neurological diseases. Antioxidants reduce the cell-damaging effects of free radicals. Besides numerous scientifically compelling studies addressing various health benefits of antioxidant supplementation in athletic training with the simple addition of a good antioxidant complex, there are numerous studies addressing the various health benefits of antioxidant supplementation in athletic training with the simple addition of a good antioxidant complex. The brain is uniquely vulnerable to oxidative injury. Due to its high metabolic rate and elevated levels of polyunsaturated lipids, it is the target of lipid peroxidation. Consequently, antioxidants are commonly used as medications to treat various forms of brain injury. Antioxidants are also being investigated as possible treatments for neurodegenerative diseases (NDs) such as Alzheimer's disease, Parkinson's disease, and amyotrophic lateral sclerosis and as a way to prevent noise-induced hearing loss [15]. People take antioxidant supplements directly from fresh fruits and vegetables. Fruits and vegetables contain a large amount of flavonoids and antioxidant supplements that contribute to protection against different types of cancers and cardiovascular health problems (Figure 1.5) [18].

1.1.4 Natural Antioxidants as Nutraceuticals in the Prevention of Oxidative Stress-Associated Diseases (with Suitable Examples)

Antioxidants are substances that retard or prevent deterioration, damage, or destruction by oxidation. Oxidative stress arises by the generation of ROS and ROS are generated in case of a lack of antioxidant defense in the body. Free radicals are a major cause of many degenerative diseases. Many chronic diseases such as diabetes, cardiovascular diseases, NDs, and carcinogenic ailments are related to free radicals [19]. Antioxidants form an integral part of the nutraceutical market. They are the major part of the most commonly occurring nutrients. Antioxidants are quite large in

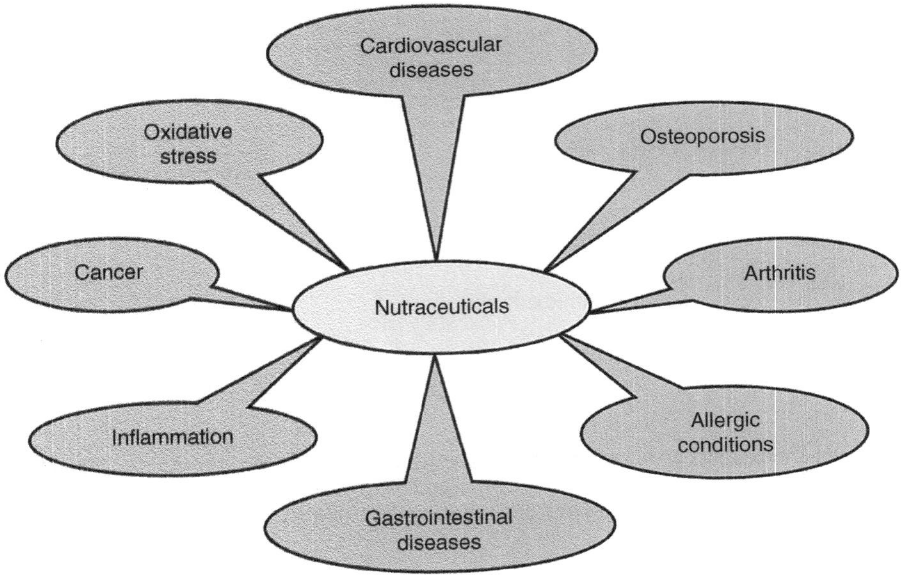

Figure 1.6 Role of nutraceuticals in the prevention of illness. *Source:* Jain and Ramawat [25]/ with permission of Springer Nature.

number and diverse in nature and prevent oxidation by neutralizing free radicals at relatively small concentrations. It can prevent the damage at the cellular level by using the following mechanisms: they may ease the energy of the free radical or subdue radical formation or repair the damage and reconstitute the membrane. Dietary intake of antioxidants may exert many potential benefits [20]. Ingestion of antioxidants from fruits and vegetables or administration of synthetic antioxidants decreases certain chronic diseases of aging. For example, intake of vitamin E may prevent Parkinson's disease [21], many dietary antioxidants can inhibit low-density lipoprotein-cholesterol (LDL-c) oxidation, thus protecting the heart against diseases. Terpenes have unique antioxidant activity as they react with free radicals by partitioning themselves into fatty membranes by virtue of their long carbon side chain and help in the cure of diseases. Thus, polyphenols act as an antioxidant protectant for humans. Phenolic antioxidants such as tocopherols, polyphenols, and phytoestrogens decline oxidative cell injuries and inflammatory reactions improving the brain's health [22, 23]. Scientific focus on the identification of health-protectant components within the foodstuff and their mechanism(s) of action needed to be studied properly. Nowadays direct attention should be given to nutraceuticals rather than to the whole-food concepts implicit in a term such as functional foods [24] (Figure 1.6).

1.2 Oxidative Stress: Major Contributor of Human Diseases

1.2.1 Oxidative Stress in the Development of Human Diseases

Oxidative stress, cellular senescence, and, consequently, various protein factors are involved in several acute and chronic pathological processes, such as metabolic syndrome (MetS), NDs, macular degeneration (MD), biliary diseases, and cancer. MetS-associated risk factors (i.e. obesity, diabetes, hypertension, and atherosclerosis) are associated with the inflammatory pathway mediated by IL-1α, IL-6, and IL-8, and lead to increased cellular senescence [26, 27]. In many neurodegenerative conditions, including Alzheimer's disease (AD), brain tissue biopsies show increased levels of p16, matrix metalloproteinase (MMP), and IL-6 [28, 29]. Chronic obstructive pulmonary disease, biliary cirrhosis, cholangitis, and osteoarthritis share several damaging senescence-associated secretory phenotype (SASP) profiles including IL-6, IL-8, and MMP. The induction of epithelial to mesenchymal transition mediated by ROS promotes cancer metastasis [30]. There is a close relationship between oxidative stress, inflammation, and aging. Aging is a loss of homeostasis due to chronic oxidative stress that affects especially the regulatory systems, such as nervous, endocrine, and immune systems. The consequent activation of the immune system increases the inflammatory responses and, consequently, increases age-related morbidity and mortality [31].

1.2.1.1 Body Mass Index (BMI)

Previous studies have reported that increased BMI is a strong risk factor for MetS-associated risk factors [32]. A strong positive association found between obesity and MetS is found in both men [33] and women [34]. Obesity is mainly associated with an increased risk of developing insulin resistance and type 2 diabetes mellitus (T2DM). In obese individuals, a large amount of nonesterified fatty acid (FA) was released by adipose tissue and also other factors responsible for the development of insulin resistance were also released. When insulin resistance is followed by dysfunction of the β-cells, the downfall of insulin secretion results in an uncontrolled rise in blood glucose levels leading to T2DM. Many genes are responsible for the obesity and pathogenesis of MetS. There are many genes, which determine the range of BMI in the population. Each gene can explain a few gram differences in body weight among the population [33]. Genes that are responsible for obesity and insulin resistance interact with environmental factors such as increased fat/calorie intake and decreased physical activity resulting in the development of obesity and insulin resistance followed ultimately by the development of MetS [35].

1.2.1.2 Hyperglycemia

T2DM is one of the serious endocrine disorders and a major health problem that has been growing in most of the countries today [36]. According to the report of the WHO (2013), about 347 million people suffer from diabetes worldwide. After cancer and cardiovascular disease (CVD), diabetes has become the third "killer" of mankind due to its high prevalence, morbidity, and mortality [29]. A recent report revealed that sales of diabetic drugs have been climbing over the last few years and the maximum sales crossed ₹97 crores since December 2013 in India alone. Type 2 DM is a heterogeneous disorder characterized by a progressive decline in insulin action (insulin resistance), followed by the inability of β-cells to compensate for insulin resistance (pancreatic β-cell dysfunction). Insulin resistance is a characteristic metabolic defect that proceeds over β-cell dysfunction. The β-cells normally compensate for insulin resistance by secreting more amounts of insulin to maintain glucose homeostasis. In the course of time, however, this β-cell function gets impaired leading to deterioration in glucose homeostasis and subsequent development of impaired glucose tolerance leading to diabetes [37].

Diabetes is a chronic disease that occurs either when the pancreas does not produce enough insulin or when the body cannot effectively use the insulin it produces. There are mainly two types of diabetes mellitus (DM). Type 1 DM is immune-mediated and requires daily administration of insulin. The other common type is type 2 DM and characterized by insulin resistance or relative insulin deficiency. T2DM is the most common form and comprises 90% of people with diabetes around the world. The prevalence of T2DM rates continues to increase with increasing number of patients at risk of serious diabetes-related complications. Having T2DM increases the risk of a myocardial infarction two times and the risk of suffering a stroke two to four times.

Diabetes is a worldwide health problem, due to its high prevalence and because the clinical course of the disease can lead to several impairments, such as vision loss, chronic renal disease, macro- and microvascular complications, and eventually result in physical incapacity and death [38]. The American Diabetes Association (ADA) estimates that in 2012, expenditure on the treatment of diabetes and its complications was approximately $245 billion in the United States. Inhibitors of α-glucosidase, which catalyzes the final step in the digestive process of carbohydrates, can retard the uptake of dietary carbohydrates and suppress postprandial hyperglycemia and may find use in the treatment of diabetic patients. The present generation of drugs used in the treatment of diabetes has certain limitations, and they sometimes lose their effect on glycemic control. Consequently, there is a need for the development of new drugs. Studies have shown that certain plants and phytochemicals have properties that improve the diabetic state. These properties include insulinotropic action on pancreatic β-cells [39]; enhancement of insulin signaling in the liver [40], adipose tissue [32], and muscle [33]; and reduction in intestinal glucose absorption [34, 41]. Plants containing antioxidant substances have also been shown to have beneficial effects on diabetes, as well as reduce the oxidative stress caused by hyperglycemia [42]. Several α-glucosidase inhibitors from plants have been screened and shown to be of clinical importance [43].

1.2.1.3 Hyperlipidemia

Hyperlipidemia is a heterogeneous disorder commonly characterized by an increased flux of free fatty acids (FFAs), raised triglycerides (TGs), LDL-c (bad cholesterol), and apolipoprotein B (apoB) levels, as well as by a reduced plasma high-density lipoprotein-cholesterol (HDL-c) concentration (good cholesterol) because of metabolic effects, or dietary and lifestyle habits (Figure 1.7) [44].

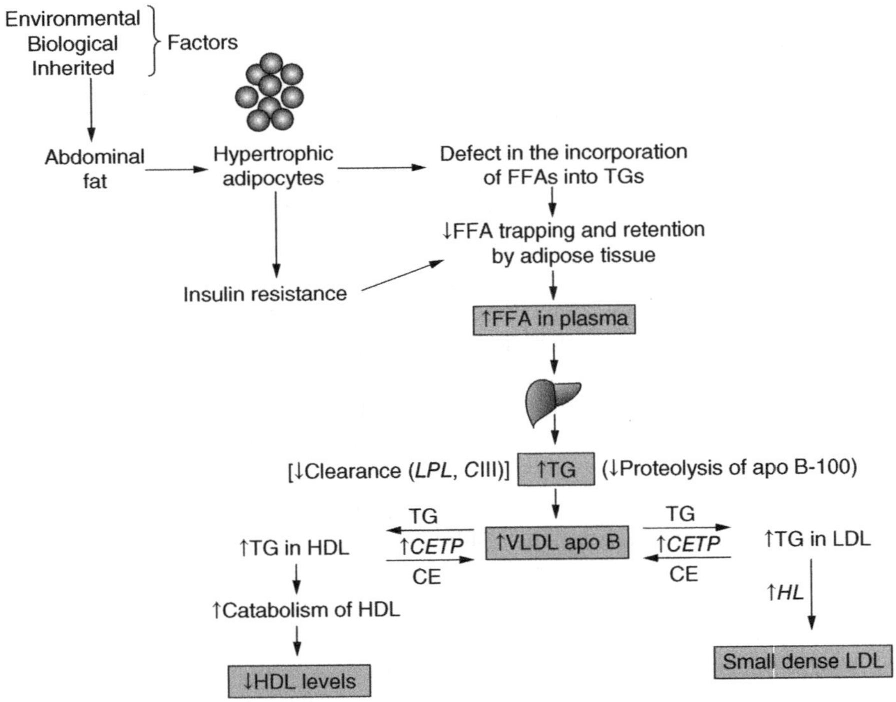

Figure 1.7 Schematic representation of dyslipidemia of metabolic syndrome.

The lipid abnormality in hyperlipidemia is an increase in circulating (nonesterified) FFAs originating from adipose tissue, and inadequate esterification and FFA metabolism [44]. The reduced retention of FAs by adipose tissue leads to an increased flux of FFA returning to the liver, which stimulates hepatic TG synthesis, promoting the production of apoB, whereby deposition and secretion of very low-density-lipoprotein-cholesterol (VLDL-c). When plasma TG concentration is subsequently increased, TG-rich HDL particles are formed and undergo catabolism. Elevated VLDL-c particles are lysed and hence fail to bind efficiently to LDL receptors (LDLr), while the exchange of cholesterol esters with TGs forms TG-rich lipoproteins, resulting in the formation of small dense LDL-c particles [45]. A strong association exists between elevated LDL-c levels and increased incidence of coronary artery disease [46]. The development of atherosclerotic plaques is associated with elevated levels of LDL-c, reduced receptor-mediated clearance, increased arterial wall retention, and an increased susceptibility [47]. Cardiovascular risk factors such as hyperlipidemia, hypertension, and thrombosis contribute to the underlying mechanisms of atherosclerotic disease, promoting endothelial dysfunction, oxidative stress, and proinflammatory pathways to peroxidation [47]. Lipid guidelines from the National Heart Foundation of Australia place great emphasis on LDL-c and HDL-c as atherogenic and antiatherogenic components, respectively. Indeed, high LDL-cholesterolemia is considered one of the major modifiable risk factors for coronary heart disease, which continues to be the leading cause of death and morbidity in the United States [44]. Conversely to the Australian lipid guidelines, the Adult Treatment Panel III (ATP III) guidelines of the US National Cholesterol Education Program place greater emphasis on TG levels [44]. According to the National Health and Nutrition Examination Survey III, 24% of individuals aged >20 years had MetS [48]. Dyslipidemia is an independent risk factor for cardiovascular disease [49, 50]. Low HDL-c and hypertriglyceridemia have been found to be independently and significantly related to myocardial infarction/stroke in patients with MetS [51]. The combination of high fasting glucose and low HDL-c were shown to have primary predictive ability for coronary heart disease [52]. Dyslipidemia may be caused by a combination of overproduction of VLDL-c, apoB-100, decreased catabolism of apoB-containing particles, and increased catabolism of HDL-apoA-I particles. Insulin resistance may be the consequence of this abnormality [53]. Dyslipidemia may arise from genetic components (e.g. mutated LDLr, mutated apoB-100, mutated proprotein convertase subtilisin/kexintype-9) [54], with or without environmental components (e.g. improper diet, familial history of hypercholesterolemia, hyperlipidemia, and/or hypertriglyceridemia) [55]. Causes of secondary hyperlipidemia include diabetes, hypothyroidism, obstructive liver disease, chronic renal failure, and drugs that increase LDL-c and decrease HDL-c, such as progestins and corticosteroids [56].

1.2.1.4 Hypertension

Previous literature and case-control studies have revealed that hypertension progression is also a predictor of MetS [57]. There is an association existing between T2DM and hypertension. Endothelial dysfunction could be one of the common pathways explaining the strong association between blood pressure and incident MetS. Studies suggested that endothelial dysfunction is associated with the onset of diabetes [58], and are closely related to blood pressure and hypertension [58]. Also, inflammatory markers such as C-reactive protein have been consistently related to the incident of T2DM [59], and to increasing blood pressure levels, suggesting that inflammation may be another marker for the association between blood pressure, MetS, and incident T2DM [59]. Thus, insulin resistance may be another link between blood pressure levels and the pathogenesis of MetS [59]. Studies also suggested that blood pressure is linked to MetS. Hence, blood pressure increases with increasing BMI, the potential risk of T2DM associated with hypertension. A causal relationship between hypertension and T2DM is further strengthened by a recent randomized clinical trial study showing a 14% reduction in risk of diabetes in subjects with glucose intolerance by allocation to five-year treatment with valsartan, an angiotensin II blocker with antihypertensive properties [60].

1.2.1.5 Physical Activity

Studies reported that physical inactivity is a strong risk factor for hyperlipidemia and diabetes. A sedentary lifestyle has been reported to be associated with MetS-related risk factors like diabetes and obesity for both men and women [61], while moderate and vigorous physical exercise is associated with a lower risk of MetS [62]. Evidence from clinical trials which included physical activity as an integral part of lifestyle interventions suggested that the onset of T2DM can be prevented or delayed as a result of successful lifestyle interventions that included physical activity as a part of these interventions [63]. Physical activity plays an important role in delaying or preventing the development of MetS in those at risk both directly by improving insulin sensitivity and reducing insulin resistance, and indirectly by beneficial changes in body mass and body composition [64].

1.2.1.6 Dietary Pattern

Dietary habit is an important lifestyle risk factor associated with the development of MetS-related risk factors. There is a positive correlation existing between dietary intake and the risk of developing MetS [65]. Previous literature showed that a higher dietary glycemic index has been consistently associated with an elevated risk of T2DM, which is a risk factor of MetS [66]. Western countries over the past years have shifted toward a more sedentary lifestyle by taking a high-calorie diet, which is mainly rich in fructose and saturated fatty acid (SFA). Dietary fructose and fat are associated with different MetS-related risk factors like obesity, insulin resistance, and T2DM. It is of two types, namely SFA and polyunsaturated fatty acid (PUFA) and monounsaturated fatty acid (MUFA). SFA is deleterious, while PUFA and MUFA have beneficial effects [67]. A study suggested that a diet rich in SFA caused insulin resistance, whereas a diet rich in ω-6-PUFA resulted in insulin sensitivity [68]. However, a PUFA linoleic acid also prevented the development of insulin resistance and negatively correlated with T2D risk and cardiovascular risk in diabetes [69]. The exact mechanism by which unsaturated FAs exert their effects on insulin sensitivity is yet unclear. Besides, linoleic acid intake may not have an effect on body weight. However, it can decrease inflammation in humans [62]. An explanation for this may be that PUFAs are precursors for leukotrienes and prostaglandins that act as proinflammatory and lipoksines and resolvines that act as anti-inflammatory, in addition, docosanoides [68] and linoleic acid is a precursor for anti-inflammatory substances [68]. In addition, the beneficial effects of PUFAs might be explained by their role in the process of cellular lipogenesis, which is responsible for preventing FA-related disorders.

1.2.1.7 Oxidative Stress and Cancer

Many studies have demonstrated the direct relationship between inflammatory responses and carcinogenesis [70]. The main chemical mediators of the inflammatory response are free radical species derived from oxidative stress. ROS may directly or indirectly damage other chemical or structural components in target cells or recruit other inflammatory cells leading to additional ROS production that amplifies the damage [71]. In particular, ROS and inflammatory cytokines, such as TNFα, activate the transcription factor NFκB, which induces the expression of genes involved in cell proliferation, apoptosis, and carcinogenesis [26]. Chronic inflammation is also associated with angiogenesis, another characteristic of cancer because ROS can also increase the expression of transcriptional factors involved in neoplastic transformation and enhancement of cancer angiogenesis [71]. Moreover, macrophages, platelets, fibroblasts, and cancer cells are a major source of angiogenic factors (i.e. fibroblast growth factor, vascular endothelial growth factor, and prostaglandins – E1 and E2), which increase the production of ROS and, subsequently, increase the risk of cancer [70]. The mutagenic/carcinogenic potential of ROS is due to their capability of reacting with DNA and chemically modifying it. In particular, ROS-induced DNA damages can result in transcriptional arrest or induction/replication errors, or genomic instability, which are all associated with carcinogenesis [72]. Among oxidized DNA products, 8-oxodG and 8-nitroguanine are considered biomarkers for inflammation-induced carcinogenesis [70]. There is a dramatic age-dependent escalation in cancer risk, and increased oxidative stress and oxidative stress and its damages over a lifetime may be responsible for this phenomenon. In fact, there is an accumulation of ROS-induced DNA damages with age, which is confirmed by the progressive and statistically significant increase in the levels of 7,8-dihydro8-oxo-2′-deoxyguanosine observed with aging [70]. On the basis of these considerations, chronic inflammation and oxidative stress should be considered high-risk factors for cancer, especially in elderly people [73]. Based on these analogies, it should be recommended that elderly people should consume higher antioxidant compounds, but the efficacy of antioxidant therapy in the prevention of carcinogenesis is currently under investigation.

1.2.2 Endogenous Antioxidant Defense and Oxidative Stress

The human body is equipped with a variety of antioxidants that are synthesized denovo, which serve to counterbalance the effect of ROS. For all practical purposes, these can be divided into two categories: enzymatic and nonenzymatic.

1.2.2.1 Enzymatic Antioxidants

The major enzymatic antioxidants of the body are superoxide dismutase (SOD), catalase, and glutathione peroxidase (GSH)-Px. In addition to these major enzymes, other antioxidants, including heme oxygenase-1, and redox proteins, such as thioredoxins, peroxiredoxins, and glutaredoxins, have also been found to play crucial roles in the antioxidant defenses.

Since superoxide is the primary ROS produced from a variety of sources, its dismutation by SOD is of primary importance for each cell. There are three forms of SOD, namely CuZn-SOD, Mn-SOD, and EC-SOD, which are widely expressed in the human lung. Mn-SOD is confined in the mitochondria matrix, and EC-SOD is primarily localized in the extracellular matrix, especially in areas containing high amounts of type I collagen fibers and around pulmonary and systemic vessels. It has also been detected in the bronchial epithelium, alveolar epithelium, and alveolar macrophages [74, 75]. CuZn-SOD and Mn-SOD act as bulk scavengers of superoxide radicals. The relatively high EC-SOD level in the lung with its specific binding to the extracellular matrix components may represent a fundamental component of lung matrix protection [76].

H_2O_2 that is produced by the action of SODs or the action of oxidases, such as xanthine oxidase, is reduced to water by catalase and the GSH-Px. Catalase is a tetramer composed of four identical monomers each of which contains a heme group at the active site. Degradation of H_2O_2 is accomplished via the conversion between two conformations of catalase–ferricatalase (iron coordinated to water) and compound I (iron complexed with an oxygen atom).

In addition, disposal of H_2O_2 is closely associated with several thiol-containing enzymes, namely TRXs (TRX1 and TRX2), thioredoxin reductases (TRRs), PRXs (which are thioredoxin peroxidases), and glutaredoxins [77]. Two TRXs and TRRs have been characterized in human cells, existing in both cytosol and mitochondria. In the lung, TRX and TRR are expressed in bronchial and alveolar epithelium and macrophages. PRX V has recently been found to function as a peroxynitrite reductase [78], which means that it may function as a potential protective compound in the development of ROS-mediated lung injury [79].

Another antioxidant enzyme family, GSTs inactivate secondary metabolites, such as unsaturated aldehydes, epoxides, and hydroperoxides. Three major families of GSTs have been described: cytosolic GST, mitochondrial GST [80], and membrane-associated microsomal GST, which has a role in eicosanoid and GSH metabolism. Seven classes of cytosolic GST are identified in mammalians, designated Alpha, Mu, Pi, Sigma, Theta, Omega, and Zeta [80].

1.2.2.2 Nonenzymatic Antioxidants

Nonenzymatic antioxidants include low-molecular-weight compounds, such as vitamins (vitamins C and E), β-carotene, uric acid, and GSH, a tripeptide.

1.2.2.2.1 Vitamin C (Ascorbic Acid) Water-soluble vitamin C (ascorbic acid) provides intracellular and extracellular aqueous-phase antioxidant capacity primarily by scavenging oxygen free radicals. It converts vitamin E free radicals back to vitamin E [81].

1.2.2.2.2 Vitamin E (α-Tocopherol) Lipid-soluble vitamin E is concentrated in the hydrophobic interior site of the cell membrane and is the principal defense against oxidant-induced membrane injury. Vitamin E donates an electron to peroxyl radical, which is produced during lipid peroxidation. α-Tocopherol is the most active form of vitamin E and the major membrane-bound antioxidant in a cell. Vitamin E triggers apoptosis of cancer cells and inhibits free radical formations [53].

1.2.2.2.3 Glutathione Another antioxidant molecule GSH is responsible for the reduction of H_2O_2 and lipid hydroperoxides (generated as a result of membrane lipid peroxidation) [82]. It is a tetrameric molecule that contains the unique amino acid selenocysteine within the active sites and uses low-molecular-weight thiols. Four GSH-Pxs have been described, encoded by different genes: GSH-Px-1 (cellular GSH-Px) is ubiquitous and reduces H_2O_2 and fatty acid peroxides, but not esterified peroxyl lipids [83]. GSH-Px-2 (gastrointestinal GSH-Px) is localized in gastrointestinal epithelial cells where it serves to reduce dietary peroxides. GSH-Px-3 (extracellular GSH-Px) is the only member of the GSH-Px family that resides in the extracellular compartment and is believed to be one of the most important extracellular antioxidant enzymes in mammals. Esterified lipids are reduced by membrane-bound GSH-Px-4 (phospholipid hydroperoxide GSH-Px), which can use several different low-molecular-weight thiols as reducing equivalents [80].

GSH is a cofactor for several detoxifying enzymes. It has a role in converting vitamins C and E back to their active forms. GSH protects cells against apoptosis by interacting with proapoptotic and antiapoptotic signaling pathways. It also regulates and activates several transcription factors, such as AP-1, NF-κB, and Sp-1 [81].

1.2.2.2.4 Carotenoids (β-Carotene) Another nonenzymatic plant-derived antioxidant is a carotenoid pigment. Primarily, β-carotene reacts with peroxyl (ROO·), hydroxyl (·OH), and superoxide ($O_2^{\cdot-}$) radicals [84]. Carotenoids show their antioxidant effects in low oxygen partial pressure but may have pro-oxidant effects at higher oxygen concentrations [85]. Both carotenoids and retinoic acids (RAs) are capable of regulating transcription factors. β-carotene inhibits the oxidant-induced NF-κB activation and interleukin (IL)-6 and tumor necrosis factor-α production. Carotenoids also affect the apoptosis of cells.

1.2.3 Modulation of ROS-Mediated Oxidative Stress by Antioxidant Small Molecules

In order to maintain proper cell signaling, it is likely that a number of radical scavenging enzymes maintain the level of ROS inside the cell. However, when the level of ROS exceeds this threshold, a rise in ROS production may lead to excessive signals to the cell, in addition to direct damage to key components in signaling pathways. Oxidative stress occurs when there is a balance between ROS formation and detoxification favors an increase in ROS levels, leading to disturbed cellular functioning. ROS can also irreversibly damage essential macromolecules such as DNA, protein, and lipids, which may initiate carcinogenesis [74]. Therefore, the concentration of ROS has to be controlled by several defense mechanisms, which also involve a number of antioxidants and detoxifying enzymes.

An antioxidant is a molecule, which has the ability to prevent or slow down the oxidation of macromolecules. The role of antioxidants is to decrease or terminate this chain of reactions by eliminating these free radicals or inhibiting other oxidation reactions by being oxidized themselves. So, antioxidants are often considered as the reducing agents such as polyphenols or thiols [86].

The antioxidant systems in the human body consist of powerful nonenzymatic and enzymatic antioxidants [87], such as vitamins C and E and glutathione, as well as different enzymatic systems which catalyze the oxidation reactions such as catalase, superoxide dismutase (SOD), and peroxidases. The defects in or inhibition of these antioxidant enzymes will lead to oxidative stress and may damage and lyse the cells [88]. The mechanisms followed by the antioxidant defense are (i) blocking of free radical production of (ii) oxidant scavenging, (iii) converting toxic-free radicals into less toxic substances, (iv) blocking the production of secondary toxic metabolites and mediators of inflammation, (v) blocking of the chain propagation of the secondary oxidants, (vi) repairing the injured molecules, (vii) initiation and enhancement of the endogenous antioxidant defense system. All of these defense mechanisms act hand in hand for the protection of the body from oxidative stress and thus from free radicals.

The antioxidant enzymes in all body cells consist of three major categories, which are the catalases, SODs, and glutathione peroxidases (GPx). All of these play crucial roles in maintaining the body's homeostasis. The induction of these enzymes reflects a specific response to oxidative stress [89]. The role of SOD is to scavenge superoxide radicals and convert them to H_2O_2. The role of GPx is achieved by the reduction of hydrogen peroxide, lipid hydroperoxides, and other organic hydroperoxides. Glutathione-*S*-transferases (GST) represent a major group of detoxifying enzymes, which form a family of multifunctional proteins involved in the cellular detoxification of cytotoxic and genotoxic compounds and in the protection of tissues against oxidative damage [90]. Besides certain roles in the endogenous metabolism, these enzymes are associated with the detoxification of xenobiotics such as drugs, carcinogens, and environmental pollutants in humans and animals, and with pesticide and herbicide resistance in insects and plants. Some studies have shown that GPx and GST are inactivated by hydroperoxides, which exert their toxicity either directly by oxidation of SH groups of proteins or indirectly by hydroxyl radical formation [91]. It is known from the literature that a significant number of GST isoenzymes also exhibit GPx activity and catalyze the reduction of organic hydroperoxides to their corresponding alcohols. SOD catalyzes the breakdown of the superoxide anion into oxygen and hydrogen peroxide. It removes O_2 by catalyzing a dismutation reaction. In the absence of this enzyme, this reaction becomes very slow.

Catalase is composed of 4 polypeptide chains, each chain is over 500 amino acids long, and contains 4 porphyrin heme (iron) groups allowing the enzyme to react with the H_2O_2. The enzymatic activity modes of catalase in decomposing H_2O_2 comprise the catalytic mode of activity and the peroxidatic mode of activity. The turnover rate of catalase is the highest among all of the other antioxidant enzymes. Catalase exists in both eukaryotic and prokaryotic cells. Most of them are located in an oxidative particle of all types of mammalian cells except red blood cells. Since H_2O_2 acts as a substrate for a specific reaction that generates highly reactive hydroxyl radical, it is believed that the primary role of catalase in cellular antioxidant defense mechanisms is to reduce the accumulation of H_2O_2. The overexpression of catalase makes cells more resistant to H_2O_2 toxicity and oxidative-mediated damage [53].

The glutathione system includes glutathione *S*-transferases, GPx, and glutathione reductase (GR). Glutathione *S*-transferases are another class of enzymatic antioxidants that catalyzes the breakdown of lipid peroxides. GPx shows high activity with H_2O_2 and organic hydroperoxides. These enzymes help in the detoxification mechanism [92].

GR catalyzes the reduction of oxidized glutathione (GSSG) to reduced glutathione (GSH). This enzyme enables the cell to maintain adequate levels of cellular GSH. GR is a flavoprotein that contains two flavin adenine dinucleotide (FAD) molecules as a prosthetic group, which is reducible by NADPH. GR is one of the thermostable enzymes. GR belongs to the defense system protecting the organism against chemical and oxidative stress. The deficiency of GR is characterized by hemolysis due to increased sensitivity of erythrocyte membranes to H_2O_2 and contributes to oxidative stress, which plays a key role in the pathogenesis of many diseases [93].

1.3 Protective and Restorative Roles of Natural Antioxidants in Various Oxidative Stress-Induced Diseases

1.3.1 Protective Effect of Natural Antioxidants in Type 2 Diabetes Mellitus

In relation to the antioxidant state in DM, Lortz and Tiedge reported that overexpression of the enzyme SOD and chloramphenicol acetyltransferase (CAT) could protect the pancreatic islets from ROS and maintain insulin production. Large-scale studies have shown that intensive early glucose control reduces the risk of micro- and macrovascular complications of DM [94]. Vitamins C and E and β-carotenes have traditionally been considered ideal supplements against DM-associated oxidative stress. García et al. suggested that vitamin E supplementation in DM patients can prevent myocardial infarction, stroke, and cardiovascular death [95]. Akbar et al. performed a meta-analysis of 14 studies where they found that supplementation with antioxidants does not affect plasma glucose or insulin levels. However, the HbA1c level is significantly reduced by supplementation with antioxidants, apparently due to having a protective effect on DM complications [96].

Melatonin is an active indoleamine (derived from tryptophan) component with antioxidant properties secreted mainly by pinealocytes. Melatonin also increases the expression of antioxidant enzymes (SOD, CAT, and GPx) and eliminates free radicals. Melatonin is indicated alone or in combination with other therapies for one to three weeks, where it can produce clinical improvement in patients with type 2 DM [97].

1.3.2 Protective Effect of Natural Antioxidants in Cardiovascular Diseases

As a public health problem of global concern, CVDs have attracted considerable attention. Some antioxidant natural products have been proven capable of preventing CVDs. Multiple epidemiological investigations enrolling participants from different countries, different ages, and so on, suggested that the consumption of antioxidant natural products was beneficial to reduce the risk of CVD events [98]. Moreover, results from experimental studies showed that some natural products exerted cardiovascular-protective effects via different mechanisms of action, such as reducing blood pressure, improving the lipid profile, ameliorating oxidative stress, mitigating inflammation, and regulating gut microbiota. Furthermore, clinical trials confirmed that some antioxidant natural products could prevent and treat CVDs. Supported by current evidence, some antioxidant natural products and their active compounds could be developed into functional foods or medicine for the prevention and treatment of CVDs. In the future, the effects of more antioxidant natural products on CVDs should be evaluated to find out more cardiovascular-protective natural products, and relative bioactive components should be isolated and identified. In addition, the mechanisms of action should be elucidated further. Furthermore, special attention should be paid to the safety of relative natural products and functional foods [99].

1.3.3 Preventive Effects of Natural Antioxidants in Lipotoxicity and Obesity

Food and diet play a key role in health and disease. In recent decades, the severity of the global epidemic of obesity has increased substantially. In fact, in 2016, more than 39% of adults were overweight, and 13% of these were obese. Additionally, the number of obese children has increased 10-fold in recent decades to a total estimated number of 124 million [100]. Obesity is generally associated with energy imbalance as well as lipogenesis, resulting in increased adipose tissue. Overweight adults are at major risk for type 2 diabetes, cardiovascular diseases, several types of cancers, osteoarthritis, and early death. Obesity-induced inflammatory stress also stimulates the release of cytokines, leading to tissue damage [53]. The illness occurs as the result of multiple factors, such as food consumption, general lifestyle, insufficient physical activity, metabolic factors, and genetic predispositions.

Functional foods represent potent dietary approaches to support weight management and protect against the metabolic consequences of obesity [100]. On the food market, these "healthy foods" have become extremely popular and will reach a market size of US$260 400 million by the end of 2025, showing a compound annual growth rate of 6.8% [100]. Thus, the development of functional foods has increased the number of studies on potential dietary or herbal products, either for their identification or to describe their mode of action at a molecular level.

One of the most prominent chemical groups in bioactives is polyphenols, which are found in various fruits, vegetables, and herbs. However, the preventative potential of phenolic compounds and other compounds is highly dependent on their bioavailability (absorption, transport, and metabolism). To elucidate their biological functioning, both in vitro and in vivo assays are indispensable for the identification and study of bioactive compounds in regard to lipid accumulation, adipogenesis, stimulation of thermogenesis, or tissue inflammation.

1.3.4 Protective Effect of Natural Antioxidants in Hepatoxicity

Several naturally derived substances showed a great antioxidant activity when tested on animals with acetaminophen (APAP)-induced liver toxicity. Kumari and Kakkar claimed in their study that lupeol, a naturally occurring triterpenoid derived from olive, mango, crataeva, strawberry, and fig, reduced oxidative damage by scavenging free radicals and prevented the alteration in the antioxidant defense, inhibited depolarization of the mitochondria, prevented downregulation of Bcl-2, upregulation of Bax, and activation of caspases, and prevented DNA damage and cell death in rats [101]. Honey is another natural substance derived from the floral nectar that prevented an increase in the serum levels of hepatic enzyme markers and reduced both oxidative stress and inflammatory cytokines, thus confirming hepatoprotection [102]. Different acids derived from natural sources also have a great antioxidant activity when tested against APAP-induced hepatotoxicity. Arjunolic acid (AA) found in the bark of Terminalia arjuna potentially inhibited P450-mediated APAP bio-activation, and c-Jun N-terminal kinase (JNK)-mediated activation of mitochondrial permeabilization [103]. α-Lipoic acid, another naturally derived compound, is found to have hepatoprotective activity at a dose of 20 or 100 mg/kg [104]. Caffeic acid generally found in coffee decreases Kcap 1 expression, inhibits binding of Kcap 1 to nuclear factor erythroid 2-related factor 2 (Nrf2), which activates Nrf2, increases expression of anti-oxidative signals including heme oxygenase-1 (HO-1), NAD(P)H quinone dehydrogenase 1 (NQO1), thus protecting the liver from APAP-induced toxicity [105].

1.3.5 Preventive Role of Natural Antioxidants in Cancer

Antioxidants are being credited for their ability to protect cells from the oxidative/electrophilic damage that makes them turn into cancerous agents. A number of antioxidants have shown to inhibit the induction of cancer by a wide variety of chemical carcinogens and/or radiation at many target sites in mice, rats, and hamsters [106]. Epidemiological studies suggest that a diet rich in plant products containing natural antioxidants may be a deterrent to carcinogenicity. Many antioxidants were tested to determine if they would inhibit tumor initiation, promotion, and/or progression. Important antioxidants can be useful in the treatment of cancer, either as sole agents or as adjuncts to standard radiation and chemotherapy protocols [107]. The knowledge of antioxidants in a cancer-curing or preventing method is still in the early stages. The relationship between antioxidants and cancer prevention cannot be depicted only on the basis of the presumed mechanism of action when used. Numerous natural antioxidants appear to have beneficial health effects. There is sufficient evidence to recommend consuming food sources rich in antioxidants [108].

1.3.6 Ameliorative Role of Natural Antioxidants in Neurodegenerative Disorders

Some natural antioxidants have been used to enhance the antioxidant effects of pharmacology therapy. An experimental study reveals that the administration of ascorbic acid or rose oil can help decrease levels of oxidative damage to lipids or proteins induced by levodopa [95]. Studies show that the administration of vitamin E decreases the toxic effects of β-amyloid and improves cognitive development, decreases neuronal damage, and slows the progression of Alzheimer's disease [109]. Green tea epigallocatechin gallate (EGCG) esters have inhibitory properties of amyloidosis and β-amyloid production both in vitro and in vivo. Melatonin is another natural component that has been shown to have neuroprotective effects. In Parkinson's disease models, melatonin contributes to decreased dopamine production and decreases lipid peroxidation (LPO) and nitrites in the cytosol [110]. Melatonin has also been observed in clinical studies to improve sleep disorders in patients with Parkinson's disease, but not to improve motor symptoms [110].

1.3.7 Ameliorative Role of Natural Antioxidants in Inflammation and Arthritis

The World Health Organization (WHO) has estimated that 80% of the world's inhabitants utilized traditional medicine for their primary health care needs and the majority of this therapy requires the use of herbal extracts and their active components. Various medicinal plant bioactive extracts and their identified/isolated active constituents have shown a variety of medicinal pharmacological properties against various acute and chronic diseases/disorders [53]. Lack of antioxidant compounds in daily diet can lead to the development of degenerative diseases such as cancers, cardiovascular diseases, Alzheimer's disease, NDs, and various inflammatory illnesses [111, 112]. Incorporation of antioxidant compounds by consuming natural plant sources in the daily diet can be a suitable solution to solve human health issues. Numerous studies exhibited that flavonoids and phenolic content have contributed to the antioxidant activities of natural compounds. Besides having antioxidant activities, flavonoids and phenolic compounds also exert an effective role as anti-inflammatory factors. The anti-inflammatory activities of natural compounds have been reported in several studies and have been observed in numerous preclinical studies [113]. The findings from anti-inflammatory researches have proven that bioactive extracts and their natural compounds exert their biological properties by blocking two major signaling pathways such as NF-κB and mitogen-activated protein kinases (MAPKs), which have the main role in the production of various proinflammatory mediators.

Hypercholesterolemia is considered the main trigger for atherosclerosis. Therefore, the control of lipoprotein levels through the administration of statins is one of the main management alternatives to reduce the risk of atherosclerosis [114]. Statins antagonize the activity of the enzyme hydroxy-methylglutaryl-coenzyme A (HMG-CoA) reductase, decreasing the production of intracellular cholesterol and also decrease of liver LDLr [115]. Statins show pleiotropic effects on endothelial function, inhibition of thrombus gene activity, the stability of atherosclerosis plaques, and decreased inflammation and OS. Statins have been shown to have antioxidant effects on redox signaling of vascular and myocardial tissue by modifying NADPH oxidase activity. Statins show effects on endothelial nitric oxide synthase (eNOS) and decreased LPO [116]. Treatment of patients with simvastatin has protective effects on lipoprotein oxidation. However, the metabolism of statins generates ROS and produces toxicity in various tissues, including skeletal muscle and liver damage. The activity of simvastatin and lovastatin inhibits the complete II, III, IV, and V of the electron transport chain, whereas fluvastatin and cerivastatin only inhibit the V complex, thus causing mitochondrial dysfunction [117]. Eight weeks of simvastatin management is sufficient to cause mitochondrial respiration dysfunction in muscles [118].

1.3.8 Restorative Effect of Natural Antioxidants in the Gastrointestinal Disease

Normal physiological processes result in the generation of ROS such as peroxide, superoxide, hydroxyl radical, and peroxynitrite. The ROS occur normally in the body at very low concentrations (nanomolar to micromolar) [119]. The body needs them for survival but excess ROS might have deleterious effects because the gastrointestinal (GI) tract maintains an appropriate balance of the levels of ROS [120]. The enzyme SOD decreases the level of superoxide by catalyzing its conversion into molecular oxygen or hydrogen peroxide [121]. Similarly, catalase also catalyzes the degradation of hydrogen peroxide into oxygen and water. GPx decreases the levels of peroxides using GSH [122]. This requires maintenance of an appropriate ratio of GSH and its oxidized form (GSSG). Whereas γ-glutamylcysteine ligase and GSH synthase are pivotal for the GSH synthesis, nicotinamide adenine dinucleotide phosphate (NADPH)-dependent GSSG reductase is key to the conversion of GSSG to GSH. Ascorbate may also quench ROS and is converted into a transient unstable species and then into dehydroascorbate. Dehydroascorbate reductase uses GSH to recycle it back into ascorbate [119]. The reduced peroxiredoxin can also quench hydrogen peroxide and gets oxidized in the process. Thioredoxin can convert the peroxiredoxin back by oxidation of thioredoxin, which, in turn, is reduced by TRR and $NADPH^+$ [119]. The cysteine–cystine cycle also plays a key role in maintaining the appropriate balance of ROS [123].

1.3.9 Restorative Effect of Natural Antioxidants in Aging and Skin Health

Nutrition has been recognized to have an important impact on overall mortality and morbidity, and its role in extending life expectancy has been the object of extensive scientific research. Dietary supplementation with antioxidants has become more and more popular. However, their biochemical mechanisms of protection against oxidative stress and antiaging effects are not fully understood. Recently, several foodstuffs have been claimed as "antiaging," principally on the basis of their anti-inflammatory and antioxidative properties: berries, dark chocolate, beans, fish, vegetables, nuts, whole grains, garlic, and avocados [124].

Most recently, Bacalini et al. discussed the potential impact of the so-called "epigenetic diet" on age-related diseases, focusing on cardiovascular disease, highlighting the involvement of epigenetic modifications rather than DNA methylation, such as microRNA [125]. Epigenetic modifications may delay the aging process and impact diverse health benefits by activating numerous intracellular pathways.

While the beneficial effects of antioxidant supplements seem undoubtful in cases of antioxidant deficiencies, additional studies are warranted in order to design adapted prescriptions in antioxidant vitamins and minerals for healthy persons.

1.3.10 Role of Natural Antioxidants in Wound Healing

There is great consumer interest these days in "natural products," which include compounds derived from fruits, plants, and herbs. Indeed, these contain a rich diversity of antioxidants and are widely available and used for a multitude of cutaneous ailments [126]. Preclinical studies have shown their promise for use in wound healing because they are versatile as antioxidant and antimicrobial sources [127]. However, only two natural products, Medihoney and β-glucans, have been critically evaluated by the FDA, and only Medihoney has been evaluated specifically for wound healing. We review a few of the natural products that have been studied for wound healing that are also available over the counter.

1.3.10.1 β-glucans

β-glucans are glucose polymers found in the cell wall of yeast, fungi, and cereal plants that have previously been approved by the FDA for use against coronary heart disease. It is currently promoted as a dietary oral supplement and in topical creams and lotions for antiaging purposes and wound healing. β-glucans are proposed to have antioxidant properties and have been studied in rats for their role in burn wound healing [128].

1.3.10.2 Epigallocatechin Gallate

EGCG belongs to a group known as catechins, naturally occurring polyphenolic compounds, and is an important component of green tea. EGCG is available as an oral over-the-counter supplement and as a topical ointment marketed as Veregen.

1.3.10.3 Honey

Honey, in general, contains a rich combination of antioxidants [129] with a mixture of osmotically active agents and antibacterial components. A Cochrane evidence-based review of honey concluded that honey may be beneficial to acute wounds but not too long-standing wounds such as venous leg ulcers [130].

1.3.10.4 Proanthocyanidins

Proanthocyanidins are polyphenolic bioflavonoids that are found in many common foods including berries, grapes, apples, chocolate, and cinnamon spices [131]. The most potent sources were noted to be grape seeds and cinnamon. Proanthocyanidins are available over the counter as health supplements in powder and capsule formulations. Grape seed proanthocyanidin extract (GSPE) has been said to have redox-active properties. GSPE treatment applied topically was assessed in mice with full-thickness excisional wounds left to heal by secondary intention [132].

1.4 Protective Mechanisms of Natural Antioxidants at the Molecular Level

The protective effects of natural antioxidants have received more attention against free radical induced toxicities [133]. Flavonoids play an important role in the protection against oxidative stress, especially in the case of cancer. Flavonoids are found widely in vegetables, red wine, fruit, cocoas, and tea [90]. Flavonoids are widespread in beverages and foods, which possess a wide range of biological activities of which antioxidation has been extensively explored. Natural antioxidants strengthen the endogenous antioxidant defenses from ROS and restore the optimal balance by neutralizing reactive species [134]. The antioxidant activities of phenolics are related to a number of different mechanisms, such as free radical scavenging, hydrogen donation, singlet oxygen quenching, metal ion chelating, and acting as a substrate for radicals such as superoxide and hydroxyl. *Olea europaea* leaves and *Mentha piperita* were showed remarkable hepatic protection against hepatotoxic agents, which may be due to the antioxidant properties of these medicinal plants and herbs.

Therefore, humans exposed to hepatotoxic agents and patients with hepatic disorders should be advised to take these medicinal plants and herbs. Natural antioxidants have a variety of biochemical actions such as inhibition of the production of ROS and scavenging of free radicals [135]. Azab et al. reported that consumption of fenugreek, curcumin, peppermint, parsley, rosemary, garlic, pomegranate, sesame, and propolis was shown protective effects against renal diseases and nephrotoxic agents induced renal dysfunctions in experimental animals and humans [90]. The nephroprotective effect may be due to the inhibition of tissue lipid peroxidation and enhancement of antioxidant activity. Therefore, the study suggested that these antioxidants may be useful for persons who expose to nephrotoxic agents and patients who suffer from renal diseases. The protective effects may be due to the presence of benzoquinones, flavonoids, flavonol glycosides, alkaloids, carotenoids, catechols, glycosides, steroid glycosides, terpenoids, glycoalkaloids, monoterpenes, diterpenes, and triterpenes, saponins, polyphenols, and sterols in these medicinal plants and herbs [136].

1.4.1 Prevention of Lipid Peroxidation

Antioxidant compounds may scavenge ROS and peroxide radicals, thereby preventing or treating certain pathogenic conditions. Lipid peroxidation has been extensively used as a research model for identifying natural antioxidants as well as the studies of their mechanisms of action. Studies on antioxidants such as vitamins, polyphenols (green tea), flavones, and ginsenosides against free radical induced lipid peroxidation have been undertaken in several systems such as lipids, human red cells, human LDL, and rat liver microsomes in homogeneous solution or micelles. The antioxidant activity of these polyphenols depends significantly on the structure of the molecules, the initiation conditions and the microenvironment of the reaction medium.

Erythrocyte membranes are rich in polyunsaturated fatty acids, which are very susceptible to oxidative stress mediated by free radicals. Peroxidation of polyunsaturated lipids in red cell membranes cause quick damage and the membrane loses its integrity, leading to the release of hemoglobin (hemolysis) and intracellular Kions. When antioxidants such as curcumin are present in the system, peroxyl radicals can be converted to nonreactive species, and thereby the radical-induced lipid peroxidation and haemolysis can be inhibited.

1.4.2 Prevention of DNA Damage

The OH and ONOO radicals generated from nitric oxide and $O_2 \cdot$ can react directly with plasmid DNA macromolecules in vivo to cleave (or "nick") one DNA strand, causing oxidative DNA damage. Cell death and mutation caused by this DNA damage are implicated in neurodegenerative and cardiovascular diseases, cancer, and aging. Treating these conditions with antioxidants is of growing interest.

The degree of DNA damage is assessed using electrophoresis to separate the damaged and undamaged forms. Adding antioxidants such as selenium can inhibit ·OH-induced DNA damage. Thus, the antioxidant potential of certain compounds can be quantified and directly compared.

1.4.3 Prevention of Protein Modification

Besides lipid peroxidation and DNA damage, ROS also cause protein modification by nitration or chlorination of amino acids. Formations of various intermediate products due to the activity of ROS lead to the alteration of protein function or structure and, usually, the inhibition of enzyme activities. Proteins containing nitrotyrosine residues have been detected in different pathologic conditions, including diabetes, hypertension, and atherosclerosis, all associated with enhanced oxidative stress.

1.5 Scope of Natural Antioxidants in Therapeutic Drug Development

Natural products have been used as traditional medicines from time immemorial. Natural products and their derivatives have been invaluable as a source of therapeutic agents. The pharmacological, clinical, and chemical studies of these traditional medicines, predominantly derived from plants, established the basis for most of the early medicines like morphine, quinine, and pilocarpine [137]. Farnsworth and Morris reported that nearly 119 compounds derived from 90 plant species could be considered important drugs currently used in many countries, out of which 77% have been derivatives of plants used in traditional medicine [138]. In 1991, the analysis of bestselling pharmaceuticals revealed that half of them were either natural or their derivatives. Out of 87 approved drugs, 62% were of natural origin or modeled on natural product parents.

Infectious diseases are the second leading cause of death and there is an urgent need for the discovery of new drugs to combat them as well as new evolving drug-resistant pathogens. Numerous natural products have been identified that represent antibacterial agents displaying unique modes of action and properties. During 2001–2005, 23 naturally derived new drugs were introduced into the market for the treatment of diseases like fungal and bacterial infections, diabetes, cancer, dyslipidemia, AD, Gaucher's disease, atopic dermatitis, etc. [139].

Natural product drug discovery has several drawbacks when compared to synthetic chemical drug discovery [137]. These may include:

1) Natural products are often synthesized in small quantities and present as a mixture in extracts, due to which there is a requirement of labor-intensive and time-taking purification process.
2) There are high possibilities of rediscovering known compounds while screening natural product libraries. This is caused by a lack of dereplication (a process by which the chemical and biological characteristics of the unknown compounds are compared with chemical and biological characteristics of known compounds from databases to eliminate previously identified compounds) methodologies for both compounds in the natural product libraries and natural product sourcing.
3) Natural products are structurally complex and their modification using organic chemistry is challenging.

Although having such limitations, natural products are considered to be good drug candidates and hence technological advancements and the development of new methods have revolutionized the screening of natural products and offer a unique opportunity to re-establish natural products as the main source of drug candidates. The recent advancements may include the development of streamlined screening processes, improved natural product sourcing, and advancements in organic synthetic methods, combinatorial biosynthesis and microbial genomics. The past few years witnessed major developments also in areas like fermentation optimization, purification, and structure elucidation of natural products, which enabled much faster access to sufficient number of pure compounds.

Actinomycetes have been widely investigated and used as a major source of novel microbial metabolites but their potential seems boundless. To these, natural sources can be added to investigate the rational design of novel structure types within certain classes of microbial metabolites using genetic engineering [140].

Plants and microbes are well known to produce secondary metabolites, required for their survival and can be used as efficient drug candidates. Naturally occurring compounds like polyphenols in fruits, vegetables, beverages, and cereals are plant secondary metabolites and are generally involved in defense against aggression by pathogens and UV radiations [141]. It has been reported that long-term consumption of a polyphenol-rich diet offers protection against cancer, diabetes, cardiovascular diseases, NDs, and osteoporosis. It is well known that polyphenol-rich food and beverages may increase plasma antioxidant capacity.

Despite the success into natural products, the research has experienced a slight decline during the past decades. It is a lengthy process, usually 10 or more years starting from the initial discovery to the subsequent launch of a new drug [139]. This decrease in the discovery of natural candidates has been attributed to a number of factors which includes high throughput screening against defined molecular targets. Advances in molecular biology, genomics, and cellular biology led to the rise in the number of targets and encouraged shortened drug discovery timelines.

1.6 Conclusion

Archeological and modern genetic evidence suggested that human populations have migrated to the Indian subcontinent since the prehistoric period. The knowledge of the medicinal values of the plant and other substances and their uses go back to the time of the earliest settlers. The vast amount of medical knowledge that has come down to modern times is the result of long evolution through trial and error and the exchange of know-how between diverse communities and regions. India has seen an era of prosperity in ancient times, especially in the use of traditional knowledge in therapeutics. The use of "Gou-mutra" (cow urine), Swarna bhasma, and medicinal herbs are just a few to mention when we boast of India's richness in the ancient medical panorama. Our country has always been in a privileged position, sitting on a gold mine of well-recorded and well-practiced knowledge of traditional medicine in the form of the Vedas and folklore practice among the different tribes.

The process and assimilation continue and today traditional medical practices are obliged to accommodate the norms of modern biomedicine. However, there is a growing awareness among the scientific community and the general public about the intrinsic value of traditional medicine and, as a result, Ayurveda, Unani, and Siddha have entered the mainstream to complement biomedicine. The challenge is to integrate the best of the different healing traditions to meet the healthcare needs of society.

India is known for its traditional medicinal systems. Medical systems are found mentioned even in the ancient Vedas and other scriptures. The Ayurvedic concept appeared and developed between 2500 and 500 BCE in India. The literal meaning of Ayurveda is "science of life," because the ancient Indian system of health care focused on views of man and his illness. It has been pointed out that positive health means metabolically well-balanced human beings. Ayurveda is also called the "science of longevity" because it offers a complete system to live a long healthy life. It offers programs to rejuvenate the body through diet and nutrition. It offers treatment methods to cure many common diseases such as food allergies, which have few modern treatments. However, one should be aware that Ayurvedic nutrition is not a "magic bullet" system but requires the full participation of the patient to succeed. It is an interactive system that is user-friendly and educational. It teaches the patient to become responsible and self-empowered. Ayurveda is not a nutritional system for those seeking an escape or an excuse to further abuse their body or mind. It is a system for empowerment, a system of freedom, and long life. Food is the major source for serving nutritional needs, but with growing modernization, some traditional methods are being given up. Hence, modern food habits are affecting balanced nutrition. There is an ever-widening gap in nutrient intake due to which normal life is no longer normal. However, the affluence of the working population with changing lifestyles and reducing the affordability of sick care, in terms of time and money involved, are some of the forces that are presently driving people toward thinking about their wellness.

India sits in a privileged position in the use of traditional health resources as it covers the world's biodiversity hotspot in the northeastern states (Indo-Burma hot spot) and the Western Ghats. But, in contrast to this, developments made by India in the field of pharmaceutics are significantly less as compared to countries that are at a lower privileged position in terms of the traditional knowledge base. India has not been able to capitalize on this wealth by promoting its use in the developed world despite its renewed interest in traditional medicines. The forecast is that the global market for herbal products is expected to be $5 trillion by 2050. The total global herbal market is of size $62.0 billion. In this, India's contribution is only $1 billion. A major cause of this is the export of medicinal herbs to other countries at low revenue. However, these herbs are processed by them after proper research and sold in the Indian market at a higher price. Owing to this, India is actually not able to make proper use of its wealth. It is now high time that India should start planning strategies to overcome this situation. This can be achieved by judicious product identification based on diseases found in the developed world for which no medicine or only palliative therapy is available; such herbal medicines will find speedy access in those countries. Backward integration from market demands will pay rich dividends. In this regard, research institutes should encourage collaboration with industries to come up with the product of Indian herbal origin that could be marketed worldwide. Strategically, India should enter through those plant-based medicines which are already well accepted in Europe, the United States, and Japan. Recently, there has been a shift in the universal trend from synthetic to herbal medicine, which is termed as "Return to Nature" and our country definitely has the potential to make proper use of this situation owing to its resources.

In the northeastern states of India, a lot of these traditional medicinal practices still continue mostly among the elderly crowd. We, being in the northeast, should make an immediate effort to make proper documentation and database preparation of this knowledge before it is lost. This knowledge can then be put into proper Research and Development to finally come up as an Indian-origin drug that could be marketed throughout the world. This can have a great impact on the development of our country and be a herbal and safe contribution to world therapeutics.

References

1 Anwar, H., Hussain, G., and Mustafa, I. (2018). Antioxidants from natural sources. In: *Antioxidants Foods Its Applications*. IntechOpen https://doi.org/10.5772/intechopen.75961.

2 Cook, N.C. and Samman, S. (1996). Flavonoids – chemistry, metabolism, cardioprotective effects, and dietary sources. *J. Nutr. Biochem.* 7: 66–76. https://doi.org/10.1016/S0955-2863(95)00168-9.

3 Bravo, L., Sources, D., and Significance, N. (1998). Polyphenols: chemistry, dietary sources, metabolism, and nutritional significance. *Nutr. Rev.* 56: 317–333. https://doi.org/10.1111/j.1753-4887.1998.tb01670.x.

4 Someya, S., Yoshiki, Y., and Okubo, K. (2002). Antioxidant compounds from bananas (*Musa cavendish*). *Food Chem.* 79: 351–354. https://doi.org/10.1016/S0308-8146(02)00186-3.

5 Heim, K.E., Tagliaferro, A.R., and Bobilya, D.J. (2002). Flavonoid antioxidants: chemistry, metabolism and structure–activity relationships. *J. Nutr. Biochem.* https://doi.org/10.1016/S0955-2863(02)00208-5.

6 Middleton, E., Kandaswami, C., and Theoharides, T.C. (2000). The effects of plant flavonoids on mammalian cells: implications for inflammation, heart disease, and cancer. *Pharmacol. Rev.* 52: 673–751.

7 Havsteen, B.H. (2002). The biochemistry and medical significance of the flavonoids. *Pharmacol. Ther.* 96: 67–202. https://doi.org/10.1016/S0163-7258(02)00298-X.

8 Korkina, L.G. and Afanas'Ev, I.B. (1996). Antioxidant and chelating properties of flavonoids. *Adv. Pharmacol.* 38: 151–163. https://doi.org/10.1016/S1054-3589(08)60983-7.

9 Shui, G. and Leong, L.P. (2006). Residue from star fruit as valuable source for functional food ingredients and antioxidant nutraceuticals. *Food Chem.* 97: 277–284. https://doi.org/10.1016/j.foodchem.2005.03.048.

10 Croft, K.D. (1998). The chemistry and biological effects of flavonoids and phenolic acids. *Ann. N. Y. Acad. Sci.* 854: 435–442.

11 He, M.L., Sharma, R., Mir, P.S. et al. (2010). Feed withdrawal abate regimens lipodystrophy and metabolic syndrome symptoms, such as glucose tolerance, are associated with the diameter of retroperitoneal adipocytes in rats. *Nutr. Res.* 30: 125–133. https://doi.org/10.1016/j.nutres.2009.09.009.

12 De Almeida, E., Bernardes, E., Fonseca, H., and Amaya-farfán, J. (2013). Cholesterol reducing and bile-acid binding properties of taioba (*Xanthosoma sagittifolium*) leaf in rats fed a high-fat diet ☆, ☆☆. *Food Res. Int.* 51: 886–891. https://doi.org/10.1016/j.foodres.2013.02.017.

13 Lewis, D.A. and Shaw, G.P. (2001). A natural flavonoid and synthetic analogues protect the gastric mucosa from aspirin-induced erosions. *J. Nutr. Biochem.* 12: 95–100. https://doi.org/10.1016/S0955-2863(00)00133-9.

14 Davey, M.W., Keulemans, J., and Swennen, R. (2006). Methods for the efficient quantification of fruit provitamin A contents. *J. Chromatogr. A* 1136: 176–184. https://doi.org/10.1016/j.chroma.2006.09.077.

15 West, T., Atzeva, M., and Holtzman, D.M. (2007). Pomegranate polyphenols and resveratrol protect the neonatal brain against hypoxic-ischemic injury. *Dev. Neurosci.* https://doi.org/10.1159/000105477.

16 Kandasamy, N. and Ashokkumar, N. (2014). Protective effect of bioflavonoid myricetin enhances carbohydrate metabolic enzymes and insulin signaling molecules in streptozotocin–cadmium induced diabetic nephrotoxic rats. *Toxicol. Appl. Pharmacol.* https://doi.org/10.1016/j.taap.2014.05.014.

17 Mushtaq, M. and Wani, S.M. (2013). Polyphenols and human health – a review. *Int. J. Pharma Bio Sci.* 42: B338–B360.

18 Devasagayam, T.P.A., Tilak, J.C., Boloor, K.K. et al. (2004). Free radicals and antioxidants in human health: current status and future prospects. *J. Assoc. Phys. India* 52: 794–804.

19 Cornelli, U. (2009). Antioxidant use in nutraceuticals. *Clin. Dermatol.* https://doi.org/10.1016/j.clindermatol.2008.01.010.

20 Barba, C.V.C. and Cabrera, M.I.Z. (2008). Recommended energy and nutrient intakes for Filipinos 2002. *Asia Pac. J. Clin. Nutr.* 17: 399–404.

21 Ahmadinejad, F., Møller, S.G., Hashemzadeh-Chaleshtori, M. et al. (2017). Molecular mechanisms behind free radical scavengers function against oxidative stress. *Antioxidants* https://doi.org/10.3390/antiox6030051.

22 Youdim, K.A. and Joseph, J.A. (2001). A possible emerging role of phytochemicals in improving age-related neurological dysfunctions: a multiplicity of effects. *Free Radic. Biol. Med.* https://doi.org/10.1016/S0891-5849(00)00510-4.

23 Bates, C.J., Benton, D., Biesalski, H.K. et al. (2002). Nutrition and aging: a consensus statement. *J. Nutr. Health Aging* 6: 103–116.

24 Chhabra, S., Bakshi, A., and Kaur, R. (2021). Role of nutraceuticals in health promotion and disease prevention: a review in the Indian context. *Curr. Top. Nutraceutical Res.* 19: 139–145. https://doi.org/10.37290/ctnr2641-452x.19:139-145.

25 Jain, N. and Ramawat, K.G. (2013). Nutraceuticals and antioxidants in prevention of diseases. In: *Natural Products: Phytochemistry, Botany and Metabolism of Alkaloids, Phenolics and Terpenes* (ed. K.G. Ramawat and J.M. Merillon). Heidelberg; New York: Springer https://doi.org/10.1007/978-3-642-22144-6_70.

26 Liguori, I., Russo, G., Curcio, F. et al. (2018). Oxidative stress, aging, and diseases. *Clin. Interv. Aging* https://doi.org/10.2147/CIA.S158513.

27 Hazarika, A., Kalita, H., Kalita, M.C., and Devi, R. (2017). Withdrawal from high-carbohydrate, high-saturated-fat diet changes saturated fat distribution and improves hepatic low-density-lipoprotein receptor expression to ameliorate metabolic syndrome in rats. *Nutrition* 38: 95–101. https://doi.org/10.1016/j.nut.2017.01.005.

28 Burton, D.G.A., Matsubara, H., and Ikeda, K. (2010). Pathophysiology of vascular calcification: pivotal role of cellular senescence in vascular smooth muscle cells. *Exp. Gerontol.* https://doi.org/10.1016/j.exger.2010.07.005.

29 Hazarika, A., Sc, M., Kalita, H. et al. (2016). Pathophysiology of metabolic syndrome: the onset of natural recovery on withdrawal of a high-carbohydrate, high-fat diet. *Nutrition* 32: 1081–1091. https://doi.org/10.1016/j.nut.2016.03.005.

30 Laberge, R.M., Awad, P., Campisi, J., and Desprez, P.Y. (2012). Epithelial–mesenchymal transition induced by senescent fibroblasts. *Cancer Microenviron.* https://doi.org/10.1007/s12307-011-0069-4.

31 Fuente, M. and Miquel, J. (2009). An update of the oxidation–inflammation theory of aging: the involvement of the immune system in oxi-inflamm-aging. *Curr. Pharm. Des.* https://doi.org/10.2174/138161209789058110.

32 Skarfors, E.T., Selinus, K.I., and Lithell, H.O. (1991). Risk factors for developing non-insulin dependent diabetes: a 10 year follow up of men in Uppsala. *BMJ* 303: 755–760.

33 Manson, J.E., Nathan, D.M., Krolewski, A.S. et al. (1992). A prospective-study of exercise and incidence of diabetes among United-States male physicians. *JAMA-J. Am. Med. Assoc.* 268: 63–67. https://doi.org/10.1001/jama.268.1.63.

34 Hebebrand, J. and Hinney, A. (2009). Environmental and genetic risk factors in obesity. *Child Adolesc. Psychiatr. Clin. N. Am.* 18: 83–94. https://doi.org/10.1016/j.chc.2008.07.006.

35 Akinyanju, P.A., Qureshi, R.U., Salter, A.J., and Yudkin, J.S. (1968). Effect of an "atherogenic" diet containing starch or sucrose on the blood lipids of young men. *Nature* 218: 975–977. https://doi.org/10.1038/218975a0.

36 Dahimi, O., Hassan, M.S., Rahimm, A.A. et al. (2014). Differentiation of lard from other edible fats by gas chromatography-flame ionisation detector (GC-FID) and chemometrics. *J. Food Pharm. Sci.* 2: 27–31. https://doi.org/10.14499/jfps.

37 Singh, R.B., Rastogi, S.S., Sharma, V.K. et al. (1990). Can dietary magnesium modulate lipoprotein metabolism? *Magnes Trace Elem.* 9: 255–264.

38 Meisinger, C., Thorand, B., Schneider, A. et al. (2002). Sex differences in risk factors for incident type 2 diabetes mellitus: the MONICA Augsburg cohort study. *Arch. Intern. Med.* 162: 82–89.

39 Knowler, W.C., Pettitt, D.J., Saad, M.F. et al. (1991). Obesity in the Pima Indians: its magnitude and relationship with diabetes. *Am. J. Clin. Nutr.* 53: 1543S–1551S.

40 Almdal, T., Scharling, H., Jensen, J.S., and Vestergaard, H. (2008). Higher prevalence of risk factors for type 2 diabetes mellitus and subsequent higher incidence in men. *Eur. J. Intern. Med.* 19: 40–45. https://doi.org/10.1016/j.ejim.2007.05.011.

41 Rask-Madsen, C. and Kahn, C.R. (2012). Tissue-specific insulin signaling, metabolic syndrome, and cardiovascular disease. *Arterioscler. Thromb. Vasc. Biol.* 32: 2052–2059. https://doi.org/10.1161/ATVBAHA.111.241919.

42 Mbanya, J.C.N., Motala, A.A., Sobngwi, E. et al. (2010). Diabetes in sub-Saharan Africa. *Lancet* 375: 2254–2266. https://doi.org/10.1016/S0140-6736(10)60550-8.

43 Lebovitz, H.E. (2002). Treating hyperglycemia in type 2 diabetes: new goals and strategies. *Cleve. Clin. J. Med.* 69: 809–820. https://doi.org/10.3949/ccjm.69.10.809.

44 Kawahito, S., Kitahata, H., and Oshita, S. (2009). Problems associated with glucose toxicity: role of hyperglycemia-induced oxidative stress. *World J. Gastroenterol.* 15: 4137–4142. https://doi.org/10.3748/wjg.15.4137.

45 Anderson, P.J., Critchley, J.A., Chan, J.C. et al. (2001). Factor analysis of the metabolic syndrome: obesity vs insulin resistance as the central abnormality. *Int. J. Obes. Relat. Metab. Disord.* 25: 1782–1788. https://doi.org/10.1038/sj.ijo.0801837.

46 Carr, D.B., Utzschneider, K.M., Hull, R.L. et al. (2004). Intra-abdominal fat is a major determinant of the National Cholesterol Education Program Adult Treatment Panel III criteria for the metabolic syndrome. *Diabetes* 53: 2087–2094. https://doi.org/10.2337/diabetes.53.8.2087.

47 Sharma, B., Salunke, R., Satapati, S.K. et al. (2011). Screening of some indian medicinal plant extracts for their antihyperglycemic activities in streptozotocin-induced diabetic mice. *J. Food Biochem.* 35: 1398–1406. https://doi.org/10.1111/j.1745-4514.2010.00458.x.

48 Fujimoto, M., Shimizu, N., Kunii, K. et al. (2005). A role for iNOS in fasting hyperglycemia and impaired insulin signaling in the liver of obese diabetic mice. *Diabetes* 54: 1340–1348. https://doi.org/10.2337/diabetes.54.5.1340.

49 Lee, J., Bok, S., Jeon, S. et al. (2010). Antihyperlipidemic effects of buckwheat leaf and flower in rats fed a high-fat diet. *Food Chem.* 119: 235–240. https://doi.org/10.1016/j.foodchem.2009.06.014.

50 Kalita, H., Boruah, D.C., Deori, M. et al. (2016). Antidiabetic and antilipidemic effect of *Musa balbisiana* root extract: a potent agent for glucose homeostasis in streptozotocin-induced diabetic rat. *Front. Pharmacol.* 7: 1–11. https://doi.org/10.3389/fphar.2016.00102.

51 Kolka, C.M., Castro, A.V.B., Kirkman, E.L., and Bergman, R.N. (2015). Modest hyperglycemia prevents interstitial dispersion of insulin in skeletal muscle. *Metabolism* 64: 330–337. https://doi.org/10.1016/j.metabol.2014.10.036.

52 Alzaid, F., Cheung, H.-M., Preedy, V., and Sharp, P. (2013). Regulation of glucose transporter expression in human intestinal Caco-2 cells following exposure to an anthocyanin-rich berry extract. *PLoS One* 8: e78932. https://doi.org/10.1371/journal.pone.0078932.

53 Chiarugi, M., Galatioto, C., Panicucci, S. et al. (2016). Nutrition in clinical practice. *Nat. Publ. Gr.* 15: 1–8. https://doi.org/10.1001/jama.2010.534.

54 Aryangat, A.V. and Gerich, J.E. (2010). Type 2 diabetes: postprandial hyperglycemia and increased cardiovascular risk. *Vasc. Health Risk Manag.* 6: 145–155. https://doi.org/10.2147/VHRM.S8216.

55 Dembinska-Kiec, A., Mykkänen, O., Kiec-Wilk, B., and Mykkänen, H. (2008). Antioxidant phytochemicals against type 2 diabetes. *Br. J. Nutr.* 99: ES109–ES117. https://doi.org/10.1017/S000711450896579X.

56 Wang, M., Avula, B., Wang, Y. et al. (2014). An integrated approach utilising chemometrics and GC/MS for classification of chamomile flowers, essential oils and commercial products. *Food Chem.* 152: 391–398. https://doi.org/10.1016/j.foodchem.2013.11.118.

57 Kim, Y.M., Jeong, Y.K., Wang, M.H. et al. (2005). Inhibitory effect of pine extract on α-glucosidase activity and postprandial hyperglycemia. *Nutrition* 21: 756–761. https://doi.org/10.1016/j.nut.2004.10.014.

58 Panchal, S.K., Poudyal, H., Iyer, A. et al. (2011). High-carbohydrate, high-fat diet-induced metabolic syndrome and cardiovascular remodeling in rats. *J. Cardiovasc. Pharmacol.* 57: 611–624. https://doi.org/10.1097/FJC.0b013e31821b1379.

59 Gutin, B., Barbeau, P., Owens, S. et al. (2002). Effects of exercise intensity on cardiovascular fitness, total body composition, and visceral adiposity of obese adolescents. *Am. J. Clin. Nutr.* 75: 818–826.

60 Al-Sabaawy, O.M. (2012). The relationship between serum lipid profile and selected trace elements for adult men in Mosul City. *Oman Med. J.* 27: 300–303. https://doi.org/10.5001/omj.2012.74.

61 Mittal, M., Siddiqui, M.R., Tran, K. et al. (2014). Reactive oxygen species in inflammation and tissue injury. *Antioxid. Redox Signal.* 20: 1126–1167. https://doi.org/10.1089/ars.2012.5149.

62 Tappy, L. and Le, K.A. (2010). Metabolic effects of fructose and the worldwide increase in obesity. *Physiol. Rev.* 90: 23–46. https://doi.org/10.1152/physrev.00019.2009.

63 Hwang, I.S., Huang, W.C., Wu, J.N. et al. (1989). Effect of fructose-induced hypertension on the renin–angiotensin–aldosterone system and atrial natriuretic factor. *Am. J. Hypertens.* 2: 424–427.

64 Miller, A.W., Katakam, P.V.G., and Ujhelyi, M.R. (1999). Impaired endothelium-mediated relaxation in coronary arteries from insulin-resistant rats. *J. Vasc. Res.* 36: 385–392. https://doi.org/10.1159/000025678.

65 Rayssiguier, Y., Gueux, E., Nowacki, W. et al. (2006). High fructose consumption combined with low dietary magnesium intake may increase the incidence of the metabolic syndrome by inducing inflammation. *Magnes. Res.* 19: 237–243. https://doi.org/10.1684/mrh.2006.0068.

66 Rioux, V., Lemarchal, P., and Legrand, P. (2000). Myristic acid, unlike palmitic acid, is rapidly metabolized in cultured rat hepatocytes. *J. Nutr. Biochem.* 11: 198–207. https://doi.org/10.1016/S0955-2863(00)00065-6.

67 Kumari, M., Head, J., and Marmot, M. (2004). Prospective study of social and other risk factors for incidence of type 2 diabetes in the Whitehall II study. *Arch. Intern. Med.* 164: 1873–1880. https://doi.org/10.1001/archinte.164.17.1873.

68 Mathers, C.D., Loncar, D., Van, H.L. et al. (2006). Projections of global mortality and burden of disease from 2002 to 2030. *PLoS Med.* 3: e442. https://doi.org/10.1371/journal.pmed.0030442.

69 Blake, G.J., Rifai, N., Buring, J.E., and Ridker, P.M. (2003). Blood pressure, C-reactive protein, and risk of future cardiovascular events. *Circulation* 108: 2993–2999. https://doi.org/10.1161/01.CIR.0000104566.10178.AF.

70 Khansari, N., Shakiba, Y., and Mahmoudi, M. (2009). Chronic inflammation and oxidative stress as a major cause of age-related diseases and cancer. *Recent Pat. Inflamm. Allergy Drug Discov.* https://doi.org/10.2174/187221309787158371.

71 Federico, A., Morgillo, F., Tuccillo, C. et al. (2007). Chronic inflammation and oxidative stress in human carcinogenesis. *Int. J. Cancer* https://doi.org/10.1002/ijc.23192.

72 Marnett, L.J. (2000). Oxyradicals and DNA damage. *Carcinogenesis* https://doi.org/10.1093/carcin/21.3.361.

73 Olinski, R., Siomek, A., Rozalski, R. et al. (2007). Oxidative damage to DNA and antioxidant status in aging and age-related diseases. *Acta Biochim. Pol.* https://doi.org/10.18388/abp.2007_3265.

74 Kinnula, V.L. and Crapo, J.D. (2003). Superoxide dismutases in the lung and human lung diseases. *Am. J. Respir. Crit. Care Med.* https://doi.org/10.1164/rccm.200212-1479SO.

75 Kinnula, V.L. (2005). Production and degradation of oxygen metabolites during inflammatory states in the human lung. *Curr. Drug Targets Inflamm. Allergy* https://doi.org/10.2174/1568010054526368.

76 Zelko, I.N., Mariani, T.J., and Folz, R.J. (2002). Superoxide dismutase multigene family: a comparison of the CuZn-SOD (SOD1), Mn-SOD (SOD2), and EC-SOD (SOD3) gene structures, evolution, and expression. *Free Radic. Biol. Med.* https://doi.org/10.1016/S0891-5849(02)00905-X.

77 Zhang, K., Chen, C., Liu, Y. et al. (2014). Cellular senescence occurred widespread to multiple selective sites in the fetal tissues and organs of mice. *Clin. Exp. Pharmacol. Physiol.* 41: 965–975. https://doi.org/10.1111/1440-1681.12328.

78 Dubuisson, M., Vander Stricht, D., Clippe, A. et al. (2004). Human peroxiredoxin 5 is a peroxynitrite reductase. *FEBS Lett.* https://doi.org/10.1016/j.febslet.2004.06.080.

79 Holmgren, A. (2000). Antioxidant function of thioredoxin and glutaredoxin systems. *Antioxid. Redox Signal.* https://doi.org/10.1089/ars.2000.2.4-811.

80 Birben, E., Sahiner, U.M., Sackesen, C. et al. (2012). Oxidative stress and antioxidant defense. *World Allergy Organ J.* https://doi.org/10.1097/WOX.0b013e3182439613.

81 Mezzetti, A., Lapenna, D., Romano, F. et al. (1996). Systemic oxidative stress and its relationship with age and illness. *J. Am. Geriatr. Soc.* https://doi.org/10.1111/j.1532-5415.1996.tb03741.x.

82 Brigelius-Flohé, R. and Maiorino, M. (2013). Glutathione peroxidases. *Biochim. Biophys. Acta Gen. Subj.* https://doi.org/10.1016/j.bbagen.2012.11.020.

83 Arthur, J.R. (2000). The glutathione peroxidases. *Cell. Mol. Life Sci.* https://doi.org/10.1007/pl00000664.

84 El-Agamey, A., Lowe, G.M., McGarvey, D.J. et al. (2004). Carotenoid radical chemistry and antioxidant/pro-oxidant properties. *Arch. Biochem. Biophys.* https://doi.org/10.1016/j.abb.2004.03.007.

85 Rice-Evans, C.A., Sampson, J., Bramley, P.M., and Holloway, D.E. (1997). Why do we expect carotenoids to be antioxidants in vivo? *Free Radic. Res.* https://doi.org/10.3109/10715769709097818.

86 Duarte, T.L. and Lunec, J. (2005). Review: when is an antioxidant not an antioxidant? A review of novel actions and reactions of vitamin C. *Free Radic. Res.* https://doi.org/10.1080/10715760500104025.

87 Harwell, B. (2007). Biochemistry of oxidative stress. *Biochem. Soc. Trans.* https://doi.org/10.1042/BST0351147.

88 Sreerama, Y.N., Takahashi, Y., and Yamaki, K. (2012). Phenolic antioxidants in some vigna species of legumes and their distinct inhibitory effects on α-glucosidase and pancreatic lipase activities. *J. Food Sci.* 77: 927–933. https://doi.org/10.1111/j.1750-3841.2012.02848.x.

89 Cheung, C.C.C., Zheng, G.J., Li, A.M.Y. et al. (2001). Relationships between tissue concentrations of polycyclic aromatic hydrocarbons and antioxidative responses of marine mussels, *Perna viridis*. *Aquat. Toxicol.* https://doi.org/10.1016/S0166-445X(00)00145-4.

90 Azab, A.E., Adwas, A.A., Elsayed, A.S.I. et al. (2019). Oxidative stress and antioxidant mechanisms in human body. *J. Appl. Biotechnol. Bioeng.* https://doi.org/10.15406/jabb.2019.06.00173.

91 Pigeolet, E., Corbisier, P., Houbion, A. et al. (1990). Glutathione peroxidase, superoxide dismutase, and catalase inactivation by peroxides and oxygen derived free radicals. *Mech. Ageing Dev.* https://doi.org/10.1016/0047-6374(90)90078-T.

92 Sharma, R., Yang, Y., Sharma, A. et al. (2004). Antioxidant role of glutathione *S*-transferases: protection against oxidant toxicity and regulation of stress-mediated apoptosis. *Antioxid. Redox Signal.* https://doi.org/10.1089/152308604322899350.

93 Ulusu, N.N. and Tandoğan, B. (2007). Purification and kinetic properties of glutathione reductase from bovine liver. *Mol. Cell. Biochem.* https://doi.org/10.1007/s11010-007-9454-1.

94 Ceriello, A. and Testa, R. (2009). Antioxidant anti-inflammatory treatment in type 2 diabetes. *Diabetes Care* https://doi.org/10.2337/dc09-s316.

95 García-Sánchez, A., Miranda-Díaz, A.G., and Cardona-Muñoz, E.G. (2020). The role of oxidative stress in physiopathology and pharmacological treatment with pro- and antioxidant properties in chronic diseases. *Oxid. Med. Cell. Longev.* https://doi.org/10.1155/2020/2082145.

96 Akbar, S., Bellary, S., and Griffiths, H.R. (2011). Dietary antioxidant interventions in type 2 diabetes patients: a meta-analysis. *Br. J. Diabetes Vasc. Dis.* https://doi.org/10.1177/1474651411407558.

97 Santos, V.R., Lima, J.A., De Mendonça, A.C. et al. (2009). Effectiveness of full-mouth and partial-mouth scaling and root planing in treating chronic periodontitis in subjects with type 2 diabetes. *J. Periodontol.* https://doi.org/10.1902/jop.2009.090030.

98 Zhou, D.D., Luo, M., Shang, A. et al. (2021). Antioxidant food components for the prevention and treatment of cardiovascular diseases: effects, mechanisms, and clinical studies. *Oxidative Med. Cell. Longev.* https://doi.org/10.1155/2021/6627355.

99 Miller, V., Mente, A., Dehghan, M. et al. (2017). Fruit, vegetable, and legume intake, and cardiovascular disease and deaths in 18 countries (PURE): a prospective cohort study. *Lancet* https://doi.org/10.1016/S0140-6736(17)32253-5.

100 Leporini, M., Loizzo, M.R., Sicari, V. et al. (2020). Citrus × clementina hort. Juice enriched with its by-products (peels and leaves): chemical composition, in vitro bioactivity, and impact of processing. *Antioxidants* 9: 1–32. https://doi.org/10.3390/antiox9040298.

101 Kumari, A. and Kakkar, P. (2012). Lupeol prevents acetaminophen-induced in vivo hepatotoxicity by altering the Bax/Bcl-2 and oxidative stress-mediated mitochondrial signaling cascade. *Life Sci.* https://doi.org/10.1016/j.lfs.2012.01.012.

102 Galal, R.M., Zaki, H.F., El-Nasr, M.M.S., and Agha, A.M. (2012). Potential protective effect of honey against paracetamol-induced hepatotoxicity. *Arch. Iran Med.* 15: 674–680.

103 De, D., Chatterjee, K., Ali, K.M. et al. (2010). Antidiabetic and antioxidative effects of hydro-methanolic extract of sepals of *Salmalia malabarica* in streptozotocin induced diabetic rats. *J. Appl. Biomed.* 8: 23–33. https://doi.org/10.2478/v10136-009-0004-x.

104 Elshazly, S.M., El-Moselhy, M.A., and Barakat, W. (2014). Insights in the mechanism underlying the protective effect of α-lipoic acid against acetaminophen-hepatotoxicity. *Eur. J. Pharmacol.* https://doi.org/10.1016/j.ejphar.2014.01.042.

105 Pang, C., Zheng, Z., Shi, L. et al. (2016). Caffeic acid prevents acetaminophen-induced liver injury by activating the Keap1-Nrf2 antioxidative defense system. *Free Radic. Biol. Med.* https://doi.org/10.1016/j.freeradbiomed.2015.12.024.

106 Piskounova E, Agathocleous M, Murphy MM, et al. Oxidative stress inhibits distant metastasis by human melanoma cells. *Nature* 2015; 527:186–191. https://doi.org/10.1038/nature15726.

107 Enwonwu, C.O. and Meeks, V.I. (1995). Bionutrition and oral cancer in humans. *Crit. Rev. Oral Biol. Med.* https://doi.org/10.1177/10454411950060010401.

108 Omenn, G.S., Goodman, G., Thornquist, M. et al. (1994). The β-carotene and retinol efficacy trial (CARET) for chemoprevention of lung cancer in high risk populations: smokers and asbestos-exposed workers. *Cancer Res.* https://doi.org/10.1016/0169-5002(94)92177-6.

109 Montiel, T., Quiroz-Baez, R., Massieu, L., and Arias, C. (2006). Role of oxidative stress on β-amyloid neurotoxicity elicited during impairment of energy metabolism in the hippocampus: protection by antioxidants. *Exp. Neurol.* https://doi.org/10.1016/j.expneurol.2006.02.126.

110 Tapias, V., Escames, G., López, L.C. et al. (2009). Melatonin and its brain metabolite *N*1-acetyl-5-methoxykynuramine prevent mitochondrial nitric oxide synthase induction in Parkinsonian mice. *J. Neurosci. Res.* https://doi.org/10.1002/jnr.22123.

111 Krishnaiah, D., Sarbatly, R., and Nithyanandam, R. (2011). A review of the antioxidant potential of medicinal plant species. *Food Bioprod. Process.* https://doi.org/10.1016/j.fbp.2010.04.008.

112 Di Matteo, V. and Esposito, E. (2003). Biochemical and therapeutic effects of antioxidants in the treatment of Alzheimer's disease, Parkinson's disease, and amyotrophic lateral sclerosis. *Curr. Drug Targets CNS Neurol. Disord.* https://doi.org/10.2174/1568007033482959.

113 Ravipati, A.S., Zhang, L., Koyyalamudi, S.R. et al. (2012). Antioxidant and anti-inflammatory activities of selected Chinese medicinal plants and their relation with antioxidant content. *BMC Complement. Altern. Med.* https://doi.org/10.1186/1472-6882-12-173.

114 Singh, N., Baby, D., Rajguru, J. et al. (2019). Inflammation and cancer. *Ann. Afric. Med.* https://doi.org/10.4103/aam.aam_56_18.

115 Hussain, S.P., Hofseth, L.J., and Harris, C.C. (2003). Radical causes of cancer. *Nat. Rev. Cancer* https://doi.org/10.1038/nrc1046.

116 Goszcz, K., Deakin, S.J., Duthie, G.G. et al. (2015). Antioxidants in cardiovascular therapy: panacea or false hope? *Front. Cardiovasc. Med.* 2: 1–22. https://doi.org/10.3389/fcvm.2015.00029.

117 Ling, J., Wei, B., Lv, G. et al. (2012). Anti-hyperlipidaemic and antioxidant effects of turmeric oil in hyperlipidaemic rats. *Food Chem.* 130: 229–235. https://doi.org/10.1016/j.foodchem.2011.07.039.

118 Wen, C.C., Shyur, L.F., Jan, J.T. et al. (2011). Traditional Chinese medicine herbal extracts of *Cibotium barometz*, *Gentiana scabra*, *Dioscorea batatas*, *Cassia tora*, and *Taxillus chinensis* inhibit SARS-CoV replication. *J. Tradit. Complement. Med.* https://doi.org/10.1016/S2225-4110(16)30055-4.

119 Ionita, P. (2021). The chemistry of dpph· free radical and congeners. *Int. J. Mol. Sci.* https://doi.org/10.3390/ijms22041545.

120 Ye, P., Cheah, I.K., and Halliwell, B. (2013). Biochimica et Biophysica Acta high fat diets and pathology in the guinea pig. Atherosclerosis or liver damage? *BBA – Mol. Basis Dis.* 1832: 355–364. https://doi.org/10.1016/j.bbadis.2012.11.008.

121 Walia, M., Kwana, C.Y., and Grovera, A.K. (2003). Effects of free radicals on coronary artery. *Med. Princ. Pract.* https://doi.org/10.1159/000068159.

122 Couto, N., Wood, J., and Barber, J. (2016). The role of glutathione reductase and related enzymes on cellular redox homoeostasis network. *Free Radic. Biol. Med.* https://doi.org/10.1016/j.freeradbiomed.2016.02.028.

123 Forman, H.J., Davies, K.J.A., and Ursini, F. (2014). How do nutritional antioxidants really work: nucleophilic tone and para-hormesis versus free radical scavenging in vivo. *Free Radic. Biol. Med.* https://doi.org/10.1016/j.freeradbiomed.2013.05.045.

124 Chrysohoou, C. and Stefanadis, C. (2013). Longevity and diet. Myth or pragmatism? *Maturitas* https://doi.org/10.1016/j.maturitas.2013.09.014.

125 Bacalini, M.G., Friso, S., Olivieri, F. et al. (2014). Present and future of anti-ageing epigenetic diets. *Mech. Ageing Dev.* https://doi.org/10.1016/j.mad.2013.12.006.

126 Sivamani, R.K. and Maibach, H.I. (2009). Fruits are rich in antioxidants and ripe for topical therapy. *J. Dermatolog. Treat.* https://doi.org/10.1080/09546630802683868.

127 Davis, S.C. and Perez, R. (2009). Cosmeceuticals and natural products: wound healing. *Clin. Dermatol.* https://doi.org/10.1016/j.clindermatol.2009.05.015.

128 Toklu, H.Z., Şener, G., Jahovic, N. et al. (2006). β-Glucan protects against burn-induced oxidative organ damage in rats. *Int. Immunopharmacol.* https://doi.org/10.1016/j.intimp.2005.07.016.

129 Henriques, A., Jackson, S., Cooper, R., and Burton, N. (2006). Free radical production and quenching in honeys with wound healing potential. *J. Antimicrob. Chemother.* https://doi.org/10.1093/jac/dkl336.

130 Jull, A.B., Cullum, N., Dumville, J.C. et al. (2015). Honey as a topical treatment for wounds. *Cochrane Database Syst. Rev.* https://doi.org/10.1002/14651858.CD005083.pub4.

131 Gu, L., Kelm, M.A., Hammerstone, J.F. et al. (2004). Concentrations of proanthocyanidins in common foods and estimations of normal consumption. *J. Nutr.* https://doi.org/10.1093/jn/134.3.613.

132 Khanna, S., Venojarvi, M., Roy, S. et al. (2002). Dermal wound healing properties of redox-active grape seed proanthocyanidins. *Free Radic. Biol. Med.* https://doi.org/10.1016/S0891-5849(02)00999-1.

133 Frei, B. and Higdon, J.V. (2003). Antioxidant activity of tea polyphenols in vivo: evidence from animal studies. *J. Nutr.* https://doi.org/10.1093/jn/133.10.3275s.

134 Al-Mamary, M., Al-Meeri, A., and Al-Habori, M. (2002). Antioxidant activities and total phenolics of different types of honey. *Nutr. Res.* https://doi.org/10.1016/S0271-5317(02)00406-2.

135 Azab, A.E., Albasha, M.O., and Elsayed, A.S.I. (2017). Prevention of nephropathy by some natural sources of antioxidants. *Yangtze Med* https://doi.org/10.4236/ym.2017.14023.

136 Ahmad, A., Abuzinadah, F.M., Alkreathy, M.H. et al. (2020). A novel polyherbal formulation containing thymoquinone attenuates carbon tetrachloride-induced hepatorenal injury in a rat model. *Asian Pac. J. Trop. Biomed.* 10: 147–155.

137 Baunthiyal, M., Singh, V., and Dwivedi, S. (2017). Insights of antioxidants as molecules for drug discovery. *Int. J. Pharmacol.* https://doi.org/10.3923/ijp.2017.874.889.

138 Martinez, J.L., Muñoz-Acevedo, A., and Rai, M. (2019). *Ethnobotany: Application of Medicinal Plants*. MOJ Biology and Medicine. London: CPC Press.

139 Butler, M.S. (2004). The role of natural product chemistry in drug discovery. *J. Nat. Prod.* https://doi.org/10.1021/np040106y.

140 Cragg, G.M., Newman, D.J., and Snader, K.M. (1997). Natural products in drug discovery and development. *J. Nat. Prod.* https://doi.org/10.1021/np9604893.

141 Beckman, C.H. (2000). Phenolic-storing cells: keys to programmed cell death and periderm formation in wilt disease resistance and in general defence responses in plants? *Physiol. Mol. Plant Pathol.* https://doi.org/10.1006/pmpp.2000.0287.

2

Phytomedicines as Sources of Natural Antioxidants

Syed Mohammad Abdullah and Papiya Mitra Mazumder

Department of Pharmaceutical Sciences and Technology, Birla Institute of Technology, Mesra, Ranchi, Jharkhand, India

2.1 Introduction

Phytomedicine is a combination of two words, i.e. phyto and medicine, where "phyto" is a Greek word meaning "plant or herb" typically, which reflects either a product or an ingredient of plant origin, and "medicine" meaning "drug or formulation" used for the prevention or treatment of the disease. Phytomedicine can be simply defined as "herbal medicine," which shows curative potential against disease. According to the regulatory agencies, phytomedicines are measured as the most reliable assets next to the primary medication for patients worldwide [1]. Phytomedicines may be crude or purified products derived from dietary materials or traditional medicinal plants, and have either preventive or therapeutic potential against specific ailments. Collective evidences obtained from all around the world indicate that approximately 35,000 plants' species are presently being utilized in the form of herbal preparations or formulae [2]. Although, recently available research data reflected that phytochemical analysis and biological screening have been carried out approximately only for 10–20% of medicinal plants' constituents so far [3]. Therefore, the rest of the active constituents of medicinal plants are still needed scientific investigational efforts to evaluate their therapeutic potential.

In general, all animals have the ability to digest their food materials and gain energy after sequential cellular metabolisms. During the cellular metabolism process, acceptance of electrons by oxygen molecules in the process of reduction reaction leads to the formation of numerous reactive oxygen species (ROS) [4]. They are typically also known as oxidants or free radicals. At the cellular level, these oxidants could be evolved endogenously as by-products of mitochondrial metabolic reactions along with energy molecule (ATP) or exogenously due to UV radiation as well as environmental pollution. Oxidants may exist in different forms such as superoxide, hydrogen peroxide, hydroxyl radical, hydroxyl ion, singlet oxygen species, and nitric oxide [5]. These oxidants are extremely unstable and may react with nucleotide, unsaturated fatty acid, and amino acid molecules and also could aggravate mutation in genetic material, lipid peroxidation, and protein oxidation [6]. In normal condition, the generated oxidants are mostly absorbed or quenched by the antioxidant enzymes available in the body. The endogenous antioxidant enzyme system is the protective mechanism present ubiquitously in all living organisms including animals and plants. They have the potential to counterbalance the noxious effects of ROS [7]. However, overproduction of ROS creates oxidative stress at cellular or tissue levels due to an imbalance between oxidant and

Phytoantioxidants and Nanotherapeutics, First Edition. Edited by Mithun Rudrapal.

endogenous antioxidant ratio and such a condition may, in due course, lead to developing various types of ailments depending on the targeted organs. The existence of a persistent oxidative stress condition at a particular site creates alterations in cellular physiology and, consequently, may develop various complications (atherosclerosis, aging, cancer, cardiovascular diseases, chronic inflammation, diabetics, myocardial infarction, postischemic perfusion injury, rheumatoid arthritis, septic shock, stroke, and other neurodegenerative disorders) in humans [8]. Currently, there are various types of phytomedicines that have been identified with oxidant scavenging potential, which could play a critical role in the prevention or treatment of oxidative stress-induced diseases [9–11]. Hence, to overcome such pathological conditions, treatment with natural antioxidants as a nutritious supplement is necessarily required to compensate for the insufficiency of the endogenous antioxidant enzymes system. Medicinal plants and dietary food materials are considered the chief sources of natural antioxidants, which are also identified as phytoantioxidants. Nowadays, various types of phytoantioxidants have been recognized and they are broadly categorized as polyphenols, carotenoids, organo-sulfural compounds, vitamins, and minerals [12].

The present chapter briefly discussed the chemical structure-based classification of phytomedicines, which act as antioxidants and portray the role of oxidative stress in the pathogenesis of various diseases. It also enlightens the knowledge associated with phytomedicines including their major sources, indications, marketed formulation, bioavailability, toxicity, and long-term risk assessment.

2.2 Sources of Phytomedicines

The principal sources of phytomedicines are dietary food and medicinal plants. As the knowledge is growing about the importance of phytoantioxidants in health benefits, various dietary food and plants are drawing great attention in food science and nutrition-related research. In the last few decades, antioxidant activities have been widely evaluated in varieties of plant products including flowers, fruits, vegetables, cereal grains, and spices. However, some of the dietary foods contain a high concentration of bioactive components, which shows noticeable antioxidant activities during biochemical assessments and they are broadly designated as phytoantioxidants.

Phytoantioxidants can be divided into different major categories depending on the basis of their basic chemical structural moieties, such as polyphenols, carotenoids, organo-sulfural compounds, vitamins, and minerals. Generally, most of the phytoantioxidants belong to the polyphenol and carotenoid categories [13, 14], whereby, more specifically, polyphenols are further classified into their subtypes (flavonoids, lignans, phenolic acids, stilbenes, tannins, and terpenes) as shown in Figure 2.1.

Most of the edible fruits and vegetables are commonly rich in flavonoid content. Among the dietary food and medicinal plants, flavonoids can be present as anthocyanidins, catechin (CTH), flavanones, flavones, flavonols, and iso-flavonoids in various concentrations [15]. Particularly, citrus fruits are the main sources of flavanones, whereas orange and grapefruit are considered the richest sources of hesperidin and narirutin, and grapefruit is the main source of naringin and narirutin flavonoids [16]. In general, the most abundantly available flavonols reported in plants is quercetin, whereas onion is considered the highest quercetin containing dietary food [17]. Similarly, tea and wine are also measured as good sources of quercetin [18]. However, other types of flavonols such as kaempferol, myricetin (MC), and isorhamnetin are chiefly available in broccoli, berries, and onion, respectively [19, 20]. Tea and red wine are the best sources of CTH. Typically, it is present as aglycone or esterified form of gallic acid (GA) [14]. However, celery and red pepper are the major sources of flavones and are occurred as apigenin and luteolin [21]. Moreover, red-, blue-, or violet-colored fruits or vegetables such as eggplant, plums, and several types of berries contains cyanidin, delphinidin, and pelargonidin as

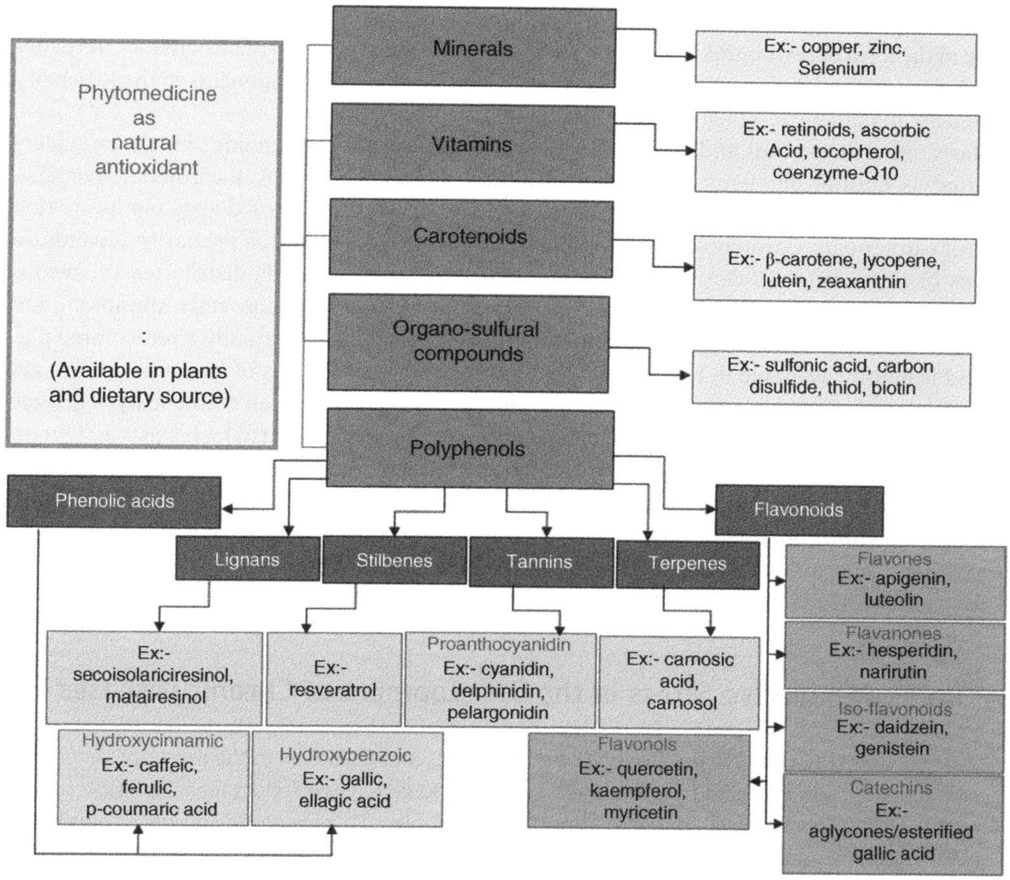

Figure 2.1 Classification of natural antioxidant obtainable from plant and dietary sources.

anthocyanin flavonoid [22]. Proanthocyanidin, a precursor of anthocyanidin, also known as condense tannin, is an oligomeric flavonoid compound that has been identified with powerful antioxidant properties primarily found in grape seeds and blueberry and executes excellent oxidant species scavenging activities [23]. Similarly, dietary foods such as legumes (soybean) present a rich source of iso-flavonoids in the form of iso-flavones mainly available as daidzein and genistein [15, 24]. Phenolic acids as another main subclass of polyphenols mostly occur either in the form of cinnamic acid or benzoic acid. Cinnamic acid usually exists as caffeic, *p*-coumaric, and ferulic, whereas benzoic acid is generally present as GA and hydroxybenzoic acids. In plants, hydroxycinnamic acids are available in higher concentrations than hydroxybenzoic acids [14]. Rich sources of hydroxycinnamic acids are edible fruits, namely apples, blueberries, cherries, kiwis, and plums, and mostly contain hydroxycinnamic acid in between 0.5 and 2 g in 1 kg of fresh fruit. In various fruits, the most available phenolic acid is caffeic acid (CA) and consists of approximately 75–100% of the total content of hydroxycinnamic acid. However, in cereal grains as phenolic acid, ferulic acid (FA) is the chief compound and consists of around 90% of the total content of polyphenol in wheat grain. In medicinal plants, hydroxybenzoic acid is generally available in negligible amounts, but certain red-colored fruits and onions show its presence. In nature, lignans are mainly present as secoisolariciresinol (major portion) and matairesinol (minor portion) in linseed [14, 24]. But, some other dietary foods such as lentils, wheat, pears, garlic, and carrots also showed lignan traces. Polyphenols such as stilbene are primarily available in red wine in the form of resveratrol. It has been reported with a wide spectrum of bioactivities. Terpenes are existing in carnosol and carnosic acid forms and are found as major components in the Rosemary plant, and are maximally responsible for its antioxidant activity [25].

Similarly, other important and abundant phytoantioxidants are carotenoids, which are widely distributed as natural pigments and specifically their main subtypes are β-carotene, lycopene, lutein, and zeaxanthin [26]. Typically, yellow-orange- and deep-green-colored leafy plants are rich sources of carotenoids. Carotenoid is only soluble in lipids and its absorption primarily depends on the types of formulation. β-carotene is the frequent form of carotenoids distributed in various plants, also known as provitamin A. It is mainly present in carrot, mango, nuts, pumpkin, and palm oil. In addition, another type of carotenoid is lycopene, which is normally a red-colored pigment and is mainly available in ripe vegetables. The chief natural sources of lycopene are tomato and its various products including juices, pulps, and peels [27]. Deep green edible leafy vegetable including broccoli, lettuce, peas, and spinach usually contains xanthophyll, which exists as a lutein compound in nature [28]. Red-colored marine algae are an extensive source of zeaxanthin, which approximately contribute 97.4% of the total natural occurrence of carotenoids [29]. The most common sources of sulfur obtained from dietary sources are broccoli, garlic, and onion [30]. Minerals such as copper, zinc, and selenium all serve as an excellent antioxidant in cytosolic defence against ROS and reactive nitrogen species (RNS) and these are chiefly available in nuts and seeds [31].

2.3 Role of Oxidative Stress in the Development of Chronic Diseases

Oxygen is essentially needed for the survival of normal cells and tissues as in the presence of oxygen, metabolic activities take place and, subsequently, generate ROS as a by-product for which host tissues are sensitive [32]. Mitochondria consume almost 80% of molecular oxygen during phosphorylation process [33]. The major portion of the total ROS, including hydroxyl free radical ($\dot{O}H$), superoxide anion ($O_2 - \dot{}$), hydrogen peroxide (H_2O_2), and peroxynitrite (ONO_2-) are generated from mitochondria during cellular respiration as their by-products of metabolic reactions [34]. In normal conditions, redox homeostasis is generally maintained by simultaneous activities of the endogenous antioxidant defence system. But, disturbance between prooxidant and antioxidant reactions leads to imbalance condition, causes consistent accumulation of oxidants (ROS/free radicals) within the cell that creates oxidative damage to the macromolecules (lipids and protein) and that leads to a gradual loss in cellular functioning and efficient processes [35]. Free radicals are quite responsible for substantial damage to the genetic materials, lipids, proteins, mitochondrial components, and other cellular machineries [36]. In disease conditions, either due to microbial infections or manifested by alterations in normal human physiology, initially produces primary cell polymorphonuclear (PMN) leukocytes or neutrophils by the host immunity in response to the pathogens or physiological changes in the body. The host immune mechanism responds by provoking inflammatory cell infiltrations to the target organ, which is accompanied by inflammation [37]. However, at the pathological site, the junctional epithelial cells are incited to produce pro-inflammatory mediators (cytokines and chemokines), which encourage to attract various phagocytic cells including monocytes, macrophages, tissue dendritic cells, and mast cells followed by neutrophils. During the cellular phagocytosis process, rapid release of oxidants also takes place, which is known as respiratory burst, catalyzed by nicotinamide adenine dinucleotide phosphate (NADPH) oxidase enzyme. The oxygen-derived free radicals having unpaired electrons are highly reactive in nature [38]. However, generated free radicals within the organ are not usually target-specific and, therefore, these cause cellular destruction in a generalized way involving localized host tissues [37]. Prolonged and uncontrolled production of ROS from macrophage due to repetitive microbial infections (respiratory burst) or imbalance in the redox homeostasis during altered physiological conditions in localized areas could finally lead to organ damage. The pathological conditions developed through such underlying mechanisms are specifically named oxidative stress-induced diseases. Based on the collective evidences, there are various chronic diseases (neurodegenerative disorder, metabolic syndrome, cancer, obesity, atherosclerosis, type 2 diabetes, etc.) which have been identified and which are usually triggered due to enhanced oxidative stress as a common underlying pathological condition [8]. The critical pathways involved in the development of oxidative stress and further leading to established chronic diseases have been portrayed in Figure 2.2.

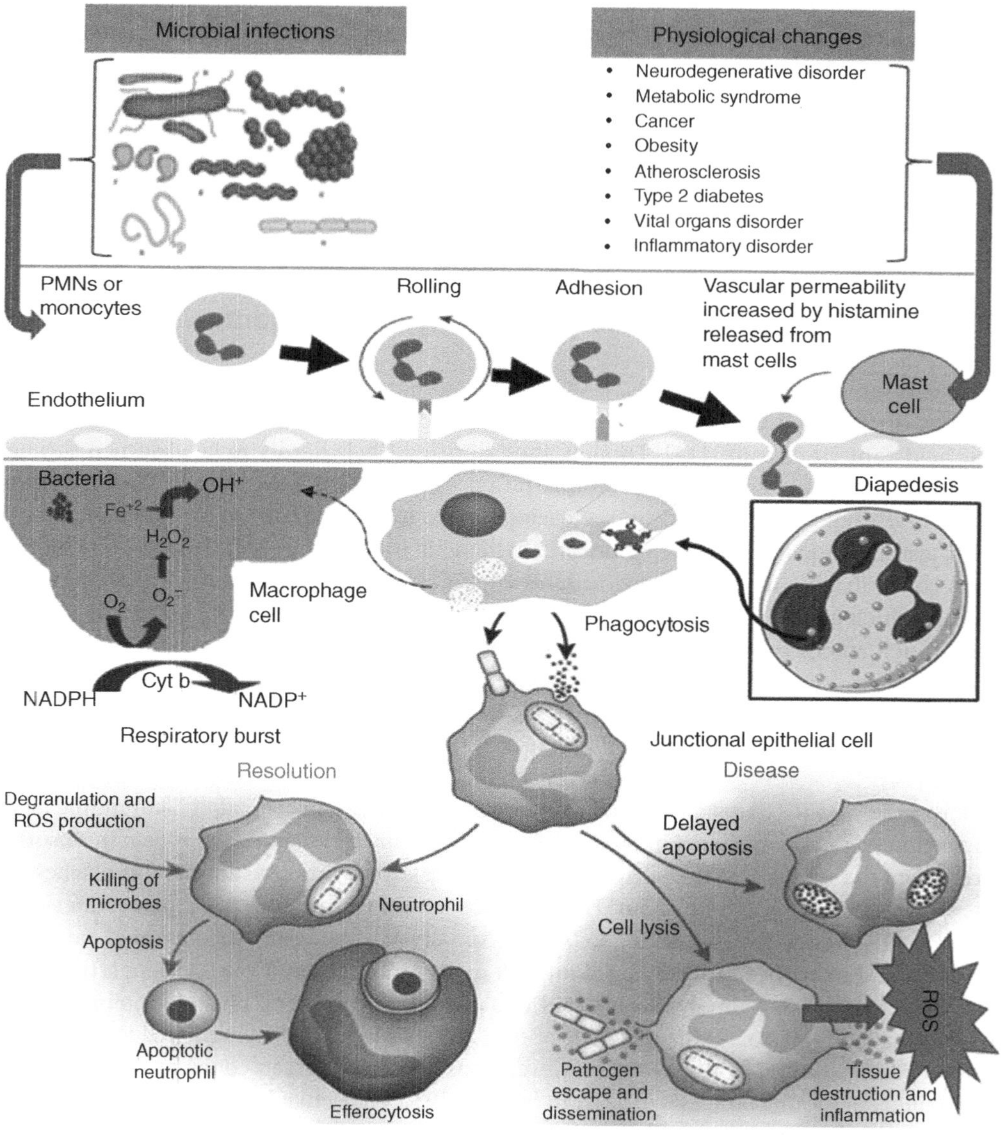

Figure 2.2 Process involved in the development of oxidative stress leads to various diseases.

2.4 Phytomedicines Along with Antioxidant Potential

Natural antioxidants derived from the plants and dietary sources usually interfere with biological processes by various mechanisms such as the scavenging of free radicals, modulating in redox signaling pathways as well as by controlling the transcription factors, and also sometimes by the formation of very complex metabolite fingerprints. However, all such mechanisms are involved in antioxidant activity and are quite interrelated at multiple stages.

Regular intake of certain fruits and vegetables as a daily diet in a balanced amount is usually recognized to diminish the risk of chronic illnesses. Many fresh fruits and vegetables contain a large quantity of natural antioxidants which evidently assist in the strengthening of defence mechanisms against various health problems [39]. Previously, several studies suggested that a diet comprising unprocessed vegetables and fruits generally has a good quantity of antioxidants and it ascertains a positive effect on health at a large scale [40]. Till now, a wide range of medicinal plants and dietary food materials have been verified with a substantial amount of antioxidant properties and could be utilized as phytomedicines for the treatment of various chronic illnesses mainly derived due to the redox imbalance condition. Some of the major phytoconstituents experimentally reported with antioxidant properties and their mechanisms of action have been discussed in Table 2.1.

2.5 Bioavailability and Metabolism of Phytoantioxidants

Among various types of polyphenols, CA is a very common phenolic acid available as both free as well as in esterified types, consisting of approximately 75–100% of the whole hydroxycinnamic acid content of fruits [14]. Though, the esterified CA is hardly to be absorbed through the intestine, consequently, after administration of esterified CA, initially to be cleaved through intestinal microbial flora and releasing its free form and then get absorbed maximally from intestinal mucosa [87] via monocarboxylic acid transporters (MCTs) [88]. Its highest plasma concentration usually reaches around 1-hour postoral administration and then rapidly declined, which could be maintained by two-hourly repeated doses of CA [89]. CA is metabolized through an enzymatic conjugation process including glucuronidation, methylation, and sulfation in presence of glucuronosyltransferases, methyltransferases, and sulfotransferase enzymes, respectively. After metabolism, the product becomes more hydrophilic and less toxic [88]. Due to its enhanced hydrophilic nature, 5.9–27% of CA is easily eliminated, mainly by urine [14].

Chlorogenic acid (CGA) is an intermediate moiety that lies between caffeic and quinic acids, available in excess as polyphenol in plants. After oral administration in rats, it is usually absorbed directly from the stomach in the unhydrolyzed form [90]. After oral consumption, it reaches systemic circulation within half an hour. However, absorption of the CGA takes place in a similar manner as observed in CA (via MCT). CGA is maximally absorbed from the proximal part of the gastrointestinal tract after being changed into FA moiety through MCT [68]. Through the microbial esterase action, CGA finally metabolized into *m*-coumaric acid and phenyl-propionic derivatives in the colon [91]. Approximately, 9.2% of the total dose is normally eliminated through urine after oral administration of CGA [92]. However, consumption of green coffee provides apparent bioavailability of around $33 \pm 23\%$ of CGA in humans [93]. The actual bioavailability of CGA may be increased by the addition of milk in coffee [94]. In addition, the concentration of CGAs present in coffee beverages will also influence its bioavailability [95].

Table 2.1 Phytoconstituents with antioxidant potentials.

Plant source	Phyto-constituents	Chemical structure [41] (lastly visited on 5 July 2021)	Antioxidant effects (mechanisms)
Allium sativum (garlic) [42]	Allicin	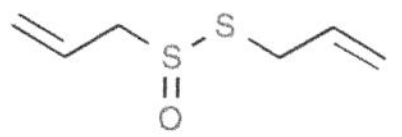	Modulating Nrf2 activation, ROS formation and increasing endogenous antioxidant enzymes [43].
	Alliin	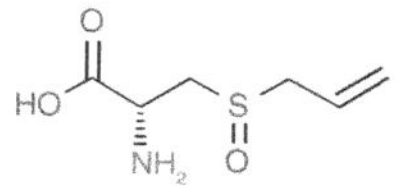	Scavenging ROS by preventing Nox1 signaling pathway [44].
	Diallyl sulfide	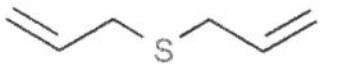	Showing Nrf2-mediated antioxidant effect by activation of ERK/p38 signaling pathways [45].
Datura fastuosa (kecubung) [46]	γ-Tocopherol	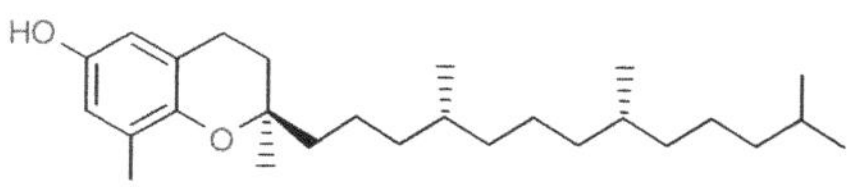	Scavenging free radicals and protecting against oxidative damage [47].
Artemisia campestris ("T" gouft) [48]	Isoquercitrin	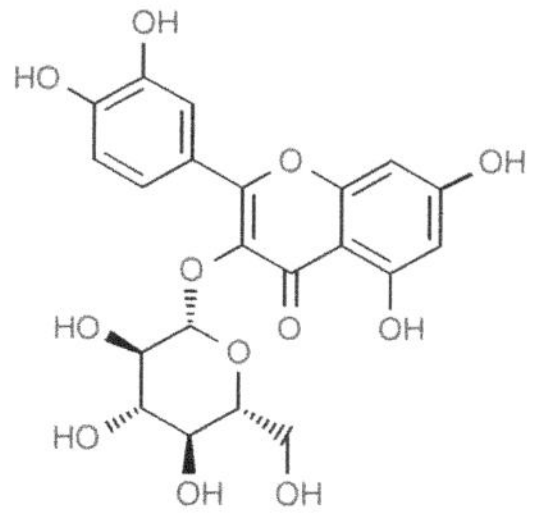	Indirectly chelating Fe^{2+} ions and scavenging ROS to protect oxidative damage. 6″-OH-group enhancing electron transfer and Fe^{2+} chelating potential and decreasing hydrogen releasing abilities by steric hindrance [49].
	Rutin	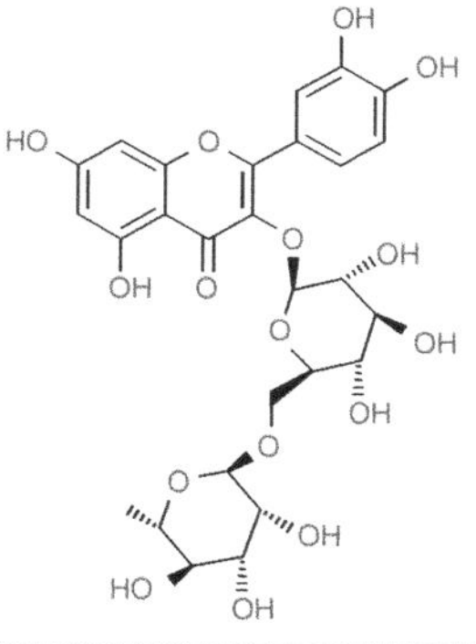	Reducing LPO along with enhancing cellular antioxidant enzyme activity [50].

(*Continued*)

Table 2.1 (Continued)

Plant source	Phyto-constituents	Chemical structure [41] (lastly visited on 5 July 2021)	Antioxidant effects (mechanisms)
Zingiber officinale (ginger) [51]	Shogaol	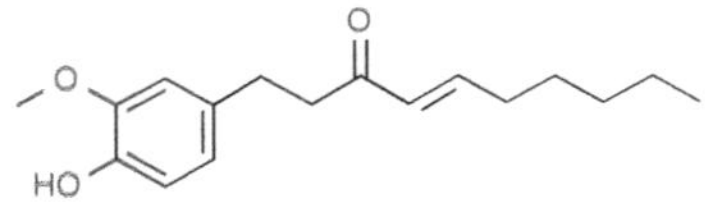	Increases intracellular GSH/GSSG ratio and decreases ROS level along with inducing Nrf2 signaling pathway [52].
	Gingerol	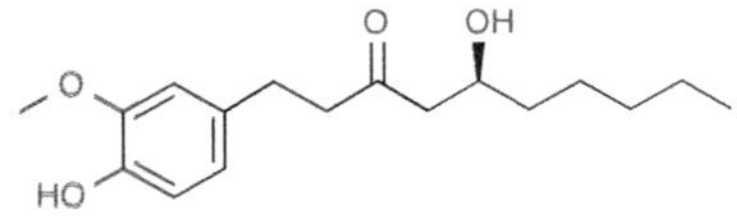	Reducing MDA level and increasing catalase enzyme activities and GSH level [53].
Anacardium occidentale (cashew) [54]	Anacardic acid	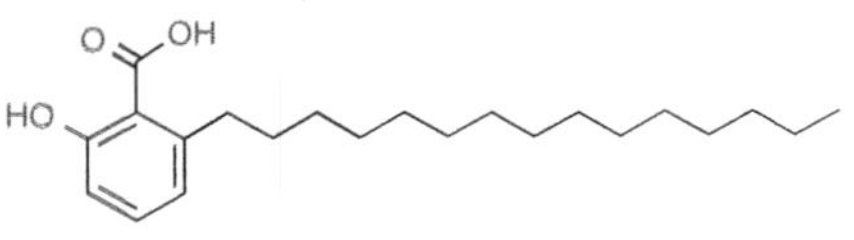	Preventing generation of superoxide radicals and inhibiting XO and LPO-1 enzymes and causes divalent metal ions (Fe^{2+} and Cu^{2+}) chelation [55].
Crocus sativus L. (saffron) [56]	Safranal	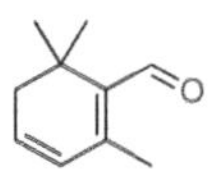	Presented free radical scavenging activities due to its H-donating potential [57].
	Crocin	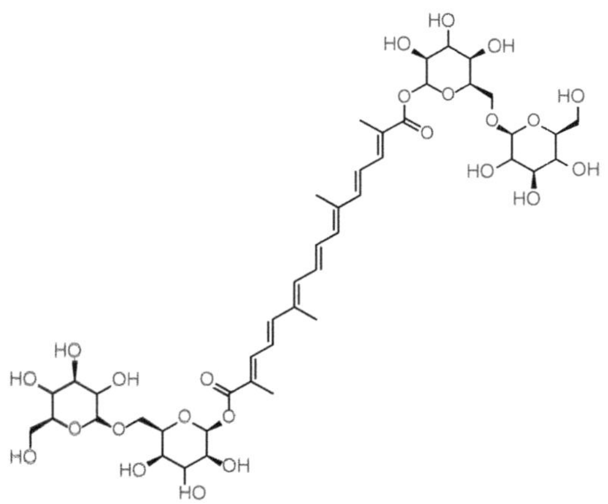	Scavenging free radical [58] and preventing LPO also assist in restoring the cellular antioxidant enzyme (SOD) activity [59].

Table 2.1 (Continued)

Plant source	Phyto-constituents	Chemical structure [41] (lastly visited on 5 July 2021)	Antioxidant effects (mechanisms)
Cannabis sativa L. (hemp) [60]	Cannabidiol	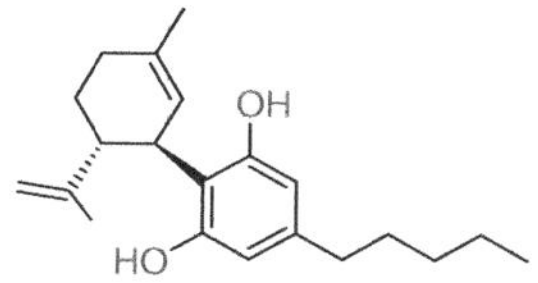	Reducing LPO and ROS production and chelating transition metal ions and modifying redox balance by changing level of antioxidants [61].
Vernonia cinerea (dandapala) [62]	Gallic acid	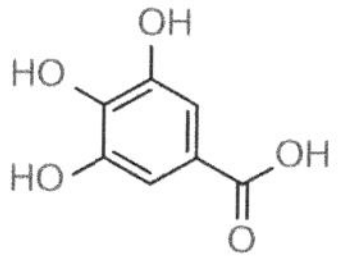	Increasing SOD, GPx and CAT activities, also GSH/GSSG ratio, and activating HO-1 an Nrf2 signaling pathway [63].
	Caffeic acid	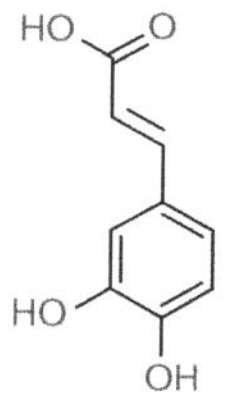	Interrupting free radical formation by inhibiting chain reaction and also acting as a chelating agent [64].
	Quercetin	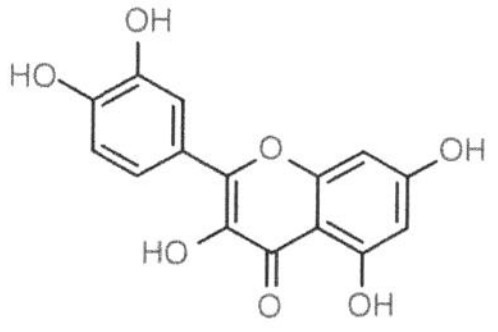	Indirectly chelating Fe^{2+} ions and scavenging ROS to protect from oxidative damage [49]. Besides, attenuating oxidative stress through activation of Nrf2/MAPK signaling cascades [65].
	Ferulic acid	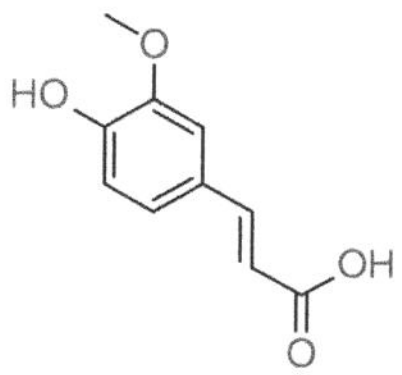	Scavenging the free radicals, binding with transition metals such as iron and copper, and prevents LPO. Also act as H-donor [66].

(*Continued*)

Table 2.1 (Continued)

Plant source	Phyto-constituents	Chemical structure [41] (lastly visited on 5 July 2021)	Antioxidant effects (mechanisms)
Ficus beecheyana (erect fig) [67]	Chlorogenic acid	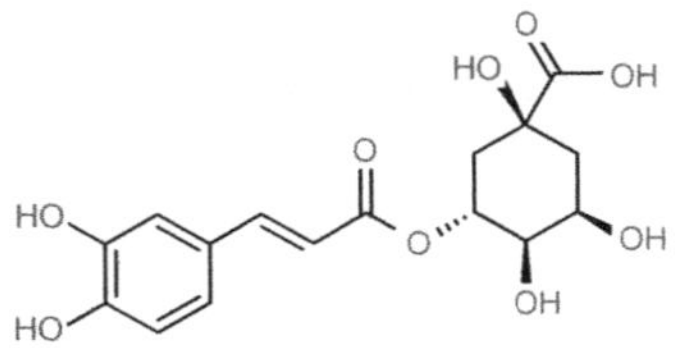	Protecting against oxidation reactions by up-regulating redox-related nuclear transcription factors involved in the antioxidant enzymes expression [68].
	p-Coumaric acid	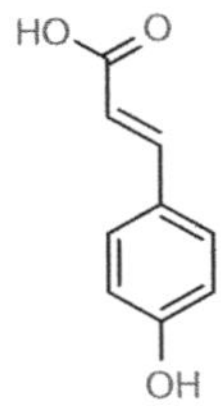	Inhibit ROS generation, scavenge free radicals. Also, enhances gene expression and antioxidant enzymes synthesis [69].
Bacopa monnieri (brahmi) [70]	Stigmasterol	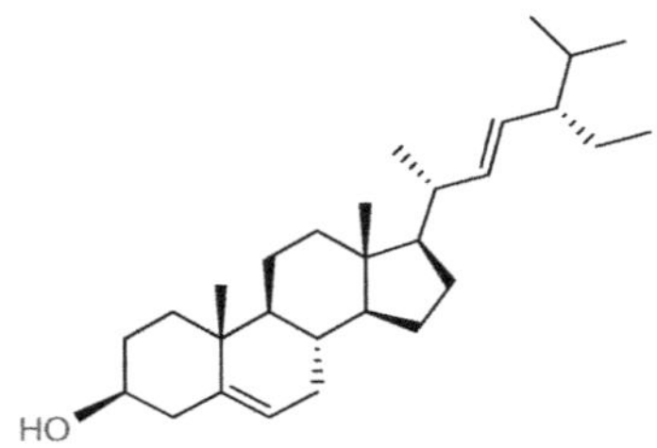	Decreasing LPO activity and increasing the level of GSH and SOD and CAT activities [71].
Aloe vera (star cactus) [72]	Catechin	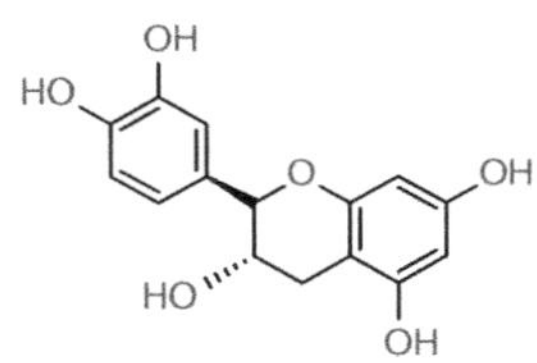	Scavenging ROS and chelating metal ions. Induces antioxidant enzymes and inhibiting prooxidant enzymes [73].
	Sinapic acid	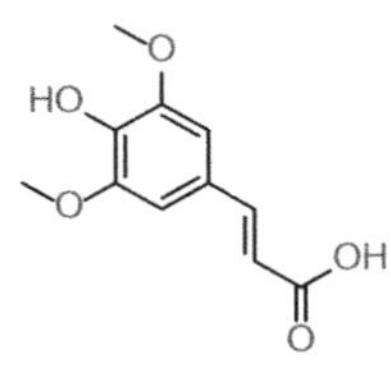	Scavenging free radicals and suppressing LPO activity [74].
	Myricetin	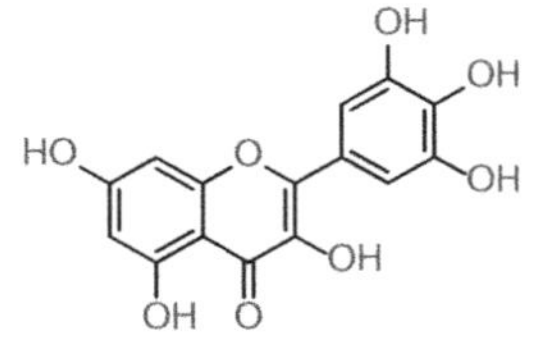	Scavenging free radicals. Decreasing the generation of MDA while increases the SOD level and GPx activity [75].
Curcuma longa (turmeric) [76]	Curcumin	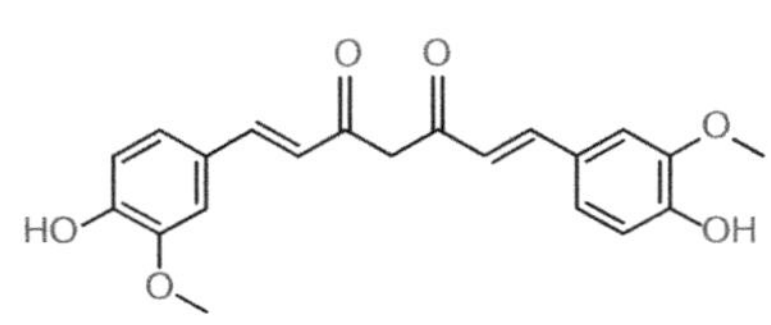	Acting by single electron transfer, radical adduct formation, H-atom transfer. Also, inhibiting LPO activity and scavenging NO [77].

Table 2.1 (Continued)

Plant source	Phyto-constituents	Chemical structure [41] (lastly visited on 5 July 2021)	Antioxidant effects (mechanisms)
Ocimum basillicum (basil) [78, 79]	Antho-cyanin	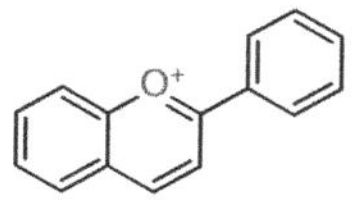	Suppress ROS formation by inhibiting cellular enzymes and chelating trace elements indulged in production of free radicals [80].
	Apigenin	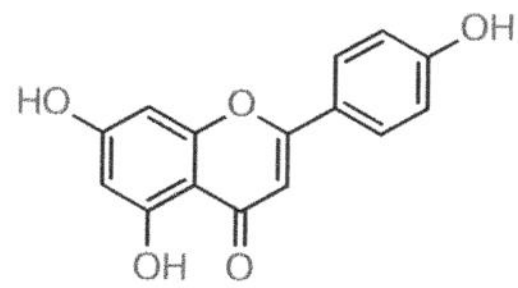	Decreasing MDA level and enhancing the CAT activity [81].
	Naringenin	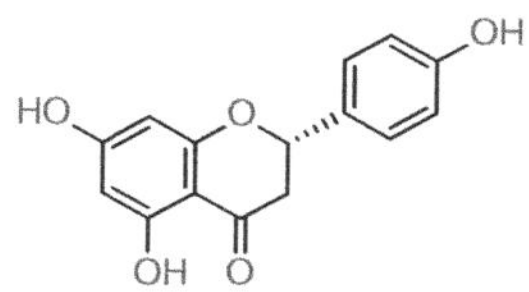	Scavenging free radicals. Also, reducing LPO activity and increasing the cellular antioxidants enzymes activities [81].
Ocimum sanctum (tulsi) [82, 83]	Rosmarinic acid	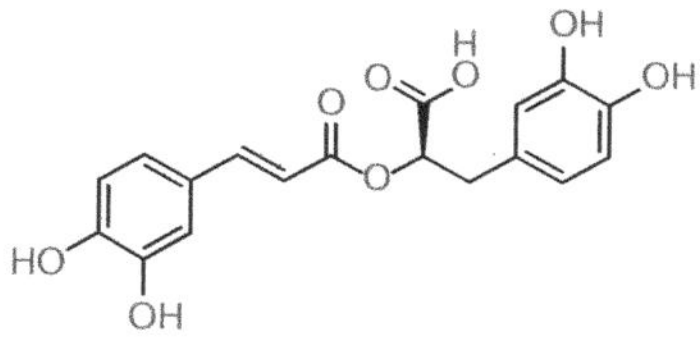	Elevating basal cellular GSH level and decreasing LPO activity [84].
	Orientin	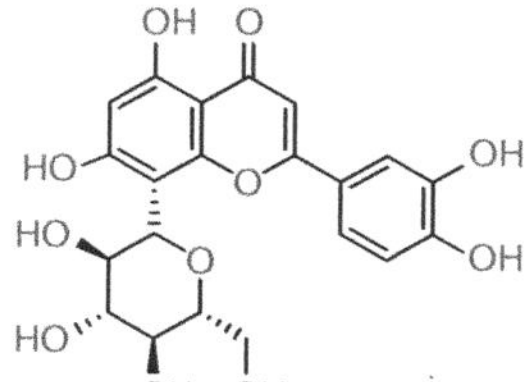	Increases the activity of anti-oxidase system by improving antioxidant capacity of SOD, CAT, and GPx [85, 86].
	Vicenin	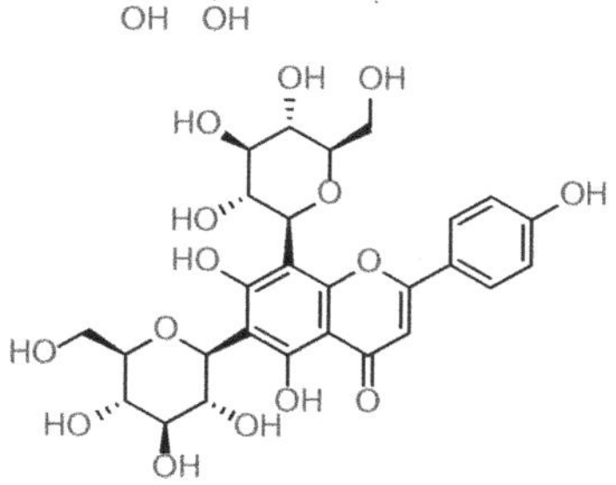	Inhibiting LPO activity and provides protection against radiation induced DNA damage [86].

CAT, catalase; ERK, extracellular signal-regulated kinase; GPx, glutathione peroxidase; GSH/GSSG, glutathione/oxidized glutathione ratio; HO-1, hemeoxygenase-1; LPO, lipid peroxidation; LPO-1, lipoxygenase-1; MAPK, mitogen-activated protein kinase; MDA, malondialdehyde; NO, nitric oxide; Nox 1, NADPH oxidase 1; Nrf2, nuclear factor erythroid 2-related factor 2; ROS, reactive oxygen species; SOD, superoxide dismutase; XO, xanthin oxidase.

Sinapic acid (SA) is a less frequent plant-derived hydroxycinnamic acid predominantly bound with albumin protein in blood due to its hydrophobic interaction and H-bonding [96]. After oral administration, SA is mainly absorbed through the small intestine by an active transport mechanism [97]. Moreover, the metabolism of SA occurs by the activity of the esterase enzyme of intestinal microflora in the small intestine [98]. After metabolism of SA, various by-products including 3-hydroxy-5-ethoxy-phenyl-propionic acid, 3-hydroxy-5-methoxycinnamic acid, methyl sinapate-glucuronide, methyl sinapate-sulfate and dihydrosinapic acid were reported in the urine of rats [99].

Another broad category of polyphenols is flavonoids distributed in various plants, vegetables, and fruits. Flavanone is an important flavonoid available mainly in citrus fruits and juices. The citrus flavonoids (flavanones) are absorbed rapidly from the intestines but are typically presented substantially low oral bioavailability that limit their efficacy [100]. After administration of citrus fruits, the peak plasma concentration of hesperidin and naringin is usually achieved at around five to six hours with relatively slight differences among various flavanone forms. They are metabolized at both hepatic as well as intestinal sites and excreted mainly with conjugation of sulfate and glucuronide moiety via urine [101].

In addition, flavonols, also being an important type of flavonoids, their pharmacokinetic (PK) profiles were assessed. In rats, a PK study of MC was carried out, it showed low bioavailability around 9.62 and 9.74% after 50 and 100 mg/kg of oral doses, respectively. Particularly, C_{max} and AUC were found to increase in a dose-dependent manner and T_{max} reached 6.4 hours after oral administration because of slow absorption of MC [102]. However, MC coadministration enhances the bioavailability of other drugs, due to the CYP-450 inhibition property [103]. Likewise, rutin (RT) is primarily absorbed through the small intestine after oral administration and enzymatically metabolizes [104]. The chief metabolites of RT are reported as 3,4-dihydroxyphenylacetic acid, 3,4-dihydroxytoluene, *m*-hydroxyphenylacetic acid, 3-methoxy-4-hydroxyphenylacetic acid/homovanillic acid, and 3,5,7,30,50-pentahydroxyflavonol/quercetin [105]. Approximately, 9% of the orally administered RT is excreted in the urine. However, the rest unchanged form is excreted through faeces [92].

In the previous study, after ingestion of raw garlic, various forms of the chemical compounds were detected in the breath of study participants including diallyl disulfide (DADS), allyl methylsulfide (AMS), diallyl sulfide (DAS), allyl methyl disulfide (AMDS), dimethyl sulfide, acetone, and diallyl trisulfide (DATS) [106]. It was observed that within two to three hours, DADS, DAS, DATS, and AMDS were achieved the peak concentrations in breath [107]. A PK study conducted in rats demonstrated that the T_{max} of alliin was achieved prior to 10 minutes and was completely eliminated from the blood after 6 hours, whereby T_{max} of allicin was obtained from 30 to 60 minutes and the total excretion (including fecal and urinary) was found 85.5% after 72 hours [108]. Normally, allicin is quickly degradable under the influence of temperature and converted into ajoenes and vinyldithiins, which are more stable than allicin [109].

2.6 Probable Use of Phytoantioxidants in Prevention and/or Mitigation of Chronic Diseases

CA compounds effectively eliminate the free radicals, prevent ROS production as well as induce DNA oxidation in cancer cells. In hepatocarcinoma, CA acts through its potent antioxidant effect, which inhibits ROS generation and reduces oxidative stress as it frequently occurs in this case [87].

Cannabidiol (CBD) on the basis of the outcome of recent studies could be utilized in the treatment of cancer, psychosis, Alzheimer's disease (AD), anxiety, epilepsy, diabetes, stroke, arrhythmia, atherosclerosis, hypertension, diabetes-related cardiomyopathy, and arthritis [110–112]. CBD is also directly involved in balancing the redox system due to its potent antioxidant property [61].

CTH plays a critical role in oxidative stress-derived disease [73] since some age-linked chronic diseases including Parkinson's disease, AD, diabetes, and cardiovascular diseases are frequently associated with an imbalance in redox homeostasis and ROS-induced cellular damages [113]. Therefore, due to the good antioxidant potential of CTH, it may be utilized in the treatment of these chronic disease conditions. Green tea is a rich source of CTH and presented profound action against various cancer and noncancer therapies [114].

CGA obtained from dietary materials showed efficacy against a wide range of chronic diseases. In diabetes patients, due to hyperglycemic conditions, generation of ROS increases, which leads to oxidative stress, which may be overcome with CGA [115]. In the environment, toxicity, which occurs by heavy metal pollution, particularly through cadmium, produces oxidative stress in the neurological system, where the CGA showed a protective effect [116].

FA reflects broad efficacy as anticancer, antidiabetic, antimicrobial, anti-inflammatory, and antioxidant due to strong antioxidant properties. It also provides a defensive effect to the cellular structure along with melanogenesis attenuation. Besides, it also has anti-wrinkles and discoloration diminishing effects, including good skin penetration ability. Additionally, it shows less toxicity to the biological system. Thus, FA has gained interest in cosmetic industries, particularly to develop facial products, which may prevent from sun-related skin changes [117].

GA efficiently modulates the prooxidant and antioxidant imbalance condition. It may also be employed for the treatment of cancer as it shows the potential to inhibit the matrix metalloproteinase (MMP) expression, which could prevent invasion and metastasis activities [118]. In addition, GA is enabled to the prevention of various metabolic syndromes including hyperglycemia, hypertriglyceridemia, and pancreatic β-cells destruction by enhancing the expression of a nuclear transcription factor (PPAR-γ) [119].

Garlic extract preparation has been widely utilized as a phytomedicine to prevent atherosclerosis, hypertension, arrhythmia, hyperlipidemia, and thrombosis due to its good antioxidant properties [120, 121]. Allicin is an important bioactive component of garlic extract, which enables to boost cellular immunity and inhibit gene expression in tumor vessels and increases apoptosis in cancerous cells due to which it has the ability to diminish pancreatic and colorectal cancers [122]. In addition, allicin also has efficacy against gastrointestinal cancer by inhibiting cell proliferation [123].

In the last few years, ginger has been recognized for efficacious antimicrobial (antibacterial, antifungal, and antiviral) effects [124]. Ginger extract was also extensively evaluated against various cancer disease models including breast, cervical, colorectal, and prostate cancers [125–127]. One of the important components of ginger, 6-shogaol shows an ability to improve cognitive dysfunction in AD-induced mice through suppressing inflammatory responses, upregulation of NGF and enhancing synaptogenesis [128]. Other components, including gingerenone A and 6-gingerol, presented an anti-obesity effect, due to mainly inhibition of adipogenesis and enhancement of fat metabolism [129].

MC, a bioactive molecule, earlier reported with antioxidant as well as anti-inflammatory actions, and showed beneficial effects in abstaining diabetes-associated conditions [130]. In the treatment of AD, MC inhibits amyloid-β fibrils and neurofibrillary tangles formation [131]. MC also presents anticancer potential by seizing the cell cycle at the G2/M phase [132].

Besides this, quercetin is a well-known antitumor phytomedicine and it acts by various mechanisms. Particularly, it has the tendency to arrest the cell cycle at the G2/M phase, promote cell apoptosis, as well as inhibit blood vessel generation in tumor cells [133].

Likewise, RT is also capable to attenuate the cancer cell growth by similar mechanisms as reported with quercetin [134]. Additionally, RT also has glycemic control effect by increasing the activity of insulin receptor kinase, which involves enhancing the insulin signaling pathway, leading to augment GLUT4 translocation and glucose uptake [135].

2.7 List of Some Marketed Preparations of Phytoantioxidants

1) Marketed product name: Curcuma plus herbal capsule
   Category: Antioxidant and anti-inflammatory
   Manufactured by: Sonark Pharmaceuticals Pvt Ltd., Maharashtra, India
   Marketed by: Nisarga Healthcare Pvt. Ltd.
   Active ingredients: *Curcuma longa* rhizome extract 400 mg + *Piper longum* fruit extract 5 mg.
2) Marketed product name: Antoxid HC capsule
   Category: Antioxidant
   Manufactured by: Dr Reddy's Laboratories Ltd., Bengaluru, India
   Active ingredients: β-Carotene (carotenoids) 30 mg + zinc sulfate 27.5 mg + selenium dioxide 200 mcg + manganese 2 mg + copper 1 mg.
3) Marketed product name: Nutrova kera strength capsule
   Category: Antioxidant
   Manufactured by: Vital Nutraceuticals Pvt. Ltd., Maharashtra, India
   Active ingredients: Grape seed extract 10 mg + green tea extract 10 mg + primrose oil 236 mg + *N*- acetyl-L-cysteine 50 mg + L-taurine 50 mg + L-methionine 30 mg + methyl sulfonyl methane 30 mg + vitamin D_3 400 IU + niacin 15 mg + vitamin B_{12} 1 mcg + folic acid 100 mcg + biotin 5 mg + choline bitartrate 10 mg + iron 8 mg + zinc 2.5 mg.
4) Marketed product name: Omega woman
   Category: Antioxidant and immune supplement
   Manufactured by: Wassen, Surrey, UK
   Active ingredients: Lycopene and resveratrol.
5) Marketed product name: Olivenol™
   Category: Antioxidant and dietary supplement
   Manufactured by: Cre Agri, Hayward, CA, USA
   Active ingredients: Natural antioxidant, hydroxytyrosol

2.8 Toxicity and Long-Term Risk Assessment of Some Phytoantioxidants

In the last few decades, a large number of active phytoconstituents were evaluated for their toxicity by using both in vitro and in vivo models. But, discussing the outcomes of toxicity studies related to all these compounds seems beyond the limit of this chapter. However, some of them have been tried to cover here.

In the animal-based study, saffron was reported as a nontoxic compound with LD_{50} value equivalent 20.7 g/kg. However, in vitro study demonstrated its noncytotoxic effects with LD_{50} value up to 200 mg/ml. During the evaluation of LD_{50} value, comparatively petals' extract was found less toxic than stigma in mice [136]. The results of LD_{50} value for petal and stigma extract of saffron were reported as 3.38 and 1.6 g/kg, respectively [137]. But, when tumor-bearing mice were treated with saffron extract, it enhanced the survival span by three times, without alteration in values of biochemical and hematological parameters during toxicity studies [138].

Similarly, an in vitro study was carried out to evaluate the toxicity of garlic-related compounds in which garlic powder or its bioactive constituent allicin was tested in the isolated rat liver where 200 mg/ml concentration of these compounds caused marked cell damage [139]. Another study reported that long-term administration of raw garlic continuously at a high dose subjected to red blood cells rupture, which usually caused anemia and loss in body weight. But, ingestion of garlic juice with a measured dose of up to 5 ml/kg creates injury in the stomach and causes death in animals [140]. Moreover, garlic powder administration of up to 50 mg/kg on daily basis has withdrawn spermatogenesis and Leydig cell functions [139]. Oral administration of dehydrated raw garlic powder triggered acute injury to gastric mucosa [141]. In humans, a high dose of garlic is generally associated with hepatotoxicity, therapeutic dose is rarely accompanied with gastrointestinal complications. However, a low dose seems safe enough [139].

Also, during the toxicity study of RT, the carcinogenic potential was assessed in hamster and the outcomes of the study confirmed that carcinogenic effects were found absent in RT [142].

Likewise, SA-related toxicity study was performed on the broiler chickens, indicating that there was no escalation in creatine kinase (CK) and lactate dehydrogenase (LDH) activity. Collectively, no harmful effects were observed related to the vital organs in the experimental animals [143].

2.9 Conclusions

Many chronic diseases are frequently associated with oxidative stress as a common pathological reason. Also, several environmental factors are generally responsible to release oxygen reactive species in our body that may lead to lipid peroxidation and genetic material impairment, concrete underlying pathways for the development of various chronic diseases. Accordingly, an attempt to create an equilibrium between oxidant and antioxidant molecules could diminish ROS-related cellular damages by using an appropriate therapeutic dose of specific phytoantioxidants. Therefore, the future of phytomedicines with antioxidant properties has great scope for developing novel therapeutic preparations against oxidative stress-induced chronic diseases.

References

1. World Health Organization (2019). *WHO Global Report on Traditional and Complementary Medicine 2019*. World Health Organization.
2. Pan, S.Y., Zhou, S.F., Gao, S.H. et al. (2013). New perspectives on how to discover drugs from herbal medicines: CAM's outstanding contribution to modern therapeutics. *Evidence-based Complementary and Alternative Medicine* 2013: 627375.
3. Altemimi, A., Lakhssassi, N., Baharlouei, A. et al. (2017). Phytochemicals: extraction, isolation, and identification of bioactive compounds from plant extracts. *Plants* 6 (4): 42.
4. Halliwell, B. (2006). Reactive species and antioxidants. Redox biology is a fundamental theme of aerobic life. *Plant Physiology* 141 (2): 312–322.
5. Zorov, D.B., Juhaszova, M., and Sollott, S.J. (2014). Mitochondrial reactive oxygen species (ROS) and ROS-induced ROS release. *Physiological Reviews* 94 (3): 909–950.
6. Sies, H. (1993). Strategies of antioxidant defense. *EJB Reviews* 1994: 101–107.
7. Poljsak, B., Suput, D., and Milisav, I. (2013). Achieving the balance between ROS and antioxidants: when to use the synthetic antioxidants. *Oxidative Medicine and Cellular Longevity* 2013: 956792.

8 Uttara, B., Singh, A.V., Zamboni, P., and Mahajan, R.T. (2009). Oxidative stress and neurodegenerative diseases: a review of upstream and downstream antioxidant therapeutic options. *Current Neuropharmacology* 7 (1): 65–74.

9 Chidambaram, U., Pachamuthu, V., Natarajan, S. et al. (2013). In vitro evaluation of free radical scavenging activity of *Codariocalyx motorius* root extract. *Asian Pacific Journal of Tropical Medicine* 6 (3): 188–194.

10 Gasca, C.A., Cabezas, F.A., Torras, L. et al. (2013). Chemical composition and antioxidant activity of the ethanol extract and purified fractions of cadillo (*Pavonias epioides*). *Free Radicals and Antioxidants* 3: S55–S61.

11 Kohen, R. and Gati, I. (2000). Skin low molecular weight antioxidants and their role in aging and in oxidative stress. *Toxicology* 148 (2–3): 149–157.

12 Zhang, Y.J., Gan, R.Y., Li, S. et al. (2015). Antioxidant phytochemicals for the prevention and treatment of chronic diseases. *Molecules* 20 (12): 21138–21156.

13 Baiano, A. and Del Nobile, M.A. (2016). Antioxidant compounds from vegetable matrices: biosynthesis, occurrence, and extraction systems. *Critical Reviews in Food Science and Nutrition* 56 (12): 2053–2068.

14 Manach, C., Scalbert, A., Morand, C. et al. (2004). Polyphenols: food sources and bioavailability. *The American Journal of Clinical Nutrition* 79 (5): 727–747.

15 Erlund, I. (2004). Review of the flavonoids quercetin, hesperetin, and naringenin. Dietary sources, bioactivities, bioavailability, and epidemiology. *Nutrition Research* 24 (10): 851–874.

16 Testai, L. and Calderone, V. (2017). Nutraceutical value of citrus flavanones and their implications in cardiovascular disease. *Nutrients* 9 (5): 502.

17 David, A.V., Arulmoli, R., and Parasuraman, S. (2016). Overviews of biological importance of quercetin: a bioactive flavonoid. *Pharmacognosy Reviews* 10 (20): 84.

18 Lakhanpal, P. and Rai, D.K. (2007). Quercetin: a versatile flavonoid. *Internet Journal of Medical Update* 2 (2): 22–37.

19 Holden, J.M., Bhagwat, S.A., and Patterson, K.Y. (2002). Development of a multi-nutrient data quality evaluation system. *Journal of Food Composition and Analysis.* 15 (4): 339–348.

20 Hakkinen, S.H., Kärenlampi, S.O., Heinonen, I.M. et al. (1999). Content of the flavonols quercetin, myricetin, and kaempferol in 25 edible berries. *Journal of Agricultural and Food Chemistry* 47 (6): 2274–2279.

21 Panche, A.N., Diwan, A.D., and Chandra, S.R. (2016). Flavonoids: an overview. *Journal of Nutritional Science* 5: e47.

22 Xu, D.P., Li, Y., Meng, X. et al. (2017). Natural antioxidants in foods and medicinal plants: extraction, assessment and resources. *International Journal of Molecular Sciences* 18 (1): 96.

23 Guo, Q., Zhao, B., Li, M. et al. (1996). Studies on protective mechanisms of four components of green tea polyphenols against lipid peroxidation in synaptosomes. *Biochimica et Biophysica Acta (BBA) – Lipids and Lipid Metabolism* 1304 (3): 210–222.

24 Liggins, J., Bluck, L.J., Runswick, S. et al. (2000). Daidzein and genistein contents of vegetables. *British Journal of Nutrition* 84 (5): 717–725.

25 Afaq, F., Adhami, V.M., and Ahmad, N. (2003). Prevention of short-term ultraviolet B radiation-mediated damages by resveratrol in SKH-1 hairless mice. *Toxicology and Applied Pharmacology* 186 (1): 28–37.

26 Mezzomo, N. and Ferreira, S.R. (2016). Carotenoids functionality, sources, and processing by supercritical technology: a review. *Journal of Chemistry* 2016: 3164312.

27 Borguini, R.G., da Silva, F., and Torres, E.A. (2009). Tomatoes and tomato products as dietary sources of antioxidants. *Food Reviews International* 25 (4): 313–325.

28 Abdel-Aal, E.S., Akhtar, H., Zaheer, K., and Ali, R. (2013). Dietary sources of lutein and zeaxanthin carotenoids and their role in eye health. *Nutrients* 5 (4): 1169–1185.

29 Raposo, M.F., De Morais, A.M., and De Morais, R.M. (2015). Carotenoids from marine microalgae: a valuable natural source for the prevention of chronic diseases. *Marine Drugs* 13 (8): 5128–5155.

30 Batchu, S.N., Chaudhary, K.R., Wiebe, G.J., and Seubert, J.M. (2012). Bioactive compounds in heart disease. *Bioactive Food as Dietary Interventions for Cardiovascular Disease: Bioactive Foods in Chronic Disease States* 22: 431.

31 Ros, E. (2010). Health benefits of nut consumption. *Nutrients* 2 (7): 652–682.

32 He, L., He, T., Farrar, S. et al. (2017). Antioxidants maintain cellular redox homeostasis by elimination of reactive oxygen species. *Cellular Physiology and Biochemistry* 44 (2): 532–553.

33 Muller, F.L., Liu, Y., and Van Remmen, H. (2004). Complex III releases superoxide to both sides of the inner mitochondrial membrane. *Journal of Biological Chemistry* 279 (47): 49064–49073.

34 Kumari, S., Badana, A.K., and Malla, R. (2018). Reactive oxygen species: a key constituent in cancer survival. *Biomarker Insights* 13: 1–9.

35 Lobo, V., Patil, A., Phatak, A., and Chandra, N. (2010). Free radicals, antioxidants and functional foods: Impact on human health. *Pharmacognosy Reviews* 4 (8): 118.

36 Droge, W. (2002). Free radicals in the physiological control of cell function. *Physiological Reviews* 82 (1): 47–95.

37 Maurya, P.K. and Dua, K. (ed.) (2020). *Role of Oxidative Stress in Pathophysiology of Diseases*. Springer.

38 Halliwell, B.A., Gutteridge, J.M., and Cross, C.E. (1992). Free radicals, antioxidants, and human disease: where are we now? *The Journal of Laboratory and Clinical Medicine* 119 (6): 598–620.

39 Hamid, A.A., Aiyelaagbe, O.O., Usman, L.A. et al. (2010). Antioxidants: its medicinal and pharmacological applications. *African Journal of Pure and Applied Chemistry* 4 (8): 142–151.

40 Bai, Z., Rui, Y., Li, J. et al. (2018). Application of several novel natural antioxidants to inhibit oxidation of tree peony seed oil. *CyTA Journal of Food* 16 (1): 1071–1078.

41 http://www.chemspider.com.

42 Elosta, A., Slevin, M., Rahman, K., and Ahmed, N. (2017). Aged garlic has more potent antiglycation and antioxidant properties compared to fresh garlic extract in vitro. *Scientific Reports* 7 (1): 1–9.

43 Zhang, M., Pan, H., Xu, Y. et al. (2017). Allicin decreases lipopolysaccharide-induced oxidative stress and inflammation in human umbilical vein endothelial cells through suppression of mitochondrial dysfunction and activation of Nrf2. *Cellular Physiology and Biochemistry* 41 (6): 2255–2267.

44 Chen, Y., Sun, J., Dou, C. et al. (2016). Alliin attenuated RANKL-induced osteoclastogenesis by scavenging reactive oxygen species through inhibiting Nox1. *International Journal of Molecular Sciences* 17 (9): 1516.

45 Ho, C.Y., Cheng, Y.T., Chau, C.F., and Yen, G.C. (2012). Effect of diallyl sulfide on in vitro and in vivo Nrf2-mediated pulmonic antioxidant enzyme expression via activation ERK/p38 signaling pathway. *Journal of Agricultural and Food Chemistry* 60 (1): 100–107.

46 Dhiman, A., La, R., Bhan, M. et al. (2012). Plebeian assessment of anti-microbial and in vitro antioxidant zest of *D. fastuosa* L. seeds. *Journal of Pharmaceutical and Scientific Innovation* 1 (4): 49–53.

47 Kurutas, E.B. (2015). The importance of antioxidants which play the role in cellular response against oxidative/nitrosative stress: current state. *Nutrition Journal* 15 (1): 1–22.

48 Akrout, A., Gonzalez, L.A., El Jani, H., and Madrid, P.C. (2011). Antioxidant and antitumor activities of *Artemisia campestris* and *Thymelaea hirsuta* from southern Tunisia. *Food and Chemical Toxicology* 49 (2): 342–347.

49 Li, X., Jiang, Q., Wang, T. et al. (2016). Comparison of the antioxidant effects of quercitrin and isoquercitrin: understanding the role of the 6″-OH group. *Molecules* 21 (9): 1246.

50 Gęgotek, A., Rybałtowska-Kawałko, P., and Skrzydlewska, E. (2017). Rutin as a mediator of lipid metabolism and cellular signaling pathways interactions in fibroblasts altered by UVA and UVB radiation. *Oxidative Medicine and Cellular Longevity* 2017: 4721352.

51 Tohma, H., Gülçin, İ., Bursal, E. et al. (2017). Antioxidant activity and phenolic compounds of ginger (*Zingiber officinale* Rosc.) determined by HPLC-MS/MS. *Journal of Food Measurement and Characterization* 11 (2): 556–566.

52 Chen, H., Fu, J., Chen, H. et al. (2014). Ginger compound [6]-shogaol and its cysteine-conjugated metabolite (M2) activate Nrf2 in colon epithelial cells in vitro and in vivo. *Chemical Research in Toxicology* 27 (9): 1575–1585.

53 Saiah, W., Halzoune, H., Djaziri, R. et al. (2018). Antioxidant and gastroprotective actions of butanol fraction of *Zingiber officinale* against diclofenac sodium-induced gastric damage in rats. *Journal of Food Biochemistry* 42 (1): e12456.

54 Trevisan, M.T., Pfundstein, B., Haubner, R. et al. (2006). Characterization of alkyl phenols in cashew (*Anacardium occidentale*) products and assay of their antioxidant capacity. *Food and Chemical Toxicology* 44 (2): 188–197.

55 Kubo, I., Masuoka, N., Ha, T.J., and Tsujimoto, K. (2006). Antioxidant activity of anacardic acids. *Food Chemistry* 99 (3): 555–562.

56 Hatziagapiou, K., Kakouri, E., Lambrou, G.I. et al. (2019). Antioxidant properties of *Crocus sativus* L. and its constituents and relevance to neurodegenerative diseases; focus on Alzheimer's and Parkinson's disease. *Current Neuropharmacology* 17 (4): 377–402.

57 Assimopoulou, A.N., Sinakos, Z., and Papageorgiou, V. (2005). Radical scavenging activity of *Crocus sativus* L. extract and its bioactive constituents. *Phytotherapy Research: An International Journal Devoted to Pharmacological and Toxicological Evaluation of Natural Product Derivatives* 19 (11): 997–1000.

58 Nanda, S. and Madan, K. (2021). The role of safranal and saffron stigma extracts in oxidative stress, diseases and photoaging: a systematic review. *Heliyon* 7 (2): e06117.

59 Yaribeygi, H., Noroozadeh, A., Mohammadi, M.T. et al. (2019). Crocin improves oxidative stress by potentiating intrinsic anti-oxidant defense systems in pancreatic cells during uncontrolled hyperglycemia. *Journal of Pharmacopuncture* 22 (2): 83.

60 Yan, X., Tang, J., dos Santos, P.C. et al. (2015). Characterization of lignanamides from hemp (*Cannabis sativa* L.) seed and their antioxidant and acetylcholinesterase inhibitory activities. *Journal of Agricultural and Food Chemistry* 63 (49): 10611–10619.

61 Atalay, S., Jarocka-Karpowicz, I., and Skrzydlewska, E. (2020). Antioxidative and anti-inflammatory properties of cannabidiol. *Antioxidants* 9 (1): 21.

62 Sonibare, M.A., Aremu, O.T., and Okorie, P.N. (2016). Antioxidant and antimicrobial activities of solvent fractions of *Vernonia cinerea* (L.) less leaf extract. *African Health Sciences* 16 (2): 629–639.

63 Yeh, C.T., Ching, L.C., and Yen, G.C. (2009). Inducing gene expression of cardiac antioxidant enzymes by dietary phenolic acids in rats. *The Journal of Nutritional Biochemistry* 20 (3): 163–171.

64 Genaro-Mattos, T.C., Maurício, Â.Q., Rettori, D. et al. (2015). Antioxidant activity of caffeic acid against iron-induced free radical generation – a chemical approach. *PLoS One* 10 (6): e0129963.

65 Sun, L., Xu, G., Dong, Y. et al. (2020). Quercetin protects against lipopolysaccharide-induced intestinal oxidative stress in broiler chickens through activation of Nrf2 pathway. *Molecules* 25 (5): 1053.

66 Srinivasan, M., Sudheer, A.R., and Menon, V.P. (2007). Ferulic acid: therapeutic potential through its antioxidant property. *Journal of Clinical Biochemistry and Nutrition* 40 (2): 92–100.

67 Yen, G.C., Chen, C.S., Chang, W.T. et al. (2018). Antioxidant activity and anticancer effect of ethanolic and aqueous extracts of the roots of *Ficus beecheyana* and their phenolic components. *Journal of Food and Drug Analysis* 26 (1): 182–192.

68 Liang, N. and Kitts, D.D. (2016). Role of chlorogenic acids in controlling oxidative and inflammatory stress conditions. *Nutrients* 8 (1): 16.

69 Shen, Y., Song, X., Li, L. et al. (2019). Protective effects of *p*-coumaric acid against oxidant and hyperlipidemia – an in vitro and in vivo evaluation. *Biomedicine & Pharmacotherapy* 111: 579–587.

70 Ghosh, T., Maity, T.K., and Singh, J. (2011). Evaluation of antitumor activity of stigmasterol, a constituent isolated from *Bacopa monnieri* Linn aerial parts against Ehrlich ascites carcinoma in mice. *Oriental Pharmacy and Experimental Medicine* 11 (1): 41–49.

71 Liang, Q., Yang, J., He, J. et al. (2020). Stigmasterol alleviates cerebral ischemia/reperfusion injury by attenuating inflammation and improving antioxidant defenses in rats. *Bioscience Reports* 40 (4): BSR20192133.

72 Lopez, A., De Tangil, M.S., Vega-Orellana, O. et al. (2013). Phenolic constituents, antioxidant and preliminary antimycoplasmic activities of leaf skin and flowers of *Aloe vera* (L.) Burm. f. (syn. *A. barbadensis* Mill.) from the Canary Islands (Spain). *Molecules* 18 (5): 4942–4954.

73 Bernatoniene, J. and Kopustinskiene, D.M. (2018). The role of catechins in cellular responses to oxidative stress. *Molecules* 23 (4): 965.

74 Chen, C. (2016). Sinapic acid and its derivatives as medicine in oxidative stress-induced diseases and aging. *Oxidative Medicine and Cellular Longevity* 2016: 3571614.

75 Wang, Z.H., Kang, K.A., Zhang, R. et al. (2010). Myricetin suppresses oxidative stress-induced cell damage via both direct and indirect antioxidant action. *Environmental Toxicology and Pharmacology* 29 (1): 12–18.

76 Maizura, M., Aminah, A., and Wan Aida, W.M. (2011). Total phenolic content and antioxidant activity of kesum (*Polygonum minus*), ginger (*Zingiber officinale*) and turmeric (*Curcuma longa*) extract. *International Food Research Journal* 18 (2).

77 Guo, Y.L., Li, X.Z., and Kuang, C.T. (2011). Antioxidant pathways and chemical mechanism of curcumin. *Advanced Materials Research* 236: 2311–2314.

78 Prinsi, B., Morgutti, S., Negrini, N. et al. (2020). Insight into composition of bioactive phenolic compounds in leaves and flowers of green and purple basil. *Plants* 9 (1): 22.

79 Jayasinghe, C., Gotoh, N., Aoki, T., and Wada, S. (2003). Phenolics composition and antioxidant activity of sweet basil (*Ocimum basilicum* L.). *Journal of Agricultural and Food Chemistry* 51 (15): 4442–4449.

80 Reis, J.F., Monteiro, V.V., de Souza, G.R. et al. (2016). Action mechanism and cardiovascular effect of anthocyanins: a systematic review of animal and human studies. *Journal of Translational Medicine* 14 (1): 1–6.

81 Ren, B., Qin, W., Wu, F. et al. (2016). Apigenin and naringenin regulate glucose and lipid metabolism, and ameliorate vascular dysfunction in type 2 diabetic rats. *European Journal of Pharmacology* 773: 13–23.

82 Pattanayak, P., Behera, P., Das, D. et al. (2010). A reservoir plant for therapeutic applications: an overview. *Pharmacognosy Reviews* 4 (7): 95.

83 Zahran, E.M., Abdelmohsen, U.R., Khalil, H.E. et al. (2020). Diversity, phytochemical and medicinal potential of the genus *Ocimum* L.(Lamiaceae). *Phytochemistry Reviews* 19: 907–953.

84 Adomako-Bonsu, A.G., Chan, S.L., Pratten, M., and Fry, J.R. (2017). Antioxidant activity of rosmarinic acid and its principal metabolites in chemical and cellular systems: importance of physico-chemical characteristics. *Toxicology in vitro* 40: 248–255.

85 An, F., Yang, G., Tian, J., and Wang, S. (2012). Antioxidant effects of the orientin and vitexin in *Trollius chinensis* Bunge in d-galactose-aged mice. *Neural Regeneration Research* 7 (33): 2565.

86 Satyamitra, M., Mantena, S., Nair, C.K. et al. (2014). The antioxidant flavonoids, orientin and vicenin enhance repair of radiation-induced damage. *SAJ Pharmacy and Pharmacology* 1 (1): 1.

87 Espíndola, K.M., Ferreira, R.G., Narvaez, L.E. et al. (2019). Chemical and pharmacological aspects of caffeic acid and its activity in hepatocarcinoma. *Frontiers in Oncology* 9: 541.

88 de Oliveira, D.M. and Bastos, D.H.M. (2011). Biodisponibilidade de acidos fenolicos. *Quim Nova* 34: 1051–1056.

89 Scalbert, A. and Williamson, G. (2000). Chocolate: modern science investigates an ancient medicine. *Journal of Medicinal Food* 3: 121–125.

90 Lafay, S., Gil-Izquierdo, A., Manach, C. et al. (2006). Chlorogenic acid is absorbed in its intact form in the stomach of rats. *The Journal of Nutrition* 136 (5): 1192–1197.

91 Gonthier, M.P., Verny, M.A., Besson, C. et al. (2003). Chlorogenic acid bioavailability largely depends on its metabolism by the gut microflora in rats. *The Journal of Nutrition* 133 (6): 1853–1859.

92 Choudhury, R., Srai, S.K., Debnam, E., and Rice-Evans, C.A. (1999). Urinary excretion of hydroxycinnamates and flavonoids after oral and intravenous administration. *Free Radical Biology and Medicine* 27 (3–4): 278–286.

93 Farah, A., Monteiro, M., Donangelo, C.M., and Lafay, S. (2008). Chlorogenic acids from green coffee extract are highly bioavailable in humans. *The Journal of Nutrition* 138 (12): 2309–2315.

94 Tagliazucchi, D., Helal, A., Verzelloni, E., and Conte, A. (2012). The type and concentration of milk increase the in vitro bioaccessibility of coffee chlorogenic acids. *Journal of Agricultural and Food Chemistry* 60 (44): 11056–11064.

95 Stalmach, A., Williamson, G., and Crozier, A. (2014). Impact of dose on the bioavailability of coffee chlorogenic acids in humans. *Food & Function* 5 (8): 1727–1737.

96 Bountagkidou, O.G., Ordoudi, S.A., and Tsimidou, M.Z. (2010). Structure–antioxidant activity relationship study of natural hydroxybenzaldehydes using in vitro assays. *Food Research International* 43 (8): 2014–2019.

97 Ader, P., Grenacher, B., Langguth, P. et al. (1996). Cinnamate uptake by rat small intestine: transport kinetics and transepithelial transfer. *Experimental Physiology: Translation and Integration* 81 (6): 943–955.

98 Kern, S.M., Bennett, R.N., Mellon, F.A. et al. (2003). Absorption of hydroxycinnamates in humans after high-bran cereal consumption. *Journal of Agricultural and Food Chemistry* 51 (20): 6050–6055.

99 Griffiths, L.A. (1969). Metabolism of sinapic acid and related compounds in the rat. *Biochemical Journal* 113 (4): 603–609.

100 Kanaze, F.I., Bounartzi, M.I., Georgarakis, M., and Niopas, I. (2007). Pharmacokinetics of the citrus flavanone aglycones hesperetin and naringenin after single oral administration in human subjects. *European Journal of Clinical Nutrition* 61 (4): 472–477.

101 Manach, C. and Donovan, J.L. (2004). Pharmacokinetics and metabolism of dietary flavonoids in humans. *Free Radical Research* 38 (8): 771–786.

102 Dang, Y., Lin, G., Xie, Y. et al. (2014). Quantitative determination of myricetin in rat plasma by ultra-performance liquid chromatography tandem mass spectrometry and its absolute bioavailability. *Drug Research* 64 (10): 516–522.

103 Choi, D.H., Li, C., and Choi, J.S. (2010). Effects of myricetin, an antioxidant, on the pharmacokinetics of losartan and its active metabolite, EXP-3174, in rats: possible role of cytochrome P450 3A4, cytochrome P450 2C9 and P-glycoprotein inhibition by myricetin. *Journal of Pharmacy and Pharmacology* 62 (7): 908–914.

104 Andlauer, W., Stumpf, C., and Furst, P. (2001). Intestinal absorption of rutin in free and conjugated forms. *Biochemical Pharmacology* 62 (3): 369–374.

105 Pashikanti, S., de Alba, D.R., Boissonneault, G.A., and Cervantes-Laurean, D. (2010). Rutin metabolites: novel inhibitors of nonoxidative advanced glycation end products. *Free Radical Biology and Medicine* 48 (5): 656–663.

106 de Rooij, B.M., Boogaard, P.J., Rijksen, D.A. et al. (1996). Urinary excretion of *N*-acetyl-*S*-allyl-l-cysteine upon garlic consumption by human volunteers. *Archives of Toxicology* 70 (10): 635–639.

107 Freeman, F. and Kodera, Y. (1995). Garlic chemistry: stability of *S*-(2-propenyl)-2-propene-1-sulfinothioate (allicin) in blood, solvents, and simulated physiological fluids. *Journal of Agricultural and Food Chemistry* 43 (9): 2332–2338.

108 El-Saber Batiha, G., Magdy Beshbishy, A., Wasef, L.G. et al. (2020). Chemical constituents and pharmacological activities of garlic (*Allium sativum* L.): a review. *Nutrients* 12 (3): 872.

109 Thomson, M. and Ali, M. (2003). Garlic [*Allium sativum*]: a review of its potential use as an anti-cancer agent. *Current Cancer Drug Targets* 3 (1): 67–81.

110 da Silva, V.K., de Freitas, B.S., Garcia, R.C. et al. (2018). Antiapoptotic effects of cannabidiol in an experimental model of cognitive decline induced by brain iron overload. *Translational Psychiatry* 8 (1): 1–8.

111 Yang, L., Rozenfeld, R., Wu, D. et al. (2014). Cannabidiol protects liver from binge alcohol-induced steatosis by mechanisms including inhibition of oxidative stress and increase in autophagy. *Free Radical Biology and Medicine* 68: 260–267.

112 Hammell, D.C., Zhang, L.P., Ma, F. et al. (2016). Transdermal cannabidiol reduces inflammation and pain-related behaviours in a rat model of arthritis. *European Journal of Pain* 20 (6): 936–948.

113 Liguori, I., Russo, G., Curcio, F. et al. (2018). Oxidative stress, aging, and diseases. *Clinical Interventions in Aging* 13: 757.

114 Zaveri, N.T. (2006). Green tea and its polyphenolic catechins: medicinal uses in cancer and noncancer applications. *Life Sciences* 78 (18): 2073–2080.

115 Bonnefont-Rousselot, D. (2002). Glucose and reactive oxygen species. *Current Opinion in Clinical Nutrition and Metabolic Care* 5 (5): 561–568.

116 Kasprzak, K.S. (2002). Oxidative DNA and protein damage in metal-induced toxicity and carcinogenesis. *Free Radical Biology and Medicine* 32 (10): 958–967.

117 Zdunska, K., Dana, A., Kolodziejczak, A., and Rotsztejn, H. (2018). Antioxidant properties of ferulic acid and its possible application. *Skin Pharmacology and Physiology* 31 (6): 332–336.

118 Giftson, J.S., Jayanthi, S., and Nalini, N. (2010). Chemopreventive efficacy of GA, an antioxidant and anticarcinogenic polyphenol, against 1,2-dimethyl hydrazine induced rat colon carcinogenesis. *Investigational New Drugs* 28 (3): 251–259.

119 Gandhi, G.R., Jothi, G., Antony, P.J. et al. (2014). GA attenuates high-fat diet fed-streptozotocin-induced insulin resistance via partial agonism of PPARγ in experimental type 2 diabetic rats and enhances glucose uptake through translocation and activation of GLUT4 in PI3K/p-Akt signaling pathway. *European Journal of Pharmacology* 745: 201–216.

120 Rahman, K. (2001). Historical perspective on garlic and cardiovascular disease. *The Journal of Nutrition* 131 (3): 977S–979S.

121 Drobiova, H., Thomson, M., Al-Qattan, K. et al. (2011). Garlic increases antioxidant levels in diabetic and hypertensive rats determined by a modified peroxidase method. *Evidence-based Complementary and Alternative Medicine* 2011: 703049.

122 Chhabria, S.V., Akbarsha, M.A., Li, A.P. et al. (2015). In situ allicin generation using targeted alliinase delivery for inhibition of MIA PaCa-2 cells via epigenetic changes, oxidative stress and cyclin-dependent kinase inhibitor (CDKI) expression. *Apoptosis* 20 (10): 1388–1409.

123 Zhang, X., Zhu, Y., Duan, W. et al. (2015). Allicin induces apoptosis of the MGC-803 human gastric carcinoma cell line through the p38 mitogen-activated protein kinase/caspase-3 signaling pathway. *Molecular Medicine Reports* 11 (4): 2755–2760.

124 Nassan, M.A. and Mohamed, E.H. (2014). Immunopathological and antimicrobial effect of black pepper, ginger and thyme extracts on experimental model of acute hematogenous pyelonephritis in albino rats. *International Journal of Immunopathology and Pharmacology* 27 (4): 531–541.

125 Zhang, M., Viennois, E., Prasad, M. et al. (2016). Edible ginger-derived nanoparticles: a novel therapeutic approach for the prevention and treatment of inflammatory bowel disease and colitis-associated cancer. *Biomaterials* 101: 321–340.

126 Saha, A., Blando, J., Silver, E. et al. (2014). 6-Shogaol from dried ginger inhibits growth of prostate cancer cells both in vitro and in vivo through inhibition of STAT3 and NF-κB signaling. *Cancer Prevention Research* 7 (6): 627–638.

127 El-Ashmawy, N.E., Khedr, N.F., El-Bahrawy, H.A., and Mansour, H.E. (2018). Ginger extract adjuvant to doxorubicin in mammary carcinoma: study of some molecular mechanisms. *European Journal of Nutrition* 57 (3): 981–989.

128 Moon, M., Kim, H.G., Choi, J.G. et al. (2014). 6-Shogaol, an active constituent of ginger, attenuates neuroinflammation and cognitive deficits in animal models of dementia. *Biochemical and Biophysical Research Communications* 449 (1): 8–13.

129 Suk, S., Kwon, G.T., Lee, E. et al. (2017). Gingerenone A, a polyphenol present in ginger, suppresses obesity and adipose tissue inflammation in high-fat diet-fed mice. *Molecular Nutrition & Food Research* 61 (10): 1700139.

130 Pandey, K.B., Mishra, N., and Rizvi, S.I. (2009). Myricetin may provide protection against oxidative stress in type 2 diabetic erythrocytes. *Zeitschrift für Naturforschung. Section C* 64 (9–10): 626–630.

131 Park, K.S., Chong, Y., and Kim, M.K. (2016). Myricetin: biological activity related to human health. *Applied Biological Chemistry* 59 (2): 259–269.

132 Zhang, X.H., Zou, Z.Q., Xu, C.W. et al. (2011). Myricetin induces G2/M phase arrest in HepG2 cells by inhibiting the activity of the cyclin B/Cdc2 complex. *Molecular Medicine Reports* 4 (2): 273–277.

133 Lee, T.J., Kim, O.H., Kim, Y.H. et al. (2006). Quercetin arrests G2/M phase and induces caspase-dependent cell death in U937 cells. *Cancer Letters* 240 (2): 234–242.

134 Araújo, J.R., Gonçalves, P., and Martel, F. (2011). Chemopreventive effect of dietary polyphenols in colorectal cancer cell lines. *Nutrition Research* 31 (2): 77–87.

135 Hsu, C.Y., Shih, H.Y., Chia, Y.C. et al. (2014). Rutin potentiates insulin receptor kinase to enhance insulin-dependent glucose transporter 4 translocation. *Molecular Nutrition & Food Research* 58 (6): 1168–1176.

136 Khazdair, M.R., Boskabady, M.H., Hosseini, M. et al. (2015). The effects of *Crocus sativus* (saffron) and its constituents on nervous system: a review. *Avicenna Journal of Phytomedicine* 5 (5): 376.

137 Srivastava, R., Ahmed, H., and Dixit, R.K. (2010). *Crocus sativus* L.: a comprehensive review. *Pharmacognosy Reviews* 4 (8): 200.

138 Nair, S.C., Pannikar, B., and Panikkar, K.R. (1991). Antitumour activity of saffron (*Crocus sativus*). *Cancer Letters* 57 (2): 109–114.

139 Rana, S.V., Pal, R., Vaiphei, K. et al. (2011). Garlic in health and disease. *Nutrition Research Reviews* 24 (1): 60–71.

140 Mathew, B.C. and Biju, R.S. (2008). Neuroprotective effects of garlic a review. *Libyan Journal of Medicine* 3 (1): 23–33.

141 Yuncu, M., Eralp, A., and Celok, A. (2006). Effect of aged garlic extract against methotrexate-induced damage to the small intestine in rats. *Phytotherapy Research: An International Journal Devoted to Pharmacological and Toxicological Evaluation of Natural Product Derivatives* 20 (6): 504–510.

142 Morino, K., Matsukura, N., Kawachi, T. et al. (1982). Carcinogenicity test of quercetin and rutin in golden hamsters by oral administration. *Carcinogenesis* 3 (1): 93–97.

143 Qiao, H.Y., Dahiya, J.P., and Classen, H.L. (2008). Nutritional and physiological effects of dietary sinapic acid (4-hydroxy-3,5-dimethoxy-cinnamic acid) in broiler chickens and its metabolism in the digestive tract. *Poultry Science* 87 (4): 719–726.

3

Herbs, Spices, and Dietary Constituents as Sources of Phytoantioxidants

N. Shaista Jabeen[1], V. Kiruthiga[1], A. Vinodhini[1], and Mithun Rudrapal[2]

[1] *Department of Zoology, Dhnabagyam Krishnaswamy Mudaliar College for Women (Autonomous), Vellore, Tamil Nadu, India*
[2] *Department of Pharmaceutical Chemistry, Rasiklal M. Dhariwal Institute of Pharmaceutical Education and Research, Pune, Maharashtra, India*

3.1 Introduction

Herbs and spices are traditionally predominant in each and every kitchen all over the globe. People use herbs and spices in their food for their aroma, flavor, and color and they also impart nutrition to the body and maintain human health. The bioactive compounds in herbs and spices possess antioxidant properties. Antioxidants are chemical substances that prevent oxidative stress. Oxidative stress damages body cells and tissues. Eventually it leads to occurrence of various diseases. Davidson [1], Hacskaylo [2], and Smith and Winder [3] commented that herbs and spices are parts of a plant used in foods for their color, flavor, aroma, and nutritive value. Herbs and spices also are used in processed food and beverages as preservatives and alternative medicine to resist against various pathogens.

Based on the statement of Velioglu et al. [4], Larson [5], and Kähkönen [6], herbs and spices are a repository of phytochemicals that bears various biological properties including antioxidant activity. Burt [7] stated that the phytochemicals of herbs and spices perform various biological activities such as antimicrobial activity, anti-inflammatory activity, antioxidant activity, and antitumor activity based on the scientific evidences in documented form.

For the past three decades, extensive research has been carried out on medicinal plants in view of the efficacy of their chemical components on human health. The scientific community explored the antioxidant properties of bioactive components of herbs and spices. The chemical compounds or secondary metabolites are produced by the plant cell in order to protect the plant from disease causing pathogens.

McCormick [8] and Peter [9] commented that herbs and spices consist of antioxidants in large amount, which includes flavonoids, alkaloids, tannins, phenolic compounds, phenolic diterpenes, and vitamins. These compounds possess antioxidant activities. For instance flavonoids are able to scavenge free radicals and can bind with metal irons to inactivate the reactive oxygen species (ROS) and reactive nitrogen species (RNS). These two species are the main causes of oxidative stress. Oxidative stress is caused by high concentration of free radicals in cells and tissues. Documented evidences proved that certain herbs and spices such as sage, rosemary, and oregano are excellent sources of antioxidants containing remarkable amount of phenolic compounds.

Phytoantioxidants and Nanotherapeutics, First Edition. Edited by Mithun Rudrapal.

3.1.1 Reactive Oxygen Species and Reactive Nitrogen Species

ROS is defined as oxygen containing reactive species. ROS is generated in the cellular organ system located in the plasma membrane, cytosol, endoplasmic reticulum, and mitochondrial membrane of a cell in living beings. According to Yesiloglu et al. [10] and Gerschman et al. [11] RNS is derived from nitric oxide (+NO) and superoxide (O_2^-) produced by the action of the enzyme nitric oxide synthase 2 (NOS2) and NADPH oxide sequentially. RNS and ROS act together to damage cells causing nitrosative stress. RNS is produced in animals by the action of nitric oxide (+NO) with superoxide (O_2^-) to form peroxynitrite (ONOO). Peroxynitrite directly reacts with proteins; therefore, it can modify proteins such as hemoglobin, myoglobin, and cytochrome C by oxidizing ferrous into ferric forms. The enzyme nitric oxide synthetase converts L-arginine into L-citrulline and NO* by oxidizing guanidine nitrogen of L-arginine. Yu [12] reported that ROS is one of the key factors in tissue injury (Table 3.1).

3.1.2 Definition of Herbs, Spices, Dietary Antioxidants, and Phytoantioxidants

Herbs and spices are a repository of therapeutic chemical compounds. The term "herbs and spices" refers to plants and various parts of plants. They add flavors to any kind of dish. Herbs and spices are used worldwide for culinary purposes, in medicines due to their sweet aroma, in cosmetics, and in perfumes. Herbs and spices are harvested mainly for culinary purposes and have medicinal and economic value too.

As reported by Patch [13] and Jiang [14], many herbs and spices are the hub of antioxidants and are extensively used in foodstuffs pertaining to their health benefits encompassing from the ancient era. The health benefits includes digestive protective, anti-inflammatory, antioxidant, cardioprotective, antimicrobial, immunomodulatory, antiviral, antidiabetic, antiallergic, antiaging, anticarcinogenic, antitumor, and antiproliferative activities. On account of these benefits, they are used in many acute and chronic diseases such as hypertension, cardiovascular diseases, cancer, diabetes, neurodegenerative diseases, arthritis, asthma, metabolic disorders, obesity, etc. Besides these benefits, herbs and spices are used in cuisines as seasonings in order to enhance the flavor, aroma, and color of foods. They are used also as preservatives in processed foods and beverages to keep the food nutrients intact for certain period of time.

Table 3.1 Reactive oxygen species (ROS) and reactive nitrogen species (RNS).

Reactive oxygen species	Reactive nitrogen species
Hydroxyl radical (*OH)	Nitric oxide (NO*)
Superoxide (O_2^-)	Peroxynitrite ($ONOO^-$)
Hydrogen peroxide (H_2O_2)	Nitrogen dioxide radical ($NO_2^{\cdot}$)
Singlet oxygen ($^1O_2$)	
Peroxyl radical (LOO*)	
Alkoxyl radical (LO*)	
Lipid hydroperoxide (LOOH)	
Peroxynitrite ($ONOO^-$)	
Hypochlorous acid (HOCl)	
Ozone (O_3)	

Herbs and spices are classified into medicinal herbs, culinary herbs, aromatic herbs, ornamental herbs, astringent herbs, bitter herbs, mucilaginous herbs, and nutritive herbs. Herbs and spices are categorized into seven distinct groups: annuals, biennials, perennials, shrubs, trees, climbers, and grasses. The growth range of herbs and spices are varied. The growing conditions and habitat are from tropical conditions.

Herbs are distinguished as biennials, perennials, and annuals because the growing span of fresh herbs tends to be dependent on their stipulated period and seasons.

3.1.2.1 Biennial Herbs

Biennial plants need two years to complete their whole life cycle from seed to plant and plant to seed. Only few herbs are biennials. They are as follows:

- Sage, stevia, and parsley

3.1.2.2 Perennials

The perennials grow each year. They continually grow for many seasons. The apex part of the plant dies in winter, the crown becomes dormant, and again blooms in the spring. Most of the culinary herbs are perennials. The herbs under this category are as follows:

- Oregano, mint, fennel, tarragon, thyme, bay leaves, chives, winter savory, lavender, rosemary, holy basil, ginger, turmeric, and lemongrass

3.1.2.3 Annuals

Plants that grow their whole life cycle from seed to flower and again to seed in a single growing season are categorized as annuals. Once the whole life cycle completes, the stems and leaves of annual plants will die, and collecting the seeds of such plants can give another growing cycle. The plants under this category are as follows:

- Basil, cilantro, marjoram, chervil, summer savory, coriander, dill, black cumin, cumin, fenugreek, mustard seeds, and red chilly (Table 3.2)

Spices used for flavor in food are coriander, cinnamon, clove, fennel, mustard, saffron, ginger, garlic, asafetida, bay leaf, cumin seeds, fenugreek, turmeric, star anise, black pepper, cardamom, etc., which are added as zest to food. Traditionally, folklore medicines are based on medicinal plants; they have been used since millennium. The efficacy of medicinal plants is also mentioned in many holy books such as the holy Quran and the holy Bible.

3.1.3 Antioxidants in Herbs, Spices, and Dietary Products

Fresh and dried herbs and spices contain a vast amount of antioxidants that impart color, flavor, and aroma to foods to enhance their organoleptic qualities. Dietary products include vegetables, fruits, dry fruits, whole grains, pulses, cereals, etc. Previously, Yashin et al. [15] reported that antioxidant components in herbs and spices are diverse in plants and plant parts commonly found in ginger, garlic, cloves, turmeric, and cayenne, which are used in daily diet.

As reported by Demmig-Adams and Adams [16], plants possess numerous antioxidant compounds such as polyphenols, flavonoid tocopherols, carotenoids, glutathione, ascorbic acids, and antioxidant enzymes against redox reactions that protect the plants from oxidative damages. Consumption of antioxidant-rich plants in the form of diet includes carotenoids, phenolic compounds like benzoic acid derivatives, flavonoids, proanthocyanidins, stilbenes, coumarins, lignins, and lignans (Table 3.3).

Table 3.2 Classification of herbs.

Classes	Common herbs
Medicinal herbs	Ginger, garlic, gingko, lavender, mint, milk thistle, Saint John's wort, calendula
Culinary herbs	Mint, parsley, thyme, cilantro, oregano, bay leaf, rosemary, sage, chervil, dill marjora, lavender, chives
Aromatic herbs	Coriander, tarragon, fennel, lavender, clove pink, rose, scented geranium, lemon verbena
Ornamental herbs	Oregano, thyme, chamomile, sage, catmint rosemary, cardoon
Astringent herbs	Wild strawberry, witch hazel, lady's mantle, Shepherd's purse, yarrow, bethroot, horsetail
Bitter herbs	Mugwort, peppermint, wormwood, neem tree, chicory gentian, angelica, common sorrel yarrow, peppermint, valerian
Mucilaginous herbs	Fenugreek, plantain, psyllium, comfrey, slippery elm, marshmallow
Nutritive herbs	Nettles, calendula, raspberries, red cloves, horsetail, chickweed

Table 3.3 Major chemical compounds in herbs and spices.

Herbs and spices	Active antioxidant compounds	Therapeutic effects
Ajwain (*Trachyspermum ammi* L.) family Apiaceae (Umbelliferae)	Carvacrol, α-pinene, *p*-cymene, limonene, α-terpinene, essential oil, thymol (phenols)	Hypolipidemic, antimicrobial, antihypertensive, antispasmodic, antilithiasis and diuretic, antitussive, nematicidal, antihelminthic, antifilarial
Allspice (*Pimenta dioica* L.) Merr family Myrtaceae (Myrtle family)	Essential oil (eugenol) phenolic acids, flavonoids, catechins, phenyl propanoids, methyl eugenol, β-caryophyllene, humulene, terpinen-4-ol	Indigestion, flatulence, diarrhea, rheumatism, arthritis, stiffness, chills, congested coughs, bronchitis, neuralgia, anesthetic, analgesic, antiseptic, carminative, muscle relaxant, rubefacient, stimulant, purgative properties, bactericidal, fungicidal, antioxidant
Angelica (*Angelica archangelica* L.) family Apiaceae (Umbelliferae)	Essential oil, pinene, α- and β-phellandrene, limonene, sabinene, δ-3-carene, myrcene, *A. archangelica* oil, monoterpene hydrocarbons, α-pinene, δ-3-carene, *p*-cymene, imonene, sabinene	Hepatic, nervine, stimulant, stomachic, tonic, antispasmodic, aphrodisiac, anticoagulant, bactericidal, carminative, diaphoretic, digestive, diuretic, emmenagogue, expectorant, febrifuge, antiproliferative activity
Anise (*Pimpinella anisum* L.) family Apiaceae (Umbelliferae)	*Trans*-(*E*)-anethole, bergapten, umbelliprenine, umbelliferone, scopoletin (coumarins) quercetin-3-glucuronide, rutin, luteolin-7-glucoside, isoorientin, isovitexin, apigenin-7-glucoside (flavonoid glycosides)	Carminatives, antiseptics, diuretics, digestives, aphrodisiacs, insomnia, constipation, dyspepsia, abdominal colic, nausea, epilepsy, seizures, expectorant actions, antiviral, immunostimulator, measles virus

Table 3.3 (Continued)

Herbs and spices	Active antioxidant compounds	Therapeutic effects
Black pepper (*Piper nigrum* L.) family Piperaceae	Piperinic acid exists in four isomeric forms: piperine, isopiperine, isochavicine, and chavicine. Quercetin, isoquercetin, isorhamnetin 3-β-δ-rutinoside, kaempferol 3-arabinoside, kaempferol-3-*O*-β-galactoside, and quercetin-3-*O*-β-δ-rutinoside are the major flavonols, ignans-cubebin, tocopherols, polyphenols	Stomachic, carminative, antioxidant, antibacterial, antimicrobial, immunomodulatory, larvicidal, antibiotic, anti-inflammatory, antitumor, antipyretic, diaphoretic properties, anticancer, pancreatic protective
Bay leaf (*Laurus nobilis*) family Lauraceae	Quercetin, kaempferol sesquiterpen, pinene, eugenol, phellandrene, pinene, cineole, lauric acid, alcohols, sesquiterpenes, cinimic acid, esters, terpene hydrocarbons	Antidiabetic, hypolipidemic, antihypercholesterolemic, anti-inflammatory, antiulcerogenic, antimicrobial, antifungal, hypoglycemic properties (in the control of diabetes)
Caraway (*Carum carvi* L.) family Apiaceae (Umbelliferae)	Limonene, volatile oil and its oleoresins, flavonoids, carvone, essential oil, flavonoids (quercetin-3-glucuronide, isoquercetin), quercetin 3-*O*-caffeyl glucoside, kaempferol 3-glucoside, phenolic compounds	Antioxidant, antibacterial, antiviral, carminative, antispasmodic, antihistaminic, antidiabetic, anticarcinogenic, antimicrobial, insecticidal antifungal, antibacterial activity, antidiabetic
Chervil (*Anthriscus cerefolium* L.) family Apiaceae (Umbelliferae)	Essential oil, methylchavicol, apiin, luteolin, glucoside, fixed oil	Antilipoperoxidation activity, expectorant, stimulant, diuretic, eczema, antihypertensive, gout, kidney stones, pleurisy, dropsy, menstrual problems
Chives (*Allium schoenoprasum* L.) family Liliaceae	DL-α-Tocopherol and quercetin (superoxide dismutase, catalase, peroxidase, glutathione peroxidase), malonyldialdehyde, superoxide, hydroxyl radicals, reduced glutathione (total flavonoids) α and β-carotenoids	Antihypertensive, aids in digestion, antiproliferative, gastric cancer, antinematodal activities, antimicrobial, anticarcinogenic (prostate cancer), membrane protective, antioxidant activity
Cilantro (*Coriandrum sativum*) family Apiaceae	Phenolic compounds linalool, caffeic acid, glycitin terpinene, quercetin, kaempferol, ferulic, *n*-coumaric and vanillic acids, ascorbic acid, pyrogallol, flavonoids	Radical scavenging activity, hydroxyl radical scavenging activity, glutathione reduction, antilipidperoxidation, hepatoprotective, antidiabetic urethritis, cystitis, urinary tract infection, urticaria, rash, burns, sore throat, vomiting, indigestion, nosebleed, cough, allergies, hay fever, dizziness, amebic dysentery
Cloves (*Syzygium aromaticum* L.) family Myrtaceae	Phenolic acids (gallic acid), flavonol, glucosides, phenolic volatile oils (eugenol, acetyl eugenol), tannins	Antiflatulence, analgesic, stimulant, carminative and antinauseant, hypolipidemic activity

(*Continued*)

Table 3.3 (Continued)

Herbs and spices	Active antioxidant compounds	Therapeutic effects
Cinnamon (*Cinnamomum verum*) family Lauraceae	Cinnamaldehyde, cinnamic acid, cinnamate, eugenol polyphenols, catechin, epicatechin, procyanidin, quercetin, kaempferol, flavonoids, proanthocyanidins, oligomers	Antibacterial, antifungal, anti-inflammatory activity, cardiovascular protective, antihypertensive, antidiabetic, spleen, breast, uterus, liver, stomach tumors, stimulant, carminative, astringent, aphrodisiac
Cardamom (*Elettaria cardamomum* L.) family Zingiberaceae	α-Terpinyl acetate, 1-80-cineole, linalool volatile oil, α-terpinyl acetate, 1,8-cineole, limonene, linalyl acetate, linalool	Antidepressive, carminative, appetizer, diuretic, immunomodulator, antitumor activity, gastroprotective
Capsicum (*Capsicum annuum* L.) family Solanaceae	Capsaicinoids (alkaloid) capsaicin, dihydrocapsaicin capsinoids, capsiate, dihydrocapsiate, β-carotene, capsanthin, quercetin, and luteolin	Carminative, diarrhea, cramps, colic, rheumatism, arthritis anti-inflammatory, antitumor activity, antihyperglycemia, antihypertensive, antitumor activity
Dill (*Anethum graveolens* L.) family Apiaceae (Umbelliferae)	Flavonol glucuronides, chlorogenic acid, quercetin, isorhamnetin	Antidiabetic, hypolipidemic, carminative, stomachic, spasmodic, sedative, lactagogue, diuretic disorders, gastrointestinal, kidney, UTI
Fenugreek (*Trigonella foenum-graecum* L.) family Fabaceae (Leguminosae)	Sesquiterpenes, aromatic aldehydes, terpenes, flavonoids (vitex and vitexin glucoside), β-pinene, camphor, β-caryophyllene, neryl acetate	Antipyretic, abdominal pain, kidney ailments, hernia, arthritis, beriberi, anthelmintic, antihypertensive, cardioprotective, antilithogenic
Fennel (*Foenicuum vulgare)* family Apiaceae	Fenchone, estragole, anise aldehyde, *trans*-anethole, essential oil, quercetin (flavonoid)	Antibacterial, antifungal, antioxidant, hepatoprotective, antidiabetic, antineoplastic, anti-inflammatory, hepatoprotective, anticancer ear-edema, antihyperlipidemic
Ginger (*Zingiber officinale*) family Zingiberaceae	Shogoal, gingerol, linolic acid, pyrogallol *p*-hydroxybenzoic acid, ferulic acid, *p*-coumaric acid	Carminative, anti-nauseant, diuretic, antiflatulence, antihistaminic, aphrodisiac, cholesterol lowering
Garlic (*Allium sativum* L.) family *Alliaceae*	Caffeic, vanillic, *p*-hydroxybenzoic, *p*-coumaric acids, allicin, volatile oil	Antimicrobial, diuretic, diaphoretic, antiflatulence, cholesterol lowering, anti-inflammatory, fevers, diabetes, rheumatism, intestinal worms, colic, flatulence, dysentery, liver disorders, tuberculosis, facial paralysis, high blood pressure, bronchitis
Mustard (*Brassica nigra*), family Brassicaceae	β-Carotene, thiamine, riboflavin, niacin, essential oil, glucosinolate sinigrin, allyl isothiocyanate	Laxative, congestion, neuralgia, spasm, cytoprotective, anticancer activity
Nutmeg (*Myristica fragrans* Houtt.) family Myristicaceae	Polyphenols, total tannins, true tannins, sabinene, α-pinene, β-pinene, myristicin (essential oil)	Antimicrobial, anti-inflammatory, arthritis, rheumatism, spasmolytic hallucinogenic effects, anti-cholinergic, antibacterial, hepatoprotective

Table 3.3 (Continued)

Herbs and spices	Active antioxidant compounds	Therapeutic effects
Oregano (*Origanum vulgaris*) family Lamiaceae	Camphene, carvacol, γ-terpinene, thymol, terpinen-4-ol, myricene, linalyl acetate	Antifungal, antibacterial, antiviral, spasmolytic, anti-inflammatory, cough, headache, menstruation, seasickness
Poppy (*Papaver somniferum* L.) family Papaveraceae	Linoleic acid, oleic acid, linolenic acid, sitosterol, stigmasterol	Anticarcinogenic, antitussive, angiogenesis property, hyperproliferative, skin disorder
Peppermint (*Mentha ×piperita* L.) family Lamiaceae (Labiatae)	Essential oil, flavonoids, phytols, tocopherols, azulenes, rosmarinic acid, carotenoids, tannins, luteolin, hesperidin, rutin, caffeic acid, chlorogenic acid, α- and γ-tocopherols, α- and β-carotenes, menthol, menthone, menthyl acetate, isomenthone, menthofuran, piperitone	Digestive disorders, headaches, common colds, sore throats, insomnia, fever, antimicrobial, antiviral, anti-inflammatory, carminative, spasmolytic, choleretic activity, dyspepsia, flatulence, intestinal colic, antispasmodic, antioxidant, anthelmintic, antiulcer, cytoprotective, hepatoprotective, cholagogue, chemopreventive, antidiabetogenic
Rosemary (*Rosmarinus officinalis* L.) family Lamiaceae	Carnosol, carnosic acid, phenolic diterpene, phenolic acid, rosmanol, epirosmanol, rosmarinic acid, caffeic acid, ester, tosemaridiphenol, 3-(3,4 dihydroxyphenyl), lactic acid, flavonoids (apigenin, diosmin, luteolin), tannincineole, pinene, camphors (essential oil)	Antidiabetic, hepatoprotective, antiangiogenic, anti-inflammatory, antioxidant, antimicrobial, antihypertensive, antithrombotic, antiproliferative, anticarcinogenic activity, hypolipidemic
Star anise (*Illicium verum*) family Illiciaceae (Magnoliaceae)	Essential oil, *trans*-anethole, essential oil, sesquiterpenes, phenylpropanoids, lignans, flavonoids, α-pinene, phellandrene, *p*-cymene, 1,4-cineole, limonene, δ-terpineol	Antibacterial, carminative, diuretic, stomachic, expectorant, anticolinesterase, abdomen ache, antifungal, antioxidant, insecticidal, analgesic, sedative, anticarcinogenic, antitumor activity
Saffron (Zaffran/ Kesar) (Iridaceae)	Crocetine (carotenoid) crocin, safranal, α,β-carotene, lycopene, zeaxanthin	Hepatoprotective, neuroprotective, anti-inflammatory, cardioprotective, bladder protective, arthritis, chemotherapeutic agent, atherosclerosis, antitumorigenic effect, pancreatic cancer, anticancer agent, antiproliferative, radical scavenging activity
Sage (*Salvia officinalis*) family Lamiacea	Carnosol, carnosic acid, rosmanol (phenolic components), rosmarinic acid (phenolic compound)	Antihyperlipidemic, antidiabetic, anti-inflammatory

(Continued)

Table 3.3 (Continued)

Herbs and spices	Active antioxidant compounds	Therapeutic effects
Turmeric (*Curcuma longa* L.) family Zingiberaceae (Ginger family)	Curcumin, (curcuminoids), demethoxylated curcumins, oleoresin, bisdemethoxycurcumin, volatile oil, α-turmerone, β-turmerone, terpinolene, α-phellandrene, curcumadiol, carotenoid, bisacumol, bisacurone, curcumenol, isoprocurcumenol, zedoaronediol, procurcumenol, epiprocurcumenol, germacrone-1, linoleic acid	Anti-inflammatory, hypocholesteremic, choleretic, antimicrobial, antirheumatic, antibacterial, antiviral, cytotoxic, spasmolytic, hypersensitive, antidiabetic, antihepatotoxic, arthritis, anticarcinogenic, atherosclerosis, anti-inflammatory, antioxidant, antiproliferative, pro-apoptotic, chemopreventive, chemotherapeutic, wound healing, antinociceptive, antiparasitic, antimalarial, anti-invasive, antiangiogenic, wound healing, diabetes, Alzheimer's disease, Parkinson's disease, cardiovascular disease, pulmonary disease, ulcerative colitis, colon cancer, pancreatic cancer, pancreatitis, psoriasis, chronic anterior uveitis
Thyme (*Thymus vulgaris* L.) family Lamiaceae (Labiatae)	Thymol, (apigenin, luteolin, diosmetin, naringenin, kaempferol, quercetin, hesperidin, caffeic acid, rosmarinic acid) (flavonoid and phenolic acids)	Antibacterial, antimicrobial, antifungal activities, anti-inflammatory activity, spasmolytic

3.2 Chemical Composition of Herbs and Spices as Potent Antioxidants

Bhattacharjee and Sengupta [17] and Allwyn Sundar Raj et al. [18] reported that the antioxidant activity of herbs, spices, and dietary products is mainly due to their chemical contents such as flavonoids and polyphenolic compounds and other biological constituents. Table 3.2 lists out potent antioxidants and their biologically active compounds present in culinary herbs and spices including flavonoids, phenolic acids, lignans, essential oils, volatile oils, and alkaloids, which are mentioned concisely in many publications.

3.3 Antioxidant Properties of Herbs and Spices

Antioxidant components are found in plant products. Its chemical components demonstrated the radical scavenging activities. In addition to other constituents of plants such as vitamins, minerals, ascorbate, zinc, carotenoids, and α-tocopherols, herbal medicines also consist of flavonoids, polyphenols, and flavoproteins. Some herbs and spices exert antioxidant properties by scavenging free radicals like superoxide dismutase actions in various cellular and tissue regions. Some fruits, vegetables, and commonly used herbs and spices bear cancer protective agents apart from antioxidants. Based on their chemical structure, the bioactive compounds are divided into polyphenols, carotenoids, flavonoids, tocopherols, terpenes, retinoids, isothiocyanates, and indoles.

The antioxidant compounds such as flavonoids, polyphenolics, terpenoids, lignans, sulfides, carotenoids, saponins, coumarins, plant sterols, curcumins, and phthalides present in spices are mainly responsible for their antioxidant properties

3.4 Role of Flavonoids in Culinary Herbs and Spices

Flavonoid and phenolic compounds are polyphenols in herbs and spices that possess antioxidant activities. Flavonoids are widely distributed secondary metabolites with a wide range of functions in plants. So far there are 10 000 flavonoids across the plant kingdom. Flavonoids are the derivatives of glycosylated compounds like apigenine, and luteolins are the components of aromatic herbs and spices. A high content of flavonoids is found in celery, parsley, oregano, saffron, dill, fennel, and pepper. Therefore, a diet rich in culinary herbs and spices provides a significant ratio of plant antioxidants. According to Winkel-Shirley [19], flavonoids are classified into six major subgroups: chalcones, flavones, flavonols, flavandiols, anthocyanins, and proanthocyanidins or condensed tannins (Figure 3.1).

Flavonoids are plant components attributed to many health benefits. There are six different types of flavonoids, each with distinct health promoters. They are as follows:

Different types of flavonoids					
Flavonols	Flavones	Flavon3ols	Flavonones	Anthocyanidins	Isoflavones

3.5 Medical Importance of Flavonoids

Flavonoids are polyphenolic compounds that include flavonols (onions, broccoli, tea, and fruits), flavones (parsley, celery, chamomile tea), flanones (citrus fruits), flavonols, catechins and procyanidins (cocoa apple, grape, red wine, tea), anthocyanidins (colored berries, red wine), and isolavones (soy).

A diet rich in plant foods provides phytonutrients that protect and prevent a body from disease causing pathogens. The phytonutrients like flavonoids possess anti-inflammatory properties that protect cells from oxidative damage that leads to generation of disease. These dietary antioxidants

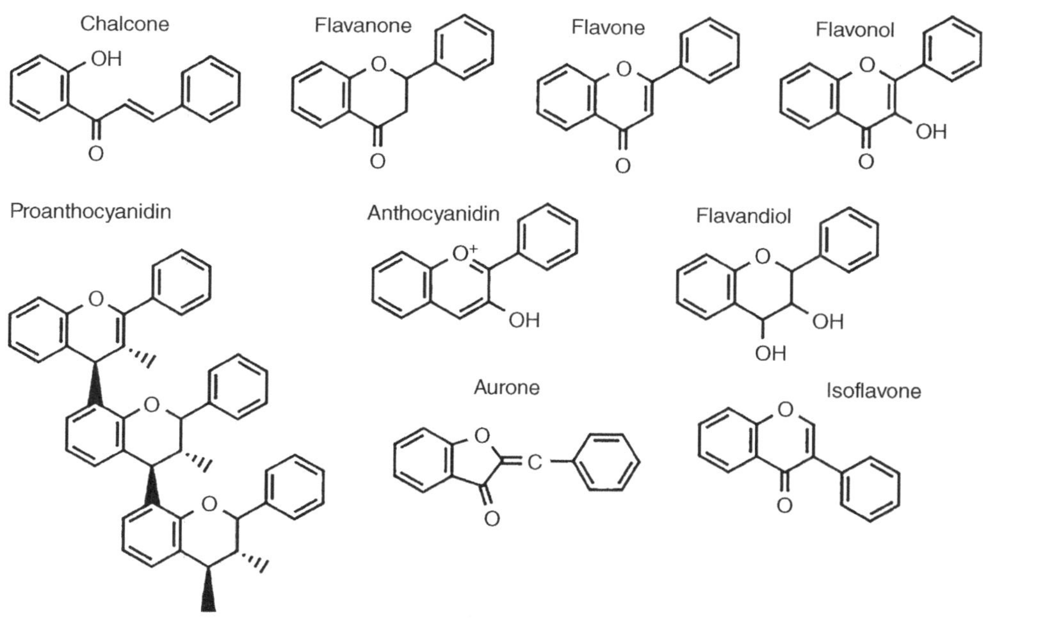

Figure 3.1 Different compounds of flavonoids.

can prevent the development of diseases such as cardiovascular disease, diabetes, cancer, and cognitive diseases such as Alzheimer's disease, Parkinson's disease, and dementia.

3.5.1 Pharmacology of Flavonoids

Flavonoids have been reported to be a potent secondary metabolite that exerts a wide range of biological activities:

- Antioxidant activity
- Cardioprotective effect
- Anticarcinogenic effect
- Anti-inflammatory
- Antimicrobial effect

3.5.1.1 Cancer Prevention

Consumption of flavonoids containing food prevents the risk of different types of cancers such as breast, prostate, and colon cancer. Many of them have been reported to be anti-carcinogenic and their action have been studied separately.

3.5.1.2 Quercetin

Torres et al. [20] stated that constituents of herbs and spices are tannins, alkaloids, phenolic diterpenes, organic sulfur, polyphenol, polyphenolpropanoid, flavonoids, vitamins, and anthocyanins; all these compounds have protective activity against oxidizing agents. All these protective effects are primarily due to the presence of "flavonoids" and "phenolic contents" in herbs and spices. Previously Slimestad et al. [21] reported that diet rich in flavonoid content decreases the risk of chronic diseases. Flavonoid content acts as defense mechanism against oxidative stress. It is also a scavenger of free radicals.

Report of Yashin et al. [15] stated that phenolic compounds undergo redox reaction. The content of phenolic compounds is a significant component that bears antioxidant activity in herbs and spices (Figure 3.2).

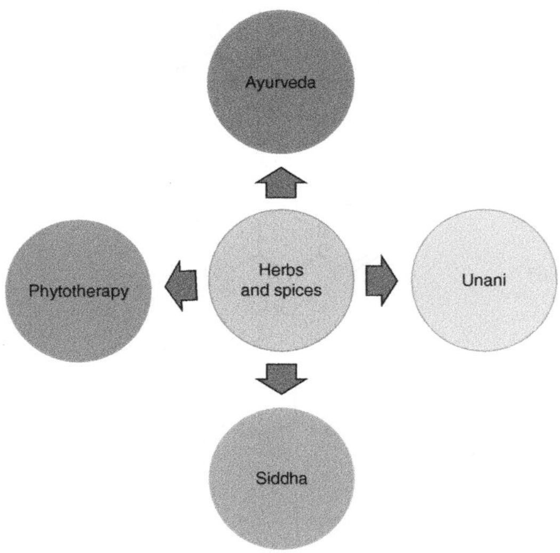

Figure 3.2 Application of herbs and spices.

3.6 Natural Antioxidants

Rice-Evans [22] commented that medicinal plants are natural sources of antioxidants. Previously Prior and Cao [23] reported that natural antioxidants enhance the antioxidant activity by reducing the risk of certain chronic diseases such cancer, stroke, and cardiac diseases. As per Mathew and Abraham [24], the bioactive compounds of plants such as flavonoids and phenolic compounds are potent to scavenge free radicals. They are found in all parts of a plant such as roots, bark, stem, leaves, fruits, and seeds. As per the report of Ito et al. [25], synthetic antioxidants cause several side effects such as hepatic damage and carcinogenesis in acclimatized laboratory animals. However, Gao et al. [26], Williams et al. [27], and Osawa and Namiki [28] stated that there is a need for harmless, nontoxic, and easily available antioxidants. Such kind of antioxidants are found in medicinal plants that are effective, harmless, and with no side effects on their consumption.

3.7 Role of Indigenous, Culinary Herbs and Spices in Daily Diet

Indian food is most popular for its wide range of herbs and spices. The variety of dishes native to India make Indian food one of the most wholesome foods all around the globe. Unfortunately studies reported that Indian foods contain lot of carbohydrates and sugars linked to high cholesterol, which is unhealthy. However, traditional Indian foods are considered to be very healthy and immensely tasty with a lot healthy ingredients that has a myriad benefits to human health such as it boosts immunity, controls blood sugar, suppresses inflammation, and improves mental health and other functions of the human body. Because traditional foods are blended with a lot of healthy ingredients such as whole grains (white rice, whole wheat flour, millet, spelt, brown rice, whole grain breads, corn) and lentils such as peas, beans, and dal varieties and a wide range of black-, green-, red-, yellow-, and brown-colored lentils, they are rich in dietary fiber and possess many health benefits such as maintaining bowel health, controlling cholesterol and blood sugar levels, and managing weight, which helps to lead a longer healthy life. Legumes are different types of beans with many nutritious values abundant with protein and fibers. Different types of vegetables, herbs, and spices naturally enhance the nutritional quality of traditional Indian foods. These ingredients of Indian diet comprise high antioxidant activity. This activity protects against cytotoxicity and prevents from various acute and chronic illnesses.

3.7.1 Basil leaves (*Ocimum basilicum*, family Lamiaceae)

- Basil leaves along with other herbs are used in Indian herbal tea.
- Fresh or dried basil leaves are added to soups.
- They are added to fried vegetables.
- They are sprinkled as seasoning on garlic bread.

Traditionally basil seeds are used in medicines to cure stomach spasm, gastritis, loss of appetite, and kidney diseases. In Indian culture basil plant is kept in houses to prevent from negativity. It is a sacred plant according to Indian tradition. Therefore, it is termed as "holy basil."

3.7.1.1 Health Benefits

- Studies have proved that holy basil is resistant against infection.
- It boosts immunity.

- Consumption of basil leaves early in the morning or after meals reduces blood sugar level.
- It reduces anxiety and depression.

3.7.2 Clove (*Syzygium aromaticum*, family Myrtaceae)

It can be used as whole or in pods in sweet dishes and curries, preferably for its rich aroma and spicy odor.

- It is commonly used in rice, meat, and vegetable curry.
- Ground clove is used in garam masala.
- It has a wide range of health benefits too. Cloves are a great source of β-carotene responsible for their rich brown color. Carotene compounds are good antioxidant substance.

Health benefits:

- Chewing clove enhances oxygen level and keeps the microbes at bay.
- The carotene pigments in cloves convert into vitamin A, which is essential for healthy eyes.
- Eugenol is an active component of clove that is an anti-inflammatory compound and a potent antioxidant. Cloves possess several antioxidant compounds that prevent the risk of developing chronic diseases such as heart disease, diabetes, cancer, and arthritis, and it is one of the best remedy to relive toothache.
- Cloves improve liver functions and prevent liver cirrhosis and is also a good remedy for stomach ulcers.

3.7.3 Ginger (*Zingiber officinale*, family Zingiberaceae)

It is a most popular and widely used spice. Ginger-garlic paste is almost always used in all types of Indian curries, and ginger is also used in herbal tea. Ginger is one of the best remedy for nausea and is used to make candies.

Health benefits:

- Ginger possesses strong anti-inflammatory properties that can reduce any kind of pain.
- Intake of ginger extract reduces colon inflammation.
- It is the best remedy for nausea.
- It reduces blood glucose levels, obesity, and prevents osteoarthritis.

3.7.4 Peppermint (*Mentha piperita*, family Lamiaceae)

The scientific name for peppermint (*Mentha × piperita*) is derived from the name *Mintha*, a Greek mythological nymph. Mint leaves are almost always used in all types of Indian cuisines such as vegetables, non-vegetarian curries, salads, desserts, and beverages such as mint tea, lime-mint juice, etc., especially for their strong flavor, aroma, and color and they also possess medicinal benefits.

Pharmacology of peppermint:

- Rich in nutrients
- Improves irritable bowel syndrome
- Relives indigestion

- Good for brain function
- Reduces lactating pain
- Improves cold symptoms
- Discards bad breath

3.7.5 Coriander (*Coriandrum sativum*, family Apiaceae)

Coriander belongs to family Apiaceae. It is also called cilantro or Chinese parsley. It is used as both fresh and dried spice in various cuisines. Coriander is used in Indian, American, and Chinese dishes such as curries, sausages, and liqueurs. Primarily it used for culinary purposes, and it has medicinal value.

3.7.5.1 Pharmacology of Coriander

Coriander is reported to have several pharmacological benefits, including antioxidant, antimutagenic, antihelminthic, antidiabetic, diuretic, antifungal, antibacterial, anticancer, anticonvulsant, anti-ulcerative, and hepatoprotective.

3.7.6 Cardamom (*Elettaria cardamomum*, family Zingiberaceae)

Cardamom consists of whole dried fruits or seeds. The seeds have pungent taste and are strongly aromatic in flavor. Cardamom is used in tea, desserts, and vegetable and mutton curries. On the whole it is used in for flavoring purpose in foods and as a carminative agent.

3.7.6.1 Pharmacology of Cardamom

Traditionally cardamom is used to treat asthma, constipation, colic, diarrhea, dyspepsia, hypertension, and epilepsy. According to Kapoor [29], cardamom is potent as antibacterial, antifungal, antiviral, carminative, diuretic, and stomachic agent.

3.8 Role of Antioxidants in Dietary Herbs and Spices – Indian Diet

Ame et al. [30] reported that ROS production is the root cause of generation of chronic diseases such as cancer, cardiovascular disease, neurodegeneration, and ageing process. As per the reports of Demmig-Adams and Adams [16], dietary antioxidants alleviate ROS generation. It is widely accepted that high intake of plant-based diet reduces the oxidative damage leading to the formation of diseases. Plants and plant products comprise high antioxidant chemical substances such as polyphenols, carotenoids, coumarins, tocopherols, glutathione, ascorbic acids, and antioxidant enzymes with antioxidant properties. There are other antioxidant components in plant-based diet such as carotenoids, flavonoids, phenolic compounds (benzoic acid derivatives), coumarins, proanthocyanidins, stilbenes, lignans, and lignins. Primary antioxidants such as phenolic compounds (e.g. α-tocopherols) scavenge ROS, whereas secondary antioxidants act by a number of mechanisms such as scavenging free radicals, ROS, binding of metal ions, and conversion of hydrogen peroxide to nonradical species. Some major antioxidant compounds found in food products are summarized in Table 3.4.

Table 3.4 Major antioxidants of herbs and spices.

Spices and herbs	Major antioxidant compounds
Allspice	Eugenol, flavonoids, proanthocyanidins
Basil	α-Tocopherol, histidine-containing dipeptides, ubiquinone, glutathione carnosine, anserine
Bay leaves	Flavonols (kaempferol, myricetin, quercetin), flavones (apigenin, luteolin), glycosylated flavonoids, sesquiterpenes, lactones, monoterpenes, germacrane alcohols
Cloves	Phenolic acids (gallic acid), flavonol glucosides, phenolic volatile oils (eugenol, acetyl eugenol), tannins
Cardamom	Limonene, 1,8-cineole, terpinolene, myrcene, caffeic acid, quercetin, kaempferol, luteolin, pelargonidin
Cinnamon	Eugenol, limonene, terpineol, catechins, proanthocyanidins, tannins, linalool, safrole, pinene, methyleugenol, benzaldehyde
Dill	Flavonoids, carotenoids, terpenoids, phenolic acids, ascorbic acids
Fennel	Phenylpropanoid derivatives, monoterpenes, *trans*-anithole, fenchone, α-pinene, β-pinene
Ginger	Shogoal, gingerol
Garlic	Allyl cysteine, alliin, allicin, allyl disulfide
Marjoram	β-Carotene, β-sitosterol, caffeic acid, carvacrol, eugenol, hydroquinone, linalyl acetate plant 3–17, myrcene, rosmarinic acid, terpinen-4-ol
Nutmeg	Myristphenone, phenolic volatile oils, phenolic acid (caffeic acid), flavanols (catechin)
Oregano	Caffeic acid, *p*-coumaric acid, rosmarinic acid, caffeoyl derivatives, carvacrol, flavonoids
Black pepper	Total flavonoids, total phenolic compounds, piperine
Rosemary	Carnosol, carnosic acid, phenolic diterpene, phenolic acid, rosmanol, epirosmanol, rosmarinic acid, caffeic acid ester, tosemaridiphenol, 3-(3,4-dihydroxyphenyl), lactic acid, flavonoids (apigenin, diosmin, luteolin), tannins
Red pepper	β-Carotene fruit, β-sitosterol plant, caffeic acid campesterol, camphene fruit, capsaicin fruit, capsanthin fruit, chlorogenic acid fruit, eugenol fruit, γ-terpinene fruit, myristic acid, hesperidin fruit
Sage	Rosmanol, epirosmanol, phenolic acids (rosmarinic acid), phenolic diterpenes (carnosic acid), flavonoids
Thyme	Phenolic acids (gallic acid, caffeic acid, rosmarinic acid), thymol, phenolic diterpenes, flavonoids

3.9 Dietary Phytonutrients Phytoantioxidants – Chemical Structure

Herbs and spices, vegetables, fruits, whole grains, lentils, and legumes are dietary constituents. Culinary herbs and spices are the most significant ingredients of foods. They are the rich sources of phytochemicals also termed "phytonutrients" or "plant bioactive compounds." The compounds are sulfur containing compounds, such as phenolic compounds, phenolic diterpenes, alkaloids, tannins, and more importantly flavonoids and polyphenols. Plant substances prevent diseases like cancer, cardiac diseases, diabetes, etc. Scientists have detected 3000 phytochemicals with possible health benefits. Based on the Neveu et al. [31] report, there is ample evidence proving that spices and herbs possess abundant dietary constituents that possess antioxidative, anti-inflammatory, antitumorigenic, anticarcinogenic, and blood glucose and LDL levels mitigating activities, which affect cognition and are actively used in preclinical, clinical, and therapeutic trials and their investigation in novel techniques for treatment is ongoing. Spices and herbs possess active constituents that are proven to be good antioxidants in scavenging free radicals (Tables 3.5 and 3.6).

Table 3.5 Major antioxidants and therapeutic effects of herbs and spices.

Herbs and spices	Major phytoantioxidants and chemical structure	Therapeutic effects on human health
Allspice 	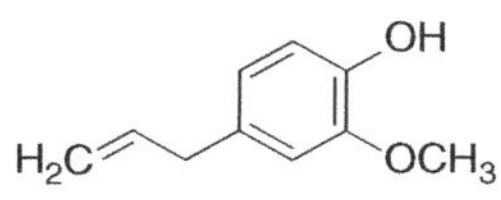 Eugenol Molecular formula – $C_{10}H_{12}O_2$	Anti-inflammatory, antimicrobial, antifungal, antibacterial, analgesic, anticancer agent
Black pepper 	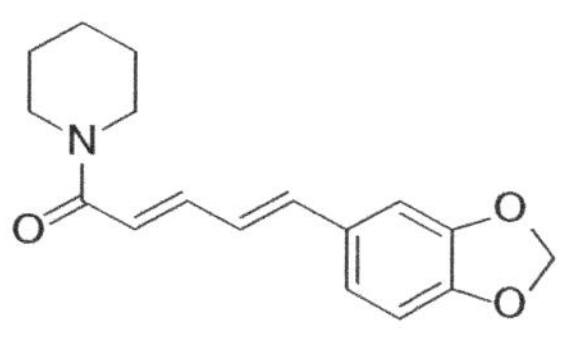 Piperine Molecular formula – $C_{17}H_{19}NO_3$	Immunomodulatory, anticarcinogenic, stimulatory, hepatoprotective, anti-inflammatory, antiulcer, antimicrobial activities
Bay leaves 	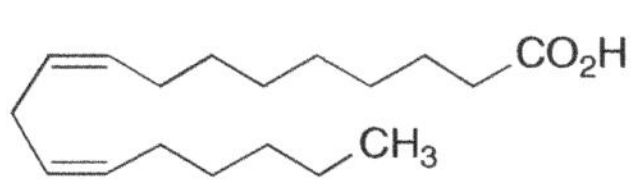 Lenolic acid Molecular formula – $C_{18}H_{32}O_2$	Diuretic, antifungal, antibacterial, hemorrhoid, rheumatic, wound healing, antimutagenic, antioxidant, anticonvulsant, analgesic, anti-inflammatory, immunostimulant, anticholinergic, insect repellent activities
Cardamom 	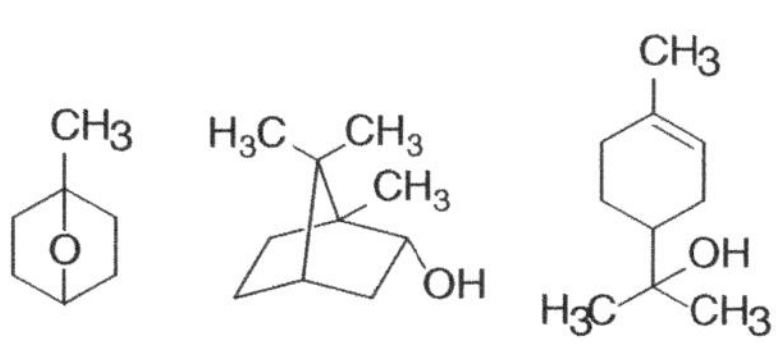 Cineole Borneol α-Terpineol	Bronchodilator, asthma, nausea, diarrhea, gastritis, carminative, stomachic
Clove 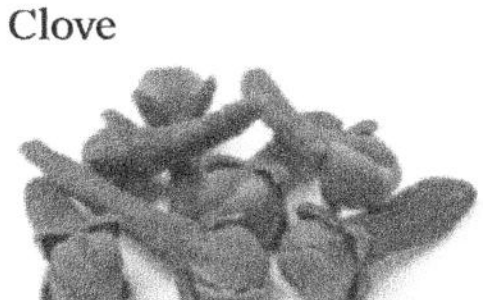	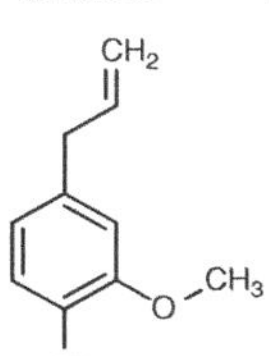 Eugenol Molecular formula – $C_{10}H_{12}O_2$	Anti-infective agent, antiallergen, sensitizer, respiratory disorders, toothache

(*Continued*)

Table 3.5 (Continued)

Herbs and spices	Major phytoantioxidants and chemical structure	Therapeutic effects on human health
Cinnamon 	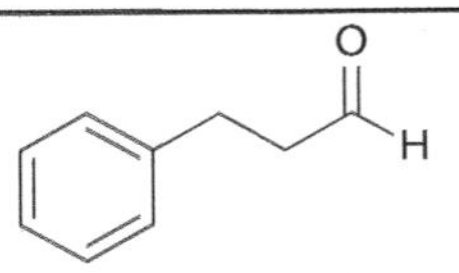 Molecular formula – $C_6H_5CH{=}CHCHO$	Antimicrobial agent, urinary issues, hepatoprotective, headache, antidiabetic, abdominal ache
Ginger 	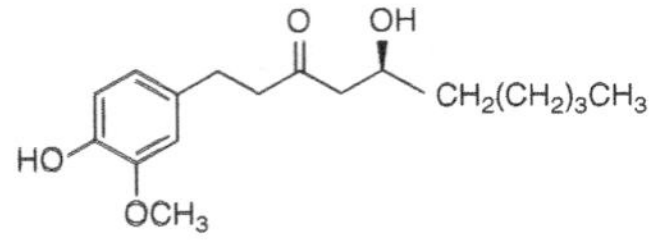 Molecular formula – $C_{17}H_{26}O_4$	Anti-inflammatory, antitumor, antioxidant property, anticancer
Garlic 	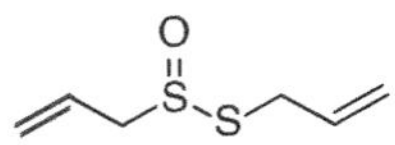 Allicin 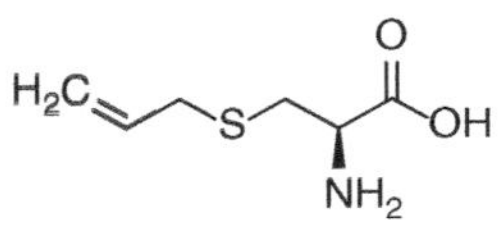 *S*-allyl cysteine Molecular formula of allicin – $C_6H_{10}OS_2$ Molecular formula of *S*-allyl cysteine – $C_6H_{11}NO_2S$	Antioxidant, antidiabetic, cardioprotective, antimicrobial, anti-cytotoxic, antihyperlipidemia
Marjoram 	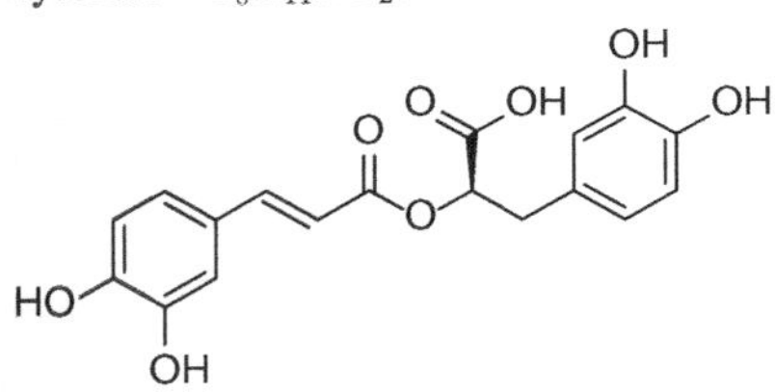 Romarinic acid Molecular formula – $C_{18}H_{16}O_8$	Anti-inflammatory, antioxidant activity, antispasmodic, colic, dysmenorrhea, respiratory disorders, hair growth, hepatoprotective, antitumorigenic
Nutmeg 	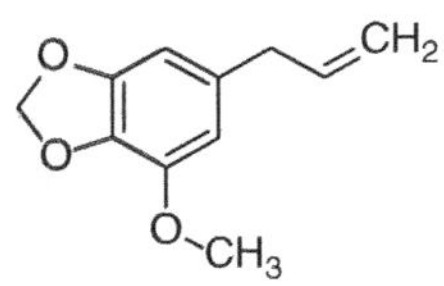 Myristicin Molecular formula – $C_{11}H_{12}O_3$	Hepatoprotective activity, insecticide, hallucinogenic effect

Table 3.5 (Continued)

Herbs and spices	Major phytoantioxidants and chemical structure	Therapeutic effects on human health
Onion 	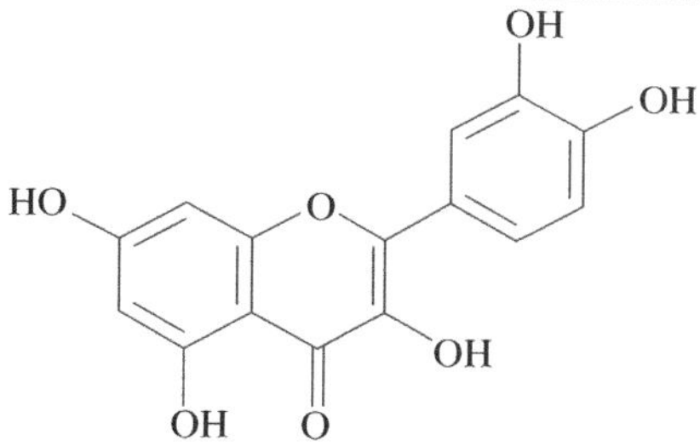 Quercetin Molecular formula – $C_{15}H_{10}O_7$	Antihypertensive, antioxidant, phytoestrogenic activity, antibacterial, antineoplastic agent, protein kinase inhibitor, cardioprotective, antidiabetic, bone density, digestive protective
Rosemary 	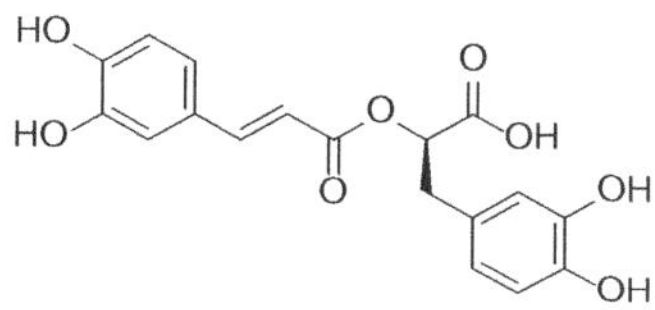 Rosmarinic acid Molecular formula – $C_{18}H_{16}O_8$	Anti-inflammatory, antioxidant, analgesic, antipyretic, platelet inhibitor, antiviral, herpes simplex lesions, nephroprotective, nonsteroidal activity
Red pepper (cayenne) 	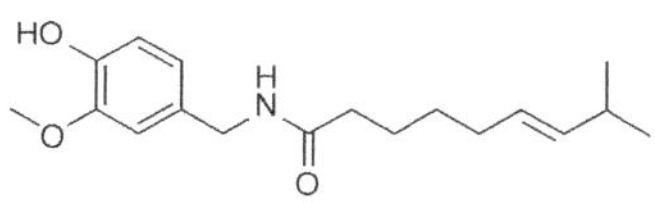 Capsaicin Molecular formula – $C_{18}H_{27}NO_3$	Antipruritic, cardiometabolic protective, obesity, hypertension, dyslipidemia, diabetes, atherosclerosis, cancer, rheumatism, stiff joints, bronchitis, headache, arthritis, heart, arrhythmias, stomachic
Sage 	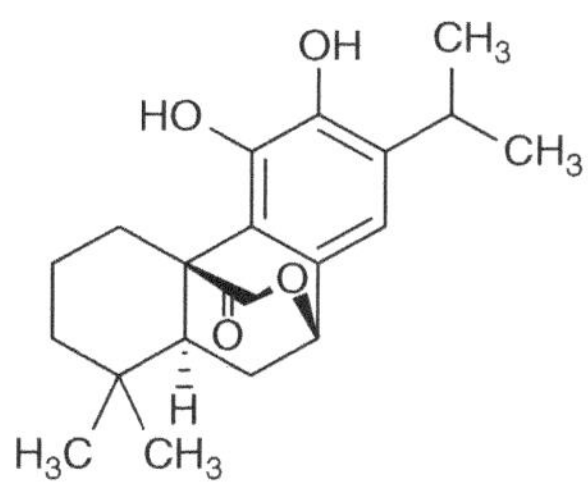 Carnosol Molecular formula – $C_{20}H_{26}O_4$	Antioxidant, antimicrobial, anti-inflammatory, anticarcinogenic, antibacterial

(Continued)

Table 3.5 (Continued)

Herbs and spices	Major phytoantioxidants and chemical structure	Therapeutic effects on human health
Saffron 	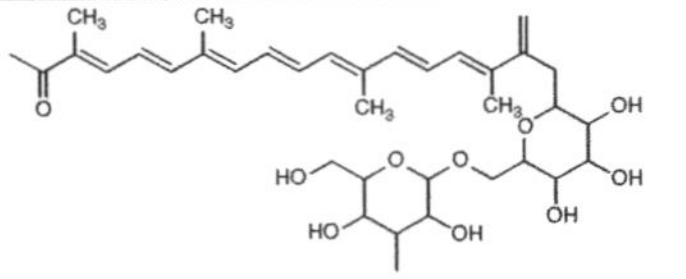 Crocin, picrocrocin Molecular structure – $C_{44}H_{64}O_{24}$	Hyperglycemia, metabolic syndrome, hypertriglyceridemia, hypercholesterolemia, antineurodegenerator, antioxidant activity
Turmeric 	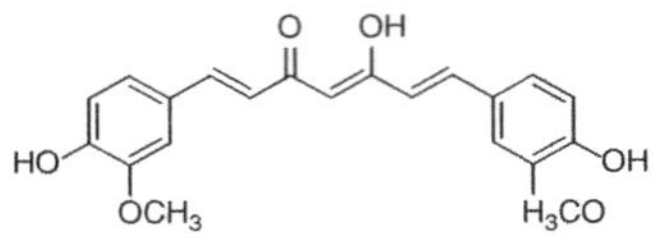 Curcumin Molecular formula – $C_{21}H_{20}O_6$	Anti-inflammatory agent, antineoplastic agent, hepatoprotective, nutraceutical, antifungal agent, neuroprotective agent, antidiabetic, immunomodulator, iron chelator
Thyme 	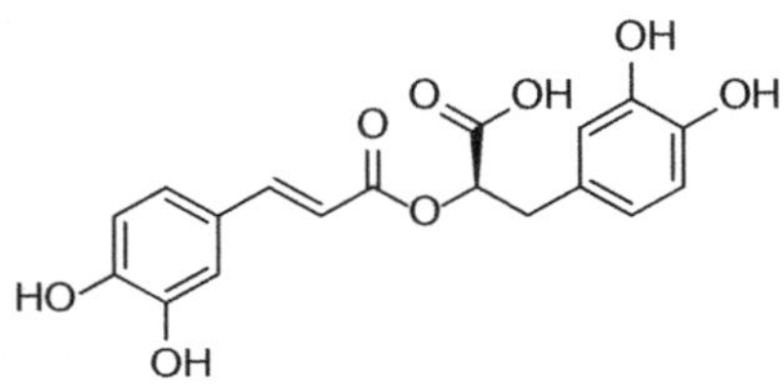 Rosmarinic acid Molecular formula – $C_{18}H_{16}O_8$	Growth modulator, fertility, antioxidant activity, immunomodulator, anti-inflammatory, anti-allergenic, antiangiogenic, chemopreventive, chemoprotective, antitumorigenic

Table 3.6 Dietary constituents.

Food items	Phytonutrients	Disease prevention
Blueberries (*Vaccinum corymbosum*)	Anthocyanin	Cardioprotective, antihypertensive, prevents high levels of LDL
Dark chocolate	Cocoa, fiber, iron, potassium, magnesium, copper, manganese, zinc, selenium, phosphorus	Antihypertensive, cardioprotective, anti-inflammatory, antilipidemic
Strawberries	Anthocyanin	Cardioprotective, antilipidemic
Raspberries	Dietary fibers, vitamin C, manganese, anthocyanins, ellagitannins, polyphenols	Antitumor activity, anti-inflammatory, cardioprotective
Artichokes	Phenolic compounds, dietary antioxidants	Anticancer activity, cardioprotective, osteoporosis, neurodegenerative, antidiabetic, hepatoprotective, antimicrobial
Pecans	Calcium, magnesium, potassium, omega 3 fatty acids, dietary fibers	Antilipidemic, cardioprotective, antidiabetic, antihypertensive, antiarthritis, Alzheimer's and Parkinson's disease

Table 3.6 (Continued)

Food items	Phytonutrients	Disease prevention
Red cabbage	Anthocyanin, vitamins A, C, K	Anti-inflammatory, anticancer, antidiabetic, cardioprotective
Beans	Kaempferol	Anticancer activity, prevents lung, breast, kidney, and bladder cancers
Purple or red grapes	Anthocyanin, proanthocyanin, vitamin C, selenium	Anticancer, cardioprotective
Spinach	Zeaxanthin, lutein	Antiaging, muscular degeneration, antiopthalmic
Beetroot	Betalain, dietary fiber, potassium, iron, folate	Anticancer, anti-inflammatory
Kale	Anthocyanins, vitamins A, C, and K	
Orange vegetables (sweet potatoes, carrots, tomatoes, pumpkins, winter squash, red peppers)	β-Carotene, lycopene, vitamins A, C, and K, zeaxanthin, lutein	Anticancer, antihypertensive, cataract

3.10 Health Benefits of Herbs and Spices

It is also used as medicine since times immemorial to date. It has been used to treat a wide range of diseases.

- In modern era garlic is used to treat conditions such as atherosclerosis, hypertension, coronary heart disease, and heart attack. Garlic prevents lung and brain cancer, hip osteoarthritis, prostate cancer, premature delivery, and controls high levels of cholesterol. Garlic is very effective against common cold.
- Fenugreek regulates blood glucose levels, lowers LDL, and regulates insulin activity (the same benefits are shown by cinnamon, flax seeds, and linseed).
- Herbs such as cloves, cinnamon, oregano, and thyme are rich sources of antioxidants, and they alleviate LDL level linked to various heart ailments.
- As reported by Topic et al. [32], ginger decreases age-related oxidative markers and reduce hepatotoxicity by suppressing oxidative stress. The bioactive compound of ginger inhibits cancer cells, asthma, dementia, diabetes, etc.
- Mint leaves are tender herbs; apart from imparting flavor and texture to foods, they soothe indigestion. Mint oil is a remedy for gut-related issues. Mint works against microbial activity. It has antiallergic effect due to the presence of a bioactive compound rosmarinic acid.
- Oregano is used as a pizza topper, and it works against harmful bacteria that cause infection such as staph. Loaded with antioxidants, it prevents cell damage.
- Turmeric is a major ingredient in most of the curries.

3.11 Traditional and Commercial Value of Herbs and Spices

Herbs and spices have traditional importance and are one of the sources of wealth since times immemorial. The Arabs were the pioneers in the trading of spices and herbs. Later the Egyptians and the Greeks expanded the trade to Asian, American, and European countries. Europe is a

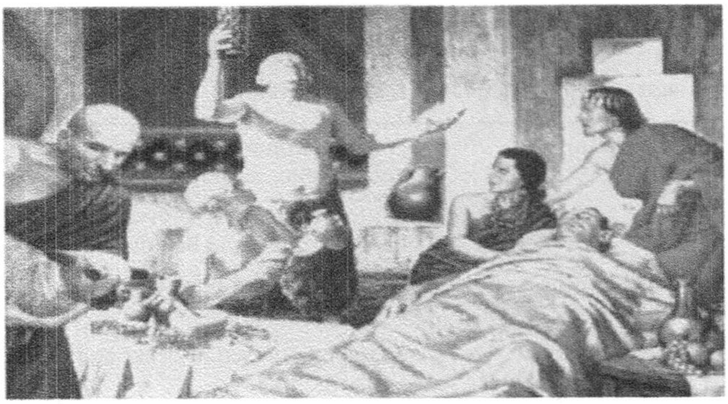

Figure 3.3 Medicinal plants used in ancient Babylonia. *Source:* Pharmapproach/Pharmapproach Limited.

leading importer of herbs and spices. Europe is the focused market for developing countries and exporters. The cultivation of herbs and spices fetches a strategic position in the global market and also raises the economy of a country.

In early medieval period, herbs and spices were used as medicines, to make holy oils, ointments, and aphrodisiacs. Priests applied them as charms to drive evil spirits. Aromatic herbs were used to impart fragrance in homes. Ancient countries like Egypt, China, Greece, Syria, and Rome explored the use of spice and herbs in the treatment of various diseases. Renowned physicians such as Hippocrates, Galen, and Pedanius Dioscorides were pioneers in using herbs and spices in healing illnesses. There are nearly 40–50 spices and herbs attributed to global economy and culinary significance (Figure 3.3).

3.12 Methods for Screening and Determination of Antioxidants

Antioxidants are chemical compounds or phytonutrients of medicinal plants including herbs, spices, and dietary products. There are several techniques used to extract the active compounds from various herbs and spices used for culinary purposes. Xu et al. [33] reported that the techniques used to extract active compounds from herbs and spices are liquid-phase extraction method, solid-phase extraction, supercritical fluid extraction, etc. There are different assays used to determine the antioxidant activity of compounds such as DPPH assay (1,1-diphenyl-2-picrylhydrazyl), ABTS assay ((2,2′-azinobis-(3-ethylbenzothiazoline-6-sulfonic acid)), FRAP assay (ferric reducing antioxidant power assay), superoxide dismutase (SOD), ORAC (oxygen radical absorbance capacity) assay, and TEAC (trolox equivalent antioxidant capacity) assay. All these assays are used in the detection of free radical scavenging activaty. These assays measure reducing ability of antioxidants toward the free radicals.

3.13 Conclusion

According to literature and studies, we know that herbs and spices not only enhance the aroma, flavor, and color of foods but also are involved in treating several chronic diseases due to their active components. The active compounds are meant to possess high antioxidant activities. The therapeutic effects of certain compounds are included in nonclinical, Clinical, and therapeutic studies. The spices that consist of antineoplastic agents are *Nigella sativa*, turmeric, clove, black pepper, ginger, garlic, etc. Plant components that are highly effective against cancer are thymoquinone, apigenine, luteolin, vinristine, vinblastin, curcumin, and quercetin.

References

1 Davidson, A. (1999). *The Oxford Companion to Food*. Oxford, UK: Oxford University Press.
2 Hacskaylo, M.M. (1996). Culinary garden. In: *The National Herb Garden Guidebook* (ed. R. Ober), 79–93. Springfield, VA: The Herb Society of America.
3 Smith, R.J. and Winder, M.L. (1996). Medicinal garden. In: *The National Herb Garden Guidebook* (ed. R. Ober), 61–71. Springfield, VA: The Herb Society of America.
4 Velioglu, Y.S., Mazza, G., Gao, L., and Oomah, B.D. (1998). Antioxidant activity and total phenolics in selected fruits, vegetables, and grain products. *J. Agric. Food Chem.* 46: 4113–4117.
5 Larson, R.A. (1988). The antioxidants of higher plants. *Phytochemistry* 27: 969–978.
6 Kähkönen, M.P., Hopia, A.I., Vuorela, H.J. et al. (1999). Antioxidant activity of plant extracts containing phenolic compounds. *J. Agric. Food Chem.* 47: 3954–3962.
7 Burt, S. (2004). Essential oils: their antimicrobial properties and potential applications in foods: a review. *Int. J. Food Microbiol.* 94 (3): 223–253.
8 McCormick (2017). The history of spices. http://www.mccormickscienceinstitute.com/resources/history-of-spices (accessed 30 May 2017).
9 Peter, K.V. (2001). *Handbook of Herbs and Spices*, vol. 1. Cambridge, UK: Woodhead Publishing Limited and CRC Press LLS.
10 Yesiloglu, Y., Audin, H., and Kilic, I. (2013). In vitro antioxidant activity of various extracts of ginger seed. *Asian J. Chem.* 25: 3573–3578.
11 Gerschman, R., Gilbert, D.L., Nye, S.W. et al. (1954). Oxygen poisoning and X-irradiation: a mechanism in common. *Science* 119 (3097): 623–626.
12 Yu, B.P. (1994). Cellular defenses against damage from reactive oxygen species. *Physiol. Rev.* 74 (1): 139–162.
13 Patch, C.S. (2006). Health benefits of herbs and spices: the past, the present, the future. *Med. J. Aust.* 185: S1–S24.
14 Jiang, T.A. (2019). Health benefits of culinary herbs and spices. *Jiang J. AOAC Int.* 102: 395–411. https://doi.org/10.5740/jaoacint.18-0418.
15 Yashin, A., Yashin, Y., Xia, X., and Nemzer, B. (2017). Antioxidant activity of spices and their impact on human health. A review. *Antioxidants* 6: 70.
16 Demmig-Adams, B. and Adams, W.W. (2002). Antioxidants in photosynthesis and human nutrition. *Science* 298: 2149–2153.
17 Bhattacharjee, S. and Sengupta, A. (2008). Spices in cancer prevention – an overview. *J. Nutr. Wellness* 7: 1–16.
18 Allwyn Sundar Raj, A., Aaron, S., Seinenbalg, S.S. et al. (2014). Review on recent trends in isolation of antioxidants from spices and its biological effects of essential oils. *J. Eng. Res. Appl.* 4: 75–84.
19 Winkel-Shirley, B. (2006). The biosynthesis of flavonoids. In: *The Science of Flavonoids* (ed. E. Grotewold), 71–95. New York, NY: Springer.
20 Torres, J.E.T., Gassara, F., Kouassi, A.P. et al. (2015). Spice use in food: properties and benefits. *Crit. Rev. Food Sci. Nutr.* https://doi.org/10.1080/10408398.2013.858235.

21 Slimestad, R., Fossen, T., and Brede, C. (2019). Flavonoids and other phenolics in herbs commonly used in Norwegian commercial kitchens. *Food Chem.* https://doi.org/10.1016/j.foodchem.2019.12578.

22 Rice-Evans, C., Miller, N., and Paganga, G. (1997). Antioxidant properties of phenolic compounds. *Trends Plant Sci.* 2: 152–159.

23 Prior, R.L. and Cao, G. (2000). Antioxidant phytochemicals in fruits and vegetables. Diet and health implications. *Hortic. Sci.* 35: 588–592.

24 Mathew, S. and Abraham, T.E. (2006). In vitro antioxidant activity and scavenging effects of *Cinnamomum verum* leaf extract assayed by different methodologies. *Food Chem. Toxicol.* 44: 198–206.

25 Ito, N., Fukushima, S., Hagiwara, A. et al. (1983). Carcinogenicity of butylated hydroxyanisole in F344 rats. *J. Natl. Cancer Inst.* 70: 343–347.

26 Gao, J.J., Igalashi, K., and Nukina, M. (1999). Radical scavenging activity of phenylpropanoid glycosides in *Caryopteris incana*. *Biosci. Biotechnol. Biochem.* 63: 983–988.

27 Williams, G.M., Iatropoulos, M.J., and Whysner, J. (1999). Safety assessment of butylated hydroxyanisole and butylated hydroxyltoluene as antioxidant food additives. *Food Chem. Toxicol.* 37: 1027–1038.

28 Osawa, T. and Namiki, M. (1981). A novel type of antioxidant isolated from leaf wax of Eucalyptus leaves. *Agric. Biol. Chem.* 45: 735–739.

29 Kapoor, L.D. (1990). *Handbook of Ayurvedic Medicinal Plants*, 172. Boca Raton: CRC Press.

30 Ame, B.N., Shigenaga, M.K., and Hagen, T.M. (1993). Oxidants, antioxidants, and the degenerative diseases of aging. *PNAS Proc. Natl. Acad. Sci. USA* 90: 7915–7922.

31 Neveu, V., Perez-Jiménez, J., Vos, F. et al. (2010). Phenol-explorer: an online comprehensive database on polyphenol contents in foods. *Database J. Biol. Databases Curation* 2010: bap024. https://doi.org/10.1093/database.bap024.

32 Topic, B., Tani, E., Tsiakitzis, K. et al. (2002). Enhanced maze performance and reduced oxidative stress by combined extracts of *Zingiber officinale* and *Ginkgo biloba* in the aged rat. *Neurobiol. Aging* 23 (1): 135–143.

33 Xu, D., Li, Y., Meng, X. et al. (2017). Nautral antioxidants in foods and medicinal plants: extraction, assessment and resources. *Int. J. Mol. Sci.* 18: 96.

4

Phytoantioxidants and Their Role in Cellular Oxidative Stress

Swarnali Bhattacharjee[1], *Prashanta Kumar Deb*[1,2], *and Rajlakshmi Devi*[1]

[1] *Life Sciences Division, Institute of Advanced Study in Science and Technology, Guwahati, Assam, India*
[2] *Department of Pharmaceutical Chemistry, School of Pharmaceutical Sciences, Shoolini University, Solan, Himachal Pradesh, India*

4.1 Introduction

In cells, aerobic metabolism under normal physiological conditions leads to the production of free radicals, especially in the form of reactive oxygen species, also called ROS. Oxygen, which is of utmost necessity for sustaining the living system, becomes a threat to it in the form of reactive free radicals, which in definition, are atoms or molecules (chemical species) with one or more unpaired electrons, which often include the derivatives of oxygen as well as nitrogen concerning the living system [1]. ROS is often released as a side-product of cells undergoing metabolism, even in normal conditions where they aid in cellular signaling as well as in other physiological processes and a very delicate balance works within the cellular environment that maintains the production of ROS and its elimination, which is in the form of redox equilibrium or oxidative balance. Whenever this equilibrium is thrown off balance, it leads to the accumulation of excessive ROS within the cellular environment, ensuing in many undesirable and harmful changes within it. This condition is known as oxidative stress, which now has established pathological implications in various diseases like hypertension, cancer, atherosclerosis, diabetes, etc. [2, 3]. Substances, which help in combating these ill effects of ROS accumulation or oxidative stress, are called antioxidants [4, 5]. The cellular endogenous antioxidant system comprises enzymes like catalase and glutathione peroxidase, along with nonenzymatic antioxidants like ascorbate, tocopherol, carotenoid, etc. The main sources of exogenous antioxidants are dietary components like fruits, vegetables, nuts, fish, meat, etc., and some of the commonly found antioxidants from such sources include β-carotene, lutein, lycopene, retinol, vitamin C, α-tocopherol, etc. [5, 6]. A variety of secondary metabolites from plants are reported, which are an important part of their development and defense mechanism, including polyphenols, anthocyanins, flavones, flavonols, and carotenes along with alkaloids and terpenoids. Many of these compounds have been found to have a very potent antioxidant activity, with accompanying therapeutic effects in various pathological conditions, thus making plant-derived antioxidants or phytoantioxidants important contributors to dealing with oxidative stress and its detrimental effects [7, 8].

This chapter travels through the crosstalk between ROS and endogenous antioxidant defense mechanism while traversing the mechanism behind oxidative stress and its associated disease

Phytoantioxidants and Nanotherapeutics, First Edition. Edited by Mithun Rudrapal.

pathophysiology, in addition to the role of phytoantioxidants in ameliorating the disease physiology while taking a deeper look at the future prospects of the use of phytoantioxidants as phytopharmaceuticals.

4.2 Oxidative Stress

Oxidative stress is a concept concerning the oxidation-reduction mechanisms and perturbed balance between oxidants and antioxidants, with an inclination toward the oxidants leading to some detrimental effects on biomolecules as well as the cell signaling pathways [9].

4.2.1 Molecular Events of Cellular Redox Balance

Superoxide ($O_2^{-\bullet}$), hydroxyl ($OH^{\bullet}$), hydrogen peroxide (H_2O_2), peroxynitrile ($ONOO^{-\bullet}$), and nitric oxide ($NO^{\bullet}$) are the main ROS/RNS radicals that are generated during different biochemical reactions in the cells, where a balance is maintained in a cyclic manner through their continuous elimination. The endogenous antioxidant system comprising superoxide dismutase (SOD), catalase, glutathione peroxidase, ascorbic acid, α-tocopherol, glutathione, and carotenoids springs to action whenever the radicals are produced in overwhelming amounts, thereby protecting the cells from oxidative damage [10]. The detailed role of oxidants in development of oxidative stress is mentioned in Table 4.1.

In addition to these, many other species of free radicals, like hydroxyl ($OH^{\bullet}$), alkoxyl ($RO^{\bullet}$), and peroxyl ($ROO^{\bullet}$), may cause extensive oxidative damage to DNA, proteins, lipids, etc., in tissues, all while generating other reactive oxidants like singlet oxygen, aggravating the extent of oxidative damage [14]. In the light of newfound evidences, all in favor of ROS, being involved in cellular differentiation, immune response mediation, signal transduction pathways, post-translational modifications, an understanding of ROS as an integral part of normal cellular metabolism is becoming more evident than it just being a nuisance in cellular metabolism [15]. In addition to the classical antioxidants present in the body, a number of metabolites like transferrin, albumin, lactoferrin, ferritin, melatonin, bilirubin, etc., also act as nonenzymatic antioxidants, taking a proactive role in the prevention of oxidative stress [16]. Not only that, ROS has been found to play a pivotal

Table 4.1 List of oxidants and their role in the development of cellular oxidative stress [3, 11–14].

Oxidant species	Radical generation	Role in redox balance
Superoxide anion (O_2^{-})	Mitochondrial ETS, via the action of xanthine oxidase or NAD(P)H oxidase.	Responsible for the bactericidal properties of leukocytes and macrophages.
Hydrogen peroxide (H_2O_2)	Via the action of SODs, NAD(P)H oxidase, from molecular oxygen in peroxisomes.	Responsible for the obliterating engulfed pathogens in granulocytes.
Hydroxyl radical ($OH^{\bullet}$)	By Haber–Weiss reaction, Fenton reaction.	May cause extensive oxidative damage to biomolecules of cells.
Nitric oxide ($NO^{\bullet}$)	By the action of nitric oxide synthase (NOS)	Aids in vasodilation, immune response, second messenger molecule in cellular signaling pathways.

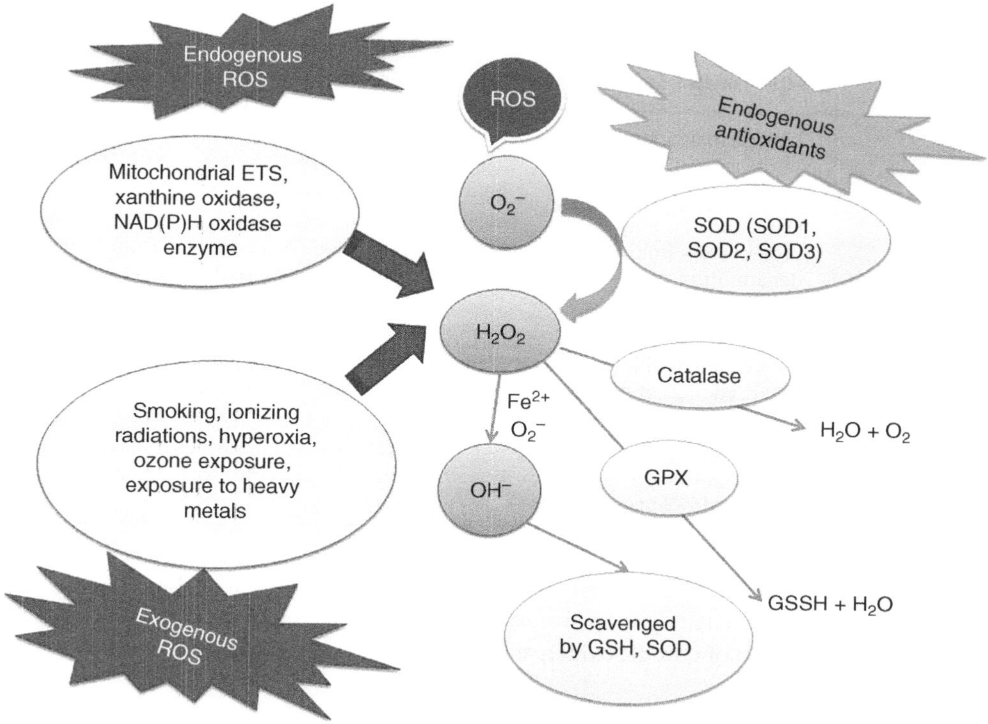

Figure 4.1 Cellular events of endogenous antioxidant defense mechanism.

role in the regulation of molecular pathways like MAPK in cellular differentiation, immune response, apoptosis, growth factor-mediated PI3K signaling in cellular proliferation, along with the antioxidant countermeasure mounting pathways via regulation of Nrf2 and Ref1, to mention a few [17]. Figure 4.1 illustrates the basic mechanism of the generation of ROS and the role of endogenous antioxidants in maintaining the redox balance.

4.2.2 ROS as the Main Inducer of Oxidative Stress

When the indigenous cellular redox balance and its defense mechanism against the ROS-mediated oxidative damage are compromised, it often leads to the situation of oxidative stress [13, 18]. Various cellular biomarkers like malonaldehyde (MDA), protein carbonyl content, 8-oxoguanine are quantified to assess the extent of ROS-mediated damage to various biomolecules like lipid, protein, or DNA. Based on the extent of modifications brought about by the ROS molecules, they may be classified as basal, low, intermediate, and high-intensity oxidative stress, where the amount of oxidative end products increases with the increasing stress and the antioxidant defense becomes intensified at first, but decreases later on when the stress becomes too large to handle [13]. The ROS produced may be endogenous or contributed through exogenous sources like smoking cigarettes, gamma rays, ultraviolet, or X-ray irradiation, exposure to high levels of oxygen, heavy metals like arsenic, lead, cadmium, etc., which may lead to the development of unwanted pathological conditions [3]. Upon exposure to an oxidant, the cells may experience acute oxidative stress, and the ROS levels may be normalized through antioxidants, but in many health conditions like

Table 4.2 ROS-mediated damage to biomolecules and their leading to cellular dysfunctions [14, 19–22].

Biomolecule	ROS-mediated damage	Underlying cellular dysfunction
Lipids	Lipid peroxidation	May trigger apoptosis via both intrinsic and extrinsic pathways, cellular autophagy.
Proteins	Oxidation of amino acids, protein cross-linking, denaturation, etc.	Impaired cellular signaling, metabolic enzymatic reactions, and formation of pernicious oxidized clusters like carbonylated proteins, may even mark them for proteolysis.
DNA	Single- and double-strand break, adducts of nitrogenous bases are formed.	May result in mutagenic lesions or even carcinogenic mutations.
RNA	Altered structure, oxidized nitrogenous bases (7,8-dihydro-8-oxo-guanosine)	Is implicated in various disease pathologies, defective translation.

diabetes mellitus, cancer, and cardiovascular diseases, the elevated ROS levels seldom return to the normal basal level. They are acted upon by antioxidants, which make them settle at a new "quasi" level, an exemplary situation of chronic oxidative stress linked to disease pathology [13].

4.2.3 Consequences of Oxidative Stress in the Cellular Level

In oxidative stress, the free radical molecules may cause the reduction of disulfide bonds in proteins changing their spatial arrangements, lipid peroxidation, single- or double-strand breaks or oxidative lesions in DNA molecules, etc., altering the normal metabolomics of the cell itself [14]. The details of oxidative assault on biomolecules and their underlying effects on cellular functioning are listed in Table 4.2.

4.2.4 Cell Signaling Pathways Involved in Oxidative Stress

Apoptotic pathways are often initiated during oxidative stress through lipid peroxidation with the underlying molecular signaling via activation of NF-κβ leading to inactivation of Bcl-2, activation of MAPKs, protein kinase C that leads to downstream activation of caspases and also interacts with AMPK/mTORC, and JNK-Bcl-2/Beclin 1 signaling, thereby implicating the autophagic pathway as well [19]. Other important signaling pathways underlying the oxidative stress include activation of Nrf2/Keap1, which enhances the antioxidant response; MAPKs which may culminate into apoptosis; NF-κβ which again acts to mount antiapoptotic and antioxidant response, along with PKC, PPARγ, and STAT3 as well [23, 24]. Figure 4.2 depicts the extensive oxidative damage to biomolecules and their related cellular dysfunctions while Figure 4.3 highlights the key molecular pathways underlying this oxidative damage.

4.2.5 Role of Antioxidants to Counter Oxidative Stress

Incriminating evidence links oxidative stress to cancer, diabetes, as well as neurodegenerative diseases, which also highlights the requirement of antioxidants as they play a significant role in ameliorating oxidative stress. The balance of antioxidants and prooxidants is somewhat inclined

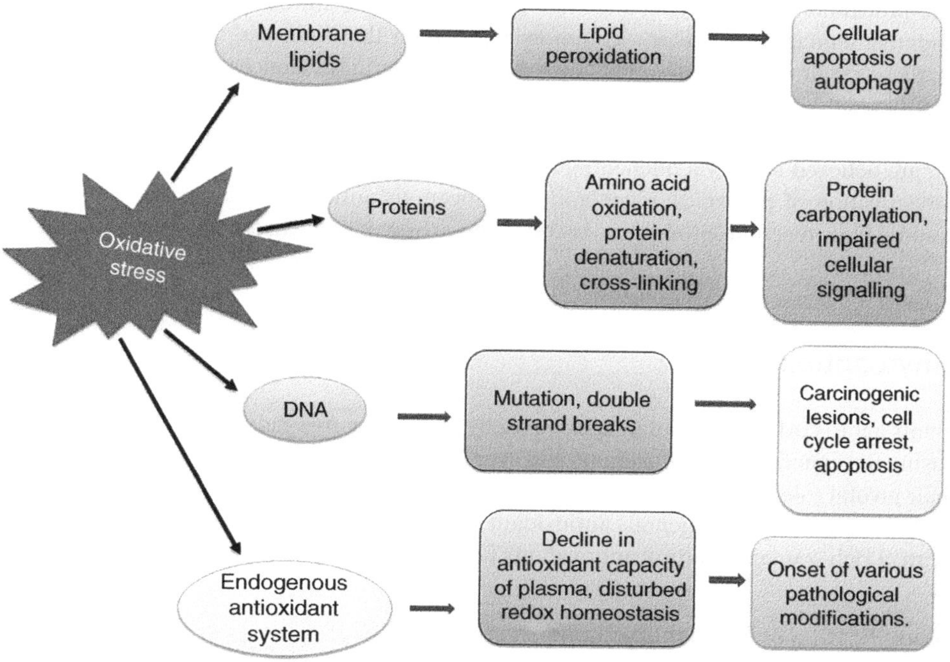

Figure 4.2 ROS-mediated assault on biomolecules and their related cellular dysfunctions.

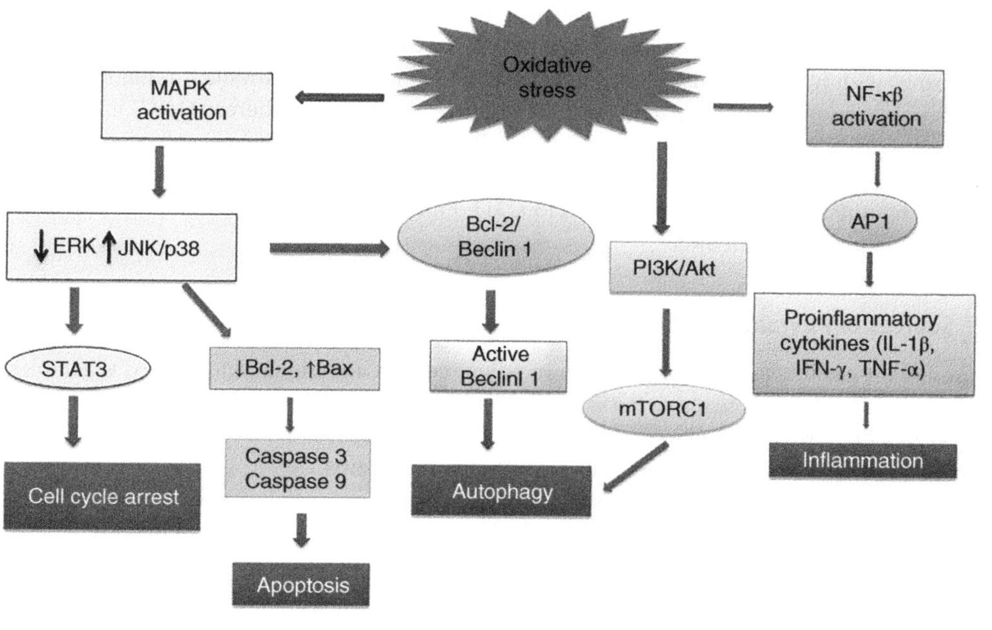

Figure 4.3 Cell signaling pathways involved in oxidative stress.

toward the prooxidants in the living system, where endogenous antioxidants act to combat and repair oxidative damages [25]. It is a common observation that with age or disease conditions setting in, these repair mechanisms often falter, and accumulating evidences suggest that naturally available antioxidants, especially from dietary sources, complement these systems. Vitamins C, E, carotenoids, and phenolic and flavonoid compounds have been acclaimed as exogenous antioxidants and are believed to act via quenching free radicals directly, chelating with transition metal cations, inactivation of peroxides, regenerating endogenous antioxidants by acting as hydrogen donors, and chain-reaction inhibitors, thereby imparting protective effect against oxidative stress [4, 26].

4.3 Phytoantioxidants

Antioxidants act to curb oxidative damage caused by rampaging free radicals by scavenging them or decreasing the concentration of oxidants and even by inhibiting their auto-oxidative chain reactions. Their pivotal role in preventing the risk of diseases has created a hype encompassing antioxidants, as they support the endogenous antioxidant system and it has been found that plant-based products are a rich source of natural antioxidants [27, 28].

4.3.1 Classification of Phytoantioxidants

A diverse array of phytochemicals has been reported from plants (fruits, vegetables, seeds, peels, and leaves), as well as medicinal plants with distinct antioxidant properties. Of these, polyphenolic compounds enclosing phenolic acids, tannins, stilbenes, terpenes, etc., along with carotenoids have been found to have potent antioxidant properties and are the main classes of phytoantioxidants [29]. Polyphenolic compounds are secondary metabolites, which are synthesized by the plants as a defensive response to biotic or abiotic stress and in structure resemble phenol with aromatic rings with hydroxyl substituent essentially present in them. Flavonoids, a major subclass of polyphenols, comprise dual benzene rings with a tri-carbon chain, whereas C_6–C_1 is the basic structure of hydroxybenzoic acids. Based on their solubility in water or organic solvents, phenolic compounds may also be classified as hydrophilic (e.g. flavonoids and phenolic acids), and hydrophobic (e.g. tannins and lignins) compounds [30, 31]. Figure 4.4 is a schematic representation of major classes of phytoantioxidants.

4.3.2 Sources of Phytoantioxidants

Plants and plant products are essentially rich in polyphenolics, thereby making fruits, vegetables, leaves, legumes, seeds, and even tree barks a rich source of antioxidants and have also been found to have therapeutic properties, playing a beneficial role in different disease conditions [32]. Listed in Table 4.3 are a few sources of different dietary phytoantioxidants.

Although the endogenous antioxidant system maintains the redox balance, sometimes it needs exogenous antioxidants to balance out overwhelming amounts of oxidants which are often obtained from food, especially from plant sources in the form of phytoantioxidants, where they might be helpful in preventing disease-related pathologies, thereby helping in maintaining a healthy life. This procedure of utilizing plant-derived antioxidants in treatment or as a preventive measure is called “phytotherapy” [50].

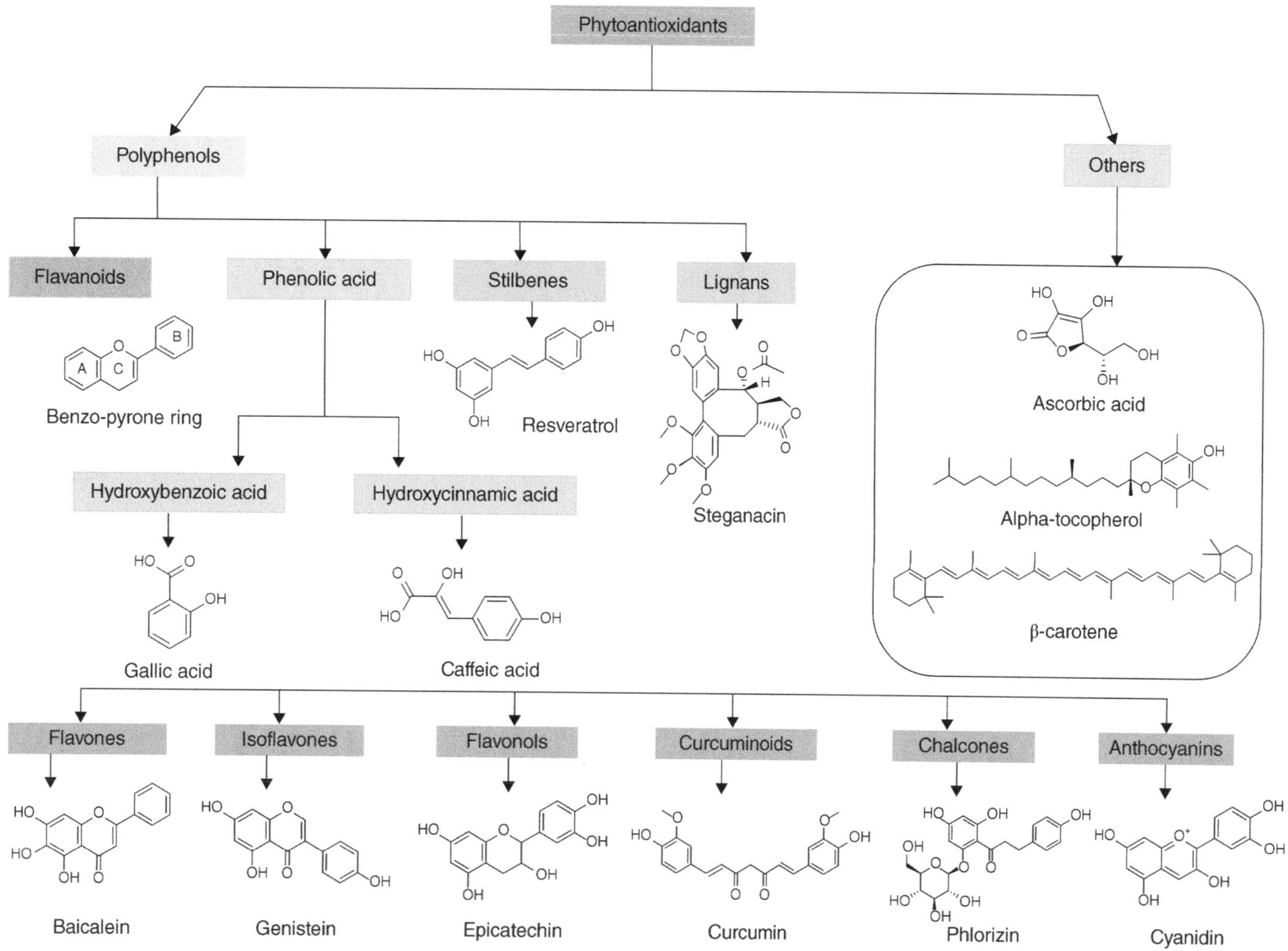

Figure 4.4 Classification of phytoantioxidants along with their structural details.

Table 4.3 Dietary sources of different classes of phytoantioxidants [33–49].

S. no.	Phytoantioxidants	Sources
1.	Carotenoids (lutein, α-carotene, β-carotene, and lycopene)	Apricot, carrot, spinach, tomato, collard greens, broccoli, papaya, microalgae
2.	Ascorbic acid/Vitamin C	Strawberries, crowberries, citrus fruits, red pepper, vegetables belonging to *Brassica* family
3.	Tocopherol/Vitamin E	Vegetable oils, rice bran oil, seed oil, nuts, meat, fish.
4.	Phenolic acids (hydroxycinnamic acid and derivatives, e.g. caffeic and ferulic acids, *p*-coumaric, sinapic acids)	Coffee beans, tea, propolis, wine, cereals, fruits (apple, blueberry, cherry, grapes, oranges, plum, strawberries, etc.); vegetables (lettuce, potatoes, spinach, sunflower seeds, turmeric) as well as spices (rosemary cinnamon, sage, thyme), wheat barn, oats.
5.	Phenolic acid (hydroxybenzoic acid and derivatives, e.g. gallic, ellagic, protocatechuic, salicylic, syringic, and vanillic acids)	Fruits (cranberries, grapes, mangoes, blackberries, pomegranate, raspberries, strawberries) and their juices. Other sources include tea, red and white wines along with nuts (peanuts, chestnuts, walnuts)
6.	Flavones (luteolin, apigenin, tangeritin, chrysin)	Celery, beans, onions, parsley, broccoli, tomatoes, red peppers, chamomile, mint, citrus fruit peels. Apple, cherries, grapes, tea, wine
7.	Flavanols [(+)-catechin (−)-epicatechin, epigallocatechin]	Bananas, apples, blueberries, peaches, pears, tea, green tea, pecan, cashews, buckwheat, etc.
8.	Flavanones (naringenin, taxifolinnaringin, hesperidin)	Citrus fruits (lemons, oranges, grapes)
9.	Isoflavonoids (genistin, biochanin A, formononetin, daidzin)	Leguminous plants including soybean, soy products, lentils, beans, peas, etc.
10.	Anthocyanidins (cyanidin, pelargonidin, malvidin, apigenidin)	Fruits (blueberries, black currant, red grapes, raspberry, pomegranate, strawberries, bilberries) and vegetables (red cabbage, purple sweet potato)
11.	Curcuminoids (demethoxycurcumin, cyclo-curcumin, bis-demethoxycurcumin, curcumin)	Zingiberaceae family, especially in the rhizomes of turmeric
12.	Chalcones (chalconaringenin, phloretin, phloridzin)	Citrus fruits, apples, strawberries, pears, tomatoes, bean sprouts
13.	Stilbenes (resveratrol)	Skin of red grapes, red wine
14.	Terpenes	Tea, thyme, cannabis, Spanish sage, rosemary
15.	Tannins	Sorghum, millets, barley, legumes, peas, apples, bananas, blackberries, dates, grapes, plums, raspberries, strawberries
16.	Lignans	Flaxseed, sesame seeds, tea, coffee, pumpkin, potato, berries, etc.
17.	Organo-sulfur compounds (indole, thiosulfonates, isothiocyanates, phytoalexins, alliins)	Onion, garlic, cabbage, cauliflower, broccoli

4.3.3 Polyphenols as Major Phytoantioxidants

Polyphenols form an enriched portion of phytochemicals present in fruits and vegetables and have a multitude of beneficial effects on human health exhibiting anticancer, hepatoprotective, antidiabetic as well as anti-inflammatory activities. These activities may be attributed to the presence of hydroxyl group substituent on the phenolic compounds, which make them suitable quenchers of ROS or RNS generated upon exposure to xenobiotic or toxic substances [51]. Various studies have effectively concluded that a diet rich in polyphenols is at a minimized risk of suffering from diabetes, obesity, or cardiovascular dysfunction. In nature, phenolic compounds often exist as glycosides and also form an important part of the pharmaceutical industry. With the associated health benefits, these compounds have intrigued keen interest of the pharmacological industries as a source of new drugs. Some of the most studied phenolics include resveratrol, rutin, kaempferol, and catechins, which have well-established ameliorative effects in a number of disease pathologies [52]. Some of the notable phenolic compounds along with their role in various diseases or pathological conditions have been listed in Table 4.4.

Table 4.4 Some dietary phytoantioxidants with their reported bioactivities [52–60].

S. no.	Polyphenolic compounds	Bioactivity
1.	Quercetin	Antioxidant, anti-inflammatory, cardioprotective, anti-ulcer, antiallergic activities.
2.	Rutin	Anticancer, anti-obesity, anti-hypercholesterolemia, increases HDL, lowers LDL, and VLDL, and triglycerides can even be used to treat hypertension.
3.	Catechins, epigallocatechin gallate (EGCG), epicatechin gallate (ECG)	Protective against coronary heart disease, neurodegenerative disease, antibacterial activity, antioxidant activity, anti-inflammatory, antimutagenic effects.
4.	Kaempferol	Minimizes the risk of cancer, cardiovascular diseases, along with having antioxidant, anti-inflammatory, analgesic, as well as antidiabetic properties.
5.	Resveratrol	Potent antioxidant anti-inflammatory, renders a protective effect against cancer, cardiovascular, diabetes, and neurodegenerative diseases.
6.	Curcumin	Potent antiangiogenesis, antitumoral, anticancer, antioxidant, anti-inflammatory activities along with protective effects in cardiovascular diseases, osteoarthritis, as well as in metabolic syndrome.
7.	Phenolic acids	Therapeutic effects in metabolic syndrome, cancer, as well as having antioxidant, anti-inflammatory activities.
8.	Naringenin	Antioxidant, hepatoprotective, anti-inflammatory, immunity booster, anticancer, antiatherogenic activities as well.
9.	Carotenoids	Source of vitamin A, improves the health of eyes, skin, bones, heart, and has cancer-preventive activity.
10.	Ascorbic acid	Brings relief in common cold, aids in wound healing, atherosclerosis, preventive action against cancer, antioxidant property.

4.3.4 Protective Mechanisms of Phytoantioxidants During Oxidative Stress

The various beneficial effects of phytoantioxidants are the result of various underlying molecular signaling pathways that help them modulate body physiology. This subsection mainly deals with pathways pursued by different phytoantioxidants for exerting their effects.

4.3.4.1 Phytoantioxidants as Free Radical Scavengers

Free radicals often have an unpaired valance electron, which often snatches electrons from the surrounding stable molecules and sets off a chain reaction. Polyphenols in a general mechanism act as hydrogen atom donors to the free radicals, quenching them. Thermodynamically, the involvement of π-electrons, bond dissociation enthalpy, hydrogen atom transfer or proton-coupled electron transfer, or electron loss-proton transfer in a sequence of biochemical reactions or stabilization by resonance can aptly be justified as the mechanism behind the radical scavenging activities of the polyphenolic compounds. With the transfer of electron or hydrogen, the reaction vigor of the radical is thereby reduced, while lower bond dissociation energy or ionization potential also contributes toward free radical scavenging activity. Often metal chelating activity of polyphenols also helps in preventing the Fe- or Cu-catalyzed free radical generation in the biological system [61–63].

4.3.4.2 Phytoantioxidants as Activators of Endogenous Antioxidant Pathways

The endogenous antioxidant enzymes of cellular redox defense mechanism as mentioned earlier include SOD, GSH-Px, catalase, and hemoxygenase-1 are often induced or activated upon interaction with different phytoantioxidants like curcumin, resveratrol, lycopene acting via Nrf2-Keap1 pathway, which further targets ARE, EpRE, or their coactivators like CREB-binding protein, or p300, and enhances the transcription of antioxidant enzyme encoding genes. Also, curcumin, catechins, and isoflavones from soy have been found to act via Akt, EKR1/2, and p38 MAPK signaling pathways, which also contribute toward the endogenous antioxidant enzyme production enhancement [64].

4.3.4.3 Phytoantioxidants as Cellular Oxidative Stress Signaling Modulator

Polyphenols have also been seen to have intricate interactions with the arachidonic acid pathway, involving phospholipase A_2 (PLA_2), cyclooxygenase (COX), and lipoxygenase (LOX) and act by inhibiting them. This also forms the crux of the anti-inflammatory effects of polyphenols like curcumin, resveratrol, ECGC, etc. Another important transcription factor is NF-$\kappa\beta$, which regulates a multitude of genes involved in inflammation, apoptosis, cell cycle regulation, angiogenesis as well as immunomodulatory functions. Phenolics and flavonoid compounds often act via inhibiting the activation of NF-$\kappa\beta$, thereby restricting the expression of proinflammatory cytokines, COX as well as NOS expression. Also, cellular proliferation, growth, and death are regulated via MAPK pathway, which, in turn, involves ERK, JNK, p-38 MAP kinase, which are often activated under stress and polyphenols act via inactivating them at various stages, which, in turn, reduces TNFα, IL-1β, IL-6, IL-8, etc. [65, 66]. Figure 4.5 illustrates the various molecular pathways involved in the protective effects of phytoantioxidantsOther Mechanisms

Many a time, polyphenols act via inhibiting crucial enzymes xanthine oxidase, aldose reductase, nitric oxide synthases, and NADPH-oxidase, which often actively contribute to ROS production [62]. Polyphenols may interact with estrogen receptors as well, acting as agonists or antagonists, modulating their function as well. They may often act via inhibiting carbohydrate and lipid digesting enzymes (pancreatic lipase and α-amylase, α-glucosidase) interfering with their

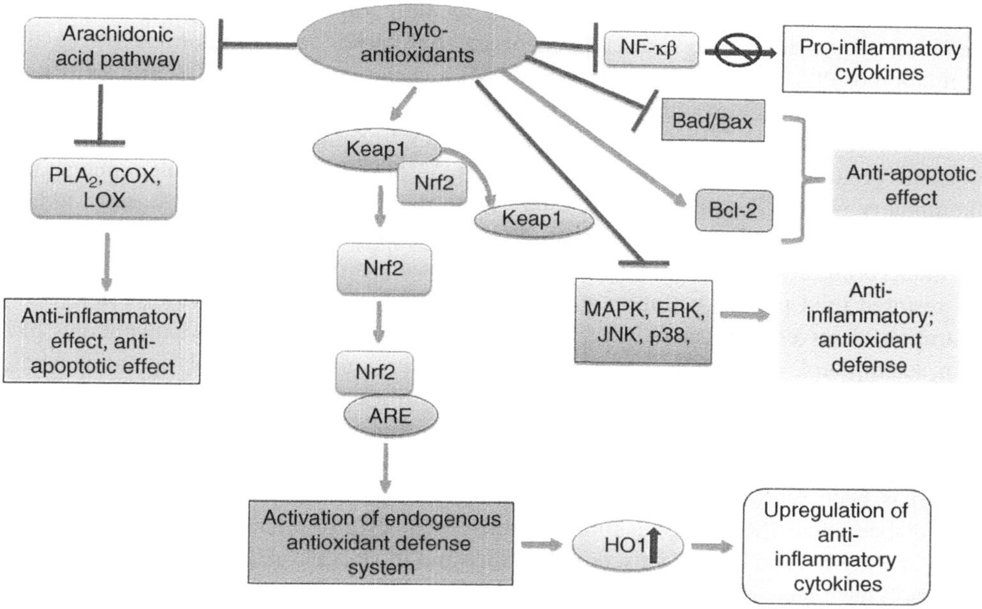

Figure 4.5 Molecular pathways involved in phytoantioxidant-mediated amelioration of oxidative stress-related cellular dysfunction.

digestion and absorption from the gastrointestinal lining restricting the calorie availability. They also interfere with glucose uptake via scrambling GLUT4 and SGLT1 transporters and lipid absorption as they suppress chylomicron formation as well as cholesterol uptake via NPC1L1 transporter. Polyphenols have also been found to regulate GI tract hormones like ghrelin, GLP1, cholecystokinin, along with other hormones like cortisol, adipokines, leptin, resistin, etc. [67].

4.4 Role of Phytoantioxidants in Oxidative Stress-Induced Disease

Oxidative stress has been linked to the pathophysiology of various human diseases like Alzheimer's, Parkinson's disease, cancer, rheumatoid arthritis, cardiovascular disease, diabetes, etc. Imbalanced excessive nutritional intake in long-term may also lead to oxidative stress and inflammation, especially in metabolic syndrome, and obesity, which, in turn, may develop further into serious pathological conditions and phytoantioxidants have been found to have a therapeutic effect in combating such situations [50, 68].

4.4.1 Phytoantioxidants in Diabetes and Metabolic Disorders

Metabolic syndrome encompasses many critical risk factors like hypertension, obesity, dyslipidemia, and insulin resistance, where oxidative stress and inflammation have a crucial interplay in the development of these conditions [69]. With recent research, a fact that dietary modification with ample fruits and vegetables significantly added to the antioxidant capacity of serum-relieving oxidative stress, while having anti-inflammatory effects as well was brought to light. Green tea catechins are known to have anti-obesity effects as they suppress adipocyte proliferation, differentiation of pre-adipocyte, restrict lipogenesis, triglyceride accumulation, following the downregulation of PPARγ and C/EBPα along with Cdk2, cyclin D1, ACC, FOXO1, and SREBP-1c genes and transcription factors and enhance the oxidation of fats via supporting the β-oxidation pathway. Similar results were also documented from resveratrol and curcumin, like upregulation of AMPK, SIRT1, and PGC-1α, and may even induce apoptosis of adipocytes, while reducing the proinflammatory NF-κβ, IL-6, and TNF-α, establishing the anti-inflammatory effect as studied in 3 T3-L1, Raw 264.7 cell lines. Animal studies on ddY mice, ob/ob mice, C57BL/6 mice, HF-diet-induced obesity model, etc., also showed improved glucose homeostasis, reduced blood glucose, restored lipid profiles acting in a coordinated manner with the cell culture data upon administration of resveratrol, catechins, and curcumin [70]. A large group of polyphenols like quercetin, kaempferol, naringenin, baicalein, limonene, curcumin, resveratrol, cinnamic acids, catechins, and mangiferin have effectively demonstrated antihyperglycemic, lowering lipid peroxidation, increased endogenous antioxidant enzymes, managing glucose uptake, increased insulin sensitivity, ameliorating dyslipidemia, etc., in STZ-induced diabetes model in rhodents [71]. Curcumin has been seen to effectively combat and diminish important pathological markers linked to diabetes like ER stress, insulin resistance, adipocyte proliferation, proinflammatory cytokines, in HFD-fed rhodent models, or *ob/ob* mice and in HUVEC cell line [72]. The majority of polyphenols have been found to act via modifying NF-κβ, CEBP α/β, AMPK, sirtuins, and PPARs signaling pathways in metabolic disorders [67]. Resveratrol has also been found to improve glucose homeostasis in rat insulinoma cell line (INS-1E) via GLUT2 translocation and SIRT1 upregulation, while increased insulin secretion was reported from HP62 and β-Min6 pancreatic cell lines. Insulin sensitivity, insulin resistance, and glucose uptake are important pathological markers in type-II diabetes and many phenolic and flavonoid compounds have been found to regulate various genes including c-AMP, PDE, p-JNK, Caspase-3, CAD, XBP-1, p-AKT, FOXO-1, etc., involved in pancreatic β-cell survival, ER stress and insulin secretion as reported from various studies performed on INS-1E, HP62, β-Min6, RINm5f, NIT-1, and β-TC6 cell lines. Also, they have been associated with the modulation of genes like GLUT4, IRS1, PI3K, and DPP4, which are involved in insulin sensitivity, AKT2, etc., in in vitro models for insulin sensitivity, glucose uptake in C2C12, L6 myoblast, HK-2, and 3T3-L1 cell lines as well as in skeletal muscles of rats. Cinnamic acid and proanthocyanidins have been reported to act by inhibiting gluconeogenetic pathways, thereby regulating blood glucose via downregulating PEPCK and G6pase enzymes, in the H4IIE cell line while EGCG promotes insulin signaling through activating the Akt/PI3K pathway, thereby upregulating IR-β, IRS-1, and IGF-1R. Similar mechanisms were also demonstrated in C57BL/KsJ-db/db mice, STZ-induced Sprague-Dawley rats. All these mechanisms added to ameliorative effects of various polyphenolic compounds in diabetes and its related complications [73].

Results from multiple human clinical trials indicated that consumption of diets rich in flavonoids, rather than single compounds, helps in minimizing the risk of diabetes, catechins were capable of lowering fasting blood sugar, resveratrol as well as anthocyanins increased insulin sensititvity, and lowered blood sugar in type II diabetic patients along with improving pancreatic β-cell functions [74].

4.4.2 Phytoantioxidants in Cardiovascular Diseases

Exposure to excessive ROS has been a determining cause in the development of cardiovascular complications like heart failure, stroke, coronary artery disease, and atherosclerosis and is also associated with diabetes, dyslipidemia and obesity [75]. The ill effects of oxidative damage to cardiac cells include damage to DNA, both nuclear and mitochondrial, protein denaturation, and peroxidation of lipids, causing either apoptosis or necrosis of the cells. It has been observed that the methylation status of DNA is one of the epigenetic markers associated with atherosclerotic lesions, and polyphenols have been capable of suppressing this methylation as well as acetylation of histone proteins, exerting their protective effect. Moreover, flavonoids like quercetin, rhamentin, and kaempferol have been reported to exert their antioxidant effects via activating NQO-1, HO-1, thioredoxin reductase, and GST, thereby inducing the endogenous antioxidant defense system, reducing the peroxidation of lipids, while restricting the activation of MAPK pathway as studied in multiple chemical-induced cardiotoxicity models in H9c2 cardiomyocyte cell line. A major trend was observed while comparing the protective activity of a myriad of compounds belonging to flavones, catechins, anthocyanins, flavanones, as well as phenolic acid in cardiac complications, which mainly involved them promoting cellular viability, strengthening the endogenous antioxidant pathways, quenching free radicals, minimizing apoptosis and necrosis, as per reports from various in vitro and in vivo experimental models [76]. Oxidized LDL is one of the major contributing factors in the development of atherosclerosis, and polyphenols have been successfully associated with lowering plasma LDL, its oxidation, while recovering HDL levels, and also minimizing the risk of cardiovascular complications and atherosclerosis as reported by a number of clinical trials over the years [77]. The ability to disrupt the process of platelet adhesion, via regulating the expression of integrin, collagen, or glycoprotein receptors, or by inhibiting phospholipase C (PLC) or the movement of calcium ions, increasing cAMP and cGTP, adds to the anti-aggregative effect of polyphenols. Resveratrol and quercetin inhibit Ca^{2+}-ATPase pump, expression of endothelin-1 and enhance eNOS activity, thereby promoting nitric oxide production in the endothelial lining, whereas isoflavonoids may induce eNOS via modulation of PI3K/Akt and ERK 1/2 pathways all aiding in their vasodilating effect. A reduction in the expression of VCAM-1, ICAM-1, E-selectin, inhibiting proinflammatory cytokines (NF-κβ, IL-6, TNF-α) along with the suppression of MCP and MIP-1α, together contributes toward the antiatherogenic effect of resveratrol. Thus, the antithrombotic, antiatherogenic, antiapoptotic, and anti-inflammatory pathways are the main regulatory pathways for polyphenols to exert their protective effect in cardiovascular complications [78–80]. Green tea polyphenols protected against ischemic injury as studied in the acute myocardial ischemia LAD-rat model via Akt/GSK-3β/caveolae pathway [81]. In spontaneously hypertensive rats as well as double transgenic rats with RAAS system, resveratrol was recorded to lower blood pressure via suppression of angiotensin II type 1 receptors and activation of SIRT1, and, in turn, inactivation of RAAS while quercetin exhibits hypotensive effect in multiple animal models without involving RAAS or oxidative stress pathways. Resveratrol and quercetin also alleviated the stroke-related complications via reducing lipid peroxidation, infracting volume in the brain, and restoring mitochondrial ATP production and respiratory complexes, as reported from various stroke models in animals. In high-fat diet-fed mice, swine with hypercholesterolemia, and Zucker obese rats, resveratrol has been seen to restore the lipid profile, reduce BMI, and have anti-inflammatory effects, along with enhancing glucose uptake as well [82]. Phenolic compounds, in general, are capable of asserting their cardioprotective role via multiple targets and gene modulation involved in major pathways like MAPK/ERK/JNK/p38 impairment, modulation of PI3K-Akt-eNOS and STAT3 pathway, AMPK-mTOR pathway via calcium-calmodulin signaling, NF-κβ pathway, and upregulation of SIRT1, VEGF-B, phosphorylated Akt, GSK3β, Bcl-2, and downregulation of COX-2, FAK, ET-1, Caspase 9, Bax genes which are crucial in shielding against cardiac hypertrophy, atherosclerotic inflammation and adhesion of platelets, apoptosis, necrosis, hypoxia, ischemia, hypertension, as well as myocardial infarction [83].

4.4.3 Phytoantioxidants in Inflammation and Arthritis

The activation of inflammatory responses is a crucial occurrence in oxidative stress, or disease pathologies, having a cause-and-effect relationship, where foreign antigen, exogenous allergen, or toxicants, as well as endogenous factors like oxidized proteins, lipids, and PRDX2 act as inflammatory signaling molecules, which, in turn, activate TNF-α and other proinflammatory cytokines and oxidative stress-related inflammation have been linked to several diseases like cardiovascular diseases, type I and type II diabetes, neurodegenerative diseases, as well as cancer. And polyphenols act via multiple mechanisms to improve the situation with the inflammatory reactions and maintain the cellular homeostasis [84]. The inhibition of the arachidonic acid pathway involving enzymes like COX, LOX, and PLA_2, along with NOS and NF-κβ are some of the notable targets for the polyphenols to act upon like catechins from green tea have been found to inhibit IL-1β, IL-8 acting through NF-κβ signaling pathway; resveratrol ameliorates LPS-induced oxidative damage via minimizing NO production, suppressed iNOS expression, and inhibition of NF-κβ [85]. Inhibition of iNOS and COX-2 has been reported from various cellular models like RAW264.7, human chondrocytes, and rat hepatocytes under the action of various polyphenolic compounds like apigenin, rutin, quercetin, kaempferol, luteolin, etc. Moreover, they also diminished the production of pro-inflammatory cytokines like TNF-α, IL-6, IL-1β, IL-8, IL-10, and IL-4 as studied through various cell and animal models like RBL-2HR, peripheral blood mononuclear cells, human leucocytes, and rat model of rheumatoid arthritis while in various studies, it has also been reported that quercetin, kaempferol suppresses various cell adhesion molecules like VCAM-1, ICAM-1, and E-selectin along with other transcription factors like AP-1, which gradually impart the anti-inflammatory effect as credited to the polyphenolic compounds along with some immunomodulatory activity as well. Similar effects have also been reported from various epidemiological studies, where the subjects were treated with grape juice, pomegranates, fruit and berry juice, blueberries, etc., enriched with phenolic compounds having strong antioxidant potentials, highlighting the beneficial role of polyphenolic compounds in inflammatory conditions [86]. Another mechanistic insight into the anti-inflammatory action of dietary polyphenolic compounds involves the capable inhibition of AGE accumulation and MMP expression, along with radical scavenging, and endogenous antioxidant-boosting potential [87]. Inflammation has been linked to all forms of chronic diseases including asthma, allergic rhinitis, atherosclerotic CVD, type II diabetes, rheumatoid arthritis, and cancer where multiple clinical trials and fruit polyphenols were successful in minimizing the clinical symptoms related to the disease [86].

Rheumatoid arthritis, an autoimmune disease and osteoarthritis, a degenerative disease of joints have inflammation and unrestricted oxidative stress as two major factors contributing to their progression. Although they are incurable at this point, polyphenols help in better management of both the diseases and hence have been found to be beneficial in these conditions. In various studies, it has been reported that polyphenols like naringenin, hesperidin derivatives, blueberries, raspberries, and pomegranate extracts have delivered improved results like decreased chondrocyte damage, decreased inflammation, and an overall improvement in the clinical symptoms of arthritis in a rhodent model of collagen-induced arthritis, surgically induced osteoarthritis, etc. Multiple clinical studies including human subjects with arthritis and their consumption of polyphenol-enriched fruits and extracts also depicted similar outcomes like a sharp decrease in MMP, TNF-α, IL-6, and IL-1β, like pro-inflammatory factors in serum, while overall alleviation of the disease symptoms in the subjects [88]. Curcumin, resveratrol, genistein, EGCG, kaempferol, have been found mainly to act via some common pathways like inhibition of TNF-α, IL-1β, IFNγ, MMPs, and VEGF decreasing the proinflammatory effect, inducing apoptosis in rheumatoid arthritis, fibroblast-like synoviocytes, suppressing MAPK phosphorylation, NOS, NO production, osteoclast differentiation, rendering their protective effect in rheumatoid arthritis [89]. Osteoarthritis is more skewed toward maintaining chondrocyte health, where compounds like EGCG, hydroxytyrosol, and curcumin displayed a significant protective effect through lowering ROS, CK, CKMB, proinflammatory markers, COX-2, iNOS, and other disease-associated pathological markers [90].

4.4.4 Polyphenols in Neurodegenerative Disorders

Neurodegenerative diseases, including Parkinson's and Alzheimer's, are often associated with the central nervous system dysfunction, where atrophic changes in the brain are a hallmark change in Alzheimer's, while loss of dopaminergic neurons from the midbrain essentially affects muscular movements causing loss of balance and walking difficulties. Alzheimer's disease is linked to the improperly folded proteins' (β-amyloid and tau proteins) accumulation around the neurons in the form of plaques, while overaccumulation and modification of α-synuclein in Parkinson's disease leads to the loss of neurons [91]. Both the diseases have been linked to high levels of expression of pro-inflammatory markers, mitochondrial dysfunction, and oxidative stress, accompanied by malfunctioning of proteins [92].

Upon treatment with curcumin, N27 dopaminergic neurons, homocysteine-induced neurotoxicity in rats, it was observed that the lipid peroxidation and free radical levels were reduced, and it also prevented the mitochondrial complex I dysfunction, improved memory, cognitive, motor functions in the rats. EGCG was able to ameliorate the neurodegenerative conditions as studied in neurotoxicity models in H 19-7, HT22 cell lines and in animal models like C57BL/6 mice with transient global cerebral ischemia, SOD1-G93A transgenic mice, etc., via boosting the endogenous antioxidant system, restoring mitochondrial functions, minimizing MDA, NF-κβ activation, caspase 3 cleavage, while activating HIF-1α pathway, along with the modulation of some common signaling pathways like PI3K/MAPK, Akt, ERK1/2, and PKC regulating the cell survival and apoptosis. The effect of resveratrol has also been studied in various in vitro and rhodent models for neurotoxicity, ischemic cerebral stroke, neurodegenerative conditions, where it prevented ROS accumulation, induces endogenous antioxidants, preserves mitochondrial functions, neurons, along with regulating pathways like ERK, SIRT-1, PI3K/Akt, Nrf-2/ARE, and MAPK as well. Polyphenols also prevent tau amyloidiopathies by regulating α, β, γ secretase activity, inhibiting β-amyloid protein oligomerization, reducing its concentration and deposition in the brain, exhibiting marked ameliorative effect on both β-amyloid protein-induced neurotoxicity and neuronal inflammation, as reported in myricitin, quercetin, kaempferol, apigenin, EGCG, and polyphenols from red wine [93]. EGCG has been reported to activate Keap1/ARF/Nrf-2 pathway, while some other polyphenols act via promoting the expression of neurotrophic factors like BDNF, GDNF, and NGF, along with Bcl-2, Bcl-xL via tyrosine receptor kinase pathway, and suppression of TNF-α and IL-6, all converging to their neuroprotective effect. The diminished expression of TLR4, along with IRF3, and MyD88 may just be another pathway for curcumin to be of benefit in Parkinson's disease [94].

Various clinical trials concerning Alzheimer's disease and the use of polyphenols like green tea, fortified fruit juice, cocoa, etc., reported better cognitive functions, increased antioxidant capacity of plasma, and better visual memory in treated patients as compared to placebo groups, enforcing the health benefits of polyphenols in neurodegenerative diseases [95].

4.4.5 Phytoantioxidants in Cancer

Cancer therapy often includes drugs with massive side effects, which led to the increased interest in safer alternatives and plant polyphenols showed promising perspective as they were reported to have potent antioxidant, chemopreventive, and anti-inflammatory activity along with anticancer [96]. Polyphenols are known to intricately act on various processes related to cancer progression including tumerogenesis, angiogenesis, metastasis, as well as chemotherapy and its side effects as well. For example, curcumin has been used as a test candidate in multiple clinical trials involving non-small cell lung cancer, prostate cancer, bladder cancer, breast cancer, multiple myeloma, adenocarcinoma, for its chemopreventive effect and molecular targets of curcumin include a vast collection of growth factors (FGF, VEGF, and TGF), transcription factors (NF-κβ, STATs, and Notch 1), inflammatory cytokines (interleukins, TNF, INF, COX-2, and MCP-1,); apoptotic markers (cytochrome C, caspases), and many more. In breast cancer cells, curcumin has been found to inhibit cellular proliferation via inhibiting NF-κβ, HER2, and PI3K/Akt signaling pathways, whereas in lung cancer, it has been found to inhibit JAK2/STAT3 pathway along with NF-κβ as studied in A549 cell lines while targeting PI3K/Akt/FoxO1/Bim in LLC cell line. In lung adenocarcinoma cell lines, curcumin induced apoptosis via inhibition of COX-2, EGFR, and ERK signaling pathways and cell cycle arrest in non-small cell lung cancer cells via inactivation of MTA1/Wnt/β-catenin pathway. These are a few pathways for the preventive effect of curcumin documented in various cancers [97, 98].

Another important polyphenol with notable chemopreventive activity is resveratrol as it regulates crucial events in tumerogenesis, apoptosis, inflammation as well as oxidative stress. Resveratrol has been found to suppress tumor initiation via regulating the cytochrome P450 enzymes and boosts the activity of UDP-glucuronyltransferase, hemeoxygenase and other detoxification enzymes for clearing out xenobiotics. It may also act as a prooxidant and initiate apoptosis in neoplastic cells. In multiple experimental cancer cell line models, resveratrol has inhibited cellular proliferation by halting DNA replication, exerting cell cycle arrest, initiating apoptosis via activation of Fas, Bax, Bak, PUMA, Bim, and Noxa and suppression of Bcl-2, Bcl-xL, along with activation of p53 gene. Resveratrol also inhibits angiogenesis via inhibition of MMPs, HIF-1α, and VEGF. Resveratrol has also been tried in many clinical studies as a potential anticancer therapeutic agent [99].

Anthocyanins, like delphinidin, were capable of suppressing NF-κβ, while black-rice anthocyanins were found to have initiated apoptosis, and halted cellular proliferation in HER2-positive breast cancer cells. EGCG, on the other hand, inhibited MMP-2 expression along with cell cycle arrest in the CL1-5 cell line, while induced apoptosis in gastric cancer cell lines. Other targets of EGCG as an anticancer agent include Akt, ERK 1/2, p38 MAPK pathways, DNA methyltransferases, as well as histone deacetylases. Similar chemopreventive effects were also reported from hesperitin, apigenin, luteolin, quercetin, kaempferol, etc. [100].

There are multiple signaling pathways whose cross talk between themselves are acted upon by polyphenols. For example, ErbB, NF-κβ, and HH/GLI may actually be an important pathway for neoplastic changes in cells, which is often seen to be regulated by polyphenols, exerting their chemopreventive effects. As a novel strategy in clinical trial, a new developing trend of utilizing these polyphenols with a chemopreventive drug to be used in combination for better therapeutic effect in the disease has also been reported [96, 101].

4.5 Future Scope of Phytoantioxidants as Phytopharmaceuticals

"Let your food be thy medicine," a quote by Hippocrates, underlines the significance of the bioactive compounds present in food itself, which now has been proven through scientific evidence. Although, prominent evidences support the fact that diets rich in fruits, vegetables, legumes, and grains are capable of decreasing the risk of high prevalence diseases, there lies huge lacunae concerning the quality and quantity of polyphenolic contents in various foods and beverages, along with their accurate biological action. Although, phytoantioxidant compounds including phenolics and flavonoids are present in plants in surplus amounts, a major problem is their bioavailability in the biological system, which mainly depends on their digestion, absorption, quantity of consumption, interaction with gut microbiota as well as their modification after cooking. Although in-depth understanding of the mode of actions of various polyphenols is underway, a major bridging is still needed to upgrade mere bioactive compounds to the status of phytopharmaceuticals [102]. The beneficial and therapeutic effects of polyphenols in cardiovascular diseases, obesity, neurodegenerative diseases, cancer, etc., are well established, which lead to the development of marketing strategies promoting food enriched with polyphenols as functional foods, and making a standard dietary allowance for these is very complicated mainly because of their excessive variety, lack of knowledge of their distribution or quantity present in different food particles, lack of clinical trial data, all of which pose a major problem in upgrading these as pharmaceutical products. Moreover, as they cannot be assigned any recommended dietary intake amount, their overconsumption may pose a risk to health itself and not many clinical studies are there to give evidence of their safety or toxicity. Hence, a lot remains to be covered so as to upgrade the polyphenolic compounds from just some beneficial research interest to the status of pharmaceuticals [53].

4.6 Conclusion

ROS in overwhelming amounts may develop into oxidative stress in the biological system, but it is not always harmful and is an integral part of normal cellular metabolism. It rather plays a significant role in cellular signaling and maintains proper homeostasis in the body, when produced in controlled amounts. In various pathological conditions or in presence of xenobiotics, when the balance between ROS production and its elimination via the endogenous antioxidant system is disturbed, it may lead to the development of various diseases, where the antioxidant defense system present in the cell may often not be enough to counterbalance the oxidative assault. This is where the external support in the form of phyto-antioxidants becomes useful. Like their name, these are a large class of bioactive compounds, which have very potent antioxidant, anticancer, antidiabetic, renoprotective, and neuroprotective activities to name a few and act via modulation of multiple signal transduction pathways. Since they are derived from plants, they are believed to have very minimal side effects. Although their protective, beneficial, and therapeutic effects in biological systems have been studied via various cellular and animal models, still they are far away from being used as a medicine or phytopharmaceutical. They are mostly used as neutraceuticals, but for their proper commercialization, many more clinical study references are needed as evidence of their mode of action, and ameliorative effect on various diseases so as to get approval as a drug. Also, much research lacunae are remaining on the efficacy of polyphenols as single compounds, their bioavailability, synergistic effect, their application in combination, all of which need to be verified via experimental models as well as in human clinical trials. Hence, it can be concluded that the role of phytoantioxidants in balancing cellular oxidative stress is of much importance as it plays a critical role in protecting against various disease pathologies and has the potential to be used as a therapeutic drug someday, with much research and development.

References

1 Sivanandham, V. (2011). Free radicals in health and diseases – a mini review. *Pharmacol. Online* 1: 1062–1077.

2 Zhang, J., Wang, X., Vikash, V. et al. (2016). ROS and ROS-mediated cellular signaling. *Oxidative Med. Cell. Longev.* 2016: 4350965.

3 Birben, E., Sahiner, U.M., Sackesen, C. et al. (2012). Oxidative stress and antioxidant defense. *World Allergy Org. J.* 5 (1): 9–19.

4 Sindhi, V., Gupta, V., Sharma, K. et al. (2013). Potential applications of antioxidants – a review. *J. Pharm. Res.* 7 (9): 828–835.

5 Hamid, A.A., Aiyelaagbe, O.O., Usman, L.A. et al. (2010). Antioxidants: its medicinal and pharmacological applications. *Afr. J. Pure Appl. Chem.* 4 (8): 142–151.

6 He, L., He, T., Farrar, S. et al. (2017). Antioxidants maintain cellular redox homeostasis by elimination of reactive oxygen species. *Cell. Physiol. Biochem.* 44 (2): 532–553.

7 Kasote, D.M., Katyare, S.S., Hegde, M.V., and Bae, H. (2015). Significance of antioxidant potential of plants and its relevance to therapeutic applications. *Int. J. Biol. Sci.* 11 (8): 982.

8 Forni, C., Facchiano, F., Bartoli, M. et al. (2019). Beneficial role of phytochemicals on oxidative stress and age-related diseases. *Biomed. Res. Int.* 2019: 8748253.

9 Sies, H. (2016). The concept of oxidative stress after 30 years. In: *Biochemistry of Oxidative Stress* (ed. R.J. Gelpi, A. Boveris and J.J. Poderoso), 3–11. Cham: Springer.

10 Trachootham, D., Lu, W., Ogasawara, M.A. et al. (2008). Redox regulation of cell survival. *Antioxid. Redox Signal* 10 (8): 1343–1374.

11 Das, K.C. and Das, C.K. (2000). Thioredoxin, a singlet oxygen quencher and hydroxyl radical scavenger: redox independent functions. *Biochem. Biophys. Res. Commun.* 277 (2): 443–447.

12 Ahmadinejad, F., GeirMøller, S., Hashemzadeh-Chaleshtori, M. et al. (2017). Molecular mechanisms behind free radical scavengers function against oxidative stress. *Antioxidants* 6 (3): 51.

13 Lushchak, V.I. (2014). Free radicals, reactive oxygen species, oxidative stress and its classification. *Chem. Biol. Interact.* 224: 164–175.

14 Tvrdá, E. and Benko, F. (2020). Free radicals: what they are and what they do. In: *Pathology* (ed. V.R. Preedy), 3–13. Academic Press.

15 Schieber, M. and Chandel, N.S. (2014). ROS function in redox signaling and oxidative stress. *Curr. Biol.* 24 (10): R453–R462.

16 Mirończuk-Chodakowska, I., Witkowska, A.M., and Zujko, M.E. (2018). Endogenous non-enzymatic antioxidants in the human body. *Adv. Med. Sci.* 63 (1): 68–78.

17 Ray, P.D., Huang, B.W., and Tsuji, Y. (2012). Reactive oxygen species (ROS) homeostasis and redox regulation in cellular signalling. *Cell. Signal* 24 (5): 981–990.

18 Roy, J., Galano, J.M., Durand, T. et al. (2017). Physiological role of reactive oxygen species as promoters of natural defenses. *FASEB J.* 31 (9): 3729–3745.

19 Su, L.J., Zhang, J.H., Gomez, H. et al. (2019). Reactive oxygen species-induced lipid peroxidation in apoptosis, autophagy, and ferroptosis. *Oxidative Med. Cell. Longev.* 2019: 5080843.

20 Cecarini, V., Gee, J., Fioretti, E. et al. (2007). Protein oxidation and cellular homeostasis: emphasis on metabolism. *Biochim. Biophys. Acta (BBA)* 1773 (2): 93–104.

21 Avery, S.V. (2011). Molecular targets of oxidative stress. *Biochem. J.* 434 (2): 201–210.

22 Kong, Q. and Lin, C.L. (2010). Oxidative damage to RNA: mechanisms, consequences, and diseases. *Cell. Mol. Life Sci.* 67 (11): 1817–1829.

23 Yin, J., Ren, W.K., Wu, X.S. et al. (2013). Oxidative stress-mediated signaling pathways: a review. *J. Food Agric. Environ.* 11 (2): 132–139.

24 Sies, H., Berndt, C., and Jones, D.P. (2017). Oxidative stress. *Annu. Rev. Biochem.* 86: 715–748.

25 Poljsak, B., Šuput, D., and Milisav, I. (2013). Achieving the balance between ROS and antioxidants: when to use the synthetic antioxidants. *Oxidative Med. Cell. Longev.* 2013: 956792.

26 Pisoschi, A.M. and Pop, A. (2015). The role of antioxidants in the chemistry of oxidative stress: a review. *Eur. J. Med. Chem.* 97: 55–74.

27 Sulaiman, M., Tijani, H.I., Abubakar, B.M. et al. (2013). An overview of natural plant antioxidants: analysis and evaluation. *Adv. Biochem.* 1 (4): 64–72.

28 Lourenço, S.C., Moldão-Martins, M., and Alves, V.D. (2019 Jan). Antioxidants of natural plant origins: from sources to food industry applications. *Molecules* 24 (22): 4132.

29 Zhang, Y.J., Gan, R.Y., Li, S. et al. (2015). Antioxidant phytochemicals for the prevention and treatment of chronic diseases. *Molecules* 20 (12): 21138–21156.

30 Tsao, R. (2010). Chemistry and biochemistry of dietary polyphenols. *Nutrients* 2 (12): 1231–1246.

31 Swallah, M.S., Sun, H., Affoh, R. et al. (2020). Antioxidant potential overviews of secondary metabolites (polyphenols) in fruits. *Int. J. Food Sci.* 2020: 9081686.

32 Hamad, A.M. (2019). Some natural antioxidants sources from foods and tree barks. *Int. J. Sci. Technol. Res.* 8 (3): 93–98.

33 Langi, P., Kiokias, S., Varzakas, T., and Proestos, C. (2018). Carotenoids: from plants to food and feed industries. *Methods Mol. Biol.* 1852: 57–71.

34 Sikora, E., Cieślik, E., and Topolska, K. (2008). The sources of natural antioxidants. *Acta Sci. Pol. Technol. Aliment.* 7 (1): 5–17.

35 Bramley, P.M., Elmadfa, I., Kafatos, A. et al. (2000). Reviews – vitamin E. *J. Sci. Food Agric.* 80 (7): 913–938.

36 Abramovič, H. (2015). Antioxidant properties of hydroxycinnamic acid derivatives: a focus on biochemistry, physicochemical parameters, reactive species, and biomolecular interactions. In: *Coffee in Health and Disease Prevention* (ed. V.R. Preedy), 843–852. Academic Press.

37 Martinez, K.B., Mackert, J.D., and McIntosh, M.K. (2017). Polyphenols and intestinal health. In: *Nutrition and Functional Foods for Healthy Aging* (ed. R.R. Watson), 191–210. Academic Press.

38 Garg, C. and Garg, M. (2019). Oxidants, oxidative stress and role of phytoantioxidants as photoprotectives: a review. *Indian J. Nat. Prod.* 33 (1).

39 Panche, A.N., Diwan, A.D., and Chandra, S.R. (2016). Flavonoids: an overview. *J. Nutr. Sci.* 5: e47.

40 Hall, C. (2001). Sources of natural antioxidants: oilseeds, nuts, cereals, legumes, animal products and microbial sources. Antioxidants in food. In: *Practical Applications* (ed. J. Pokorny, N.Y. Yanishlieva and M. Gordon), 1. Cambridge: CRC Woodhead Publishing Limited.

41 Kumar, S. and Pandey, A.K. (2013). Chemistry and biological activities of flavonoids: an overview. *Sci. World J.* 2013: 162750.

42 Gacek, M. (2014). Soy and legume seeds as sources of isoflavones: selected individual determinants of their consumption in a group of perimenopausal women. *Prz. Menopauzalny* 13 (1): 27.

43 Khoo, H.E., Azlan, A., Tang, S.T., and Lim, S.M. (2017). Anthocyanidins and anthocyanins: colored pigments as food, pharmaceutical ingredients, and the potential health benefits. *Food Nutr. Res.* 61 (1): 1361779.

44 Dei Cas, M. and Ghidoni, R. (2019). Dietary curcumin: correlation between bioavailability and health potential. *Nutrients* 11 (9): 2147.

45 Orlikova, B., Tasdemir, D., Golais, F. et al. (2011). Dietary chalcones with chemopreventive and chemotherapeutic potential. *Genes Nutr.* 6 (2): 125–147.

46 Cox-Georgian, D., Ramadoss, N., Dona, C., and Basu, C. (2019). Therapeutic and medicinal uses of terpenes. In: *Medicinal Plants*, 333–359. Cham: Springer.

47 Chung, K.T., Wong, T.Y., Wei, C.I. et al. (1998). Tannins and human health: a review. *Crit. Rev. Food Sci. Nutr.* 38 (6): 421–464.

48 Adlercreutz, H. (2007). Lignans and human health. *Crit. Rev. Clin. Lab. Sci.* 44 (5–6): 483–525.

49 Międkus, N., Marszałek, K., Podlacha, M. et al. (2020). Health benefits of plant-derived sulfur compounds, glucosinolates, and organosulfur compounds. *Molecules* 25 (17): 3804.

50 Mishra, T., Kondepati, A.K., Pasumarthi, S.D. et al. (2020). Phytotherapeutic antioxidants. *Asian J. Med. Sci.* 11 (2): 96–100.

51 Sarkar, B., Deb, P., and Banerjee, S. (2021). Oxidative stress, inflammasome, and cancer: polyphenols may be the answer. In: *Handbook of Oxidative Stress in Cancer: Mechanistic Aspects.* Singapore: Springer Nature https://doi.org/10.1007/978-981-15-4501-6_165-1.

52 Rasouli, H., Farzaei, M.H., and Khodarahmi, R. (2017). Polyphenols and their benefits: a review. *Int. J. Food Prop.* 20 (sup2): 1700–1741.

53 Cory, H., Passarelli, S., Szeto, J. et al. (2018). The role of polyphenols in human health and food systems: a mini-review. *Front. Nutr.* 5: 87.

54 Calderon-Montano, M.J., Burgos-Morón, E., Pérez-Guerrero, C., and López-Lázaro, M. (2011). A review on the dietary flavonoid kaempferol. *Mini-Rev. Med. Chem.* 11 (4): 298–344.

55 Hewlings, S.J. and Kalman, D.S. (2017). Curcumin: a review of its' effects on human health. *Foods* 6: 10–92.

56 Noorafshan, A. and Ashkani-Esfahani, S. (2013). A review of therapeutic effects of curcumin. *Curr. Pharm. Des.* 19 (11): 2032–2046.

57 Călinoiu, L.F. and Vodnar, D.C. (2018). Whole grains and phenolic acids: a review on bioactivity, functionality, health benefits and bioavailability. *Nutrients* 10 (11): 1615.

58 VenkateswaraRao, P., Kiran, S.D., Rohini, P., and Bhagyasree, P. (2017). Flavonoid: a review on naringenin. *J. Pharmacogn. Phytochem.* 6: 2778–2783.

59 Eggersdorfer, M. and Wyss, A. (2018). Carotenoids in human nutrition and health. *Arch. Biochem. Biophys.* 652: 18–26.

60 Naidu, K.A. (2003). Vitamin C in human health and disease is still a mystery? An overview. *Nutr. J.* 2 (1): 1–10.

61 Di Meo, F., Lemaur, V., Cornil, J. et al. (2013). Free radical scavenging by natural polyphenols: atom versus electron transfer. *J. Phys. Chem. A* 117 (10): 2082–2092.

62 Fraga, C.G., Galleano, M., Verstraeten, S.V., and Oteiza, P.I. (2010). Basic biochemical mechanisms behind the health benefits of polyphenols. *Mol. Asp. Med.* 31 (6): 435–445.

63 Leopoldini, M., Russo, N., and Toscano, M. (2011). The molecular basis of working mechanism of natural polyphenolic antioxidants. *Food Chem.* 125 (2): 288–306.

64 Upadhyay, S. and Dixit, M. (2015). Role of polyphenols and other phytochemicals on molecular signaling. *Oxidative Med. Cell. Longev.* 2015: 1–15.

65 Santangelo, C., Varì, R., Scazzocchio, B. et al. (2007). Polyphenols, intracellular signalling and inflammation. *Annali-Istitutosuperiore Di Sanita* 43 (4): 394.

66 Seifried, H.E., Anderson, D.E., Fisher, E.I., and Milner, J.A. (2007). A review of the interaction among dietary antioxidants and reactive oxygen species. *J. Nutr. Biochem.* 18 (9): 567–579.

67 García-Conesa, M.T. (2017). Dietary polyphenols against metabolic disorders: how far have we progressed in the understanding of the molecular mechanisms of action of these compounds? *Crit. Rev. Food Sci. Nutr.* 57 (9): 1769–1786.

68 Tan, B.L., Norhaizan, M.E., and Liew, W.P. (2018). Nutrients and oxidative stress: friend or foe? *Oxidative Med. Cell. Longev.* 2018: 1–24.

69 Roberts, C.K. and Sindhu, K.K. (2009). Oxidative stress and metabolic syndrome. *Life Sci.* 84 (21–22): 705–712.

70 Wang, S., Moustaid-Moussa, N., Chen, L. et al. (2014). Novel insights of dietary polyphenols and obesity. *J. Nutr. Biochem.* 25 (1): 1–8.

71 Bacanli, M., Dilsiz, S.A., Başaran, N., and Başaran, A.A. (2019). Effects of phytochemicals against diabetes. *Adv. Food Nutr. Res.* 89: 209–238.

72 Jin, T., Song, Z., Weng, J., and Fantus, I.G. (2018). Curcumin and other dietary polyphenols: potential mechanisms of metabolic actions and therapy for diabetes and obesity. *Am. J. Physiol. Endocrinol. Metab.* 314 (3): E201–E205.

73 Kang, G.G., Francis, N., Hill, R. et al. (2020). Dietary polyphenols and gene expression in molecular pathways associated with type 2 diabetes mellitus: a review. *Int. J. Mol. Sci.* 21 (1): 140.

74 Cao, H., Ou, J., Chen, L. et al. (2019). Dietary polyphenols and type 2 diabetes: human study and clinical trial. *Crit. Rev. Food Sci. Nutr.* 59 (20): 3371–3379.

75 Csányi, G. and Miller, F. Jr. (2014). Oxidative stress in cardiovascular disease. *Int. J. Mol. Sci.* [Internet]. MDPI AG 15 (4): 6002–6008.

76 Mattera, R., Benvenuto, M., Giganti, M.G. et al. (2017). Effects of polyphenols on oxidative stress-mediated injury in cardiomyocytes. *Nutrients* 9 (5): 523.

77 Manna, E. and Maiti, S. (2016). Cardio-protecting effect of natural bioactive compound (polyphenol) by inhibiting ldl oxidation with the scavenging of reactive oxygen species (ROS). *J. Clin. Exp. Cardiol.* 7 (453): 2.

78 Tangney, C.C. and Rasmussen, H.E. (2013). Polyphenols, inflammation, and cardiovascular disease. *Curr. Atheroscler. Rep.* 15 (5): 324.

79 Quiñones, M., Miguel, M., and Aleixandre, A. (2013). Beneficial effects of polyphenols on cardiovascular disease. *Pharmacol. Res.* 68 (1): 125–131.

80 Michalska, M., Gluba, A., Mikhailidis, D.P. et al. (2010). The role of polyphenols in cardiovascular disease. *Med. Sci. Monit.* 16 (5): RA110–RA119.

81 Hsieh, S.R., Cheng, W.C., Su, Y.M. et al. (2014). Molecular targets for anti-oxidative protection of green tea polyphenols against myocardial ischemic injury. *Biomedicine* 4 (4): 7–16.

82 Andriantsitohaina, R., Auger, C., Chataigneau, T. et al. (2012). Molecular mechanisms of the cardiovascular protective effects of polyphenols. *Br. J. Nutr.* 108 (9): 1532–1549.

83 Otręba, M., Kośmider, L., and Rzepecka-Stojko, A. (2021). Polyphenols' cardioprotective potential: review of rat fibroblasts as well as rat and human cardiomyocyte cell lines research. *Molecules* 26 (4): 774.

84 Hussain, T., Tan, B., Yin, Y. et al. (2016). Oxidative stress and inflammation: what polyphenols can do for us? *Oxidative Med. Cell. Longev.* 2016: 1–9.

85 Biesalski, H.K. (2007). Polyphenols and inflammation: basic interactions. *Curr. Opin. Clin. Nutr. Metab. Care* 10 (6): 724–728.

86 González-Gallego, J., García-Mediavilla, M.V., Sánchez-Campos, S., and Tuñón, M.J. (2010). Fruit polyphenols, immunity and inflammation. *Br. J. Nutr.* 104 (S3): S15–S27.

87 Crascì, L., Lauro, M.R., Puglisi, G., and Panico, A. (2018). Natural antioxidant polyphenols on inflammation management: anti-glycation activity vs metalloproteinases inhibition. *Crit. Rev. Food Sci. Nutr.* 58 (6): 893–904.

88 Basu, A., Schell, J., and Scofield, R.H. (2018). Dietary fruits and arthritis. *Food Funct.* 9 (1): 70–77.

89 Mateen, S., Moin, S., Zafar, A., and Khan, A.Q. (2016). Redox signaling in rheumatoid arthritis and the preventive role of polyphenols. *Clin. Chim. Acta* 463: 4–10.

90 Oliviero, F., Scanu, A., Zamudio-Cuevas, Y. et al. (2018). Anti-inflammatory effects of polyphenols in arthritis. *J. Sci. Food Agric.* 98 (5): 1653–1659.

91 Li, J., Li, W., Jiang, Z.G., and Ghanbari, H.A. (2013). Oxidative stress and neurodegenerative disorders. *Int. J. Mol. Sci.* 14 (12): 24438–24475.

92 Bhullar, K.S. and Rupasinghe, H.P. (2013). Polyphenols: multipotent therapeutic agents in neurodegenerative diseases. *Oxidative Med. Cell. Longev.* 2013: 1–18.

93 Ebrahimi, A. and Schluesener, H. (2012). Natural polyphenols against neurodegenerative disorders: potentials and pitfalls. *Ageing Res. Rev.* 11 (2): 329–345.

94 Di Meo, F., Valentino, A., Petillo, O. et al. (2020). Bioactive polyphenols and neuromodulation: molecular mechanisms in neurodegeneration. *Int. J. Mol. Sci.* 21 (7): 2564.

95 Colizzi, C. (2019). The protective effects of polyphenols on Alzheimer's disease: a systematic review. *Alzheimer's Dement. Transl. Res. Clin. Interv.* 5: 184–196.

96 Mohan, A., Narayanan, S., Sethuraman, S., and Maheswari, K.U. (2013). Combinations of plant polyphenols & anti-cancer molecules: a novel treatment strategy for cancer chemotherapy. *Anti-Cancer Agents Med. Chem. (Formerly Curr. Med. Chem.-Anti-Cancer Agents)* 13 (2): 281–295.

97 Giordano, A. and Tommonaro, G. (2019). Curcumin and cancer. *Nutrients* 11 (10): 2376.

98 Asensi, M., Ortega, A., Mena, S. et al. (2011). Natural polyphenols in cancer therapy. *Crit. Rev. Clin. Lab. Sci.* 48 (5–6): 197–216.

99 Ferraz da Costa, D.C., Fialho, E., and Silva, J.L. (2017). Cancer chemoprevention by resveratrol: the p53 tumor suppressor protein as a promising molecular target. *Molecules* 22 (6): 1014.

100 Zhou, Y., Zheng, J., Li, Y. et al. (2016). Natural polyphenols for prevention and treatment of cancer. *Nutrients* [Internet]. MDPI AG 8 (8): 515.

101 Fantini, M., Benvenuto, M., Masuelli, L. et al. (2015). In vitro and in vivo antitumoral effects of combinations of polyphenols, or polyphenols and anticancer drugs: perspectives on cancer treatment. *Int. J. Mol. Sci.* 16 (5): 9236–9282.

102 Tresserra-Rimbau, A., Lamuela-Raventos, R.M., and Moreno, J.J. (2018). Polyphenols, food and pharma. Current knowledge and directions for future research. *Biochem. Pharmacol.* 156: 186–195.

5

Bioactive Flavonoids as Phytoantioxidants

Joyani Das

Sarojini Naidu Vanita Pharmacy Maha Vidyalaya, Secunderabad, Telangana, India

5.1 Introduction

A major cause of systemic cell and tissue damage has been attributed to the formation of oxygen reactive forms as part of various oxidative pathways in a living system. Antioxidants are recruited to protect against any excess of these reactive species by undergoing oxidation themselves. For this purpose, the reinforcement of the body's endogenous antioxidant system by intake of natural antioxidants through diet has played an important role and has been the subject of intensive research for a long time [1]. Naturally obtained antioxidants may be grouped into enzymatic and nonenzymatic based on their inherent nature and mode of action (Figure 5.1). Examples of enzymatic antioxidants primarily include glutathione peroxidase (GPx), catalase, and superoxide dismutase (SOD), whereas nonenzymatic ones may be vitamins (e.g. A, E, and C), enzyme cofactors (Q10), peptides, minerals (e.g. zinc and selenium) and the most acclaimed, phytoantioxidants (polyphenols) [2].

Phytonutrients, namely polyphenols, terpenoids, flavonoids, phytosterols, phytoestrogens, and the like, are bioactive components obtained from plants, which, though originally synthesized for their own sake, provide a range of health benefits when consumed by humans [3]. Flavonoids constitute a large class of secondary metabolites, which are hypothesized to have antioxidant functions in higher plants that encounter various environmental stresses [4]. In these plants, following upregulation of enzymes for polyphenol oxidation, preferential accumulation of free radical scavenging, dihydroxy B-ring-substituted flavonoids occur in response to severe stress. Thus, changes in redox homeostasis have been hypothesized to activate the biosynthesis of flavonoids, particularly flavonol metabolism via modulation of MYB (myeloblastosis) transcription factor [5–7].

This antioxidant property of flavonoids has evoked keen interest among the scientific fraternity to find ways to extrapolate the same for human benefits. As free radical generation and oxidative stress rose to progressively occupy a major fraction in the pathophysiology of various diseases, research with flavonoids flooded the scientific literature [8]. The human diet has always played a vital role in shaping physiology and it is quite well established that a diet rich in fruits and vegetables is beneficial in preventing premature aging and its various consequences such as the onset of various organ system disorders [9]. In this regard, polyphenols have emerged to be responsible for the beneficial effects of these dietary components of which bioactive flavonoids have been the prime focus of research in the quest for the most effective phytoantioxidants.

Phytoantioxidants and Nanotherapeutics, First Edition. Edited by Mithun Rudrapal.

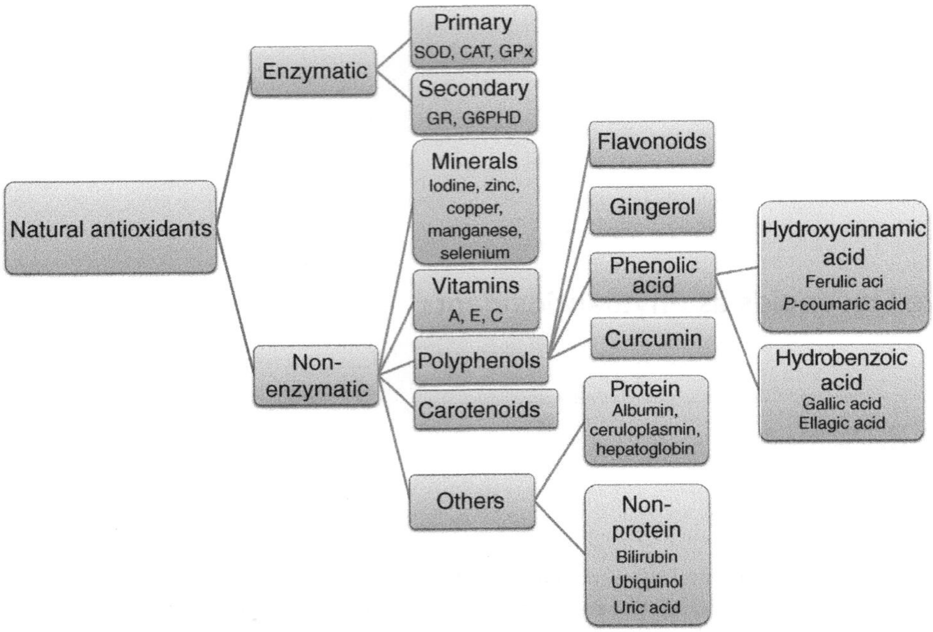

Figure 5.1 Classification of various natural antioxidants.

5.2 Structural Features and Natural Sources of Bioactive Flavonoids

Various plants have been found to contain polyphenols of which flavonoids, stilbenes, phenolic acids, phenolic alcohols, and lignans form the broad classification [10]. Foods found to be rich in flavonoids include citrus fruits, apples, berries, grapes, onions, soybeans, cowpeas, beans, and tea among various others [11]. Some very common Indian food items, namely guava, turmeric, mustard seeds, cloves, red chilis, Indian gooseberry, cumin, cardamom, betel leaf, kidney beans, ginger, coriander powder, and brinjal are of special mention in this category with respect to their quantity of consumption-wise flavonoid content [12]. Interestingly, a difference in the flavonoid content in plant sources has been noticed based on climate, the season of harvest as well as on processing and storage conditions [13].

Flavonoids may be subgrouped into flavonols, flavones, flavanones, flavan-3-ols, and anthocyanidins. The basic structure of flavonoids, as depicted in Figure 5.1, consists of two benzene rings usually connected with a three carbon chain from a pyran ring, forming a C_6-C_3-C_6 backbone [10]. Natural flavonoids contain two hydroxyl groups on A (C5, C7) and one on B ring (C3), where the bases for subclassification are C ring oxidation, C ring carbon number to which B ring attaches, and the degree of unsaturation [11]. Examples of compounds comprising various subclasses have been elaborated in Table 5.1. Structural features of the subgroups have been summarized in Figure 5.2a–e [14].

Flavonols attached to sugar moieties are known as flavonol glycosides and the ones which are not referred to as aglycone forms. Quercetin, myricetin, and isorhamnetin are major flavonol aglycones, which appear in the form of *O*-glycosides in fruits and vegetables [15]. A detailed list of various food substances along with their flavonoid contents is given in Table 5.2 [16].

Table 5.1 Subclasses of flavonoids.

Subclass	Examples
Flavonols	Isorhamnetin, Kaempferol, Myrecetin, Quercetin
Flavones	Apigenin, Luteolin
Flavanones	Eriodictyol, Hesperetin, Naringenin
Flavan-3-ols	(+)-Catechin, (+)-Gallocatechin, (−)-Epicatechin, (−)-Epigallocatechin, (−)-Epicatechin-3-gallate, (−)-Epigallocatechin-3-gallate, Theaflavin-3-gallate, Theaflavin-3′-gallate, Theaflavin-3,3′-digallate, Thearubigins
Anthocyanidins	Cyanidin, Delphidin, Malvidin, Pelargonidin, Peonidin, Petunidin

Source: Data from Ref. [13].

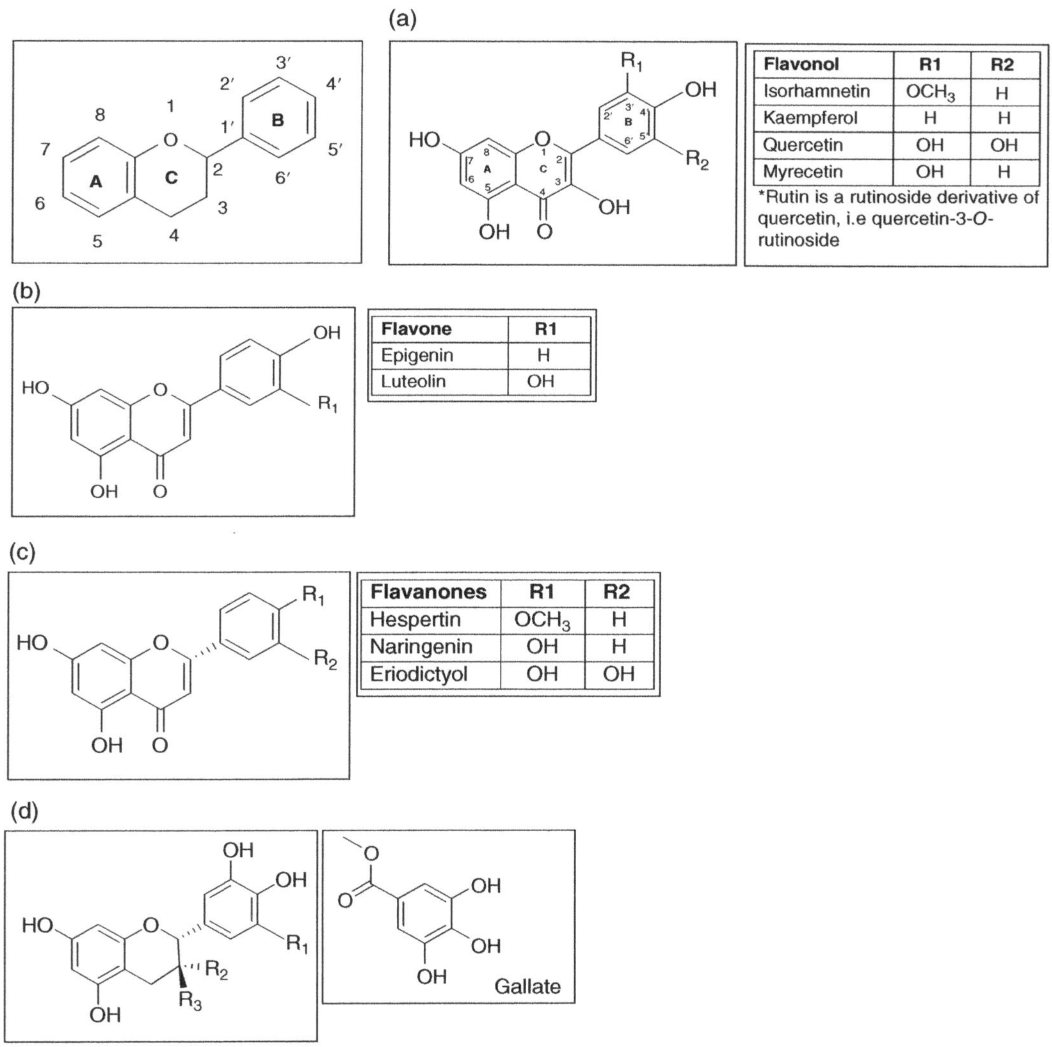

Figure 5.2 General structure of flavonoids. (a) Chemical structure of flavonols. (b) Chemical structure of flavones. (c) Chemical structure of flavanones. (d) Chemical structure of flavan-3-ols. (e) Chemical structure of theaflavins. (f) Chemical structure of anthocyanidins.

(e)

Flavan-3-ols	R1	R2	R3
(+)-Catechin	H	H	OH
(+)-Catechin-3-gallate	H	H	Gallate
(−)-Epicathechin	H	OH	H
(−)-Epicathechin-3-gallate	H	Gallate	H
(−)-Epigallocatechin	OH	OH	H
(−)-Epigallocatechin-3-gallate	OH	Gallate	H
(+)-Gallocatechin	OH	H	OH
(+)-Gallocatechin-3-gallate	OH	H	Gallate

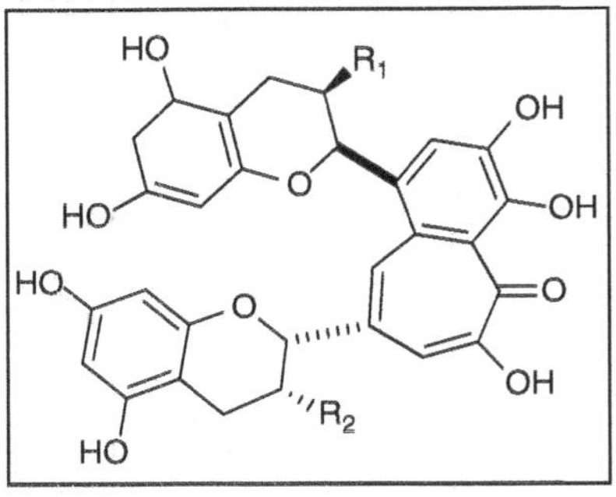

Theaflavins	R1	R2
Theaflavin	OH	OH
Theaflavin-3-gallate	Gallate	OH
Theaflavin-3′-gallate	OH	Gallate
Theaflavin-3,3′-digallate	Gallate	Gallate

(f)

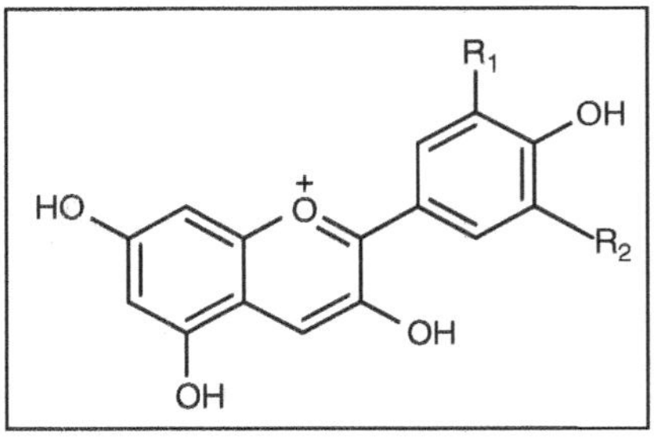

Anthocyanidin	R1	R2
Malvidin	OCH_3	OCH_3
Pelargonidin	H	H
Delphinidin	OH	OH
Cyanidin	H	OH
Peonidin	H	OCH_3
Petunidin	OH	OCH_3

Figure 5.2 (Continued)

5.3 Mechanism of Action as Antioxidants

The health benefits of flavonoids have been recognized in traditional medicine much before their isolation. Subsequently, Hungarian biochemist, Albert Szent-Gyorgyi (Nobel Prize in Physiology and Medicine, 1937) and coworkers reported the potential of citrin and eriodictin obtained from the lemon peel in treating abnormal capillary permeability and fragility [17]. These substances, identified as flavonoids, came to be classified as vitamin P but the term was dropped in 1950. A surge in flavonoid research got restarted in the 1990s wherein the potential of flavonoids against the progression of a number of chronic degenerative disorders was demonstrated and various hypotheses about their mechanism of action were proposed [18, 19].

Reactive oxygen species (ROS) and reactive nitrogen species (RNS), namely hydroxyl (OH·), superoxide (O_2·), nitric oxide (NO·), nitrogen dioxide (NO_2·), peroxyl (ROO·), and the like are

Table 5.2 Dietary sources of some common bioflavonoids with their approximate amounts.

Flavonoid class with common compounds	Common food sources and content (mg/100 g)			
Flavan-3-ols	Source	(+)-Catechin	(−)-Epicatechin	Procyanidin B2
(+)-Catechin	Apples (with skin)	2.00	9.83	15.12
(−)-Epicatechin	Apricots (raw)	3.67	4.74	1.33
Procyanidin B2	Peaches (raw)	4.92	2.34	12.24
	Pears (raw)	0.27	3.76	2.73
	Strawberries (raw)	6.65	1.56	5.26
	Black tea (brewed)	1.51	2.13	3.74
	Blueberries (raw)	5.29	0.62	5.71
	Cranberries (raw)	0.39	4.37	25.93
	Cocoa (dry powder)	64.82	196.43	183.49
	Grapes (black/red)	0.82	0.96	2.38
	Red wine	7.12	3.76	20.49
Flavonols	Source	Kaempferol	Myricetin	Quercetin
Kaempferol	Blueberries	1.66	1.26	7.67
Myricetin	Garlic	0.26	1.61	1.74
Quercetin	Onions	0.63	0.03	21.40
	Kale	46.80	0. 00	22.58
	Broccoli	7.84	0.06	3.26
	Spinach	15.75	—	5.75
	Black tea (brewed)	1.31	0.45	1.99
Isoflavones	Source	Daidzein	Genistein	Glycitein
Daidzein	Tofu (regular, raw)	8.56	12.99	1.98
Genistein Glycitein	Soyabean (raw, mature seeds)	61.33	86.33	13.33
	Tempeh	22.66	36.15	3.82
	Beans	0.29	0.30	0.00
Flavanones	Source	Eriodictyol	Hisperetin	Naringenin
Eriodictyol Hisperetin	Grapefruit (juice, white)	0.65	2.35	18.23
Naringenin	Lemon (juice)	4.88	14.74	1.38
	Orange (juice)	0.17	20.39	3.27
	Peppermint	30.92	9.52	—
Flavones	Source	Apigenin	Leuteolin	
Apigenin	Celery	2.85	1.05	
Luteolin	Celery seed (spice)	83.70	811.41	
	Parsley	215.46	1.09	
	Green peppers	0.00	4.71	
	Peppermint	8.71	11.33	
	Common in leafy plants (esp. Apiaceae family)			

(Continued)

Table 5.2 (Continued)

Flavonoid class with common compounds	Common food sources and content (mg/100 g)				
Anthocyanins	Source	Cyanidin	Delphinidin	Malvidin	Pelargonidin
Cyanidin	Apples	1.27	0.00	0.00	0.00
Delphinidin Malvidin	Blueberries (lowbush)	17.92	34.00	54.00	2.65
Pelargonidin	Red wine	0.45	2.75	15.29	–
	Strawberries	1.63	0.31	0.01	25.69

Source: Based on Ref. [16].

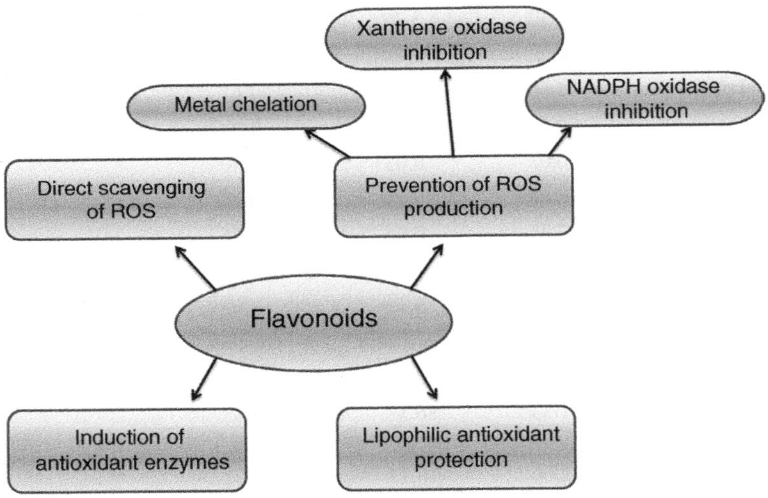

Figure 5.3 Pathways through which flavonoids exert their antioxidant action.

produced as products of physiological metabolic reactions but are rendered harmless by antioxidants like SOD, catalase, GSH, etc. The oxidative stress theory came into existence in 1985 and has, since then, been of substantial importance in the pathophysiology of chronic degenerative diseases, which are explained on the basis of impairment of this delicate equilibrium between prooxidants and endogenous antioxidants [20–22].

Several mechanisms (Figure 5.3) have been proposed as the functional basis of flavonoids. These compounds have been found to be capable of direct quenching of the free radicals or scavenging ROS and RNS by single electron transfer (SET) or hydrogen (H) atom transfer (HAT) resulting in the stabilization of free radicals. Additionally, flavonoids have demonstrated the metal-chelating effect and activation of natural antioxidant defense coupled with oxidase inhibition [23, 24]. The schematic depiction of these mechanisms has been made in Figure 5.4.

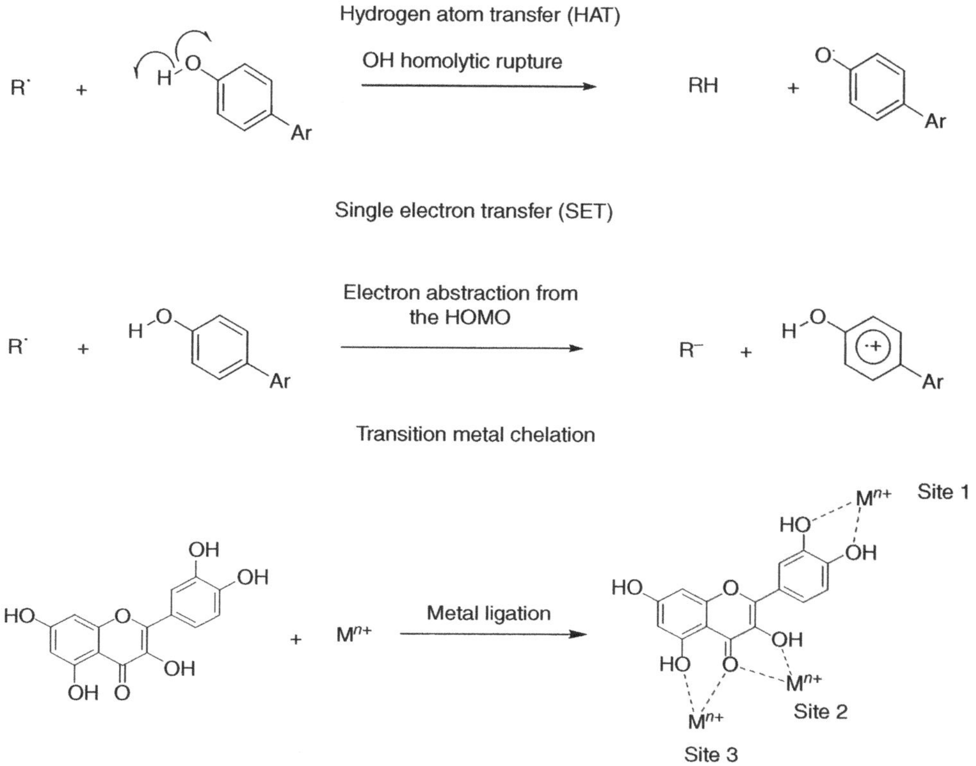

Figure 5.4 Mechanisms of antioxidant effect of flavonoids.

The various mechanisms by which flavonoids show their antioxidant effect are discussed in detail as follows:

Free radical scavenging action is carried out by these compounds by direct donation of a hydrogen atom and is proportional to the total number of hydroxyl groups and dependent on the functional groups present and the substitution pattern. When an electron is donated by the substituent to the phenolic (OH) oxygen, it causes a weakening of the O—H bond, making it easier to donate H to a radical [25]. Compounds with a B-ring catechol group have been found to be more potent free radical scavengers due to their readily available hydrogen, thereby forming a relatively stable aroxyl radical [26]. The presence of 2,3-double bond in conjugation with a 4-oxo functional group (in C-ring) along with 3-(and 5-)hydroxyl group has been found to confer optimal scavenging activity as they provide a stable radical by extensive delocalization of electrons over the A, B, C ring structure [27]. Highlights of this feature have been depicted in Figure 5.5.

The **metal ion chelation effect** is the next aspect of antioxidant effect of flavonoids in mitigating oxidative stress related to metal overload. Flavonoids are capable of chelating iron and copper ions along with other transition metals such as cobalt and manganese.

These metal ions are capable of catalyzing the production of oxidant species, which, in turn, may oxidize lipids, proteins, or even DNA. The most reactive, hydroxyl radical production occurs when hydrogen peroxide reacts with Fe(II) ions via Fenton reaction [28]. Whereas the formation of both hydroxyl (OH·) and superoxide ion (O_2·) occurs when the reaction is catalyzed by Cu(II) ions. This Fenton reaction has been found to contribute to oxidative damage of neuronal cells where

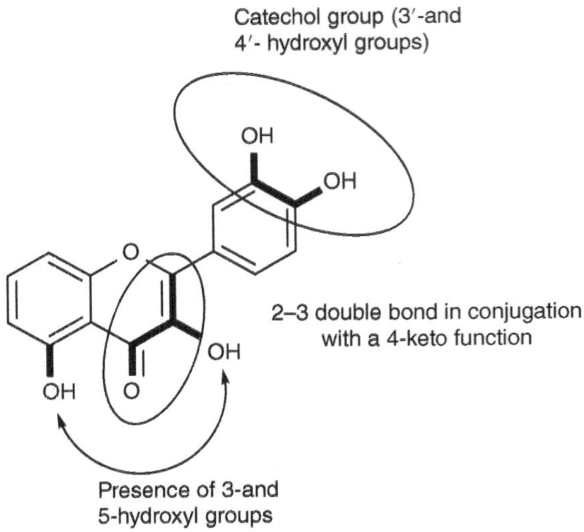

Figure 5.5 Structural requirements associated with antioxidant activities of bioflavonoids.

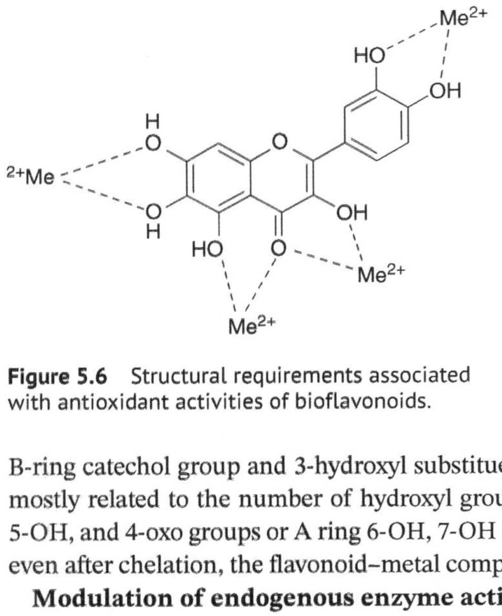

Figure 5.6 Structural requirements associated with antioxidant activities of bioflavonoids.

physiological dopamine catabolism forms some amounts of hydrogen peroxide. This mechanism has been found to be consistent with reports from several authors who have mentioned it to be an etiological factor for neurodegenerative disorders such as Alzheimer's disease (increased ferritin iron content in basal ganglia), Parkinson's disease (higher percentage of iron found in substantia nigra pars compacta) and Huntington's chorea (significantly increased iron concentrations in parts of brain) [29–31].

The reaction with iron ions has been found to be favorable in flavonoids having structural features, namely 2,3-double bond and those with both B-ring catechol group and 3-hydroxyl substituent. Whereas, reactivity with copper was found to be mostly related to the number of hydroxyl groups [28]. Metal complexation usually involves 3-OH, 5-OH, and 4-oxo groups or A ring 6-OH, 7-OH or B ring 3′-OH and 4′-OH (Figure 5.6). Additionally, even after chelation, the flavonoid–metal complex is capable of scavenging ROS [32–34].

Modulation of endogenous enzyme activities adds to the spectrum of flavonoid activity. As mentioned earlier, enzymes such as SOD, GPx, and catalase, as well as some antioxidant metabolites are responsible for preventing oxidative damage to cellular components under normal physiological conditions [35]. Flavonoids have been found to activate these enzymes and also to induce phase II detoxifying enzymes, namely NADPH-quinone reductase, UDP-glucuronosyl transferase, glutathione-*S*-transferase, etc. Such modulation has been related to alterations in gene expression of phase II genes comprising cis-acting regulatory elements known as antioxidant response elements (AREs) or electrophile-responsive elements (EpREs). An example of such a modulation is via regulation of nuclear factor erythroid 2-related factor 2 (Nrf2) which acts as a transcription factor for ARE-regulated genes [36]. This action has been demonstrated by quercetin and myricetin, which contains a free hydroxyl group at 3 position of C-ring [23]. Inhibition of prooxidant enzymes also adds to the antioxidant efficacy of flavonoids. Xanthine dehydrogenase, which is converted to xanthine oxidase under ischemic conditions, is a source of free radicals. Flavonoids such as quercetin, silibin, and luteolin were reported to inhibit xanthine oxidase activity, thereby protecting against oxidative injury [37].

Flavonoids have also been reported to interfere with inducible nitric oxide synthase activity [38]. Though initial nitric oxide release is very important in maintaining dilatation of blood vessels, its excess due to prolonged induction of nitric oxide synthase enzyme in macrophages may result in superoxide radical generation along with higher amounts of nitric oxide which then reacts with free radicals producing highly reactive peroxynitrite further precipitating oxidative tissue injury. Peroxynitrite is capable of oxidizing LDLs causing irreversible damage to cellular membranes. Besides modulating nitric oxide synthase, flavonoids that directly scavenge free radicals make them less available to react with the excess nitric oxide and additionally may scavenge nitric oxide directly [39–41]. This activity is again more prominent in flavonoids with 3′,4′-catechol arrangement along with an unsubstituted 3-OH group [42].

Lipophilic antioxidant protection by flavonoids deals with their effect on lipid peroxidation. In vivo preclinical studies have demonstrated that consuming tea stimulates hepatic UDP-glucuronyl transferase, which, in turn, protects against kidney lipid peroxidation [43]. These observations may be attributed to theaflavins formed during tea processing due to enzymatic oxidation of flavonols, catechin, and gallocatechin via polyphenol oxidase. In vitro evidence of theaflavins possessing antilipid peroxidation activity in erythrocyte membranes and microsomes is also reported [44, 45].

Flavonoids possess α-tocopherol-regenerating properties. α-tocopherol is an important lipid-soluble antioxidant present in the blood, cell membrane, and low-density lipoprotein and functions in the prevention of a number of degenerative disorders. These compounds have the ability to interact with polar head groups of cellular membranes, resulting in their increased concentration at the water–lipid interface of membranes [46]. The α-tocopheroxyl radical formed after reaction and quenching of free radicals has been found to be regenerated at these water–lipid interfaces by dietary flavonoids as well as catechins [47]. In summary, the LDL protective effect may be carried out by direct reduction of free radicals, metal ion chelation, or sparing effect and regeneration of α-tocopherol where promising results have been obtained from rutin, quercetin, and myricetin [48, 49].

5.4 Bioavailability and Metabolism

Absorption determines the bioavailability of flavonoids, which generally reach the human system via the oral route. It depends on a number of physicochemical properties, namely molecular size, configuration, solubility, lipophilicity, and pK_a. Absorption of flavonoid glycosides (flavonoids except for catechins generally occur in their β-glycosidic form in food) has been found to take place in the small intestine along with aglycones [50]. Here, an important aspect to be discussed is the effect of cooking on the flavonoid content. Some authors have reported the detrimental effect of frying on bioactive and antioxidant activity in leafy vegetables where steaming or boiling showed minimum such effects. Additionally, cooking in the microwave has been shown to be related to better retention of antioxidant effects [51–53]. Though hydrophilic in nature, the absorption of glycosides from the small intestine is hypothesized to be due to the action of the enzyme lactase-phlorizin hydrolase (LPH), which happens to be a β-galactosidase enzyme located on the outside of the brush border membrane of the small intestine. LPH releases the aglycone, which may then be absorbed from the small intestine [54]. However, cyanidin-3-glucoside (anthocyanin) was found to be an exception to this hypothesis. The glycosides may also be directly transported into the epithelium via sodium-dependent glucose cotransporter (SGLT1) to be subsequently hydrolyzed by intracellular glucosidases such as cytosolic β-glucosidases [54, 55]. The data obtained from the aforementioned studies strongly suggest that absorption of dietary flavonoids is determined by the sugar moieties attached to them.

Catechins are present in aglycone and galloylated forms in food both of which have been found to be absorbed from the small intestine. Additionally, catechins also occur as oligomers (up to 17 catechin units) in certain foods, which may be hydrolyzed to monomers and dimmers in the acidic environment of the stomach.

Glycosides that are unabsorbed in the small intestine move to the colon where they are hydrolyzed by gut bacteria but this will also simultaneously degrade the liberated aglycone [56].

Flavonoids undergo extensive hepatic metabolism as well as in the small intestine and colon and hence encounter biotransformation reactions, namely methylation, glucuronidation, and sulfation along with degradation into phenolic or other compounds. Biotransformation is an important cause of alteration in biological activity. Flavonoid metabolism has been grouped under three heads, namely phase I biotransformation (hydroxylation and demethylation), conjugate formation, and metabolism in the colon.

5.4.1 Phase I Flavonoidal Biotransformation

Flavonoid metabolism involves the action of liver microsomal enzymes mediated by cytochrome P450. Hydroxylation in flavones (glycosides and aglycones alike) has been found to favor at 4, 3, 6, and 8 positions when the B ring contains single or no hydroxyl groups, whereas demethylation occurs at methoxy groups located at 6, 7, or 4 positions. Flavonols generally show hydroxylation at 3 and 4 positions and those having a 4 methoxy group encounter O-demethylation to form corresponding hydroxylated analogs. Reactivity toward hydroxylation for isoflavones is targeted at 3, 6, and 8 positions when the B ring has a single or no hydroxyl group and 4 and 6 positions are often demethylated when these positions are carrying a hydroxyl moiety. Chalcones get hydroxylated at 3′ and 8′ positions where the latter has been found to be more favored [57].

Some examples include the conversion of daidzein (obtained from soy) to monohydroxylated and dihydroxylated metabolites, where the aromatic monohydroxylated products, 5,6,7,4-tetrahydroxyisoflavone, 5,7,8,4-tetrahydroxyisoflavone, and 5,7,3,4-tetrahydroxyisoflavone were found to be the main metabolites [58]. Similarly, microsomal enzymes have been found responsible for hydroxylation and demethylation, respectively, of galangin, structurally, 3,5,7-trihydroxyflavone and kaempferide, that is, 3,5,7-trihydroxy-4-methoxyflavone, to produce kaempferol (3,5,7,4-tetrahydroxyflavone) [59]. Cytochrome P450-mediated demethylation has been shown to be related mostly to polymethoxyflavones (flavones bearing four or more methoxy groups) where the B ring 3′ and 4′ positions are the primary site for biotransformation. However, polymethoxyflavone metabolism is greatly influenced by the number and position of B ring methoxy and hydroxyl functional groups [60].

5.4.2 Phase II Conjugation

Flavonoids are efficiently conjugated in humans and this may be proven by the fact that flavonoids are found in conjugated form in plasma and urine samples and that aglycone forms are rarely detected in the plasma. Transformations include methylation and the major conjugates are formed by sulfate and glucuronide. Given that flavonoid structure resembles that of catechol and pyrogallol, the primary metabolic products are respectively of 3′-*O*-methyl and 4′-*O*-methyl derivatives. Glucuronidation has been found to generally occur at 6,7,3′,4′-hydroxyl group for flavones, 7,3,3′ and 4′-hydroxyl group for flavonols and 4 or 5-hydroxyl moiety for isoflavones [57]. Glucuronidation occurs more in flavonoids containing catechol or pyrogallol B ring compared to those having monohydroxylated B ring [61]. Interactions may occur with glutathione and *N*-acetylcysteine as well, for example, glutathione–quercetin conjugates have been demonstrated in vivo. This is particularly important as bile enzymes (or intestinal bacteria) have been found to be capable of sequentially removing glutamic acid and glycine from glutathione–quercetin conjugate forming cysteine quercetin, which can then be reabsorbed in the gastrointestinal tract and eventually excreted via urine after addition of an acetyl group in the kidneys. The transient and unstable nature of these adducts rules out any possible toxicity [62]. The existence of flavonoid–thiol conjugate pathway has also been documented where quinine forms of quercetin may be generated followed by the acquiring of a sulfhydryl or glutathione adduct. The event of being oxidized to quinine form during metabolism has been reported particularly in flavonoids having a di-hydroxylated or catechol B-ring [62, 63].

5.4.3 Metabolism in the Colon

Early experimentation to reveal flavonoid metabolism hinted at the existence of enterohepatic cycling of these compounds [64]. Colonic metabolism involves the effect of gut microbiota on flavonoids, their glycosides as well as some of their metabolites to yield degradation products and presents as a rate-limiting step in the elimination of these compounds. Flavonoid glycosides together with flavonoid glucuronides and sulfates have been shown to undergo hydrolysis in presence of colonic microflora. Additionally, the microbiota is capable of causing the heterocyclic oxygen ring to split forming phenolic and carboxylic acid derivatives. Irrespective of the various classes of flavonoids, phenyl acetic acid along with its hydroxylated forms has been demonstrated as the main fermentation product [65]. These phenylcarboxylic acids undergo subsequent bacterial degradation and enzymatic biotransformations in various body tissues; for example, oxidation of phenylpropionic acids to benzoic acid. Further, the presence of these products in urine proves their systemic absorption [66].

Flavonoids show low urinary excretion, which proves that maximum of the intact flavonoids gets metabolized in the human body. Quercetin glycosides are about 50% absorbed but urinary excretion is extremely low. Available data suggests isoflavones the most excreted in urine among flavonoid classes, whereas flavonol glucosides demonstrate higher bioavailability coupled to lower urinary excretion [67]. The beneficial activities obtained from these compounds are attributed to the metabolites as well as in the case of anthocyanins where protocatechuic acid metabolite of anthocyanin accounts for their activity [68]. Some metabolic pathways even yield more active products compared to the parent compounds [60]. A schematic metabolic pathway for bioactive flavonoids has been shown through Figures 5.7 and 5.8.

5.4.4 Attempts to Improve Bioavailability of Flavonoids

Improvement in intestinal absorption of flavonoids may be achieved by the usage of enhancers. Substances such as fatty acids, bile salts, surfactants, cyclodextrin, chitosan, and some chelating agents have successfully improved the absorption of biomolecules like peptides, proteins, etc. [69].

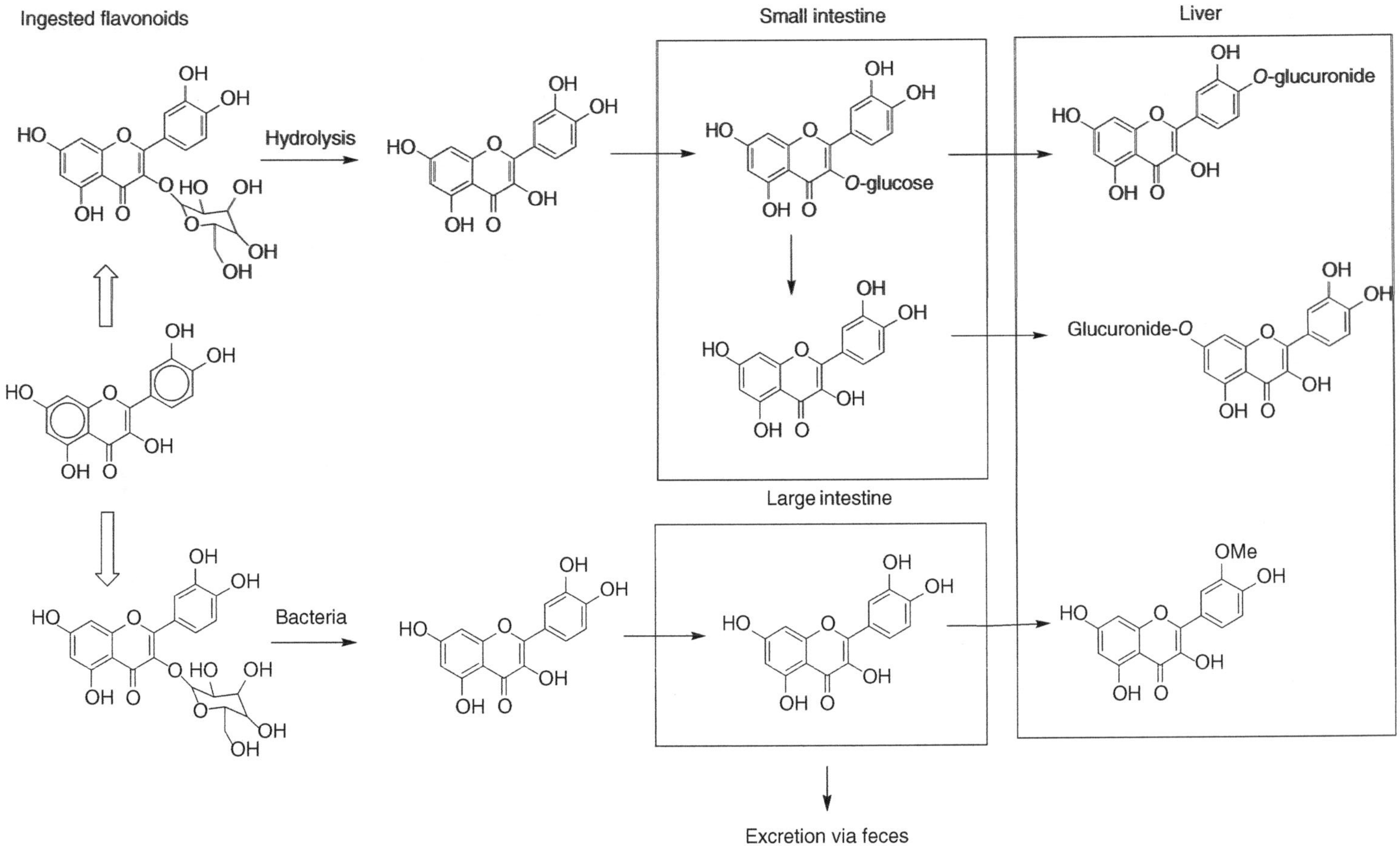

Figure 5.7 Metabolism of bioactive flavonoids.

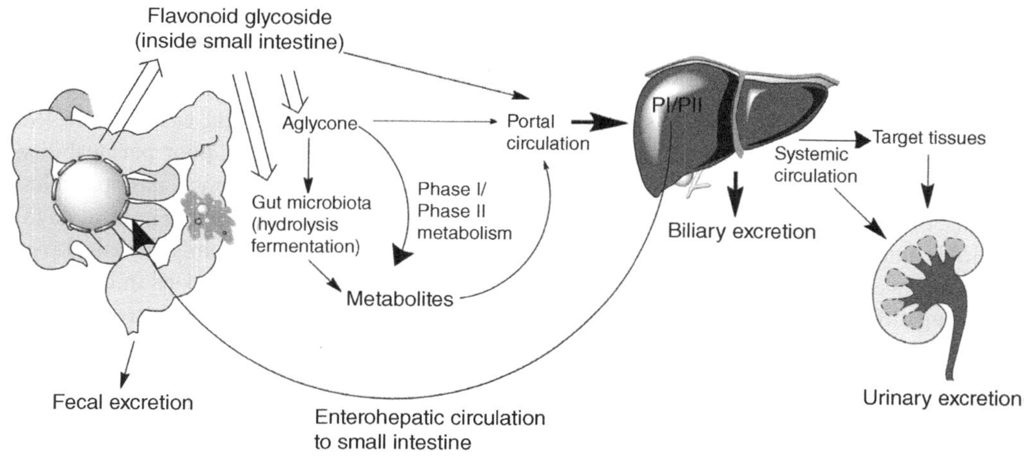

Figure 5.8 Enterohepatic circulation in the metabolism of bioactive flavonoids.

Quercetin bioavailability was found to be improved in rats on being coadministered with apple pectin in which the effect of this natural substance was attributed to modulation of absorption capacity in the small intestine by improving some of its morphological and physiological properties [70]. Similarly, coadministration with phytic acid showed improved bioavailability of isorhamnetin, quercetin, and kaempferol by improving water solubility and permeability under preclinical settings [71].

Structural alterations indicate another aspect of improving the bioavailability of flavonoids. Prodrug strategy has been proven useful for natural compounds like triterpenoids where ester, phosphate, sulfonate, amide, and glycoside prodrugs were tested for enhanced absorption and, hence, bioavailability. The same may be adopted in the case of flavonoids [72].

Nanotechnology is an emerging method that involves the control of the size and shape of material on a nanometer scale and which is increasingly becoming popular in the design of delivery systems [73]. Evaluation of solid lipid nanoparticles in oral delivery of poorly water-soluble flavonoid, quercetin, has revealed enhanced gastrointestinal absorption of the compound from the formulation [74]. Similar results have been obtained from quercetin nanocrystals produced by media milling technology [75], nano-suspension of naringin [76] as well as baicalin nanoemulsion [77]. Though most of the experiments have been performed in preclinical settings, the results hold the potential to be extrapolated to clinical studies.

Some other systems have also produced promising outcomes in enhancing the bioavailability of flavonoids. Cyclodextrins are degradation products of starch, which serve as encapsulation media for a number of natural compounds as they are capable of providing environmental protection to polyphenols as well as improving solubility and bioavailability [78]. Naringin solid dispersion has been produced by solvent evaporation and kneading methods and in vivo studies performed, which demonstrated higher solubility and dissolution compared to pure naringin [79]. Phytophospholipid complexation as a method to improve availability as well as efficacy has been experimented on using luteolin and has yielded positive results in vivo [80]. Evaluation of luteolin-isoniazid cocrystals and luteolin–caffeine cocrystals have demonstrated increased solubility to about three times compared to luteolin monohydrate in the rat model [81].

5.5 Semisynthetic Alterations in Bioactive Flavonoids

Flavonoids exhibit anti-inflammatory, antiallergic, antiviral, anticancer, cardioprotective, and a variety of beneficial effects by virtue of their antioxidant effect by protecting against ROS [8]. Natural bioactive flavonoids often face the problem of poor water solubility, poor permeability, and, hence, poor bioavailability. Various types of formulations, namely liposomes, nanoparticles, cocrystals, or micelles have been designed to improve the bioavailability of isolated natural flavonoids [69]. Attempts have been made to alter certain natural flavonoids semisynthetically to achieve improved features. Mukherjee et al. have reported semisynthetic alterations in quercetin by installing different substitutions at C-3′ and C-5′, which resulted in improved in vitro cytotoxicity due to possible improved water solubility [82].

A test conducted on transesterified rutin derivatives by Mecenas et al. showed that monoacetate and diacetate derivatives of the flavonoid had improved antioxidant potential (2.33 ± 1.08 μM Trolox/g) when compared to rutin standard (0.53 ± 0.08 μM Trolox/g) as evaluated by the ORAC method. Additionally, these derivatives demonstrated reduced production of ROS in RAW macrophages [83]. Es-Safi et al. semisynthesized modified (+)-catechin derivatives and compared their antioxidant efficacies against quercetin, BHT, and (+)-catechin in which the C8-substituted (+)-catechin derivatives showed improved antiradical activity [84]. Li et al. reported the design and synthesis of 4-hydroxyl aurone derivatives primarily as a treatment against Alzheimer's disease. Few of these compounds displayed high antioxidant values as well.

In an approach to improve anti-inflammatory activity, Freitas et al. targeted improving antioxidant properties of natural flavonoids by chlorinating flavone derivatives, where 8-chloro-3′,4′,5,7-tetrahydroxyflavone was found to have maximum efficacy. These derivatives mimicked the neutrophil oxidative burst-modulating activity of chlorinated metabolites of natural compounds (flavones, luteolin, and quercetin) by induction of caspase-dependent neutrophil apoptosis. The preclinical investigation was conducted by Mrvová et al. where a comparison of the protective role in neuroinflammation was studied between quercetin and its synthetically modified derivative, 3′-*O*-(3-chloropivaloyl) quercetin (CPQ). Here, CPQ was found to provide better suppression in phorbol myristate acetate (PMA)-induced superoxide generation compared to quercetin, proving its superior antioxidant effect [85].

5.6 Application in Therapeutics

The antioxidant property of flavonoids has by far emerged as a major contributor in the beneficial health effects produced by this class of compounds. These compounds have demonstrated a wide array of activities including anticancer, cardioprotective, anti-inflammatory, neuroprotective, and antiaging as a whole but here the therapeutic applications most closely related to the antioxidant action of flavonoids shall be discussed [86]. The antioxidant potential of flavonoids has been found to be comparable to or even more than well-known antioxidants, namely vitamin C and vitamin E [23]. Improvement in bone health has been found to be related to antioxidant and anti-inflammatory actions of a diet rich in flavonoids and this fact may be utilized in fields of bone tissue engineering by incorporating these compounds in bone biomaterials [87]. A lot of scientific evidence is currently resent to relate endothelial inflammation as well as oxidative stress to atherogenesis, which has been considered a major causative factor in almost all cardiogenic issues [88]. Research spread across many decades has demonstrated the relation between cardioprotective activity of flavonoids such as quercetin, rutin, kaempferol, luteolin, and isorhamnetin, and their antioxidant-anti-inflammatory potential. There has been evidence of protection being extended against drug-induced cardiotoxicity (e.g. chemotherapeutic agent, doxorubicin) as well [89–91]. Beneficial effects on oxidative skin damage due to UV irradiation have been achieved with polyphenols, especially flavonoids. The photoprotective effect has been demonstrated by apigenin, which is found in medicinal plants and beverages such as red wine and chamomile tea. Similarly, topically applied quercetin has shown promising UV-damage protection in preclinical studies. Other examples include silymarin, genistein, isoflavonoid metabolite, equol, and soy milk isoflavones, which have been found to repair skin damage due to UV-irradiation [91]. Direct antidiabetic effects have been achieved by modulation of various pathways affecting glucose homeostasis and/or insulin resistance. Oxidative stress or generation of ROS is a major outcome of chronic type II diabetes, which has been held responsible for a number of secondary complications such as neuropathy and cardiac dysfunctions, seen as long-term effects of this progressive metabolic disorder. Flavonoids, namely genistein, luteolin, quercetin, morin, diosmin, rutin, hesperidin, naringin, catechin, apigenin, and daidzein have the ability to effectively quench ROS or stimulate endogenous antioxidant enzymes, have been found effective in prolonging the onset of such complications in chronic diabetes [92–95].

5.7 Conclusion

At a time when neutraceuticals are becoming extremely popular, the use of bioactive flavonoids for their antioxidant and many other beneficial activities is on the rise. These agents are marketed in different combinations and different formulations such as capsules, tablets, creams, lotions, etc. Their benefits are undoubtedly immense but over-the-counter availability or lack of regulatory limits may lead to long-term chronic usage at high doses, which may not be favorable as the supplements provide much higher amounts than what is obtained from a regular balanced diet. There might be instances of unusual iron chelation resulting in a deficiency in people already having marginal iron contents. Additionally, polyphenols are known inhibitors of CYP3A4 and might be responsible for interfering with the metabolism of other drugs. Hence, to reap the maximal benefits of these phytoantioxidants, their judicious use is highly recommended.

References

1 Brewer, M.S. (2011). Natural antioxidants: sources, compounds, mechanisms of action, and potential applications. *Comprehensive Reviews in Food Science and Food Safety* 10: 221–247. https://doi.org/10.1111/j.1541-4337.2011.00156.x.

2 Anwar, H., Hussain, G., and Mustafa, I. (2018). Antioxidants from natural sources. In: *Antioxidants in Foods and Its Applications* (ed. E. Shalaby and G.M. Azzam). InTech https://doi.org/10.5772/intechopen.75961.

3 Gupta, C. and Prakash, D. (2014). Phytonutrients as therapeutic agents. *Journal of Complementary and Integrative Medicine* 11. https://doi.org/10.1515/jcim-2013-0021.

4 Agati, G., Azzarello, E., Pollastri, S., and Tattini, M. (2012). Flavonoids as antioxidants in plants: location and functional significance. *Plant Science* 196: 67–76. https://doi.org/10.1016/j.plantsci.2012.07.014.

5 Lillo, C., Lea, U.S., and Ruoff, P. (2008). Nutrient depletion as a key factor for manipulating gene expression and product formation in different branches of the flavonoid pathway. *Plant, Cell & Environment* 31: 587–601. https://doi.org/10.1111/j.1365-3040.2007.01748.x.

6 Pourcel, L., Routaboul, J., Cheynier, V. et al. (2007). Flavonoid oxidation in plants: from biochemical properties to physiological functions. *Trends in Plant Science* 12: 29–36. https://doi.org/10.1016/j.tplants.2006.11.006.

7 Tattini, M., Galardi, C., Pinelli, P. et al. (2004). Differential accumulation of flavonoids and hydroxycinnamates in leaves of *Ligustrum vulgare* under excess light and drought stress. *The New Phytologist* 163: 547–561. https://doi.org/10.1111/j.1469-8137.2004.01126.x.

8 Liguori, I., Russo, G., Curcio, F. et al. (2018). Oxidative stress, aging, and diseases. *Clinical Interventions in Aging* 13: 757–772. https://doi.org/10.2147/CIA.S158513.

9 Fraga, C.G. (2007). Plant polyphenols: how to translate their in vitro antioxidant actions to in vivo conditions. *IUBMB Life* 59: 308–315. https://doi.org/10.1080/15216540701230529.

10 Abbas, M., Saeed, F., Anjum, F.M. et al. (2017). Natural polyphenols: an overview. *International Journal of Food Properties* 20: 1689–1699. https://doi.org/10.1080/10942912.2016.1220393.

11 Kozłowska, A. and Szostak-Węgierek, D. (2018). Flavonoids – food sources, health benefits, and mechanisms involved. In: *Sweeteners, Reference Series in Phytochemistry* (ed. J.-M. Merillon and K.G. Ramawat), 1–27. Cham: Springer International Publishing https://doi.org/10.1007/978-3-319-54528-8_54-1.

12 Pal, D. and Nayak, A.K. (2021). *Bioactive Natural Products for Pharmaceutical Applications*. Springer.

13 Bhagwat, S., Haytowitz, D.B., Holden, J.M. (2013). USDA database for the flavonoid content of selected foods.

14 Bhagwat, S. and Haytowitz, D.B. (2015). USDA database for the flavonoid content of selected foods.

15 Kamiloglu, S., Tomas, M., and Capanoglu, E. (2020). Dietary flavonols and O-glycosides. In: *Handbook of Dietary Phytochemicals* (ed. J. Xiao, S.D. Sarker and Y. Asakawa), 1–40. Singapore: Springer Singapore https://doi.org/10.1007/978-981-13-1745-3_4-1.

16 Thilakarathna, S.H. and Rupasinghe, H.P.V. (2013). Flavonoid bioavailability and attempts for bioavailability enhancement. *Nutrients* 5: 3367–3387. https://doi.org/10.3390/nu5093367.

17 Bentsáth, A., Rusznyák, S., and Szent-Györgyi, A. (1936). Vitamin nature of flavones. *Nature* 138: 798–798. https://doi.org/10.1038/138798a0.

18 Hoensch, H.P. and Oertel, R. (2015). The value of flavonoids for the human nutrition: short review and perspectives. *Clinical Nutrition Experimental* 3: 8–14. https://doi.org/10.1016/j.yclnex.2015.09.001.

19 Lin, J.-K. and Weng, M.-S. (2006). Flavonoids as nutraceuticals. In: *The Science of Flavonoids* (ed. E. Grotewold), 213–238. New York, NY: Springer https://doi.org/10.1007/978-0-387-28822-2_8.

20 González-Paramás, A.M., Ayuda-Durán, B., Martínez, S. et al. (2019). The mechanisms behind the biological activity of flavonoids. *Current Medicinal Chemistry* 26: 6976–6990. https://doi.org/10.2174/0929867325666180706104829.

21 Sies, H. (2015). Oxidative stress: a concept in redox biology and medicine. *Redox Biology* 4: 180–183. https://doi.org/10.1016/j.redox.2015.01.002.

22 Sies, H. (ed.) (1985). *Oxidative Stress*. London, Orlando: Academic Press.

23 Procházková, D., Boušová, I., and Wilhelmová, N. (2011). Antioxidant and prooxidant properties of flavonoids. *Fitoterapia* 82: 513–523. https://doi.org/10.1016/j.fitote.2011.01.018.

24 Treml, J. and Šmejkal, K. (2016). Flavonoids as potent scavengers of hydroxyl radicals: flavonoids versus hydroxyl radical. *Comprehensive Reviews in Food Science and Food Safety* 15: 720–738. https://doi.org/10.1111/1541-4337.12204.

25 Haenen, G.R.M.M., Arts, M.J.T.J., Bast, A., and Coleman, M.D. (2006). Structure and activity in assessing antioxidant activity in vitro and in vivo. *Environmental Toxicology and Pharmacology* 21: 191–198. https://doi.org/10.1016/j.etap.2005.07.010.

26 Leopoldini, M., Russo, N., and Toscano, M. (2011). The molecular basis of working mechanism of natural polyphenolic antioxidants. *Food Chemistry* 125: 288–306. https://doi.org/10.1016/j.foodchem.2010.08.012.

27 Bors, W., Heller, W., Michel, C., and Saran, M. (1990). Flavonoids as antioxidants: determination of radical-scavenging efficiencies. In: *Methods in Enzymology* (ed. L. Packer and A.N. Glazer), 343–355. Elsevier https://doi.org/10.1016/0076-6879(90)86128-I.

28 Mira, L., Tereza Fernandez, M., Santos, M. et al. (2002). Interactions of flavonoids with iron and copper ions: a mechanism for their antioxidant activity. *Free Radical Research* 36: 1199–1208. https://doi.org/10.1080/1071576021000016463.

29 Gerlach, M., Ben-Shachar, D., Riederer, P., and Youdim, M.B.H. (2002). Altered brain metabolism of iron as a cause of neurodegenerative diseases? *Journal of Neurochemistry* 63: 793–807. https://doi.org/10.1046/j.1471-4159.1994.63030793.x.

30 Hirsch, E.C. and Faucheux, B.A. (1998). Iron metabolism and Parkinson's disease. *Movement Disorders* 13 (Suppl 1): 39–45.

31 Youdim, M.B.H., Ben-Shachar, D., and Riederer, P. (1993). The possible role of iron in the etiopathology of Parkinson's disease. *Movement Disorders* 8: 1–12. https://doi.org/10.1002/mds.870080102.

32 Cheng, I.F. and Breen, K. (2000). On the ability of four flavonoids, baicilein, luteolin, naringenin, and quercetin, to suppress the fenton reaction of the iron–ATP complex. *BioMetals* 13: 77–83. https://doi.org/10.1023/A:1009229429250.

33 Pietta, P.-G. (2000). Flavonoids as antioxidants. *Journal of Natural Products* 63: 1035–1042. https://doi.org/10.1021/np9904509.

34 Deng, W., Fang, X., and Jilan, W. (1997). Flavonoids function as antioxidants: by scavenging reactive oxygen species or by chelating iron? *Radiation Physics and Chemistry* 50: 271–276. https://doi.org/10.1016/S0969-806X(97)00037-6.

35 Davies, K.J.A. (1995). Oxidative stress: the paradox of aerobic life. *Biochemical Society Symposia* 61: 1–31. https://doi.org/10.1042/bss0610001.

36 Mastaloudis, A. and Wood, S.M. (2012). Age-related changes in cellular protection, purification, and inflammation-related gene expression: role of dietary phytonutrients: phytonutrients and age-related gene expression. *Annals of the New York Academy of Sciences* 1259: 112–120. https://doi.org/10.1111/j.1749-6632.2012.06610.x.

37 Nijveldt, R.J., van Nood, E., van Hoorn, D.E. et al. (2001). Flavonoids: a review of probable mechanisms of action and potential applications. *The American Journal of Clinical Nutrition* 74: 418–425. https://doi.org/10.1093/ajcn/74.4.418.

38 Shoskes, D.A. (1998). Effect of bioflavonoids quercetin and curcumin on ischemic renal injury: a new class of renoprotective agents. *Transplantation* 66: 147–152. https://doi.org/10.1097/00007890-199807270-00001.

39 Huk, I., Brovkovych, V., Nanobash Vili, J. et al. (1998). Bioflavonoid quercetin scavenges superoxide and increases nitric oxide concentration in ischaemia–reperfusion injury: an experimental study. *The British Journal of Surgery* 85: 1080–1085. https://doi.org/10.1046/j.1365-2168.1998.00787.x.

40 Shutenko, Z., Henry, Y., Pinard, E. et al. (1999). Influence of the antioxidant quercetin in vivo on the level of nitric oxide determined by electron paramagnetic resonance in rat brain during global ischemia and reperfusion. *Biochemical Pharmacology* 57: 199–208. https://doi.org/10.1016/S0006-2952(98)00296-2.

41 Vanacker, S.A.B.E., Tromp, M.N.J.L., Haenen, G.R.M.M. et al. (1995). Flavonoids as scavengers of nitric oxide radical. *Biochemical and Biophysical Research Communications* 214: 755–759. https://doi.org/10.1006/bbrc.1995.2350.

42 Choi, J.S., Chung, H.Y., Kang, S.S. et al. (2002). The structure–activity relationship of flavonoids as scavengers of peroxynitrite. *Phytotherapy Research* 16: 232–235. https://doi.org/10.1002/ptr.828.

43 Bu-Abbas, A., Clifford, M.N., Ioannides, C., and Walker, R. (1995). Stimulation of rat hepatic UDP-glucuronosyl transferase activity following treatment with green tea. *Food and Chemical Toxicology* 33: 27–30. https://doi.org/10.1016/0278-6915(95)80244-4.

44 Rice-Evans, C., Miller, N., and Paganga, G. (1997). Antioxidant properties of phenolic compounds. *Trends in Plant Science* 2: 152–159. https://doi.org/10.1016/S1360-1385(97)01018-2.

45 Yang, C.S. and Wang, Z.-Y. (1993). Tea and cancer. *JNCI Journal of the National Cancer Institute* 85: 1038–1049. https://doi.org/10.1093/jnci/85.13.1038.

46 Erlejman, A.G., Verstraeten, S.V., Fraga, C.G., and Oteiza, P.I. (2004). The interaction of flavonoids with membranes: potential determinant of flavonoid antioxidant effects. *Free Radical Research* 38: 1311–1320. https://doi.org/10.1080/10715760400016105.

47 Chan, A.C., Tran, K., Raynor, T. et al. (1991). Regeneration of vitamin E in human platelets. *The Journal of Biological Chemistry* 266: 17290–17295.

48 Korkina, L.G. and Afanas'Ev, I.B. (1996). Antioxidant and chelating properties of flavonoids. *Advances in Pharmacology* 38, Elsevier: 151–163. https://doi.org/10.1016/S1054-3589(08)60983-7.

49 Zhu, Q.Y., Huang, Y., and Chen, Z.Y. (2000). Interaction between flavonoids and α-tocopherol in human low density lipoprotein. *The Journal of Nutritional Biochemistry* 11 (1): 14–21. https://doi.org/10.1016/S0955-2863(99)00065-0.

50 Wang, Y. and Ho, C.-T. (2009). Metabolism of flavonoids. In: *Forum of Nutrition* (ed. T. Yoshikawa), 64–74. Basel: KARGER https://doi.org/10.1159/000212739.

51 Gunathilake, K.D.P.P., Ranaweera, K.K.D.S., and Rupasinghe, H.P.V. (2018). Effect of different cooking methods on polyphenols, carotenoids and antioxidant activities of selected edible leaves. *Antioxidants (Basel)* 7. https://doi.org/10.3390/antiox7090117.

52 Ioku, K., Aoyama, Y., Tokuno, A. et al. (2001). Various cooking methods and the flavonoid content in onion. *Journal of Nutritional Science and Vitaminology* 47: 78–83. https://doi.org/10.3177/jnsv.47.78.

53 Wu, X., Zhao, Y., Haytowitz, D.B. et al. (2019). Effects of domestic cooking on flavonoids in broccoli and calculation of retention factors. *Heliyon* 5: e01310. https://doi.org/10.1016/j.heliyon.2019.e01310.

54 Day, A.J., Cañada, F.J., Díaz, J.C. et al. (2000). Dietary flavonoid and isoflavone glycosides are hydrolysed by the lactase site of lactase phlorizin hydrolase. *FEBS Letters* 468: 166–170. https://doi.org/10.1016/S0014-5793(00)01211-4.

55 Hollman, P.C.H., Bijsman, M.N.C.P., van Gameren, Y. et al. (1999). The sugar moiety is a major determinant of the absorption of dietary flavonoid glycosides in man. *Free Radical Research* 31: 569–573. https://doi.org/10.1080/10715769900301141.

56 Cassidy, A. and Minihane, A.-M. (2017). The role of metabolism (and the microbiome) in defining the clinical efficacy of dietary flavonoids. *The American Journal of Clinical Nutrition* 105: 10–22. https://doi.org/10.3945/ajcn.116.136051.

57 Xiao, J. and Hogger, P. (2013). Metabolism of dietary flavonoids in liver microsomes. *Current Drug Metabolism* 14: 381–391. https://doi.org/10.2174/1389200211314040003.

58 Kulling, S.E., Honig, D.M., and Metzler, M. (2001). Oxidative metabolism of the soy isoflavones daidzein and genistein in humans in vitro and in vivo. *Journal of Agricultural and Food Chemistry* 49: 3024–3033. https://doi.org/10.1021/jf0012695.

59 Otake, Y. and Walle, T. (2002). Oxidation of the flavonoids galangin and kaempferide by human liver microsomes and CYP1A1, CYP1A2, and CYP2C9. *Drug Metabolism and Disposition* 30: 103–105. https://doi.org/10.1124/dmd.30.2.103.

60 Li, S., Pan, M.-H., Lo, C.-Y. et al. (2009b). Chemistry and health effects of polymethoxyflavones and hydroxylated polymethoxyflavones. *Journal of Functional Foods* 1: 2–12. https://doi.org/10.1016/j.jff.2008.09.003.

61 Prior, R.L., Wu, X., and Gu, L. (2006). Flavonoid metabolism and challenges to understanding mechanisms of health effects. *Journal of the Science of Food and Agriculture* 86: 2487–2491. https://doi.org/10.1002/jsfa.2611.

62 Awad, H.M., Boersma, M.G., Boeren, S. et al. (2003). Quenching of quercetin quinone/quinone methides by different thiolate scavengers: stability and reversibility of conjugate formation. *Chemical Research in Toxicology* 16: 822–831. https://doi.org/10.1021/tx020079g.

63 Lambert, J.D., Sang, S., and Yang, C.S. (2007). Biotransformation of green tea polyphenols and the biological activities of those metabolites. *Molecular Pharmaceutics* 4: 819–825. https://doi.org/10.1021/mp700075m.

64 Griffiths, L.A. (1982). Mammalian metabolism of flavonoids. In: *The Flavonoids* (ed. J.B. Harborne and T.J. Mabry), 681–718. Boston, MA: Springer US https://doi.org/10.1007/978-1-4899-2915-0_12.

65 Serra, A., Macià, A., Romero, M.-P. et al. (2012). Metabolic pathways of the colonic metabolism of flavonoids (flavonols, flavones and flavanones) and phenolic acids. *Food Chemistry* 130: 383–393. https://doi.org/10.1016/j.foodchem.2011.07.055.

66 Hollman, P. and Katan, M. (1997). Absorption, metabolism and health effects of dietary flavonoids in man. *Biomedicine & Pharmacotherapy* 51: 305–310. https://doi.org/10.1016/S0753-3322(97)88045-6.

67 Hollman, P.C.H. (2004). Absorption, bioavailability, and metabolism of flavonoids. *Pharmaceutical Biology* 42: 74–83. https://doi.org/10.3109/13880200490893492.

68 Galvano, F., Vitaglione, P., Volti, G.L. et al. (2008). Protocatechuic acid: the missing human cyanidins' metabolite. *Molecular Nutrition & Food Research* 52: 386–387. https://doi.org/10.1002/mnfr.200890011.

69 Zhao, J., Yang, J., and Xie, Y. (2019a). Improvement strategies for the oral bioavailability of poorly water-soluble flavonoids: an overview. *International Journal of Pharmaceutics* 570: 118642. https://doi.org/10.1016/j.ijpharm.2019.118642.

70 Nishijima, T., Iwai, K., Saito, Y. et al. (2009). Chronic ingestion of apple pectin can enhance the absorption of quercetin. *Journal of Agricultural and Food Chemistry* 57: 2583–2587. https://doi.org/10.1021/jf803547h.

71 Xie, Y., Luo, H., Duan, J. et al. (2014). Phytic acid enhances the oral absorption of isorhamnetin, quercetin, and kaempferol in total flavones of *Hippophae rhamnoides* L. *Fitoterapia* 93: 216–225. https://doi.org/10.1016/j.fitote.2014.01.013.

72 Zhou, M., Zhang, R.-H., Wang, M. et al. (2017). Prodrugs of triterpenoids and their derivatives. *European Journal of Medicinal Chemistry* 131: 222–236. https://doi.org/10.1016/j.ejmech.2017.03.005.

73 Kamal, M.A., Jabir, N.R., Tabrez, S. et al. (2012). Nanotechnology-based approaches in anticancer research. *International Journal of Nanomedicine* 4391. https://doi.org/10.2147/IJN.S33838.

74 Li, H., Zhao, X., Ma, Y. et al. (2009a). Enhancement of gastrointestinal absorption of quercetin by solid lipid nanoparticles. *Journal of Controlled Release* 133: 238–244. https://doi.org/10.1016/j.jconrel.2008.10.002.

75 Lu, M., Ho, C.-T., and Huang, Q. (2017). Improving quercetin dissolution and bioaccessibility with reduced crystallite sizes through media milling technique. *Journal of Functional Foods* 37: 138–146. https://doi.org/10.1016/j.jff.2017.07.047.

76 Singh, M.K., Pooja, D., Ravuri, H.G. et al. (2018). Fabrication of surfactant-stabilized nanosuspension of naringenin to surpass its poor physiochemical properties and low oral bioavailability. *Phytomedicine* 40: 48–54. https://doi.org/10.1016/j.phymed.2017.12.021.

77 Wu, L., Bi, Y., and Wu, H. (2018). Formulation optimization and the absorption mechanisms of nanoemulsion in improving baicalin oral exposure. *Drug Development and Industrial Pharmacy* 44: 266–275. https://doi.org/10.1080/03639045.2017.1391831.

78 Pinho, E., Grootveld, M., Soares, G., and Henriques, M. (2014). Cyclodextrins as encapsulation agents for plant bioactive compounds. *Carbohydrate Polymers* 101: 121–135. https://doi.org/10.1016/j.carbpol.2013.08.078.

79 Khan, A.W., Kotta, S., Ansari, S.H. et al. (2015a). Enhanced dissolution and bioavailability of grapefruit flavonoid naringenin by solid dispersion utilizing fourth generation carrier. *Drug Development and Industrial Pharmacy* 41: 772–779. https://doi.org/10.3109/03639045.2014.902466.

80 Khan, J., Saraf, S., and Saraf, S. (2015b). Preparation and evaluation of luteolin–phospholipid complex as an effective drug delivery tool against GalN/LPS induced liver damage. *Pharmaceutical Development and Technology* 1–12. https://doi.org/10.3109/10837450.2015.1022786.

81 Luo, Y., Chen, S., Zhou, J. et al. (2019). Luteolin cocrystals: characterization, evaluation of solubility, oral bioavailability and theoretical calculation. *Journal of Drug Delivery Science and Technology* 50: 248–254. https://doi.org/10.1016/j.jddst.2019.02.004.

82 Mukherjee, A., Mishra, S., Kumar, K.N., and Krishnendu, M. (2019). Semisynthetic quercetin derivatives with potent antitumor activity in colon carcinoma. *ACS Omega* https://doi.org/10.1021/acsomega.9b00143.

83 Mecenas, A.S., Adão Malafaia, C.R., Sangenito, L.S. et al. (2018). Rutin derivatives obtained by transesterification reactions catalyzed by Novozym 435: antioxidant properties and absence of toxicity in mammalian cells. *PLoS ONE* 13: e0203159. https://doi.org/10.1371/journal.pone.0203159.

84 Es-Safi, N.-E., Ghidouche, S., and Ducrot, P. (2007). Flavonoids: hemisynthesis, reactivity, characterization and free radical scavenging activity. *Molecules* 12: 2228–2258. https://doi.org/10.3390/12092228.

85 Mrvová, N., Kuniaková, M., Knezl, V. et al. (2014). Can semi-synthetic flavonoids return old microglia to their youthful state? *Free Radical Biology and Medicine* 75: S45. https://doi.org/10.1016/j.freeradbiomed.2014.10.803.

86 Ullah, A., Munir, S., Badshah, S.L. et al. (2020). Important flavonoids and their role as a therapeutic agent. *Molecules* 25. https://doi.org/10.3390/molecules25225243.

87 Preethi Soundarya, S., Sanjay, V., Haritha Menon, A. et al. (2018). Effects of flavonoids incorporated biological macromolecules based scaffolds in bone tissue engineering. *International Journal of Biological Macromolecules* 110: 74–87. https://doi.org/10.1016/j.ijbiomac.2017.09.014.

88 Victor, V., Rocha, M., Sola, E. et al. (2009). Oxidative stress, endothelial dysfunction and atherosclerosis. *Current Pharmaceutical Design* 15: 2988–3002. https://doi.org/10.2174/138161209789058093.

89 Grassi, D., Desideri, G., and Ferri, C. (2010). Flavonoids: antioxidants against atherosclerosis. *Nutrients* 2: 889–902. https://doi.org/10.3390/nu2080889.

90 Terao, J. (2017). Factors modulating bioavailability of quercetin-related flavonoids and the consequences of their vascular function. *Biochemical Pharmacology* 139: 15–23. https://doi.org/10.1016/j.bcp.2017.03.021.

91 Tungmunnithum, D., Thongboonyou, A., Pholboon, A., and Yangsabai, A. (2018). Flavonoids and other phenolic compounds from medicinal plants for pharmaceutical and medical aspects: an overview. *Medicine* 5: 93. https://doi.org/10.3390/medicines5030093.

92 Omolola, R., Nicole, L., and Oluwafemi, O. (2014). Oxidative stress and diabetic complications: the role of antioxidant vitamins and flavonoids. In: *Antioxidant-Antidiabetic Agents and Human Health* (ed. O. Oguntibeju). InTech https://doi.org/10.5772/57282.

93 Süntar, I. and Yakıncı, Ö.F. (2020). Potential risks of phytonutrients associated with high-dose or long-term use. In: *Phytonutrients in Food*, 137–155. Elsevier https://doi.org/10.1016/B978-0-12-815354-3.00010-1.

94 Temme, E. and Van Hoydonck, P. (2002). Tea consumption and iron status. *European Journal of Clinical Nutrition* 56: 379–386. https://doi.org/10.1038/sj.ejcn.1601309.

95 Lilja, J. (2000). Duration of effect of grapefruit juice on the pharmacokinetics of the CYP3A4 substrate simvastatin. *Clinical Pharmacology & Therapeutics* 68: 384–390. https://doi.org/10.1067/mcp.2000.110216.

6

Nanoparticulate Delivery Systems for Phytoconstituents

Khomendra Kumar Sarwa[1], Dipali Patel[2], Dhaneshwar Uraon[1], Manmohan Singh Jangde[3], Vijendra Kumar Suryawanshi[4], and Swarnlata Saraf[5]

[1] *Department of Pharmacy, Government Girls Polytechnic, Raipur, Chhattisgarh, India*
[2] *Department of Pharmaceutics, Shri Rawatpura Sarkar University, Raipur, Chhattisgarh, India*
[3] *Shri Balaji College of Pharmaceutical Sciences, Sakti, Chhattisgarh, India*
[4] *Department of Pharmaceutics, M J College, Bhilai, Chhattisgarh, India*
[5] *University Institute of Pharmacy, Pt. Ravishankar Shukla University, Raipur, Chhattisgarh, India*

6.1 Natural Product as the Biggest Source of the Drug

The natural product is the biggest considerable source of medicine and is listed as a drug in various pharmacopeia, still challenging the isolation and extraction about 50% of marketed drugs derived from a natural product [1]. In the segment of a plant's natural product, a new class is added as semisynthetic, semipurified phytoconstituents due to coverage of a variety of therapeutic indications and the development of large range of chemical structure possibility in small modification. The interesting thing for phytoconstituents is that a lot of research work is going on crude extract or fraction rather than a single isolated compound. This strategy is fulfilling many tasks like the multi-targeting effect. Sometimes this action has only been taken due to the lack of identification of the principal compounds [2]. The delivery of phytoconstituents is an emerging area due to the advancement of phytochemical and phytopharmaceutical science. Sometimes economic, political, social, and cultural factors may also be responsible to enhance the acceptance of the natural product as medicine. The delivery of plant constituents is challengeable and limited, especially an issue related to immunological response, uptake efficiency at target tissue high scaling cost and environmental factors still unsolved [3].

6.1.1 Challenge for Delivery of Phytoconstituents

Phytoconstituents are diverse in their chemical composition. Therefore, each phytoconstituent suffers from some specific as well as nonspecific issues related to its delivery [4]. Treatment therapy is classified into two categories, either first-line choice or adjuvant therapy. In the development of the first-line choice, the delivery of therapeutic molecules is always challengeable. Phytoconstituents are also facing such problems as poor aqueous solubility, poor permeability in the biological site, short circulating life in the biological fluid, and high first-pass metabolism by the liver [5]. Phytoconstituents having various problems need to be solved by classical as well as novel approaches presented in Figure 6.1.

Phytoantioxidants and Nanotherapeutics, First Edition. Edited by Mithun Rudrapal.

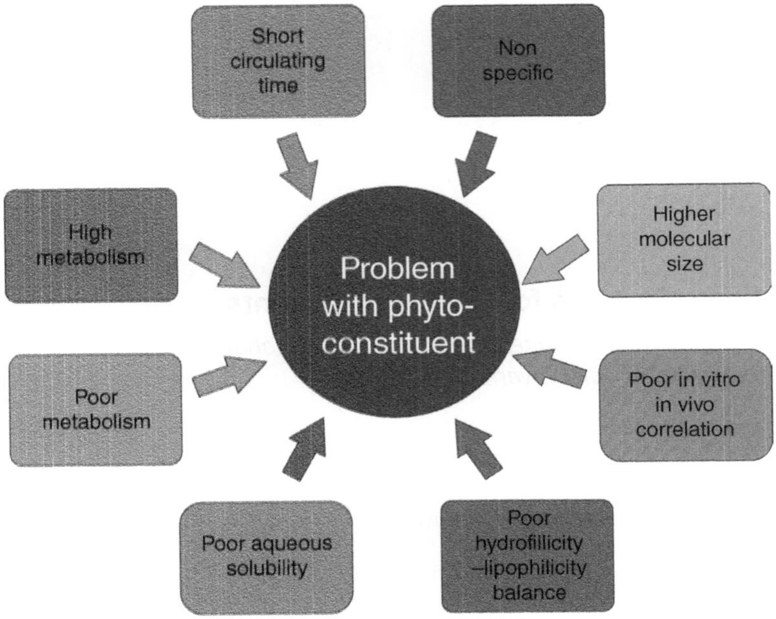

Figure 6.1 Issues with phytoconstituents need to be handled during delivery.

Most of the listed therapeutic plant constituents are plant secondary metabolic products, categorized as alkaloid, glycoside, flavonoids, terpenoids, tannin, saponin, steroids, phenolic compounds, etc.

They are high molecular weight compounds because of their inability to cross lipid membranes and their quantity in the circulation system is very limited. Therefore, metabolism needs to be prevented if delivered as a therapeutic compound. A lot of variation in in vitro–in vivo correlation data also potentiates the research for suitability studies of delivery carriers for phytoconstituents [6].

6.1.2 Difficulty in Isolation of Phytoconstituents

The modern concept of treatment is a sophisticated format of older naturally complex compound-based therapeutics where a single isolated pure compound should be preferred for formulations. This is a reason behind the requirement of long time to introduce a new therapeutic molecule into the market. Unsuccessfulness is mainly due to difficulty in the identification of lead compounds due to loss of activity, difficulty in purification, instability after isolation, etc. [7]. This difficulty may open a path to do work in a direction of considering a crude extract or semi-purified fraction as a chief source instead of a pure isolated compound. But still, the remaining work is not over to ensure quality, safety, and efficacy that could be tried to solve by using the nano-size-based delivery carrier. They act as protective vehicles to deliver in biological systems [8].

6.2 Nanotechnology as a Solution for Delivery of Phytoconstituents

A delivery of phytoconstituents via conventional delivery methodology suffers from a variety of problems like aqueous solubility, instability, and permeability issues. Plant constituent-associated problems can be easily solved by using a nanotechnological approach and resulting in safe and better efficacy and patient compliance. The active phytoconstituents are present in various parts of medicinal plants. Nanotechnology helps in the preparation of nanomedicines and nanocarriers using pure active phytoconstituents as well as semipurified extract.

The involvement of nanotechnology drastically improves the final results by reducing toxicity and adverse effects and producing more therapeutic or pharmacological effects by improving pharmacokinetic parameters. This achievement is by modifying the surface chemistry and property and tailoring as per need. An ability to target specificity, stability in biological media, better retention time, and less metabolism solve the problem related to bioavailability and therapeutic index [9]. Figure 6.2 represents the key benefit of nanotechnology in drug delivery.

The nanotechnological concept has been studied to solve issues related to activity, specificity, and bioavailability of phytoconstituents. The complexity of phytochemicals can be easily handled by nanoscale formulation by release modification, surface landing, improving solubility, decreasing loading quantity, minimizing degradation process, and enhancing absorption and bioavailability. Figure 6.3 represents the expected goal of nanodelivery carriers

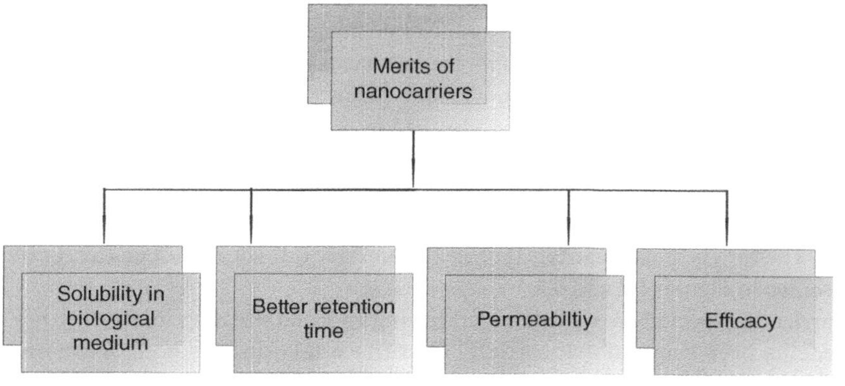

Figure 6.2 Key benefits of nanocarriers as drug-delivery vehicles.

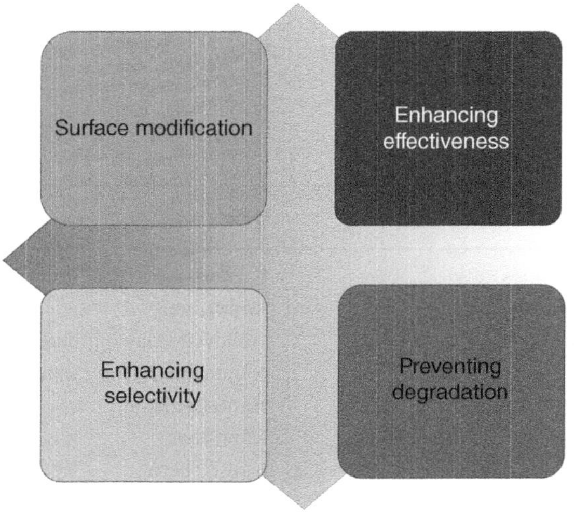

Figure 6.3 Various goals to be achieved by nanocarriers.

for phytoconstituents. The combination of nanotechnology with the delivery of phytoconstituents makes a revolution by enhancing effectiveness, selectivity, preventing degradation, and modification in the release kinetics. The possibility of altering surface property, shape, and size by selecting base vehicle from a variety of raw materials and by adopting sophisticated instruments makes very attractive carriers to deliver a large range of phytoconstituents with and without isolation [4].

A characteristic feature of the individual nanocarrier may help in the decision of the selection of carriers for particular phytoconstituents and their extract. Similarly, the properties of phytoconstituents and extracts should also be taken into consideration. Polarity and solubility of active phytoconstituents, volatility, and amount of organic solvents present in the extract are such important considerations for selection of carriers [10].

6.3 Nanocarriers for Phytoconstituents

A list of nanocarriers is too large with continuously adding new ones due to the innovations in the field of materials science. All these nanocarriers are classified into three classes, one is a nanoparticulate system, another one is a nanovesicular system, and the last one is another type of nanocarriers. In the first segment, nanoparticulate means carrier, which is nano-sized and spherical-shaped. In this carrier, phytoconstituents are entrapped and/or adsorbed into the manufacturing material. A nanovesicle means these carriers are nano-sized with a vesicular shape. In these vesicles, phytoconstituents are enclosed. Other types are nanocarriers, which have special structural components for carrying phytoconstituents [11]. In this chapter, detailed information about the nanoparticulate system is summarized. The variety of nanocarriers is presented in Table 6.1. The representative image of nanocarrier is presented in Figures 6.4 and 6.5.

Nanoparticulate carriers are specially designed carriers having a normal size range of 10–1000 nm. They have many benefits over conventional carriers like surface tailoring for both active and passive drug targeting, altering biodistribution of the therapeutic compound, and controlled release drug delivery. In a nanoparticulate system, drugs are capable of entrapped, encapsulated, bonded or attached to a matrix system open to the path for application in the delivery of a large range of phytoconstituents. The nanoparticulate carriers are broadly classified into four types based on applied core preparative material presented in Table 6.2.

Table 6.1 Various nanocarriers used for the delivery of phytoconstituents.

Nanoparticular carriers	Nanovesicular carriers	Others
Calcium carbonate nanoparticle	Cubasomes	Carbon nanotubes
Liquid crystals	Ethosomes	Dendrimer
Magnetic nanoparticle	Liposomes	Halloysite clay nanotube
Mesoporous silica nanoparticle	Micelles	Nanoemulsifying carrier
Nanocrystals	Nanocapsule	Nanoemulsion
Nanopowder	Niosomes	Nanofiber
Polymeric nanoparticles	Phytosomes	Nanosphere
Solid lipid nanoparticles	Polymersomes	Nanosuspension
	Transferosomes	

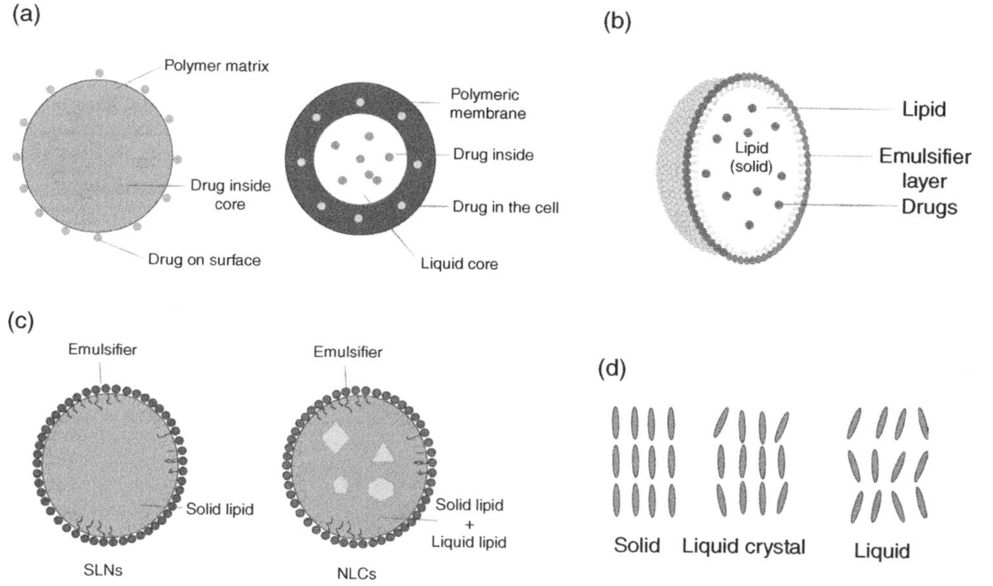

Figure 6.4 Nanocarrier (a) Polymeric nanoparticles. (b) Solid lipid nanoparticle. (c) Nanostructure lipid carrier. (d) Liquid crystals.

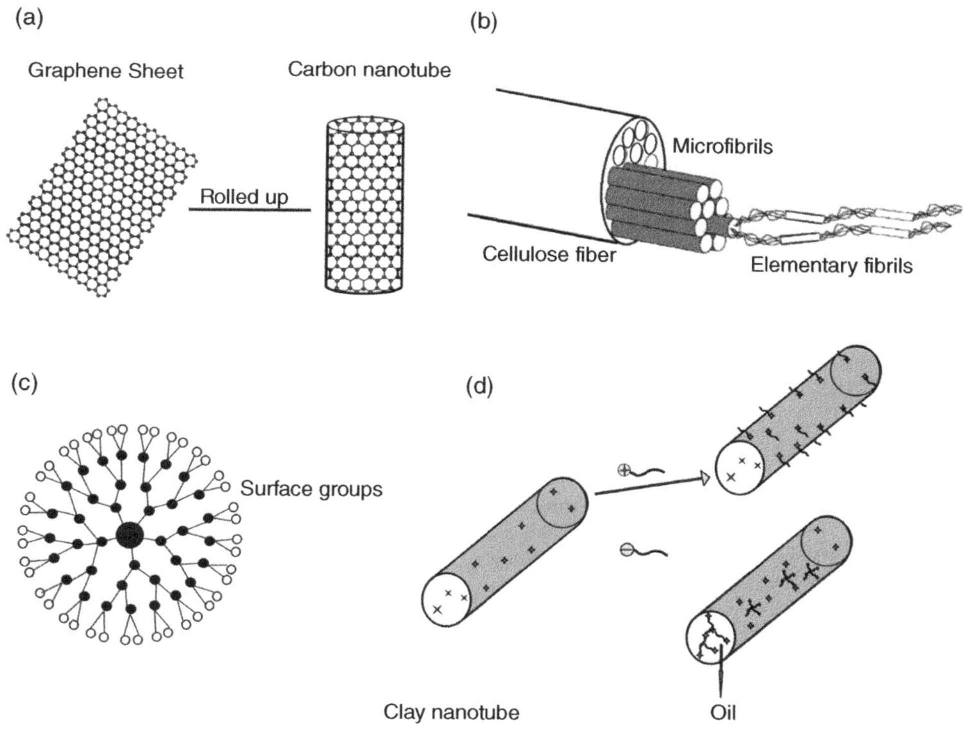

Figure 6.5 Nanocarrier (a) Carbon nanotube. (b) Nanofibers. (c) Dendrimers. (d) Halloysite clay nanotube.

Table 6.2 Classification of nanoparticular carrier based on manufacturing core material used for the preparation of nanocarrier for phytoconstituents.

Polymer base	Lipid base	Metal base	Nonmetal base
Biopolymer-based hydrogel	Linoleic and melic acid	Copper nanoparticle	Calcium carbonate nanoparticle
Biopolymer	Nanostructured lipid carriers	Gold nanoparticle	Carbon nanotube
Cyclodextrin	Solid lipid nanoparticles	Iron nanoparticles	Halloysite clay nanotube
Dendrimer	Liquid crystals	Iron nanoparticles	Mesoporous silica nanoparticle
Nanocapsules		Nanofiber	
Nanoparticle		Quantum dots	
Nanosphere		Silver nanoparticle	
		Zink nanoparticle	

6.4 Polymeric Nanoparticles

Biodegradable and biocompatible polymers are often used in the preparation of nanoparticles. Non-toxicity, non-immunogenicity, and noninflammatory are key factors for the selectivity of a polymer. The biodegradable polymer may be tailored for controlled release and targeting. Both natural and artificial polymers are used in this regard, which are summarized in Table 6.3.

Gelatin is a natural protein used as a base material for nanoparticle preparation. *Centella asiatica* (Apiaceae) extract was reported as a UV-protective function. Gelatin–*Centella asiatica* nanoparticle showed better UVA-induced matrix metalloproteinase (MMP-1) tyrosinase inhibitory activity

Table 6.3 List of polymers commonly used in the preparation of nanoparticles for delivery of phytoconstituents.

Natural	Artificial
Alginate	Cellulose acetate phthalate
Carrageenan	Isopropylacrylamide
Casein	Methacryclic acid
Cashew	Poly(alkyl cyanoacrylate)
Chitosan	Poly(alkyl methacrylate)
Collagen	Poly(ε-caprolactone)
Cyclodextrin	Poly(ε-caprolactone)
Dextran	Poly(lactic-*co*-glycolic acid)
Gallon gum	Poly(lactic acid)
Gelatin	Poly(lactic-*co*-glycolic acid)
Guar Gum	Poly(methacrylate)
Hyaluronan	Poly(styrene)
Pectin	Poly(vinylpyridine)
Serum albumin	Poly(β-hydroxybutyrate)
Silk fibroin	Poly(ε-caprolactone)
Starch	Polyethylene glycol
Xanthan gum	Polylactic acid (PLA)
	Polyvinyl acetate

compared to free extract. This difference is probably due to stabilizing effect of phytoconstituent of *Centella asiatica* from photodegradation [12].

Green tea (*Camellia sinensis*, Theaceae) is famous for its antioxidant property due to the presence of epicatechin and its derivative. Alginate and poly (ε-caprolactone) (PECL) nanoparticles were developed by encapsulating an aqueous extract of green tea to obtain control release. The developed nanoparticle confirmed more than 30% encapsulation of active extract. In a two-hour release study, nanoparticles claim only 25% release of an active principle compared to 100% release from the free extract [13].

Camptotheca acuminate Decne (Cornaceae) is a source of potent anticancer drug camptothecin, which targets intracellular topoisomerase for inhibition. Its clinical application is restricted due to unstable lactone rings and low aqueous solubility. This problem was tried to be solved by the development of a hydrophobically modified glycol chitosan-loaded nanoparticle. The lactone ring of the active principle is protected from hydrolysis by encapsulation in the hydrophobic core of the nanoparticle. Higher antitumor activity was recorded for this formulation against MDA-MB-231 human breast cancer cells in a mouse model after intravenous injection. The near-infrared study confirms the prolonged blood circulation and higher accumulation in tumor cells of the nanoparticle [14].

Phytolacca decandra (Phytolaccaceae) is a well-established chemoprotective agent extract [15] specific for lung cancer. The low bioavailability issue was solved by developing poly(lactic-*co*-glycolic acid) encapsulation of plant extract in the form of nanoparticles. Prepared formulation confirms better chemoprevention tested in vivo in mice and in vitro in a 549 cell compared to free plant extract.

Ginkgo biloba (Ginkgoaceae) herb is used in brain-related disorders. It has the ability to promote blood circulation in the brain and improve cognitive functioning by protecting neuronal damage. Its shorter shelf life is challenging for researchers. Monomethoxy polyethylene glycol and polylactide-*co*-glycolide injectable nanoparticles were developed to obtain sustained release by enhancing the shelf life of the principle active constituent of *Ginkgo biloba*. In both in vitro and in vivo tests, a successful, sustained, and synchronized release was observed. This result is due to the slow biological degradation of poly(lactic-*co*-glycolic acid) and the long circulation of PEGylated nanoparticles [16]. The application of polymeric nanoparticles for the delivery of the anticancer drug Paclitaxelis is presented in Table 6.4.

6.5 Lipid-based Nanoparticles

Lipid is one of the main constituents of our body. A compatibility issue of nanoparticle manufacturing lipid is adjusted due to similarity in the chemical composition of natural body building lipid. Another benefit of lipid as a manufacturing vehicle for nanocarrier is easily uptaken by the lymphatic circulatory system. It also acts as a penetration enhancer by solubilizing the drug on the digestive tract by avoiding first-pass metabolism by diffusion. Solid lipid nanoparticles (SLNs), nanostructured lipid carriers (NLCs), and liquid crystalline (LC) are basic carriers, which come under this nanoparticulate system. SLNs confirmed better physiochemical stability due to their solid nature at room temperature. An NLC is a second-generation system that contains a specific lipid having the ability to disorganize the liquid lipid matrix of biological membrane. NLC is composed of both solid and liquid lipids. Therefore, a large range of phytoconstituent may be loaded inside the core. SLNs suffer from low loading efficiency, which is trying to solve by mixing liquid lipid in NLC [17].

Table 6.4 Paclitaxel delivery via polymeric nanoparticles.

Phytoconstituents	Polymeric nanoparticle	Key benefits
Paclitaxel	Nanoparticle albumin-bound	Treatment of gastric cancer
Paclitaxel	Nanoparticle albumin-bound	Treatment of solid cancer
Paclitaxel	Multi-layered pluronic F127 and polyvinyl alcohol stabilized and poly-L-lysine-coated Paclitaxel-loaded poly(lactic-*co*-glycolic acid) Nanoparticle formulation Paclitaxel polymeric nanoparticles	Paclitaxel polymeric nanoparticle altered the nanomechanical properties of PanCa cells
Paclitaxel	The Paclitaxel-encapsulated core-shell nanoparticle of cetyl alcohol delivered via the oral route	Oral bioavailability was approximately 95% and chemotherapeutic efficacy was better than Taxol® and nab-Paclitaxel
Paclitaxel	Chitosan-based multifunctional nanoparticle using cationic polylysine polymer and L-cysteine delivered via the oral route	Effectively enhance the cellular uptake in Caco-2 cells. The oral bioavailability in rats was 5.63-fold higher than that of Taxol

6.5.1 Solid Lipid Nanoparticles

SLNs are nano-sized particles made of lipid. Its physical status is solid, therefore called SLNs. In this, lipophilic drugs are solubilized in a solid lipid core matrix by surfactants that emulsified them. Examples of core solid lipid materials used in the formulation of SLNs are cetyl alcohol, glyceryl monostearate, glyceryl dilaurate hydrine, stearic acid, tristearin glyceride, etc. Brain targeting is the most challenging task for formulation pharmacists due to the tight endothelial barrier, which restricts the entry of the drug into the central nervous system. Cytotoxic drug camptothecin was successfully delivered into the brain via SLNs prepared by using cetyl palmitate and surfactant polysorbate 80. In the 1,2-dimyristoyl-sn-glycero-3-phosphocholine (DMPC) membrane an artificial endothelial barrier model prepared SLN cable to change the lamellar structure of barrier and accumulated efficient quantity of camptothecin in the brain. Researchers further studied its antitumor efficiency against glioma and macrophage human cell lines and found a lower maximal inhibitory concentration (IC_{50}) after intravenous administration in the rat [18].

Quercetin oral delivery is obstructed due to poor solubility in gastrointestinal fluid and high biological degradation in gastric pH. The oral bioavailability of quercetin was tried to improve by delivery via SLNs developed by glyceryl monostearate and soy lecithin. A controlled manner drug release obtained five times better bioavailability in the in vitro test. Surprisingly, the intestine was identified as a major absorption site for SLNs loaded quercetin in the experiment [19].

Silibinin is an antitumor phytoconstituent developed in the form of SLNs using D-α-tocopherol polyethylene glycol 1000 succinate and phosphatidylcholine. Prepared SLN confirming twice cellular uptake level compared to free silibinin tested in MDA-MB-231 breast cancer cells. The result was supported by cell viability, invasion, and migration assay [20].

6.5.2 Liquid Crystals

The physical stability of the solid systems is always better than the liquid system but liquid systems have better permeation and penetration properties. This finding are helpful in the development of a new class of nanocarrier called nano liquid crystals, they are mesophasic in nature. Mesophasic means positioned between crystalline solid and isotropic liquid. They are composed of lipid which shows temperature depends interchangeable ability like cetearyl alcohol. Oily drugs are the best candidate for capsulizing in liquid crystals. Essential oil of *Carpa guanensis* Auble (Meliaceae), *Prunus persica* (Rosaceae), *Bixa orellana* (Bixaceae), and *Calendula officinalis* (Asteraceae) are delivered via liquid crystal [21].

Apigenin is a popular antioxidant compound obtained in various vegetables studied for dermal delivery via fabricating nanocrystals. In vitro results confirm twice of antioxidant potential as free apigenin by improving solubility, stability, bioavailability, and biodistribution [22]. Ursolic acid is obtained from the extraction of *Mirabilis jalapa* (Nyctaginaceae) encapsulated in nanocrystal for better dissolution rate and bioavailability. The result correlated with improved aqueous solubility and permeation as well as prolonged retention of apigenin-loaded nanocrystals [23].

6.6 Metallic Nanoparticles

Metals are minute essential components of our body. These metals are involved in redox reactions. The metallic nanoparticle is one of the most conspicuous and encouraging areas for the progress of new applications in bioactive medicine. This unique feature is extended into the delivery system by developing metallic nanoparticles. Gold, silver, platinum, zinc, copper, titanium oxide, iron, and nickel are well-reported metals used for medicinal applications. Some modified nanoparticles like quantum dots, nanowires, and nanofilms have also come under the heading of metallic nanoparticles. Gold, silver, and platinum have a unique conduction property. Similarly, iron and cadmium are reported for the preparation of magnetic nanoparticles. The large surface-volume ratio provides high encapsulation efficiency. Large surface energy provides an opportunity to surface modification as per need. There are various phytoconstituent-delivered through metallic nanoparticles for the treatment of different diseases reported in Table 6.5.

6.6.1 Magnetic Nanoparticles

Magnetic nanoparticles are also known as iron oxide/magnetic/Fe_3O_4 nanoparticles. Their unique features are targeting and tailoring. They can deliver phytoconstituents at the targeted site and are considered a replacement for post-synthesis surface modification [24]. The unique targeting ability is very helpful for cytotoxic drug delivery avoiding unnecessary exposure to healthy tissues.

Camptothecin suffers from the problem of chemical instability and poor aqueous solubility. Superparamagnetic camptothecin iron oxide nanoparticles were developed and evaluated on the H460 lung cancer line. Researchers compare the effect of both PEGylated and non-PEGylated iron oxide nanoparticles of camptothecin in blood. The antitumor effect was obtained as a percentage of apoptotic cells, 80% for both [25].

6.6.2 Silver and Gold Nanoparticles

Methanolic stem extract of *Callicarpa maingayi* (Lamiaceae) was developed as a silver nanoparticle that acts as a cap for phytoconstituents and is capable of being bound with protein's outer cell wall

Table 6.5 A list of various metallic nanoparticles-based bioactive marker compounds used for the treatment of diseases.

Phytoconstituents	Metallic nanoparticles	Key benefits
Aloin	Au metallic nanoparticle	Inhibition of human neuroectodermal tumors
Curcumin	Ag metallic nanoparticle	Synthesis and stabilization
Ellagic acids	Ag metallic nanoparticle	Effective antibacterial agent against both gram classes of bacteria
Gallic acid	Fe_3O_4–Ag magnetic metal nanoparticles	Antimicrobial activity against *Staphylococcus epidermidis*
Gallic acid	Zinc oxide nanoparticles	Antioxidant and antibacterial agents
Naringenin	Ag metallic nanoparticle	Treatment of solid cancer
Quercetin	Au metallic nanoparticle	Treatment of antimicrobial activity
Quercetin	Silver–selenium alloy nanoparticles	Anticancer and antioxidant activity
Resveratrol	Au metallic nanoparticle	Synthesis and stab ilization

of microorganisms (bacteria, fungi, and viruses). This leads to breaking the lipoprotein of the cell wall. The action finally leads to stopping cell division [26]. Diabetes mellitus is one of the diseases characterized by the metabolic dysfunction of sugar production in the body. *Cassia fistula* (Fabaceae) aqueous extract was developed as a gold nanoparticle. In vivo studies confirm the reduction of the level of various marker such as alkaline phosphate and alanine transaminase as well as uric acid and serum creatinine in diabetic mice [27].

Besides iron, silver, gold, some other metals and alloys are also popular among researchers. The delivery of quercetin and gallic acid silver-selenium alloy nanoparticle was confirming the antibacterial effect as well as the anticancer and antioxidant effects [28]. Gallic acid-coated zinc nanoparticles were assessed for antioxidant as well as antibacterial properties against methicillin-resistant *Staphylococcus aureus*. Prepared nanoparticles have shown a higher affinity toward the bacterial cell membrane [29].

6.7 Other Particulate Nanocarriers

6.7.1 Carbon Nanotubes

Carbon nanotubes are graphene-rolled cylinders that make honeycomb-like hexagonal structures of carbon. It produces a pills-like structure by looking at the end. They are available both in single-walled and multi-walled forms. Due to the structural specificity, these carry a high payload and provide an opportunity for tailoring by functionalizing at the tip and sidewall. The solubility issue of carbon nanotubes is solved by functionalization making it applicable for biological applications. In the delivery of anticancer drugs, carbon nanotubes act as targeting tools. Small diameter semiconducting carbon nanotubes are promising carriers due to the strong therapeutic infrared window. The main advantages of carbon nanotubes are the availability of a very hydrophobic hollow interior in which water-insoluble drugs can be easily loaded [30].

Table 6.6 List of drugs delivered via carbon nanotube with its therapeutic activity.

Drug	Type	Route	Therapeutics
Apigenin	Single-walled	Intravenous	Neurogenerative Alzheimer syndrome
Carvedilol-A	Multi-walled	Oral delivery	Heart failure and hypertension
Paclitaxel	Single-walled	Subcutaneous	Murine 4T1 breast cancer
Paclitaxel	Single-walled	Intravenous	Kaposi sarcoma, cervical cancer
Pirarubicin	Single-walled	Intravenous	Human bladder cancer cell line BIU-87
Quercetin	Single-walled	Intravenous	Anticancer therapy

Carbon nanotubes are less toxic and more biocompatible and have the ability to deliver the drug to the target site. Drug particles are attached to their surface. Recent studies emphasize greatly delivering chemotherapeutic agents to various cancer sites by employing carbon nanotubes. Along with delivering drugs for the treatment of cancer, carbon nanotubes are also found effective in gene delivery. These are very good nonviral vectors for gene therapy as they cross cell membranes by the endocytosis process. Moreover, due to functionalization, DNA transfer takes place without any degradation. Various phytoconstituents are delivered via carbon nanotube summarized in Table 6.6.

Folic acid is a well-known targeting ligand for delivery of anticancer drug Paclitaxel via multi-walled carbon nanotube. In the in vitro evaluation of Hella cell, using 3-(4,5-dimethylthiazol-2-yl)-2, 5-diphenyl tetrazolium bromide (MTT) assay carbon nanotube exhibited higher targeting ability with good aqueous solubility and biocompatibility [31]. Ginsenoside is an active anticancer constituent of *Panax ginseng* (Araliaceae). A multi-walled carbon nanotube was developed by the catalytic vapor deposition technique. In the MTT assay performed on MCF-3 and PANC-1 cancer cell lines, a significant decrement of the proliferation of about 73 and 60% was recorded with developed ginsenoside carbon nanotube [32].

6.7.2 Dendrimers

Dendrimers are very structural-specific nanocarriers having a tree-like structure of manufacturing compounds. Polyamidoamine is the most commonly used manufacturing material. A high-branched structure with a surface-active group is available for attachment of drug moiety and targeting ligand [33]. Dendrimers comprise three regions: polyfunctional core, branches, and multivalent surface. The high surface density of dendrimers allows them for multivalent interaction between functionalized dendrimers and biomolecules. Their unique physical properties like encapsulation ability, water solubility, and monodispersity make them drug delivery carrier. The availability of the site for functionalization make them very specific for targeting. A dendrimer is very popular for targeting an anticancer drug due to the opportunity of surface ligands by PEGylation, acetylation, glycosylation, and various amino acids [24].

Capsaicin is a pungent principle of capsicum showing divergent medicinal properties [34]. Capsaicin was formulated as a dendrimer by the esterification process. The efficiency of the capsaicin-loaded dendrimer was assessed on the VERO cell MCF-7 and HEP-2 cells line. For cytotoxicity, IC_{50} values of 1.25 and 0.62 μg/ml for normal and cancer cell lines obtained, respectively.

This was effective at IC_{50} 0.62 μg/ml on the MCF-7 cell and HEP-2 cell [35]. Gallic acid was formulated into a dendrimer using polyamidoamine. This compound provides surface modification for drug loading. The effectiveness of this dendrimer was found better than free gallic acid in MCF-7 breast cancer cells line [36].

The root of *Puerarin lobata* (Leguminosae) is a source of isoflavonoid Puerarin. Puerarin-Polyimidoamine dendrimer was developed, the prepared dendrimer showed better solubility and improved bioavailability in a sustained-release fashion. Hemolytic toxicity studies in in vitro studies in fresh rat erythrocytes confirm a low level of hemolytic activity by puerarin dendrimer formulation [37]. Anticancer compound anthocyanin was extracted from *Daucuscarota* (Apiaceae) and developed as dendrimer in silica-polyamidoamine base. The anticancer effect was assessed on the neuroblastoma (A2) cell line. The better response was might be due to improved stability, solubility, and sustained release of active principle from dendrimer [38].

6.7.3 Nanofiber

Nanofiber is a nanomaterial carrier system for a drug that has a large length diameter ratio. High porosity, small diameter, low density, large surface to volume ratio, and excellent mechanical property make it a unique drug carrier for phytoconstituents. Its applicability was extended from wound dressing and medical implants to drug delivery carriers. Its features like high loading capacity, high encapsulation efficacy, simultaneous delivery of various therapeutics, easy operation, and cost-effectiveness make it suitable for the delivery of phytoconstituents. It is made of biodegradable and biocompatible polymers like polyvinyl, pyrrolidone etc. The stimuli responsiveness of nanofibers like pH, temperature, electricity, magnate, and light make it suitable for targeting purposes [30].

Aloe vera gel-loaded nanofibers were developed to increase hydrophilicity for wound healing and the skin regeneration process. Accelerated wound healing and skin regeneration in the male BALB/c mice model of an induced wound were found [39]. *Lycium barbarum* (Solanaceae) was used for nerve cell regeneration by fabricating nanofiber. Extract-loaded nanofiber showed sustained release through diffusion. In vitro test in rat, both rat Schwann cell myelination and neurite outgrowth of dorsal root ganglion neuron obtained. Prepared nanofiber has shown better peripheral nerve regeneration and neuroprotection by better nerve growth factor promotion in a better way than free extract [40]. Copaiba oil (Copaifera plant)-loaded nanofiber confirms increased hydrophilicity and controlled drug release pattern with greater antimicrobial action against *Staphylococcus aureus* in vitro [41].

6.7.4 Halloysite Clay Nanotube

Aluminosilicate clay is a mineral compound, commonly used as a pharmaceutical excipient due to its biocompatibility and large availability, with low price. Halloysite clay nanotube is novelly carried and composed of alumina and silica. Halloystic clay nanotubes have both positive and negative charges on the surface. The surface is negatively charged due to silica oxide providing a space to load positively charged drug molecules by repelling them. The inner core of the tube is positively charged by the presence of aluminum oxide, providing an opportunity to load a negatively charged biomolecule. Halloysite is a naturally occurring biocompatible listed nanomaterial. This tube has a fine size, large surface area, and excellent dispersion property with a high aspect ratio. It supports to load of a large number of active particles and full opportunity for surface functionalization for loading and targeting. They provide faster absorption due to pores [30].

Thyme oil (*Thymus vulgans*) and rosemary oil (*Rosmaris officinalis*) were loaded in halloysite nanotube and assessed for antimicrobial activity. Both tubes confirm the sustained and controlled release of the active principle. Thyme oil nanotube showed an in vitro antimicrobial effect against *Escherichia coli* O157:H7, yeasts, and mesophilic aerobic bacteria. While rosemary oil-loaded nanotube did not show mold growth after three months at room temperature storage condition [42, 43]. Silibinin obtained from *Silybum marianum* (Asteraceae) and quercitin both are potent cytotoxic drugs formulated in a tube with surface functionalization with cyclodextrin providing improved stability, drug uptake, and sustainability. Antiproliferative activity against human anaplastic thyroid cancer cell line (8505C) confirms a positive report in vitro using MTT assay [44].

6.8 Target Specificity and Toxicity Concerns of Nano-sized Drug Carriers

There is no doubt about the application of nanomaterials in nano-pharmaceutics. This specific ability provides some exceptional characteristics such as size, size distribution, surface charge, expanded surface area, self-assembly, and stability. Researchers are attracting nano-drug carriers to solve the problem of any aspects of drug delivery. Providing the right dose of a particular drug to a specific site of disease controlled distribution, and less toxic systemic side effects, thereby bringing about better patient compliance, especially in anticancer therapy. Figure 6.6 represents the strategic approach of apigenin-loaded nano-sized delivery carriers and Table 6.7 summarizes various nanocarriers under investigation for delivery of curcumin. At the same time, the feature of nano-drug carriers possibly affects the basic adsorption, distribution, metabolism, and excretion phenomena of the human body. The specific ability of the nano-sized drug carriers may also be affect the natural process like cellular uptake, transport, and distribution of nutrients within

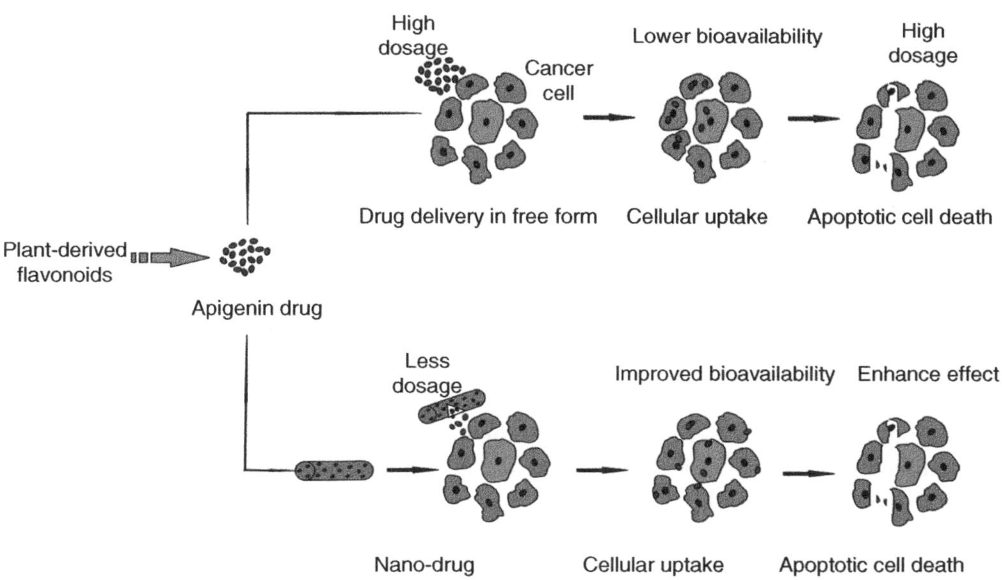

Figure 6.6 Strategy for the treatment of the nanocarriers' approach to skin cancer.

Table 6.7 Delivery of curcumin in various nanoparticulate carriers with their key features.

Nanoparticulate carrier	Specification	Advantages
Polymeric nanoparticle	Curcumin-loaded poly D,L-lactic-*co*-glycolic acid nanoparticle	Improved the oral bioavailability
	Curcumin-loaded polybutylcyanoacrylate	Enhance the transport of curcumin to the brain
	Curcumin-loaded chitosan nanoparticle	Increased cellular uptake
	Curcumin-loaded *N*-isopropylacrylamide–methacrylic acid	Treatment of breast cancer cells
Solid lipid nanoparticle	Curcumin-loaded SLN	Treatment for lung cancer
	N-carboxymethyl chitosan coating curcumin-loaded SLN	Release simulated gastric fluid (SFG)
	Curcumin-loaded cetylpalmitate SLN	Anticancer efficacy
Protein-based nanocarriers	Curcumin-loaded bovine serum albumin nanoparticle	Higher encapsulation efficiency
	Curcumin-loaded albumin nanoparticles	Enhanced penetration of albumin
	Curcumin-loaded ovalbumin	Higher antioxidant activity
	Curcumin-zein fibers	Inhibit bacterial growth
	Curcumin-loaded casein nanoparticles	Increase the solubility and bioactivity
	Curcumin-loaded silk fibroin nanoparticles	Influenced encapsulation efficiency and drug loading
Nanocrystals	Curcumin-loaded HPH nanocrystal	Enhancement of dermal penetration
	Curcumin-loaded sodium lauryl sulfate and poloxamer 188	The prevention of nanocrystal aggregation
	Curcumin-loaded hyaluronic acid nanocrystals	Increased cellular uptake
Mesoporous silica nanoparticles	Curcumin-loaded mesoporous silica nanoparticles	Enhanced cell cytotoxicity
	Folate receptor targeted Curcumin-loaded mesoporous silica nanoparticles	Tumor acidic environment responsive release
	Curcumin-loaded polyethyleneimine and folic acid	Enhanced cellular uptake and increased cytotoxicity

the human body. Another noticeable point that the nanosize of carriers enables them to cross the natural biological barriers like simple capillary endothelial barrier, simple cell membrane barrier as well as some natural low-penetration barriers like blood–brain barrier, cerebrospinal fluid barrier, placental barrier, and testis barrier. So, the possibility of unwanted accumulation of the drug can also be able to create unknown problems.

6.9 Future Prospects

The matter is that non-debated acceptance of the application of nanotechnology for the delivery of phytoconstituent from the herbal drug can potentially enhance. This can be an easily abolished issue related to the lack of suitability of the conventional dosage form. However, so many issues still need to be solved which are related to the delivery of phytoconstituents via nanocarrier. These are presented in Figure 6.7.

In the path of success of nanocarriers, the biggest aspect is the scaling technique in low-cost inventory involvement. Sometimes the use of a sophisticated instrument like an electron microscope in evaluation enhances the premarket release investment. In the discovery of drugs from herbal sources, the isolation procedure with high yield is still under investigation for so many plant sources. In the delivery of the extract, clinical viability in therapy is also under investigation. The combination of the nanotechnological approach with phytoconstituent may search out the answer to the earlier questions. South Asian countries like India, where already phytoconstituent-based treatment therapy, the Ayurveda system, is well-established, will play a lead role in the production and standardization of nano-scaled phytoconstituent formulation. They have already applied sophisticated modern techniques of standardization such as gas chromatography-mass spectrometry, liquid chromatography–mass spectrometry, high performance liquid chromatography, and high-performance thin-layer chromatography.

So many pharma companies are working in the nanocarrier-based formulation development like cytimune, invitrogen, antaria, nanoledge, and nanoces, indicating the bright feature of this combination in the drug delivery segment. A few nanocarrier-based formulations are under the consideration of clinical and preclinical trials and some of them got approval from the Food and Drug Administration (FDA) confirming their safety for human application. The concept of surface modification will be playing a key role in the development of inexpensive, safe, tolerable, and approachable first-line medication through the synergistic benefit of phytoconstituents with nanotechnology.

Figure 6.7 Unsolved issues with nanocarriers.

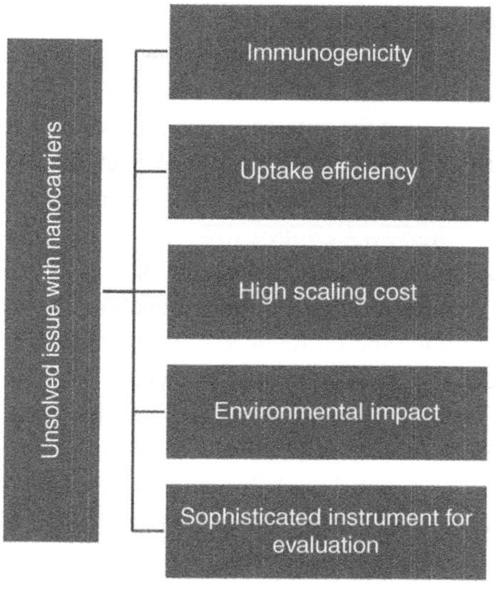

References

1 Sarwa, K.K., Das, P.J., and Mazumder, B. (2014). A nanovesicle topical formulation of Bhut Jolokia (hottest capsicum): a potential anti-arthritic medicine. *Expert Opin. Drug Deliv.* 11: 661–676.

2 Rudrapal, M., Sarwa, K., and Bhattacharya, S. (2020). Standardization and quality evaluation of botanicals with special reference to marker components. *J. Adv. Lab. Res. Biol.* 2: 405–426.

3 Sarwa, K., Rudrapal, M., and Debnath, M. (2013). Extraction of green tea leaves: the use of different methods, their optimization and comparative evaluation. *Biosci. Biotechnol. Res. Asia* 10: 383–386.

4 Sarwa, K.K., Mazumder, B., Rudrapal, M., and Verma, V.K. (2015). Potential of capsaicin-loaded transfersomes in arthritic rats. *Drug Deliv.* 22: 638–646.

5 Rauta, P.R. (2019). *Nanotechnology in Biology and Medicine: Research Advancements & Future Perspectives*. CRC Press (Taylor and Francis).

6 Jangdey, M., Gupta, A., and Sarwa, K. (2018). Apigenin and quercetin: potential therapeutic challenging effective against in Alzheimer's disease. *UK J. Pharmaceut. Biosci.* 6: 46.

7 Sarwa, K.K., Mazumder, B., Suresh, P.K., and Kaur, C.D. (2016). Topical analgesic nanolipid vesicles formulation of capsaicinoids extract of bhut jolokia (*Capsicum chinense* Jacq): pharmacodynamic evaluation in rat models and acceptability studies in human volunteers. *Curr. Drug Deliv.* 13: 1325–1338.

8 Sarwa, K., Mazumder, B., and Rudrapal, M. (2017). Effect of extraction methods and long term storage on capsaicinoids content of Bhut Jolokia fruits. *Indian J. Nat. Prod. Resour.* 8: 69–77.

9 Kumar Sarwa, K., Rudrapal, M., and Mazumder, B. (2015). Topical ethosomal capsaicin attenuates edema and nociception in arthritic rats. *Drug Deliv.* 22: 1043–1052.

10 Zorzi, G.K., Carvalho, E.L.S., Von Poser, G.L., and Teixeira, H.F. (2015). On the use of nanotechnology-based strategies for association of complex matrices from plant extracts. *Rev. Bras* 25: 426–436.

11 Verma, V.K., Zaman, M.K., Verma, S. et al. (2020). Role of semi-purified andrographolide from Andrographis paniculata extract as nano-phytovesicular carrier for enhancing oral absorption and hypoglycemic activity. *Chin. Herb. Med.* 12: 142–155.

12 Kwon, M.C., Choi, W.Y., Seo, Y.C. et al. (2012). Enhancement of the skin-protective activities of *Centella asiatica* L. urban by a nano-encapsulation process. *J. Biotechnol.* 157: 100–106.

13 Sanna, V., Lubinu, G., Madau, P. et al. (2015). Polymeric nanoparticles encapsulating white tea extract for nutraceutical application. *J. Agric. Food Chem.* 63: 2026–2032.

14 Min, K.H., Park, K., Kim, Y.S. et al. (2008). Hydrophobically modified glycol chitosan nanoparticles-encapsulated camptothecin enhance the drug stability and tumor targeting in cancer therapy. *J. Control. Release* 127: 208–218.

15 Das, J., Das, S., Samadder, A. et al. (2012). Poly(lactide-*co*-glycolide) encapsulated extract of *Phytolacca decandra* demonstrates better intervention against induced lung adenocarcinoma in mice and on A549 cells. *Eur. J. Pharm. Sci.* 47: 313–324.

16 Han, L., Fu, Y., Cole, A.J. et al. (2012). Co-encapsulation and sustained-release of four components in ginkgo terpenes from injectable PELGE nanoparticles. *Fitoterapia* 83: 721–731.

17 Bonifácio, B.V., Silva, P.B., Ramos, M.A. et al. (2014). Nanotechnology-based drug delivery systems and herbal medicines: a review. *Int. J. Nanomedicine* 9: 1–15.

18 Martins, S.M., Sarmento, B., Nunes, C. et al. (2013). Brain targeting effect of camptothecin-loaded solid lipid nanoparticles in rat after intravenous administration. *Eur. J. Pharm. Biopharm.* 85: 488–502.

19 Li, H., Zhao, X., Ma, Y. et al. (2009). Enhancement of gastrointestinal absorption of quercetin by solid lipid nanoparticles. *J. Control. Release* 133: 238–244.

20 Xu, P., Yin, Q., Shen, J. et al. (2013). Synergistic inhibition of breast cancer metastasis by silibinin-loaded lipid nanoparticles containing TPGS. *Int. J. Pharm.* 454: 21–30.

21 Silvestrini, A.V.P., Caron, A.L., Viegas, J. et al. (2020). Advances in lyotropic liquid crystal systems for skin drug delivery. *Expert Opin. Drug Deliv.* 17: 1781–1805.

22 Al Shaal, L., Shegokar, R., and Müller, R.H. (2011). Production and characterization of antioxidant apigenin nanocrystals as a novel UV skin protective formulation. *Int. J. Pharm.* 420: 133–140.

23 Pi, J., Liu, Z., Wang, H. et al. (2016). Ursolic acid nanocrystals for dissolution rate and bioavailability enhancement: influence of different particle size. *Curr. Drug Deliv.* 13: 1358–1366.

24 Khan, T. and Gurav, P. (2017). Phytonanotechnology: enhancing delivery of plant based anti-cancer drugs. *Front. Pharmacol.* 8: 1002.

25 Castillo, P.M., de la Mata, M., Casula, M.F. et al. (2014). PEGylated versus non-PEGylated magnetic nanoparticles as camptothecin delivery system. *Beilstein J. Nanotechnol.* 5: 1312–1319.

26 Shameli, K., Bin Ahmad, M., Jaffar Al-Mulla, E.A. et al. (2012). Green biosynthesis of silver nanoparticles using *Callicarpa maingayi* stem bark extraction. *Molecules* 17: 8506–8517.

27 Daisy, P. and Saipriya, K. (2012). Biochemical analysis of *Cassia fistula* aqueous extract and phytochemically synthesized gold nanoparticles as hypoglycemic treatment for diabetes mellitus. *Int. J. Nanomed.* 7: 1189–1202.

28 Mittal, A.K., Kumar, S., and Banerjee, U.C. (2014). Quercetin and gallic acid mediated synthesis of bimetallic (silver and selenium) nanoparticles and their antitumor and antimicrobial potential. *J. Colloid Interface Sci.* 431: 194–199.

29 Lee, J., Choi, K.H., Min, J. et al. (2017). Functionalized ZnO nanoparticles with gallic acid for antioxidant and antibacterial activity against methicillin-resistant *S. aureus*. *Nanomaterials (Basel)* 7: 365.

30 Rahman, H.S., Othman, H.H., Hammadi, N.I. et al. (2020). Novel drug delivery systems for loading of natural plant extracts and their biomedical applications. *Int. J. Nanomed.* 15: 2439–2483.

31 Tian, Z., Shi, Y., Yin, M. et al. (2011). Functionalized multiwalled carbon nanotubes-anticancer drug carriers: synthesis, targeting ability and antitumor activity. *Nano Biomed. Eng.* 3: 157–162.

32 Lahiani, M.H., Eassa, S., Parnell, C. et al. (2017). Carbon nanotubes as carriers of *Panax ginseng* metabolites and enhancers of ginsenosides Rb1 and Rg1 anti-cancer activity. *Nanotechnology* 28: 015101.

33 Chauhan, A.S. (2018). Dendrimers for drug delivery. *Molecules* 23: 938.

34 Sarwa, K., Mazumder, B., Rudrapal, M. et al. (2013). Capsaicinoids content of some indigenous capsicum varieties of Assam, India. *J. Nat. Sci. Res.* 3: 112–116.

35 Malar, C. and Bavanilathamuthiah (2015). Dendrosomal capsaicin nanoformulation for the invitro anti-cancer effect on HEp2 and MCF-7 cell lines. *Int. J. Appl. Bioeng.* 9: 30–35.

36 Sharma, A., Gautam, S.P., and Gupta, A.K. (2011). Surface modified dendrimers: synthesis and characterization for cancer targeted drug delivery. *Bioorg. Med. Chem.* 19: 3341–3346.

37 Gu, L., Wu, Z., Qi, X. et al. (2013). Polyamidomine dendrimers: an excellent drug carrier for improving the solubility and bioavailability of puerarin. *Pharm. Dev. Technol.* 18: 1051–1057.

38 Yesil-Celiktas, O., Pala, C., Cetin-Uyanikgil, E.O., and Sevimli-Gur, C. (2017). Synthesis of silica-PAMAM dendrimer nanoparticles as promising carriers in Neuro blastoma cells. *Anal. Biochem.* 519: 1–7.

39 Jouybar, A., Seyedjafari, E., Ardeshirylajimi, A. et al. (2017). Enhanced skin regeneration by herbal extract-coated poly-l-lactic acid nanofibrous scaffold. *Artif. Organs* 41: E296–E307.

40 Wang, J., Tian, L., He, L. et al. (2018). *Lycium barbarum* polysaccharide encapsulated poly lactic-*co*-glycolic acid Nanofibers: cost effective herbal medicine for potential application in peripheral nerve tissue engineering. *Sci. Rep.* 8: 8669.

41 Bonan, R.F., Bonan, P.R., Batista, A.U. et al. (2015). In vitro antimicrobial activity of solution blow spun poly(lactic acid)/polyvinylpyrrolidone nanofibers loaded with Copaiba (*Copaifera* sp.) oil. *Mater. Sci. Eng. C Mater. Biol. Appl.* 48: 372–377.

42 Cavallaro, G., Lazzara, G., Massaro, M. et al. (2015). Biocompatible poly(*N*-isopropylacrylamide)-halloysite nanotubes for thermoresponsive curcumin release. *J. Phys. Chem. C* 119: 8944–8951.

43 Lee, M.H., Seo, H.S., and Park, H.J. (2017). Thyme oil encapsulated in halloysite nanotubes for antimicrobial packaging system. *J. Food Sci.* 82: 922–932.

44 Riela, S., Massaro, M., Colletti, C. et al. (2015). Multicavity halloysite-amphiphilic cyclodextrin hybrids for co-delivery of natural drugs into thyroid cancer cells. *J. Mater. Chem. B* 3: 4074–4081.

7

Nanodelivery of Herbal and Phyto-Antioxidants

Gülsen Kaya[1], Merve Keskin[2], Saban Keskin[2], and Mithun Rudrapal[3]

[1] *Scientific and Technological Research Center, Inonu University, Malatya, Turkey*
[2] *Vocational School of Health Services, Bilecik Seyh Edebali University, Bilecik, Turkey*
[3] *Department of Pharmaceutical Chemistry, Rasiklal M. Dhariwal Institute of Pharmaceutical Education and Research, Pune, Maharashtra, India*

7.1 Introduction

Food-grade antioxidants derived from plant-derived extracts are the most common ingredients in antioxidant supplements. Phyto-bioactive compounds such as curcumin, resveratrol, catechins, and quercetin are among the most widely used natural compounds used as potential modulators of free radical-induced cellular damage. However, the therapeutic potential of these compounds is limited by their low bioavailability associated with poor solubility, stability, and absorbance in the gastrointestinal tract. Recently, new nanotechnology-based systems have been developed for the therapeutic delivery of natural antioxidants with improved bioavailability and, consequently, efficacy in clinical practice. Such systems provide many benefits in preclinical research over conventional preparations, including superior solubility and stability, extended half-life, improved epithelial permeability and bioavailability, increased tissue targeting, and minimized side effects.

7.2 Natural Antioxidants and Herbal/Phyto-Antioxidants

Natural antioxidants are widely found in food and herbs. These natural antioxidants, especially polyphenols and carotenoids, exhibit a wide variety of biological effects, including anti-inflammatory, antiaging, anti-atherosclerosis, and anticancer. Antioxidant phytochemicals are widely found in fruits, vegetables, grains, edible macrofungi, microalgae, and medicinal plants [1–3]. Common fruits such as berries, grapes, Chinese dates, pomegranates, guava, sugar bowls, dates, Chinese wampee, and plums are rich in antioxidant phytochemicals [4–6]. In addition, wild fruits such as *Eucalyptus robusta*, *Eurya nitida*, *Melastoma sanguineum*, *Melaleuca leucadendron*, *Lagerstroemia indica*, *Caryota mitis*, *Lagerstroemia speciosa*, and *Gordonia axillar* also have high antioxidant capacities and total phenolic content [7]. Also, fruit waste (peel and seed) contains high amounts of antioxidant phytochemicals, including catechin, cyanidin 3-glucoside, epicatechin, gallic acid, kaempferol, and chlorogenic acid [8]. Some vegetables such as Chinese

Phytoantioxidants and Nanotherapeutics, First Edition. Edited by Mithun Rudrapal.

toon, cereal, penis leaf, cowpea, cumin, lotus root, sweet potato leaf, soybean (green), chili leaf, ginseng leaf, chives, and broccoli have been found to have high antioxidant capacities. High amounts of antioxidant phytochemicals (flavones and tannins) are found among cereal grains such as black rice, red rice, and purple rice [9]. Many flowers, including edible and wild ones, also contain high amounts of antioxidant phytochemicals [10].

Polyphenols and carotenoids are the two main types of antioxidant phytochemicals, and they contribute the most to the antioxidant properties of foods/herbs. For example, carotene, quercetin, myricetin, and kaempferol are the major antioxidant phytochemicals found in Cape gooseberry [11], and anthocyanins and ellagitannins are the major antioxidant compounds among strawberry phytochemicals [12]. Additionally, flavonoids isolated from Euterpe oleracea pulp offer a significant antioxidant activity measured by oxygen radical absorbance capacity [13]. When fruits contain higher total phenolic content, they have stronger antioxidant activity [14]. For example, the scavenging activity of grape seed extract against 2,2′-azino-bis(3-ethylbenzoth iazoline-6-sulfonic acid) (ABTS) radical was strongly associated with the level of phenolic compounds [15]. These natural antioxidants, especially polyphenols and carotenoids, exhibit a wide variety of biological effects, including anti-inflammatory, antiaging, anti-atherosclerosis, and anticancer. Therefore, nano release of herbal and phyto-antioxidants and studies in this area are very important.

7.3 Types of Nanodelivery Systems (Nanoparticles and Others) for Phyto-Antioxidants

Many nanodelivery systems loaded with plant-based bioactive compounds have been shown to be effective in modulating anti-inflammatory, antiaging, anti-atherosclerosis, anticancer, and related chronic inflammation mediating many of the disorders associated with oxidative stress. These nanodelivery systems are described in the following section.

7.3.1 Nano-Resveratrol

Resveratrol (3,5,4′-*trans*-trihydroxystylbene) is a polyphenol compound found in grape skins and seeds and, to a lesser extent, in some other plant sources [16]. Recently, some nano-sized resveratrol-loaded formulations have been investigated for their potential clinical use. The bioavailability of orally administered trans-resveratrol loaded on lipid core nanocapsules was found to be twice that of free trans-resveratrol in the brain, kidney, and liver of male Wistar rats [17]. Nanodelivery has also improved gastrointestinal safety in this model. It has also been shown that the bioavailability of resveratrol nanoparticles encapsulated with folate conjugated human serum albumin is sixfold higher following intravenous administration compared to natural resveratrol [18].

Many studies have demonstrated the antioxidant properties of nano-encapsulated resveratrol. For example, according to Chen et al., resveratrol exhibited higher antioxidant capacity in the nano-encapsulated, self-microemulsified drug delivery system and lower toxicity compared to free resveratrol [19]. High reactive oxygen species (ROS) clearance efficacy has also been demonstrated for the vitamin E-loaded resveratrol nanoemulsion (mean globule diameter of about 100 nm) in patients with Parkinson's disease [20]. It has been shown that the activities of endogenous antioxidant enzymes, including SOD and glutathione (GSH) levels, were significantly higher and malondialdehyde levels were significantly lower in the resveratrol nanoemulsion group.

7.3.2 Nano-Curcumin

Curcumin [1,7-bis(4-hydroxy-3-methoxyphenyl)-1,6-heptadiene-3,5-dione] is a polyphenol obtained from the rhizome of the turmeric plant *Curcuma longa*. Besides its antioxidant properties, this polyphenol compound exhibits anti-inflammatory, anti-neurodegenerative, and anticancer activities [21]. In the last decade, the health-promoting potential of this compound has been extensively investigated in clinical trials [22]. However, its bioavailability is limited due to its low water solubility and gastrointestinal stability [23]. Innovative nano-deployment strategies are currently being developed to overcome these limitations [24, 25]. Both in vitro and in vivo evidence suggest that nanocurcumin formulations have better health-promoting properties than traditional (natural) curcumin. The solubility and bioavailability of this compound in water have been shown to be significantly increased by nano-encapsulation with lipid or polymeric nanoparticles, nanogels, and dendrimers as well as conjugation to metal oxide nanoparticles [26]. Specifically, the oral bioavailability of polylactic-*co*-glycolic acid (PLGA) curcumin nanoformulation has been shown to be 22 times higher than that of natural curcumin [27].

7.3.3 Nano-Quercetin

Quercetin (3,5,7,3′,4′-pentahydroxyflavone) is a bioactive flavonoid with powerful antioxidant properties. In particular, it is known to have the capacity to prevent the oxidation of low-density lipoproteins by scavenging free radicals and chelating transition metal ions. Quercetin's antioxidant ability is due to the presence of phenolic hydroxyl groups accessible to oxidizing agents. In addition to its antioxidant properties, quercetin is known to exhibit anti-inflammatory, anti-obesity, anti-diabetic, anti-atherosclerotic, anti-hypercholesterolemic, and antihypertensive activities [28]. However, these activities are limited due to low bioavailability (<2% of the administered dose). In recent years, innovative nanotechnology-based approaches have been developed to increase quercetin bioavailability. Quercetin-loaded nanoparticles have also been tested on animal models that can develop antioxidant defense mechanisms. For example, in the streptozotocin-induced diabetic rat model, nanoparticles of quercetin-loaded PLGA, approximately 180 nm in size, showed antioxidant properties similar to that of free quercetin [29]. Similar effects were also observed for catalase (CAT) and superoxide dismutase (SOD) activities of the pancreas and kidneys.

7.3.4 Nano-Genistein

Genistein [4′,5,7-trihydroxyisoflavone or 5,7-dihydroxy-3-(4-hydroxyphenyl) chromen-4-one] is phytoestrogenic and is a powerful isoflavone-derived antioxidant with strong antioxidant activity [30]. The clinical use of this compound is generally limited due to its low bioavailability. Moreover, when used in high doses, it may cause endocrine-disrupting and toxic effects [31]. To overcome these potential side effects, innovative nanotechnological solutions have recently been proposed. Improved oral bioavailability has been demonstrated in polymeric nanomicelles compared to Genistein. This effect is due to the high solubility and gastrointestinal tract.

7.3.5 Nano-Epigallocatechin-3-Gallate

Epigallocatechin-3-gallate [EGCG, (2*R*,3*R*)-5,7-dihydroxy-2-(3,4,5-trihydroxyphenyl)chroman-3-yl 3,4,5-trihydroxybenzoate] is one of the major polyphenols and is used in green tea Epidemiological studies to investigate EGCG are controversial and often conflicting findings with in vitro findings are encountered. This contradiction can be attributed to poor bioavailability. Therefore, in improving the bioavailability of EGCG, innovative nanodelivery systems are widely used [32]. EGCG-loaded nanoparticles show good bioavailability and strong antioxidant properties.

7.4 Nanodelivery of Phyto-Antioxidant and Therapeutic Applications

Drug delivery with nanocarriers can generally have a variety of therapeutic advantages. These advantages include, inter alia, increased half-life, extended circulation time, improved mean residence time, and improved pharmacokinetic clearance of these agents [33]. Currently, nano-sized delivery systems for many oral applications, 95% of phytotherapeutics have reached the clinical trial stage and are increasingly applied in clinical practice. Successful nanodelivery systems are characterized by optimum properties for loading and unloading and the therapeutic agent is known for its long storage life as well as its high therapeutic side effects or minimized efficacy [34, 35].

7.5 Challenges and Future Perspectives

The use of nanocarrier-based systems increases the solubility, stability, and gastrointestinal absorption of released bioactive phytochemicals and protects against early enzymatic degradation. Concluding evidence suggests that the bioavailability of nanoparticle-encapsulated phytochemicals may be 5–10 times higher than natural formulations [36, 37]. Moreover, such an oral mode of administration may result in a prolonged circulation time of these compounds and thus a decrease in their potential toxicity and side effects. Applying nanotherapeutic approaches will likely provide an opportunity to overcome many shortcomings. However, whether the applied nanomaterials can bioaccumulate in the human body, whether they can be converted into potentially hazardous metabolites, can be easily removed from the body after the therapeutic agent is administered, and whether they can become dangerous after disposal are important challenges.

7.6 Conclusion

For nanodelivery of herbal/phyto-antioxidants, several key challenges still exist. The nanocarriers used to encapsulate phytotherapeutic agents should be extensively studied to determine whether these nanomaterials themselves can exhibit adverse health effects, especially when used by patients for long periods. While several nanoantioxidant formulations may cause concerns about drug release cellular toxicity, a slower drug release, conversely, may lead to inadequate therapeutic efficacy. Therefore, the development of nanoformulations with release profiles optimized according to the physicochemical properties of the transported therapeutic agents is an important research challenge that requires further investigation.

References

1 Deng, G.F., Xu, X.R., Guo, Y.J. et al. (2012). Determination of antioxidant property and their lipophilic and hydrophilic phenolic contents in cereal grains. *J. Funct. Foods* 4: 906–914. https://doi.org/10.1016/j.jff.2012.06.008.

2 Guo, Y.J., Deng, G.F., Xu, X.R. et al. (2012). Antioxidant capacities, phenolic compounds and polysaccharide contents of 49 edible macro-fungi. *Food Funct.* 3: 1195–1205. https://doi.org/10.1039/c2fo30110e.

3 Wu, S., Li, S., Xu, X.R. et al. (2012). Sources and bioactivities of astaxanthin. *Int. J. Mod. Biol. Med.* 1: 96–107.

4 Fu, L., Xu, B.T., Xu, X.R. et al. (2011). Antioxidant capacities and total phenolic contents of 62 fruits. *Food Chem.* 129: 345–350. https://doi.org/10.1016/j.foodchem.2011.04.079.

5 Manganaris, G.A., Goulas, V., Vicente, A.R., and Terry, L.A. (2014). Berry antioxidants: small fruits providing large benefits. *J. Sci. Food Agric.* 94: 825–833. https://doi.org/10.1002/jsfa.6432.

6 Xia, E.Q., Deng, G.F., Guo, Y.J., and Li, H.B. (2010). Biological activities of polyphenols from grapes. *Int. J. Mol. Sci.* 11: 622–646. https://doi.org/10.3390/ijms11020622.

7 Fu, L., Xu, B.T., Xu, X.R. et al. (2010). Antioxidant capacities and total phenolic contents of 56 wild fruits from south China. *Molecules* 15: 8602–8617. https://doi.org/10.3390/molecules15128602.

8 Deng, G.F., Shen, C., Xu, X.R. et al. (2012). Potential of fruit wastes as natural resources of bioactive compounds. *Int. J. Mol. Sci.* 13: 8308–8323. https://doi.org/10.3390/ijms13078308.

9 Deng, G.F., Xu, X.R., Zhang, Y. et al. (2013). Phenolic compounds and bioactivities of pigmented rice. *Crit. Rev. Food Sci. Nutr.* 53: 296–306. https://doi.org/10.1080/10408398.2010.529624.

10 Li, A.N., Li, S., Li, H.B. et al. (2014). Total phenolic contents and antioxidant capacities of 51 edible and wild flowers. *J. Funct. Foods* 6: 319–330. https://doi.org/10.1016/j.jff.2013.10.022.

11 Zhang, Y.J., Deng, G.F., Xu, X.R. et al. (2013). Chemical components and bioactivities of cape gooseberry (*Physalis peruviana*). *Int. J. Food Nutr. Saf.* 3: 15–24.

12 Giampieri, F., Tulipani, S., Alvarez-Suarez, J.M. et al. (2012). The strawberry: composition, nutritional quality, and impact on human health. *Nutrition* 28: 9–19. https://doi.org/10.1016/j.nut.2011.08.009.

13 Costa, A., Garcia-Diaz, D.F., Jimenez, P., and Silva, P.I. (2013). Bioactive compounds and health benefits of exotic tropical red-black berries. *J. Funct. Foods* 5: 539–549. https://doi.org/10.1016/j.jff.2013.01.029.

14 Sun, J., Chu, Y.F., Wu, X.Z., and Liu, R.H. (2002). Antioxidant and antiproliferative activities of common fruits. *J. Agric. Food Chem.* 50: 7449–7454. https://doi.org/10.1021/jf0207530.

15 Sung, J. and Lee, J. (2010). Antioxidant and antiproliferative activities of grape seeds from different cultivars. *Food Sci. Biotechnol.* 19: 321–326. https://doi.org/10.1007/s10068-010-0046-6.

16 Salehi, B., Mishra, A.P., Nigam, M. et al. (2018). Resveratrol: a double-edged sword in health benefits. *Biomedicine* 6: E91. https://doi.org/10.3390/biomedicines6030091.

17 Frozza, R.L., Bernardi, A., Paese, K. et al. (2010). Characterization of trans-resveratrol-loaded lipid-core nanocapsules and tissue distribution studies in rats. *J. Biomed. Nanotechnol.* 6: 694–703. https://doi.org/10.1166/jbn.2010.1161.

18 Lian, B., Wu, M., Feng, Z. et al. (2019). Folate-conjugated human serum albumin-encapsulated resveratrol nanoparticles: preparation, characterization, bioavailability and targeting of liver tumors. *Artif. Cells Nanomed. Biotech.* 47: 154–165. https://doi.org/10.1080/21691401.2018.1548468.

19 Chen, Y., Zhang, H., Yang, J., and Sun, H. (2015). Improved antioxidant capacity of optimization of a self-microemulsifying drug delivery system for resveratrol. *Molecules* 20: 21167–21177. https://doi.org/10.3390/molecules201219750.

20 Pangeni, R., Sharma, S., Mustafa, G. et al. (2014). Vitamin E loaded resveratrol nanoemulsion for brain targeting for the treatment of Parkinson's disease by reducing oxidative stress. *Nanotechnology* 25: 485102. https://doi.org/10.1088/0957-4484/25/48/485102.

21 Sarker, M.R. and Franks, S.F. (2018). Efficacy of curcumin for age-associated cognitive decline: a narrative review of preclinical and clinical studies. *Geroscience* 40: 73–95. https://doi.org/10.1007/s11357-018-0017-z.

22 Salehi, B., Stojanović-Radić, Z., Matejić, J. et al. (2019). The therapeutic potential of curcumin: a review of clinical trials. *Eur. J. Med. Chem.* 163: 527–545. https://doi.org/10.1016/j.ejmech.2018.12.016.

23 Kumar, A., Ahuja, A., Ali, J., and Baboota, S. (2010). Conundrum and therapeutic potential of curcumin in drug delivery. *Crit. Rev. Ther. Drug Carrier Syst.* 27: 279–312. https://doi.org/10.1615/CritRevTherDrugCarrierSyst.v27.i4.10.

24 Flora, G., Gupta, D., and Tiwari, A. (2013). Nanocurcumin: a promising therapeutic advancement over native curcumin. *Crit. Rev. Ther. Drug Carrier Syst.* 30: 331–368. https://doi.org/10.1615/CritRevTherDrugCarrierSyst.2013007236.

25 Ahmad, M.Z., Akhter, S., Mohsin, N. et al. (2014). Transformation of curcumin from food additive to multifunctional medicine: nanotechnology bridging the gap. *Curr. Drug Discov. Technol.* 11: 197–213. https://doi.org/10.2174/1570163811666140616153436.

26 Shome, S., Talukdar, A.D., Choudhury, M.D. et al. (2016). Curcumin as potential therapeutic natural product: a nanobiotechnological perspective. *J. Pharm. Pharmacol.* 68: 1481–1500. https://doi.org/10.1111/jphp.12611.

27 Tsai, Y.M., Jan, W.C., Chien, C.F. et al. (2011). Optimised nano-formulation on the bioavailability of hydrophobic polyphenol, curcumin, in freely-moving rats. *Food Chem.* 127: 918–925. https://doi.org/10.1016/j.foodchem.2011.01.059.

28 Anand David, A.V., Arulmoli, R., and Parasuraman, S. (2016). Overviews of biological importance of quercetin: a bioactive flavonoid. *Pharmacogn. Rev.* 10: 84–89. https://doi.org/10.4103/0973-7847.194044.

29 Chitkara, D., Nikalaje, S.K., Mittal, A. et al. (2012). Development of quercetin nanoformulation and in vivo evaluation using streptozotocin induced diabetic rat model. *Drug Deliv. Transl. Res.* 2: 112–123. https://doi.org/10.1007/s13346-012-0063-5.

30 Spagnuolo, C., Russo, G.L., Orhan, I.E. et al. (2015). Genistein and cancer: current status, challenges, and future directions. *Adv. Nutr.* 6: 408–419. https://doi.org/10.3945/an.114.008052.

31 Patisaul, H.B. (2017). Endocrine disruption by dietary phyto-oestrogens: impacton dimorphic sexual systems and behaviours. *Proc. Nutr. Soc.* 76: 130–144. https://doi.org/10.1017/S0029665116000677.

32 Granja, A., Frias, I., Rute Neves, A. et al. (2017). Therapeutic potential of epigallocatechin gallate nanodelivery systems. *Biomed. Res. Int.* 5813793: https://doi.org/10.1155/2017/5813793.

33 Ravindran, S., Suthar, J.K., Rokade, R. et al. (2018). Pharmacokinetics, metabolism, distribution and permeability of nanomedicine. *Curr. Drug Metab.* 19: 327–334. https://doi.org/10.2174/1389200219666180305154119.

34 Bilia, A.R., Piazzini, V., Guccione, C. et al. (2017). Improving on nature: the role of nanomedicine in the development of clinical natural drugs. *Planta Med.* 83: 366–381. https://doi.org/10.1055/s-0043-102949.

35 Piazzini, V., Lemmi, B., D'Ambrosio, M. et al. (2018). Nanostructured lipid carriers as promising delivery systems for plant extracts: the case of silymarin. *Appl. Sci.* 8: 1163. https://doi.org/10.3390/app8071163.

36 Ganesan, P., Karthivashan, G., Park, S.Y. et al. (2018a). Microfluidization trends in the development of nanodelivery systems and applications in chronic disease treatments. *Int. J. Nanomedicine* 13: 6109–6121. https://doi.org/10.2147/IJN.S178077.

37 Ganesan, P., Ramalingam, P., Karthivashan, G. et al. (2018b). Recent developments in solid lipid nanoparticle and surface-modified solid lipid nanoparticle delivery systems for oral delivery of phyto-bioactive compounds in various chronic diseases. *Int. J. Nanomedicine* 13: 1569–1583. https://doi.org/10.2147/IJN.S155593.

8

Nanodelivery of Antioxidant Herbal Extracts, Spices, and Dietary Constituents

Shiv Kumar Prajapati[1], Dolly Jain[2,3], Sabnam Parveen[4], Siddhartha Maji[1], and Prashanta Kumar Deb[5]

[1] *Institute of Pharmacy, Ram-Eesh Institute of Vocational and Technical Education, Greater Noida, Uttar Pradesh, India*
[2] *Oriental College of Pharmacy and Research, Oriental University, Indore, Madhya Pradesh, India*
[3] *Adina College of Pharmacy, Sagar, Madhya Pradesh, India*
[4] *Birbhum Pharmacy School, Birbhum, West Bengal, India*
[5] *Department of Pharmaceutical Chemistry, School of Pharmaceutical Sciences, Shoolini University, Solan, Himachal Pradesh, India*

8.1 Introduction

During the last couple of decades, herbs and spices were extensively studied in the field of phytochemistry not only because of their availability or their flavor for the food but also their organoleptic properties, which process tremendous antioxidant activity and that makes them unique in the field of medicine. Antioxidants are chemicals that protect other molecules from oxidizing [1]. The major distinction between spices and herbs is that spices are derived from portions of a plant other than the leaves, whereas herbs are derived from the leaves. Herbs and spices, on the other hand, have been employed in a variety of ways. Spices and culinary herbs have been used to enhance flavor and increase organoleptic characteristics in food since ancient times. Spices and herbs have long been used as medicine and preservation. Spices and herbs, in addition to fruits and vegetables, might offer us additional natural antioxidants in our diet [2, 3]. Flavonoids, phenolic compounds, sulfur-containing compounds, tannins, alkaloids, phenolic diterpenes, and vitamins are among the many bioactive chemicals found in spices. These chemicals have various antioxidant properties. Flavonoids, for instance, can scavenge free radicals and form complexes with catalytic metal ions, making them inactive [4–6]. Spices and herbs strong in phenolic compounds, such as clove, turmeric, ginger, garlic, curcumin, coriander, black pepper, and oregano, have been demonstrated to be great sources of antioxidants in studies [7]. The use of nanocarriers for antioxidant delivery has attracted attention due to their ability to deliver various therapeutic agents, herbal extracts, etc. The nanocarriers show distinguished properties such as high loading capacity, targeting ability, the requirement of reduced dose, biocompatibility, and higher stability. Various nanocarriers show antioxidant activities or enhance the antioxidant activity by loading them into it. The chapter describes the various antioxidant herbs, spices, and mechanism of antioxidants.

Phytoantioxidants and Nanotherapeutics, First Edition. Edited by Mithun Rudrapal.

8.1.1 What Is an Antioxidant?

Antioxidant capability refers to chemicals that can protect a biological system from the potentially detrimental effects of reactive oxygen and nitrogen species during processes or reactions (ROS and RNS). Antioxidants' protective benefits have gotten a lot of interest in the biological, medicinal, nutritional, and agrochemical domains, which has necessitated the development of simple, convenient, and accurate antioxidant capacity measurement techniques [8, 9]. Natural antioxidants are in more demand than synthetic antioxidants, owing to safety concerns about synthetic antioxidants, as well as a growing customer desire for natural goods, clear labels, and reduced use of food additives in food items. Due to the intrinsic antioxidant components, spices and herbs are good sources of antioxidants for food preservation. Using natural antioxidants from spices and herbs has added benefits. These include their health advantages as well as the fact that they are easily absorbed by the body [10, 11]. Generally, the mechanism of natural antioxidants reacts with lipid radicals and converts them to more stable molecules leading to the breaking of the antioxidants chain. The majority of their structures are phenolic. Table 8.1 describes the mode of action of the antioxidant of an individual compound. Synthetic antioxidants include some phenolic compounds, which capture free radicals and prevent chain reactions. Such compounds are [12]:

1) Butylated hydroxyanisole (BHA)
2) Butylated hydroxytoluene (BHT)
3) Propyl gallate (PG)
4) Nordihydroguaiaretic acid (NDGA)
5) Tertiary butylhydroquinone (TBHQ)

In the literature, several techniques have been devised and evaluated that differ in terms of reaction processes, oxidant and target/probe species, reaction settings, and result in expression.

8.1.1.1 DPPH Assay

DPPH is a chemical and its IUPAC name is 1,1-diphenyl-2-picrylhydrazyl. It can capture the free electron donated by an antioxidant (Figure 8.1). This most widely reported DPPH assay method is based on the reduction of methanolic solution of colored free radical DPPH by a free radical scavenger. The procedure involves measurement of the decrease in absorbance of DPPH at its absorption maxima of 516 nm, which is proportional to the concentration of free radical scavenger added to the DPPH reagent solution. The activity is expressed as effective concentration EC_{50}. DPPH radical was calculated by the following equation [13, 14]:

$$\text{DPPH radical scavenging activity}(\%): 100-\left[\left(\text{Abs}_{\text{sample}}-\text{Abs}_{\text{blank}}\right)\times 100\right]/\text{Abs}_{\text{control}}$$

where, $\text{Abs}_{\text{sample}}$, absorbance of DPPH radical; $\text{Abs}_{\text{blank}}$, absorbance of sample; $\text{Abs}_{\text{control}}$, absorbance of DPPH radical.

8.1.1.2 ABTS Method

ABTS is also a chemical (2,2-azino-bis(3-ethyl benzothiazoline-6-sulfonic acid) diammonium salt). This is a measure of antioxidant activity. It also permits the distinction between additive and synergistic effects. The assay is based on the interaction between antioxidant and $ABTS^{+}$ radical cation which has a characteristic color showing maxima at 645, 734, and 815 nm. ABTS radical was calculated by the following equation [15, 16]:

$$\text{ABTS radical scavenging activity } (\%): [(\text{Abs}_{\text{control}}-\text{Abs}_{\text{sample}})/\text{Abs}_{\text{control}}]\times 100$$

where, $\text{Abs}_{\text{control}}$, the absorbance of ABTS; $\text{Abs}_{\text{sample}}$, the absorbance of sample + ABTS.

Table 8.1 Different herbs/spices with antioxidant compounds.

Spice/herb	Scientific name	Antioxidant compounds	Mode of action
Black pepper	*Piper nigrum* L.	Kaempferol, rhamnetin, quercetin	Free radical scavenger
Chili pepper	*Capsicum frutescence* L.	Capsaicin, capsaicinol	Free radical scavenger
Clove	*Eugenia caryophyllus*	Flavonol glucosides, gallic acid, volatile oil eugenol, acetyl eugenol, tannins	Free radical scavenger, metal chelator
Cumin	*Cumimum cymimum*	Cuminal, γ-terpinene, pinocarveol, linalool, 1-methyl-2-(1-methylethyl)benzene, carotol	Free radical scavenger, metal chelator
Ginger	*Zingiber officinale*	Gingerol, shogoal	Free radical scavenger
Mace (outer part of nutmeg)	*Myristica fragrans*	Myristphenone	Free radical scavenger
Marjoram	*Majorana hortensis*	β-Carotene, β-sitosterol, caffeic acid, carvacrol, eugenol, hydroquinone, myrcene, rosmarinic acid	Free radical scavenger
Nutmeg	*Myristica fragrans*	Myristphenone, phenolic volatile oils, caffeic acid, flavanols (catechin)	Free radical scavenger
Oregano	*Origanum vulgaris*	Caffeic acid, *p*-coumaric acid, rosmarinic acid, cavacrol, flavonoids	Free radical scavenger
Red pepper	*Capsicum annuum*	β-Carotene fruit, β-sitosterol plant, caffeic acid campesterol, capsaicin fruit, capsanthin fruit, eugenol fruit, hesperidin fruit, myristic acid	Free radical scavenger
Rosemary	*Rosemarinus officinalis*	Caffeic acid, carnosol, rosmanol, caffeoyl derivatives, carnosol, rosmarinic acid, phenolic diterpenes (carnosic acid), epirosmanol, flavonoids	Scavenge superoxide radicals, lipid antioxidants and metal chelator
Sage	*Salvia officinalis* L.	Rosmanol, epirosmanol, rosmarinic acid, phenolic diterpenes (carnosic acid), flavonoids	Free radical scavenger
Sesame seed	*Sesamum indicum*	Sesaminol, α-tocopherol, sesamol	Free radical scavenger
Turmeric	*Curcuma domestica* L.	Curcumin, 4-hydroxycinnamoylmethane	Free radical scavenger
Thyme	*Thymus vulgaris* L.	Phenolic acids (gallic acid, caffeic acid, rosmarinic acid), thymol, phenolic diterpenes, flavonoids	Free radical scavenger

8.1.1.3 Conjugated Diene Assay

By detecting UV absorbance at 234 nm, this approach permits dynamic measurement of conjugated dienes because of early polyunsaturated fatty acid (PUFA) oxidation. The double bonds in linoleic acid are transformed into conjugated double bonds during oxidation, which has a significant UV absorbance at 234 nm and is used in this experiment. The inhibitory concentration is used to measure the activity (IC_{50}) [17].

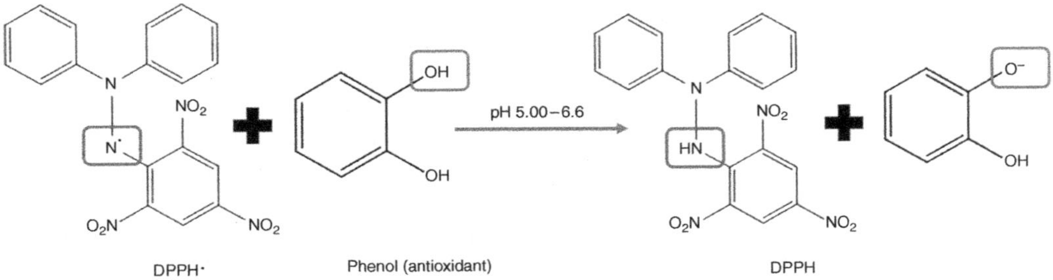

Figure 8.1 Reaction mechanism of DPPH with antioxidants.

8.1.1.4 Hydroxyl Radical Scavenging Activity

This approach includes employing the Fenton reaction to generate hydroxyl radicals in vitro using the Fe^{3+}/ascorbate/EDTA/H_2O_2 system. The amount of hydroxyl radical that is scavenged in the presence of an antioxidant is assessed. The hydroxyl radicals produced by oxidation are produced to react with dimethyl sulfoxide (DMSO) to create formaldehyde in one of the techniques. With Nash reagent, the produced formaldehyde displays a bright yellow hue (2 M ammonium acetate with 0.05 M acetic acid and 0.02 M acetylacetone in distilled water). The intensity of the yellow hue produced by that reaction is spectrophotometrically quantified at 412 nm against a blank reagent. The activity is measured in hydroxyl radical scavenging percentages [18].

8.2 Antioxidant Properties of Spices

8.2.1 Ginger

The subterranean stems or rhizomes of *Zingiber officinale*, which belong to the Zingiberaceae family, are used to make the spice ginger. Ginger rhizome is commonly used to flavor tea as a fresh paste and dried powder. Since ancient times, this plant's subterranean stem or rhizome has been utilized as a medicine in Asian, Indian, and Arabic herbal traditions [19]. Ginger tea has long been used throughout the world to treat coughs and asthma caused by allergies or inflammation along with upset stomach, rheumatic disorders [20], fevers, motion sickness, bronchitis, diabetes, cancer, etc. Ginger is a rubefacient and has traditionally been used to treat headaches and toothaches. Ginger has anti-inflammatory and anti-aggregator properties, and it lowers the risk of heart attack or stroke by lowering blood platelet clumping. The antioxidant properties of ginger are attributed to the components zingiberene, the primary terpenoid, and 6-gingerol (Table 8.2), the pungent alkaloid, according to research. Besides that, it has zingerone, shogaol derivatives, camphene, R-terpineol, farnesene, *p*-cineole, β-mycrene, pentadecanoic acid, and geranyl isobutyrate. The total phenols of the alcohol extract were found to be 870.1 mg/g dry [21] due to its capacity as donors of hydrogen atoms or electrons and to capture the free radicals. DPPH analysis is one of the tests used to prove the ability of the components of the ginger extract to act as donors of hydrogen atoms. The DPPH method showed the total antioxidant activity for Ginger reaches 90.1% of that of butylated hydroxytoluene (BHT). IC_{50} value for DPPH inhibition was found to be 0.641 g/ml [21].

8.2.2 Turmeric

Turmeric (*Curcuma longa*) is a Zingiberaceae rhizomatous medicinal plant native to India. Turmeric has been utilized in traditional medicine since the dawn of civilization. Turmeric and its value-added derivatives are produced, marketed, and exported in large quantities by India.

Table 8.2 The major chemical constituents found in spices and herbs which are biologically active.

Herbs and spices	Chemical constituents	Most active compounds	Structure
Ginger	Zingiberene, gingerol, zingerone, shogaol	Gingerol	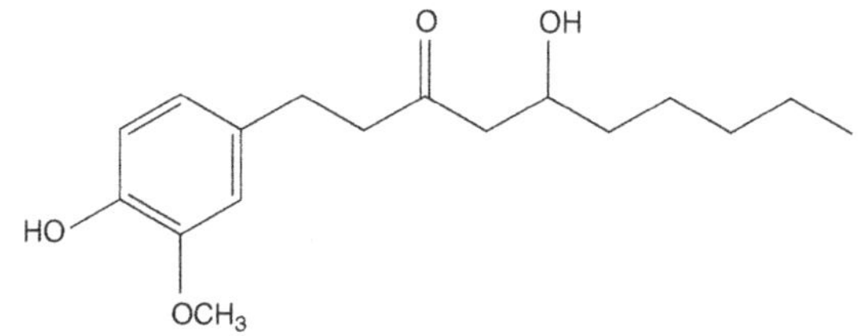
Garlic	Allicin, *S*-allyl-cysteine, diallyl sulphides	Allicin, *S*-allyl cysteine	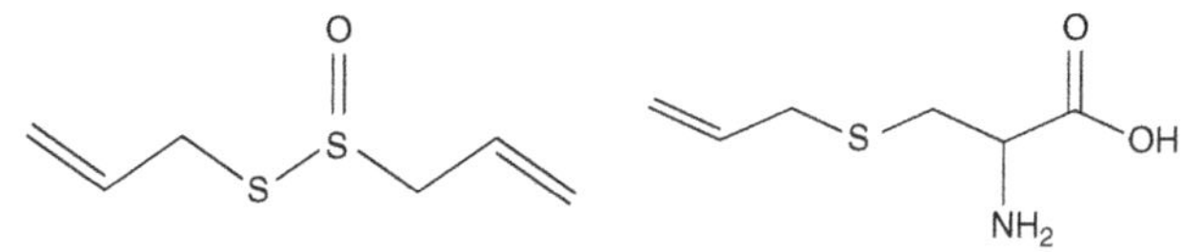
Turmeric	Curcumins, *p*-coumaric, ascorbic acid, syringic, vanillic acid	Curcumin	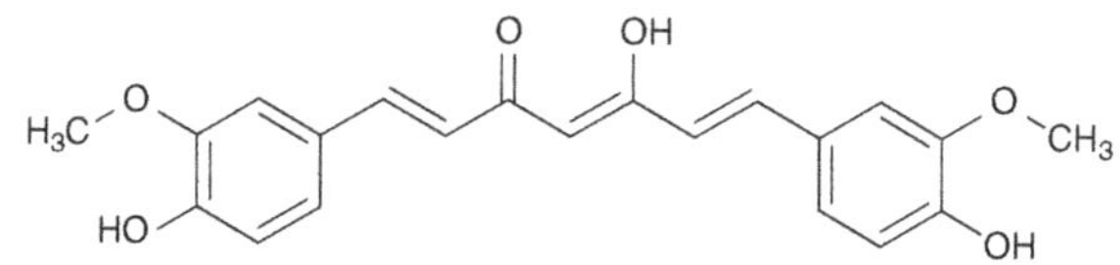
Black pepper	Piperine, pinene, camphene, terpenes, piperidine	Piperine	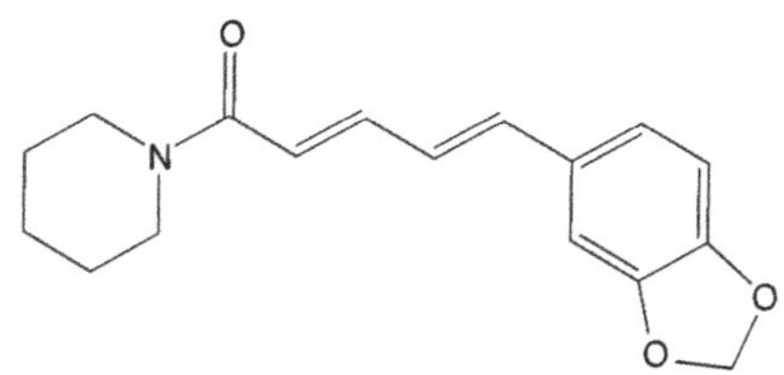
Coriander	Linalool, borneol, geraniol, terpineol, terpinene, quercetin	Inalool	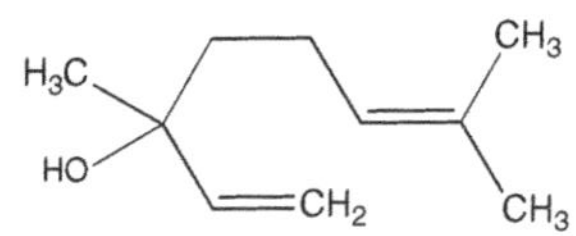

(*Continued*)

Table 8.2 (Continued)

Herbs and spices	Chemical constituents	Most active compounds	Structure
Onion	Quercetin, apigenin, dipyridyl disulfide, rutin, quercetin-4-glucoside	Quercetin	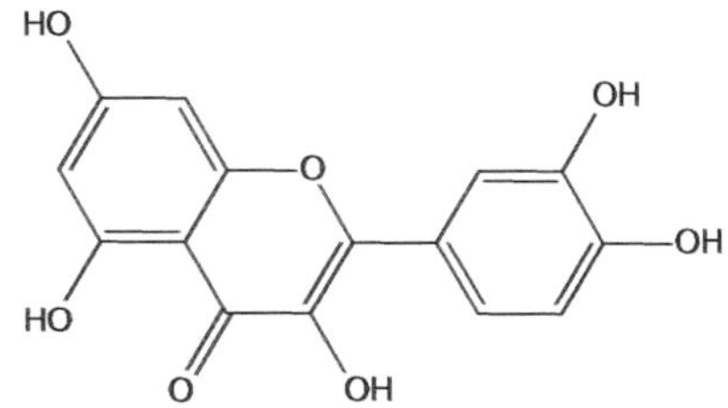
Clove	Euginol, caryophyllene, isoeugenol, eugenol acetate, 3-allyl-6-methoxyphenol	Euginol	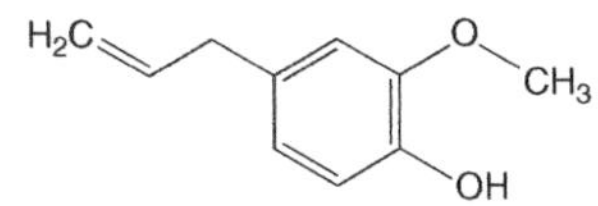
Oregano	Luteolin, quercetin, myricetin, apigenin, eriodictyol, carvacrol	Luteolin	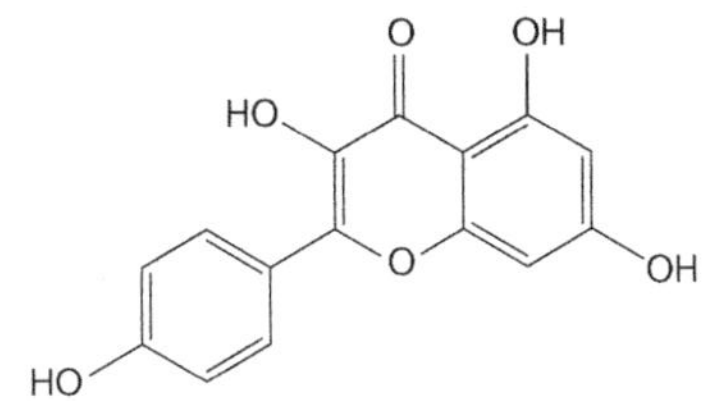
Cumin	Pinene, cymene, terpinene, cumin aldehyde, oleoresin	Cumin aldehyde	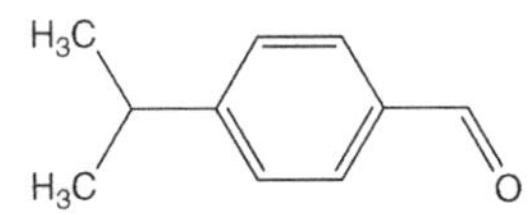
Cardamom	Cineole, α-terpineol, terpinyl acetate, D-limonene, sabinene, borneol	Cineole	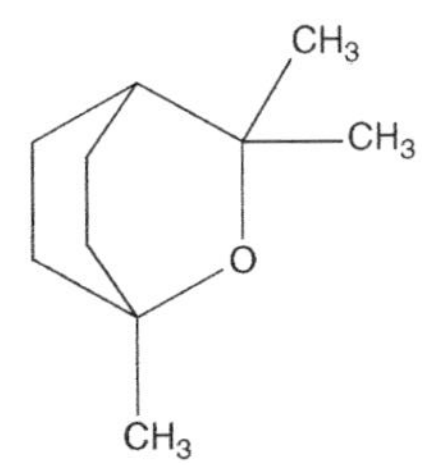

Turmeric is increasingly being used in India to treat anorexia, liver issues, cough, diabetic sores, rheumatism, and sinusitis. Turmeric extract was investigated for anticarcinogenic [22] and antimutagenic effects [22]. From in vivo experiments, it was found that turmeric reduced the number of tumors in mice [23]. Preventing cancer now receives the attention it has long deserved. Laboratory experiments have established that the active principle of turmeric (curcumin) is a potent antimutagenic agent by the theory of free-radical inactivation. Mainly curcumin (Table 8.2) is responsible for antioxidant activity, anti-inflammatory properties, anticancer activity, antimetabolic syndrome activities, neuroprotective activity, antimicrobial effects, anti-arthritic effects, antiviral effects, anti-asthma, antidiabetic effects, anti-obesity, cardio and liver toxicity protection activities, and anti-depression and anxiety activities. Turmeric has been widely used as a typical household treatment for cough, sore throat, and respiratory ailments and could be an effective immunity booster against SARS-CoV-2 therapy during the ongoing pandemic situation [24–26]. Another study concluded that since curcumin is a potent antioxidant, it may scavenge the epoxides and prevent binding to macromolecules. In other words, this spice's cell-protective properties are similar to nutrient antioxidants, vitamins C and E, which inhibit free radical reactions. This type of herb is known as a nonsteroidal anti-inflammatory drug (NSAID). Curcumin inhibits cyclooxygenase and lipoxygenase enzymes. It has three main mechanisms of action: (i) antioxidant activity; (ii) lipoxygenase inhibitor; and (iii) cyclooxygenase inhibition. The fresh juice from the turmeric rhizome will reduce swelling in recent bruises, wounds, and insect bites; and the dried powdered root kills parasites and relieves head colds and arthritic aches. A pharmacological review in 1991 confirmed many turmerics upregulate the rate of wound healing, gastric mucosa protection, antispasmodic activity, reduction of intestinal gas formation, protection of liver cells, increasing bile production, diminishing platelet aggregation [24] (i.e. blood clumping), lowering serum cholesterol (at very high doses), antibacterial properties, antifungal properties, and potential antitumor activity [27]. All these activities are due to the presence of curcumins, essential oils, eugenol, carotene, ascorbic acid, caffeic, *p*-coumaric, protocatechuic, syringic, vanillic acid in turmeric as a chemical constituent, which is a rich source of flavonoids, phenolic acids, lignans, essential oils, and alkaloids [28]. The total phenol content was found to be around 83 mg/g with total antioxidant activity of 9.6% via the DPPH method.

8.2.3 Onion

For the last couple of centuries, onion has been a well-known spice throughout the world and it has become medicinally important due to the presence of quercetin (Table 8.2), which is the most common flavonoid in higher plants. Besides this, a few glycosides such as rutin, isoquercitrin, quercitrin, hyperin, and quercimeritrin are also rich sources of flavonoids, present in onion as major chemical constituents. Quercetin is a powerful antioxidant that decreases the concentration of superoxide anions in enzyme and nonenzyme systems. A recent animal study demonstrated its antiulcer and gastro-protective effects, especially against ethanol injury [24]. The cytoprotective activity was affected through several interacting pathways involving stimulation of prostaglandin and inhibition of leukotriene production and through quercetin's antioxidant medicines. It possesses 4.6–74.1 mg/g GAE of total phenolic contents. In the year 2007, Panpatil et al. [28] performed an experiment to determine the antioxidant activity of various layers of an onion and the value is ranging from 13.6 to 84.1%. It has been concluded that the outermost layer of the onion, which goes to waste, is having most of the polyphenols, and is responsible for giving the highest amount of antioxidants. The free radical scavenging property of quercetin helps inhibit gastric carcinoma, and along with it reducing platelet aggregation, blood cholesterol level, and enhancing blood fibrinolytic activity [29, 30].

8.2.4 Garlic

Garlic (*Allium sativum* L.) originated from western Asia and the Mediterranean coast. Garlic has long been used as a spice and as a natural remedy. Garlic, which includes allicin, *S*-allyl-cysteine, diallyl sulfides, and other sulfur compounds, is known to have natural antioxidants that can eliminate reactive oxygen species (ROS) and decrease lipid peroxides and low-density lipoprotein (LDL) oxidation. Garlic also has numerous physiological effects and activities in numerous metabolic pathways since it includes allicin (Table 8.2), *S*-allyl-cysteine, and diallyl sulfides. Plant extracts' antioxidant properties are largely intertwined with phenolic-type chemicals [31–33]. In the year 2018, Jang et al. compared the qualitative test of aged and nonaged garlic with different solvents and conditions and concluded that regardless of solvents, flavonoid and phenolic contents of aged groups were significantly higher compared to those of nonaged groups. The flavonoid concentration of water extract from aged garlic was roughly 4.5-fold greater than that of nonaged garlic, which is why aged garlic, also known as black garlic, is catching customers' attention these days [34].

8.2.5 Coriander

Coriander (*Coriandrum sativum* L.) is a widely used medicinal herb with nutritional and therapeutic characteristics that belongs to the Apiaceae family. Coriander extracts and essential oils offer antibacterial, antioxidant, free radical, antidiabetic, anticancer, and antimutagenic properties [35]. Although all portions of the coriander plant are edible, their taste and purposes differ. The aerial portions of *C. sativum* have been the subject of the most reported investigations on antioxidant activity. In this investigation, the predominant volatile constituent of coriander essential oil was linalool (Table 8.2), which accounted for 66.07% of the total along with camphor 8.34%, geranyl acetate 6.91%, and cymene 6.35%. A couple of research papers based on the chromatographic techniques have reported the presence of chemical constituents, like linalool, terpinene, pinene, camphor, decanal geranyl acetate, limonene, geraniol, camphene, and D-limonene to be the main volatile compounds for seeds, while (*E*)-2-decenal, linalool, (*E*)-2-dodecenal, (*E*)-2-tetradecenal, 2-decen-1-ol, (*E*)-2-undecenal, dodecanal, (*E*)-2-tridecenal, (*E*)-2-hexadecenal, pentadecenal and pinene were the main compounds of the coriander essential oils obtained from leaves. According to Shahwar et al., the antioxidant activity of the coriander essential oil determined by DPPH was 66.48% [36–38].

8.2.6 Black Pepper

Black pepper is a popular spice that is prized for its scent as well as its pungency. Black pepper was said to have digestive properties, enhance appetite, and heal colds, coughs, dyspnoea, throat disorders, intermittent fever, colic, dysentery, worms, and piles. It also has broad-spectrum antibacterial properties, like many spices. Analgesic, antipyretic, and anti-inflammatory properties are also discussed, with piperine (Table 8.2) being one of the active molecules in these circumstances. Neither pepper nor piperine produces any hepatotoxicity. In contrast, a protective impact on key liver enzymes has been reported [39]. Pepper passed the Ames test with flying colors in terms of mutagenic and carcinogenic qualities. In Turkish folk medicine, black pepper seeds were also utilized as an appetizer medication. Piperine and its structural counterparts including isopiperine, chavicine, isochavicine, piperic acid, piperanine, piperettine, piperylin A, piperolein B, and pipericine, which have 47–53% of fiber, 11–14% of protein, and 10–13.5% of starch, are believed to be responsible for these features of black pepper [40, 41]. Besides, pepper contains a variety of chemical constituents, such as piperolides, propenylphenols, amides, neolignans, lignans, flavonoids, terpenes, and steroids, which are the main reasons to give antioxidant activity [42]. The radical scavenging activity was measured using the DPPH technique, and the total antioxidant was found to be in the range of 48–79% [43].

8.2.7 Clove

Clove (*Eugenia caryophyllata*) is the aromatic flower bud of a tree that belongs to the family Myrtaceae. It is native to Indonesia but nowadays it is found in many tropical plant countries. It is cultivated throughout the year according to weather conditions. Clove is generally used as a spice in African, Mediterranean, Asian, and Middle East countries for adding flavor to meats and curries [44]. During the last decades, researchers started taking interest in clove along with some other spices because they got to know that spice helps preserve food. During World War II, Clove earned increasing popularity as a powerful means for extending the shelf-life of products and for decreasing the nutritional losses by preventing or slowing the oxidation process, which was eventually used to help military ration to be useable for a longer period [45]. Besides all these, humans have used clove oil for centuries, as an anesthetic for toothaches, headaches, and joints pain [46]. Due to the presence of various biochemical constituents. Clove oil is used as a mosquito and ant repellent. Recently, it has been published that clove has been used against cancer and all because of eugenol (Table 8.2), caryophyllene, isoeugenol, eugenol acetate, and 3-allyl-6-methoxyphenol, which are the major chemical constituents, which are rich sources of flavonoids and polyphenols, and provide powerful antioxidant property than any other spices. Few other chemical constituents like sesquiterpene, pinene, vanillin, and gallic acid are also important for the antioxidant activity [47, 48]. Clove has the highest amount of flavonoids (366.5 mg/100 g) [48]. Clove oil was found to have the greatest radical scavenging inhibitory activity via the DPPH method, 91.2% at 0.5 μg/ml. It also showed 93.7% hydroxyl radical scavenging activity [49].

8.2.8 Oregano

Oregano (*Origanum vulgare*) belongs to the Lamiaceae family and is one of the most renowned spices in the world. Oregano is derived from the Greek ancient words "*Oros*" and "*Ganos*," which mean "mountain beauty." Oregano has been used since the dawn of time, dating back to 50 000–70 000 BCE [50]. It is primarily used as a flavoring herb in many cuisines around the world, especially in the Mediterranean, where it grows naturally. Throughout history, however, each culture has used it for a variety of purposes, including wedding decorations, cures for different illnesses, perfumes, and snuff additives. The first scientific paper on oregano bioactivity was published in 1945, according to the PubMed database [51]. Since then, there has been a significant increase in interest in oregano bioactivity, which has coincided with advances in phytochemical studies that have culminated in the isolation and identification of various bioactive substances from Origanum. European oregano is used in salads, stews, sauces, and soups, as well as in meat and meat-like items such as sausages. The leaves and essential oils are also used in fruit, cosmetics, and alcoholic liqueurs. Mexican and Indian oreganos are also used as food flavorings and are considered medicinal plants as antimicrobial, antitumour, antifungal, and anti-inflammatory drugs, antidiabetic agents, and for antispasmodic activities [52, 53]. All these activities of oregano are because of its powerful chemical constituents like luteolin (Table 8.2), quercetin, myricetin, apigenin, eriodictyol, carvacrol, thymol, diosmetin, protocatechuic acid, caffeic, and rosmarinic, which are the rich sources of flavonoid in which luteolin is present in the highest quantity 1028.7 mg/100 g, naringenin 372.0 mg/g, eriodictyol 85.3 mg/g, quercetin 42.0 mg/g, and apigenin 17.7 mg/g. In oregano, the amount of rosmeric acid was found to be 2562.7 mg/100 g, which is total phenol contain helps for the antioxidant activity and oxygen radical absorption capacity (ORAC). The total antioxidant activity of oregano was found to be around 42% via the DPPH radical scavenging inhibitory activity method [48, 53, 54].

8.2.9 Cumin

Cumin (*Cuminum cyminum*) belongs to the family Apiaceae. Cumin seeds are used as popular aromatic herbs and culinary spices, used for the preparation of food. Cumin is a native spice of Egypt, Turkistan, and the East Mediterranean, and it is widely produced in Iran, China, India, Morocco, South Russia, Japan, Indonesia, Algeria, and Turkey [55]. The cumin seeds have a strong aromatic scent and a warm bitterish taste. In India, cumin seeds are utilized in curry powder and spice mixtures. Cumin (*Cuminum cyminum*), black cumin (*Nigella sativa*), and bitter cumin are the three types of cumin grown in India (*Cinnamomum nigrum*). Cumin (*C. cyminum*) is belonging to the family of Umbelliferae, mostly cultivated in Arabia, India, and China. The Ranunculaceae family includes black cumin (*N. sativa*). It is extensively dispersed in Mediterranean, Central European, and Western Asian countries. The Apiaceae family includes bitter cumin (*C. nigrum*). Central Asia, Iran, Pakistan, and northern India are among the countries where it is grown. Bitter cumin seeds are brownish-black in color and form, yet they have a bitter flavor. In India, black cumin is known as "shahi jeera" [56]. Cumin seeds include fixed oil, volatile oils, acids, essential oils, protein, and other components, according to proximate analysis. Pinene, cymene, terpinene, cumin aldehyde, oleoresin, thymol, cuminic alcohol, and other compounds found in cumin have been demonstrated to be effective against a variety of ailments [57]. It provides energy, boosts the immune system, and protects against a variety of ailments. Besides giving flavor to food, cumin seeds have also been utilized in traditional medicine for a variety of ailments. Toothaches, dyspepsia, diarrhea, epilepsy, and jaundice are just a few of the ailments and diseases that may be treated with it. According to recent studies, cumin has been used to treat a variety of severe diseases, including cancer, hypolipidemia, and diabetes [58, 59]. Cumin aldehyde (Table 8.2) is a powerful antioxidant, along with γ-terpinene, pinocarveol, linalool, 1-methyl-2-(1-methylethyl) benzene, and carotol, which also show an antioxidant property. The total antioxidant activity of oregano was found to be around 32.7% via the DPPH radical scavenging inhibitory activity method and the total phenol content was found to be 49.5 mg/g [1].

8.2.10 Cardamom

Cardamom is a remarkable spice made from the seeds of a perennial plant (*Elettaria cardamomum*). Cardamom is a spice that comes from India's coast. Guatemala, Tanzania, Sri Lanka, Vietnam, Laos, and Cambodia are now growing it. Dried cardamom is mostly exported from India. The spice cardamom is referred to as the "Queen of Spices" [60]. It is one of the world's most expensive and unusual spices. It is a tropical perennial herb plant that belongs to the Zingiberaceae family of ginger plants. Generally, there are two types of cardamoms, one is small green cardamom (*Eletteria cardamomum*) and the other one is large red/black cardamom (*Amomum subulatum* Roxb). In India, cardamom is known as "elaichi" [60]. The seeds of cardamom have a lovely scent as well as a warm, somewhat spicy flavor. Cardamom is mostly used in home cooking and medicine across the world. The volatile oil and the chemical constituents in cardamom are responsible for its aromaticity and therapeutic effects present in it like cineole (Table 8.2), α-terpineol, terpinyl acetate, D-limonene, sabinene, and borneol. Limonene, terpinolene, myrcene, caffeic acid, quercetin, kaempferol, luteolin, and pelargonidin are potential flavonoids and polyphenols which help to protect against cellular oxidation by scavenging free radicals. The IC_{50} value for DPPH scavenging activity was found to be 1.483 μg/ml [1, 61].

8.3 Nanodelivery of Herbal Antioxidant Extracts, Spices, and Dietary Constituents

Nanocarriers have been acclaimed as a novel method for delivering herbal antioxidants with improved properties. Because some of the extracts are volatile in nature [62] (e.g. essential oils), encapsulating them in nanocarriers reduces the rate of evaporation while simultaneously increasing antioxidant activity by preserving them from oxidation and heating [63, 64]. Various nano-drug delivery vehicles such as polymeric NPs, metallic NPs, CNTs, micelles, nanoemulsions, etc. [65–67] are being explored for the delivery of extracts of the herbs, spices, and dietary constituents. NPs are functionalized with herbal antioxidants where NPs act as the carrier or antioxidant delivery carrier to boost the antioxidant activity and provide targeted delivery of antioxidants that show poor permeation across cell membranes and cell internalization. Various nanocarriers for the delivery of herbal antioxidant extracts, spices, and dietary constituents are displayed in Figure 8.2.

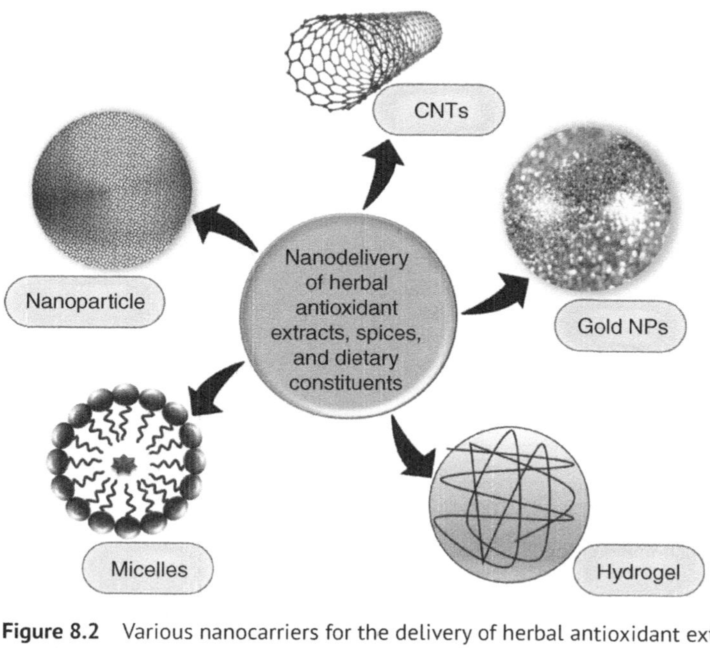

Figure 8.2 Various nanocarriers for the delivery of herbal antioxidant extracts, spices, and dietary constituents.

8.3.1 Nanoparticles

Oliveira et al. [68] successfully developed biodegradable NPs of poly(ε-caprolactone) (PCL). Quercetin and biapigenin were extracted to encapsulate into PCL NPs to assess their antioxidant potential. The values of free radical scavenging activity for quercetin–biapigenin extract (5.95 ± 0.97 μg/ml) and quercetin–biapigenin-PCL NPs (5.73 ± 1.20 μg/ml) were almost identical. While the iron-chelating activity of quercetin–biapigenin-PCL NPs was significantly higher (23.50 ± 0.55 μg/ml) than that of the quercetin–biapigenin extract (11.56 ± 0.44 μg/ml). The long-term liver toxicity was evaluated on HepG2 cells, the outcomes of the study revealed that the developed nanoformulation significantly reduced the intrinsic toxicity of the compound even at higher concentrations while in the case of extract, the cell viability was found to be decreased by 30 and 80% at the concentration of 25 and 100 g/ml, respectively. The data obtained confirm the synergistic antioxidant potential of nanoformulation [68]. Hadidi et al. [69] developed chitosan NPs for the delivery of clove essential oil (CEO) to evaluate their antioxidant and antimicrobial efficacies. Chitosan NPs showed good retention of CEO ranging from 55.8 to 73.4% confirming the good loading efficiency of NPs. The antioxidant activity of CEO-loaded chitosan nanoparticles was higher than free CEO. The 2,2-diphenylpicrylhydrazyl (DPPH) radical scavenging assay was performed to evaluate the antioxidant activity. The scavenging activity of free CEO was from 15.4 to 60.4%, and for encapsulated CEO, DPPH radicals ranged from 5.9 to 71.8% (Figure 8.3). The chitosan NPs showed significantly higher antioxidant activity compared to free CEO [69].

8.3.2 Metallic Nanoparticles

Metallic NPs have been explored for the delivery of not only for drug and gene delivery but also for delivery of herbal and dietary constituents containing antioxidant molecules and are considered a novel approach of antimicrobial therapeutic agents for the prevention and reduction of deterioration of food and pathogenic microorganisms. Physiochemical properties, high stability, high

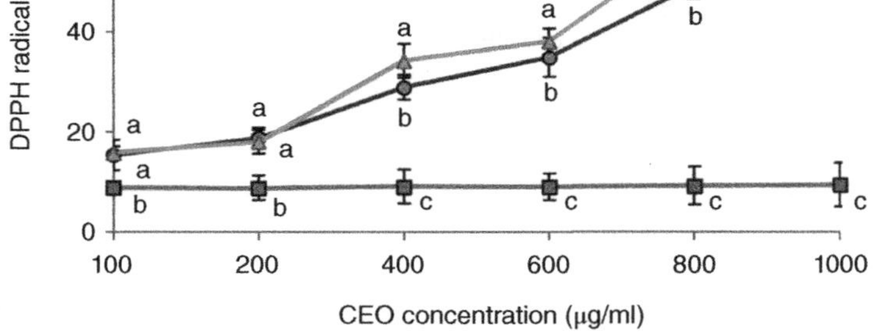

Figure 8.3 DPPH radical scavenging activity (%) of chitosan nanoparticles, CEO-loaded chitosan nanoparticles with chitosan/CEO ratio of 1 : 0.5, and CEO. *Source:* Hadidi et al. [69]/with permission of Elsevier.

reactivity, and photothermal and plasmonic properties of metallic NPs make them potent carriers as therapeutic agents [70]. Novel metal nanoparticles (Au, Ag, and Pt) and transition metal oxide (CuO and NiO) are commonly used and tested for their antioxidant activity. However, the antioxidant properties of these nanoparticles depend on their nature, chemical composition, surface charge, particle size, surface-to-volume ratio, and surface coating. The conjugation of desired/target molecules to metals, especially NPs in aqueous media, might provide an enhanced activity, increased half-life, enhanced stability, and possible resistance to metabolic processes. Singh et al. [71] worked to develop AuNPs by direct reduction of chloroauric acid ($HAuCl_4$) and functionalized with curcumin. The results showed that the curcumin function as a capping and reducing agent stabilized the gold solution. The antioxidant property is assessed by free radical DPPH. The deionized water was used as a control and did not show any quenching effect. The results showed that the antioxidant property was found to be increased when AuNPs were developed by curcumin and the antioxidant activity was progressively enhanced. The antioxidant activity of AuNPs-curcumin was found greater than curcumin [71]. Otunola and Afolayan [72] developed the AgNPs via the green synthesis method using a blend of three spice extracts (garlic, ginger, and cayenne pepper) in equal ratios. The color of the AgNPs changed from yellow to brown within an hour confirming the formation of NPs. The AgNPs displayed higher potential of inhibition on DPPH radical with IC50 < 68.75 and 31.25 μg/ml for ABTS when compared with rutin (control). The significant radical scavenging activity was in a dose-dependent manner. The higher inhibition of DPPH radicals was observed by AgNPs than the ABTS. The improved antioxidant behavior was due to the functional group expression by the use of spice blend [72]. The metallic NPs of silver (Ag), copper (Cu), iron (Fe), and zinc (Zn) were developed by El-Refai et al. [73] using ethanolic (E) and water (W) extract of ginger (Gin) and garlic (Gar) to evaluate their antimicrobial and antioxidant potential. The antioxidant potential of NPs was evaluated by free radical scavenging $ABTS^+$ and DPPH assays. The NPs of ginger extract (GinE, GinE+AgNp, GinE+ZnNp, GinE+FeNp, GinW, and GinW+AgNp) showed significant antioxidant efficacy than that of the garlic extract NPs. Among all these, GinE+ZnNp displayed radical scavenging activity 77.31% via DPPH assay and 63.90% via ABTS assays. The radical scavenging activity of garlic extract formulations (GarE, GarE+FeNp, GarW, GarW+ZnNp, and GinW+FeNp) was in the range of 56.09–69.60% and 40.61–57.89% through DPPH assay and ABTS assay, respectively. NPs of both extracts showed the antioxidant effect, but the outstanding potential of ginger ZnNPs might be due to the presence of higher phenolic and flavonoid concentrations. The values of each NPs formulation are displayed in Table 8.3 [73].

8.3.3 Micelles

Gorinova et al. [74] preferred Pluronic® P 123 or Pluronic® F 127 triblock copolymers for the development of micelles. The curcumin was encapsulated in the micelles and the higher encapsulation was observed in micelles of Pluronic® P 123. So, these were preferred for further antioxidant and neuroprotective investigation. Fe^{2+}/AA-induced lipid peroxidation model was selected to assess the in vitro antioxidant potential of curcumin loaded into micelles. After Fe^{2+}/AA treatment, a remarkable threefold raise was observed in the amount of malondialdehyde (MDA) vs. untreated controls. MDA is a naturally occurring marker for oxidative stress. Fe^{2+}/AA-induced lipid peroxidation in rat liver microsomes was observed when an equal concentration of both curcumin and curcumin micelles was used. Naked curcumin significantly diminished the Fe^{2+}/AA-induced lipid damage by 39%. Curcumin-loaded micelles exhibited identical protection against lipid damage; especially it dampened the MDA formation by 47%. Likewise, curcumin loaded in Pluronic® F

Table 8.3 DPPH and $ABTS^{+•}$ antioxidant assays for garlic and ginger extracts and their nanoparticles.

Extract and NPs	Total polyphenols (mgGAE/g)	Total flavonoids (mgQE/g)	Radical scavenging activity (RSA) (%)	
			DPPH	$ABTS^{+•}$
GarE	52.55^{g}	17.21^{d}	67.5^{h}	55.98^{o}
GarE+AgNps	25.36^{p}	11.56^{j}	57.77^{n}	45.13^{t}
GarE+ZnNps	40.39^{j}	12.81^{i}	55.52^{p}	42.09^{u}
GarE+CuNps	13.46^{r}	10.68^{k}	45.4^{t}	35.89^{y}
GarE+FeNps	39.54^{k}	16.53^{e}	69.6^{f}	57.89^{n}
GarW	24.44^{q}	12.32^{i}	62.54^{j}	50.47^{r}
GarW+AgNps	34.38^{m}	13.4^{h}	49.91^{r}	39.99^{w}
GarW+ZnNps	32.22^{n}	12.76^{i}	56.09^{o}	40.61^{w}
GarW+CuNps	35.55^{l}	12.11^{i}	55.21^{p}	41.81^{v}
GarW+FeNps	29.41^{o}	12.29^{i}	48.05^{s}	38.3^{x}
GinE	56.9^{d}	15.21^{f}	72.14^{d}	60.2^{l}
GinE+AgNps	55.14^{e}	16.3^{e}	74.18^{b}	62.05^{j}
GinE+ZnNps	63.44^{b}	21.31^{a}	77.31^{a}	63.9^{i}
GinE+CuNps	45.07^{i}	12.68^{i}	50.62^{r}	40.63^{w}
GinE+FeNps	54.22^{f}	18.44^{c}	73.11^{c}	61.08^{k}
GinW	67.09^{a}	19.31^{b}	74.19^{b}	62.06^{j}
GinW+AgNps	52.26^{g}	16.31^{e}	71.78^{e}	59.87^{m}
GinW+ZnNps	49.84^{h}	12.59^{i}	60.55^{l}	48.66^{s}
GinW+CuNps	55.39^{e}	12.82^{i}	62.61^{j}	51.53^{q}
GinW+FeNps	61.8^{c}	14.63^{g}	68.29^{g}	56.7^{o}

"a" represents the highest values and descending to "z" represents the lowest values.
Source: El-Refai et al. [73]/Springer Nature/CC BY 4.0.

127 micelles reduced the formation of MDA by 54%. Thus, the results confirm the potential of micelles to preserve the antioxidant activity of curcumin [74]. To enhance the therapeutic potential of curcumin, Akbar et al. [75] developed the polymeric micelles using chitosan. The formulation was evaluated for its anticancer, antioxidant, and antimicrobial potential. DPPH scavenging activity of various micellar formulations was evaluated, where it was evident that curcumin-loaded polymeric micelles encapsulating curcumin unveiled notable antioxidant potential more than pure curcumin. The MTT assay of formulations on HeLa cells showed significant anticancer potential even at low concentrations. Similarly, the antimicrobial potential of curcumin was not as significantly effective as that of the micellar formulation [75]. Caffeic is a naturally occurring chemical compound and is obtained from various plans. It shows broad biological activity such as antioxidant, anticancer, and anti-inflammatory. For the delivery of caffeic acid phenethyl ester (CAPE), Yoncheva et al. [76] formulated micelles using block copolymer poly(ethylene oxide)-b-poly(ε-caprolactone)-b-poly(ethylene oxide) (PEO-b-PCL-b-PEO). The encapsulation of CAPE was done by the dialysis method. The activity of the developed formulation was evaluated on

HepG2 and SH-SY5Y cells. The micellar formulation showed better protective activity against oxidative damage at a very low concentration (0.1 μg/ml) when compared with CAPE alone. When micellar CAPE (71 μg/ml) (Figure 8.4) was incubated at the highest concentration with the neuronal SH-SY5Y and HepG2 cells, it showed moderate antioxidant activity with 56 and 35% increase in cell viability [76].

8.3.4 Carbon Nanotubes

Carbon nanotubes (CNTs) have been explored for herbal drug delivery. CNTs are considered unique delivery carriers because of properties such as high surface functionality, high encapsulation potential, and high aspect ratio [77]. It has been confirmed that CNTs are tremendous free-radical scavengers and can escalate the thermostability and photooxidative stability of polymer matrices [78, 79]. The radical scavenging activity of CNTs is associated with their peculiar structure, and the CNTs are good acceptors and elector donors and are very reactive toward free radicles due to their paradigm of conjugated double bonds [78]. Even though CNTs are able to donate or accept electrons via the mechanism of electron transfer, their acceptor-like electron properties due to the lattice defect make CNTs for good antioxidant activity [80]. The nano-hybrid of quercetin and CNTs was developed by Arrigo et al. [81]. The studies exhibited that the CNTs showed notable stabilizing action attributable to firm physical interaction between quercetin and CNTs. Certain carbon atoms change the hybridization from sp2 to sp3 on the surface of CNTs and cause lattice defects. Moreover, upon immobilization of quercetin on CNT surface, the quercetin fragments exerted antioxidant action at the interface of CNT and polymer ultra-high-molecular-weight polyethylene (UHMWPE) interface and that interface reflected the critical zone of degradation and thus enhanced the contact between quercetin and UHMWPE, resulting in improved usefulness of antioxidant potential of quercetin [81]. Nichita and Stamatin [82] successfully functionalized flavonoids quercetin (Q) and rutin (R) to hydroxylated single-walled CNTs (SWCNTs-OH) and carboxylate SWCNTs (SWCNTs-COOH). The chemiluminescence assay evidenced that the CNTs have the potential antioxidant capacity (as electron donor or proton donor) in the redox transition reaction with free radicals. The biohybrids, i.e. SWCNT-COOH-R, SWCNT-COOH-Q, SWCNT-OH-R, and SWCNT-OH-Q showed 3- to 4-folds more antioxidant activity compared to single flavonoids. The scavenging rate of different formulations is shown in Figure 8.5. The order of antioxidant activity of blank CNTs was SWCNT < SWCNT-OH < SWCNT-COOH [82].

8.3.5 Solid Lipid Nanoparticles

SLNs are submicron carrier systems ranging from 50 to 1000 nm in size. These are generally fabricated of biocompatible and biodegradable lipids, which enable the encapsulation of lipophilic and lipophobic drugs. Singh et al. [83] worked on to fabricate Sesamol (S)-loaded SLNs (S-SLN) for the delivery of sesamol to evaluate it's subacute hepatotoxicity. The S-SLN was prepared by microemulsification method. Hepatoprotective ability was found to be improved in the case of S-SLN compared to other treatment groups. The MDA level was found to be decreased by 1.77 folds when treated with S-SLN. SILY showed more reduction because it was administered in three times higher doses. Superoxide dismutase (SOD), an enzymatic antioxidant assay, showed that the CCl4 and Bla-SLNs CCl4 group significantly reduced the level of SOD. Significant upregulation in SOD level was seen in the S-SLN-treated group than FS and SILY groups. Similarly, GSH level was increased than in other treatment groups. The results confirmed the better antioxidant activity of SLN with the improved hepatoprotective potential [83] for the delivery of rosmarinic acid, which

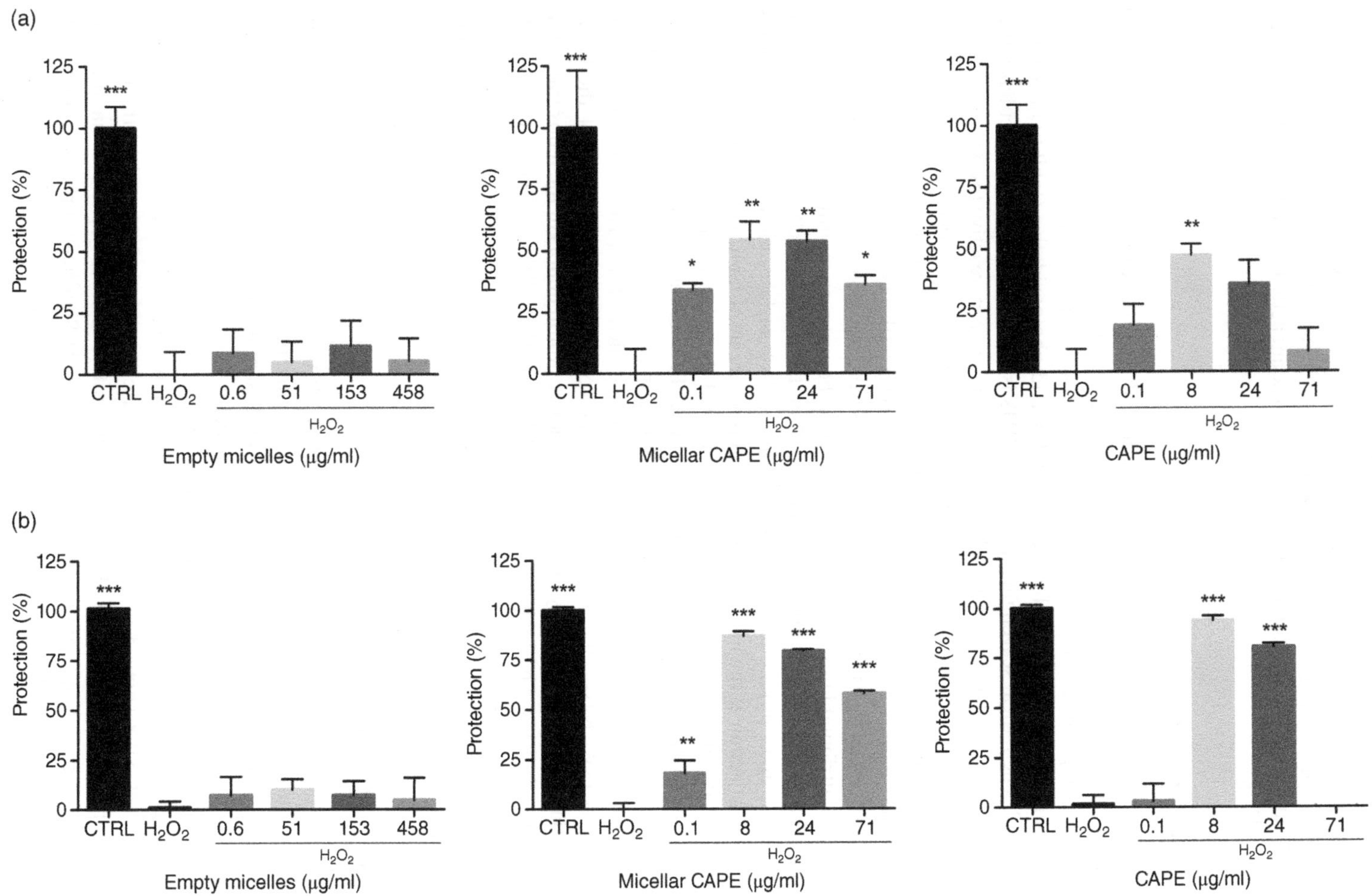

Figure 8.4 Protective effects of micellar CAPE and pure CAPE in a model of H_2O_2-induced oxidative damage in HepG2 (a) and SH-SY5Y (b) cells.
Source: Yoncheva et al. [76]/Taylor & Francis/CC BY 4.0.

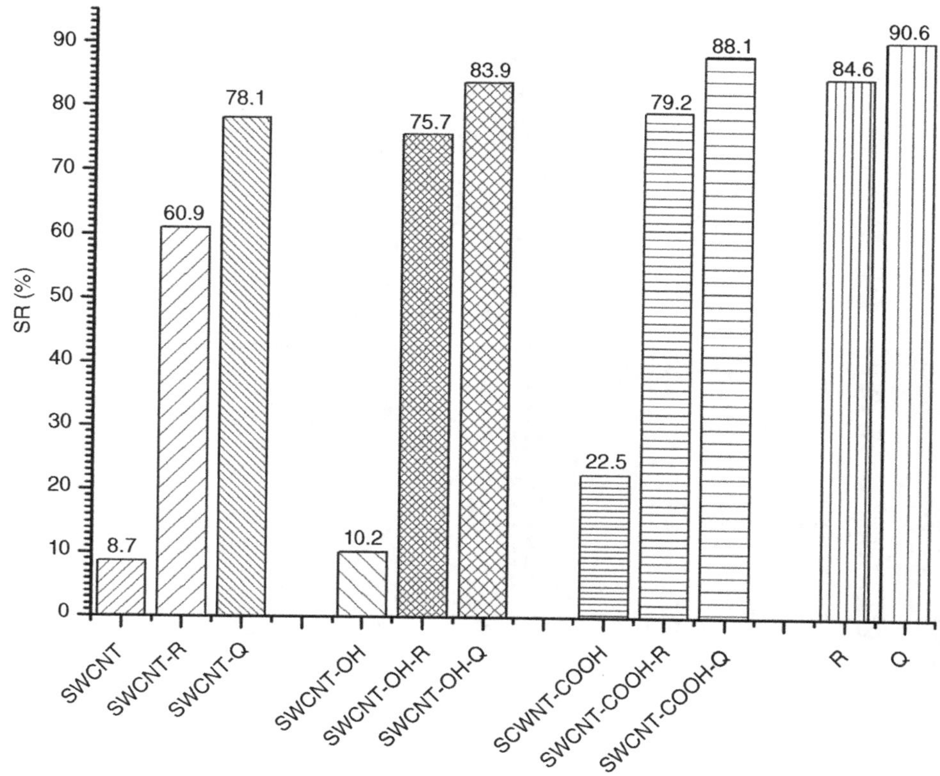

Figure 8.5 The scavenging rate for SWCNTs and biohybrids with quercetin and rutin. *Source:* Nichita and Stamantin [82], licensed under CC BY 4.0.

is rich in the sage extract was delivered by SLNs. The results of the experiment designed by Madureira et al. [84] evaluated antioxidant activity, genotoxicity, cytotoxicity, and redox status. The results concluded that rosmarinic acid or rosmarinic acid-loaded SLNs were harmless at a low dose (0.15 mg/ml) but as the dose increased (1.5 mg/ml), it led to genotoxicity (necrosis). An antioxidant effect was observed in the liver tissue, and slight variations occurred in the composition of gut bacterial microbiota and metabolism [84].

8.3.6 Nanoemulsions

Nanoemulsions are also termed submicron emulsions, miniemulsions, fine-dispersed emulsions, etc. The volume of oil in O/W-type nanoemulsions ranges from 5 to 20% w/w. To enhance the solubilization drug, the oil may be used in the combination or the cosolvent or cosurfactant may be used to simplify the process of stabilization [85, 86]. The nanoemulsions are preferred because of their characteristics such as greater transparency and stability, improved physicochemical properties, and taste and smell masking ability of the core material and so enhance organoleptic properties and help improve biological activity due to the high surface area [87, 88]. Neem oil is the deep yellow extract of Neem, which shows good antioxidant activity. Rinaldi et al. [89] worked to explore the successful delivery of neem oil via nanoemulsion. Free radical scavenging activity of formulations was evaluated by different methods, i.e. the antioxidant and

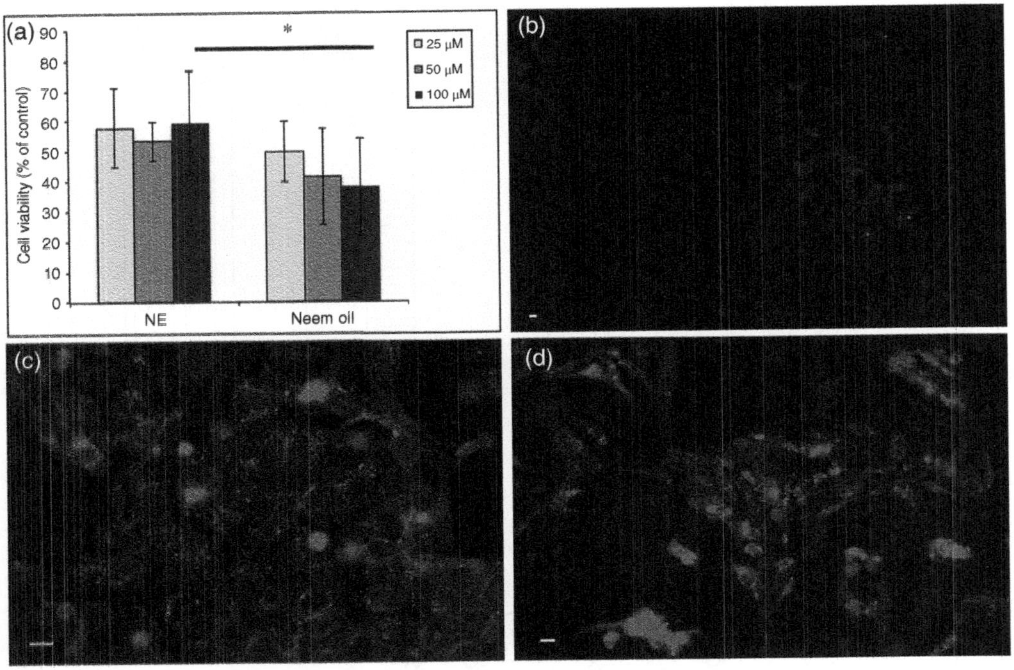

Figure 8.6 (a) Cell cytotoxicity study of neem oil and with nanoemulsion by MTT assay. Fluorescence microscopy HEp-2 control cells; (b) Free Nile red-treated; (c) Nile red-loaded NEs for seven hours; and (d) Nile red-loaded NEs up to 24 hours. *Source:* Rinaldi et al. [89]/Taylor & Francis Group/CC BY 4.0.

chelating activity was analyzed by DPPH. ABTS (2,2-azino-bis(3-ethylbenzothiazoline-6-sulfon ic acid)) radical cation scavenging activity and phosphomolybdenum assay; cupric ion-reducing (CUPRAC) method, ferric reducing antioxidant power (FRAP) method and metal chelating activity on ferrous ions. Nanoemulsion containing neem oil showed significantly high antioxidant activity than that of neem oil. Neem oil also shows its anticancer potential via caspase-dependent apoptosis. Due to this, the author accessed the cytotoxic potential and their uptake of formulation on HEp-2 human laryngeal carcinoma cells. Cell cytotoxicity study of neem oil and with nanoemulsion was evaluated by MTT assay. The remarkable outcomes were observed and the cytotoxicity was in dose-dependent manner (Figure 8.6). Similarly the cellular uptake by nanoemulsion confirmed their potential for the delivery of a bioactive compound by nanoemulsion [89]. *Achyrocline satureioides* is a species of a plant abundant with quercetin which shows good antioxidant potential. For this, Zorzi et al. [90] developed nanoemulsion by the spontaneous method using the extract (E) (NEE) of achyrocline satureioides or with quercetin (Q) (NEQ). NEE showed a higher accumulation of quercetin in porcine skin compared to NEQ. The antioxidant potential was assessed by thiobarbituric acid-reactive species (TBA-RS) assay. The protection against lipoperoxidation was found to be 8.3 and 72.6% for quercetin solution and NEQ, respectively. The production of TBA-RS was quantified and a reduction in the TBA-RS production was observed in the skin treated with NEE when compared with both NEB (Blank NE) and NEQ treatments [90]. Nanocarrier-based antioxidant potential of herbal and spice constituents is depicted in Table 8.4.

Table 8.4 Nanocarrier-based antioxidant potential of some extracts of herbs and spices.

Nanocarrier	Herb/spice/dietary constituents	Remarks	Reference
Nanoparticles	Ferulic acid	FA-PLGA NPs showed greater anticancer potential by declining cell proliferation, antioxidant behaviors, boosting ROS intracellularly, change in mitochondrial membrane potential, lipid peroxidation, and apoptosis.	[91]
Nanoparticles	Ferulic acid	FA-treated ME-180 cells showed escalated ROS levels intracellularly hanging up the cell growth, enhanced lipid peroxidation, and extremely induced cell apoptosis.	[92]
Nanoparticles	Curcumin	Curcumin-loaded NPs provocatively boosted the antioxidant and antihepatoma activities of curcumin.	[93]
Nanoemulsion	Clove essential oil	Clove oil-loaded nanoemulsion (CON) showed greater antioxidant activity than clove oil. The CON showed fourfolds more cellular toxicity compared to the clove oil	[94]
Nanoemulsion	Cumin essential oil	Cumin essential oil (CEO) nanoemulsion confirmed improvement in in vitro and in vivo antioxidant potential together with hepatoprotective ability.	[95]
Gold NPs	Lotus extract	AuNPs exhibited more significant DPPH radical scavenging potential than the flavonoid extract. Mild to low cytotoxicity on MCF7 cells was observed in a dose-dependent manner.	[96]
Gold NPs		PgAuNPs showed dose-dependent antioxidant activity. The NPs revealed good moisture retention capacity and showed potential to inhibit mushroom tyrosinase. MTT results revealed no toxicity to HDF and B16 cells and reduced melanin content.	[97]
Selenium (Se) NPs	*Withani somnifera* leaves extract	Se NPs were synthesized by green synthesis using extract of *Withania somnifera* leaves, showed significant antioxidant activity and excellent A549 cells growth control.	[98]
SLNs (HSLNs)	Rosmarinic acid	Rosmarinic acid-loaded SLNs showed improved nose-to-brain delivery to manage Huntington's disease. SLNPRT (rosmarinic acid-loaded SLN) markedly lessened lipid peroxidation, nitrite concentration, and reinstated antioxidant enzyme (catalase and GSH) activities.	[99]
SLNs	Myricitrin	In vitro valuation concluded that myricitrin-SLNs amended the antioxidant defense, glycogen, and cellular survival of the hyperglycemic ailment.	[100]

(Continued)

Table 8.4 (Continued)

Nanocarrier	Herb/spice/dietary constituents	Remarks	Reference
SLN and nanostructured lipid carriers (NLCs)	Ferulic acid (FA)	The cells treated with FA-SLN disclosed a significant reduction in ROS production when compared with the free FA confirming the potential of FA-SLNs, a defensive ability against oxidative stress in neurons than free FA confirming their excellency for Alzheimer's disease.	[101]
Noisome	Gallic acid, curcumin, and quercetin	Co-encapsulations of gallic acid/curcumin and ascorbic acid/quercetin showed a great impact on physicochemical properties; also, the antioxidants resulted in an encouraging capability to reduce free radicals.	[102]
Hydrogel	Puerarin	The developed hydrogel exhibited improved mechanical and antioxidant behavior, which is advantageous for wound dressing.	[103]
Nanofibers	Quercetin	The nanoformulation revealed improved antioxidant potential than raw quercetin. Besides, sustained quercetin release was observed.	[104]
CNTs	Ginsenosides of *Panax ginseng*	Detection of CNT conjugates using Raman spectroscopy together with the observed enhancement of the anticancer activity of ginsenosides successfully conjugated to CNT showed potential to overcome cell barriers and delivered antioxidant complex molecules, ginsenosides (Rb1and Rg1).	[105]

8.3.7 Patents

Attempts have been made to design and develop the NPs for the delivery of herbs, spices, or dietary constituents or to develop the NPs by green synthesis methods. Most of the extracts of herbs, spices, and their constituents exhibit antioxidant properties or enhance the antioxidant behavior of NPs, which assist in the treatment of fatal ailments such as Alzheimer's, cancer, diabetes, etc. Some of the patents based are described in Table 8.5.

8.4 Conclusion

After having an exhaustive literature review, we know that spices are not only meant for their color, flavor, taste, and aromaticity but also have the ability to protect individuals from acute and chronic illnesses due to their strong antioxidant activity. This chapter shares the information related to the chemical constituents with total flavonoid and polyphenol contents. It has also been found that spices hold strong antioxidant properties, have extremely high biological activity, and are frequently employed in preclinical, clinical, and therapeutic studies to find novel ways to cure illnesses. Nanocarriers, which have amply benefitted in developing a multidisciplinary approach to drug development and various NPs themselves show antioxidant potential. So, for the delivery

Table 8.5 Some patents on the NPs-based delivery of extracts of herbs and spices showing antioxidant potential.

Patent number	Remarks	Reference
US9381252B2	Zein matrix and basic amino acid NPs were developed by encapsulating antioxidant quercetin. The developed NPs are applicable to pharmaceuticals, cosmetics, and in nanotechnology.	[106]
US8784881B2	Curcumin liposome complex was developed for the management of various kinds of diseases.	[107]
US8563053B2	NPs of *Lepidium* extracts were developed with calcium, vitamin D, and antioxidants like flavonoids, lycopene, green tea extract, and pomegranate extract, for the treatment of osteoporosis in animals.	[108]
US20100104646A1	Invented herbal nanoparticles to prevent hair loss and to improve hair growth by encouraging hair metabolism with antioxidant activity on the scalp.	[109]
WO/2013/078477	Curcumin NPs were invented for an antioxidant, anti-inflammatory, antiviral, antibacterial, and antifungal effects.	[110]

of antioxidant herbs and spices, NPs have been widely studied, though their potential application and their toxicological profile should be well studied clinically. This chapter also provided information regarding the history and ongoing preclinical and clinical studies of possible antioxidant drug delivery options using nanocarriers.

References

1 Embuscado, M.E. (2015). Spices and herbs: natural sources of antioxidants – a mini review. *J. Funct. Foods* https://doi.org/10.1016/j.jff.2015.03.005.

2 Decker, E.A., Elias, R.J., and McClements, D.J. (2010). *Oxidation in Foods and Beverages and Antioxidant Applications*, 1e. Woodhead Publishing.

3 Dwivedi, S., Vasavada, M.N., and Cornforth, D. (2006). Evaluation of antioxidant effects and sensory attributes of Chinese 5-spice ingredients in cooked ground beef. *J. Food Sci.* 71: C12–C17. https://doi.org/10.1111/j.1365-2621.2006.tb12381.x.

4 Peter, K.V. (2012). *Handbook of Herbs and Spices*, vol. 2. Woodhead Publishing.

5 Peter, K.V. (2001). *Handbook of Herbs and Spice*, 1e. Woodhead Publishing.

6 Xiao, Q., Zhang, S., Wang, Z. et al. (2016). Onion-like glycodendrimersomes from sequence-defined Janus glycodendrimers and influence of architecture on reactivity to a lectin. *PNAS* 113: 1162–1167.

7 McCormick Science Institute (2021). The history of spices [WWW Document]. http://www.mccormickscienceinstitute.com/resources/history-of-spices (accessed 20 May 2021).

8 Halliwell, B., Murcia, M.A., Chirico, S., and Aruoma, O.I. (1995). Free radicals and antioxidants in food and in vivo: what they do and how they work. *Crit. Rev. Food Sci. Nutr.* 35: 7–20. https://doi.org/10.1080/10408399509527682.

9 Huang, D., Boxin, O.U., and Prior, R.L. (2005). The chemistry behind antioxidant capacity assays. *J. Agric. Food Chem.* https://doi.org/10.1021/jf030723c.

10 Patra, K., Jana, S., Mandal, D.P., and Bhattacharjee, S. (2016). Evaluation of the antioxidant activity of extracts and active principles of commonly consumed Indian spices. *J. Environ. Pathol. Toxicol. Oncol.* 35: 299–315. https://doi.org/10.1615/JEnvironPatholToxicolOncol.2016016387.

11 Serafini, M. and Peluso, I. (2017). Functional foods for health: the interrelated antioxidant and anti-inflammatory role of fruits, vegetables, herbs, spices and cocoa in humans. *Curr. Pharm. Des.* 22: 6701–6715. https://doi.org/10.2174/1381612823666161123094235.

12 Hurrell, R.F. (2003). Influence of vegetable protein sources on trace element and mineral bioavailability. *J. Nutr. Am. Inst. Nutr.* https://doi.org/10.1093/jn/133.9.2973s.

13 Mensor, L.L., Menezes, F.S., Leitão, G.G. et al. (2001). Screening of Brazilian plant extracts for antioxidant activity by the use of DPPH free radical method. *Phyther. Res.* 15: 127–130. https://doi.org/10.1002/ptr.687.

14 Sánchez Moreno, C., Larrauri, A.J., and Saura-Calixto, F. (1999). Free radical scavenging capacity of selected red, rosé and white wines. *J. Sci. Food Agric.* 79: 1301–1304. https://doi.org/10.1002/(SICI)1097-0010(19990715)79:10.

15 Choi, H.R., Choi, J.S., Han, Y.N. et al. (2002). Peroxynitrite scavenging activity of herb extracts. *Phyther. Res.* 16: 364–367. https://doi.org/10.1002/ptr.904.

16 Re, R., Pellegrini, N., Proteggente, A. et al. (1999). Antioxidant activity applying an improved ABTS radical cation decolorization assay. *Free Radic. Biol. Med.* 26: 1231–1237. https://doi.org/10.1016/S0891-5849(98)00315-3.

17 Bostwick, D.G., Alexander, E.E., Singh, R. et al. (2000). Antioxidant enzyme expression and reactive oxygen species damage in prostatic intraepithelial neoplasia and cancer. *Cancer* 89 (1): 123–134. https://doi.org/10.1002/1097-0142(20000701)89:1.

18 Babu, B.H., Shylesh, B.S., and Padikkala, J. (2001). Antioxidant and hepatoprotective effect of *Acanthus ilicifolius*. *Fitoterapia* 72: 272–277. https://doi.org/10.1016/S0367-326X(00)00300-2.

19 El-Ghorab, A.H., Nauman, M., Anjum, F.M. et al. (2010). A comparative study on chemical composition and antioxidant activity of ginger (*Zingiber officinale*) and cumin (*Cuminum cyminum*). *J. Agric. Food Chem.* 58: 8231–8237. https://doi.org/10.1021/jf101202x.

20 Altman, R.D. and Marcussen, K.C. (2001). Effects of a ginger extract on knee pain in patients with osteoarthritis. *Arthr. Rheum.* https://doi.org/10.1002/1529-0131.

21 Stoilova, I., Krastanov, A., Stoyanova, A. et al. (2007). Antioxidant activity of a ginger extract (*Zingiber officinale*). *Food Chem.* 102: 764–770. https://doi.org/10.1016/j.foodchem.2006.06.023.

22 Basnet, P. and Skalko-Basnet, N. (2011). Curcumin: an anti-inflammatory molecule from a curry spice on the path to cancer treatment. *Molecules* https://doi.org/10.3390/molecules16064567.

23 Kuttan, R., Bhanumathy, P., Nirmala, K., and George, M.C. (1985). Potential anticancer activity of turmeric (*Curcuma longa*). *Cancer Lett.* 29: 197–202. https://doi.org/10.1016/0304-3835(85)90159-4.

24 Al-Suhaimi, E.A., Al-Riziza, N.A., and Al-Essa, R.A. (2011). Physiological and therapeutical roles of ginger and turmeric on endocrine functions. *Am. J. Chin. Med.* 39: 215–231. https://doi.org/10.1142/S0192415X11008762.

25 Sahoo, J.P., Behera, L., Praveena, J. et al. (2021). The golden spice turmeric (*Curcuma longa*) and its feasible benefits in prospering human health – a review. *Am. J. Plant Sci.* 12: 455–475. https://doi.org/10.4236/ajps.2021.123030.

26 Sahoo, J.P., Upasana, M., Ambika, P.M., and Kailash, C. (2020). Turmeric (Haldi) – a strapping strategy for enhancing the immune system to reduce the effect of SARS-CoV-2. *Food Sci. Rep.* 1: 10–12.

27 Vasanthi, H.R. and Parameswari, R.P. (2010). Indian spices for healthy heart – an overview. *Curr. Cardiol. Rev.* 6: 274–279. https://doi.org/10.2174/157340310793566172.

28 Panpatil, V.V., Tattari, S., Kota, N. et al. (2013). In vitro evaluation on antioxidant and antimicrobial activity of spice extracts of ginger, turmeric and garlic. *J. Pharmacogn. Phytochem.* 2: 143–148.

29 Dorant, E., Van den Brandt, P.A., Goldbohm, R.A., and Sturmans, F. (1996). Consumption of onions and a reduced risk of stomach carcinoma. *Gastroenterology* 110: 12–20. https://doi.org/10.1053/gast.1996.v110.pm8536847.

30 Kleijnen, J., Knipschild, P., and Ter Riet, G. (1989). Garlic, onions and cardiovascular risk factors. A review of the evidence from human experiments with emphasis on commercially available preparations [see comments]. *Br. J. Clin. Pharmacol.* 28: 535–544. https://doi.org/10.1111/j.1365-2125.1989.tb03539.x.

31 Ide, N. and Lau, B.H.S. (1997). Garlic compounds protect vascular endothelial cells from oxidized low density lipoprotein-induced injury. *J. Pharm. Pharmacol.* 49: 908–911. https://doi.org/10.1111/j.2042-7158.1997.tb06134.x.

32 Rose, P., Whiteman, M., Moore, P.K., and Yi, Z.Z. (2005). Bioactive *S*-alk(en)yl cysteine sulfoxide metabolites in the genus *Allium*: The chemistry of potential therapeutic agents. *Nat. Prod. Rep.* https://doi.org/10.1039/b417639c.

33 Saravanan, G. and Prakash, J. (2004). Effect of garlic (*Allium sativum*) on lipid peroxidation in experimental myocardial infarction in rats. *J. Ethnopharmacol.* 94: 155–158. https://doi.org/10.1016/j.jep.2004.04.029.

34 Jang, H.J., Lee, H.J., Yoon, D.K. et al. (2018). Antioxidant and antimicrobial activities of fresh garlic and aged garlic by-products extracted with different solvents. *Food Sci. Biotechnol.* 27: 219–225. https://doi.org/10.1007/s10068-017-0246-4.

35 Kačániová, M., Galovičová, L., Ivanišová, E. et al. (2020). Antioxidant, antimicrobial and antibiofilm activity of coriander (*Coriandrum sativum* L.) essential oil for its application in foods. *Foods* 9: 282. https://doi.org/10.3390/foods9030282.

36 Marangoni, C. and de Moura, N.F. (2011). Atividade antioxidante do óleo essencial de *Coriandrum sativum* L. em salame Italiano. *Food Sci. Technol.* 31: 124–128. https://doi.org/10.1590/S0101-20612011000100017.

37 Shahwar, M.K., El-Ghorab, A.H., Anjum, F.M. et al. (2012). Characterization of coriander (*Coriandrum sativum* L.) seeds and leaves: volatile and non volatile extracts. *Int. J. Food Prop.* 15: 736–747. https://doi.org/10.1080/10942912.2010.500068.

38 Singh, K., Rani, R., Bansal, P. et al. (2015). Antioxidant activity of essential oil of *Coriandrum sativum* and standardization of HPTLC method for the estimation of major phytomarkers. *J. Anal. Chem.* 70: 220–224. https://doi.org/10.1134/S1061934815020094.

39 Deans, S.G. (2001). Black pepper, *Piper nigrum*. Medicinal and aromatic plants/industrial profiles. *Phytochemistry* 58: 827–829.

40 Govindarajan, V.S. and Stahl, W.H. (1977). Pepper – chemistry, technology, and quality evaluation. *CRC Crit. Rev. Food Sci. Nutr.* 9: 115–225. https://doi.org/10.1080/10408397709527233.

41 Lee, J.-G., Chae, Y., Shin, Y., and Kim, Y.-J. (2020). Chemical composition and antioxidant capacity of black pepper pericarp. *Appl. Biol. Chem.* 63: 35. https://doi.org/10.1186/s13765-020-00521-1.

42 Jayashree, E., Zachariah, T.J., and Gobinath, P. (2009). Physico-chemical properties of black pepper from selected varieties in relation to market grades. *J. Food Sci. Technol.* 46: 263–265.

43 Gülçin, I. (2005). The antioxidant and radical scavenging activities of black pepper (*Piper nigrum*) seeds. *Int. J. Food Sci. Nutr.* 56: 491–499. https://doi.org/10.1080/09637480500450248.

44 Kamatou, G.P., Vermaak, I., and Viljoen, A.M. (2012). Eugenol – from the remote Maluku Islands to the international market place: a review of a remarkable and versatile molecule. *Molecules* https://doi.org/10.3390/molecules17066953.

45 Kramer, R.E. (1985). Antioxidants in clove. *J. Am. Oil Chem. Soc.* 62: 111–113. https://doi.org/10.1007/BF02541505.

46 Soto, C.G. and Burhanuddin (1995). Clove oil as a fish anaesthetic for measuring length and weight of rabbitfish (*Siganus lineatus*). *Aquaculture* 136: 149–152. https://doi.org/10.1016/0044-8486(95)01051-3.

47 Pandey, K., Sharma, S., and Lamsal, R. (2014). Antioxidant properties of various fruits, herbs, spices and vegetable: a review. *World J. Pharm. Pharm. Sci.* 3: 1101–1109.

48 Yashin, A., Yashin, Y., Xia, X., and Nemzer, B. (2017). Antioxidant activity of spices and their impact on human health: a review. *Antioxidants.* https://doi.org/10.3390/antiox6030070.

49 Gülçin, I., Elmastaş, M., and Aboul-Enein, H.Y. (2012). Antioxidant activity of clove oil – a powerful antioxidant source. *Arab. J. Chem.* 5: 489–499. https://doi.org/10.1016/j.arabjc.2010.09.016.

50 Kintzios, S.E. (ed.) (2002). *Oregano: The Genera Origanum and Lippia*, 1e. CRC Press/Taylor & Francis.

51 Franz, C., Baser, K.H.C., and Windisch, W. (2010). Essential oils and aromatic plants in animal feeding – a European perspective. A review. *Flavour Fragr. J.* https://doi.org/10.1002/ffj.1967.

52 Giannenas, I., Bonos, E., Christaki, E., and Florou-Paneri, P. (2018). Oregano: a feed additive with functional properties. In: *Therapeutic Foods* (ed. G. Ilias), 179–208. Elsevier https://doi.org/10.1016/b978-0-12-811517-6.00006-4.

53 Karadayı, M., Yıldırım, V., and Güllüce, M. (2020). Antimicrobial activity and other biological properties of oregano essential oil and carvacrol. *Anatol. J. Biol.* 2: 52–68.

54 Spiridon, I., Colceru, S., Anghel, N. et al. (2011). Antioxidant capacity and total phenolic contents of oregano (*Origanum vulgare*), lavender (*Lavandula angustifolia*) and lemon balm (*Melissa officinalis*) from Romania. *Nat. Prod. Res.* 25: 1657–1661. https://doi.org/10.1080/14786419.2010.521502.

55 Tuncturk, R. and Tuncturk, M. (2006). Effects of different phosphorus levels on the yield and quality components of cumin (*Cuminum cyminum* L.)|AVESİS. *Res. J. Agric. Biol. Sci.* 2: 336–340.

56 Hajlaoui, H., Mighri, H., Noumi, E. et al. (2010). Chemical composition and biological activities of Tunisian *Cuminum cyminum* L. essential oil: a high effectiveness against *Vibrio* spp. strains. *Food Chem. Toxicol.* 48: 2186–2192. https://doi.org/10.1016/j.fct.2010.05.044.

57 Fatima, T., Beenish, Naseer, B. et al. (2018). Antioxidant potential and health benefits of cumin. *J. Med. Plants Stud.* 6: 232–236.

58 Mnif, S. and Aifa, S. (2015). Cumin (*Cuminum cyminum* L.) from traditional uses to potential biomedical applications. *Chem. Biodivers.* 12: 733–742. https://doi.org/10.1002/cbdv.201400305.

59 Nostro, A., Cellini, L., Di Bartolomeo, S. et al. (2005). Antibacterial effect of plant extracts against *Helicobacter pylori*. *Phyther. Res.* 19: 198–202. https://doi.org/10.1002/ptr.1640.

60 Guenther, E. (1975). *The Essential Oils*. Robert E. Krieger Publishing.

61 Amma, K.P.P., Sasidharan, I., Sreekumar, M.M. et al. (2015). Total antioxidant capacity and change in phytochemicals of four major varieties of cardamom oils during decortication. *Int. J. Food Prop.* 18: 1317–1325. https://doi.org/10.1080/10942912.2011.587621.

62 Prajapati, S.K., Mishra, G., Malaiya, A. et al. (2021). Antimicrobial application potential of phytoconstituents from turmeric and garlic. In: *Advanced Structured Materials*, 409–435. Springer Science and Business Media Deutschland GmbH https://doi.org/10.1007/978-3-030-54027-2_12.

63 Pan, L., Qiu, H., Dou, C. et al. (2010). Conducting polymer nanostructures: template synthesis and applications in energy storage. *Int. J. Mol. Sci.* https://doi.org/10.3390/ijms11072636.

64 Woranuch, S. and Yoksan, R. (2013). Eugenol-loaded chitosan nanoparticles: I. Thermal stability improvement of eugenol through encapsulation. *Carbohydr. Polym.* 96: 578–585. https://doi.org/10.1016/j.carbpol.2012.08.117.

65 Prajapati, S.K., Jain, A., Shrivastava, C., and Jain, A.K. (2019a). Hyaluronic acid conjugated multi-walled carbon nanotubes for colon cancer targeting. *Int. J. Biol. Macromol.* 123: 691–703. https://doi.org/10.1016/j.ijbiomac.2018.11.116.

66 Prajapati, S.K., Jain, A., Jain, A., and Jain, S. (2019b). Biodegradable polymers and constructs: a novel approach in drug delivery. *Eur. Polym. J.* https://doi.org/10.1016/j.eurpolymj.2019.08.018.

67 Prajapati, S.K., Malaiya, A., Kesharwani, P. et al. (2020a). Biomedical applications and toxicities of carbon nanotubes. *Drug Chem. Toxicol.* 1–16. https://doi.org/10.1080/01480545.2019.1709492.

68 Oliveira, A.I., Pinho, C., Fonte, P. et al. (2018). Development, characterization, antioxidant and hepatoprotective properties of poly(Ɛ-caprolactone) nanoparticles loaded with a neuroprotective fraction of *Hypericum perforatum*. *Int. J. Biol. Macromol.* 110: 185–196. https://doi.org/10.1016/j.ijbiomac.2017.10.103.

69 Hadidi, M., Pouramin, S., Adinepour, F. et al. (2020). Chitosan nanoparticles loaded with clove essential oil: characterization, antioxidant and antibacterial activities. *Carbohydr. Polym.* 236: 116075. https://doi.org/10.1016/j.carbpol.2020.116075.

70 Sardoiwala, M.N., Kaundal, B., and Choudhury, S.R. (2018). Development of engineered nanoparticles expediting diagnostic and therapeutic applications across blood–brain barrier. In: *Handbook of Nanomaterials for Industrial Applications*, 696–709. INST https://doi.org/10.1016/B978-0-12-813351-4.00038-9.

71 Singh, D.K., Jagannathan, R., Khandelwal, P. et al. (2013). In situ synthesis and surface functionalization of gold nanoparticles with curcumin and their antioxidant properties: an experimental and density functional theory investigation. *Nanoscale* 5: 1882–1893. https://doi.org/10.1039/c2nr33776b.

72 Otunola, G.A. and Afolayan, A.J. (2018). In vitro antibacterial, antioxidant and toxicity profile of silver nanoparticles green-synthesized and characterized from aqueous extract of a spice blend formulation. *Biotechnol. Biotechnol. Equip.* 32: 724–733. https://doi.org/10.1080/13102818.2018.1448301.

73 El-Refai, A.A., Ghoniem, G.A., El-Khateeb, A.Y., and Hassaan, M.M. (2018). Eco-friendly synthesis of metal nanoparticles using ginger and garlic extracts as biocompatible novel antioxidant and antimicrobial agents. *J. Nanostructure Chem.* 8: 71–81. https://doi.org/10.1007/s40097-018-0255-8.

74 Gorinova, C., Aluani, D., Yordanov, Y. et al. (2016). In vitro evaluation of antioxidant and neuroprotective effects of curcumin loaded in Pluronic micelles. *Biotechnol. Biotechnol. Equip.* 30: 991–997. https://doi.org/10.1080/13102818.2016.1186500.

75 Akbar, M.U., Zia, K.M., Nazir, A. et al. (2018). Pluronic-based mixed polymeric micelles enhance the therapeutic potential of curcumin. *AAPS PharmSciTech* 19: 2719–2739. https://doi.org/10.1208/s12249-018-1098-9.

76 Yoncheva, K., Tzankova, V., Yordanov, Y. et al. (2019). Evaluation of antioxidant activity of caffeic acid phenethyl ester loaded block copolymer micelles. *Biotechnol. Biotechnol. Equip.* 33: 64–74. https://doi.org/10.1080/13102818.2018.1537753.

77 Prajapati, S.K., Malaiya, A., Kesharwani, P. et al. (2020b). Biomedical applications and toxicities of carbon nanotubes. *Drug Chem. Toxicol.* https://doi.org/10.1080/01480545.2019.1709492.

78 Galano, A. (2010). Carbon nanotubes: promising agents against free radicals. *Nanoscale* 2: 373–380. https://doi.org/10.1039/b9nr00364a.

79 Shi, X., Jiang, B., Wang, J., and Yang, Y. (2012). Influence of wall number and surface functionalization of carbon nanotubes on their antioxidant behavior in high density polyethylene. *Carbon N. Y.* 50: 1005–1013. https://doi.org/10.1016/j.carbon.2011.10.003.

80 Watts, P.C.P., Fearon, P.K., Hsu, W.K. et al. (2003). Carbon nanotubes as polymer antioxidants. *J. Mater. Chem.* 13: 491–495. https://doi.org/10.1039/b211328g.

81 Arrigo, R., Dintcheva, N.T., Guenzi, M., and Gambarotti, C. (2016). Nano-hybrids based on quercetin and carbon nanotubes with excellent anti-oxidant activity. *Mater. Lett.* 180: 7–10. https://doi.org/10.1016/j.matlet.2016.05.096.

82 Nichita, C. and Stamantin, I. (2013). The antioxidant activity of the biohybrides based on carboxylated/hydroxylated carbon nanotubes-flavonoid compounds. *Dig. J. Nanomater. Biostruct.* 8: 445–455.

83 Singh, N., Khullar, N., Kakkar, V., and Kaur, I.P. (2016). Hepatoprotective effects of sesamol loaded solid lipid nanoparticles in carbon tetrachloride induced sub-chronic hepatotoxicity in rats. *Environ. Toxicol.* 31: 520–532. https://doi.org/10.1002/tox.22064.

84 Madureira, A.R., Nunes, S., Campos, D.A. et al. (2016). Safety profile of solid lipid nanoparticles loaded with rosmarinic acid for oral use: In vitro and animal approaches. *Int. J. Nanomed.* 11: 3621–3640. https://doi.org/10.2147/IJN.S104623.

85 Anton, N., Benoit, J.P., and Saulnier, P. (2008). Design and production of nanoparticles formulated from nano-emulsion templates – a review. *J. Control. Release* https://doi.org/10.1016/j.jconrel.2008.02.007.

86 Sharma, N., Bansal, M., Visht, S. et al. (2010). Nanoemulsion: a new concept of delivery system. *Chron. Young Sci.* 1: 2–2.

87 McClements, D.J. (2011). Edible nanoemulsions: fabrication, properties, and functional performance. *Soft Matter.* https://doi.org/10.1039/c0sm00549e.

88 McClements, D.J. and Rao, J. (2011). Food-grade nanoemulsions: formulation, fabrication, properties, performance, biological fate, and potential toxicity. *Crit. Rev. Food Sci. Nutr.* https://doi.org/10.1080/10408398.2011.559558.

89 Rinaldi, F., Hanieh, P.N., Longhi, C. et al. (2017). Neem oil nanoemulsions: characterisation and antioxidant activity. *J. Enzyme Inhib. Med. Chem.* 32: 1265–1273. https://doi.org/10.1080/14756366.2017.1378190.

90 Zorzi, G.K., Caregnato, F., Moreira, J.C.F. et al. (2016). Antioxidant effect of nanoemulsions containing extract of *Achyrocline satureioides* (Lam) D.C. – asteraceae. *AAPS PharmSciTech* 17: 844–850. https://doi.org/10.1208/s12249-015-0408-8.

91 Merlin, J.J.P., Rajendra Prasad, N., Shibli, S.M.A., and Sebeela, M. (2012). Ferulic acid loaded poly-d,l-lactide-*co*-glycolide nanoparticles: systematic study of particle size, drug encapsulation efficiency and anticancer effect in non-small cell lung carcinoma cell line in vitro. *Biomed. Prev. Nutr.* 2: 69–76. https://doi.org/10.1016/j.bionut.2011.12.007.

92 Panwar, R., Sharma, A.K., Kaloti, M. et al. (2016). Characterization and anticancer potential of ferulic acid-loaded chitosan nanoparticles against ME-180 human cervical cancer cell lines. *Appl. Nanosci.* 6: 803–813. https://doi.org/10.1007/s13204-015-0502-y.

93 Yen, F.L., Wu, T.H., Tzeng, C.W. et al. (2010). Curcumin nanoparticles improve the physicochemical properties of curcumin and effectively enhance its antioxidant and antihepatoma activities. *J. Agric. Food Chem.* 58: 7376–7382. https://doi.org/10.1021/jf100135h.

94 Nagaraju, P.G., Sengupta, P., Chicgovinda, P.P., and Rao, P.J. (2021). Nanoencapsulation of clove oil and study of physicochemical properties, cytotoxic, hemolytic, and antioxidant activities. *J. Food Process Eng.* 44: e13645. https://doi.org/10.1111/jfpe.13645.

95 Mostafa, D.M., Kassem, A.A., Asfour, M.H. et al. (2015). Transdermal cumin essential oil nanoemulsions with potent antioxidant and hepatoprotective activities: in-vitro and in-vivo evaluation. *J. Mol. Liq.* 212: 6–15. https://doi.org/10.1016/j.molliq.2015.08.047.

96 Oueslati, M.H., Tahar, L.B., and Harrath, A.H. (2020). Catalytic, antioxidant and anticancer activities of gold nanoparticles synthesized by kaempferol glucoside from *Lotus leguminosae*. *Arab. J. Chem.* 13: 3112–3122. https://doi.org/10.1016/j.arabjc.2018.09.003.

97 Jiménez-Pérez, Z.E., Singh, P., Kim, Y.J. et al. (2018). Applications of *Panax ginseng* leaves-mediated gold nanoparticles in cosmetics relation to antioxidant, moisture retention, and whitening effect on B16BL6 cells. *J. Ginseng Res.* 42: 327–333. https://doi.org/10.1016/j.jgr.2017.04.003.

98 Alagesan, V. and Venugopal, S. (2019). Green synthesis of selenium nanoparticle using leaves extract of *Withania somnifera* and its biological applications and photocatalytic activities. *Bionanoscience* 9: 105–116. https://doi.org/10.1007/s12668-018-0566-8.

99 Bhatt, R., Singh, D., Prakash, A., and Mishra, N. (2015). Development, characterization and nasal delivery of rosmarinic acid-loaded solid lipid nanoparticles for the effective management of Huntingtons disease. *Drug Deliv.* 22: 931–939. https://doi.org/10.3109/10717544.2014.880860.

100 Ahangarpour, A., Oroojan, A.A., Khorsandi, L. et al. (2018). Solid lipid nanoparticles of myricitrin have antioxidant and antidiabetic effects on streptozotocin–nicotinamide-induced diabetic model and myotube cell of male mouse. *Oxidative Med. Cell. Longev.* 2018: 7496936.

101 Bondi, M., Montana, G., Craparo, E. et al. (2009). Ferulic acid-loaded lipid nanostructures as drug delivery systems for alzheimers disease: preparation, characterization and cytotoxicity studies. *Curr. Nanosci.* 5: 26–32. https://doi.org/10.2174/157341309787314656.

102 Tavano, L., Muzzalupo, R., Picci, N., and De Cindio, B. (2014). Co-encapsulation of antioxidants into niosomal carriers: gastrointestinal release studies for nutraceutical applications. *Colloids Surf. B Biointerfaces* 114: 82–88. https://doi.org/10.1016/j.colsurfb.2013.09.058.

103 Qianmin, O., Zhang, S., Fu, C. et al. (2021). More natural more better: triple natural anti-oxidant puerarin/ferulic acid/polydopamine incorporated hydrogel for wound healing. *J. Nanobiotechnol.* https://doi.org/10.21203/RS.3.RS-460628/V1.

104 Li, X., Liu, Y., Yu, Y. et al. (2019). Nanoformulations of quercetin and cellulose nanofibers as healthcare supplements with sustained antioxidant activity. *Carbohydr. Polym.* 207: 160–168. https://doi.org/10.1016/j.carbpol.2018.11.084.

105 Lahiani, M.H., Eassa, S., Parnell, C. et al. (2017). Carbon nanotubes as carriers of *Panax ginseng* metabolites and enhancers of ginsenosides Rb1 and Rg1 anti-cancer activity. *Nanotechnology* 28: 015101. https://doi.org/10.1088/0957-4484/28/1/015101.

106 Bazo, M.A., Catalán, I.E., Carolina G.-F., *et al.* (2016). Nanoparticles for encapsulation of compounds, the production and uses thereof. US Patent, WO/2012/007628, date published 19 January 2012.

107 Kurzrock, R., Li, L., Mehta, K., et al. (2014). Liposomal curcumin for treatment of diseases – Google Patents. US Patent 8784881B2.

108 Mousa, S., Qari, M., Ardawi, M. (2013). Compositions and methods of natural products in nanoformulations for the prevention and treatment of osteoporosis. US Patent 8563053B2.

109 Kim, J. (2010). Nanoparticle composition for prevention of hair loss and promotion of hair growth. US Patent 20100104646A1.

110 Payne, A., Kurien, J., Biji, T. (2010). Solubilization of curcuminoid compounds and products thereof. WO Patent 2013078477A3.

9

Nanophytomedicine in Disease and Therapy

Sonal Upadhyay[1], *Ravi Bhushan*[1], *and Mithun Rudrapal*[2]

[1] *Centre for Genetic Disorder, Institute of Science, Banaras Hindu University, Varanasi, Uttar Pradesh, India*
[2] *Department of Pharmaceutical Chemistry, Rasiklal M. Dhariwal Institute of Pharmaceutical Education & Research, Pune, Maharashtra, India*

9.1 Introduction

Nanotechnology is an emerging field in science and engineering that involves the design, synthesis, characterization, and application of devices and materials whose smallest functional organization in at least one dimension is on the nanometer scale (one-billionth of a meter) [1]. The prefix "nano" (Greek word) means "dwarf." One nanometer (nm) is equal to one-billionth of a meter, or about the width of 6 carbon atoms, or 10 water molecules, or three orders of magnitude smaller than a micron, roughly the size scale of a molecule itself (e.g. a DNA molecule is about 2.5 nm long while a sodium atom is about 0.2 nm) [2]. The concept of nanotechnologies was first laid out in 1959 by physicist Richard Feynman in his lecture, "There's plenty of room at the bottom." The term "nanotechnology" was first used in 1974 by Norio Taniguchi, a researcher at the University of Tokyo, to refer to the ability to engineer materials precisely at the nanometer level. Nanotechnology is a multidisciplinary field that covers a diverse and vast array of devices and materials derived from engineering, biology, chemistry, and physics [1]. Thus, nanotechnology is not in itself a single emerging scientific discipline but rather a meeting of traditional sciences such as chemistry, physics, materials science, and biology to bring together the required collective expertise needed to develop these novel technologies [2].

In the context of biology and medicine, nanotechnology has tremendous potential for advancing medical science with great opportunities in the field of drug-delivery vehicles, contrast agents, and diagnostic devices. Many of them have already been approved or under clinical investigation by the Food and Drug Administration (FDA) for use in humans (Table 9.1). Recent advances in nanotechnology in the field of medical science include microelectromechanical systems (MEMS) and biocompatible electronic devices that will be a game changer in the diagnosis and treatment of many disorders. One example of such an application includes novel drug delivery systems (specifically for the blood–brain barrier in some cases) using nanoparticles (NPs) or highly porous self-assembling bilayer tubule systems. Chemically functionalized dendrimers are another class of applications being developed. It is a highly branched molecule with a "tree-like" branching structure. It can be used as molecular building blocks for gene therapy

Phytoantioxidants and Nanotherapeutics, First Edition. Edited by Mithun Rudrapal.

Table 9.1 List of FDA-approved nano-medicines.

Disease	Nano-medicine	Year of approval
Severe combined immunodeficiency disease (SCID)	Adagen®/pegademase bovine (Sigma-Tau Pharmaceuticals)	1990
Crohn's disease	Cimzia®/certolizumab pegol (UCB)	2008
Rheumatoid arthritis		2009
Psoriatic arthritis		2013
Ankylosing spondylitis		2013
Multiple sclerosis	Copaxone®/Glatopa (Teva)	1996
Prostate cancer	Eligard® (Tolmar)	2002
Macular degeneration, neovascular age-related	Macugen®/Pegaptanib (Bausch & Lomb)	2004
Anemia associated with chronic kidney disease	Mircera®/Methoxy polyethylene glycolepoetin beta (Hoffman-La Roche)	2007
Neutropenia, chemotherapy-induced	Neulasta®/pegfilgrastim (Amgen)	2002
Hepatitis B; Hepatitis C	Pegasys® (Genentech)	2002
Hepatitis C	PegIntron® (Merck)	2001
Chronic kidney disease	Renagel® [sevelamer hydrochloride]/ Renagel® [sevelamer carbonate] (Sanofi)	2000
Acromegaly	Somavert®/pegvisomant (Pfizer)	2003
Acute lymphoblastic leukemia	Oncaspar®/pegaspargase (Enzon Pharmaceuticals)	1994
Chronic gout	Krystexxa®/pegloticase (Horizon)	2010
Multiple sclerosis 2014	Plegridy® (Biogen)	2014
Hemophilia	ADYNOVATE (Baxalta)	2015
Osteoarthritis (OA) of the knee	Zilretta	2017
Control and prevention of bleeding episodes and prevention of bleeding in the perioperative setting for hemophilia B patients	Rebinyn	2017

agents or as magnetic resonance imaging (MRI) contrast agents. For gene therapy applications, dendrimers are attractive molecules as they are a nonviral delivery vehicle for DNA (an important clinical consideration), which tends to wrap itself in the dendrimer branches. In addition, much research is going into biologically inspired functional nanodevices, such as the development of DNA/polymerase chain reaction or protein-based molecular computers, which encode information in the nucleotide sequences of DNA molecules or in the tertiary structures of proteins and carry out computations by following different biochemical reaction pathways. Another area of interest is biomimetic self-assembling molecular motors such as the flagella of bacteria or the mechanical forces produced by RNA polymerase during protein transcription. These applications represent just a few examples of emerging nanotechnology research aimed at medicine and physiology [2].

For applications to medicine and physiology, these materials and devices can be designed to interact with cells and tissues at a molecular (i.e. subcellular) level with a high degree of functional specificity, thus allowing a degree of integration between technology and biological systems not previously attainable.

Nanomedicine is an emerging field that uses the knowledge and techniques of nanoscience in medical biology for the prevention and remediation of disease. Nanomedicine has great potential to treat or repair (regenerative medicine, not just killing of diseased cells) the tissues and organs within individually targeted cells, cell by cell. It typically uses molecular biosensors to provide feedback control of treatment and repair. Nanomedicine can target and adjust the use of drugs appropriately for individual cell treatment at the proper dose for each cell (single-cell medicine). Nanomedicine utilizes nano-dimensional materials including NPs [3], nanorobots [4], nanosensors for diagnosis [5], delivery [6],sensory purposes [7], and actuate materials in live cells.

9.2 Nanodelivery/Nanoparticulate Systems in Disease and Therapy

Various nanotools such as NPs, nanotubes, nanobiosensors, nanochannels, nanopores, nanowires (NWs), and nanopatterned surfaces are becoming of great importance in the study of biological events at the nanoscale.

9.2.1 Nanoparticles

NPs include a wide class of materials in the range of 1–100 nm. These are among the most currently developed nanotools with a wide variety of applications such as cell trackers, biomarkers, and diagnostics. Currently, various types of NPs are in use in the diagnostics field, which may be inorganic NPs, metallic NPs, protein, and polysaccharide NPs. The most currently used are gold NPs, magnetic NPs, and quantum dots (QDs). Due to their ability to be detected by a variety of analytical techniques, these NPs serve as good labels for sensors. Magnetic NPs are also used as labeling molecules for bio-screening. Magnetic NPs usually consist of a magnetic core coated by a polymeric layer functionalized with antibodies for capturing cells or specific biomolecules. For example, perfluorocarbon NPs are magnetic NPs detected by MRI. In this technology, an MRI scanner is tuned to the specific frequency of the fluorine compound in the NPs enabling the detection of only those cells that contain the perfluorocarbon NPs. QDs are inorganic fluorophores and are considered good fluorescent markers over conventional markers due to their much slower photobleaching rate [8]. High sensitivity, broad excitation spectra, stable fluorescence with simple excitation, and no need for lasers are unique properties of QDs, which make them the most suitable markers. QDs have wide applicability and have been proved to be promising tools for diagnostics, in particular, in viral diagnosis and in some diseases such as cancer. Gold NPs can assemble onto a sensor surface in the presence of a complementary target.

9.2.2 Liposomes

Liposomes are single or multiple concentric lipid bilayers encapsulating an aqueous compartment, which may include natural and/or synthetic lipids and surfactants. It was discovered by Alec D. Bangham in the 1960s at the Babraham Institute, University of Cambridge. It consists of phospholipids and amphiphilic molecules that have a hydrophilic head and two apolar hydrophobic chains and thus has the capability of entrapping lipophilic agents in the lipid membrane while hydrophilic agents in the aqueous core. The size of the liposomes can range from a few nanometers to several micrometers. However, liposomes used in the medical field range between 50 and 450 nm. The morphological similarity to cellular membranes and ability to incorporate various substances makes liposomes an almost ideal drug-carrier system.

Liposomes can be classified on the basis of the preparation method (reverse-phase evaporation vesicles or vesicle-extruded technique), size (small, intermediate, or large), and lamellarity (uni-, oligo-, and multilamellar vesicles) [9]. Due to their biocompatibility and biodegradability nature, liposomes are, in fact, considered safe nanocarriers. In addition to being biocompatible, all liposomes have in common a structure that gives them the ability to contain both hydrophilic and hydrophobic drugs. The encapsulation of the active form of a drug into the lipid bilayer protects it against naturally occurring phenomena, such as enzymatic degradation and immunologic and chemical inactivation. Therefore, liposomes prevent a drug from being metabolized prior to reaching target tissues, and simultaneously they minimize exposure of healthy tissue to the encapsulated drug during its circulation in the blood. Both of these effects contribute to increasing the therapeutic index. Liposomes should store, protect, and transfer substantial quantities of drugs while being well tolerated in patients receiving the cure. These unique characteristics could provide for an improved biopharmaceutical profile through reduced toxicity, favorable pharmacokinetic behavior, and an enhanced therapeutic index in comparison to the free-form drug.

9.2.3 Carbon Nanotubes (CNTs)

Carbon nanotubes (CNTs) are cylindrical molecules made of graphene sheets (carbon atoms). CNTs (seamless cylinders) were discovered in 1991 by Iijima. It can be open-ended or capped, with a high aspect ratio of at least 1 nm diameter and a length of several micrometers. CNTs may consist of a single graphene sheet resulting in single-walled nanotubes (SWNT) or several graphene sheets making up multiwalled CNTs (MWNTs). CNTs have been investigated as potential nanocarriers for the delivery of drugs, genes, and proteins. CNTs are capable of adsorbing or conjugating with a wide variety of therapeutic molecules due to their high surface area. Their unique needle-like shapes enable them to adsorb or covalently link to a wide variety of therapeutic materials ranging from small drug molecules to macromolecules such as genes and proteins and internalize them into the target cell.

CNTs have proved to serve as safer and more effective alternatives to previous drug delivery systems [10]. Nanoscale dimensions of CNTs allow them to head toward specific targets via crossing the cell membrane. CNTs can be used to carry therapeutic drugs, vaccines, peptides, and nucleic acids to the specific targets inside the cell. CNTs can also be used as reinforcement in composite materials, as narrow conduits for flow-based assays, or as tips in AFM for high-resolution imaging. Other compounds like boron nitride, gallium nitride, boron carbide, and some organic polymers have also been used in the fabrication of nanotubes for multiple applications [11]. Though CNTs seem to be a very promising tool and have several advantages over conventional drug delivery systems, their in vivo behavior and elimination are very controversial and doubtful.

9.2.4 Nanobiosensors

A nanobiosensor is an analytical device that detects, transmits, and records the information regarding an analyte consisting of a biological component with a physicochemical detection system. It consists of a biologically active element with an appropriate physical transducer, which generates a measurable signal proportional to the concentration of chemical species in any type of sample. Nanobiosensor acts on the nanoscale region.

There are different types of nanobiosensors – optical biosensors, electrical biosensors, electrochemical biosensors, NW biosensors, nanotube-based biosensors, viral nanobiosensors, and nanoshell biosensors.

Nanobiosensors have wide applications in the field of biology. Listed here are some of the biological applications of nanobiosensors:

1) As a DNA sensor: genetic monitoring of diseases,
2) As immunosensors: HIV, hepatitis, other viral diseases, drug testing, environmental monitoring,
3) Cell-based sensors: functional sensors, drug testing,
4) Point-of-care sensors: blood, urine, electrolytes, gases, steroids, drugs, hormones, proteins, and others,
5) Bacteria (*E. coli*, *Streptococcus*, other) sensors: food industry, medicine, environmental, and others,
6) Enzyme sensors: diabetics, drug testing, and others [12].

Nanobiosensors can be used for the detection of chemical or biological signals at the nanoscale. These sensors can be electronically tuned to react to the binding of a single molecule. Some prototype sensors have been demonstrated to detect nucleic acids, proteins, and ions. These sensors can operate in the liquid or gas phase, opening up an enormous variety of downstream applications. They are inexpensive to manufacture and are portable. It may even be possible to develop implantable detection and monitoring devices on the basis of these detectors [13].

QDs and silicon NP sensors are two major nanobiosensors. QDs, as the name suggests, are very small dot-like structures that confine electrons in three dimensions and the smallest objects that can be synthesized on the nanoscale. QDs may be like pyramids, cylinders, lens shapes, and spheres. Depending on its application, the total diameter of a QD varies between 3 and 60 nm [12].

9.2.5 Nanowires

NWs are single-crystalline, highly anisotropic, semiconducting, insulating, and/or metallic nanostructures that result from rapid growth along one direction. NWs are typically one-dimensional (1D) nanostructures that represent an important and broad class at the forefront of nanoscience and nanotechnology. The cross section of NWs is uniform and much smaller than the length. The shape of NWs may be cylindrical, hexagonal, square, or triangular in cross section with a large anisotropy in dimensions. Compared to other nanoscale building blocks like CNTs, NWs represent one of the best defined and controlled classes of building blocks. NWs can spontaneously assemble into nano springs [14], nanobots [15], nanorings [16], and nano helices [25], which are candidates for nanoscale transducers, actuators, and sensors [17].

9.3 Nanomedicine as Biomedical Devices for Disease Treatment

Nanomedicine may include nanoantibiotics, nanorobotic microbivores, nanosponges, nanorobots, biological sensors, nanogenerators, etc.

9.3.1 Nanorobots

A nanorobot, also known as nanorobots or nanoids, is an extremely small robot designed to perform a specific task with precision at nanoscale. Nanorobots can effectively convert diverse energy sources into movement and force. The field of nanotechnology, which deals with the design and construction of robotic devices at an atomic, molecular, or cellular level, is called nanorobotics. Nanorobotics integrates the advantages of nanotechnologies and robotic sciences. Over the past decade, nanorobotics has emerged as a novel and versatile platform [18]. Nanorobotics deals with the study of designing, programming, manufacturing, and controlling robots at the nanoscale. The design of nanorobots is inspired by biological models of bacteria. The concept of injecting small robots into the human body was given by K. Eric Drexler in 1986. A nanorobot can be made of mechanical parts such as bearing, gears, motors, etc. The outer shell of a nanorobot is likely to be constructed using diamondoid material due to its inert properties, high thermal conductivity, and strength. The super-smooth surfaces could reduce the chances of triggering the body's immune system. The nanoscale gears and other components designed for special purposes could be constructed using elements like hydrogen, sulfur, oxygen, nitrogen, silicon, etc. Medical nanorobots like respirocytes (artificial mechanical red blood cells), microbivores (artificial mechanical white blood cell), and clottocytes (artificial platelets) that could transverse the human body were designed by Robert A. Freitas. The components involved in the construction of a nanorobot involve onboard power supply, sensors, nanocomputers, pumps, manipulators, and pressure tanks [18]. Desirable characteristics to be found in a nanorobot include nano-information processing and programmability, self-assembly and replication, swarm intelligence, and nano to macro world interface architecture. A nanorobot could be introduced into the human body either via a natural [19] opening or surgically. Once inside the body, it could act as a homing device, searching for pathology, diagnosing it, and even curing it by nanomanipulation. Nanorobots can communicate with computers controlling them by encoding the messages to acoustic signals at a wave frequency ranging from 1 to 100 MHz. Furthermore, these nanorobots can be easily retrieved through the usual human excretory channels or can also be removed by active scavenger systems. Various nanorobots in medicine include respirocyte, microbivore, clottocyte, pharmacytes (for nanorobotic drug delivery), dentifrobots (dental nanorobots), and vasculoids (as an artificial nanomechanical vascular system). Medical nanorobots can perform a wide range of tasks in diagnosis, monitoring, and treating vital diseases, for example, Femtolaser, a nanorobot, acts like a pair of nano scissors and vaporizes diseased tissue locally while leaving adjacent tissue unharmed. This surgery has already been performed on individual chromosomes [20].

The potential applications of nanorobots include:

1) Drug delivery
2) Body surveillance
3) Dentistry
4) In surgery
5) Cancer detection and treatment
6) Diagnosis and treatment of diabetes
7) Delicate surgeries
8) Gene therapy

9.3.2 Nanogenerators

Nanogenerators are generators at the nanoscale level, which convert mechanical energy from body movement, muscle stretching, or water flow into electricity. This is another exciting area and could be used to create a new class of self-powered implantable medical devices, sensors, and portable electronics. This novel technology will convert mechanical energy generated by movement within the body into electrical energy to run the several medical devices like pacemakers or ICDs obviating the need for external batteries, thus leading to permanent "self-powered medical devices" [20].

9.3.3 Nanoantibiotics

Nanoantibiotics are nanomaterials that either show antimicrobial activity by themselves or elevate the effectiveness and safety of antibiotics administration [21]. Unlike many antimicrobial drugs currently being used, antimicrobial NPs may not pose direct and acute adverse effects, although potential toxicity upon long-term exposure is questionable. Most importantly, antimicrobial NPs target multiple biological pathways found in broad species of microbes. Furthermore, antimicrobial NPs could be cost-effective, compared with antibiotics synthesis, and they are quite stable enough for long-term storage with a prolonged shelf-life [22]. In addition, some NPs can withstand harsh conditions, such as high-temperature sterilization, under which conventional antibiotics are inactivated. Antibiotics delivery using nanomaterials offer multiple advantages: (i) controllable and relatively uniform distribution in the target tissue, (ii) improved solubility, (iii) sustained and controlled release, (iv) improved patient compliance, (v) minimized side effects, and (vi) enhanced cellular internalization [21].

9.3.4 Nanosponges

Nanosponges are tiny sponges with a porous structure that can circulate in the body to reach the specific site and bind on the surface to release the drug in a controlled and predictable manner. The porous structure of nanosponges has the unique ability to entrap the drug moieties and offers the merit of the desired release. Nanosponges are made up of tiny particles with a narrow cavity of a few nanometers, which can be filled with various types of substances. Nanosponges can be administered by oral as well as intravenous (IV) route. The nanosponges have a three-dimensional scaffold (backbone) or network of polyester that is biodegradable. It is about the size of a virus with a "backbone" (a scaffold structure) of naturally degradable polyester. Nanosponges are prepared by cross-linking the cyclodextrin with carbonyl or dicarboxylate (crosslinkers) [23].

Nanosponges have several advantages like increasing aqueous solubility of the poorly water-soluble drug, can release the drug molecules in a predictable fashion, because of their tiny pore size (0.25 μm), bacteria cannot penetrate the nanosponges, and they act like a self-sterilizer, nanosponges drug delivery system are nonirritating, nonmutagenic, and nontoxic. It helps remove the toxic and venom substance from the body. Nanosponges drug delivery system minimize side effect, increase formulation stability, enhance the flexibility of the formulation, reduce dosing frequency, and have better patient compliance. Nanosponges complexes are stable over a wide range of pH (i.e. 1–11) and a temperature of 130 °C [17, 19–21, 24, 25]. However, it has disadvantages too, for example, it is not suitable for larger molecules and may cause dose dumping at times [23, 26–29].

9.4 Nanophytomedicine as Therapeutics

9.4.1 Nanophytomedicine Against Infectious Diseases

Infectious diseases causing pathogens like bacteria, viruses, fungi, or parasites are one of the major causes of death, accounting for approximately 15 million annual deaths worldwide. Other infectious diseases like tuberculosis (TB), diarrheal diseases, pneumonia, and malaria are also major causes of death, accounting for 1.5, 2.2, 1.8, and 1.2 million deaths, respectively [19, 21]. Despite great efforts to develop effective pharmaceuticals and new technologies to produce drugs at low cost, the remarkable increase in drug resistance of infectious agents prevents efficient treatment of these diseases. However, overuse or misuse of antibiotics or incomplete therapy may result in the emergence of resistant infectious bacteria. Infections caused by drug-resistant microorganisms often fail to respond to conventional treatment, resulting in a prolonged infection and increased risk of death. In addition, patients infected with such microorganisms remain infectious for a long time, increasing the risk of spreading the disease to others. Despite the discovery and development of new therapeutic strategies for treatment, complete prevention and eradication of infectious diseases have not been achieved. Certain therapeutic compounds have serious side effects and therefore cannot be widely applied; for example, nerve damage caused by the DPT vaccine (diphtheria, pertussis, and tetanus) and long-term pain, numbness, infertility, and paralysis due to the HPV vaccine in Japan. In addition to the difficulty in developing new therapeutics, the emergence of drug resistance of existing pathogens that escape from current treatment strategies is another significant problem in infectious diseases.

Consequently, there is an increase in the demand for new strategies, pharmaceuticals, and devices to diagnose and treat diseases accurately, easily, and efficiently. Emerging nanotechnology-based techniques have recently attracted interest as an approach that may overcome the problems of current diagnostic techniques through their specific mode of actions and unique physical properties (i.e. shape, size, surface charge, and dimension). A number of nanotechnology-based materials have been studied with the aim of effective control and prevention of infectious diseases. Nanotechnology can aid in the diagnosis and treatment of infectious diseases.

9.4.2 Nanophytomedicine Against Cancer

Cancer, also called malignancy, is a population of erratically growing abnormal cells, which can metastasize to other tissue organs of the body. Cancer is one of the most common causes of morbidity and mortality worldwide and the second leading cause of death globally making it a major public health issue. It caused around 8.8 million deaths in 2015. Approximately, 70% of deaths due to cancer occur in low- and middle-income countries [30]. According to the International Agency for Research on Cancer (IARC), 18.1 million new cancer cases were found in 2018 while mortality reached 9.6 million, projecting a rise of cancer prevalence to 23.6 million by 2030 around the world. Till date, more than 100 types of cancers have been detected of which lung, colorectal, stomach, liver, and breast cancers are the most prevalent and the foremost cause of cancer death worldwide [31].

Almost all existing anticancer drugs show severe side effects including severe damage to the surrounding normal cells, anemia due to bone marrow suppression, diarrhea/constipation, digestive tract infection, inflammation, immunosuppression, and multiorgan toxicity [31]. Most of the clinically used anticancer drugs (formulated as conventional delivery systems) suffer some limitations like low solubility in water, unsuitability for oral administration, short half-life in the body, and poor specificity associated with severe side effects [30, 32]. Nanotechnology has the potential to revolutionize cancer diagnosis and therapy. Nanotechnologically improved drugs and drug delivery systems are increasing day by day in cancer treatment and some are already in clinical use with good results. Nanotechnology may help in better cancer diagnosis, more efficient drug delivery to tumor cells, and molecular targeted cancer therapy that improves the therapeutic management of cancer patients [32–34]. Nanotechnology has opened a new window to improvise drug delivery using new concepts and carriers that conventional technologies have never achieved. The majority of current nanotechnology platforms for chemotherapy repackage the traditional anticancer agents into various forms of nanometer-sized delivery vehicles. Often there is a defective, leaky vascular architecture associated with a tumor as a result of the poorly regulated nature of tumor angiogenesis. The interstitial fluid within a tumor is drained by a poorly formed lymphatic system. As a result, submicron-sized particulate matter can easily extravasate into the tumor and be retained resulting in an "enhanced permeability and retention" (EPR) effect. This EPR effect is advantageous for a properly designed nanoparticulate system for passive targeting that allows nanocarriers loaded with cytotoxic agents to accumulate in the tumor tissues. This contributes to an increased amount of drug reaching the targeted tumor sites with minimum systemic drug toxicity. Several studies have reported the advantages of NPs and other nano-drug delivery systems in the case of cancer.

Chen et al. co-encapsulated the VCR (vincristine) and VRP in poly(lactic-*co*-glycolic acid) (PLGA) NPs to reduce toxicity. The in vitro and in vivo studies revealed that VCR-VRP encapsulation into PLGA NPs at a synergistic ratio exhibited good antitumor efficacy in MDR MCF-7/ADR human breast tumor xenograft models and enhance antitumor activity in multidrug-resistant breast [35]. Das et al. studied the effect of apigenin-loaded PLGA nanoparticles (Nap) against ultraviolet B (UVB) and benzo(a)pyrene (BaP)-induced skin tumor and mitochondrial dysfunction in mice. The results showed a superior effect for the Nap vs. apigenin alone attributed due to their smaller size, and faster mobility [36]. Similarly, Siddiqui et al. prepared PEG–PLA NPs of epigallocatechin-3-gallate (EGCG) and evaluated them on Mel 928 human melanoma cells (in vitro and in vivo studies). The results indicated an eightfold increase in efficacy for the NPs as compared to free EGCG [37]. In another study, Lopes et al. prepared liposomes of ursolic acid using the lipid hydration method and evaluated them on MDA-MB-231 breast cancer cells. The results indicated that the liposomes had better anticancer activity than free ursolic acid indicating the improved anticancer activity of the nano-liposomes [38]. Tian et al. synthesized and functionalized multiwalled CNTs (MWNTs) as anticancer drug carriers. In vitro cytotoxicity studies performed on HeLa cell line using MTT assay showed that these functionalized MWNTs exhibited good aqueous solubility, biocompatibility, and high targeting ability. The results showed significant enhancement in the cytotoxic capability, thus improving the antitumor activity of the drug [39].

9.4.3 Nanophytomedicine Against Arthritis and Inflammatory Disorders

Inflammation is an essential and fundamental process designed to protect the human body by eliminating both the causes and consequences of injury, such as microbes, toxins, and necrotic cells, thereby enabling survival by maintaining tissue homeostasis. Thus, inflammation is a mechanism of the body's defense that originates from multiple causes such as infectious agents (viruses, bacteria), physical agents, radical oxygen species (ROS), and metabolic stress (hypoxia). Inflammation can either be acute or chronic. Acute inflammation has a rapid onset, and is usually caused by stimuli such as infection and foreign bodies and lasts for a few hours to a few days. However, chronic inflammation, on the other hand, has a duration of weeks to several months, causing degeneration of connective tissue in one or more body parts [40]. In some conditions, the inflammatory process becomes detrimental, contributing to the pathogenesis of a disease. It is now known that a large spectrum of diseases including sepsis, asthma, obesity and type 2 diabetes, neurodegenerative diseases, cardiovascular diseases (CVDs) (e.g. atherosclerosis, cardiac ischemia/reperfusion), cancer, bowel disease, Crohn's disease, rheumatoid polyarthritis, etc., are accompanied by an uncontrolled inflammatory response.

Today, therapeutic strategies for inflammation include the use of steroidal or nonsteroidal anti-inflammatory drugs. Corticosteroids, particularly glucocorticoids, a class of steroid hormones involved in a wide range of physiological processes, including stress response, immune response, and regulation of inflammation, are the main steroidal drugs administered for inflammatory disease. Hydrocortisone and derivatives, prednisolone, dexamethasone, or budesonide are some examples of the mostly used glucocorticoids for the treatment of inflammation. Besides these steroid drugs, two additional nonsteroidal-based therapeutic approaches have recently emerged, which include the use of anti-leukotrienes (inhibit the recruitment of inflammatory cells) and pro-inflammatory cytokine inhibitors, like anti-TNF alpha or anti-IL-1 monoclonal antibodies. However, steroid drugs as well as nonsteroid drugs are sometimes not sufficient to obtain an optimal pharmacological activity. Limitations of these drugs include nonspecific biodistribution, low bioavailability, and/or short half-life in the body. Moreover, high dosages are needed to be administered to patients, which cause off-target side effects with only modest success in controlling inflammatory disease symptoms [32]. These limitations can be avoided by the use of NPs, with a size from a few tenths to a few hundred nanometers. Nowadays biocompatible nanomedicines (i.e. NPs loaded with active principle) with a highly controlled shape, size, and surface charge have gained increasing interest. The formulation of NPs regulating the expression of pro- and anti-inflammatory molecules and targeting inflammatory sensors or macrophages through phagocytosis can be a game changer in the treatment of inflammatory diseases. In addition, specific targeting of effector cells (antigen-presenting cells) using nanocarriers could be of great value to promote cellular response or immune tolerance. Furthermore, encapsulation of steroids using nanocarriers too can increase the drug efficacy. For example, prednisolone (glucocorticoids) nanocapsules, composed of a polymer wall surrounding an oily core, and coated with a pH-responsive enteric polymer (Eudragit S100), increase the selectivity and efficacy of prednisolone for the treatment of inflammatory bowel diseases (IBD) or other colonic diseases in healthy rats [41].

Rheumatoid arthritis (RA) is a chronic systemic inflammatory disease which affects around 1% of the general population worldwide. It is characterized by the destruction of articular cartilage and bone of the affected joints due to persistent polyarticular inflammation of the synovial tissues. Current therapeutic strategies to treat RA are based on anti-inflammatory drugs, which may be steroidal or nonsteroidal (nonsteroidal anti-inflammatory drugs, NSAIDs). NSAIDs possess adverse side effects along with limited effectiveness and the inability to modify disease courses in the long term. Disease-modifying antirheumatic drug (DMARD) is another popular method to treat RA; however, it also has a 70% failure rate. Anti-inflammatory drugs like glucocorticoids (GCs) along with other DMARDs are also widely used in active RA as co-therapy. However long-term use of GCs leads to a number of unwanted side effects, including increased risks of CVDs, osteoporosis, infections, and impaired glucose metabolism. Thus, RA remains a challenging disease as currently recommended drugs are often associated with the development of drug resistance and adverse events.

Hence, the development of a targeted nanomedicine/nano phytomedicine or nanocarrier system for RA is highly desirable. These nanocarrier systems will enhance the solubility of certain drugs and protect them against degradation in circulation, further increasing their local bioavailability. Therefore, the use of nanocarriers promises to increase drug specificity and bioavailability while reducing unwanted off-target side effects. The current nanotechnological approaches to treat RA include coated solid lipid nanoparticles (SLNs), lipid-microspheres, and dendrimers. The current approaches aim at decreasing adverse effects related to NSAIDs through site-specific delivery and controlled release. For lipophilic drugs such as NSAIDs, several groups of investigators found that lipid microspheres (LMs), lipid-based preparations with an internal oil phase surrounded by a phospholipid monolayer, offer better loading capacity compared with conventional liposomes. Early studies with LM preparations encapsulating the NSAID indomethacin indicated that encapsulation improved the anti-inflammatory activity of NSAIDs while lowering the gastrointestinal side effects [37]. However, LMs are rapidly cleared by the RES. The incorporation of polyethylene glycol (PEG) to LM delayed their uptake by the RES system threefold, prolonging their circulation time and increasing the bioavailability of the encapsulated NSAIDs. Systemic administration of PEG-LM encapsulating the NSAID indomethacin led to a significantly higher accumulation of the drug in the paws of arthritic rats, 7.5-fold higher compared with conventional LM [42]. Several reports of PEG-liposomes containing different GC preparations so far look promising in rodent models of experimental arthritis. By encapsulating the GC prednisolone into small PEG-liposomes (~100 nm in size), Metselaar et al. showed that these nanocarriers remained in the circulation with a half-life of 50 hours and a single systemic administration of this preparation led to a complete reversal of paw inflammation within 2 days of injection, with the effect lasting for 2 weeks [43].

9.4.4 Nanophytomedicine Against Diabetes and Metabolic Disorders

Metabolic syndrome (MetS) includes a series of metabolic abnormalities that lead to diabetes mellitus and CVDs. MetS, also known as "syndrome X" and "insulin-resistance syndrome," is characterized by several metabolic abnormalities, including insulin resistance, type 2 diabetes, obesity, hypertension, and dyslipidemia. About 20–30% of the world population is diagnosed with MetS, which makes the disease a global health issue.

MetS is the result of a series of genetic and environmental factors; however, the exact etiology is not yet understood. The underlying mechanisms encompass insulin resistance, elevated plasma free fatty acids, chronic inflammation, and oxidative stress [44]. Insulin resistance is the most widely accepted hypothesis for the underlying pathophysiology of MetS. The increased level of free fatty acids results in suppression of insulin clearance and is closely associated with insulin resistance in obese individuals. To overcome the resistance, the pancreas secretes more insulin, leading to hyperinsulinemia. Free fatty acids cause induction and suppression of protein kinase in the liver and the muscle cells, respectively, which subsequently increases gluconeogenesis in the liver and diminishes glucose uptake in muscles. Chronic inflammation is implicated in visceral obesity and exacerbates insulin resistance, which is characterized by the abnormal production of adipocytokines such as tumor necrosis factor-a (TNF-a), interleukin-1 (IL-1), IL-6, leptin, and plasminogen activator inhibitor-1 (PAI-1). Adiponectin is an important anti-inflammatory and antiatherogenic adipokine and is considered a protective factor against the development and progression of chronic diseases related to metabolic disorders and oxidative stress including diabetes, hypertension, and CVDs [45]. Oxidative stress induces insulin resistance and also abrogates adiponectin production by adipocytes [46].

Currently, the management of MetS involves lifestyle modification, which consists of particular recommendations on physical activity and dietary interventions to achieve a normal weight, modulation of glycemic and lipid profile, as well as a decrease in blood pressure [47]. The conventional treatment guideline for MetS involves the drugs that lower the blood glucose level, TGs, and blood pressure along with improvement of the patient's lifestyle. These common medications exert unpleasant adverse effects such as flatulence and other related side effects in the early stage of treatment. Furthermore, the long-term application of medications leads to a weak tolerance of the patient. Several studies have shown the beneficial therapeutic effects like anti-inflammatory, antioxidant, and insulin-sensitizing properties of medicinal plants and their isolated compounds [48]. Oral administration of plant extracts is shown to reduce visceral obesity, systolic and diastolic blood pressure, and blood glucose. Additionally, medicinal plants enhance insulin secretion and cardiovascular function, while suppressing gluconeogenesis, inflammation, and oxidative stress [48, 49]. However, the bioactivity of these plant extracts is diminished due to their low bioavailability.

Thus, in order to decrease the side effects of these medications along with their increased efficacy and bioavailability, nanoformulation of natural and synthetic agents is a promising option. Nanosized drug carriers, consisting of phytochemicals from conventional medicines endowed with developed pharmacodynamics and pharmacokinetic characteristics, are a novel therapeutic approach. The nano-vehicles provide unique properties including enhanced bioavailability and solubility of the drug, diminished systemic adverse effects, elongated circulation time, and privileged accumulation in the precise target organ as compared to conventional drug delivery systems, which are often accompanied by some critical limitations such as high dosage, low efficacy, low bioavailability, lack of target specificity, and dose-dependent side effects [50, 51]. Therefore, different phytochemicals including curcumin, berberine (BBR), oleoresin capsicum, naringenin, emodin, gymnemic acid, resveratrol, quercetin (QUE), scutellarin, stevioside, silybin, baicalin, and others have been incorporated into diverse nanoformulation structures including polymeric-based polyesters, polyanhydride, poly(alkyl cyanoacrylate), etc., and natural ones (including cellulose, collagen, albumin, chitosan, dextran, etc.) for treatment and alleviation of MetS. Lipid-based drug delivery systems such as micelles, liposomes, SLNs, nanostructured lipid carriers (NLCs), and other related nanoformulations like nanoemulsions and polymeric NPs have also been included. In vitro and in vivo investigations showed that nano-drug delivery systems such as nanomicelles, liposomes, and hydrogel-based nanocomposites can provide drug targeting to a specific site [52, 53]. In the case of diabetes and diabetes-associated dysfunctions, effective delivery of insulin via oral route by these nanoformulations is highly preferred compared to the available parenteral preparations, due to higher patient compliance [54, 55].

Nanostructured formulations of herbal extracts can potentially enhance their antidiabetic properties by enhancing their pharmacokinetics and bioavailability [56]. Several studies have shown these nanoformulations based on herbal extracts.

Marrubiin, a diterpenoid in the *Leonotis leonurus* (L.) R. Br., has a preventive and therapeutic effect on diabetes. Marrubiin has the ability to induce insulin secretion, increase insulin sensitivity, and upregulate the glucose transporter type (GLUT)-2 gene [57]. But this phytochemical is poorly soluble in water. NLCs based on acetonic leaf extract of *L. leonurus* stimulated insulin release in INS1 pancreatic b cells and elevated glucose uptake in Chang liver cells compared to the extract alone.

Ficus religiosa L., commonly known as the peepal tree, possesses several pharmacological activities including antioxidant, anti-inflammatory, antidiabetic, and neuroprotective activity. SLNs prepared from ethanolic stem bark extract of *F. religiosa* showed pronounced hypoglycemic and insulin-sensitizing effects compared with the extract suspension. The nanoformulations based on SLNs provided an initial burst release followed by sustained release [58] and, hence, can be the subject of future studies.

Syzygium cumini (L.) Skeels [Synonym: *Syzygium jambolanum* (Lam.) DC.] is an Indian medicinal plant with previously demonstrated antidiabetic activity [59]. High content of polyphenols in leaf extract of *S. cumini* is responsible for antidiabetic and anti-inflammatory effects [60]. The encapsulation in polymeric NPs can augment the ameliorative effects of *S. cumini*. PLGA is a biodegradable and biocompatible polymer made from lactic acid and glycolic acid monomers [61]. The *S. cumini*-based nanoformulation caused a significant increase in glucose uptake, glucokinase activity, and GLUT4 protein expression in L6 rat skeletal muscle cells.

Curcumin is a derivative of turmeric (*Curcuma longa*) and belongs to the curcuminoid subgroup of polyphenols. Curcumin has numerous pharmacological activities such as antitumor, antioxidant, anti-inflammatory, hypolipidemic, and anticarcinogenic effects [64, 66] and has also been shown to possess therapeutic effects regarding MetS. SNEDDS is a branch of SEDDS (self-emulsifying drug delivery system), which contains an isotropic mixture of drug substance, surfactant, co-surfactant, and oil, resulting in the formation of a nanoemulsion just after consumption in the gastrointestinal tract. SNEDDS based on a curcumin formulation has been developed for alleviating diabetic neuropathy and found to be an efficient tool for relieving neuroinflammation, and modifying antioxidant defense in diabetic neuropathy [62].

Argyreia nervosa (Burm. f.) Bojer, from the family Convolvulaceae has been used in traditional Indian medicine for many therapeutic aspects such as antidiabetic, anti-inflammatory, and diabetic wound healing [63]. Silver NPs (AgNPs) prepared using an aqueous leaf extract of *A. nervosa* showed in vitro inhibitory effects on α-amylase and α-glucosidase, which are important enzymes in carbohydrate metabolism with IC50 values of 55.5 and 51.7 mg/ml, respectively.

Costus speciosus (J. Koenig) Sm. (synonym *Cheilocostus speciosus* (J. Koenig) C. D. Specht) has shown an antidiabetic effect via induction of insulin secretion and improvement in insulin sensitivity [64]. Ethanolic leaf extract of *C. speciosus* was loaded in PLGA NPs to enhance bioactivity and provide sustained release of active constituents. PLGA NPs increased the expression of insulin (I and II), and GLUT4 genes; while decreasing the expression of the GLUT2 gene. In addition, the nanoformulation diminished blood sugar, enhanced high-density lipoprotein cholesterol (HDL-C) and decreased total cholesterol (TC), triacylglycerol (TAG), and low-density lipoprotein cholesterol (LDL-C) in STZ-induced diabetic rats. The nanoformulated extract was more effective in controlling the lipid profile and blood glucose in comparison to the simple extract [65].

Marsilea quadrifolia L. is known as an "aquatic fern," with antidiabetic and antioxidant activities. Biosynthesized AuNPs with methanolic extract of *M. quadrifolia* elevated glucose utilization in 3T3 L1 adipocytes, as an in vitro model of diabetes, in a dose-dependent manner. Polyphenols and flavonoids of the extract attached to the surface of AuNPs improved bioavailability, induced transmitted GLUT4 vesicle to the cell membrane, and probably increased glucose uptake in adipocytes [66].

The ginger extracted from the *Zingiber officinale* is considered an antidiabetic agent through interfering with the NF-kB cascade pathway and displaying anti-inflammatory and antioxidant properties. The application of nanotechnology provides a useful delivery system for ginger compounds and enhances their biological activities [67]. Nano transdermal delivery for a ginger ingredient, gingerol, increased its bioavailability ratio. In an in vivo study performed by fabricated AgNPs using ethanolic extract of ginger rhizome exhibited notable hypoglycemic properties.

Chamaecostus cuspidatus (Nees & Mart.) C. D. Specht and D. W. Stev. is known as an "insulin plant" with antidiabetic and antioxidant activities [68]. AuNPs green-synthesized using aqueous leaf extract of *C. cuspidatus* elevated body weight, decreased TC, and abolished lipid peroxidation, hydroxyl radicals, and nitric oxide in STZ-induced diabetic rats. Additionally, AuNPs slightly decreased blood glucose and increased glycogen and insulin level.

Altogether, several findings revealed that nanostructured formulations of plant extracts were able to overcome the drawbacks of chemical counterparts, affecting the promising effects of plant extracts in diabetes. Considering the critical role of herbal-based nanoformulations in combating inflammation and oxidative damages, future studies should focus on other aspects of their uses in the management of oxidative and inflammation-related diseases.

9.4.5 Nanophytomedcine in CVDs and Heart Diseases, Hypertension

CVDs are a major cause of disability and they are currently among the most significant and ever-growing concerns during the last several decades. CVDs are the most lethal diseases worldwide and are responsible for a significant number of deaths in a large percentage of the world population. In 2015, around 17.7 million deaths were associated with CVDs worldwide, which corresponds to about 31% of all global deaths. According to a World Health Organization (WHO) report, nearly 7.4 and 6.7 million deaths are caused by coronary heart diseases and stroke, respectively [69, 70]. Hypertension and hypercholesterolemia are two key risk factors for the emergence and development of CVDs like acute myocardial infarction, stroke, thrombosis, etc. [71]. The most common controllable risk factors for CVDs include a high degree of psychosocial factors, overconsumption of alcohol, the elevated value of low-density lipoprotein (LDL) and cholesterol, hypertension, insufficient consumption of fruits and vegetables, diabetes mellitus, abdominal obesity, smoking, and the lack of regular physical activity. Plaques may form due to the accumulation of cholesterol, fat, calcium, and other harmful substances, leading to atherosclerosis or hardening of the arteries, cerebrovascular disease, coronary artery disease, cardiac arrhythmias, and the subsequent development of heart failure can occur [72, 73].

Multiple drug therapies based on the use of synthetic active ingredients are known for the treatment of CVD. However, the use of these drugs is frequently associated with the occurrence of adverse effects. Furthermore, most of them are not much efficient to prevent or significantly reduce CVD's progression and prevalence. In this context, it is appealing to implement nanotechnology and phytochemical-based nanoformulations as an attractive and innovative tool for the treatment of CVDs. Encapsulation of natural products and their derivatives provides unique advantages, including reduced systemic adverse side effects, enhanced biosafety, high solubility and bioavailability of drugs, extended circulation time, and restricted accumulation in target organs. Diverse organic and inorganic nanostructures, including NPs, have been developed to deliver phytochemicals such as curcumin, emodin, gymnemic acid, tilianin, puerarin (PUE), BBR, QUE, scutellarin, magnolol, breviscapine, resveratrol, baicalin, naringenin, and so on [74, 75].

9.4.5.1 Curcumin

Curcumin, an antioxidant, is functional against oxidative stress-induced damages. Rachmawati et al. evaluated the antihypercholesterolemic and antihypertensive effects of curcumin nanoemulsion in vitro using acetylcholinesterase (ACE) and 3-hydroxy-3-methylglutaryl coenzyme A reductase (HMGR) assay kits. The results revealed that nanoemulsified curcumin slightly increased its inhibitory effect on ACE compared with pure curcumin. In another study, the encapsulation of curcumin into multipolymer poly(γ-benzyl-L-glutamate)–poly(ethylene glycol)–poly(γ-benzyl-L-glutamate), known as PBLG–PEG–PBLG, improved its water solubility, biological activity, and dispersion in aqueous solutions in a diabetic cardiomyopathy model. In vivo administration of curcumin/P enhanced the serum levels of H_2S and Ca_{2+i} content in myocardial cells and upregulated the expression of cystathionine-γ-lyase, calmodulin, and calcium-sensing receptors. Cardioprotective effects of curcumin NPs were also shown in ISO-induced AMI in diabetic rats, attributed to their antioxidant and anti-inflammatory features [76]. Treatment with curcumin NPs (50 mg/kg ip) attenuated the development of right ventricular hypertrophy, decreased right ventricle weight to the bodyweight ratio, decreased right ventricle mRNA expression of TNF-α and IL-1β, and diminished oxidative stress [77, 78]. Administration of curcumin capped gold-loaded poly (lactic-coglycolic acid) NPs (CAu-PLGA Nps) inhibited cardiac hypertrophy via increasing solubility, bioavailability, and absorption, improving resistance to enzymatic degradation and drug delivery efficiency in a rat model.

9.4.5.2 Ginsenoside Rg3

Ginsenoside Rg3 (Rg3), an active constituent of *Panax ginseng*, has anti-inflammatory, immunomodulatory, antioxidant, and antiaging activities in different diseases [71, 79, 80]. Rg3 loaded in Pluronic F127 micelles (P-Rg3) effectively reduced doxorubicin-induced mortality, preserved the peak aortic blood velocity, and decreased the levels of atrial natriuretic peptide and B-type natriuretic peptide in C57/BL mice. P-Rg3 was more effective than Rg3 at protecting the heart against doxorubicin-induced cardiotoxicity and prevented depolarization of the mitochondrial membrane in H9C2 cells. Thus, co-administration of the polymeric micellar encapsulated natural products with chemotherapeutic agents increased their efficacy, decreased their toxicity, and were more effective than drug administration alone [71].

9.4.5.3 Panax Notoginsenoside

Panax notoginsenoside (PNS) is the primary active component of *P. notoginseng* [71]. PNS is poorly absorbed after oral administration [81]. Combination of PNS with a liposomal system [encapsulation of methyl ether poly(ethylene glycol)–poly(lactide-*co*-glycolide) (mPEG-PLGA)-based NPs] enhanced the entrapment efficiency. Treatment with PNS-core–shell hybrid liposomal vesicles (HLV) significantly decreased infarction volume and improved biochemical parameters by improving its physicochemical properties [82].

9.4.5.4 Berberine

BBR, derived from *Berberis* spp., is a small fluorescent isoquinoline quaternary alkaloid. Despite the potent anti-inflammatory, antioxidative, and cardioprotective activities of BBR, its application is limited due to low solubility in aqueous buffers, short half-life in blood circulation, and low absorption rate in the intestine [83, 84]. In a mouse model of myocardial infarction, liposomal encapsulation of BBR improved its solubility, reduced adverse cardiac remodeling, and preserved left ventricular ejection. Furthermore, BBR encapsulation into long-circulating liposomes enhanced the accumulation of the compound in infarcted heart tissue and improved its local delivery and treatment efficacy [71].

9.4.5.5 Tilianin

Tilianin, a natural flavonoid compound extracted from *Dracocephalum moldavica*, has been used to treat cardiovascular conditions such as coronary heart disease, hypertension, atherosclerosis, and myocardial ischemia [85]. Pretreatment of H/R-induced H9c2 cardiomyocytes with tilianin-loaded micelles (TLMs), reduced the LDH levels, preserved the ROS production, increased cell viability, and significantly reduced the MDA level in the cardiomyocytes.

9.4.5.6 Puerarin

PUE, a primary active ingredient derived from Radix puerariae, acts as an antioxidant and has been widely used to improve CVDs [86–88]. PUE-loaded PEG-PE (1,2-Distearoyl-sn-glycero-3-phosphoethanolamine-*N*-[methoxy(polyethylene glycol)-2000]) micelles improved the bioavailability of PUE and extended its circulation time. Furthermore, administration of PUE-loaded TPP/PEG-PE (triphenylphosphonium [TPP] cation) enhanced the anti-apoptosis effect of PUE in ISO-induced H9c2 cells. PUE@TPP/PEG-PE micelles enhanced the intracellular uptake of the desired drug in the ischemic zone. The system reduced the ROS level, Bax expression, and caspase-3 activity while increasing the Bcl-2 expression. Overall, PUE@TPP/PEG-PE micelles were suggested as a potent nanocarrier system to treat acute myocardial infarction.

9.4.5.7 Quercetin

QUE, a flavonoid isolated from various citrus fruits, tea, onions, and apples, exerts significant antioxidant, anti-inflammatory, and antihypertensive activities [89, 90].

QUE-loaded phosphatidylcholine liposomes (PCLs) were found protective against myocardial injuries induced by peroxynitrite by directly acting as both the scavenger and decomposer of endogenously formed peroxynitrite ions. In another study, polymeric micelles loaded with a combination of resveratrol and QUE were used to assess their cardioprotective actions against doxorubicin-induced cardiotoxicity. This combination was effective in attenuating the cardiotoxicity in both in vivo and in vitro experiments [91]. Furthermore, Giannoulia et al. fabricated QUE-loaded polymeric PLGA NPs by electrohydrodynamic atomization process to improve its drawbacks and prevent atherosclerosis. QUE demonstrated excellent potential for preventing atherosclerosis and other CVDs, attributed to its biphasic release (24 hours and 59 days, respectively) from NPs [92].

9.4.5.8 Baicalin

Baicalin is one of the significant bioactive flavone glucuronides derived from Scutellaria Radix. It has been used to alleviate diabetes and CVDs [93]. Various nanosized formulations have been used to overcome low hydrophilicity, short half-life, and poor absorption of baicalin following oral administration. Baicalin-loaded PEGylated NLCs (BN-PEG-NLCs) prolonged their plasma circulation time, enhanced their delivery to the ischemic area of the heart, and reduced the infarct size in the acute myocardial infarction model [94].

9.4.5.9 Resveratrol

Resveratrol is a polyphenol commonly found in berries, grapes, and peanuts with a broad spectrum of pharmacological activities such as cardioprotective. Administration of resveratrol nanocapsules significantly reduced both the systolic and diastolic blood pressures and alleviated insulin resistance in mice due to reduced whole-body fat [95]. In another study, administration of resveratrol-loaded SLNs significantly increased the heart rate, ejection fractions, and fractional shortening in doxorubicin-induced cardiotoxicity mice. The myocardial fibers were also arranged regularly, and a few vacuole degenerations were seen in the myocardial cells [96].

9.4.6 Nanomedicine in Neurological Disorders

Central nervous system (CNS) disorders, especially neurodegenerative disorders, are a major challenge for public health. Neurological disorders are among the highest reasons for causing death and disability in the world. To date, the diagnosis and treatment of the neurological disorders such as Alzheimer's disease (AD), Parkinson's disease (PD), Huntington's disease (HD), head trauma, brain tumor, and epilepsy are still a challenging task [97]. A wide spectrum of potential drugs are in use to treat several neurological disorders but they face a range of challenges and hence have limited therapeutic success. One of the major and most common challenges is the difficulty in delivering agents such as drugs, nucleic acids, proteins, imaging agents and other macromolecules to the CNS across the peripheral barriers, namely blood–brain barrier (BBB) and blood–cerebrospinal fluid barrier (BCSFB), particularly the BBB [98, 99]. Thus, the current neurotherapeutics are associated with two major limitations, i.e. before entering into brain cells, they are posed to limited entry via BBB as an insufficient entry in the brain as well as limited access to immune cells to the brain [100].

The inability of drugs to cross the BBB highlights the need for developing NPs-based strategies targeted toward drug delivery. Several NPs have been used for crossing the BBB. The NPs penetrate through the tight junctions between the endothelial cells of the vessels and enable the drug to pass through the BBB. Following are the nanomaterials used for the transfusion from BBB.

9.4.6.1 Lipid Nanoparticles (NPs)

In 1961, Müller and Gasco discovered lipid NPs. Lipid NPs mainly include SLNs, NLCs, and liposomes. Encapsulation of drugs in lipid NPs [101] elongates the time of drug distribution in the bloodstream, decreases the side effects of the medication, and increases its therapeutic effects on CNS disorders. These characteristics make lipid NPs very valuable for drug delivery to the CNS [102]. Liposomes are rapidly absorbed by the macrophages, microglia, and steroids in the CNS. Gurthork et al. introduced this nanosystem to be effective for drug delivery to the CNS by passing the BBB. They designed a liposomal drug delivery system for transferring levodopa (used for the treatment of Parkinson's disease) through the BBB.

SLNs are aqueous colloidal nanocarrier systems, which are composed of physiological lipid (triglycerides, fatty acids, steroids, waxes, etc.), dispersed in water or in an aqueous surfactant solution. Sterylamine-based SLNs containing clozapine, an antipsychotic drug, have been fabricated and demonstrated to deliver the drug successfully inside the brain after IV and intraduodenal administration [103].

9.4.6.2 Polymeric Nanoparticles (NPs)

Polymeric NPs are solid colloidal dispersion of biodegradable and biocompatible polymers such as poly(alkylcyanoacrylate) (PACA), polyesters such as poly(lactide) (PLA), poly(D,L-lactide-*co*-glycolic acid) (PLGA), and several others. Polycyanoacrylate NPs have been evaluated for the delivery of several drugs to the CNS based on polyethylene glycol (PEG). These NPs are coated with PEG-containing surfactant to increase drug diffusion. Polystyrene nanospheres (20–200 nm) showed suitable properties for drug delivery to the CNS. It has been established that coating of poly(*n*-butyl cyanoacrylate) (PBCA) NPs with 1% polysorbate 80 (PS80) amplified the concentration of rivastigmine or tacrine drug inside the brain as compared with free drug and selectively targeted to the CNS for AD reducing the hepatic or gastrointestinal side effects coupled with conventional treatment approach [104].

9.4.6.3 Micelles

Polymeric micelles have been developed for drug delivery in recent years [105, 106]. These micelles form spontaneously in amphiphilic copolymer solutions and show shell-core structures, which consist of hydrophobic block polymers (e.g. L,D-lactine polycaprolactone) as a core and hydrophilic block polymer (usually PEGs) as a shell. These structures improve drug stability and bioavailability [107]. The micelle shell protects drugs from interacting with serum proteins and nontarget cells.

9.4.6.4 Dendrimers

Dendrimers are the emerging polymeric structures having 3-D symmetrical architecture possessing an inner core from which a number of hyperbranches, known as "generations," having various chemical functional groups at the peripheral terminal surface arise. Nowadays, as a drug carrier, dendrimers are given more preference over traditional polymers due to various features such as high aqueous solubility, biocompatibility, polyvalency, and precise molecular weight which make them amenable for targeted drug delivery in the brain [108]. Researchers synthesized poly(propylene imine) dendrimers core and maltose-histidine shell (G4HisMal) and checked their biocompatibility and ability to penetrate BBB. They tested them against AD and found their high ability to cross BBB with significant biocompatibility showing memory protection in transgenic mice model of Alzheimer's disease via synapse protection.

9.4.6.5 Nanoemulsions

Nanoemulsions are either oil-in-water (O/W) or water-in-oil (W/O) colloidal particulate systems that are composed of edible oils, surface-active agents (surfactants), and water, having a size range of 100–500 nm [109, 110]. The fabrication of nanoemulsions with oils that are rich in omega-3 polyunsaturated fatty acids (PUFA) imparts specific properties to nanocarriers for overcoming a biological barrier, including the BBB, therefore helping in achieving the rapid distribution of drugs to peripheral sites, especially the brain [111].

9.4.6.6 Nanogels

Nanogels are networks of nano-sized polymers, which form ionic or non-ionic chains, such as polyethylene amine and PEG, or polyacrylic acid and pluronics. The modification of nanogel surface with transferrin and insulin enhances distribution through BBB [112]. According to the in vivo studies, nanogels increase the absorption of oligonucleotides in the brain while reducing their absorption in the liver and spleen [113]. Nanogels are generally promising transporters for drug delivery to the CNS.

9.5 Nanophytomedicine as Nutraceuticals/Functional Foods

Foods contain an array of phytochemicals that are associated with health benefits and offer an excellent opportunity to improve public health and well-being. These phytochemicals are glucosinolates, sulfur-containing compounds of the Alliaceae, terpenoids (carotenoids, monoterpenes, and phytosterols), and various groups of polyphenols (anthocyanins, flavones, isoflavones, stilbenoids, ellagic acid, etc.). This wide range of products cannot be truly classified as "food" and a new hybrid term between nutrients and pharmaceuticals, "nutraceuticals," has been coined to designate them. The term "nutraceutical" was invented in 1989 by Stephen L. Defelice, who established The Foundation for Innovation in Medicine in 1976 [114]. There is no internationally recognized definition of a nutraceutical, and various confusing and contradictory definitions have appeared. In general, nutraceuticals are more than food ("nutrition") but less than "pharmaceuticals." According to the US Nutraceutical Research and Education Act, which was presented to the House of Representatives in the first session of the 106th Congress of 1999–2000, "Nutraceuticals" are a dietary supplement, food, or medical food . . . that (i) has a benefit which prevents or reduces the risk of a disease or health condition, including the management of a disease or health condition or the improvement of health; and (ii) is safe for human consumption in the quantity, and with the frequency required to realize such properties [115].

On the contrary, functional foods are different from nutraceuticals. Functional foods are whole, fortified, enriched, or enhanced foods that provide health benefits more than vitamins and minerals when they are consumed as part of a varied diet on a regular basis. Like nutraceuticals, functional foods have no universally accepted definition. It can be considered similar to conventional foods, which are consumed as part of a usual diet but are known to improve health status beyond primary nutritional function.

Basic differences between nutraceuticals and functional foods are that nutraceuticals are the products produced from foods but sold in medicinal forms of either a capsule, tablet, powder, solution, or potion, which is not generally associated with the food and have demonstrated physiological benefits and/or provide protection against chronic diseases while functional foods are whole, fortified, enriched, or enhanced foods.

9.5.1 Benefits of Nutraceuticals

Nutraceuticals have shown several beneficial effects on human health over the years; for example, they have been used to treat inflammation, arthritis [116], cholesterol [117], diabetes [114], and many conditions [118]. A mixture of nutraceuticals (chlorogenic acid, vitamin C, bergamot extract, and phytosterols), administered to overweight subjects with dyslipidemia, was able to increase the metabolism of lipids and glucose compared with overweight subjects treated with placebo [119, 120]. Numerous scientific works in the literature demonstrated how the use of nutraceuticals is related to the control of blood cholesterol levels in order to reduce the risk of CVD [121, 122].

9.5.2 Nanotechnology in Nutraceuticals

Nanotechnology, based on structures with sizes in the order of nanometers, is a revolutionary technology that has allowed researchers to overcome numerous limitations related to the use of nutraceuticals following their encapsulation into these structures, such as their stability, low solubility, and poor bioavailability. It is noteworthy that the bioavailability, that is the fraction of a taken compound, which is absorbed and available for physiological functions, is a key aspect of nutraceutical compounds because their effectiveness is strictly related to their bioavailability [123, 124]. Unfortunately, different endogenous and exogenous factors can compromise the bioavailability of nutraceuticals, such as the biochemical transformations they may undergo in the epithelial cells, their physicochemical features, the food storage, and so on [125]. For this reason, many innovative strategies have been thought to exert their beneficial effects when introduced into the organism. Among them, different nanoformulations have been designed to enhance the beneficial effects of nutraceuticals. The nanoencapsulation technique provides the possibility to protect the chemical structure of nutraceuticals from environmental agents such as pH, light, temperature, radicals, or oxygen [126, 127]; increases their bioavailability; allows specific delivery to target sites [128, 129]; and allows a controlled release of the encapsulated compound [130]. Consequently, nanoencapsulation of nutraceuticals could enhance their positive effects on human health, thus reducing their side effects.

9.5.3 Nanophytoantioxidants/Nanoflavanoids

The antioxidant activity of natural compounds is known to have a lot of beneficial effects on human health [131–133]. Antioxidants protect the organism, acting as free radical scavengers, thus preventing oxidative damage. In fact, their efficacy in the prevention of CVDs and cancer, their anti-inflammatory activity, and their ability in the prevention and management of Alzheimer's disease are well known [124]. However, almost all antioxidants are insoluble in water and hence biologically less available to the body. In order to increase their bioavailability, encapsulation is needed and here comes the role of nanotechnology.

Curcumin, a polyphenol with antioxidant and anti-inflammatory properties, is hydrophobic in nature. Soy-conglycinin (a storage globulin) was employed for the development of NPs for the encapsulation of hydrophobic curcumin. -conglycinin, which is the vicilin storage protein of soybeans, was performed using urea and without adding any organic solvent; the obtained NPs, produced with this new technology, turn out to be more natural and are characterized by a good solubility and encapsulation efficiency (around 80%) greater than that obtained in previous works.

The bioaccessibility of curcumin was found to be around 40% (while that of free curcumin was found to be around 20%).

Resveratrol is another nutraceutical with antioxidant properties but like curcumin it is also insoluble in water. To overcome this difficulty, the encapsulation into cyclodextrins was recently proposed; in particular, it was evidenced that complexation produced a consistent improvement in the solubility of resveratrol in water and consequently a significant improvement in the anti-cancer activity of resveratrol on several cell lines. Another approach to improving the oral bioavailability of resveratrol was the encapsulation into zein NPs. The NPs were administered to human volunteers in aqueous suspension, obtaining satisfactory plasma levels of resveratrol compared with those found in literature data, in which free resveratrol was employed. This is probably related to the increase in resveratrol oral bioavailability when administered in zein NPs [134].

SLNs and NLCs are two lipid based nanocarrier systems [135], widely applied in food and nutraceuticals fields. SLNs and NLCs proved to be suitable carriers for oral administration of resveratrol.

In fact, in the experimental work of Neves and collaborators, these two lipidic nanocarriers allowed an encapsulation efficiency of around 80%, proved to protect resveratrol and to be well tolerated at the level of intestinal mucosa; furthermore, NLCs, and to a lesser extent SLNs, improved resveratrol gastrointestinal permeability, since they are able to cross the intestinal barrier, as demonstrated by the *in vitro* tests on Caco-2 cells.

QUE and linseed oil are two bioactive agents with extraordinary therapeutic properties, such as anti-cancer, anti-inflammatory, and antioxidant, but, like all lipophilic nutraceuticals, their use is limited by their poor water solubility and their instability with regard to environmental conditions. The co-encapsulation of QUE and linseed oil in NLCs solved these problems.

Oleuropein, tyrosol, and hydroxytyrosol are three compounds extracted from various parts of *Olea europaea* L., a typical tree of the Mediterranean region, largely known for its antioxidant activities [136]. These nutraceuticals were encapsulated in liposomes in order to protect them from degradation phenomena and to improve their bioavailability in human chondrocyte cells.

Flavonoids are very sensitive to environmental conditions and are therefore very subject to degradation phenomena. Luteolin, QUE, and kaempferol were successfully encapsulated in liposomes.

9.6 Opportunities, Scopes, and Challenges

Nanomedicine is one of the fastest-growing areas in nanotechnology and is poised to revolutionize healthcare and medicine through transformative new diagnostic and therapeutic tools. Nowadays, nanotechnology is considered a worldwide priority. The emerging and potential commercial applications of nanotechnologies clearly have great potential to significantly advance and even potentially revolutionize various aspects of medical practice and medical product development. Nanotechnology is already touching upon many aspects of medicine, including drug delivery, diagnostic imaging, clinical diagnostics, nanomedicines, and the use of nanomaterials in medical devices. Nanotechnology is also addressing many unmet needs in the pharmaceutical industry, including the reformulation of drugs to improve their bioavailability or toxicity profiles.

Nanomedicine has the potential advantages to overcome biological barriers, effectively deliver hydrophobic drugs and biologics, and preferentially target sites of disease [137]. However, only a relatively small number of NP-based medicines have been approved for clinical use because of numerous challenges and hurdles at different stages of development [138]. The first FDA-approved nano-drug is Doxil® in 1995, which can prolong drug circulation time and is avoidance of the RES due to the use of PEGylated nano-liposomes.

Successfully translating nanomedicine from preclinical proof of concept to demonstration of therapeutic value in the clinic is still challenging. Following are the major challenges:

1) **Biological barriers and specific targeting**
   Before the drugs successfully reach the disease sites, the nano-formulation has to help them cross multiple biological barriers. For example, the blood–brain barrier (BBB) restricts the diffusion of 99% large or hydrophilic molecules into the cerebrospinal fluid and is a major obstacle for brain targeting delivery.
2) **Characterization of nano-formulations**
   Due to variable properties, it is very difficult to characterize nanomedicine products. For example, biodegradable properties of nanomaterials would also alter the properties of nanomedicine products during the storage either in the solutions or in a lyophilized powder form. Second, nanomedicine products with the same average size but different polydispersities (PDs) may differ in secondary properties such as targeting properties, drug release rate, biocompatibility, toxicity, and in vivo behaviors.
3) **Manufacturing**
   The key point for successful manufacturing of any medicine is controlling and keeping constant stability of physicochemical properties on a batch-to-batch basis. In the case of nanomedicine, it is very difficult to control the polydispersity of nanomaterial, which could cause batch-to-batch variations of physical and chemical properties.
4) **Toxicity**
   The nanoscale dimension of nanomedicine products can simulate the intracellular organelles or biomolecules such as protein, polysaccharide, and enzyme involved in cell signaling, which may lead to detrimental biological interactions leading to toxicity. Nonetheless, it remains difficult to evaluate the toxicity of nanomaterials, especially in vivo and long-term toxicity. At present, the assessment of NP toxicity using classical drug toxicity assays may be inadequate. And there is no standard list of required tests.
5) **Regulatory challenges**
   There are no specifically implemented guidelines for drug products containing soft materials by FDA, EMA, and other regulatory bodies yet. Regulatory guidelines for the examination of nanomedicine and nanomedicine therapeutics are lacking.

9.7 Conclusion

In the past few decades, a diverse range of therapeutic formulations has been developed for combating all kinds of diseases, whether it is genetic or infectious or metabolic. These medications have greatly improved global health and have reduced the disease burden. However, it is well known that despite significant improvements in treating diseases, these formulations have adverse effects on human health, which cannot be neglected. Most of the currently available allopathic medications have less/severe side effects on human health. Hence in recent years, to overcome all these limitations, drawbacks, and side effects of allopathic medicine, the concept of nanomedicine and nano phytomedicine has emerged with dramatic response in curing disease without any harmful effects. Seeing the advantage of nanomedicine over traditional therapeutic formulations, there is an exponential increase in the field of nanomedicine. In the coming future, there are limitless opportunities for nanomedicine. There is a need to advance nanomedicine, which can only be achieved by expanding cross-disciplinary training for researchers, medical care providers, and public health professionals working in the industry, government, and academia. Research and development in the field of novel nanomedicine applications will transform the way of treatment along with human health across the world.

References

1 Sahoo, S.K., Parveen, S., and Panda, J.J. (2007). The present and future of nanotechnology in human health care. *Nanomedicine: Nanotechnology, Biology and Medicine* 3 (1): 20–31.

2 Silva, G.A. (2004). Introduction to nanotechnology and its applications to medicine. *Surgical Neurology* 61 (3): 216–220.

3 McNamara, K. and Tofail, S.A.M. (2015). Nanosystems: the use of nanoalloys, metallic, bimetallic, and magnetic nanoparticles in biomedical applications. *Physical Chemistry Chemical Physics* 17 (42): 27981–27995.

4 Saadeh, Y. and Vyas, D. (2014). Nanorobotic applications in medicine: current proposals and designs. *American Journal of Robotic Surgery* 1 (1): 4–11.

5 Oliveira, O.N. Jr., Iost, M.R., Siqueira, J.R. Jr. et al. (2014). Nanomaterials for diagnosis: challenges and applications in smart devices based on molecular recognition. *ACS Applied Materials & Interfaces* 6 (17): 14745–14766.

6 DeJong, J. (2003). *Making an Impact in HIV and AIDS*. London: ITDG Publishing.

7 Holzinger, M., Le Goff, A., and Cosnier, S. (2014). Nanomaterials for biosensing applications: a review. *Frontiers in Chemistry* 2 (2014): 63.

8 Chan, W.C.W., Maxwell, D.J., Gao, X. et al. (2002). Luminescent quantum dots for multiplexed biological detection and imaging. *Current Opinion in Biotechnology* 13 (1): 40–46.

9 Euliss, L.E., DuPont, J.A., Gratton, S. et al. (2006). Imparting size, shape, and composition control of materials for nanomedicine. *Chemical Society Reviews* 35 (11): 1095–1104.

10 Lin, Y., Taylor, S., Li, H. et al. (2004). Advances toward bioapplications of carbon nanotubes. *Journal of Materials Chemistry* 14 (4): 527–541.

11 Ciofani, G., Raffa, V., Menciassi, A. et al. (2009). Boron nitride nanotubes: an innovative tool for nanomedicine. *Nano Today* 4 (1): 8–10.

12 Ghoshal, S., Mitra, D., Roy, S. et al. (2010). Biosensors and biochips for nanomedical applications: a review. *Sensors & Transducers* 113 (2): 1.

13 Navarro, M. and Planell, J.A. (2012). Is nanotechnology the key to unravel and engineer biological processes? *Nanotechnology in Regenerative Medicine* 811: 1–16.

14 (a) Pan, Z.W., Dai, Z.R., and Wang, Z.L. (2001). Nanobelts of semiconducting oxides. *Science* 291: 1947; (b) Kong, X.Y. and Wang, Z.L. (2003). Spontaneous polarization-induced nanohelixes, nanosprings, and nanorings of piezoelectric nanobelts. *Nano Letters* 3: 1625.

15 Hughes, W.L. and Wang, Z.L. (2004). Formation of piezoelectric single-crystal nanorings and nanobows. *Journal of the American Chemical Society* 126 (21): 6703–6709.

16 Kong, X.Y., Din, Y., Yang, R., and Wang, Z.L. (2004). Single-crystal nanorings formed by epitaxial self-coiling of polar nanobelts. *Science* 303: 1348.

17 Lieber, C.M. and Wang, Z.L. (2007). Functional nanowires. *MRS Bulletin* 32 (2): 99–108.

18 Li, J., De Ávila, B.E.F., Gao, W. et al. (2017). Micro/nanorobots for biomedicine: delivery, surgery, sensing, and detoxification. *Science Robotics* 2 (4): eaam6431.

19 Manjunath, A. and Kishore, V. (2014). The promising future in medicine: nanorobots. *Biomedical Science and Engineering* 2 (2): 42–47.

20 Emerich, D.F. and Thanos, C.G. (2003). Nanotechnology and medicine. *Expert Opinion on Biological Therapy* 3 (4): 655–663.

21 Huh, A.J. and Kwon, Y.J. (2011). Nanoantibiotics: a new paradigm for treating infectious diseases using nanomaterials in the antibiotics resistant era. *Journal of Controlled Release* 156: 128–145.

22 Khurana, C. and Chudasama, B. (2018). Nanoantibiotics: strategic assets in the fight against drug-resistant superbugs. *International Journal of Nanomedicine* 13 (T-NANO 2014 Abstracts): 3.

23 Selvamuthukumar, S., Anandam, S., Krishnamoorthy, K. et al. (2012). Nanosponges: a novel class of drug delivery system-review. *Journal of Pharmacy & Pharmaceutical Sciences* 15 (1): 103–111.

24 Bozzuto, G. and Molinari, A. (2015). Liposomes as nanomedical devices. *International Journal of Nanomedicine* 10: 975.

25 Elhissi, A.M.A., Ahmed, W., Hassan, I.U. et al. (2012). Carbon nanotubes in cancer therapy and drug delivery. *Journal of Drug Delivery* 2012: 837327.

26 Shringirishi, M., Prajapati, S.K., Mahor, A. et al. (2014). Nanosponges: a potential nanocarrier for novel drug delivery – a review. *Asian Pacific Journal of Tropical Disease* 4: S519–S526.

27 Zhao, Y.S., Di, C., Yang, W. et al. (2006). Photoluminescence and electroluminescence from tris(8-hydroxyquinoline) aluminum nanowires prepared by adsorbent-assisted physical vapor deposition. *Advanced Functional Materials* 16 (15): 1985–1991.

28 Srivastava, A., Seth, A., and Katiyar, K. (2021). Microrobots and nanorobots in the refinement of modern healthcare practices. In: *Robotic Technologies in Biomedical and Healthcare Engineering*, 13–37. CRC Press.

29 Bhowmik, A., Jaggi, S., Varghese, C. et al. (2012). Block designs balanced for second order interference effects from neighbouring experimental units. *Statistics and Applications* 10: 1–12.

30 Khan, T. and Gurav, P. (2018). PhytoNanotechnology: enhancing delivery of plant based anti-cancer drugs. *Frontiers in Pharmacology* 8: 1002.

31 Rizwanullah, M., Amin, S., Mir, S.R. et al. (2018). Phytochemical based nanomedicines against cancer: current status and future prospects. *Journal of Drug Targeting* 26 (9): 731–752.

32 Ali, I. (2011). Nano anti-cancer drugs: pros and cons and future perspectives. *Current Cancer Drug Targets* 11 (2): 131–134.

33 Mansoori, G.A., Mohazzabi, P., Percival, M. et al. (2007). Nanotechnology in cancer prevention, detection and treatment: bright future lies ahead. *World Review of Science, Technology and Sustainable Development* 4 (2–3): 226–257.

34 Chidambaram, M., Manavalan, R., and Kathiresan, K. (2011). Nanotherapeutics to overcome conventional cancer chemotherapy limitations. *Journal of Pharmacy & Pharmaceutical Sciences* 14 (1): 67–77.

35 Chen, Y., Zheng, X.L., Fang, D.L. et al. (2014). Dual agent loaded PLGA nanoparticles enhanced antitumor activity in a multidrug-resistant breast tumor xenograft model. *International Journal of Molecular Sciences* 15 (2): 2761–2772.

36 Das, S., Das, J., Samadder, A. et al. (2013). Efficacy of PLGA-loaded apigenin nanoparticles in benzo[a]pyrene and ultraviolet-B induced skin cancer of mice: mitochondria mediated apoptotic signalling cascades. *Food and Chemical Toxicology* 62: 670–680.

37 Siddiqui, I.A., Bharali, D.J., Nihalet, M. et al. (2014). Excellent anti-proliferative and pro-apoptotic effects of (−)-epigallocatechin-3-gallate encapsulated in chitosan nanoparticles on human melanoma cell growth both in vitro and in vivo. *Nanomedicine: Nanotechnology, Biology and Medicine* 10 (8): 1619–1626.

38 Lopes, S.C.A., Novais, M.V.M., Teixeira, C.S. et al. (2013). Preparation, physicochemical characterization, and cell viability evaluation of long-circulating and pH-sensitive liposomes containing ursolic acid. *BioMed Research International* 2013: 467147.

39 Tian, Z., Shi, Y., Yin, M. et al. (2011). Functionalized multiwalled carbon nanotubes-anticancer drug carriers: synthesis, targeting ability and antitumor activity. *Nano Biomedicine & Engineering* 3 (3): 157.

40 Kumar, V., Abbas, A., Aster, Jon, et al., Robbins and Cotran pathologic basis of disease, in Kumar V, Abbas A, Aster J Robbins and Cotran Pathologic Basis of Disease. 2010. p. 1450–1450.

41 Brusini, R., Varna, M., and Couvreur, P. (2020). Advanced nanomedicines for the treatment of inflammatory diseases. *Advanced Drug Delivery Reviews* 157: 161–178.

42 Changasi, A.H., Shams, T.A., Pouget, J.G. et al. (2014). Genetics of antipsychotic drug outcome and implications for the clinician: into the limelight. *Translational Developmental Psychiatry* 2 (1): 24663.

43 Lobatto, M.E., Calcagno, C., Otten, M.J. et al. (2015). Pharmaceutical development and preclinical evaluation of a GMP-grade anti-inflammatory nanotherapy. *Nanomedicine: Nanotechnology, Biology and Medicine* 11 (5): 1133–1140.

44 Bergman, R.N. and Mittelman, S.D. (2001). Central role of the adipocyte in the metabolic syndrome. *Journal of Investigative Medicine* 49 (1): 119–126.

45 Rokling-Andersen, M.H., Reseland, J.E., Veierød, M.B. et al. (2007). Effects of long-term exercise and diet intervention on plasma adipokine concentrations. *The American Journal of Clinical Nutrition* 86 (5): 1293–1301.

46 Furukawa, S., Fujita, T., Shimabukuro, M. et al. (2004). Increased oxidative stress in obesity and its impact on metabolic syndrome. *Journal of Clinical Investigation* 487 (114): 1752–1761.

47 Grundy, S.M. (2016). Metabolic syndrome update. *Trends in Cardiovascular Medicine* 26 (4): 364–373.

48 Naseri, R., Farzaei, F., Haratipour, P. et al. (2018). Anthocyanins in the management of metabolic syndrome: a pharmacological and biopharmaceutical review. *Frontiers in Pharmacology* 9: 1310.

49 Payab, M., Ranjbar, S.H., Shahbal, N. et al. (2020). Effect of the herbal medicines in obesity and metabolic syndrome: a systematic review and meta-analysis of clinical trials. *Phytotherapy Research* 34 (3): 526–545.

50 do Nascimento, T., Todeschini, A.R., Oliveira, R.S. et al. (2020). Trends in nanomedicines for cancer treatment. *Current Pharmaceutical Design* 26 (29): 3579–3600.

51 Gupta, R. (2017). Diabetes treatment by nanotechnology. *Journal of Biotechnology and Biomaterials* 7 (3): 268.

52 Lin, Z.-L., Ding, J., Sun, G.P. et al. (2020). Application of paclitaxel-loaded EGFR peptide-conjugated magnetic polymeric liposomes for liver cancer therapy. *Current Medical Science* 40 (1): 145–154.

53 Kesharwani, P., Gorain, B., Low, S.Y. et al. (2018). Nanotechnology based approaches for anti-diabetic drugs delivery. *Diabetes Research and Clinical Practice* 136: 52–77.

54 Maity, S., Mukhopadhyay, P., Kundu, P.P. et al. (2017). Alginate coated chitosan core–shell nanoparticles for efficient oral delivery of naringenin in diabetic animals – an in vitro and in vivo approach. *Carbohydrate Polymers* 170: 124–132.

55 Fangueiro, J.F., Silva, A.M., Garcia, M.L. et al. (2015). Current nanotechnology approaches for the treatment and management of diabetic retinopathy. *European Journal of Pharmaceutics and Biopharmaceutics* 95: 307–322.

56 Ganesan, P., Arulselvan, P., and Choi, D.-K. (2017). Phytobioactive compound-based nanodelivery systems for the treatment of type 2 diabetes mellitus – current status. *International Journal of Nanomedicine* 12: 1097.

57 Mnonopi, N., Levendal, R.A., Mzilikazi, N. et al. (2012). Marrubiin, a constituent of *Leonotis leonurus*, alleviates diabetic symptoms. *Phytomedicine* 19 (6): 488–493.

58 Priyanka, K., Sahu, P.L., and Singh, S. (2018). Optimization of processing parameters for the development of *Ficus religiosa* L. extract loaded solid lipid nanoparticles using central composite design and evaluation of antidiabetic efficacy. *Journal of Drug Delivery Science and Technology* 43: 94–102.

59 Samadder, A., Das, S., Das, J. et al. (2012). Ameliorative effects of *Syzygium jambolanum* extract and its poly(lactic-*co*-glycolic) acid nano-encapsulated form on arsenic-induced hyperglycemic stress: a multi-parametric evaluation. *Journal of Acupuncture and Meridian Studies* 5 (6): 310–318.

60 Ajiboye, B.O., Ojo, O.A., Akuboh, O.S. et al. (2018). Anti-hyperglycemic and anti-inflammatory activities of polyphenolic-rich extract of *Syzygium cumini* Linn leaves in alloxan-induced diabetic rats. *Journal of Evidence-Based Integrative Medicine* 23: 2515690X18770630.

61 Dong, Y., Su, H., Jiang, H. et al. (2017). Experimental study on the influence of low-frequency and low-intensity ultrasound on the permeability of the *Mycobacterium smegmatis* cytoderm and potentiation with levofloxacin. *Ultrasonics Sonochemistry* 37: 1–8.

62 Joshi, R.P., Negi, G., Joshi, A.K. et al. (2013). SNEDDS curcumin formulation leads to enhanced protection from pain and functional deficits associated with diabetic neuropathy: an insight into its mechanism for neuroprotection. *Nanomedicine: Nanotechnology, Biology and Medicine* 9 (6): 776–785.

63 Singhal, A.K., Gupta, H., and Bhati, V.S. (2011). Wound healing activity of *Argyreia nervosa* leaves extract. *International Journal of Applied and Basic Medical Research* 1 (1): 36.

64 Ali, H.A., Almaghrabi, O.A., and Afifi, M.E. (2014). Molecular mechanisms of anti-hyperglycemic effects of *Costus speciosus* extract in streptozotocin-induced diabetic rats. *Saudi Medical Journal* 35 (12): 1501.

65 Vergès, B. (2015). Pathophysiology of diabetic dyslipidaemia: where are we? *Diabetologia* 58 (5): 886–899.

66 Chowdhury, A., Kunjiappan, S., Bhattacharjee, C. et al. (2017). Biogenic synthesis of *Marsilea quadrifolia* gold nanoparticles: a study of improved glucose utilization efficiency on 3T3-L1 adipocytes. *In vitro Cellular and Developmental Biology – Animal* 53 (6): 483–493.

67 Zhang, M., Xu, C., Liu, D. et al. (2018). Oral delivery of nanoparticles loaded with ginger active compound, 6-shogaol, attenuates ulcerative colitis and promotes wound healing in a murine model of ulcerative colitis. *Journal of Crohn's and Colitis* 12 (2): 217–229.

68 Ponnanikajamideen, M., Rajeshkumar, S., and Annadurai, G. (2016). In vivo antidiabetic and in vitro antioxidant and antimicrobial activity of aqueous leaves extract of *Chamaecostus cuspidatus*. *Research Journal of Pharmacy and Technology* 9 (8): 1204–1210.

69 Martín Giménez, V.M., Kassuha, D.E., and Manucha, W. (2017). Nanomedicine applied to cardiovascular diseases: latest developments. *Therapeutic Advances in Cardiovascular Disease* 11 (4): 133–142.

70 Mir, M.A., Bhat, B.A., Sheikh, B.A. et al. (2021). Nanomedicine in human health therapeutics and drug delivery: nanobiotechnology and nanobiomedicine. In: *Applications of Nanomaterials in Agriculture, Food Science, and Medicine*, 229–251. IGI Global.

71 Hesari, M., Mohammadi, P., Khademi, F. et al. (2021). Current advances in the use of nanophytomedicine therapies for human cardiovascular diseases. *International Journal of Nanomedicine* 16: 3293.

72 Dimmeler, S. (2011). Cardiovascular disease review series. *EMBO Molecular Medicine* 3 (12): 697–697.

73 Stewart, J., Manmathan, G., and Wilkinson, P. (2017). Primary prevention of cardiovascular disease: a review of contemporary guidance and literature. *JRSM Cardiovascular Disease* 6: 2048004016687211. Epub 2017/03/14. https://doi.org/10.1177/2048004016687211. PMID: 28286646.

74 Taghipour, Y.D., Hajialyani, M., Naseri, R. et al. (2019). Nanoformulations of natural products for management of metabolic syndrome. *International Journal of Nanomedicine* 14: 5303.

75 Chakravorty, A. (2020). Nano approach: indian spices as antimicrobial agents. In: *Ethnopharmacological Investigation of Indian Spices*, 205–241. IGI Global.

76 Boarescu, P.-M., Boarescu, I., Bocşan, I.C. et al. (2019). Antioxidant and anti-inflammatory effects of curcumin nanoparticles on drug-induced acute myocardial infarction in diabetic rats. *Antioxidants* 8 (10): 504.

77 Rice, K.M., Manne, N.D.P.K., Kolli, M.B. et al. (2016). Curcumin nanoparticles attenuate cardiac remodeling due to pulmonary arterial hypertension. *Artificial Cells, Nanomedicine, and Biotechnology* 44 (8): 1909–1916.

78 Rather, G.A., Bhat, B.A., Majeed, I. et al. (2021). Routes of synthesis and characterizations of nanoparticles. In: *Applications of Nanomaterials in Agriculture, Food Science, and Medicine*, 288–309. IGI Global.

79 Karthik, L., Kumar, G., Keswani, T. et al. (2014). Protease inhibitors from marine actinobacteria as a potential source for antimalarial compound. *PLoS One* 9 (3): e90972.

80 Yang, L.Q., Wang, B., Gan, H. et al. (2012). Enhanced oral bioavailability and anti-tumour effect of paclitaxel by 20 (s)-ginsenoside Rg3 in vivo. *Biopharmaceutics & Drug Disposition* 33 (8): 425–436.

81 Lee, J., Lee, E., Kim, D. et al. (2009). Studies on absorption, distribution and metabolism of ginseng in humans after oral administration. *Journal of Ethnopharmacology* 122 (1): 143–148.

82 Zhang, J., Han, X., Li, X. et al. (2012). Core–shell hybrid liposomal vesicles loaded with panax notoginsenoside: preparation, characterization and protective effects on global cerebral ischemia/reperfusion injury and acute myocardial ischemia in rats. *International Journal of Nanomedicine* 7: 4299.

83 Gautam, R. and Jachak, S.M. (2009). Recent developments in anti-inflammatory natural products. *Medicinal Research Reviews* 29 (5): 767–820.

84 Huang, Z., Han, Z., Ye, B. et al. (2015). Berberine alleviates cardiac ischemia/reperfusion injury by inhibiting excessive autophagy in cardiomyocytes. *European Journal of Pharmacology* 762: 1–10.

85 Tian, L., Cao, W., Yue, R. et al. (2017). Tilianin pretreatment prevents myocardial ischemia–reperfusion injury via preservation of mitochondrial function in rat heart. *Phytomedicine* 34: 106–114.

86 Fokker, P.A., Visser, K., Peters, E. et al. (2012). Inversion of surface subsidence data to quantify reservoir compartmentalization: a field study. *Journal of Petroleum Science and Engineering* 96: 10–21.

87 Wong, K.H., Li, G.Q., Li, K.M. et al. (2011). Kudzu root: traditional uses and potential medicinal benefits in diabetes and cardiovascular diseases. *Journal of Ethnopharmacology* 134 (3): 584–607.

88 Vickers, N.J. (2017). Animal communication: when i'm calling you, will you answer too? *Current Biology* 27 (14): R713–R715.

89 Larson, A., Witman, M.A.H., Guo, Y. et al. (2012). Acute, quercetin-induced reductions in blood pressure in hypertensive individuals are not secondary to lower plasma angiotensin-converting enzyme activity or endothelin-1: nitric oxide. *Nutrition Research* 32 (8): 557–564.

90 Anand, P., Kunnumakkara, A.B., Newman, R.A. et al. (2007). Bioavailability of curcumin: problems and promises. *Molecular Pharmaceutics* 4 (6): 807–818.

91 Cote, B., Carlson, L.J., Rao, D.A. et al. (2015). Combinatorial resveratrol and quercetin polymeric micelles mitigate doxorubicin induced cardiotoxicity in vitro and in vivo. *Journal of Controlled Release* 213: 128–133.

92 Marano, F., Hussain, S., Lima, F.R. et al. (2011). Nanoparticles: molecular targets and cell signalling. *Archives of Toxicology* 85 (7): 733–741.

93 Kong, F., Luan, Y., Zhang, Z.H. et al. (2014). Baicalin protects the myocardium from reperfusion-induced damage in isolated rat hearts via the antioxidant and paracrine effect. *Experimental and Therapeutic Medicine* 7 (1): 254–259.

94 Zhang, S., Wang, J., and Pan, J. (2016). Baicalin-loaded PEGylated lipid nanoparticles: characterization, pharmacokinetics, and protective effects on acute myocardial ischemia in rats. *Drug Delivery* 23 (9): 3696–3703.

95 Shahraki, A., Bahadorikhalili, S., Hashemzaei, M. et al. (2017). Resveratrol nanocapsule as an efficient tool for blood pressure regulation: a study on metabolic syndrome induced mice. *Bioscience Biotechnology Research Commununications* 10 (4): 623–630.

96 Zhang, L., Zhu, K., Zeng, H. et al. (2019). Resveratrol solid lipid nanoparticles to trigger credible inhibition of doxorubicin cardiotoxicity. *International Journal of Nanomedicine* 14: 6061.

97 Barchet, T.M. and Amiji, M.M. (2009). Challenges and opportunities in CNS delivery of therapeutics for neurodegenerative diseases. *Expert Opinion on Drug Delivery* 6 (3): 211–225.

98 Pathan, S.A., Iqbal, Z., Zaidi, S.M.A. et al. (2009). CNS drug delivery systems: novel approaches. *Recent Patents on Drug Delivery & Formulation* 3 (1): 71–89.

99 Lee, H.J., Liu, Y., Zhao, J. et al. (2013). In vitro and in vivo mapping of drug release after laser ablation thermal therapy with doxorubicin-loaded hollow gold nanoshells using fluorescence and photoacoustic imaging. *Journal of Controlled Release* 172 (1): 152–158.

100 Nagpal, K., Singh, S.K., and Mishra, D.N. (2013). Drug targeting to brain: a systematic approach to study the factors, parameters and approaches for prediction of permeability of drugs across BBB. *Expert Opinion on Drug Delivery* 10 (7): 927–955.

101 Abbott, N.J. (2002). Astrocyte–endothelial interactions and blood–brain barrier permeability. *Journal of Anatomy* 200 (5): 523–534.

102 Murakami, K., Kondo, T., Kawase, M. et al. (1998). The development of a new mouse model of global ischemia: focus on the relationships between ischemia duration, anesthesia, cerebral vasculature, and neuronal injury following global ischemia in mice. *Brain Research* 780 (2): 304–310.

103 Wang, K., Zhu, X., Yu, E. et al. (2020). Therapeutic nanomaterials for neurological diseases and Cancer therapy. *Journal of Nanomaterials* 2020: 2047379.

104 Wilson, B., Samanta, M.K., Santhi, K. et al. (2008). Poly(*n*-butylcyanoacrylate) nanoparticles coated with polysorbate 80 for the targeted delivery of rivastigmine into the brain to treat Alzheimer's disease. *Brain Research* 1200: 159–168.

105 Zhang, Y., Huang, Y., and Li, S. (2014). Polymeric micelles: nanocarriers for cancer-targeted drug delivery. *AAPS PharmSciTech* 15 (4): 862–871.

106 Kedar, U., Phutane, P., Shidhaye, S. et al. (2010). Advances in polymeric micelles for drug delivery and tumor targeting. *Nanomedicine: Nanotechnology, Biology and Medicine* 6 (6): 714–729.

107 Xu, W., Ling, P., and Zhang, T. (2013). Polymeric micelles, a promising drug delivery system to enhance bioavailability of poorly water-soluble drugs. *Journal of Drug Delivery* 2013: 340315.

108 Duncan, R. and Izzo, L. (2005). Dendrimer biocompatibility and toxicity. *Advanced Drug Delivery Reviews* 57 (15): 2215–2237.

109 Sarker, D.K. (2005). Engineering of nanoemulsions for drug delivery. *Current Drug Delivery* 2 (4): 297–310.

110 Lovelyn, C. and Attama, A.A. (2011). Current state of nanoemulsions in drug delivery. *Journal of Biomaterials and Nanobiotechnology* 2 (5): 626.

111 Shah, L., Yadav, S., and Amiji, M. (2013). Nanotechnology for CNS delivery of bio-therapeutic agents. *Drug Delivery and Translational Research* 3 (4): 336–351.

112 Vinogradov, S. (2004). The second annual symposium on nanomedicine and drug delivery: exploring recent developments and assessing major advances. *Expert Opinion on Drug Delivery* 1 (1): 181–184.

113 Vinogradov, S.V., Batrakova, E.V., and Kabanov, A.V. (2004). Nanogels for oligonucleotide delivery to the brain. *Bioconjugate Chemistry* 15 (1): 50–60.

114 Kolodziejczyk-Czepas, J., Bijak, M., Saluk, J. et al. (2015). Radical scavenging and antioxidant effects of *Matricaria chamomilla* polyphenolic–polysaccharide conjugates. *International Journal of Biological Macromolecules* 72: 1152–1158.

115 Basu, A., Betts, N.M., Ortiz, J. et al. (2011). Low-energy cranberry juice decreases lipid oxidation and increases plasma antioxidant capacity in women with metabolic syndrome. *Nutrition Research* 31 (3): 190–196.

116 Silva, R.O., Damasceno, S.R.B., Brito, T.V. et al. (2015). Polysaccharide fraction isolated from *P. assiflora* edulis inhibits the inflammatory response and the oxidative stress in mice. *Journal of Pharmacy and Pharmacology* 67 (7): 1017–1027.

117 Drummond, E.M., Harbourne, N., Marete, E. et al. (2013). An in vivo study examining the antiinflammatory effects of chamomile, meadowsweet, and willow bark in a novel functional beverage. *Journal of Dietary Supplements* 10 (4): 370–380.

118 Georgiev, V., Ananga, A., and Tsolova, V. (2014). Recent advances and uses of grape flavonoids as nutraceuticals. *Nutrients* 6 (1): 391–415.

119 Ghosh, S., Banerjee, S., and Sil, P.C. (2015). The beneficial role of curcumin on inflammation, diabetes and neurodegenerative disease: a recent update. *Food and Chemical Toxicology* 83: 111–124.

120 Goya, L., Martín, M.A., Sarriá, B. et al. (2016). Effect of cocoa and its flavonoids on biomarkers of inflammation: studies of cell culture, animals and humans. *Nutrients* 8 (4): 212.

121 Magrone, T., Heredia, F.P., Jirillo, E. et al. (2013). Functional foods and nutraceuticals as therapeutic tools for the treatment of diet-related diseases. *Canadian Journal of Physiology and Pharmacology* 91 (6): 387–396.

122 Potter, A.S., Foroudi, S., Stamatikos, A. et al. (2011). Drinking carrot juice increases total antioxidant status and decreases lipid peroxidation in adults. *Nutrition Journal* 10 (1): 1–6.

123 Farokhzad, O.C. and Langer, R. (2009). Impact of nanotechnology on drug delivery. *ACS Nano* 3 (1): 16–20.

124 Paolino, D., Mancuso, A., Cristiano, M.C. et al. (2021). Nanonutraceuticals: the new frontier of supplementary food. *Nanomaterials (Basel)* 11 (3): 792.

125 Dima, C., Assadpour, E., Dima, S. et al. (2020). Bioavailability of nutraceuticals: role of the food matrix, processing conditions, the gastrointestinal tract, and nanodelivery systems. *Comprehensive Reviews in Food Science and Food Safety* 19 (3): 954–994.

126 Cristiano, M.C., Froiio, F., Mancuso, A. et al. (2021). Oleuropein-laded ufasomes improve the nutraceutical efficacy. *Nanomaterials* 11 (1): 105.

127 Rasti, B., Erfanian, A., and Selamat, J. (2017). Novel nanoliposomal encapsulated omega-3 fatty acids and their applications in food. *Food Chemistry* 230: 690–696.

128 Meenambal, R. and Bharath, M.M.S. (2020). Nanocarriers for effective nutraceutical delivery to the brain. *Neurochemistry International* 140: 104851.

129 Fresta, M., Mancuso, A., Cristiano, M.C. et al. (2020). Targeting of the pilosebaceous follicle by liquid crystal nanocarriers: in vitro and in vivo effects of the entrapped minoxidil. *Pharmaceutics* 12 (11): 1127.

130 Assadpour, E. and Mahdi Jafari, S. (2019). A systematic review on nanoencapsulation of food bioactive ingredients and nutraceuticals by various nanocarriers. *Critical Reviews in Food Science and Nutrition* 59 (19): 3129–3151.

131 Deng, L.Z., Mujumdar, A.S., Zhang, Q. et al. (2019). Chemical and physical pretreatments of fruits and vegetables: effects on drying characteristics and quality attributes – a comprehensive review. *Critical Reviews in Food Science and Nutrition* 59: 1408.

132 Imran, M., Ghorat, F., Ul-Haq, I. et al. (2020). Lycopene as a natural antioxidant used to prevent human health disorders. *Antioxidants* 9 (8): 706.

133 Borodina, I., Kenny, L.C., McCarthy, C.M. et al. (2020). The biology of ergothioneine, an antioxidant nutraceutical. *Nutrition Research Reviews* 33 (2): 190–217.

134 Brotons-Canto, A., Navarro, C.J.G., Javier, G. et al. (2020). Zein nanoparticles improve the oral bioavailability of resveratrol in humans. *Journal of Drug Delivery Science and Technology* 57: 101704.

135 Paolino, D., Cosco, D., Cilurzo, F. et al. (2007). Innovative drug delivery systems for the administration of natural compounds. *Current Bioactive Compounds* 3: 262–277.

136 Larussa, T., Oliverio, M., Suraci, E. et al. (2017). Oleuropein decreases cyclooxygenase-2 and interleukin-17 expression and attenuates inflammatory damage in colonic samples from ulcerative colitis patients. *Nutrients* 9 (4): 391.

137 Pignatello, R. (2011). *Biomaterials: Applications for Nanomedicine*. BoD – Books on Demand.

138 Gaspar, R. and Duncan, R. (2009). Polymeric carriers: preclinical safety and the regulatory implications for design and development of polymer therapeutics. *Advanced Drug Delivery Reviews* 61 (13): 1220–1231.

10

Biochemical and Therapeutic Targets for Nanophytomedicines

Prithviraj Chakraborty[1], *Debarupa Dutta Chakraborty*[2], *and Mainak Chakraborty*[3]

[1] *Department of Pharmaceutics, Mata Gujri College of Pharmacy, Mata Gujri University, Kishanganj, Bihar, India*
[2] *Department of Pharmaceutical Chemistry, Mata Gujri College of Pharmacy, Mata Gujri University, Kishanganj, Bihar, India*
[3] *Departments of Pharmaceutical Technology and Pharmacology, NSHM Knowledge Campus, Kolkata, West Bengal, India*

10.1 Introduction

Chemically reactive molecules containing oxygen, referred to as reactive oxygen species (ROS), are vital participants in both the beneficial and detrimental aspects of the immune response [1, 2].

ROS are incredibly reactive ions and free radicals, comprising superoxide (O_2^-), hydroxyl radical (·OH), hypochlorite ion (OCl^-), hydrogen peroxide (H_2O_2), singlet oxygen ($^1O_2$), and so forth. The three most significant sites of endogenous ROS production are mitochondria, endoplasmic reticulum (ER), and NADPH oxidase, which produce O_2^-. Furthermore, superoxide dismutases (SODs) and transition metals also produce H_2O_2 and ·OH by reacting with O_2^- and H_2O_2, respectively. ROS is produced to perform a key part in physiological features involving the modulation of protein functions, hormone productions, cell signaling regulation, inflammation mediation, and pathogen elimination. These versatile roles of ROS are governed by their amount and duration in addition to localization. Broadly speaking, low concentrations of ROS control cell-signaling pathways and stimulate cell proliferation, while elevated levels induce the nonspecific damage of proteins, lipids, and DNA and obliterate the bacteria and pathogens.

Consequently, a deficit or an overabundance of ROS can cause numerous illnesses involving autoimmune, cardiovascular, neurodegenerative diseases, etc. Precisely, the rise of ROS levels is at risk of mutating the cellular DNA, which is firmly linked to the development of numerous cancer cells [3–15].

ROS may exert distinct impacts dependent on their nature, localization, and levels. An imbalance in both production of ROS and scavenging can lead to a prolonged surge of the ROS levels and to a pro-oxidant state that has been related to a broad range of pathological illnesses involving cancer. ROS participate in all facets of carcinogenesis, in adjunct to nonsurgical cancer treatments. ROS modulation caused by conventional radio- and chemotherapy can have an impact on tumor outcome, albeit with poor selectivity and high toxicity. Therefore, new treatments are sought after to address these concerns. Consequently, research on treatments of cancer based on ROS modulation has grown up in the last decade and, in current years, the study on nano-materials for medical applications added to producing novel nanostructures with redox-active properties that might augment the oncologic diagnosis and therapies [16–19].

Phytoantioxidants and Nanotherapeutics, First Edition. Edited by Mithun Rudrapal.

ROS perform a decisive role in numerous physiological processes, like cellular growth, differentiation, proliferation, signaling, and senescence, in addition to pathological processes. ROS is principally produced endogenously in mitochondria via cellular respiration. Oxidase accepts electrons liberated from an organic substrate via membrane carriers, like the ubiquinone and cytochrome c, forming superoxide anion. The superoxide anion is subsequently converted quickly to hydrogen peroxide (H_2O_2) in the extracellular space in addition to the mitochondria and cytosol employing SODs or xanthine oxidase. H_2O_2 may also be produced without superoxide anion formation via the NADPH and monoamine oxidase reactions. Other ROS production sources are via β-oxidation of fatty acids and flavin oxidase activity in peroxisomes. Protein oxidation in the ER, in addition, could also produce ROS. Exogenously, ROS may be generated by an environmental toxin, for instance, cigarette smoke and ultraviolet light exposure.

Amongst ROS, H_2O_2 has a relatively longer biological life span and is stable comparatively. H_2O_2 is additionally recognized to be diffusible easily within and between cells. Even though H_2O_2 overproduction all through oxidative stress conditions can be lethal, H_2O_2 at a deficient physiological level has been demonstrated to be vital for cellular signaling for normal physiological cellular functions. In humans, H_2O_2 concentration can be more flexible because nonpolar H_2O_2 can spontaneously diffuse all through the membrane. The specialized aquaporin channels further assist the membrane transport of H_2O_2. Excessive H_2O_2 production is also the vital pathogenesis of oxidative stress-induced tissue injury by the release of pro-inflammatory cytokines and initiating apoptosis (Figure 10.1) [20–27].

Oxidative stress results from the imbalance between ROS production and biological systems' capacity to detoxify the reactive intermediates or fix the subsequent damages, affecting all cell components, involving proteins, lipids, and DNA. Moreover, reactive nitrogen species (RNS) might form through a combination of nitric oxide ($NO^{\cdot}$) with $O_2^{\cdot-}$ to produce peroxynitrite ($ONOO^-$) and behave along with ROS to harm cells, initiating nitrosative stress [28].

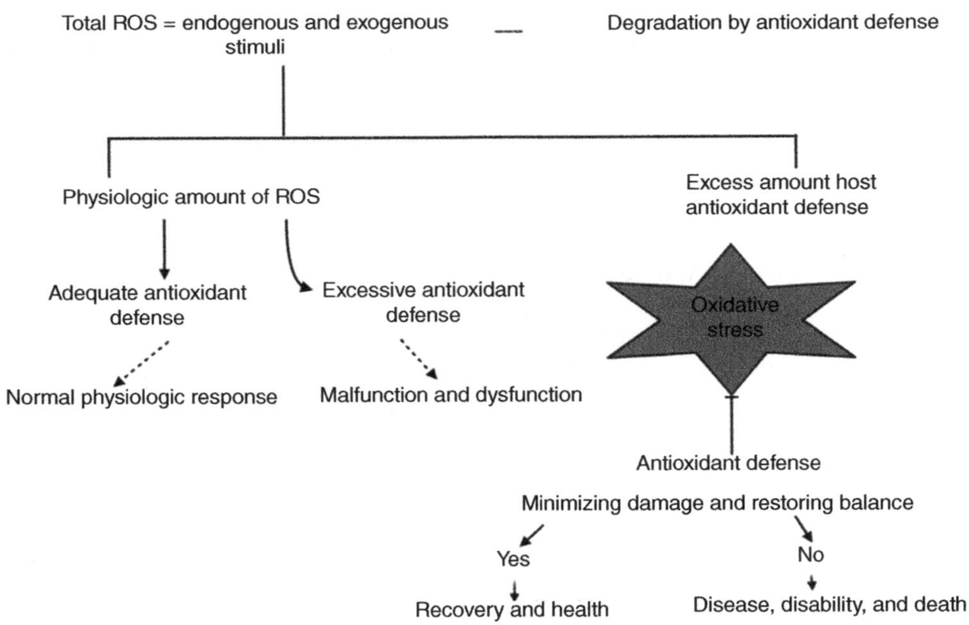

Figure 10.1 The broad concept of physiologic and pathologic reactive oxygen species/oxidative stress. *Source:* Kim et al. [20]/with permission of Future Science Group.

10.1.1 Functions of ROS in Disease Conditions

Oxidative stress injuries are observed in a wide range of clinical environments. The quantity of ROS that surpasses local antioxidant capacity defines the propensity for oxidative damage. Innumerable disease conditions might be linked to either acute overproduction or chronic rise in ROS.

10.1.2 Diseases Correlated with Acute Excess ROS Production

Acute overproduction of ROS induces oxidative stress that outcomes in substantial cellular dysfunction, cell death, and tissue loss leading to organ dysfunction. Perhaps the utmost well-recognized illnesses related to acute ROS overproduction are cardiovascular diseases, such as myocardial infarction, ischemia/reperfusion (I/R) injury, and heart failure. Inhibition of ROS overproduction in I/R utilizing different antioxidants has been displayed to successfully obstruct the lethal impacts of ROS, like apoptosis, in experimental settings.

Like cardiovascular diseases, ROS has been involved in a cascade of ischemic and reperfusion injuries of the brain, like stroke, which is by far the most widely recognized explanation of adult disability worldwide.

ROS could be generated both exogenously and endogenously within the respiratory system, like pollutants, cigarette smoking, and high-altitude hypoxia. ROS might cause immediate, acute, and chronic injuries via immune-mediated inflammation.

10.1.3 Diseases Correlated with Chronic Increase in ROS

ROS is a vitally important signaling molecule to sustain cellular homeostasis and normal physiological function. But, if there is a chronic rise in the production of ROS, normal homeostasis is interrupted, and disease could arise. In general, aging results in enhanced ROS production and endogenous scavenging capacity reduction that presents more considerable potential hazards of creating oxidative stress-implicated diseases, like, for instance, Alzheimer's disease.

Oxidative stress has been known to play a significant role in cancer cells' commencement and advancement. Oxidative stress management is essential in avoiding tumor development and enabling response to anticancer therapy. ROS may contribute to tumor initiation at low-to-moderate levels either by functioning as signaling molecules or through promoting genomic DNA mutation [18, 20, 23, 29–36].

10.2 Biochemical Basis and Metabolic Interrelations of Redox Regulation/Chemistry of ROS-Based Nanoplatforms

Enormous efforts have been committed to improving drug delivery systems (DDSs) that can effectively convey restorative therapeutic agents into disease destinations. Notwithstanding, restorative efficiency is regularly hampered by untimely medication delivery and quick body clearance, which requires a considerable portion of medication and conveys undesirable fundamental toxicity. On-request drug delivery is subsequently of most extreme significance in accomplishing site-specific delivery with diminished side effects. Taking this into consideration, advanced responsive DDSs have acquired expanding fame owing to their capacity to deliver payload in light of a specific boost related to certain infection conditions. Typical stimuli-responsive DDSs incorporate endogenous (e.g. ROS, redox, pH, thermal, and enzyme) and exogenous (e.g. light, temperature, magnetic field, and ultrasound) ones.

Specifically noteworthy are the ROS-responsive DDSs because ROS overproduction has extraordinary ramifications on an assortment of sicknesses, and arising ROS-responsive materials hold incredible guarantee in the improvement of cutting edge nanomedicines. An effective DDS depends on both a strong comprehension of physiological states of the sickness locales and a sane plan of improvements responsive medication transporter that can go through sharp synthetic or physical changes due to the upgrades to consider freight escape. Multiple-stimuli-responsive DDSs better designate several obsessive conditions with numerous upgrades to empower further developed medication efficacy [37].

ROS are exceptionally receptive particles and free radicals, including superoxide (O_2^-), hydroxyl radical (·OH), hypochlorite ion (OCl^-), hydrogen peroxide (H_2O_2), singlet oxygen ($^1O_2$), etc. The significant three locales of endogenous ROS generation are mitochondria, ER, and NADPH oxidase (NOX), which produce O_2^-. Likewise, SODs and transition metals additionally partake in the generation of H_2O_2 and ·OH by responding with O_2^- and H_2O_2, separately. The generated ROS assume a fundamental part in physiological capacities, including modulation of functions of proteins, production of several hormones, regulation of cell signaling, mediation of inflammation, and elimination of pathogens [3].

10.2.1 ROS-Prompted Drug Release from Polymeric Nanoparticles

Polymeric nanoparticles, framed through unconstrained self-assembly of amphiphilic block copolymers, have been broadly perceived as productive transporters for conveying a wide scope of imaging and therapeutic agents [38–41]. By regulating the hydrophilic–lipophilic equilibrium of block copolymers and the condition of solution-state self-assembly, we can, without much of a stretch, tune their physical properties, for example, morphology and size, which incredibly impact in vivo biodistribution. Their centre shell nano-sized construction empowers polymeric nanoparticles to collect at the tumor site through enhanced permeation and retention (EPR) impact following foundational organization. It is likewise conceivable to design their surface focusing on ligands for more effective intracellular conveyance through receptor-intervened endocytosis pathways. Notwithstanding these focusing on capacities, incorporating polymeric micelles with stimuli-responsive units holds an incredible guarantee since it permits us to tweak discharge qualities of the stacked medications in a spatiotemporally controlled style at the objective site. However, a few improvements, responsive units that can react to components like pH, temperature, light, chemical, or reducible climate have effectively been investigated in polymeric nanocarriers, the fuse of ROS-responsive units in polymeric nanoplatforms for tunable medication discharge is just acquiring consideration as of late, as expanding proof proposes that few pathogenic cycles are ensnared with raised ROS [3].

10.2.2 ROS-Receptive Chemical Structures

There are two significant ROS-responsive components for the generally utilized ROS-responsive polymers, specifically ROS-instigated non-cleavable hydrophobic–hydrophilic transition and ROS-prompted structural cleavage (Table 10.1). On the one hand, ROS can oxidize chalcogen components (like S, Se, and Te), changing their valence from +2 to +4 or +6. During this cycle, oxygen particles structure covalent bonds with the chalcogen atoms and the polarized groups empower hydrogen-holding arrangement with the ecological water molecules, hence initiating the hydrophobic–hydrophilic transition of the polymer backbone without cleaving its synthetic construction. Then again, ROS can respond with substance designs, for example, thioketals (TKs), phenylboronic acids/esters (PBAs/PBEs), vinyl dithioethers, and proline oligomers, prompting cleavage of these constructions.

Table 10.1 A brief summary of certain ROS-responsive structures and mechanisms.

Functional group	**ROS-responsive reactions**
Sulfides	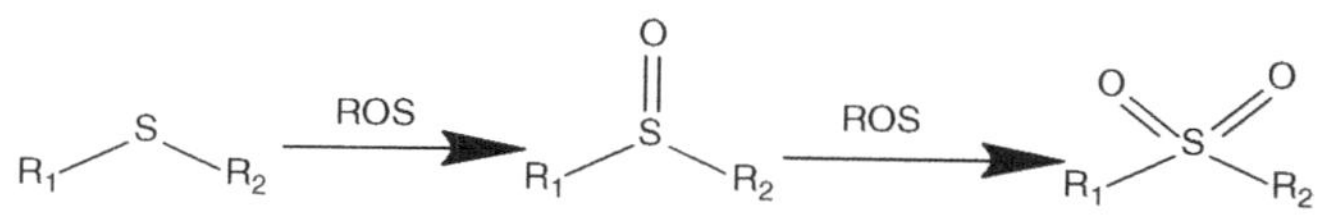
Selenide	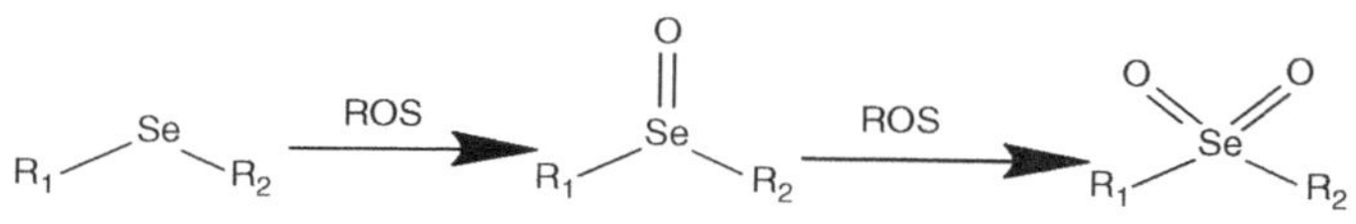
Thioketal	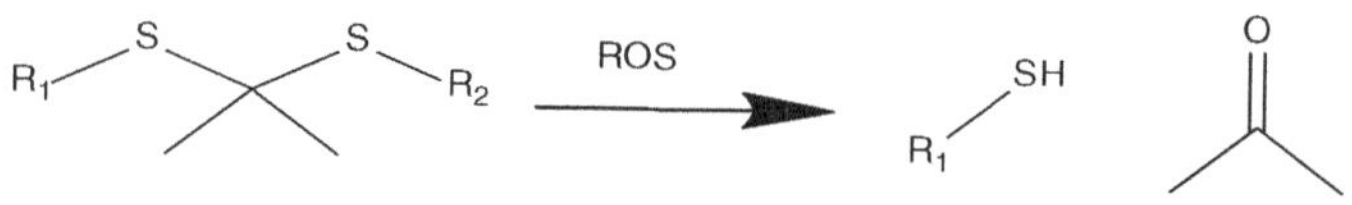
Boronic ester group	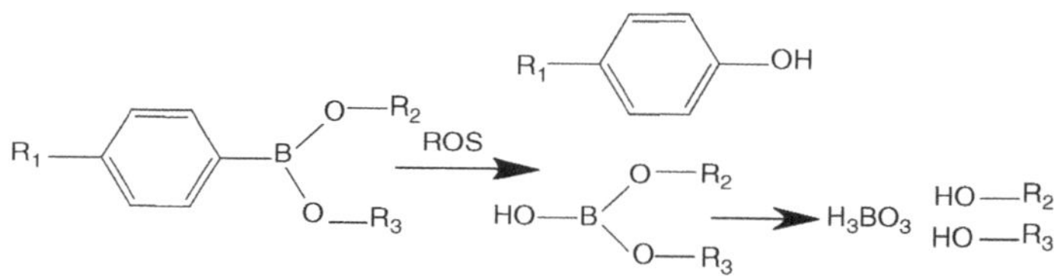
Polyoxalate	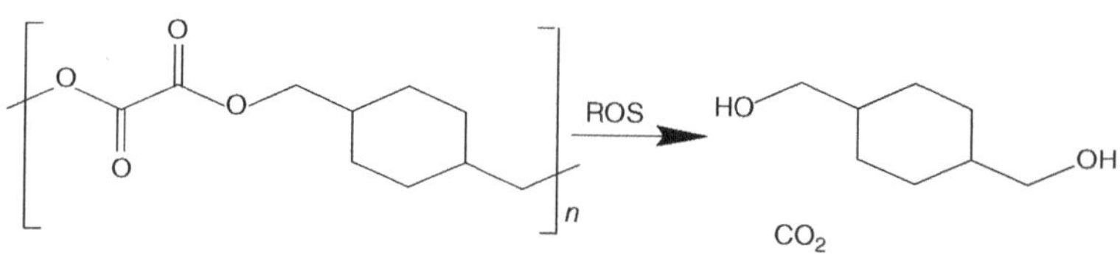

Source: Based on Ref. [42].

10.2.3 Certain ROS-Targeting Nano-Materials

10.2.3.1 Polypropylene Sulfide

The oxidative reaction of polypropylene sulfide (PPS) depends on the change of hydrophobic sulfide to hydrophilic gatherings relying upon the oxidant [43]. When presented to $H_2O_2^-$ rich climate, it oxidizes to hydrophilic polysulfoxides and sulfones. Most typically, PPS-based materials are created by amphiphilic block copolymers through the formation of hydrophilic PEG subsidiaries. They have been known as self-assembled that convey hydrophobic medications inside that enlarge when uncovered in an oxidant-rich climate. For DDS, it takes advantage of these solvency changes to convey hydrophobic medications or imaging agents through, under any circumstances, troublesome biologic hindrances. For aqueous stability, hydrophilic ends are by and large added, making up micelle-shaping A-B-A block copolymer. A few varieties of PPS-based materials incorporate deblock PEG-bPPS, PPS-b-poly dimethyl acrylamide, small PPS-PEG (20, 45, 100 nm), and polymers conjugated to model antigen. Oxidation-sensitive PPS-based materials can be additionally made delicate to decrease in one grid (transporter framework) contingent on the microenvironment by adding disulfide linkage. PPS-PEG micelles will set off the arrival of hydrophobic load alongside the capacity to change superoxide over to more steady H_2O_2 by adding SOD and catalase mimetics have been likewise evolved [20]. The advantages of PPS-based polymeric specialists are simple combination and flexibility. Of note, nevertheless, oxidation of PPS in profoundly oxidative conditions, for example, within sight of natural peroxy acids, there is a propensity to create sulfones. PPS-based materials could be constructive in ROS (prevalently H_2O_2)-ensnared infections, like malignancy and inflammatory diseases.

10.2.3.2 Selenium- or Tellurium-Containing Polymers

Similar to sulfur-containing polymers, compounds containing selenium (Se) and tellurium (Te), which also are from the chalcogen group, were exploited for drug delivery applications also. The organoselenium and organotellurium compounds are often oxidized from divalent to tetravalent states, which makes them desirable ROS scavengers. ROS oxidation of monoselenium and monotellurium compounds may cause phase change from hydrophobic to hydrophilic, which may be capitalized in the ROS-responsive drug carriers' construction. Initial in 2010, Xu and Zhang produced an amphiphilic block copolymer (PEG-PUSe-PEG) with a hydrophobic mono selenide-containing block polymer and two hydrophilic blocks of PEG [44]. The polymer self-assembles into micelles with a mean diameter of 71 nm. On oxidation, polymer micelles undergo a hydrophobic-to-hydrophilic phase change that causes the disassembly of the micelles. It was discovered that the PEG-PUSe-PEG polymer is more sensitive to oxidation stimuli than its sulfur analogue PEG-PUS-PEG: almost complete conversion was attained for the previous exposure to 0.1% H_2O_2 for five hours while only ~30% was converted for the latter. These polymer micelles were effectively displayed for drug release using doxorubicin (DOX) because of the model drug. Liu et al., in 2013, described an ROS-responsive nanocarrier employing an Se-containing amphiphilic hyperbranched polymer micelle [37, 45].

Compounds containing tellurium are considered to be more sensitive than their selenium [44, 46] counterparts because of both their lower electronegativity and toxicity, which make them desirable drug carriers. The acid uses cyclic voltammetry, suggesting that telluride is a superior candidate drug carrier [47]. Due to the reversible redox properties of Te-containing polymers, the components can be oxidized with dilute H_2O_2 solution and reduced with acid, ascorbic acid, in presence of 100 mM H_2O_2. However, ROS oxidation did not cause substantial morphological changes. According to reports, polymer micelle systems containing Te also react with H_2O_2 and 2 Gy gamma radiation, causing NPs to swell and dissociate [48].

10.2.3.3 Thioketal

Thioketals may be oxidized to ketones and natural thiols or disulfide through ROS (usually through superoxide anion and H_2O_2) [20]. Upon oxidation, thioketal-primarily-based substances have been designed to launch the loaded therapeutics through conversion to sulfoxides and sulfones, that could be hydrophilic – this growth in polarity and water affinity consequences in morphologic disruption to launch the cargo. Thioketal hyperlink within the polymeric subunits has been cleaved through the intracellular ROS found in most cancer cells. Thioketal-primarily-based totally, degradable polymers have been implemented for focused gene shipping to prostate most cancer cells that are wealthy in ROS. Shim et al. confirmed ROS-responsive polyaminothioketal (PATK) having better performance in gene shipping and functionalization in comparison with a nondegradable polyamine or the generally used transfection agent, branched polyethyleneimine [48]. Murthy's organization additionally confirmed some other use of thioketal-primarily-based substances totally in intestinal inflammatory disease [49].

Thioketal-primarily-based fabric devised from a polymer, poly-(1,4-phenyleneaceton dimethylene thioketal) (PPADT) degraded selectively in reaction to extra ROS produced through activated macrophages. However, it becomes strong through physiologic, acidic, primary, and protease ample environment, inclusive of the GI tract. PPADT derives ROS (usually superoxide) sensitivity from the thioketal linkages, and PPADT is cleaved to acetone and 4-(mercaptomethyl) phenyl methanethiol in an oxidizing environment.

10.2.3.4 Boronic Ester Groups

Much just like the thioketal bond, boronic esters go through cleavage because of oxidation [1, 50]. Therefore, nanoparticles using this bond are properly acceptable for the burst launch of medicine in reaction to inflammation [51–55]. Indeed, this chemistry has additionally been used to deal with strokes that have improved quantities of ROS.

Boronic acids can engage with Lewis bases (vulnerable bases) to generate boronated molecules. They can additionally attach with diol gadgets to shape cyclic boronated esters. When saccharides are introduced to boronic acids, pH drops, and the innate acidity of boronic acid is stronger while 1.2-, 1.3-, or 1.4-diols react with boronic acid to shape cyclic boronic esters (5.6.7-membered rings). In the aqueous media, those interactions shape the diol-primarily-based totally receptors that can be used as sensors or separation systems.

Although the boronic ester bond is that of a covalent bond, its formation is reversible below certain conditions [56]. Almutari's group created polymers with each monomer combining ROS-degradable aryl boronic esters and adipic acid. Two polymer systems they evolved differ by the linkage among the boronic ester companies and the polymeric spine, one without delay and the alternative via way of means of ether linkage. Both had been solid in aqueous answer and at physiologic pH. Upon publicity to H_2O_2 consisting of 50–100 mM conc. range, they underwent backbone degradation and launched the payload [57].

10.2.3.5 Polyoxalate

Lee and colleagues [58] were among the first to report the reaction of peroxalate-based nanoparticles with H_2O_2 [20]. These nanoparticles contain a large amount of aromatic peroxalate, making them very unstable and degrading rapidly during the hydrolysis process [59, 60]. Therefore, the therapeutic use of these nanoparticles is minimal. To eliminate this instability, H_2O_2-sensitive nanoparticles based on polyoxalate have been recently established [61, 62]. These nanoparticles are based on 4-hydroxybenzyl alcohol (HBA) or 4-vanillyl alcohol (VA) and are referred to collectively as HBA-containing copolyoxalate (HPOX) or VA-containing copolyoxalate (PVAX).

These two kinds of nanoparticles are rapidly activated by H_2O_2 at the ROS generation site, and H_2O_2 is removed in compliance with the concentration [61, 62]. PVAX and HPOX also show several clinically meaningful therapeutic effects on I-induced lesions, such as anti-oxidation, anti-inflammatory, and anti-apoptotic effects. These therapeutic effects may be explained by the synergistic effects of H_2O_2, ester bonds, and the antioxidant and anti-inflammatory effects of HBA and VA. The blend of fluorophore with these particles also makes it possible to plot the overproduction of H_2O_2 within a range in the body from 0.25 μM to 10 mM.

Another possible application of these nanoparticles is in DDS. They can be loaded with small molecules of interest, such as drugs or imaging agents.

Drug-loaded nanoparticles react with high concentrations of H_2O_2 at the oxidative stress site, thereby "activating" the nanoparticles and then releasing the encapsulated drug for therapeutic and diagnostic purposes [59].

10.3 Nanomedicines in Drug Delivery Systems: Biochemical Pathways and Therapeutic Targets

10.3.1 Nanomedicine in Cancer

There are two primary sources for free radicals generation. Those are mainly endogenous and exogenous sources. The exogenous sources for free radicals production include inflammatory cytokines like interleukins (IL-1 and IL-6), pathogens, xenobiotics, radiation, chlorinated compounds and, on the other hand, endogenous sources include inflammation that includes NF-κB pathway that involves IL-6, C-reactive protein (CRP), TNF-α, STAT3, and HIF1, oxidative phosphorylation, peroxisome metabolism, P-450 metabolism (Table 10.2) [60, 66].

The exogenous and endogenous pathways result in the generation of ROS. ROS leads to redox adaptation and, on the other hand, it leads to mitochondrial malfunction. This increases the oxidative stress, which resulted in H_2O_2 formation, leading to the mitochondrial-DNA damage in presence of mito-ETC, which finally leads to mitochondrial malfunctioning. A rise in oxidative stress in presence of redox-sensitive kinase causes damage to DNA, RNA, and protein, which, in turn, leads to cancer [67].

Free radicals cause carcinogen activation and persistent oxidative stress. Carcinogen activation causes DNA damage by structural alteration and gene mutation. This leads to inheritable mutation. Further, persistent oxidative stress leads to abnormal gene expression and enzyme activity. This leads to decreased apoptosis and ultimately causes clonal expansion. Persistent oxidative stress generates resistance against chemotherapy, which leads to metastasis and invasion (Figure 10.2) [68].

10.3.2 Nanomedicines in Wound Healing

In the wound-healing process, oxygen plays a very crucial role. The vital roles of oxygen in wound healing include angiogenesis, collagen synthesis, epithelialization, oxidative bacterial killing, etc. [69–71]. Thus, the wound-healing process is hindered under the condition of hypoxia. Though, the exact role of oxygen in the wound-healing process is not brought to light yet. Energy is produced

Table 10.2 AGEs-RAGE axis-induced cytokines and their possible after-effects in cancer models.

Induced cytokines	Effects	Mechanism
Interleukin-1 and -6 (IL-1 and IL-6)	Production of a large number of pro-inflammatory mediators	Activate immune cells of lymphoid and myeloid origin and trigger signaling pathway
Interferon (INF) and tumor necrosis factor α (TNF-α)	Generation of ROS, oxidation of HMGB1, and delivery of tolerogenic signals to dampen immune activation	Upregulate co-stimulatory molecules such as CD80, CD86, and expand effector T cells
Interleukin 1 beta (IL-1β)	Activate TLR4 and induce inflammation by HMG B1	Signals through IL-1R/IL-1RAcP complex or associates with lipopolysaccharides
Interleukin-12 (IL-12) and interferon-γ (IFN-γ)	Stimulation of allogeneic T cells and enhancement of pro-inflammatory activity and apoptosis	HMGB 1 helps in the maturation of immature DCs (iDCs) with upregulation of co-stimulatory molecules and MHC Class II

Source: Based on Ref. [63–65]

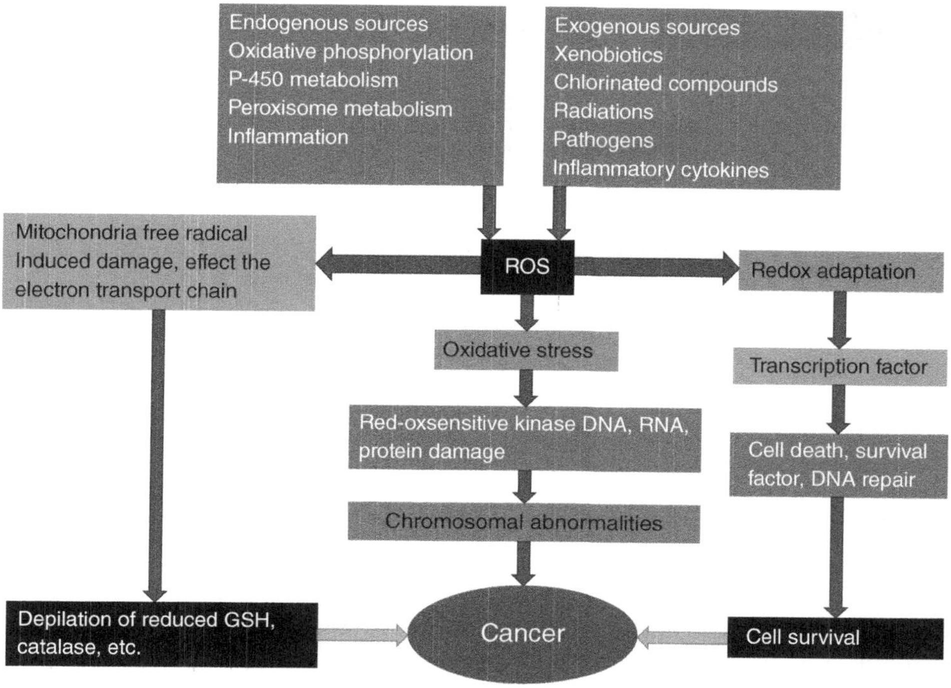

Figure 10.2 Biochemical pathway for free radicals in cancer.

via oxidative phosphorylation by the utilization of oxygen. By the utilization of oxygen, ROS is produced, and it causes oxidative injury. Though ROS supports various kinds of physiological processes still, it often exerts life-threatening damage. Majorly, the pathological role of ROS in the inflammatory phase is very well known.

All through the inflammatory phase, neutrophils and macrophages arrive at the wound area and start secreting a large amount of ROS along with pro-inflammatory cytokines and proteolytic enzymes like the matrix metalloproteinase (MMP). During high levels in the plasma membranes of inflammatory cells, NADPH oxidase (NOX2) is expressed and triggered through the process of phagocytosis, which leads to the production of a large number of superoxide radical anions. The ROS that has been produced directly attacks annexing pathogens and ultimately kills them to succour phagocytosis. Superoxide (when excessively produced, damages the surrounding tissues) is dismutated to hydrogen peroxide (H_2O_2) and molecular oxygen either by SOD or a spontaneous reaction. H_2O_2 is then detoxified using peroxidases like peroxiredoxin (PRDX), catalase, glutathione peroxidase (GPx), to circumvent the Fenton reaction, which occurs in presence of transition metal ions such as iron or copper and that results in the production of hydroxyl radicals, which is the main damaging ROS [71–76].

The ROS generated through other cells is of a much lower amount compared to ROS generated by immune cells. Physiologically, ROS signaling is at a low level in response to stimuli. The H_2O_2 levels that are moderate increase the vascular endothelial growth factor (VEGF) production, which results in an acceleration in the process of angiogenesis. NOX2 and NOX4 both play a vital role in modulating vascular endothelial cell proliferation. ROS are additionally entailed in re-epithelialization. H_2O_2 enhances the stimulation of receptors for epidermal growth factor (EGF) and the keratinocyte growth factor (KGF) and directs to the TGFα production (a member of EGF)

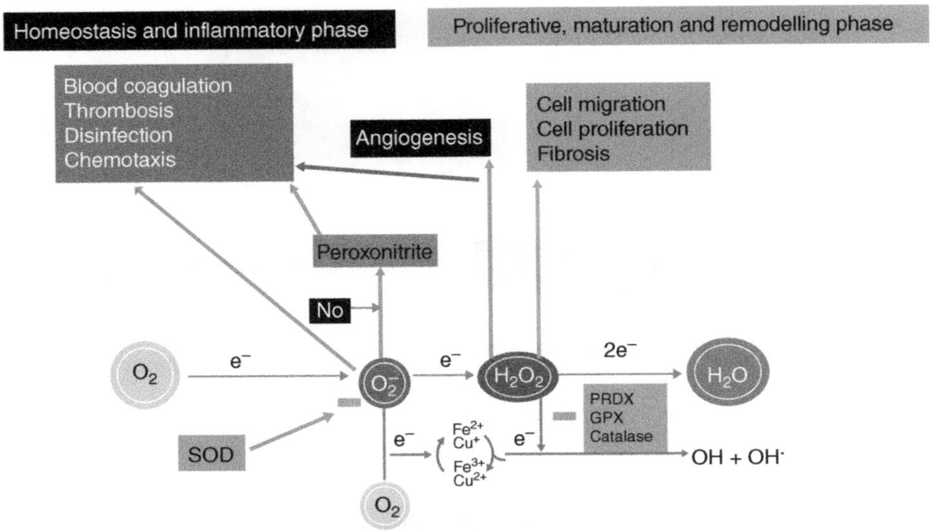

Figure 10.3 Biochemical pathway showing ROS in wound healing. *Source:* Based on Ref. [71].

in fibroblasts [71, 77]. The origin of H_2O_2 is imputed to dual oxidase (DUOX) activity that occurs in epithelial cells. Knockdown of DUOX1 leads to a decrease of H_2O_2 production around the wound site and concomitantly hinders the recruitment of leukocytes to the wound site. Certain enzymes involved in the signaling pathways, such as phosphotyrosine phosphatase and sulfhydryl residues, act as the catalytic centre, are majorly sensitive to oxidative modification and tend to suffer from oxidative inactivation. Therefore, excess ROS strain signaling networks create imbalanced redox homeostasis, resulting in impaired wound healing. Delayed wound healing results in immunodeficiency, diabetes, malnutrition, and aging. Under these pathological conditions, redox imbalance occurs, leading to elevated oxidative injury in the body (Figure 10.3).

10.3.3 Nanomedicines in Ischemia–Reperfusion Injury: Biochemical Pathway

Ischemia–reperfusion (IR) injury is commonly known as tissue damage, which is the result of the process of ischemia followed by the further troubles induced due to the restoration of blood supply to the ischemic tissues [78]. Reperfusion results in an inflammatory response resulting from the induction of various kinds of cytokines and chemokines, which, in turn, increases oxidative stress, resulting in microvascular dysfunction in the ischemic tissues and organs [78–80].

As IR injury results in the inflammatory response, it triggers the inflammatory signaling pathways (Figure 10.4). The trigger in the inflammatory signaling pathway leads to free radical-mediated cell damage, which leads to the generation of ROS, production of cytokines and chemokines, complement activation, and changes in endothelial nitric oxide synthase (eNOS) and induced NO synthase activity [78].

10.3.4 Nanomedicines in Alzheimer's Disease: Impact of Free Radicals

Research has suggested an important pathophysiological role of increased reactive oxygen species (ROS) production leading to oxidative stress [81] as well as the increase of RNS in older people [82] and in neurodegenerative disorders [83], which includes Alzheimer's disease [84]. Alzheimer's disease is designated by synaptic loss, nerve cell loss, extracellular deposition of β-amyloid (Aβ)

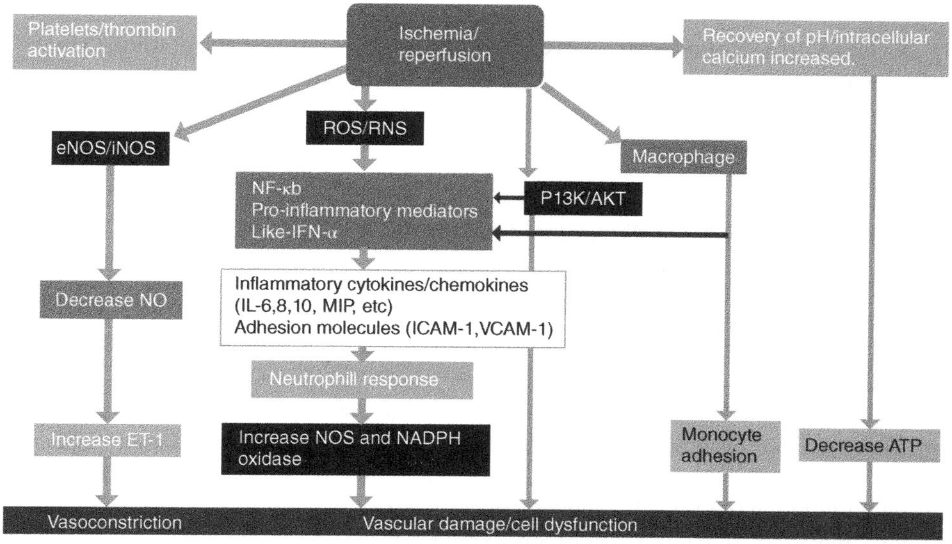

Figure 10.4 Mechanisms involved in ischemia–reperfusion injury and targeting pathways of statins. Ca^{2+}, calcium ion; ROS, reactive oxygen species; RNS, reactive nitrogen species; PI3K, phosphoinositide 3-kinase; Akt, protein kinase B; eNOS, endothelial nitric oxide synthase; iNOS, induced nitric oxide synthase; NFκB, nuclear factor κB; INF, interferon; MCP, monocyte chemotactic protein; HO, heme oxygenase; IL, interleukin; MIP, macrophage inflammatory proteins; ATP, adenosine triphosphate; ICAM, intracellular cell adhesion molecule; VCAM, vascular cell adhesion molecule; CO, carbon monoxide; NO, nitric oxide; ET, endothelin; NOS, nitric oxide synthase; NADPH, nicotinamide adenine dinucleotide phosphate. *Source:* Based on Ref. [78].

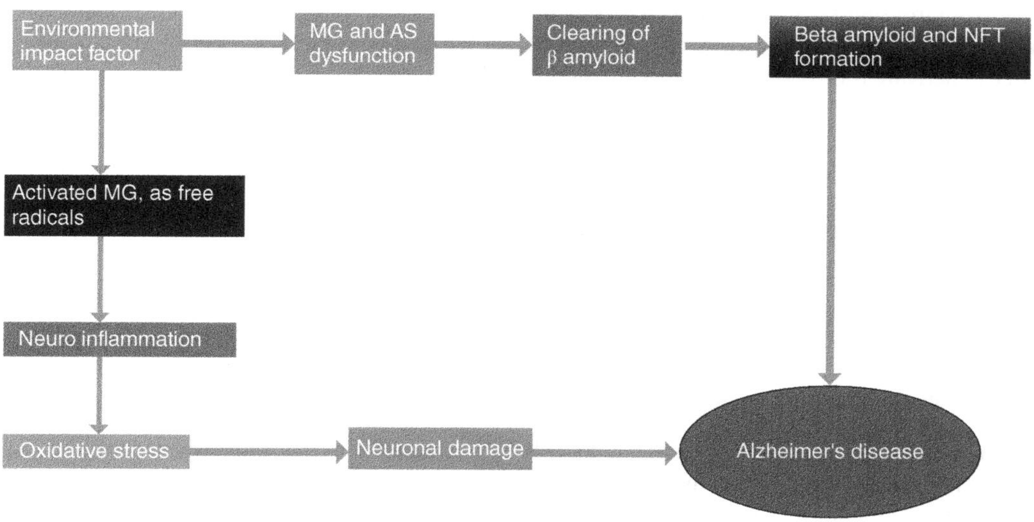

Figure 10.5 Pathways of free radicles in Alzheimer's disease. *Source:* Based on Ref. [85].

protein (forming senile plaques), and intracellular precipitation of hyperphosphorylated τ-protein (Figure 10.5).

Due to environmental factors and outer sources, the ROS and RNS increase, which leads to activation of free radicals leading to neuroinflammation leading to an increase in oxidative stress and leading to programmed cell death resulting in Alzheimer's disease.

Some environmental factors cause dysfunction in microglia and astrocyte causing unusual accumulation of amyloid β-protein, which causes disruption of cell function and neurofibrillary tangles (NFTs), which aggregates of hyperphosphorylated τ-protein that leads to Alzheimer's disease.

Neuroinflammation and oxidative stress might cause accumulation of A-β protein and NFT, leading to Alzheimer's.

10.3.5 Nanomedicines' Targets in Rheumatoid Arthritis

Rheumatoid arthritis (RA) is a persistent systemic inflammatory disorder that might affect many tissues and organs but attacks synovial joints. The process causes an inflammatory response of the synovium (synovitis) secondary to hyperplasia of synovial cells, causes an excess of synovial fluid, and the development of pannus in the synovium [86].

The highly reactive species do have the potential to damage lipids, proteins, and DNA in joint tissues. Under normal circumstances, ROS production is controlled by various antioxidant defence systems present in the body. The nonenzymatic antioxidant defence might include vitamins A and C, reduced glutathione (GSH), while enzymatic antioxidant activity includes SOD, catalase, GPx, glutathione reductase (GR) and glutathione-*S*-transferase (GST). The imbalance between oxidants and antioxidants is due to enhancement of chemical reaction or insufficient antioxidant defence system resulting in oxidative stress [87]. These ROS, if not scavenged properly, may damage biological macromolecules [85, 88].

10.3.6 Nanomedicine Targets in Inflammation

Due to environmental and external factors, body receives stimuli and increases the ROS and activates the inflammatory pathway of tyrosine kinase, tyrosine phosphatase, and MAPK pathway and can activate the gene transcription related to inflammation in two ways: (a) via transcription factors, such as NF-jB, AP-1, and ARE-binding proteins (ARE-BP) that can interact directly with specific DNA motifs on promoters of target genes, or (b) via activation of MAPK cascades, which, in turn, start transcription factors that trigger target gene transcription such as nuclear factor-kappa B; activator protein-1; a hypoxia-inducible factor which will lead to inflammation (Figure 10.6) [89].

10.4 Natural Nanotherapeutics on ROS-Based Platform

10.4.1 Natural Nanotherapeutics in Cancer

Elevated ROS levels are a hallmark of cancer cells and are related to the aggressiveness of tumors; however, this phenomenon creates a physical mismatch between the actual carcinogenic potential of ROS and our general belief that excessive ROS can cause accidental oxidative damage. Why can cancer cells maintain high ROS levels while maintaining their carcinogenic ability to reproduce? In cancer cells, elevated ROS levels can also activate powerful endogenous cellular antioxidant defence mechanisms, leading to increased antioxidant activity and redox kinetics shifting from normal redox homeostasis to new ones with a high ROS formation rate. Stabilize the equilibrium state and remove. This keeps the total ROS value below the toxicity threshold [5]. These endogenous redox adaptation mechanisms, such as the activation of redox transcription factors (such as KB and Nrf2), the expression of ROS capture molecules (such as SOD and GSH) or through various

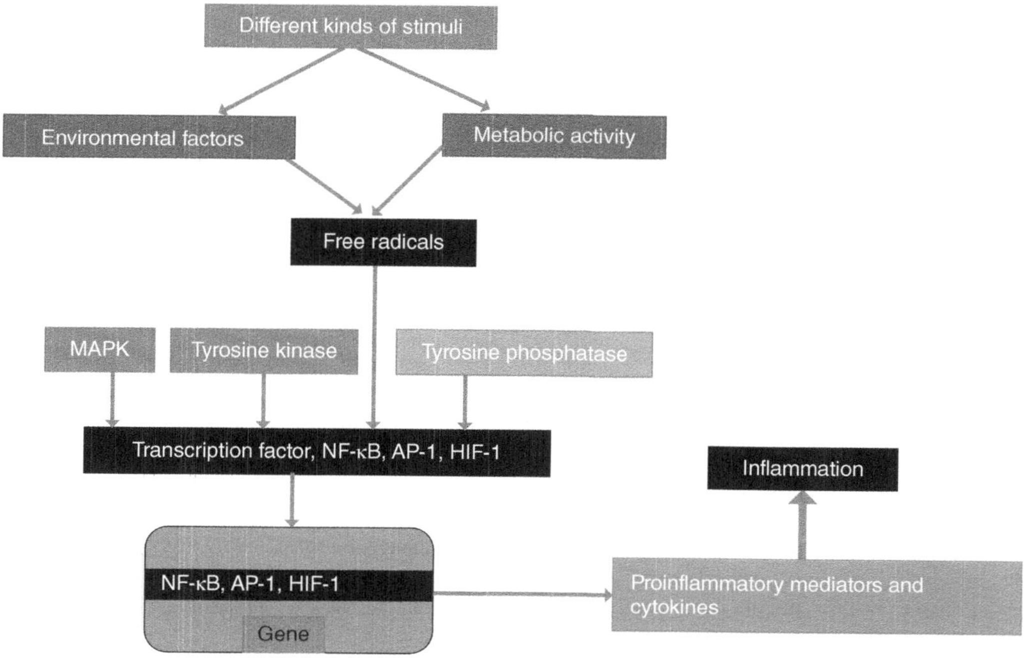

Figure 10.6 Pathways of free radicles in Alzheimer's disease. *Source:* Based on Ref. [89].

mechanisms (such as genetic or epigenetic pathways) genetic modification, which can counteract the potentially toxic effects of excessive ROS, and enable cancer cells to avoid oxidative damage and survive under high redox stress conditions [5].

Therefore, cancer cells will rely more on the endogenous antioxidant system to maintain the newly formed redox homeostasis and are more susceptible to further aggravation of oxidative stress by exogenous ROS generators. When the antioxidant system is overwhelmed, interventions such as ROS-producing nano-materials can cause ROS levels to be higher than the cell tolerance threshold. This may be an attractive but feasible strategy for preferentially emptying cancer cells, producing critical therapeutic effects. Based on the redox biology of cancer, a material-driven and ROS-mediated treatment method is provided to destroy the adaptive mechanism of cancer cells to achieve therapeutic effects for antineoplastic therapy.

On the other hand, the redox-elevating feature of cancer cells reminds us of another possible or complementary cancer treatment. These excess ROS in cancer cells can act as a "biochemical engine," activating ROS-sensitive nanomedicine drugs for on-demand administration. In addition to the above strategies, the characteristics of materials in response to ROS can also be considered to enhance the therapeutic effect [90].

The field of antioxidants is undoubtedly undergoing a renaissance, largely because oxidative stress is closely related to the development of many chronic diseases, including diabetes, cancer, and various central nervous system and age-related diseases [91].

Naturally occurring antioxidants include flavonoids, carotenoids, and carnosine, enzymes, vitamins, chelating agents, and other phenolic and polyphenolic compounds. The antitumor properties of lycopene, resveratrol, ascorbic acid, tocopherol, capsaicin, curcumin, garlic, and quercetin have been fully demonstrated (Table 10.3) [91].

Table 10.3 Some phytonutrients used in cancer nanomedicines [92].

Phytochemical	Application target	Delivery system
β-Lapachone	Colon cancer cells, lung, prostate, breast cancer cells	PEG-PLA polymer micelles
Curcumin	Colon cancer cells, lung, prostate, breast cancer cells Brain, leukemia, colon, breast, prostate, cervical, pancreatic cancer cells, neuroblastoma	PEG-PLA polymer micelles PLGA, PLA-vitamin E TPGS copolymer, alginate NPs, soy protein NPs, PVP conjugate micelle, α-CD derivatives, thermosensitive polymer, NP, nanoprecipitation, liposomal formulation, magnetic NP, hollow capsules, albumin nanosuspension
Daidzein	Cardiovascular system, breast cancer	SLN with PEGylated phospholipid
Dibenzoylmethane	Cervical cancer cells, hepatic cancer, prostate cancer, lung cancer, osteosarcoma	PLA NP
Dihydroartemisinin	Breast cancer, ovarian cancer, esophageal cancer	Magnetic NP
Ellagic acid	Breast cancer, prostate cancer, colorectal cancer, melanoma, ovarian cancer, non-small cell lung cancer (NSCLC), bladder cancer	PLGA NP, PEG, mesoporous silica NP
Epigallocatechin gallate (EGCG)	Prostate cancer cells, pancreatic cancer cells, breast cancer, ovarian cancer, endometrial cancer, renal cancer, and colon cancer	Lipid NP, polymeric NP (PLA and PLGA), liposomes, gold NP, selenium nanocarriers, PEG
Eugenol	Colon and liver cancer, breast cancer	Nanoemulsions, magnetic NP
Ferulic acid	Hepatocellular cancer, colon cancer, pancreatic cancer	NP, chitosan-coated SLN
Gambogic acid	Breast cancer, pancreatic cancer	Glycol chitosan NP, NP coated with red blood cell membranes, carbon nanotube, and graphene nanodelivery, magnetic NPs
Genistein	Breast cancer, prostate cancer, colon cancer, osteosarcoma, human gastric carcinoma, neuroblastoma, bladder cancer, lung cancer, cervical cancer	Biodegradable TPGS-b-PCL NP, PEGylated silica NP, hybrid nano-material
Honokiol	Liver cancer, breast cancer, colorectal cancer, lymphoid malignant cells, melanoma, thyroid cancer, glioblastoma	NP, nanosome, QD
Naringenin	Cervical cancer, glioblastoma, lung cancer, pancreatic cancer	Silk fibroin NP, naringenin-loaded PCL NP, multi-walled carbon nanotubes, naringenin-loaded PLGA NP
Nobiletin	Breast cancer, ovarian cancer, prostate cancer, colon cancer, liver cancer, hepatocellular carcinomas, glioblastoma, gastric cancer, lung cancer, nasopharyngeal cancer	Nano-emulsion, micelles

Table 10.3 (Continued)

Phytochemical	Application target	Delivery system
Quercetin	Breast cancer, ovarian cancer, cervical cancer, prostate cancer, colorectal cancer, gastrointestinal cancer, liver cancer, thyroid cancer, lung cancer, pancreatic cancer, lymphomas	Quercetin encapsulated in SLN, gold NP-quercetin into PLGA, graphene oxide nanocarrier
Resveratrol	Skin cancer, breast cancer, prostate cancer, pancreatic cancer	Gold NP, SLN
Thymoquinone	Glioblastoma, breast cancer, ovarian cancer, osteosarcoma, colorectal cancer, adenocarcinoma, pancreatic carcinoma, myeloblastic leukemia	Silica NP core loaded with thymoquinone, thymoquinone-loaded nanostructured lipid carriers, PLGA, and PEG, thymoquinone-encapsulated chitosan NP, 1,2-dipalmitoyl-sn-glycerol-3-phosphocholine liposomal system, micelles
Triptolide	Breast cancer, liver cancer, lung cancer, pancreatic cancer	Triptolide-loaded cationic liposomes, triptolide coupled to vitamin E using dithiodiglycolic acid and co-dissolved with PEG2000-linoleic acid, nucleolin-specific aptamer-mediated polymeric nanocarrier
Taxifolin	Colorectal cancer, breast cancer	Zinc oxide NP, NP by liquid antisolvent precipitation
Ursolic acid	Cervical cancer, breast cancer, prostate cancer, lung cancer, hepatocellular carcinoma, gallbladder carcinoma, melanoma	Gold-ursolic acid into PLGA NP, and long-circulating and pH-sensitive liposomes. PEG-modified liposome, ursolic acid-loaded chitosan NP, pH-sensitive mesoporous silica NP

CD, cyclodextrin; NP, nanoparticle; PCL, polycaprolactone; PEG, polyethylene glycol; PLA, poly(lactide); PLGA, poly lactic-co-glycolic acid; PVP, poly(vinyl pyrrolidone); QD, quantum dot; SLN, solid lipid NP; TPGS, tocopheryl polyethylene glycol 1000 succinate.

Although the use of bioactive food ingredients for chemoprevention in preclinical settings has achieved remarkable success, its "lab to bedside" applicability to humans has only achieved limited success.

Some of the reasons that are considered to cause the disappointment in chemoprevention clinical trials are: (i) different genetic origin; (ii) different dietary habits; most importantly, (iii) systemic administration is ineffective, and the drug's bioavailability is poor. Given that the FDA is considering more and more nanoparticle formulations and the number of patent applications for new formulations is increasing exponentially, the prospects of nanoparticle systems in cancer treatment are as follows. So far, various related nanodevices have been used in chemoprevention research, including injectable nanocarriers, such as liposomes and new biological targeting nanoparticle conjugates [93, 94].

10.4.2 Natural Nanotherapeutics in Wound Healing

The wound-healing process is a series of well-organized biochemical and cellular phenomena designed to restore the skin and subcutaneous tissue integrity [95]. Various plant extracts and their plant components are considered promising alternatives to wound healing agents due to their different active ingredients, easy accessibility, and low side effects.

In addition to the advantages of medicinal plants in wound healing, one of the most promising ways to improve effectiveness is to use them in nanoscale processes or integrate them into nanostructures. Due to their nano-scale size and high surface area to volume ratio, nano-materials have unique properties, and nano-scale medicinal plants can be produced from changes in their physical and chemical properties. Oxygen is an essential local factor, which is vital to the wound-healing process because it plays a vital role in all stages of wound healing, mediating angiogenesis, enhancing epithelial reformation, and increasing collagen synthesis and fibroblast proliferation. Active oxygen substances such as hydrogen peroxide (H_2O_2) and superoxide (O_2-) are by-products of human oxygen metabolism and are essential regulators of wound healing. In order to maintain the level of free radicals at the desired level, medicinal plants are a good choice because of their inherent antioxidant potential, which can regulate the SOD enzymes GPx and catalase. Herbal compounds, plant extracts, and essential oils in wound dressings, such as curcumin, genistein, cellulose (from citrus reticulate extract), asiaticoside, *Chromola enaodorata* extract, Fuchs, black cumin oil, and wheat germ oil can inhibit oxidative stress and show antioxidant effect [95–100].

As we all know, curcumin is a powerful antioxidant, and this nano-stable botanical drug can be prepared without significantly changing its intrinsic antioxidant activity. Lee and others have proved this. Compared with the unmodified curcumin (natural form), the antioxidant activity of nano curcumin was reduced by <1%, and added to the chitosan/alginate hydrogel [101].

In another study, AgNP was biologically conjugated with *Catharanthus roseus* leaf extract. Due to the adhesion of functional groups in *C. roseus* extract, traditional AgNP produces robust antioxidant activity [102].

In in vitro and in vivo studies, *Fraxinus angustifolia*'s bark and leaf extracts showed antioxidant effects on H_2O_2-induced oxidative stress, which is related to high quercetin and tannins, the two most potent antioxidants. The ability of these extracts to chelate ferrous ions has been proven, inhibiting their ability to convert peroxides into free radicals. Its high chelating activity may be another reason for its antioxidant capacity in wound tissues. The higher cellular uptake of the nanoformulation samples results in these structures having a higher antioxidant activity and a better wound-healing function (Table 10.4) [103].

The formation of biocompatible metal-based nanoparticles using medicinal plants and their extracts, which are environmentally friendly and low-cost, makes it possible to obtain a set of nanostructures with excellent antimicrobial and antibacterial activities. According to the literature, impregnating plant compounds with nanoparticles, nanoemulsions, nanoliposomes, hydrogels, etc., is one of the most effective methods to improve their availability by controlling their release to the site, in the form of a sustainable delivery system, of wounds and increase the permeability of these treatments to the bottom layer of the skin, which is necessary for the healing process.

10.4.3 Natural Nanotherapeutics in Ischemia–Reperfusion Injury

IR injury is a term used to describe the functional and structural changes that occur when blood flow is restored after a period of ischemia. Consequently, destructive effects, such as the necrosis of irreversibly damaged cells, severe swelling of cells, and the restoration of uneven blood flow in various parts of the tissue occur. This disordered restoration of tissue blood flow, a reflux phenomenon, results from a vicious circle of vascular, endothelial, and mitochondrial blood flow dysfunction and local blood flow reduction, severe dysfunction, edema, and many other [104] consequences. At the systemic level, reperfusion injury first mobilizes neutrophils through the chemotaxis and

Table 10.4 List of herbal nanoformulations in wound management.

Plant	Fraction	Model for wound healing
Astragali dadix	Hydroethanolic extract	In vivo on Sprague Dawley rats
Tragacanth gum	—	In vivo on Sprague Dawley rats
Calendula officinalis	NM	In vivo on Sprague Dawley rats
Henna	Leaves ethanolic extract	In vivo on Wistar rats
Fenugreek	Seed extract	In vivo on Wistar rats
Soybean	Protein isolates	In vitro against primary HDFs
Spartium junceum L.	Vegetable branch extracts	In vitro against human keratinocytes (HaCaT)
Tecomella undulata	Bark methanolic extract	In vitro
Centella asiatica	Crude extracts	In vitro on NHDF
Cassia roxburghii	Aqueous extract	In vivo (male Wistar albino rats)
Arnebia nobilis	Root aqueous extract	In vivo on Wistar albino rats
Biophytum sensitivum	Aqueous extract	In vitro on L929 fibroblast
Cellulose gum	—	In vivo on Sprague Dawley rats
Coleus forskohlii	Aqueous root extract	In vivo on albino Wistar rats
Drosera binata	Leaf chloroform extract	In vitro on human keratinocytes
Ficus religiosa	Leaf aqueous extract	In vivo on Wistar albino rats
Naringi crenulata	Aqueous extracellular leaf extract	In vivo on Wistar albino rats
Citrus reticulate	Toluene–ethanol extract	In vitro
Moringa oleifera	Aqueous leaf extract	In vivo on male albino rats
Potato starch	—	In vivo on Swiss albino rats
Guar gum	—	In vivo on Wistar rats
Nyctanthes arbor-tristis L.	Ethanol leaf extract	In vitro
Turmeric	Acetone extract	In vivo albino rats
Bryonia laciniosa	Aqueous leaf extract	In vivo on Wistar rats
Cassia auriculata	Aqueous leaf extract	In vivo on Wistar albino rats
Lansium domesticum	Fruit peel aqueous extract	In vivo Sprague Dawley rats
Phytophthora infestans	Aqueous leaf extract	In vivo on albino rats
Azadirachta indica	Aqueous leaf extract	In vivo on rabbits
Piper nigrum	Aqueous leaf extract	In vitro
Catharanthus roseus	Methanolic leaf extract	In vivo on albino rats
Propolis	Propolis–gelucire aqueous extract	In vivo on Wistar rats
Orchidantha chinensis	Endophytic fungus OC-11 isolate	In vivo Sprague Dawley rats
Linseed	Linseed hydrogels obtained from aqueous extract	In vitro
Momordica charantia	Fruit ethanolic extract	In vitro on L929 fibroblast
Eucalyptus	Eucalyptus essential oil	In vivo on Sprague Dawley rats
Aloe vera	Aqueous extract	In vitro against human fibroblast cells

(*Continued*)

Table 10.4 (Continued)

Plant	Fraction	Model for wound healing
Calendula officinalis	Aqueous extract	In vitro against conjunctival epithelial cells
Dendrocalamus hamiltonii	NM	In vivo on Swiss albino mice
Bambusa bambos	Leaf extract	In vivo on Swiss albino mice
Danggui buxue	Aqueous extract	In vivo on Sprague Dawley rats
Black seed	Essential oil	In vitro against HaCaT cells
Wheat germ	Essential oil	In vitro against HaCaT cells
Pluchea indica	Leaf ethanolic extract	In vitro against HO-1-N-1 buccal mucosa carcinoma-derived cells
Tragacanth gum	—	In vitro against human fibroblast cells
Silybum marianum L.	Isolated silymarin content	In vivo on male BALB/c mice
Fraxinus angustifolia	Leaf ethanolic extract	In vivo on CD-1 mice
Gum arabic	—	In vivo on C57BL6 mice

HDF, human dermal fibroblast; NHDF, normal human dermal fibroblast; NM, not mentioned.
Source: Hajialyani et al. [95]/Dove Press Ltd.

adhesion of endothelium, CD4+T lymphocytes, and circulating platelets in the vascular space. Neutrophils stimulate the production of ROS, tumor necrosis factor-alpha (TNF-α), and local inflammatory mediators, thereby exacerbating tissue damage. CD4+T cells produce macrophage-stimulating factors, interferon-gamma, and TNF-α, thereby increasing local macrophage activation and cytokine release [104]. In addition, reoxygenation will increase the amount of free oxygen free radicals that penetrate the parenchyma of the lesion, endothelial cells, and lymphocytes. The formation of superoxide anion is due to the incomplete recovery of oxygen by damaged mitochondria and the effect of neutrophils and endothelial cells. These processes can lead to the accumulation of free radicals. These unstable molecules can destroy the stability of inorganic and organic chemicals and cause cell damage [105–107].

At the systemic level, increased levels of circulating ROS and pro-inflammatory factors can lead to oxidative stress and increased endothelial adhesion molecules in distant organs. In addition, the level of NO decreases, and there is an imbalance between the production of endothelin-1 and NO through-NO synthase (NOS), which leads to vasoconstriction, increased expression of binding adhesion molecules, and promotes platelet uptake and neutrophils in the local vascular structure, microphage activation, and release of ROS and inflammatory cytokines [108].

Oxidative stress is caused by the imbalance between oxidants and the immune system; cells in a physiological state have low levels of oxygen free radicals and oxidants, which regulate cell homeostasis, survival, mitosis, differentiation, and signal transduction. Overproduction of ROS is generally toxic to cells and destroys all cellular components such as protein, DNA, and lipids. As we all know, oxidative stress is a vital damage mechanism, involving the pathogenesis and progression of many diseases, including ischemic myocardial injury [104].

In recent years, studies have confirmed the beneficial effects of curcumin as a powerful antioxidant and have shown that curcumin has the ability to prevent or reduce IR damage. Turmeric has antioxidant activity and protects cells from oxidative damage, mainly by eliminating various ROS [109].

Sreejayan and Rao [110] found that the presence of phenolic groups in the structure of curcumin is responsible for its antioxidant activity and ability to remove ROS from cells. They can remove hydroxyl radicals, nitrogen dioxide, and NO [109, 111, 112]. Curcumin reduces the formation of superoxide radicals [113]. Curcumin can reduce myocardial insufficiency after IR by activating the silent information regulator 1 (SIRT1) signal, upregulating Bcl-2 and downregulating Bax. Curcumin can also maintain mitochondrial redox potential, significantly increase the activity of mitochondrial SOD, and reduce the production of mitochondrial hydrogen peroxide and malondialdehyde [114, 115].

Nanocarriers contain various antioxidant and neuroprotective compounds for the treatment of ischemic tissue damage [116]. Free radical scavengers have shown promise in animal models of stroke, but none of them has been approved for clinical use in the United States. Compounds that failed clinical trials include Tiliraza, Citicoline, Disufenton Sodium (NXY059), and edavarone [113, 117, 118]. Various strategies to increase the level of ascorbic acid (vitamin C) and tocopherol (vitamin E) or change the level of NO in the blood have also not been implemented [113]. The half-lives of ascorbic acid and tocopherol are only 30 and 12 minutes, respectively [119, 120]. In addition, excessive use of these compounds can induce oxidative stress in cells and lead to cell death [121]. It is logical that the composition of nanoparticles provides a longer cycle time and lower free radical scavenger toxicity. However, in the context of stroke or other reperfusion injury, nanoparticles containing the above-mentioned scavengers are rarely studied. A method for producing controlled-release nanoparticles containing tocopherol and/or ascorbic acid using polylactic acid–glycolic acid copolymer [122–124] and liposome preparation [125] has been reported. *N*-acetylcysteine [126, 127], melanin [128], and many other conjugates [125, 129] have also been evaluated in vivo and in vitro [130].

Most of the published reports on the use of nanoparticles to treat stroke involve introducing antioxidant enzymes, which, unlike small molecule scavengers, can reduce more than one free radical at a time. Early attempts made it possible to increase the cycle time by combining polyethylene glycol from 5 minutes to 25 hours [131]. The enzyme is also packaged in solid polymer particles [132] and liposomes [133] to improve blood stability. In addition, Kabanov and his colleagues reported a series of so-called "nanozymes" composed of SOD, which were formed by electrostatic interaction [134] with new nanoparticles with covalent cross-linking [135] and linear polymers complex. SOD is also encoded by catalase in polylactide–glycolic acid nanoparticles [136], although experiments were not designed to show whether this effect is additive or synergistic. Compositions sensitive to SOD or catalase payloads include mesoporous silica [137] and iron oxide [138, 139]. It has been reported that some of these formulations can reduce infarct volume (necrotic brain tissue) in rodent stroke models by more than 50% compared with free enzymes, as well as other recovery measures [98]. Lee D and coworkers proposed a novel IR-focused nano-therapeutic agents primarily based totally on molecularly engineered PVAX nanoparticles, organized from bioactive VA, a lively pharmaceutical element in *Gastrodia elata* Blume, a herbal agent for brain ischemic damage and coronary heart diseases, and it exerts antioxidant, anti-inflammatory and anti-nociceptive interest, which are sensitive and precise to H_2O_2. PVAX nanoparticles show off great intrinsic antioxidant, anti-inflammatory, and anti-apoptotic activities in each in vitro and in vivo models of IR damage [62].

10.5 Conclusion and Future Prospects

The cap potential blessings of the scientific trials of the nanoscale studies are widely recognized and are pretty apparent because the small size lets in fantastic access to target areas withinside the body. For instance, the assay of minimal quantities of bioanalyte, remedy or restore internal diseased cells and tissues, nanostructures, and imaging devices are more excellent sensitive prognoses for nanomedicine. The swiftly developing interest in nanotechnology is for the motive that the character of a few materials is altered in surprising methods as size is reduced, called "quantum effects," thereby generating modifications in properties including color, reactivity, electric conductivity, and strength [140]. So far, there are numerous applications concerned with nanomedicine; however, handiest few pieces of research were accomplished in the field of oxidative stress. Antioxidants are compounds that shield cells from reactive oxygen cells or free radicals withinside the body. Although they are created as a part of the body's regular metabolic functions, free radicals react with different cells and intrude on their capacity to function. Free radicals are unstable atoms or molecules with free outer electrons. This makes them fantastically reactive since free electrons usually try to shape a stable bond. This stabilization entails gaining an electron from any other molecule, triggering a series reaction. Such reactions are omnipresent withinside the human body; however, beneath neath, certain circumstances can harm biomolecules. Whether nanoparticles are intracellularly taken up leading to the activation of free radical manufacturing is presently being discussed. Various in vitro and in vivo research display that free radical formation may be prompted through nanoparticles, viz. fullerenes, carbon nanotubes, quantum dots, and emission particles. Nanoparticles may be taken up actively (phagocytosis) through specific cells (macrophages) and provoke ROS formation. Ongoing research is investigating whether the quantity of free radicals formed on the surface of nanoparticles is sufficient to result in cellular effects. This file offers a top-level view approximately what free radicals are, the manner by which they generate, why we need them, how they are neutralized, and what we understand roughly the connection between nanoparticles and free radical assembling.

References

1 Ballance, W.C., Qin, E.C., Chung, H.J. et al. (2019). Reactive oxygen species-responsive drug delivery systems for the treatment of neurodegenerative diseases. *Biomaterials* 217: 119292.

2 Meyer, U., Buchter, A., Wiesmann, H.P. et al. (2005). Basic reactions of osteoblasts on structured material surfaces. *European Cells & Materials* 9: 39–49.

3 Saravanakumar, G., Kim, J., and Kim, W.J. (2017). Reactive-oxygen-species-responsive drug delivery systems: promises and challenges. *Advanced Science* 4 (1): 1600124.

4 Lee, S.H., Gupta, M.K., Bang, J.B. et al. (2013). Current progress in reactive oxygen species (ROS)-responsive materials for biomedical applications. *Advanced Healthcare Materials* 2 (6): 908–915.

5 Trachootham, D., Alexandre, J., and Huang, P. (2009). Targeting cancer cells by ROS-mediated mechanisms: a radical therapeutic approach? *Nature Reviews Drug Discovery* 8 (7): 579–591.

6 Brieger, K., Schiavone, S., Miller, F.J., and Krause, K.H. (2012). Reactive oxygen species: from health to disease. *Swiss Medical Weekly* 142: w13659.

7 Schumacker, P.T. (2015). Reactive oxygen species in cancer: a dance with the devil. *Cancer Cell* 27 (2): 156–157.

8 Song, C.C., Du, F.S., and Li, Z.C. (2014). Oxidation-responsive polymers for biomedical applications. *Journal of Materials Chemistry B* 2 (22): 3413–3426.

9 Loukogeorgakis, S.P., van den Berg, M.J., Sofat, R. et al. (2010). Role of NADPH oxidase in endothelial ischemia/reperfusion injury in humans. *Circulation* 121 (21): 2310–2316.

10 Sorce, S. and Krause, K.H. (2009). NOX enzymes in the central nervous system: from signaling to disease. *Antioxidants & Redox Signaling* 11 (10): 2481–2504.

11 Szatrowski, T.P. and Nathan, C.F. (1991). Production of large amounts of hydrogen peroxide by human tumor cells. *Cancer Research* 51 (3): 794–798.

12 Kawanishi, S. and Hiraku, Y. (2006). Oxidative and nitrative DNA damage as biomarker for carcinogenesis with special reference to inflammation. *Antioxidants & Redox Signaling* 8 (5–6): 1047–1058.

13 Nishikawa, M. (2008). Reactive oxygen species in tumor metastasis. *Cancer Letters* 266 (1): 53–59.

14 Pelicano, H., Carney, D., and Huang, P. (2004). ROS stress in cancer cells and therapeutic implications. *Drug Resistance Updates* 7 (2): 97–110.

15 Waris, G. and Ahsan, H. (2006). Reactive oxygen species: role in the development of cancer and various chronic conditions. *Journal of Carcinogenesis* 5: 14.

16 Ibañez, I.L., Notcovich, C., Catalano, P.N. et al. (2015). The redox-active nanomaterial toolbox for cancer therapy. *Cancer Letters* 359 (1): 9–19.

17 Davies, K.J. (2000). An overview of oxidative stress. *IUBMB Life* 50 (4–5): 241–244.

18 Valko, M., Leibfritz, D., Moncol, J. et al. (2007). Free radicals and antioxidants in normal physiological functions and human disease. *The International Journal of Biochemistry & Cell Biology* 39 (1): 44–84.

19 Spyratou, E., Makropoulou, M., Mourelatou, E.A., and Demetzos, C. (2012). Biophotonic techniques for manipulation and characterization of drug delivery nanosystems in cancer therapy. *Cancer Letters* 327 (1–2): 111–122.

20 Kim, K.S., Lee, D., Song, C.G., and Kang, P.M. (2015). Reactive oxygen species-activated nanomaterials as theranostic agents. *Nanomedicine* 10 (17): 2709–2723.

21 Colavitti, R., Pani, G., Bedogni, B. et al. (2002). Reactive oxygen species as downstream mediators of angiogenic signaling by vascular endothelial growth factor receptor-2/KDR. *Journal of Biological Chemistry* 277 (5): 3101–3108.

22 Bienert, G.P., Schjoerring, J.K., and Jahn, T.P. (2006). Membrane transport of hydrogen peroxide. *Biochimica et Biophysica Acta (BBA) – Biomembranes* 1758 (8): 994–1003.

23 Droge, W. (2002). Free radicals in the physiological control of cell function. *Physiological Reviews* 82 (1): 47–95.

24 Veal, E.A., Day, A.M., and Morgan, B.A. (2007). Hydrogen peroxide sensing and signaling. *Molecular Cell* 26 (1): 1–14.

25 Bienert, G.P., Møller, A.L., Kristiansen, K.A. et al. (2007). Specific aquaporins facilitate the diffusion of hydrogen peroxide across membranes. *Journal of Biological Chemistry* 282 (2): 1183–1192.

26 Darley-Usmar, V. and White, R. (1997). Disruption of vascular signalling by the reaction of nitric oxide with superoxide: implications for cardiovascular disease. *Experimental Physiology: Translation and Integration* 82 (2): 305–316.

27 Ramachandran, A., Levonen, A.L., Brookes, P.S. et al. (2002). Mitochondria, nitric oxide, and cardiovascular dysfunction. *Free Radical Biology and Medicine* 33 (11): 1465–1474.

28 Le Lay, S., Simard, G., Martinez, M.C., and Andriantsitohaina, R. (2014). Oxidative stress and metabolic pathologies: from an adipocentric point of view. *Oxidative Medicine and Cellular Longevity* 2014: 908539.

29 Gottlieb, R.A., Burleson, K.O., Kloner, R.A. et al. (1994). Reperfusion injury induces apoptosis in rabbit cardiomyocytes. *The Journal of Clinical Investigation* 94 (4): 1621–1628.

30 Logue, S.E., Gustafsson, Å.B., Samali, A., and Gottlieb, R.A. (2005). Ischemia/reperfusion injury at the intersection with cell death. *Journal of Molecular and Cellular Cardiology* 38 (1): 21–33.

31 Whelan, R.S., Kaplinskiy, V., and Kitsis, R.N. (2010). Cell death in the pathogenesis of heart disease: mechanisms and significance. *Annual Review of Physiology* 72: 19–44.

32 Zweier, J.L. (1988). Measurement of superoxide-derived free radicals in the reperfused heart. Evidence for a free radical mechanism of reperfusion injury. *Journal of Biological Chemistry* 263 (3): 1353–1357.

33 Kang, P.M. and Izumo, S. (2000). Apoptosis and heart failure: a critical review of the literature. *Circulation Research* 86 (11): 1107–1113.

34 Yaoita, H., Ogawa, K., Maehara, K., and Maruyama, Y. (1998). Attenuation of ischemia/reperfusion injury in rats by a caspase inhibitor. *Circulation* 97 (3): 276–281.

35 Go, A.S., Mozaffarian, D., Roger, V.L. et al. (2014). Heart disease and stroke statistics – 2014 update: a report from the American Heart Association. *Circulation* 129 (3): e28–e292.

36 Domej, W., Oettl, K., and Renner, W. (2014). Oxidative stress and free radicals in COPD – implications and relevance for treatment. *International Journal of Chronic Obstructive Pulmonary Disease* 9: 1207.

37 Liang, J. and Liu, B. (2016). ROS-responsive drug delivery systems. *Bioengineering & Translational Medicine* 1 (3): 239–251.

38 Kataoka, K., Harada, A., and Nagasaki, Y. (2001). Block copolymer with cross-linking core was also highlighted here. micelles for drug delivery: design, characterization and Since the thin hydrogel layer made by alternate biological significance. *Advanced Drug Delivery Reviews* 47: 113–131.

39 Oerlemans, C., Bult, W., Bos, M. et al. (2010). Polymeric micelles in anticancer therapy: targeting, imaging and triggered release. *Pharmaceutical Research* 27 (12): 2569–2589.

40 Gong, J., Chen, M., Zheng, Y. et al. (2012). Polymeric micelles drug delivery system in oncology. *Journal of Controlled Release* 159 (3): 312–323.

41 Torchilin, V.P. (2007). Micellar nanocarriers: pharmaceutical perspectives. *Pharmaceutical Research* 24 (1): 1–16.

42 Ye, H., Zhou, Y., Liu, X. et al. (2019). Recent advances on reactive oxygen species-responsive delivery and diagnosis system. *Biomacromolecules* 20 (7): 2441–2463.

43 Hu, P. and Tirelli, N. (2012). Scavenging ROS: superoxide dismutase/catalase mimetics by the use of an oxidation-sensitive nanocarrier/enzyme conjugate. *Bioconjugate Chemistry* 23 (3): 438–449.

44 Engman, L. (1985). Synthetic applications of organotellurium chemistry. *Accounts of Chemical Research* 18 (9): 274–279.

45 Liu, J., Pang, Y., Zhu, Z. et al. (2013). Therapeutic nanocarriers with hydrogen peroxide-triggered drug release for cancer treatment. *Biomacromolecules* 14 (5): 1627–1636.

46 Fang, R., Xu, H., Cao, W. et al. (2015). Reactive oxygen species (ROS)-responsive tellurium-containing hyperbranched polymer. *Polymer Chemistry* 6 (15): 2817–2821.

47 Cao, W., Gu, Y., Li, T., and Xu, H. (2015). Ultra-sensitive ROS-responsive tellurium-containing polymers. *Chemical Communications* 51 (32): 7069–7071.

48 Shim, M.S. and Xia, Y. (2013). A reactive oxygen species (ROS)-responsive polymer for safe, efficient, and targeted gene delivery in cancer cells. *Angewandte Chemie* 125 (27): 7064–7067.

49 Wilson, D.S., Dalmasso, G., Wang, L. et al. (2010). Orally delivered thioketal nanoparticles loaded with TNF-α-siRNA target inflammation and inhibit gene expression in the intestines. *Nature Materials* 9 (11): 923–928.

50 Wang, C., Wang, J., Zhang, X. et al. (2018). In situ formed reactive oxygen species-responsive scaffold with gemcitabine and checkpoint inhibitor for combination therapy. *Science Translational Medicine* 10 (429): eaan3682.

51 Zhang, Q., Zhang, F., Chen, Y. et al. (2017). Structure–property correlations of reactive oxygen species-responsive and hydrogen peroxide-eliminating materials with anti-oxidant and anti-inflammatory activities. *Chemistry of Materials* 29 (19): 8221–8238.

52 Qiao, C., Yang, J., Shen, Q. et al. (2018). Traceable nanoparticles with dual targeting and ROS response for RNAi-based immunochemotherapy of intracranial glioblastoma treatment. *Advanced Materials* 30 (18): 1705054.

53 Deng, Z., Qian, Y., Yu, Y. et al. (2016). Engineering intracellular delivery nanocarriers and nanoreactors from oxidation-responsive polymersomes via synchronized bilayer cross-linking and permeabilizing inside live cells. *Journal of the American Chemical Society* 138 (33): 10452–10466.

54 Bandyopadhyay, A. and Gao, J. (2016). Iminoboronate-based peptide cyclization that responds to pH, oxidation, and small molecule modulators. *Journal of the American Chemical Society* 138 (7): 2098–2101.

55 Granzhan, A., Schouwey, C., Riis-Johannessen, T. et al. (2011). Connection of metallamacrocycles via dynamic covalent chemistry: a versatile method for the synthesis of molecular cages. *Journal of the American Chemical Society* 133 (18): 7106–7115.

56 Bull, S.D., Davidson, M.G., Van den Elsen, J.M. et al. (2013). Exploiting the reversible covalent bonding of boronic acids: recognition, sensing, and assembly. *Accounts of Chemical Research* 46 (2): 312–326.

57 de Gracia Lux, C., Joshi-Barr, S., Nguyen, T. et al. (2012). Biocompatible polymeric nanoparticles degrade and release cargo in response to biologically relevant levels of hydrogen peroxide. *Journal of the American Chemical Society* 134 (38): 15758–15764.

58 Lee, D., Khaja, S., Velasquez-Castano, J.C. et al. (2007). In vivo imaging of hydrogen peroxide with chemiluminescent nanoparticles. *Nature Materials* 6 (10): 765–769.

59 Lee, D., Bae, S., Ke, Q. et al. (2013). Hydrogen peroxide-responsive copolyoxalate nanoparticles for detection and therapy of ischemia–reperfusion injury. *Journal of Controlled Release* 172 (3): 1102–1110.

60 Karin, M. and Greten, F.R. (2005). NF-κB: linking inflammation and immunity to cancer development and progression. *Nature Reviews Immunology* 5 (10): 749–759.

61 Lee, D., Erigala, V.R., Dasari, M. et al. (2008). Detection of hydrogen peroxide with chemiluminescent micelles. *International Journal of Nanomedicine* 3 (4): 471.

62 Lee, D., Bae, S., Hong, D. et al. (2013). H_2O_2-responsive molecularly engineered polymer nanoparticles as ischemia/reperfusion-targeted nanotherapeutic agents. *Scientific Reports* 3 (1): 1–8.

63 Basta, G. (2008). Receptor for advanced glycation endproducts and atherosclerosis: from basic mechanisms to clinical implications. *Atherosclerosis* 196 (1): 9–21.

64 Brinkley, T.E., Leng, X., Nicklas, B.J. et al. (2017). Racial differences in circulating levels of the soluble receptor for advanced glycation endproducts in middle-aged and older adults. *Metabolism* 70: 98–106.

65 Li, T., Qin, W., Liu, Y. et al. (2017). Effect of RAGE gene polymorphisms and circulating sRAGE levels on susceptibility to gastric cancer: a case–control study. *Cancer Cell International* 17 (1): 1–10.

66 Meng, H.Z., Zhang, W.L., Liu, F., and Yang, M.W. (2015). Advanced glycation end products affect osteoblast proliferation and function by modulating autophagy via the receptor of advanced glycation end products/Raf protein/mitogen-activated protein kinase/extracellular signal-regulated kinase kinase/extracellular signal-regulated kinase (RAGE/Raf/MEK/ERK) pathway. *Journal of Biological Chemistry* 290 (47): 28189–28199.

67 Ahmad, S., Akhter, F., Shahab, U. et al. (2018). Do all roads lead to the Rome? The glycation perspective! *Seminars in Cancer Biology* 49: 9–19. Academic Press.

68 García-Gasca, T., Martínez-Dávila, I., Reyes, P., and Mejia, C. *Apoptosis: The Role of Free Radicals in Physiological and Pathological States*. Signal Post Research Publish.

69 Sen, C.K. (2009). Wound healing essentials: let there be oxygen. *Wound Repair and Regeneration* 17 (1): 1–18.

70 Schreml, S., Szeimies, R.M., Prantl, L. et al. (2010). Oxygen in acute and chronic wound healing. *British Journal of Dermatology* 163 (2): 257–268.

71 Kurahashi, T. and Fujii, J. (2015). Roles of antioxidative enzymes in wound healing. *Journal of Developmental Biology* 3 (2): 57–70.

72 Schäfer, M. and Werner, S. (2008). Oxidative stress in normal and impaired wound repair. *Pharmacological Research* 58 (2): 165–171.

73 Rhee, S.G. (2006). H_2O_2, a necessary evil for cell signaling. *Science* 312 (5782): 1882–1883.

74 Breton-Romero, R. and Lamas, S. (2014). Hydrogen peroxide signaling in vascular endothelial cells. *Redox Biology* 2: 529–534.

75 Marinho, H.S., Real, C., Cyrne, L. et al. (2014). Hydrogen peroxide sensing, signaling and regulation of transcription factors. *Redox Biology* 2: 535–562.

76 Sen, C.K., Khanna, S., Babior, B.M. et al. (2002). Oxidant-induced vascular endothelial growth factor expression in human keratinocytes and cutaneous wound healing. *Journal of Biological Chemistry* 277 (36): 33284–33290.

77 Vivekananda, J., Lin, A., Coalson, J.J., and King, R.J. (1994). Acute inflammatory injury in the lung precipitated by oxidant stress induces fibroblasts to synthesize and release transforming growth factor-alpha. *Journal of Biological Chemistry* 269 (40): 25057–25061.

78 Maeda, K. and Ruel, M. (2014). Prevention of ischemia–reperfusion injury in cardiac surgery: therapeutic strategies targeting signaling pathways. *The Journal of Thoracic and Cardiovascular Surgery* 149 (3): 910–911.

79 Murata, T., Kinoshita, K., Hori, M. et al. (2005). Statin protects endothelial nitric oxide synthase activity in hypoxia-induced pulmonary hypertension. *Arteriosclerosis, Thrombosis, and Vascular Biology* 25 (11): 2335–2342.

80 Chouchani, E.T., Pell, V.R., Gaude, E. et al. (2014). Ischaemic accumulation of succinate controls reperfusion injury through mitochondrial ROS. *Nature* 515 (7527): 431–435.

81 Sies, H. (1997). Oxidative stress: oxidants and antioxidants. *Experimental Physiology: Translation and Integration* 82 (2): 291–295.

82 Polidori, M.C., Cherubini, A., Senin, U., and Mecocci, P. (2001). Peripheral non-enzymatic antioxidant changes with human aging: a selective status report. *Biogerontology* 2 (2): 99–104.

83 Calabrese, V., Scapagnini, G., Stella, A.G. et al. (2001). Mitochondrial involvement in brain function and dysfunction: relevance to aging, neurodegenerative disorders and longevity. *Neurochemical Research* 26 (6): 739–764.

84 Beal, M.F. (2000). Oxidative metabolism. *Annals of the New York Academy of Sciences* 924 (1): 164–169.

85 Mahajan, A. and Tandon, V.R. (2004). Antioxidants and rheumatoid arthritis. *Journal of Indian Rheumatology Association* 12: 139–142.

86 Majithia, V. and Geraci, S.A. (2007). Rheumatoid arthritis: diagnosis and management. *The American Journal of Medicine* 120 (11): 936–939.

87 Öztürk, H.S., Çimen, M.B., Çimen, Ö.B. et al. (1999). Oxidant/antioxidant status of plasma samples from patients with rheumatoid arthritis. *Rheumatology International* 19 (1–2): 35–37.

88 Leonavičienė, L., Bradūnaitė, R., Vaitkienė, D. et al. (2008). Collagen-induced arthritis and pro-/antioxidant status in Wistar and Lewis rats. *Biologija* 54 (4): 290–300.

89 Ranneh, Y., Ali, F., Akim, A.M. et al. (2017). Crosstalk between reactive oxygen species and pro-inflammatory markers in developing various chronic diseases: a review. *Applied Biological Chemistry* 60 (3): 327–338.

90 Yang, B., Chen, Y., and Shi, J. (2019). Reactive oxygen species (ROS)-based nanomedicine. *Chemical Reviews* 119 (8): 4881–4985.

91 Hunyadi, A. (2019). The mechanism(s) of action of antioxidants: from scavenging reactive oxygen/nitrogen species to redox signaling and the generation of bioactive secondary metabolites. *Medicinal Research Reviews* 39 (6): 2505–2533.

92 Salama, L., Pastor, E.R., Stone, T., and Mousa, S.A. (2020). Emerging nanopharmaceuticals and nanonutraceuticals in cancer management. *Biomedicines* 8 (9): 347.

93 Khurana, R.K., Jain, A., Jain, A. et al. (2018). Administration of antioxidants in cancer: debate of the decade. *Drug Discovery Today* 23 (4): 763–770.

94 Siddiqui, I.A., Sanna, V., Ahmad, N. et al. (2015). Resveratrol nanoformulation for cancer prevention and therapy. *Annals of the New York Academy of Sciences* 1348 (1): 20–31.

95 Hajialyani, M., Tewari, D., Sobarzo-Sánchez, E. et al. (2018). Natural product-based nanomedicines for wound healing purposes: therapeutic targets and drug delivery systems. *International Journal of Nanomedicine* 13: 5023.

96 Suwantong, O., Ruktanonchai, U., and Supaphol, P. (2010). In vitro biological evaluation of electrospun cellulose acetate fiber mats containing asiaticoside or curcumin. *Journal of Biomedical Materials Research Part A* 94 (4): 1216–1225.

97 Ali, A., Haq, I.U., Akhtar, J. et al. (2017). Synthesis of Ag-NPs impregnated cellulose composite material: its possible role in wound healing and photocatalysis. *IET Nanobiotechnology* 11 (4): 477–484.

98 Naraginti, S., Kumari, P.L., Das, R.K. et al. (2016). Amelioration of excision wounds by topical application of green synthesized, formulated silver and gold nanoparticles in albino Wistar rats. *Materials Science and Engineering: C* 62: 293–300.

99 Gumus, Z.P., Guler, E., Demir, B. et al. (2015). Herbal infusions of black seed and wheat germ oil: their chemical profiles, in vitro bio-investigations and effective formulations as phyto-nanoemulsions. *Colloids and Surfaces B: Biointerfaces* 133: 73–80.

100 Kim, J.E., Lee, J., Jang, M. et al. (2015). Accelerated healing of cutaneous wounds using phytochemically stabilized gold nanoparticle deposited hydrocolloid membranes. *Biomaterials Science* 3 (3): 509–519.

101 Li, X., Chen, S., Zhang, B. et al. (2012). In situ injectable nano-composite hydrogel composed of curcumin, *N,O*-carboxymethyl chitosan and oxidized alginate for wound healing application. *International Journal of Pharmaceutics* 437 (1–2): 110–119.

102 Al-Shmgani, H.S., Mohammed, W.H., Sulaiman, G.M., and Saadoon, A.H. (2017). Biosynthesis of silver nanoparticles from *Catharanthus roseus* leaf extract and assessing their antioxidant, antimicrobial, and wound-healing activities. *Artificial Cells, Nanomedicine, and Biotechnology* 45 (6): 1234–1240.

103 Moulaoui, K., Caddeo, C., Manca, M.L. et al. (2015). Identification and nanoentrapment of polyphenolic phytocomplex from *Fraxinus angustifolia*: in vitro and in vivo wound healing potential. *European Journal of Medicinal Chemistry* 89: 179–188.

104 Mokhtari-Zaer, A., Marefati, N., Atkin, S.L. et al. (2019). The protective role of curcumin in myocardial ischemia–reperfusion injury. *Journal of Cellular Physiology* 234 (1): 214–222.

105 Nastos, C., Kalimeris, K., Papoutsidakis, N. et al. (2014). Global consequences of liver ischemia/reperfusion injury. *Oxidative Medicine and Cellular Longevity* 2014: 906965.

106 Kumar, V., Abbas, A.K., and Aster, J.C. (2015). *Robbins & Cotran Pathologic Basis of Disease*, 955. Elsevier.

107 Lee, P.J. and Choi, A.M. (2003). Pathways of cell signaling in hyperoxia. *Free Radical Biology and Medicine* 35 (4): 341–350.

108 Soares, R.O., Losada, D.M., Jordani, M.C. et al. (2019). Ischemia/reperfusion injury revisited: an overview of the latest pharmacological strategies. *International Journal of Molecular Sciences* 20 (20): 5034.

109 Rao, M.N.A. (1997). Nitric oxide scavenging by curcuminoids. *Journal of Pharmacy and Pharmacology* 49 (1): 105–107.

110 Sreejayan, N. and Rao, M.N. (1996). Free radical scavenging activity of curcuminoids. *Arzneimittel-Forschung* 46 (2): 169–171.

111 Reddy, A.C.P. and Lokesh, B.R. (1994). Studies on the inhibitory effects of curcumin and eugenol on the formation of reactive oxygen species and the oxidation of ferrous iron. *Molecular and Cellular Biochemistry* 137 (1): 1–8.

112 Unnikrishnan, M.K. and Rao, M.N.A. (1995). Curcumin inhibits nitrogen dioxide induced oxidation of hemoglobin. *Molecular and Cellular Biochemistry* 146 (1): 35–37.

113 Shirley, R., Ord, E.N., and Work, L.M. (2014). Oxidative stress and the use of antioxidants in stroke. *Antioxidants* 3 (3): 472–501.

114 Yang, Y., Duan, W., Lin, Y. et al. (2013). SIRT1 activation by curcumin pretreatment attenuates mitochondrial oxidative damage induced by myocardial ischemia reperfusion injury. *Free Radical Biology and Medicine* 65: 667–679.

115 Glantzounis, G.K., Salacinski, H.J., Yang, W. et al. (2005). The contemporary role of antioxidant therapy in attenuating liver ischemia-reperfusion injury: a review. *Liver Transplantation* 11 (9): 1031–1047.

116 Amani, H., Habibey, R., Hajmiresmail, S.J. et al. (2017). Antioxidant nanomaterials in advanced diagnoses and treatments of ischemia reperfusion injuries. *Journal of Materials Chemistry B* 5 (48): 9452–9476.

117 Cheng, Y.D., Al-Khoury, L., and Zivin, J.A. (2004). Neuroprotection for ischemic stroke: two decades of success and failure. *Neuro Rx* 1 (1): 36–45.

118 Margaill, I., Plotkine, M., and Lerouet, D. (2005). Antioxidant strategies in the treatment of stroke. *Free Radical Biology and Medicine* 39 (4): 429–443.

119 Duconge, J., Miranda-Massari, J.R., Gonzalez, M.J. et al. (2008). Pharmacokinetics of vitamin C: insights into the oral and intravenous administration of ascorbate. *Puerto Rico Health Sciences Journal* 27 (1): 7–19.

120 Bjørneboe, A., Bjørneboe, G.E.A., and Drevon, C.A. (1987). Serum half-life, distribution, hepatic uptake and biliary excretion of α-tocopherol in rats. *Biochimica et Biophysica Acta (BBA)-Lipids and Lipid Metabolism* 921 (2): 175–181.

121 Chakraborty, A. and Jana, N.R. (2017). Vitamin C-conjugated nanoparticle protects cells from oxidative stress at low doses but induces oxidative stress and cell death at high doses. *ACS Applied Materials & Interfaces* 9 (48): 41807–41817.

122 Zigoneanu, I.G., Astete, C.E., and Sabliov, C.M. (2008). Nanoparticles with entrapped α-tocopherol: synthesis, characterization, and controlled release. *Nanotechnology* 19 (10): 105606.

123 Astete, C.E., Dolliver, D., Whaley, M. et al. (2011). Antioxidant poly (lactic-co-glycolic) acid nanoparticles made with α-tocopherol–ascorbic acid surfactant. *ACS Nano* 5 (12): 9313–9325.

124 Stevanović, M., Uskoković, V., Filipović, M. et al. (2013). Composite PLGA/AgNpPGA/Asc H nanospheres with combined osteoinductive, antioxidative, and antimicrobial activities. *ACS Applied Materials & Interfaces* 5 (18): 9034–9042.

125 Lee, C.Y., Sharma, A., Uzarski, R.L. et al. (2011). Potent antioxidant dendrimers lacking pro-oxidant activity. *Free Radical Biology and Medicine* 50 (8): 918–925.

126 Kurtoglu, Y.E., Navath, R.S., Wang, B. et al. (2009). Poly(amidoamine) dendrimer–drug conjugates with disulfide linkages for intracellular drug delivery. *Biomaterials* 30 (11): 2112–2121.

127 Zarchi, A.A.K., Abbasi, S., Faramarzi, M.A. et al. (2015). Development and optimization of *N*-acetylcysteine-loaded poly(lactic-*co*-glycolic acid) nanoparticles by electrospray. *International Journal of Biological Macromolecules* 72: 764–770.

128 Liu, Y., Ai, K., Ji, X. et al. (2017). Comprehensive insights into the multi-antioxidative mechanisms of melanin nanoparticles and their application to protect brain from injury in ischemic stroke. *Journal of the American Chemical Society* 139 (2): 856–862.

129 Park, H., Kim, S., Kim, S. et al. (2010). Antioxidant and anti-inflammatory activities of hydroxybenzyl alcohol releasing biodegradable polyoxalate nanoparticles. *Biomacromolecules* 11 (8): 2103–2108.

130 Poellmann, M.J., Bu, J., and Hong, S. (2018). Would antioxidant-loaded nanoparticles present an effective treatment for ischemic stroke? *Nanomedicine* 13 (18): 2327–2340.

131 Boccu, E., Velo, G.P., and Veronese, F.M. (1982). Pharmacokinetic properties of polyethylene glycol derivatized superoxide dismutase. *Pharmacological Research Communications* 14 (2): 113–120.

132 Reddy, M.K. and Labhasetwar, V. (2009). Nanoparticle-mediated delivery of superoxide dismutase to the brain: an effective strategy to reduce ischemia-reperfusion injury. *The FASEB Journal* 23 (5): 1384–1395.

133 Yusa, T., Crapo, J.D., and Freeman, B.A. (1984). Liposome-mediated augmentation of brain SOD and catalase inhibits CNS O_2 toxicity. *Journal of Applied Physiology* 57 (6): 1674–1681.

134 Rosenbaugh, E.G., Roat, J.W., Gao, L. et al. (2010). The attenuation of central angiotensin II-dependent pressor response and intra-neuronal signaling by intracarotid injection of nanoformulated copper/zinc superoxide dismutase. *Biomaterials* 31 (19): 5218–5226.

135 Manickam, D.S., Brynskikh, A.M., Kopanic, J.L. et al. (2012). Well-defined cross-linked antioxidant nanozymes for treatment of ischemic brain injury. *Journal of Controlled Release* 162 (3): 636–645.

136 Petro, M., Jaffer, H., Yang, J. et al. (2016). Tissue plasminogen activator followed by antioxidant-loaded nanoparticle delivery promotes activation/mobilization of progenitor cells in infarcted rat brain. *Biomaterials* 81: 169–180.

137 Ambati, J., Lopez, A.M., Cochran, D. et al. (2012). Engineered silica nanocarriers as a high-payload delivery vehicle for antioxidant enzymes. *Acta Biomaterialia* 8 (6): 2096–2103.

138 Chorny, M., Hood, E., Levy, R.J., and Muzykantov, V.R. (2010). Endothelial delivery of antioxidant enzymes loaded into non-polymeric magnetic nanoparticles. *Journal of Controlled Release* 146 (1): 144–151.

139 Hood, E.D., Chorny, M., Greineder, C.F. et al. (2014). Endothelial targeting of nanocarriers loaded with antioxidant enzymes for protection against vascular oxidative stress and inflammation. *Biomaterials* 35 (11): 3708–3715.

140 Watal, G., Watal, A., Rai, P.K. et al. (2013). Biomedical applications of nano-antioxidant. *Oxidative Stress and Nanotechnology* 1028: 147–151.

11

Green Approaches for Synthesis of Nanophytopharmaceuticals/ Nanophytodelivery Systems

Gülsen Kaya[1], Merve Keskin[2], and Mithun Rudrapal[3]

[1] *Scientific and Technological Research Center, Inonu University, Malatya, Turkey*
[2] *Vocational School of Health Services, Bilecik Seyh Edebali University, Bilecik, Turkey*
[3] *Department of Pharmaceutical Chemistry, Rasiklal M. Dhariwal Institute of Pharmaceutical Education and Research, Pune, Maharashtra, India*

11.1 Introduction

Nanotechnology includes approaches that benefit from different fields such as biology, chemistry, engineering, and medicine. Any technology at the nanoscale can be defined as nanotechnology. Many applications and researches are carried out in this area. With this technology, the unknown properties of the substance can be revealed, as well as the synthesis of new featured substances, and the data obtained can be used in various applications [1].

Nanotechnology aims to design and synthesize nanostructures and to give extraordinary new properties to materials thanks to these structures. These features contribute to the solution of problems in different areas and enable them to be used by gaining new functions. Thanks to these potential developments, nanotechnology is of great importance in the development of advanced technologies. Particles of nanoscale (below 100 nm) are of particular interest in applications. It offers excellent solutions for many fields, especially nanomedicine, by direct use or modification of nanoparticles [2].

Metallic nanoparticles (gold, silver, platinum, copper, etc.), which have a wide range of uses, are used in direct contact with the human body such as shampoo, soap, toothpaste, textile, cosmetics, etc., are put into the content of the materials. They are used in pharmaceutical companies, in catalysis studies, in cancer diagnosis, and as antimicrobial agents. In addition, platinum nanoparticles have also been used as an antitumor agent and have been found to be effective [3]. They are synthesized using biological, physical, and chemical methods, but synthesis with biological sources is very advantageous because they do not contain toxic chemicals and are very cheap. Especially, synthesis studies with plant extracts are important for this [3, 4]. Since plant extracts are used both as a reducing agent and as a stabilizing agent, they also have another advantage over chemical methods [5]. Various studies examining the effects of nanoparticles on cells are of interest to nanotoxicology. It is also necessary to know the possible negative effects of nanoparticles, which are used in many applications, on the biotic environment. More and more intensive research is being conducted on approaches to their toxicological effects.

Phytoantioxidants and Nanotherapeutics, First Edition. Edited by Mithun Rudrapal.

Problems that can reach carcinogenesis may occur as a result of exposure to nanoparticles. Therefore, studies are needed to reveal the possible epigenetic effects of nanoparticles [6].

11.2 Synthesis of Nanoparticles

It is possible to synthesize nanoparticles using various methods (biological, physical, or chemical). Among these, synthesis methods with biological sources (green synthesis) attract a lot of attention. This is because biological methods include environmentally friendly approaches and do not contain toxic chemicals. Various studies emphasize that nanoparticles obtained by biological synthesis are more environmentally friendly than those produced by other methods. For this reason, applications for the use of plant resources in synthesis by biological methods have gained momentum.

11.2.1 Production Methods of Nanoparticles

It has been reported that the methods known for the synthesis of nanoparticles were introduced before the seventeenth century and that gold nanoparticles were used in the content of the drug called "Suvarna bhasma," which was used in the treatment of rheumatoid arthritis of ancient Hindus. It is known that Hindus produce gold nanoparticles with the top-down method, that is, the biological method, and Michael Faraday was the first person to try the chemical method in 1857 [7].

On the methods used in the production of nanoparticles: There are two main approaches called top-down "top down" and bottom-up "bottom up" [8]. These two approaches can be explained as follows:

11.2.2 Top-down Approach

In the methods included in the top-down approach, the volumetric material is based on the principle of separating the material into small pieces that can reach the nano-size as a result of the external mechanical and/or chemical processes. The most general examples that can be given to methods working with the top-down approach can be mechanical grinding and grinding. Since these techniques consume much more energy than conventional grinding processes, they are also called high-energy grinding or high-speed mills. In this approach, production starts from larger materials to reach the micro level. A good example of this approach is to turn wheat into flour with the dry milling method and thus increasing its water retention ability [9]. In a study, it was reported that by reducing the green tea particles to 1000 nm, the digestion and absorption of green tea were facilitated. As a result, the activities of enzymes that remove oxygen increased, and more scientifically, the antioxidant effect increased [10].

11.2.3 Bottom-up Approach

The essence of the method is based on building an organic or inorganic structure with atoms or molecules. The control of carbon nanotubes can be achieved by using the forces in nature and the self-assembly feature of biological systems such as DNA to join nanostructures. The reason for the application of this method is defined as the realization of particle formation by enlarging atomic or molecular-sized structures by chemical reactions. The gas condensation technique, which is the first method used in the production of nano metals and alloys, works with a bottom-up approach.

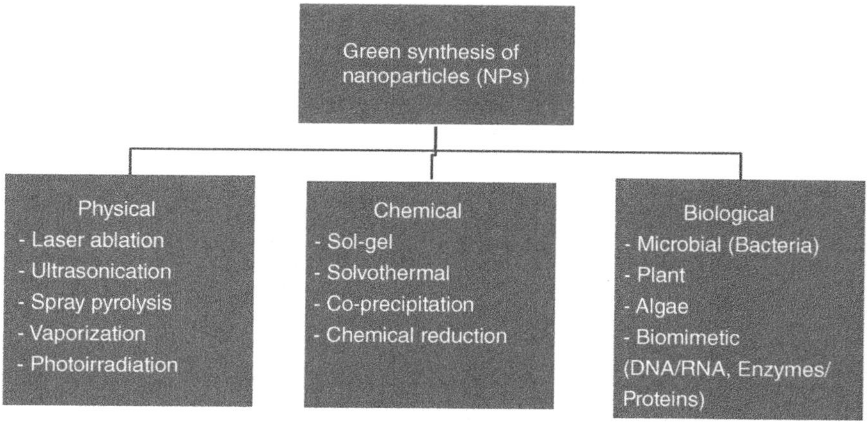

Figure 11.1 Approaches in nanoparticle synthesis.

Chemical vapor coating, chemical vapor condensation, sol gel, and spray pyrolysis methods are also the most well-known members of this approach [11, 12]. Many classical methods, physically and chemically, have been used for the synthesis of nanoparticles in a solution environment, and nowadays "green nanotechnology," which includes cheaper, environmentally friendly, nontoxic biological methods, comes to the fore [13]. At the same time, various methods have been developed in classical methods. Some of these are techniques such as electrochemical synthesis, reverse micelle/microemulsion method, hydrothermal synthesis, sonochemical precipitation, and chemical reduction. It is aimed that the nanostructures obtained in most of these studies have a certain size and morphology. The green synthesis of nanoparticles (NPs) is presented in Figure 11.1.

11.3 Green Synthesis

Thanks to the physical and chemical technologies that have been used in nanoparticle production for a long time, high resolution, desired small particles can be produced in a short time, but their toxic content is high, particle stability is not good, and the technologies used are expensive. So, new technologies should be investigated [14].

In this sense, with the start of researches, the study of materials and living things that already exist in nature, containing perfectly designed nanoscale dimensions, has inspired scientists, and the production of inorganic substances using living structures has begun to be researched. As a result of the researches, the term "green nanotechnology" has emerged, which is environmentally friendly, with low toxic substance content, and is based on the production of nanoparticles from living cells. This term is expressed as working methods that research easy-to-apply methods that reduce the problem of waste products and do not harm human health, within the scope of nanotechnology science [15]. Green technology has been developed as a viable substitute for chemical and physical techniques. Its advantages include the absence of unwanted reaction solvents. The application of biological syntheses is focused on the production of noble metal nanomaterials due to their ease of reduction using weak reducing agents [16]. For this purpose, materials such as bacteria [17], cyanobacteria [18], seaweeds [19], fungi [20], biological [21], and honey [22] are used (Table 11.1). The main benefits of synthesizing NPs from biological sources are simplicity,

Table 11.1 Synthesis of NPs using various biological entities and plant extracts.

Plant or biological entity	Reaction conditions	Average size (nm)	Shape	Biomedical application	Reference
Nigella sativa (black cumin), (seeds)	75 °C; 2 days	3.47	Spherical	Anticancer activity against human breast (MDA-MB-231) and cervical (HeLa) cancer cells_ bactericidal effect against Gram-negative and Gram-positive bacteria	[23]
Green tea (powder extract)	50 °C; 4 h	2	Spherical	Anticancer activity against two human colon cancer cell lines (SW480 and SW620)	[24]
Punica granatum (pomegranate), (crusts)	Room temperature ultrasonication 24 h	20.12	Spherical and cubic	Antiproliferation effect and enhanced apoptosis against human breast adenocarcinoma cell line (MCF-7)	[25]
Honey	100 °C for 2 and 4 h	2.2	Nanowires	—	[26]
Streptomyces species (Gram-positive bacteria)	50 °C; 24 h	20–50	Spherical	Anticancer activity against human breast cancer cell lines (MCF-7)	[27]
Padina gymnospora (brown algae)	Room temperature for 10 min	5–50	Truncated octahedral	Bactericidal activity against *Escherichia coli*, *Lactococcus lactis*, and *Klebsiella pneumoniae*	[28]
Sheep milk	Room temperature for 3 h	9	Spherical	—	[29]
Prunus × yedoensis (gum extract)	pH 8; Gum extract concentrations of 7 and 8% 30 min	10–50	Spherical	Antifungal against *Colletotrichum acutatum* and *Cladosporium fulvum*	[30]

convenience, cheapness, easy scalability, low energy requirement, environmental friendliness, and minimal use of hazardous substances and maximum efficiency of the synthesis process. Synthesis based on biological resources increases the stability of NPs in terms of size and shape and provides higher yields than other physical and chemical methods.

Nanoparticles can be used in many applications depending on their size, shape, surface components, and atomic arrangement on the surface, and these properties enable them to take place in important applications in the field of medicine (in terms of antimicrobial and anticarcinogenic effects). Some metals preferred in metallic nanoparticle synthesis: silver, gold, iron, copper, lead, palladium, cadmium, nickel, etc. (Figure 11.2).

Many studies provide an overview of the preparation of nanoparticles with green synthesis approaches, which have advantages over traditional methods involving chemicals associated with environmental toxicity. Green methods include mixed-valence polyoxometallates, polysaccharides, Tollens, irradiation, and bio-assisted synthesis methods.

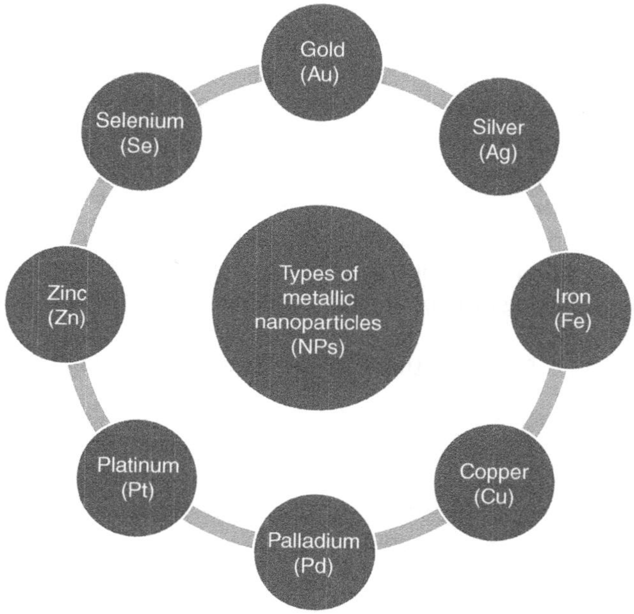

Figure 11.2 Metal nanoparticle types and biomedical applications of these types.

11.4 Nanoparticle Synthesis from Biological Materials

11.4.1 Nanoparticle Synthesis from Various Plants

Environmentally friendly bio-organisms in plant extracts contain proteins that act as both reducing and capping agents that form stable and shape-controlled Ag NPs.

Jae Yong Song and Beom Soo Kim used five plant leaf extracts (Pine, persimmon, Ginkgo, Magnolia, and Platanus) and compared them for the extracellular synthesis of metallic silver nanoparticles. Stable silver nanoparticles were formed by treating the aqueous $AgNO_3$ solution with plant leaf extracts as the reducing agent of Ag+ to Ag0. This eco-friendly method of biological silver nanoparticle production provides comparable or faster synthesis rates to chemical methods and can potentially be used in a variety of human contact areas such as cosmetics, foods, and medical applications [31].

In a study, it was determined that silver nanoparticles obtained by using *Argyeria nervosa* seed extracts have an antagonistic effect against fungus and Gram-negative and Gram-positive bacteria, which have potential applications in medicine and pharmacy, and inhibit nano-sized silver metals reducing agents [32].

Platinum nanoparticles (PtNPs) have superior physicochemical properties and great potential in biomedical applications. In the green synthesis of PtNPs, various biogenic entities including plant extracts, algae, and fungal bacteria were used and their biomedical effects were evaluated. Other biological derivatives have also been used in the synthesis of PtNPs, such as egg yolk, sheep's milk, honey, and bovine serum albumin protein [33].

PtNPs produced using *Nigella sativa* seed extract have been shown to have a spherical shape and an average size of 3.47 nm. Biosynthesized PtNPs (IC) human breast (MDA-MB-231) and neck were found to have significant anticancer activity against cancer cells (IC50: 36.86 and 19.83 μg/ml). It showed significant bactericidal activity against Gram-negative and Gram-positive bacteria at concentrations of 100 μg and 500 μg/ml [23].

PtNPs prepared using tea polyphenols had flower-shaped morphology and an average size of 30–60 nm. In addition, biosynthesized PtNPs have been reported to be effective against cervical human cancer cells (IC50 and IC75 between 18.34 and 11.4 μg/ml, respectively) [34].

Phytosynthesis of monodisperse spherical and cubic PtNPs with an average size of 20 nm was performed using ethanolic extract of pomegranate peel (*Punica granatum*). Biosynthesized PtNPs MCF-7 have been found to have significant cytotoxicity against cancer cells. After 48 hours of incubation, the IC50 was calculated as 17.84 μg/ml [25].

Spherical PtNPs, 5–20 nm, were prepared using a leaf extract of *Bacopa monnieri* (water hyssop). Biosynthesized nanoparticles were found to have pronounced antioxidant and neuroprotective activities, making them promising candidates for the potential treatment of Parkinson's disease [35].

11.4.2 Green Synthesis of NPs Using Various Biological Entity

There are few studies reporting the synthesis of NPs using biological entities other than plants.

Sheep milk was used for fast and environmentally friendly biosynthesis of spherical PtNPs with an average size of 9 nm and zeta potential −31 mV. Sheep milk was mixed with Pt(IV) ions and incubated for three hours at room temperature on a rotary shaker (200 rpm). Bio-reduction of Pt(IV) ions and formation of PtNPs were confirmed by the appearance of the UV/vis peak in the 278–284 nm range [29].

Two studies have been reported on the green synthesis of PtNPs using aqueous extract of Indian brown seaweed *Padina gymnospora*, which yields truncated octahedral and spherical PtNPs of 5–50 and 20–35 nm sizes, respectively [19, 28]. Truncated octahedral PtNPs showed maximum bactericidal activity. It was found to be biocompatible with red blood cells without any hemolytic activity observed. Spherical PtNPs had anticancer activity and remarkable cytotoxicity (IC_{50}: 19.3 μg/ml) against A549 lung carcinoma cells.

Another study used Streptomyces bacterial strains for the green synthesis of PtNPs. PtNPs were created by incubating the bacterial strain with Pt(IV) ions at 50 °C for 24 hours. It has been suggested that the biological reduction of PtNPs is mediated by NAD-bound chloride reductase (electron shuttle enzymatic metal reduction process). PtNPs exhibited significant cytotoxic activity against human breast cancer cell lines (MCF-7) with IC. Support further investigation of the potential use of biogenic PtNPs in breast cancer treatment [27].

11.5 Conclusion

Metallic nanoparticles, which have an important place in nanotechnology, find application opportunities in many fields such as physics, chemistry, electronics, food, health, and biomedical and are the subject of intensive studies in this field. NPs derived from biological entities can be used as promising antibacterial, anticancer, and antifungal agents. Despite efforts to develop biosynthesized NPs for potential use in nanomedicine, many challenges remain. Additional in vitro and animal studies are needed to understand its mechanisms, effects on the immune system and antioxidant activity, and other long-term effects.

References

1 Kuzma, J. and Susanna, P. (2010). Nanotechnology, risk, and oversight: learning lessons from related emerging technologies. *Risk Anal.* 30: 1688–1698.

2 Tomalia, D.A., Reyna, L.A., and Svenson, S. (2007). Dendrimers as multi-purpose nanodevices for oncology drug delivery and diagnostic imaging. *Biochem. Soc. Trans.* 35: 61–67.

3 Song, J.Y., Kwon, E.Y., and Beom, S.K. (2010). Biological synthesis of platinum nanoparticles using *Diopyros kaki* leaf extract. *Bioprocess Biosyst. Eng.* 33: 159–164.

4 Raghasudha, M. (2016). Green synthesis of silver nano particles and study of catalytic activity. *Int. J. Mod. Chem. Appl. Sci.* 3: 306–308.

5 Ramkumar, V.S., Pugazhendhi, A., Gopalakrishnan, K. et al. (2017). Biofabrication and characterization of silver nanoparticles using aqueous extract of seaweed *Enteromorpha compressa* and its biomedical properties. *Biotechnol. Rep.* 14: 1–7.

6 Atlı Şekeroğlu, Z. (2013). From nanotechnology to nanogenotoxicology: genotoxic effect of cobalt-chromium nanoparticles. *Turk. J. Hyg. Exp. Biol.* 70: 33–42.

7 Reddy, A. (2016). *Gold Nanoparticles: Synthesis and Applications*, 1791, and references therein Michael Faraday, 1857. London: Philosophical Transactions of the Royal Society.

8 Ravichandran, R. (2010). Nanotechnology applications in food and food processing: innovative green approaches, opportunities and uncertainities for global market. *Int. J. Green Nanotechnol. Phys. Chem.* 1: 72–96.

9 Shibata, N. (2019). The cell wall galactomannan antigen from *Malassezia furfur* and *Malassezia* detection has diagnostic potential. *Microbiology (Reading)* 155: 3420–3429.

10 Machado, S., Pinto, S.L., Grosso, J.P. et al. (2013). Green production of zero-valent iron nanoparticles using tree leaf extracts. *Sci. Total Environ.* 445–446: 1–8.

11 Wolfgang, L. (2007). Bottom-up methods for making nanotechnology products. http://www.azonano.com/details.asp?ArticleID=1079 (accessed 28 February 2021).

12 Zaki, A. (2007). Processing and synthesis techniques for the preparation of nanomaterials. http://www.azonano.com/details (accessed 15 January 2021).

13 Bar, H., Bhui, D.K., Sahoo, G.P. et al. (2009). Green synthesis of silver nanoparticles using latex of *Jatropla curas*. *Colloids Surf. A. Physicochem. Eng. Aspects* 2009 (339): 134–139.

14 Narayanan, K.B. and Sakthivel, N. (2010). Biological synthesis of metal nanoparticles by microbes. *Adv. Colloid Interf. Sci.* 156: 1–13.

15 Duncan, T.V. (2011). Applications of nanotechnology in food packaging and food safety: barrier materials, antimicrobials and sensors. *J. Colloid Interface Sci.* 363 (1): 1–24.

16 Rabeea, M.A., Owaid, M.N., Aziz, A.A. et al. (2020). Mycosynthesis of gold nanoparticles using the extract of *Flammulina velutipes*, physalacriaceae, and their efficacy for decolorization of methylene blue. *J. Environ. Chem. Eng.* 8: 03–841.

17 Martins, M., Mourato, C., Sanches, S. et al. (2017). Biogenic platinum and palladium nanoparticles as new catalysts for the removal of pharmaceutical compounds. *Water Res.* 108: 160–168.

18 Brayner, R., Barberousse, H., Hemadi, M. et al. (2007). Cyanobacteria as bioreactors for the synthesis of Au, Ag, Pd, and Pt nanoparticles via an enzyme-mediated route. *J. Nanosci. Nanotechnol.* 7: 2696–2708.

19 Shiny, P., Mukherjee, A., and Chandrasekaran, N. (2016). DNA damage and mitochondria-mediated apoptosis of A549 lung carcinoma cells induced by biosynthesised silver and platinum nanoparticles. *RSC Adv.* 6: 27775–27787.

20 Longoria, E.C., Velasquez, S.M., Nestor, A.V. et al. (2012). Production of platinum nanoparticles and nanoaggregates using *Neurospora crassa*. *J. Microbiol. Biotechnol.* 22: 1000–1004.

21 Tahir, K., Nazir, S., Ahmad, A. et al. (2017). Facile and green synthesis of phytochemicals capped platinum nanoparticles and in vitro their superior antibacterial activity. *J. Photochem. Photobiol. B* 166: 246–251.

22 Nadaroglu, H., Gungor, A.A., Ince, S., and Babagil, A. (2017). Green synthesis and characterisation of platinum nanoparticles using quail egg yolk. *Spectrochim. Acta A Mol. Biomol. Spectrosc.* 172: 43–47.

23 Ayguna, A., Gülbagcaa, F., Ozerb, L.Y. et al. (2020). Biogenic platinum nanoparticles using black cumin seed and their potential usage as antimicrobial and anticancer agent. *J. Pharm. Biomed. Anal.* 179: 112961.

24 Depciuch, J., Stec, M., Maximenko, A. et al. (2020). Synthesis method-dependent photothermal effects of colloidal solutions of platinumnanoparticles used in photothermal anticancer therapy. *Appl. Organomet. Chem.* 34: e5401.

25 Sahin, B., Aygün, A., Gündüz, H. et al. (2018). Cytotoxic effects of platinum nanoparticles obtained from pomegranate extract by the green synthesis method on the MCF-7 cell line. *Colloids Surf. B: Biointerfaces* 163: 119–124.

26 Venu, R., Ramulu, T.S., Anandakumar, S. et al. (2011). Bio-directed synthesis of platinum nanoparticles using aqueous honey solutions and their catalytic applications. *Colloid Surf. A* 384: 733–738.

27 Baskaran, B., Muthukumarasamy, A., Chidambaram, S. et al. (2017). Cytotoxic potentials of biologically fabricated platinum nanoparticles from *Streptomyces* sp. on MCF-7 breast cancer cells. *IET Nanobiotechnol.* 11: 241–246.

28 Ramkumar, V.S., Pugazhendhi, A., Prakash, S. et al. (2017). Synthesis of platinum nanoparticles using seaweed *Padina gymnospora* and their catalytic activity as PVP/PtNPs nanocomposite towards biological applications. *Biomed. Pharmacother.* 92: 479–490.

29 Gholami-Shabani, M., Gholami-Shabani, Z., Shams-Ghahfarokhi, M. et al. (2016). Biogenic approach using sheep milk for the synthesis of platinum nanoparticles: the role of milk protein in platinum reduction and stabilization. *Int. J. Nanosci. Nanotechnol.* 12: 199–206.

30 Velmurugan, P., Shim, J., Kim, K., and Oh, B.T. (2016). *Prunus yedoensis* tree gum mediated synthesis of platinumnanoparticles with antifungal activity against phytopathogens. *Mater. Lett.* 174: 61–65.

31 Song, J.Y. and Kim, B.S. (2009). Rapid biological synthesis of silver nanoparticles using plant leaf extracts. *Bioprocess Biosyst. Eng.* 32: 79–84.

32 Thombre, R., Parekh, F., and Patil, N. (2014). Green synthesis of silver nanoparticles using seed extract of *Argyreia nervosa*. *Int J Pharm. Bio. Sci* 5: P114–P119.

33 Ashraf Fahmy, S., Preis, E., Bakowsky, U., and El-Said Azzazy, H.M. (2020). Platinum nanoparticles: green synthesis and biomedical applications. *Molecules* 25 (21): 4981.

34 Alshatwi, A.A., Periasamy, J.A., and Subbarayan, V. (2015). Green synthesis of platinum nanoparticles that induce cell death and G2/M-phase cell cycle arrest in human cervical cancer cells. *J. Mater. Sci. Mater. Med.* 26: 7.

35 Nellore, J., Pauline, C., and Amarnath, K. (2013). *Bacopa monnieri* phytochemicals mediated synthesis of platinum nanoparticles and its neurorescue effect on 1-methyl 4-phenyl 1,2,3,6 tetrahydropyridine-induced experimental Parkinsonism in zebrafish. *J. Neurodegener. Dis.* 2013: 1–8.

12

Characterization of Nanophytopharmaceuticals

Ashwini Kumar Mishra and Pravat Kumar Sahoo

Department of Pharmaceutics, Delhi Institute of Pharmaceutical Sciences and Research (DIPSAR), Delhi Pharmaceutical Sciences and Research University, New Delhi, Delhi, India

12.1 Introduction

Herbal medicines have been widely utilized around the world since ancient times, and physicians and patients alike have acknowledged their superior therapeutic value because they have fewer side effects than contemporary pharmaceuticals. The herbal treatment boosts the therapeutic value of pharmaceuticals by lowering their toxicity and side effects while also increasing their absorption [1]. Nanotechnology plays a significant part in this strategy, and its usage in herbal medicine, specifically drug delivery, is expected to grow fast. Nano-herbal medication delivery systems offer the potential to improve the activity of medicinal plants while also addressing their drawbacks. As a result, herbal nanocarriers aid in the treatment of severe diseases such as cancer and diabetes. For a long time, herbal medicines were not considered for the development of novel formulations due to a lack of scientific justification and processing difficulties, but modern phytopharmaceutical research can address the scientific needs for herbal medicines to be incorporated into novel drug delivery systems (NDDSs) (such as nanoparticles, microemulsions, matrix systems, and solid dispersions). Herbal remedies have been utilized for centuries throughout the world; herbal medications are especially popular in India [2, 3]. Herbal medications have become more popular as a result of their capacity to treat a variety of ailments with fewer negative effects. To overcome several constraints such as low bioavailability, in vivo stability, water insolubility, intestinal absorption, and unspecific site of action, the development of innovative drug delivery systems (NDDS) is critical. The potential of herbal medications for treating chronic diseases such as cancer and other destructive diseases is enhanced by the integration of nanotechnology as an NDDS in traditional medicine. Nanoparticles can be synthesized using unique approaches such as polymer nanoparticle, magnetic nanoparticle, and metallic nanoparticle, be governed by the nanoparticle's characteristics. Advanced technologies will help characterize nanoparticles in order to determine their toxicity profiles based on their physical and chemical properties. Anthocyanidins, lycopenes, flavonoids, carotenoids, glucosinolates, limonoids, polyphenols, isoflavonoids, omega-3 fatty acids, phytoestrogens, resveratrol, phytosterols, and terpenoids are some of the important phytopharmaceuticals that have significant pharmacological activities, which provide therapeutic

Phytoantioxidants and Nanotherapeutics, First Edition. Edited by Mithun Rudrapal.

actions like anti-inflammatory, antiallergic, antioxidant, antimicrobial, antispasmodic, anti-aging, and antidiabetic. With the passing of time, the knowledge of healing from natural resources developed into the science of healing and was documented in Ayurveda [4, 5]. The other world event had comparable historical processes that led to the validation of traditional therapeutic knowledge, particularly the effectiveness of herbal formulations in the treatment of sickness and physical disorders. The majority of herbal formulations are taken orally and are poorly absorbed due to hydrophobic properties, resulting in lower bioavailability and higher systemic clearance. To sustain therapeutic efficacy, repeated administration or dose escalation is required, restricting the use of herbal medications [6]. Herbal medicines commonly use crude powders, extracts, and formulations based on statutory compendia, which may not be required in high quantities, have a limited method of drug administration, and may be less effective than modern therapeutic formulations. Allopathic drugs have dominated the therapeutic sector due to their rapid advancement. Solid dosage form (tablets, capsules, and powders), liquid dosage form (syrup, suspension, and emulsion provided by oral route and injectables delivered via parental route), and aerosols are allopathic drug dosage forms offered to treat illness conditions. Nausea, vomiting, and dizziness are just a few of the most prevalent side effects associated with standard allopathic dosage forms. A continuous study has been conducted in order to reduce adverse effects encountered by patients, resulting in some development to such a traditional dosage form. An NDDS is a type of drug delivery that focuses on the target particular location of action while minimizing the negative effects of traditional dose forms. There has been a shift in the global trend away from synthetic medicine and toward natural medicine, which we might call "return to nature." The active molecules contained in medicinal plant species, or herbal medicine, are important because they produce synergistic activity and, hence, improve therapeutic effectiveness. Research is ongoing to combine novel drug delivery and phytoconstituents in order to avoid any common negative effects of NDDS in the allopathic system of treatment after prolonged use [7, 8]. Herbal treatments and supplements are widely utilized around the world. The global market for phytomedicines is currently worth more than $60 billion per year. It is widely acknowledged that natural resources are used to treat 70–80% of the world's population. The American Botanical Council reported that the sale of herbal supplements climbed by 7.7% in the United States in 2017 (on the basis of the turnover of the herbal market in 2016). Europe is at the top of the list due to extensive herbal medicine R&D and a significant rise in funding for research on herbal medicinal plants. In 2017, America ranked second in the global herbal medicine market, with 16.2% of the market share. According to the National Center for Biotechnology Information, herbal supplements were primarily utilized in the United States for disorders such as heart attack (48.7%), cancer (43.1%), and arthritis (43%) in 2017 [9–11]. This chapter includes details of phytoconstituents and polymers generally employed in the fabrication of nanophytopharmaceuticals, characterization techniques of phytopharmaceuticals along with its treatment approaches for several disorders are also discussed.

12.2 Polymers Generally Applied in the Fabrication of Nanophytopharmaceuticals

The fabrication, evaluation, and manipulation of the nanosystems having a size of 1–100 nm require a combination of physics, chemistry, materials science, and biology [12]. Polymers are an ideal block for the fabrication of nanosystems as they have well-defined structures, length scale, controlled synthesis, easy processing, and can incorporate a wide range of active ingredients [13].

Polymers play an important role in the nanosystems as they have various favorable features such as flexibility, low cost, diverse functionalities, and processability [14]. In this section, we are going to study various polymers employed in the fabrication of nanophytopharmaceuticals.

12.3 Eudragit

Eudragit is prepared by the polymerization of methacrylic and acrylic acids or their esters, e.g. dimethyl aminoethyl ester or butyl esters. Eudragit was developed by Röhm and Haas GmbH in Darmstadt in 1950s. The Eudragit-based formulations had varying degrees of solubility. Formulations were designed to release a drug in the intestine while protecting it from the acid. Different variants of the Eudragit show different release times and are, thus, known as retard preparations. They are nonbiodegradable, nonabsorbable, and nontoxic [15].

Eudragit has various variants and can be combined to obtain the desired drug release profile. It enables to release the drug at the desired site at the desired time. Eudragit is used as an enteric, protective, and sustained release polymer, which protects the drug from moisture, masks the taste and odor and releases the drug at the desired site [16].

Eudragit exists in aqueous dispersion, powders, and organic solution granules and different grades can be mixed in order to adjust the dissolution pH and GI targeting of the drug. Table 12.1 shows the summary of some grades of Eudragit. Eudragit S and L grades are used to target particular areas of the intestine protecting the active constituent from gastric fluids. The grades show application with drugs poorly soluble in the gastrointestinal tract. EUDRAGIT® FS 30 D releases

Table 12.1 The summary of some grades of Eudragit.

Eudragit polymer	Year of development [15]	Physical form [16]	Dissolution property [16]	Description [15–17]	Application [18]	Drug release site
L 12,5	1954	12.5% organic solution	Dissolution above pH 5.5	Cationic, yellow in colour	Film coating	Jejunum pH 6–7
L 30 D-55	1972	30 % aqueous dispersion	Dissolution above pH 5.5	Anionic, white free-flowing powders	Enteric coatings	Duodenum pH > 5.5
L 100	1977	Powder	Dissolution above pH 6.0	Anionic, white free-flowing powders	Enteric coatings	Jejunum pH 6–7
L 100–55	1985	Powder	Dissolution above pH 5.5	Anionic, white free-flowing powders	Enteric coating	Duodenum pH > 5.5
S 12,5	1954	12.5% organic solution	Dissolution above pH 7.0	Anionic, white free-flowing powders	Enteric coatings	Ileum, colon pH > 7.0
S 100		Powder	Dissolution above pH 7.0	Anionic, white free-flowing powders	Enteric coatings	Ileum, colon pH > 7.0
FS 30 D	1999	30% aqueous dispersion	Dissolution above pH 7.0			Ileum, colon pH > 7.0

Table 12.2 Recent application of various grades of Eudragit.

Eudragit grade	Year and researcher	Active ingredient	Type of nanoparticle	Application
Eudragit® RS100	2019 and Froiio F. et al.	Bergamot oil (*Citrus bergamia* Risso)	Nanocapsules	Antimicrobial [33]
Eudragit® L100–55	2019 and Pina-Barrera A.M. et al.	Thyme oil (*Thymus vulgaris* L.)	Nanocapsules	Antioxidant [34]
Eudragit® EPO	2019 and Silva Flores P.G. et al.	Lavender oil	Nanocapsules	Antioxidant [35]

the drug in the colon with additional advantages of aqueous processing, multiarticulate tablet preparation, and flexible coating [17].

The oral drug release mechanisms of Eudragit involve dissolution, diffusion, osmotic pressure, and ion exchange. The release mechanism from Eudragit E 30 D-coated pellets was found to be diffusion through water-filled pores in the film. The propranolol HCl release from the monolithic matrix of Eudragit NE 30 D was governed by diffusion through the pores and channels. The drug release mechanism can be a combination of diffusion and erosion as seen in a polymer (Eudragit L)-wax matrices [15].

12.3.1 Applications

Eudragit-based formulations find application in ophthalmic, buccal and sublingual, gastrointestinal, intestinal, colon, transdermal, and vaginal drug delivery. It is also used for gene delivery and vaccine delivery. The characteristics such as no toxicity, controlled release profile, and positive charge make Eudragit a potential polymer for ophthalmic drug delivery [19–23]. Eudragit provides good mucoadhesion and produces elastic, flexible, and soft buccal films for buccal and sublingual drug delivery [24–29]. Different grades of Eudragit exhibit different properties such as low density for buoyancy in gastric fluid, high density to retain dosage form at the bottom of the stomach, mucoadhesion, expansion by unfolding or swelling to increased size to avoid the passage of formulation through pyloric sphincter that makes it a successful candidate for gastrointestinal drug delivery [30]. Eudragit S and L are used to develop dosage forms that could bypass the stomach to release the drug in the intestine for an extended period [31]. Eudragit S100 is used in transdermal drug delivery as it produces adhesive, wrinkle-free transparent films [32]. Some of the recent applications of various grades of Eudragit in nanophytopharmaceuticals have been shown in Table 12.2.

12.4 Poly Lactic Acid (PLA) and Polylactide-*co*-Glycolic Acid (PLGA) – The Polyesters

PLA and PLGA are synthetic aliphatic polyesters possessing promising processing and mechanical properties [36, 37]. The polymers are synthesized by fermenting glycolic and lactic acid from sugars and thus it makes them eco-friendly and safe. The polymers are biocompatible and biodegradable [38]. They undergo hydrolytic degradation to glycolic and lactic acid which finally eliminates as CO_2 and water from the body [39]. The drug release kinetics of the system depends both on the degradation of the polymer as well as the diffusion of the drug from the polymer matrix [40]. These

Table 12.3 Merits of PLA- and PLGA-based delivery systems.

Merits of PLA and PLGA as polymers for nanophytopharmaceuticals	Enhanced drug adherence
	Increased stability
	Enhanced cellular uptake of the delivery system via endocytosis
	Reduced drug blood concentration fluctuations
	Controlled drug release kinetics
	Sustained and targeted delivery
	Biodegradable and biocompatible
	Can deliver a variety of drug payloads

polymers are utilized to prepare many DDS enabling the delivery of small molecules to large proteins and nucleic acids in a targeted and sustained manner. As a result, it can be inferred that the polymers have controllable biodegradability and erosion, exceptional biocompatibility, and high safety profiles which makes them suitable for both local and systemic drug delivery [41, 42]. The polymers also play an important role in protecting the active ingredient from enzymatic degradation resulting in enhanced stability [43].

PLA is found in two forms, either L or D (PDLA or PLLA) as it has asymmetric centers in its backbone, while PLGA is the copolymer of D,L-lactic acid with glycolic acid [44].

PLLA is crystalline in nature while PLDA is amorphous with disordered polymer chains. PLA and PLGA could be easily formulated for any drug delivery system with different architectures [45]. They are soluble in various solvents such as chloroform, tetrahydrofuran, acetone, dichloromethane, and ethyl acetate but are insoluble in water [46]. The two polymers absorb the water and hydrolyze their ester linkages resulting in their degradation [47]. Incorporation of D-lactic acid in the backbones of the polymers leads to retardation of biodegradation in the human body as the human body does not have enzymes that could consume D-lactic acid and thus reduced degradability [48].

The polymers are currently used in nanophytopharmaceuticals extensively due to their merits as listed in Table 12.3. Zhao J. et al. in 2009 formulated and evaluated brucine-loaded PLA nanoparticles using the solvent diffusion method. It was found that the nanoparticles exhibited small size, enhanced entrapment efficiency of brucine, and sustained release in vitro. Kesente M. et al. in 2017 formulated PLA nanoparticles loaded with olive leaves extract and further loaded them in a cosmetic formulation. The technique used was nanoprecipitation. The resultant nanoparticles had a small size and burst effect at the first two hours. The cosmetic formulation exhibited enhanced stability [49, 50]. Meng-Tsai Y. et al. in 2011 prepared and evaluated curcumin-loaded PLGA nanoparticles using the high-pressure emulsification solvent evaporation method. The particle size reduced and the curcumin concentration in many organs like the spleen and lungs drastically increased, which resulted in enhanced AUC, $t_{1/2}$, and MRT of curcumin. The C-NPs could also cross the BBB and get concentrated in the hippocampus. The formulation showed sustained diffusion [50, 51].

12.5 Chitosan

Chitosan is a biodegradable and biocompatible polymer derived from naturally occurring sources such as crustaceans, insects, and exoskeletons of insects [52]. Chitosan is a semisynthetic amino polysaccharide with multidimensional properties, unique structure, and complex functionality [53–55]. The deacetylation of its parent compound, chitin leads to the generation of chitosan, which is a copolymer of β-(1–4)-linked 2-acetamido-2-deoxy-β-D-glucopyranose and 2-amino-2-deoxy-β-D-glycopyranose [56, 57]. Properties such as low toxicity, compatibility, and ecological safety make it an effective biomaterial with a wide range of applications in phytopharmaceuticals [58–61]. It is polycationic, hemostatic, and insoluble at high pH but these properties can be reversed by sulfating chitosan that renders the molecule anionic and water-soluble and additionally anticoagulant [62, 63].

Chitosan is a bioadhesive polymer with the excellent mucoadhesive property. It can adhere to hard and soft tissues and is an extremely safe excipient as it does not cause any allergic or inflammatory response [64, 65]. Chitosan has the ability to regulate the drug release and helps in avoiding the use of harmful and deleterious organic solvents while preparing the formulation as it is soluble in aqueous acidic solvents [66–69]. Chitosan aids in targeted drug delivery, enhances the absorption of peptides, and improves the dissolution rate of the poorly soluble drug [70–73]. The nano-based drug delivery systems incorporating chitosan can be prepared using emulsion-droplet coalescence and coacervation/precipitation methods. The drug loading into the nanoparticles can be achieved by either incorporating the drug during the preparation or after the formation of the nanosystem. Both water-soluble and insoluble drugs can be loaded into the system by the above technique as the drug is absorbed onto the surface as well as physically incorporated into the matrix [74]. The release mechanism of the drug from chitosan can be: (i) from surface of the particles, (ii) diffusion through swollen matrix, and (iii) surface erosion. The absorbed drug, when comes in contact with the release medium, gets dissolved and thus follows release from the particle surface. The diffusion mechanism occurs in three steps. The first is the penetration of water into the particulate system followed by the swelling of the matrix and, lastly, the diffusion of the drug via the matrix [75].

Yosra S.R. Elnaggar et al. in 2015 prepared monodisperse chitosan nanoparticles loaded with piperine for brain-targeted treatment of Alzheimer's disease through the intranasal route. It was concluded from the study that chitosan nanoparticles are effective, safe, and noninvasive formulation that decreased the oral dose 20-folds and could serve as a potential candidate for Alzheimer's therapy [76]. Hend Mohamed Anter et al. in 2019 prepared apocynin-loaded chitosan nanoparticles to potentiate the antiulcerogenic activity of apocynin using the gelation method [77]. In another study conducted by Kamel M. Kamel et al., the cinnamon-loaded solid lipid nanoparticles coated with chitosan were found to show decreased side effects and uninterrupted therapy in combination with 5-fluorouracil for the treatment of colon carcinoma [78].

12.6 Alginates

Alginates are natural, biodegradable, and anionic polymers derived from brown seaweed (phaeophycea) on treatment with alkali solution, typically NaOH. Its structure has homogeneous blocks of linear polymers consisting of D-mannuronic acid and L-guluronic acid residues separated by units of mannuronic and glucuronic acid [79]. The polymer undergoes time, temperature, and pH-dependent proton-catalyzed hydrolysis leading to the formation of highly viscous "acid gel." The intermolecular forces and the capillary forces lead to the water-holding capacity of the polymer. It forms soluble salts with monovalent ions whereas it forms gels or precipitates with multivalent cations (except Mg^{2+}). Alginate can form two types of pH-dependent gels, i.e. an acid gel and an inotropic gel [80, 81].

The alginates, especially their sodium and calcium salts, are biocompatible, nontoxic, and non-immunogenic and possess the property of adhesion [82]. The mechanism of drug release from

Table 12.4 Properties of alginate enabling it to be a potential polymer for nanophytopharmaceuticals.

S. No.	Property
1.	Readily available
2.	Inexpensive
3.	Naturally occurring, nontoxic
4.	Protective effect on mucous membranes of GIT
5.	Biodegradable
6.	Water-soluble, eliminates noxious solvent usage
7.	Forms formulation at room temperature, protects against drug degradation due to high temperature
8.	Mucoadhesiveness
9.	Cell affinity, mechanical strength, gelation can be altered by mixing alginate with other polymers

alginate-based systems is mainly through diffusion via matrix and surface erosion at certain pH [83]. The properties listed in the table enable alginate to serve as a potential polymer for nanosystems [84, 85]. Alginates find a wide range of applications in drug delivery, wound healing, and cell transplantation. Meenakshi Choudhary et al. in 2021 prepared calcium alginate nanoparticles hydrogel as a hemostatic agent in diabetic wound healing [86]. Qingguan Liu et al. in 2019 used sodium alginate in combination with sodium caseinate to double coat curcumin-loaded zein nanoparticles [87]. Coating with sodium alginate resulted in increased entrapment efficiency, improved water solubility, enhanced stability, and improved antioxidant property of curcumin. In a study conducted by Mingyu Ji et al., the nanoparticles formulated with cinnamaldehyde, sodium alginate, and chitosan showed sustained release with narrow PDI, small particle size, and high encapsulation efficiency [88] (Table 12.4).

12.6.1 Characterization Techniques of Phytopharmaceuticals

12.6.1.1 Scanning Electron Microscope (SEM)

SEM is an advanced technology that uses an electron beam to investigate the morphology of particles. The electron beam is radiated on the sample surface in SEM to obtain pictures of the surface from various angles. Depending on the electron beam's penetration and the type of SEM used, different interactions occur between the electron beam and the particle's surface [89]. The receiving signal detected by the various types of detectors is used to construct an image of the surface.

Various types of SEM detectors are used for material characterization, X-ray mapping, backscattered electron imaging, and electron channeling.

- The electron beam gun, which is placed at the top of the column, fires an electron beam.
- Hairpin tungsten cannon, which produces a high-diameter electron beam, resulting in high-resolution pictures.
- Electromagnetic lenses, which concentrate electron beams and generate a small, focused point on the sample surface.
- The electron beam does not scatter in a high vacuum environment.

S. Nandani et al. in a study used field emission scanning electron microscopy to examine the surface morphology of the vasaka extract-phospholipid complex. The powdered sample was placed on a metal stub of an electron microscope and coated with gold in a particle sputter.

The sample was digitally depicted by irregularly inspecting the stub at varying magnifications. The diameters of the phytosomes were measured arbitrarily using Image J software and FE-SEM images [90, 91].

12.6.1.2 Transmission Electron Microscope (TEM)

The crystal structure, shape, and content of various biomaterials and phytoconstituents are studied using transmission electron microscopy (TEM). The electron beam in a TEM was focused onto the sample after passing via a vacuum chamber and electromagnetic lenses. The sample and the electron beam interact, causing some electrons to be scattered and others to be transferred through the sample [92].

The sample must be thin enough to allow a sufficient number of electrons to pass through it. The denser the sample, the fewer electrons pass through it, resulting in a denser image.

In a study conducted by Tao Wang et.al., ginkgo and cornstarch-based nanocarriers were produced and TEM was used to analyze the morphologies and sizes of native starches and nanoparticles. A dilute SNP suspension (0.05%, w/v) drop was put on a copper grid and lyophilized for TEM examination [92].

12.6.1.3 Atomic Force Microscopy (AFM)

Atomic Force Microscopy (AFM) is a cutting-edge technology for studying nanoscale materials' surface characteristics and structural features. Researchers can utilize AFM with a variety of structures and media thanks to a large number of potential signals and the technique's flexibility. Unlike previous methods, this technique is not limited by the material and environment available for studying surface attributes [89].

12.6.1.4 FTIR (Fourier Transform Infrared Spectroscopy)

The structural characteristics and functional groups of the biological extract with the nanoparticle polymers are determined using FTIR in nano-phytopharmaceuticals. It measures the infrared intensity of light with a wavelength of 400–4000 cm^{-1}. It is utilized to see if the nanoparticle elements cause any incompatibilities or changes in the characteristics of the herbal extract. Chemical incompatibilities are demonstrated by a shift in the peaks [90].

The KBr pellets approach, in which a little amount of the sample is squeezed between the KBr pellets or ground with KBr, and then pressed as a disc at a predetermined pressure and time in hydraulic pressure, is the most frequent way to do FTIR. Transmittance (percentage) vs. wavenumber (cm^{-1}) is plotted as the results [91].

To identify the structural characteristics and nature of related functional groups of euphorbia cyathphora-loaded copper oxide nanoparticles, M. Rajendran et al. colleagues used FTIR [90].

12.6.1.5 UV–Vis-Spectroscopic Analysis

UV-visible spectroscopy (UV–Vis) analyzes the extinction of light passing through a substance (scatter + absorption). UV–Vis is a great tool for recognizing, characterizing, and investigating nanomaterials because nanoparticles have unique optical properties that are sensitive to size, shape, concentration, aggregation state, and refractive index near the nanoparticle surface.

The optical characteristics of the nanoformulation generated are determined using a UV–Vis spectrophotometer [90]. M. Rajendran et al. used UV spectroscopy to investigate the characteristic absorption peak of Euphorbia cyathophora-loaded copper oxide nanoparticles between 200 and 800 nm.

12.6.1.6 Percentage Yield

The percentage yield is obtained by using the following equation:

$$\text{Percentage yield}\ (\text{w/w}) = \frac{\text{Practical yield}}{\text{Theoretical yield}} \times 100$$

Practical yield: total weight of the prepared formulation.

Theoretical yield: sum of each constituent used for preparing formulation [91].

S. Nandhini et al. calculated the percentage yield of vasaka phospholipid nano delivery system and found it to be between 30.03 and 97.03%.

In the manufacturing of items, the percent yield is critical. Chemical producers spend a lot of time and money trying to improve their percent yield. When many separate reactions are used to make complicated molecules, one step with a low percent yield can quickly squander a lot of reactants and cost a lot of money.

Because of the reasons stated earlier, percent yields are typically less than 100%. However, if the measured result of the reaction contains impurities that cause its mass to be larger than it would be if the product were pure, percent yields greater than 100% are feasible.

12.6.1.7 Entrapment Efficiency

The direct or indirect technique can be used to calculate the entrapment effectiveness of nanocarriers. After centrifugation, the unentrapped or free amount was measured in the supernatant using the indirect approach. The difference between the initial drug amount and the free or unentrapped quantity of drug in the supernatant in relation to the total quantity included in the nanocarrier preparation was used to calculate the entrapment efficiency or percentage of the content. In the direct procedure, however, nanocarriers were dissolved in a suitable solvent and examined after filtering and dilution to determine entrapment effectiveness using an appropriate methodology.

The ultracentrifugation method is commonly used to measure entrapment efficiency. A specific rpm, duration, and temperature are applied to the formulation. Following centrifugation, the supernatant is pipetted out, diluted with the standard solvent, and analyzed for active ingredient entrapment using HPLC or UV–Vis spectrophotometer at the active ingredient's maximum concentration.

The formula used to calculate the entrapment is:

$$\text{Entrapment efficiency}\ (\%) = \text{total amount of active constituent} - \text{amount of free active constituent} * 100 / \text{total amount of active constituents}$$

S. Nandhini et al. performed the entrapment efficiency of vasaka phospholipid nano delivery system using the centrifugation method and found it to be between 20.02 and 95.88% [91].

12.6.1.8 Particle Size, Polydispersity, and Zeta Potential

The permeation and stability of a nanoparticle are determined by its particle size and surface charge, respectively. Both parameters are routinely assessed using photon correlation spectroscopy (Zetasizer), a technology based on the dynamic light scattering (DLS) concept. The samples are normally diluted and sonicated with distilled water. Triplicate measurements should always be taken [91].

The particle size distribution of nanospheres derived via dynamic laser light scattering is defined by the polydispersity index (PDI) (DLS). Its dimensionless number is extrapolated from the autocorrelation function, and PDI varies from 0.1 to 0.5 for adequate measurements and good-quality colloidal suspensions to close to one for poor-quality samples with wide size distribution and big particles or aggregates.

In a study conducted by Tao Wang et al., Ginkgo and cornstarch-based nanocarriers were produced and the sample's zeta potential was measured using a DLS analyzer at a wavelength of 633 nm. The material was diluted in deionized water before being analyzed. Without filtering or eliminating dust, the nanoparticle samples were analyzed immediately [92].

12.6.1.9 Differential Scanning Calorimetry (DSC)

DSC is a thermodynamical method for determining the direct heat energy uptake in a sample during a controlled temperature increase or decrease. The calorimetry technique is used to track changes in phase transitions [83, 84].

A single molecular transition of a molecule from one conformation to another is analyzed by DSC, which is extensively employed for the study of biochemical reactions [93]. The samples' thermal transition temperatures (Tt; melting points) are calculated in solution, solid, or mixed phases such as suspensions [84].

Energy is simultaneously supplied into a sample cell (which contains a solution containing the molecule of interest) and a reference cell (containing only the solvent) in a basic DSC experiment. Both cells' temperatures rise at the same rate over time. The amount of excess heat absorbed or released by the molecule in the sample would be the difference in the input energy required to match the temperature of the sample to that of the reference (during an endothermic or exothermic process, respectively). As the presence of the target molecule requires extra energy to raise the sample to the same temperature as the reference, the idea of excess heat comes into the picture [94–96].

Tao Wang et al. in a study conducted differential scanning calorimetry to investigate the starches' thermal properties. An identical amount of moisture was introduced to each sample in an aluminum pan. Before DSC analysis, the pan was hermetically sealed and left to stand for 12 hours. Each pan was heated at a rate of 10 °C/min from 25 to 125 °C. An empty metal pan was used as a reference for calibrating the DSC analyzer. The instrument's software was used to determine the temperature at the start (T_o), peak (T_p), and end (T_c). Meanwhile, the gelatinization enthalpy (H) was calculated by integrating the region between the thermogram and a baseline under the peak and expressed in Joules per gram of dry sample [92].

12.6.1.10 Thermogravimetric Analysis

Thermogravimetric analysis is used to determine the prepared formulation's thermal stability. The active constituent, polymers, and formulation are put into an aluminum crucible constructed specifically for TGA and heated at a specified temperature based on the extract's and constituents' melting points under a nitrogen atmosphere. The thermogram plotted between weight (percent) and temperature is used to determine the weight loss and exothermic/endothermic nature of the formulation (C) [91].

S. Nandhini et al. in a study used a thermogravimetric analyzer to evaluate the thermal stability of the improved vasaka phytosome. The extract, phospholipids, and phytosomes were put into a thermogravimetric-aluminum crucible, which was subsequently heated at a rate of 10 °C/min from 30 to 350 °C under a nitrogen environment. For TGA analysis, a blank aluminum pan was used as a reference [91].

12.6.1.11 Solubility Studies

The content of the active ingredient in the nanophytoformulation is determined by analyzing the saturation solubility of the produced formulation. It is done with a typical shake flask approach, where an excessive amount of formulation is mixed with distilled water and other solvents. The contents are sealed in glass vials and shaken for 24 hours at 100 rpm and 37 °C in a water bath. The samples were then collected and tested for active ingredient content using spectrophotometry or HPLC. Solubility studies aid in the selection of a solvent for in vitro drug release experiments [91].

During studying vasaka nanosystem, S. Nandani adopted the standard shake flask method to assess the saturation solubility of phytosomes in distilled water and various mediums. A large amount of phytosomes were introduced to distilled water and various phosphate buffers with pH 1.2 (HCl), pH 6.8, and pH 7.5. The contents were sealed in a glass vial and shaken for 24 hours at 37 °C at 100 rpm in a shaking water bath. The samples were centrifuged, and the supernatant was fed into HPLC analysis to measure the vasicine content in phytosomes.

12.6.1.12 X-Ray Diffraction Analysis

For characterizing crystalline materials, X-ray diffraction is an efficient nondestructive technique. Structures, phases, preferred crystal orientations (texture), and other structural data such as average grain size, crystallinity, strain, and crystal defects could be studied. A monochromatic beam of X-rays scattered at precise angles from each set of lattice planes in a sample produces XRD peaks by constructive interference. The atomic locations within the lattice planes determine the peak intensities. As a result, the XRD pattern serves as a fingerprint for periodic atomic configurations in each material [97].

An X-ray diffractometer with a Cu-Ka radiation source was used to create the vasaka phytosome's powder X-ray diffraction (XRD) pattern by S. Nandini et.al. The experiment was conducted at room temperature in Bragg–Brentano geometry with a scanning rate of 0.02°/min through 2 ranging from 5° to 80°. Origin software was used to determine the crystallinity of the phytosomes.

12.6.1.13 In Vivo Drug Release Studies

The amount of drug released and its therapeutic effect on the biological system are measured in an in vivo drug release research. It is used to investigate the manufactured formulation's pharmacokinetics and pharmacodynamics profile. Mice, rats, rabbits, and guinea pigs are employed as experimental animals in this study. Normally, the animals are divided into two groups: control and test. Before undertaking this study, a well-equipped animal protocol must be filed and approved by an authorized animal ethics committee.

Following the procurement of the animals, proper care and environmental conditions, such as temperature, humidity, and light, must be maintained in accordance with the experiment's needs. The animals will be given a conventional laboratory feed as well as water. Once the animals are ready, the experiment begins by generating the ailment for which the produced formulation's therapeutic efficacy is to be determined. After the disease has been developed, the animals are given the produced formulation, and the drug levels in the body, as well as its therapeutic efficacy, are measured using various methods such as hematological assay and visual inspection.

12.6.1.14 Phytopharmaceuticals in the Treatment of Diseases

Phytopharmaceuticals include natural substances that are plant-based products. Anthocyanidins, lycopenes, flavonoids, carotenoids, glucosinolates, limonoids, polyphenols, isoflavonoids, omega-3 fatty acids, phytoestrogens, resveratrol, phytosterols, and terpenoids are some of the important phytopharmaceuticals that have significant pharmacological activities, which provide therapeutic actions like anti-inflammatory, antiallergic, antioxidant, antimicrobial, antispasmodic, antiaging, antidiabetic. They are widely used as dietary supplements.

Anthocyanins are a class of water-soluble natural phenolic pigments, which are widely found in plant families like *Rosaceae, Vitaceae, Saxifragaceae, Ericaceae, Caprifoliaceae, Cruciferae*, and *Fabaceae* [98]. They have potential pharmacological activities like antioxidant, anti-inflammatory, antidiabetic, and anticancer, and are widely used in treating chronic age-related diseases like cardiovascular, neurodegenerative, and eye-related diseases. Recent in vitro studies reveal that ACNs also possess antiviral properties as they can actively prohibit the replication of several virus species such as herpes simplex, syncytial virus, parainfluenza virus, HIV, rotavirus, and adenovirus [99].

According to several epidemiological studies, dietary ACNs help in preventing cardiovascular diseases. Dietary ACNs have protective action against atherosclerosis and oxidative stress observed in ischemic reperfusion. In a study, Toufektsian et al. demonstrated that tissue damage due to ischemia in rats fed with CAN-rich corn was reduced to 39% in comparison to rats that were fed with CAN-free corn [3]. Consumption of berries rich in ACN significantly reduced the risk of myocardial infarction by 32% in middle-aged and young women and further lowered the chance of coronary heart disease and mortality associated with it [100–103].

Gao et al. in an epidemiological study suggested that regular consumption of strawberries and blueberries, which are rich in ACNs, helps in reducing the risk of Parkinson's disease [104].

Anthocyanins modulate the antioxidant defense action by activating antioxidant enzymes and improving glutathione synthesis. They chelate metal ions, such as iron and copper, thereby reducing the production of free radicals by Fenton and other reactions [105]. Studies reveal that ACNs can activate various antioxidant enzymes such as glutathione peroxidase, superoxide dismutase, and catalase and they can further augment their enzymatic activities [106, 107]. Several studies demonstrated that cell survival, proliferation, controlling apoptosis, and inflammation are the major pathways that are involved in cancer prevention and inhibition by ACNs.

Carotenoids are lipid-soluble colorful pigments, which are found in plants, bacteria, fungi, algae, and in various foods such as fruits, vegetables, fish, etc. More than 600 carotenoids are known, which are subdivided into xanthophylls, carotenes, and lycopene on the basis of structural variations [108, 109]. Various studies indicate that consuming a diet rich in carotenoids lowers the risk of cardiovascular diseases, cancer, diabetes, osteoporosis, cataract, macular degeneration which is related to age, and other infectious diseases like HIV [110, 111].

The plasma concentration of carotenoids in HIV patients is usually low as all the carotenoids in these patients are destroyed, which increases the risk of mortality. Studies suggest that HIV patients administering 60 mg of β-carotene per day showed an increase in the CD4+ and CD8+ lymphocytes and the symptoms of the disease depreciated in 24–36 months [14]. Several biological actions associated with carotenoids have been explored which include antioxidant activity, immunity enhancement, reduction in malignant lesions, and mutagenesis suppression [110].

Carotenoids have significant antioxidant properties and they act as a scavenger of reactive oxygen species. They actively scavenge radicals in three steps, namely electron transfer, hydrogen abstraction, and addition [112]. Studies reveal that β-carotene is an efficient chain-breaking antioxidant and the amalgamation of two antioxidants that are lipophilic in nature such as β-carotene with vitamins E and C, produces a synergistic effect. Upadhyaya et al. in their study demonstrated that the consumption of apricots, carrots, kale, mangoes, red pepper, spinach, and broccoli that are rich in β-carotene, can significantly help in fighting leukemia by delaying the cell cycle in the G1 phase and inducing apoptosis [113]. Several other studies also revealed that β-carotene is highly beneficial against colon cancer, gastric cancer, adenocarcinoma, neuroblastoma, hepatocarcinoma, cervical cancers, and prostate cancer [114–116]. Moeller et al., in a clinical study, suggested that lutein/zeaxanthin helps in protecting women who are younger than 75 years of age from macular degeneration [117].

Lycopene is a red-colored pigment, which is commonly found in tomatoes, apricots, guavas, watermelons, and papayas. It is an important aid in treating several oral diseases such as leukoplakia, oral submucous fibrosis, lichen planus, oral squamous cell carcinoma, etc. [118].

Sahin et al. performed an in vivo study on female Wistar rats for studying the effect of lycopene in breast cancer and he affirmed that lycopene in a dose of 20 mg/kg hindered the growth of tumor and expression of apoptosis-associated proteins [119]. Literature indicates that lycopene plays a crucial role in the prevention of atherogenesis and carcinogenesis as it protects cellular biomolecules, which include lipoproteins, proteins, and DNA [120]. Aung et al. reported that consuming a daily dose of 8 mg of lycopene is highly competent in treating oral leukoplakia and this efficacy of lycopene is attributed to its antioxidant property.

Flavonoids are one of the important classes of phytopharmaceuticals, which are widely distributed in plant-based food and beverages like vegetables, fruits, tea, wine, cocoa, etc.

Flavonoids actively inhibit acetylcholinesterase, which is the major enzyme of the central nervous system. Inhibition of acetylcholinesterase augments the neural acetylcholine levels, which provides symptomatic relief from mild to moderate Alzheimer's disease [121]. In vitro studies performed on various flavonoids such as quercetin, macluraxanthone, rutin, and kaempferol 3-*O*-β-D-galactoside revealed that quercetin and macluraxanthone had inhibitory action against acetylcholinesterase and butyrylcholinesterase, which was concentration-dependent [122]. Mello et al. performed a study for exploring the binding modes of different flavonoids with COX-2 enzyme, using in silico methods and the results of the study revealed that flavones and flavonols with 2,3-double bond present in it act as potential inhibitors for COX 2 [123]. A study evaluated genistein, scopoletin, silibinin, galangin, hesperitin, taxifolin, naringenin, daidzein, celecoxib, and esculatin for their COX-inhibitory activity and the results affirmed that the flavonoids had a higher binding energy range as compared to the standard, which helped in developing potential and efficacious COX inhibitors for treating inflammation [124]. In a study, Ganugapati et al. demonstrated that flavonoids present in green tea are effective insulin mimetics. The study suggested that Epicatechin in green tea is an active insulin receptor activator, which reduces the adverse effects associated with diabetes [28]. Clinical trials suggest that genistein, daidzein, and equol possess oestrogenic activity, which are being significantly explored for designing treatment regimens for cardiovascular disease, cancer, and osteoporosis. Genistein is one of the most efficacious flavonoids for preventing postmenopausal bone loss in females [125–127].

Flavonoids show antimicrobial, antiviral, and antifungal activities, which make them extensively useful in different fields like health, medicines, nutrition, and food sciences. Wang et al. in his study explored the antiviral activity of flavonoids in the treatment of viral infections. A variable degree of antiviral activities is present in quercetin, hesperetin, naringin, and catechin. Quercetin and apigenin possess antibacterial activity. Quercetin extracted from leaves of lotus showed significant potential in treating periodontitis [128–130].

Resveratrol is a plant compound, which is found in red wine, grapes, peanuts, and berries. It is a potent antioxidant, which is beneficial in lowering blood pressure and suppressing cancer cells. It was reported that resveratrol suppresses the gastric cancer cells in G1 phase, which results in senescence. In nude mice, administration of 40 mg/kg/day of resveratrol actively inhibits the development of gastric cancer [131]. Resveratrol is an efficacious chemopreventive agent against liver cancer.

Moussa et al. reported that patients who administered resveratrol 1 g/day for 52 weeks showed a depreciation in MMP-9 levels (an inflammatory marker associated with Alzheimer's disease) when compared to a placebo group. Patients who were treated with resveratrol had less Aβ accumulation inside the brain which indicated that resveratrol rejuvenates the central nervous system by hindering the permeation of MMP-9 and reduces its inflammatory activity [132, 133]. Resveratrol has a potential neuroprotective action, which makes this polyphenol an important compound for treating neurodegenerative diseases like Alzheimer's disease, Parkinson's disease, and amyotrophic lateral sclerosis. It is effective in modulating blood glucose levels, reducing resistance to insulin, augments blood lipid profiles, attenuates diabetic hypertension, and hinders oxidative stress [134].

Another class of phytopharmaceuticals called limonoids has significant pharmacological activities against several diseases. These are tetranortriterpenoids, which are extensively found in *Rutaceae* and *Meliaceae* families. Citrus fruits such as oranges, lemons, and grapes are rich in limonoids. They possess multiple therapeutic actions like antihyperglycemic, antitumor, neuroprotective, insecticidal, antioxidative, anti-inflammatory, immunomodulatory, antibacterial, anti-obesity, and antiviral. Studies demonstrate that limonoids acquired from citrus plants inhibit the proliferation of hepatocellular, colorectal, pancreatic, gastric, breast, and prostate cancers, neuroblastoma, and melanoma [135–137]. Jayaprakasha et al. in a study explored the inhibitory effect of isolimonic acid and ichanexic acid obtained from *Citrus aurantium* L. on colon cancer cells of humans and the result demonstrated that these limonoids play a potential role in inhibiting the growth of the cancerous colon cells [138].

Glucosinolates are another class of active phytopharmaceuticals that are known to protect from serious disease conditions like myocardial infarction, prostate cancer, breast cancer, and colorectal cancer. Cruciferous vegetables like arugula, broccoli, Brussels sprouts, cabbage, kale, turnip, and cauliflower are rich sources of glucosinolates. They are primarily prominent in genus *Brasica*. Glucosinolates are sulfur- and nitrogen-containing compounds and are extracted from glucose and amino acids. Inside the human gut, these glucosinolates change into isothiocyanates through hydrolysis. Indole products and isothiocyanates are the breakdown products of glucosinolates, which significantly induce apoptosis, reduce the growth of tumor cells by inhibiting the phase 1 enzymes, modifying the estrogen metabolism, and arresting the cell cycle. They act as potential anti-inflammatory agents by inhibiting the histamine release, which is responsible for inflammation [139].

A study reported that extract obtained from sprouts of *Raphanus sativus* L. induces cytotoxicity and apoptosis in human breast cancer cells and hepatocarcinoma cells [140, 141]. Raphasatin obtained from radish sprouts has been identified as a potent chemopreventive compound. Miao et al. suggested that sprout extracts obtained from Chinese kale are having significant antioxidant activities [142].

Clinical studies were conducted to evaluate the glucosinolate mechanism of action. In a pilot study, Broccoli sprouts that are rich in sulforaphane were administered to melanoma patients and the result affirmed their chemopreventive action [143] and demonstrated their ability to reduce serum insulin. It also augmented insulin resistance in patients with type 2 diabetes, which improved glycemic control reducing the risk of cardiovascular diseases [144].

It was reported that consumption of Broccoli soup reduces the progression of prostate cancer and daily intake of free sulforaphane helps in managing biochemical recurrence of prostate cancer in patients who have undergone radical prostatectomy. Another study revealed that sulforaphane augments bronchoconstrictor activity in asthmatic patients which is helpful in improving mild asthma in patients with schizophrenia. Consumption of broccoli sprouts helps in managing cognitive functions in such patients.

Fish is a rich source of omega-3 fatty acids, which contain docosahexenoic acid (DHA) and eicosapentenoic acid (EPA). Other dietary sources of omega-3 fatty acids include vegetable oils, seeds, and nuts. Omega-3 fatty acids that are derived from marine sources are having a great potential in preventing cardiovascular diseases. They exert a cardioprotective action through varied mechanisms, which include reduced arrhythmias and altered production of prostaglandins, which helps in reducing inflammation and enhances the endothelial and platelet functions. A study recommended consumption of 200–400 g of fatty fish two times a week and intake of food that is rich in α-linolenic acid for protection from cardiovascular diseases. Another research suggested that consumption of 200–400 g of fatty fish or fish oil supplementation containing 900 mg of eicosapentenoic acid and docosahexenoic acid every day along with a diet rich in α-linolenic acid is good for patients who are suffering from cardiovascular disease or congestive heart failure [145]. Omega-3 fatty acids are found to be crucial for the functioning of the central nervous system. Ongoing research is developing a correlation between the use of Omega-3 fatty acids and serious depressive disorders. Evidences from various experimental, epidemiological, and clinical studies reveal that Omega-3 fatty acids can influence the outcomes and vulnerability of depressive diseases. It was suggested that Omega-3 fatty acids supplementation has significant potential for treating depressive disorders and its deficiency can lead to mood disorders.

12.6.1.15 Phytopharmaceutical Formulations

The Drugs and Cosmetics Act of 1940 provided guidelines for the use of a refined and filtered fraction of the phytochemical components of a drug ingredient or part thereof, the interior or exterior use of humans or animals to diagnose, treat, reduce, or prevent any disease, but exempted the management of parental trauma. New phytopharmaceutical drugs have not been used in the country at a critical level under prescribed conditions and labeled, and not considered effective and safe by the licensing authority referred to under rule 21 in the proposed applications. Phytopharmaceutical medicine differs from Ayurvedic medicine, Siddha, and Unani and includes all interior or exterior uses for diagnostic, therapeutic, mitigation, or prevention of diseases in humans only by the formulas prescribed in the Ayurvedic, Siddha, and Unani formulas. The phytopharmaceutical drugs are part of the raw material and separated by purification and regulation (Zulfa Nooren).

12.6.1.16 Clinical Aspects of Phytopharmaceutical Formulations

It is essential to interact with the clinical aspect of phytopharmaceutical formulations to achieve better safety and efficacy over the conventional system of herbal drugs. Clinical research ASU drug/patent drug on patients or non-patient volunteers human subjects for the diagnosis or validation of clinical, pharmacological including pharmacokinetic/pharmacodynamic and side effects/adverse effects to determine their safety and efficacy. For the clinical study of phytopharmaceutical drugs, data requirements are described in Appendix IB of Schedule Y of D&C rules. The clinical trial of phytopharmaceutical drugs is to be performed by applicable laws and strategies for new drugs. For all phytopharmaceutical drugs, details from phase I (determining the greatest tolerance and related toxicity) and agreements will be submitted before conducting studies. The results of quantitative research and the goals will be presented earlier in the study. The phytopharmaceutical drugs have already been sold for five years or where there is sufficiently published evidence regarding the safety of phytopharmaceutical drugs, studies can be shortened, modified, or free (John Dike Nwabueze Ogbonna).

12.6.1.17 Clinical Trial of Phytopharmaceutical Formulation for the Treatment of Cancer

Food and drug administration comprehensive, the clinical data for phytopharmaceuticals (Madhusmita Dhupal, Seca AML).

Plants	Phytochemical	Activity	Clinical trials
Dysoxylum binectariferum (Meliaceae family)	Alvocidib (Flavopiridol)	G1 phase cell cycle bound, successful in apoptosis produces cell fatality in chemotherapy	Phase II, avlocidib clinical study followed by for the treatment of acute myeloid leukemia, prostate cancer, and kidney cancer
Camptotheca acuminata Decne (Nyssaceae)	Topotecan (Hycamtin)	It is damaged DNA, disrupts cell phase and apoptosis to cell death	Clinical trial on ovarian, diminutive cell lung cancers and CNS tumors
Camptotheca acuminata Decne (Nyssaceae)	Belotecan hydrochloride	It is damaged DNA replication and consequentially apoptotic tumor cell death	Phase III, a clinical trial for the comparison of the effect of hebelotecan/cisplatin treatment and general treatment of etoposide/cisplatin in diminutive cell lung carcinoma Presently in a clinical study for the treatment of cervical cancer, diminutive cell lung cancer, lung cancer, and ovarian cancer
Camptotheca acuminata Decne (Nyssaceae)	Irinotecan HCl (Topotecin)	The cell cycle is arrested at S-G2 phase and consequently generates apoptosis. It acts as a topoisomerase inhibitor and inhibited of DNA synthesis	Phase III, the combination or alone therapy an intravenous infusion in the treatment of colorectal cancer, diminutive cell lung cancer, and stomach cancer
Camptotheca acuminata Decne (Nyssaceae)	10-Hydroxy camptothecin (AR67)	It is damaged DNA and apoptosis of proliferative cell	Numerous clinical studies were enduring on glioblastoma multiforme and patients with metastatic solid malignancy continued to confirm the efficacy of AR67
Vinca rosea Linn. (Apocynaceae)	Vincristine sulfate (Kyocristine)	The S and M phases of cancer cells arrested by this phytochemical agent	It is approved by the US FDA, in 1963, like Oncovin, for the treatment of leukemias such as lymphoblastic lymphoma, Burkitt lymphoma, Philadelphia chromosome-negative acute lymphoblastic leukemia, and solid tumor. Major clinical trials that continue the ongoing development of this drug
Vinca rosea Linn. (Apocynaceae)	Vinorelbine tartrate	It is inhibited microtubule polymerization and results in apoptosis	Enduring clinical trials using vinorelbine as a chief chemotherapeutic agent in patients with various types of cancer such as metastatic breast, esophageal squamous cell carcinoma, and lung cancer

Plants	Phytochemical	Activity	Clinical trials
Vinca rosea Linn. (Apocynaceae)	Vinflunine ditartrate (Javlor)	It is arrested by the cell cycle at the G2/M phase; resultant apoptosis precedes cell death of proliferative cells.	Clinical trials in the treatment of a range of cancer such as hormone-refractory prostate cancer (stage II), relapsed extensive SCLC (stage II), recurrent or metastatic head and neck cancers (stage III)
Combretum caffrum (African bush willow)	Fosbretabulin tromethamine (CA4P)	It is depolymerized microtubule compound, disintegrates mitosis, disintegrates cancer vasculature, and assists apoptosis in cancer cells	Phase II clinical trials undergoing epithelial, ovarian, neuroendocrine tumor, fallopian tube, and lung cancer.
Combretum caffrum (Combretaceae)	Ombrabulin	It inhibits tubulin polymerization and reduces mitotic binding and apoptosis in endothelial proliferative cells	Phase I trials that study the treatment of sarcoma, neoplasms, solid tumors, and malignant cancer cells
Glycin max Linn. (Fabaceae)	Genistein	It is inhibited topoisomerase-II and proteintyrosine kinase activity, foremost to cell cycle arrest at the G2/M stage and leading to apoptosis. It has an antioxidant, anti-inflammatory and antiangiogenic activities	Metastatic colorectal malignancy (stage I/II), early-stage prostate growth (stage II), stage IV breast malignancy (stage II) undergoing different clinical study
Euphorbia peplus (Euphorbiaceae)	Ingenol mebutate	It regulates the cell cycle that causes apoptosis in a variety of cancer cells	Clinical study of basal cell growth (stage I/II) and noninvasive lentigo malignancy of the face (stage II)
Betula pubescens (Betulaceae)	Betulinic acid	It has anti-inflammatory and antineoplastic activities. It is a DNA topoisomerase inhibitor, which promotes apoptosis through mitochondrial disturbance	Phase-I trial continues as an ointment applied topically for the treatment of cutaneous metastatic melanoma
Maytenus ovatus (Celastraceae)	Maytansine	It is an anti-tubulin, antimitotic drug used in the combination with an antibody (ADC) called trastuzumab emtansine approved by FDA	Clinical trials on various types of cancer such as metastatic non-small cell lung cancer (NSCLC) (stage II), advanced breast cancer (stage I/II), and colorectal neoplasm
Brassica oleracea (Brassicaceae)	Sulforaphane	It has antioxidant, antimitotic, and antiangiogenic activities	Phase II, study on prostate cancer, lung cancer, and metastatic breast cancer

(*Continued*)

Plants	Phytochemical	Activity	Clinical trials
Camellia sinensis (Theaceae)	Epigallocatechin gallate (EGCG)	It is an antineoplastic and anti-Hsp90 agent having antioxidant properties. It also produces neuroprotective actions	Colorectal and prostate cancer surgery (phase I) and lung cancer treatment (phase II)
Curcuma longa (Zingiberaceae)	Curcumin	It has antifungal, antibacterial, and antineoplastic activities	Study on breast, pancreatic, and prostate cancer treatment
Gossypium herbaceum (Malvaceae)	*R*-(−)-Gossypol acetic acid	G0/G1 phase is arrested by this phytochemical to cause apoptosis, therefore restricting proliferation of tumor cell	Phase I/II, clinical study on myeloma, lung cancer, prostate cancer, and solid tumor treatment
Tabebuia avellanedae (Bignoniaceae)	β-Lapachone (ARQ 501)	It is spoiled DNA and caused apoptosis by arresting cell cycle	Clinical study on carcinoma, pancreatic cancer, solid tumors, head and neck neoplasm, and lymphoma
Blue algae Cyanobacteria Nostocaceae	Cryptophycins 52 (LY355703)	It disrupts microtubule, restricted cell cycle and resulting in cancer cell apoptosis	Phase-II clinical study has been completed on platinum resistant ovarian cancer and advanced non-small cell lung cancer treatment
Oyster mushroom *Pleurotus ostreatus* (Pleurotaceae)	Lovastatin	It acts as a HMG-CoA reductase inhibitor and antitumor	Ongoing clinical study on melanoma, breast cancer treatment
Coixlacryma jobi (Cramineae)	Kanglaite (*Coix* seed oil)	It is arrested cell cycle at the G2/M phase, resulting in cancer cells apoptosis	Clinical study on pancreatic cancer, non-small cell lung cancer, and cachexia treatment
Fugus *Aspergillus terreus* (Trichocomaceae)	Simvastatin	It is prevented the growth of cancer cell at the G0/G1 phase, resulting apoptotic in cancer cell	Clinical study to cure metastatic breast cancer, adenocarcinoma of the stomach and rectum, pancreatic cancer, and prostate cancer
Taxus brevifolia (Taxaceae)	Paclitaxel (Taxol)	Cells proliferation and cell division arrested by blocking G2/M phase of cell cycle which move to apoptosis in cancer cells	Clinical study on non-small cell lung cancer and metastatic melanoma treatment
Bleekeria vitensis (Apocynaceae)	Elliptinium acetate (Celiptium)	It is intercalating agent, inhibited the topoisomerase(II) enzyme to prompt the breakages of DNA, thus restricting transcription and translation of proteins	Phase II for retrieve treatment (weekly dosage) of advanced breast cancer.
Cephalotaxus harringtonia var. drupacea (Cephalotaxaceae)	Omacetaxine mepesuccinate (Synribo)	It is protein translation inhibitor and prevent tyrosine kinase receptor antagonists in myeloid leukemia	Clinical study to cure different types of cancer such as leukemia and the solid tumors

12.6.1.18 Phytopharmaceutical Market in India

In modern pharmacy, the era may have preeminence of synthetic drugs but phytopharmaceuticals are making a powerful and theatrical response. In India, the Ayush (Ayurveda, Yoga, Unani, Siddha, and Homoeopathy) approach to remedy is rapidly growing. The Chinese traditional medicine has been reputed as the peak level of the herbal market. Phytopharmaceutical class is a new class of drugs system beginning from a traditional herbal system of medicine. The outlook of the Indian market, the phytopharmaceutical system is a raised area. The markets of ASU (Ayurveda, Siddha, Unani) pharmacies are not acquainted with the up-and-coming market for phytopharmaceuticals system. The standardized and validated extracts consequent from *Silybum marianum*, *Ginkgo biloba*, *Commiphora guggul*, and *Zingiber officinale* are only some examples to represent the phytopharmaceutical class of drugs in the Indian market. These extracts of phytopharmaceuticals are widely used by allopathic agents in a variety of disease conditions. The market for the composition of phytopharmaceuticals is based on the prescription of allopathic and Ayush practitioners. However, it has been noted that a large proportion of the traditional proprietary and over-the-counter medications are prescribed by Ayush staff members working in the clinical practice. The use of standardized validated extracts of plant-based phytopharmaceutical formulations is very narrow just because of less awareness and also some expensiveness. The estimated values of 1300 medicinal plants are used for the preparation of phytopharmaceutical formulations. According to Ayurveda itself, many drugs have been established for the treatment of liver diseases, such as the extract of *Andrographis paniculata*, *Solanum nigrum*, *Picrorrhizakurroa*, *Cicohrium intybus*, *Achillea millefolium*, *Capparis spinosa*, *Cassia occidentalis*, *Tamarix gallica*, *Terminalia arjuna*, and *Eclipta alba* are used in phytopharmaceutical formulations (Ex-Liv-52, Livergen, etc.) for the treatment of liver's disease. It is time to bring out the exposure of herbal drugs under the umbrella of the phytopharmaceutical system. Additionally, it is predictable that the phytopharmaceutical market shall be governed by standardized and validated extracts due to enhanced bioavailability, fewer side effects, and improved effect of the dosage form. Nanophytopharmaceutical is also knocking on the door of the phytopharmaceutical industries to achieve a better bioavailability, less toxicity, and improved the effect of phytopharmaceutical formulations. From the future point of view, nanophytopharmaceuticals may be seen as a dynamic force in the growing Indian phytopharmaceutical market (Vani Mamillapalli).

Marketed formulation of phytopharmaceuticals.

S. No.	Name	Description	Uses
1.	Nanoparticles of *Cuscuta chinensis*	Flavonoids and lignans containing nanostructured formulations to get better water solubility administered orally	Hepatoprotective and antioxidant effect
2.	Artemisinin nanocapsule	Artemisinin containing nanostructured sustained drug release formulations administered by IV route	Anticancer
3.	Adix salvia miltiorrhiza nanocapsule	*R. salvia* miltiorrhiza containing nanostructured formulations to improve the bioavailability administered by IV route	Coronary heart diseases, angina pectoris, and myocardial infarction
4.	Taxel-loaded nanoparticles	Taxel-loaded nanostructured formulations to enhance the bioavailability and sustained effect administered by IV route	Anticancer

(Continued)

S. No.	Name	Description	Uses
5.	Berberine-loaded nanoparticles	Berberine-loaded nanostructured Sustained drug release formulation administered by IV route	Anticancer
6.	Nano herbal sunscreen	Sunscreen ultraviolet (UV) filters are used in the form of nano instead of bulk form to make the sunscreen more translucent than white. They are also reported to be very effective when used in nano form. Nanoparticles of lipids can act as a UV blocker.	Protect skin form UV light
7.	*Pueraria mirifica* nano breast serum	*Pueraria mirifica*-loaded niosomes expand the cellular structure and increase the lobules and alveoli of the breasts	Skin moisturizing and nourishment
8.	Nanocluster hair shampoo	Shampoo and conditioner (nanoceutical citrus mint) are made with nanoclusters to give glowing hairs. It contains an extract mixture of nettle leaf, blackberry, and chamomile with citrus and mint oils.	Antidandruff, anti-inflammatory for hair follicles

References

1 Kesarwani, K. and Gupta, R. (2013). Bioavailability enhancers of herbal origin: an overview. *Asian Pac. J. Trop. Biomed.* 3 (4): 253–266.

2 Goyal, A., Kumar, S., Nagpal, M. et al. (2011). Potential of novel drug delivery systems for herbal drugs. *Indian J. Pharm. Educ. Res.* 45: 225–235.

3 Sachan, A.K. and Gupta, A. (2015). A review on nanotized herbal drugs. *Int. J. Pharm. Sci. Res.* 6 (3): 961.

4 Bonifácio, B.V., Silva, P.B., Aparecido, M. et al. (2014). Nanotechnology-based drug delivery systems and herbal medicines: a review. *Int. J. Nanomed.* 9: 1–15.

5 Chauhan, N.S., Rajan, G., and Gopalakrishna, B. (2009). Phytosomes: potential phyto-phospholipid carriers for herbal drug delivery. *J. Pharm. Res.* 2 (7): 1267–1270.

6 Rana, M., Kumar, A., and Rana, A.J. (2020). Drug delivery through targeted approach with special references to phytosomes. *Role Novel Drug Deliv. Vehicles Nanobiomed.* 22: 125.

7 Inamdar, N.S., Edalat, V.B., Kotwal, V.B., and Pawar, S.S. (2008). Herbal drugs in milieu of modern drugs. *Int. J. Green Pharm.* 2: 2–8.

8 Calixto, J.B. (2000). Efficacy, safety, quality control, marketing and regulatory guidelines for herbal medicines (phytotherapeutic agents). *Braz. J. Med. Biol. Res.* 33 (2): 179–189.

9 Christie, A. (2001). Herbs for health, but how safe are they. *Bull. World Health Org.* 79: 691692.

10 Herbal Medicine Market Research Report-Forecast to 2023. http://www.markertresearchfuture.com/reports/herbal-medicine-market-3250.

11 Mukherjee, P.K. and Wahile, A. (2006). Integrated approaches towards drug development from ayurveda and other Indian system of medicine. *J. Ethnopharmacol.* 103: 25–35.

12 Xia, Y., Rogers, J.A., Paul, K.E., and Whitesides, G.M. (1999). Unconventional methods for fabricating and patterning nanostructures. *Chem. Rev.* 99: 1823–1848.

13 Ito, H. (1997). Chemical amplification resists: history and development within IBM. *IBM J. Res. Dev.* 41: 69–80.

14 Binder, W.H., Kunz, M.J., and Ingolic, E. (2003). Supramolecular poly(ether ketone)–polyisobutylene pseudo-block copolymers. *J. Polym. Sci. Part A Polym. Chem.* 42 (1): 162–172.

15 http://www.scribd.com/doc/5682786/Kollicoat-MAE-grades.
16 Jiao, Y.Y., Ubrich, N., Hoffart, V. et al. (2002). Eudragit and its pharmaceutical significance. *Drug Dev. Ind. Pharm.* 28: 1033–1041.
17 Riley, S.A., Tavares, I.A., Bennett, A., and Mani, V. (1988). Delayed-release mesalazine (5-aminosalicylic acid): coat dissolution and excretion in ileostomy subjects. *Br. J. Clin. Pharmacol.* 26: 173–177.
18 Rowe, R.C., Sheskey, P.J., and Weller, P.J. *Hand Book of Pharmaceutical Excipients*, 3e. Washington DC, USA/London, UK: American Pharmaceutical Association/Pharmaceutical Press.
19 Duarte, A.R., Roy, C., Vega-Gonzalez, A. et al. (2007). Preparation of acetazolamide composite microparticles by supercritical anti-solvent techniques. *Int. J. Pharm.* 332: 132–139.
20 Pignatello, R., Bucolo, C., and Puglisi, G. (2002). Ocular tolerability of Eudragit RS100 and RL100 nanosuspensions as carriers for ophthalmic controlled drug delivery. *J. Pharm. Sci.* 91: 2636–2641.,
21 Khopade, A.J. and Jain, N.K. (1995). Self assembling nanostructures for sustained ophthalmic drug delivery. *Pharmazie* 50: 812–814.
22 Bucolo, C., Maltese, A., Maugeri, F. et al. (2004). Eudragit RL100 nanoparticle system for the ophthalmic delivery of cloricromene. *J. Pharm. Pharmacol.* 56: 841–846.
23 Galey, W.R., Lonsdale, H.K., and Nacht, S. (1976). The in vitro permeability of skin and buccal mucosa to selected drugs and tritiated water. *J. Invest. Dermat.* 67: 713–717.
24 Nair, M.K. and Chien, Y.W. (1996). Development. *Drug Dev. Ind. Pharm.* 22: 243–253.
25 Chen, W.G. and Hwang, G. (1992). The in vitro permeability of skin and buccal mucosa to selected drugs and tritiated water. *Int. J. Pharm.* 82: 61–66.
26 Hango, R., Kavimani, S., Mullaicharam, A.R., and Jayakar, B. (1997). In vitro studies on buccal strips of glibenclamide using chitosan. *Indian J. Pharm. Sci.* 59: 232–235.
27 Bremecker, K.D., Strempel, H., and Klein, G. (1984). Novel concept for a mucosal adhesive ointment. *J. Pharm. Sci.* 73: 548–552.
28 Shin, S.C., Bum, J.P., and Choi, J.S. (2000). Enhanced bioavailability by buccal administration of triamcinolone acetonide from the bioadhesive gels in rabbits. *Int. J. Pharm.* 209: 37–43.
29 Madgulkar, A., Kadam, S., and Pokharkar, V. (2008). *Asian J. Pharm.* 2: 57–60.
30 Md. Rahman, A. and Ali, J. (2008). Development and in vitro evaluation of enteric coated multiparticulate system for resistant. Tuberculosis. *Indian J. Pharm. Sci.* 70: 477–481.
31 Jain, S.K., Chourasia, M.K., and Dengre, R. (2005). Azopolymers for colon targeted drug delivery. *Indian J. Pharm. Sci.* 67: 43–50.
32 Małolepsza-Jarmołowska, K., Kubis, A.A., and Hirnle, L. (2003). Studies on gynaecological hydrophilic lactic acid preparations, part 6: use of Eudragit E-100 as lactic acid carrier in intravaginal tablets. *Pharmazie* 58 (5): 334–336.
33 Froiio, F., Ginot, L., Paolino, D. et al. (2019). Essential oils-loaded polymer particles: preparation, characterization and antimicrobial property. *Polymers (Basel)* 11 (6): 1017. https://doi.org/10.3390/polym11061017.
34 Pina-Barrera, A.M., Alvarez-Roman, R., Baez-Gonzalez, J.G. et al. (2019 Oct). Application of a multisystem coating based on polymeric nanocapsules containing essential oil of *Thymus vulgaris* L. to increase the shelf life of table grapes (*Vitis vinifera* L.). *IEEE Trans Nanobiosci.* 18 (4): 549–557.
35 Silva-Flores, P.G., Pérez-López, L.A., Rivas-Galindo, V.M. et al. (2019). Simultaneous GC-FID quantification of main components of *Rosmarinus officinalis* L. and *Lavandula dentata* essential oils in polymeric nanocapsules for antioxidant application. *J. Anal. Methods Chem.* 2019: 2837406.
36 Martins, C., Sousa, F., Araújo, F., and Sarmento, B. (2018). Functionalizing PLGA and PLGA derivatives for drug delivery and tissue regeneration applications. *Adv. Healthc. Mater.* 7: 1701035.

37 Reis, C.P., Neufeld, R.J., and Veiga, F. (2017). Preparation of drug-loaded polymeric nanoparticles. In: *Nanomedicine in Cancer* (ed. L.P. Balogh), 197–240. Singapore: Pan Stanford.
38 Juturu, V. and Wu, J.C. (2016). Microbial production of lactic acid: the latest development. *Crit. Rev. Biotechnol.* 36: 967–977.
39 Tačić, A., Ristić, I.S., Ilić, D. et al. (2017). *Polymeric Matrix Systems for Drug Delivery, Drug Delivery Approaches and Nanosystems*, vol. 1, 95–131. Waretown, NJ: Apple Academic Press.
40 Xu, Y., Kim, C.S., Saylor, D.M., and Koo, D. (2017). Polymer degradation and drug delivery in PLGA-based drug–polymer applications: a review of experiments and theories. *J. Biomed. Mater. Res. B Appl. Biomater.* 105: 1692–1716.
41 Agrahari, V., Agrahari, V., and Mitra, A.K. (2016). Nanocarrier fabrication and macromolecule drug delivery: challenges and opportunities. *Therap. Deliv.* 7: 257–278.
42 Yan, Y., Such, G.K., Johnston, A.P. et al. (2012). Engineering particles for therapeutic delivery: prospects and challenges. *ACS Nano* 6: 3663–3669.
43 Mundargi, R.C., Babu, V.R., Rangaswamy, V. et al. (2008). Nano/micro technologies for delivering macromolecular therapeutics using poly(d,l-lactide-co-glycolide) and its derivatives. *J. Control. Release* 125: 193–209.
44 Silva, A.T.C.R., Cardoso, B.C.O., MESR, E.S. et al. (2015). Synthesis, characterization, and study of PLGA copolymer in vitro degradation. *J. Biomater. Nanobiotechnol.* 6: 8.
45 Müller, A.J., Arnal, M.L., and Lorenzo, A.T. (2013). Crystallization in nano-confined polymeric systems. In: *Handbook of Polymer Crystallization* (ed. E. Piorkowska and G.C. Rutledge). New York: Wiley.
46 Mohammadi-Samani, S. and Taghipour, B. (2015). PLGA micro and nanoparticles in delivery of peptides and proteins; problems and approaches. *Pharm. Dev. Technol.* 20: 385–393.
47 Makadia, H.K. and Siegel, S.J. (2011). Polylactic-*co*-glycolic acid (PLGA) as biodegradable controlled drug delivery carrier. *Polymers* 3: 1377–1397.
48 Gorrasi, G. and Pantani, R. (2013). Effect of PLA grades and morphologies on hydrolytic degradation at composting temperature: assessment of structural modification and kinetic parameters. *Polym. Degrad. Stab.* 98: 1006–1014.
49 Kesente, M., Kavetsou, E., Roussaki, M. et al. (2017). Encapsulation of olive leaves extracts in biodegradable PLA nanoparticles for use in cosmetic formulation. *Bioengineering* 4 (3): 75.
50 Tsai, Y.-M., Chien, C.-F., Lin, L.-C., and Tsai, T.-H. (2011). Curcumin and its nano-formulation: the kinetics of tissue distribution and blood–brain barrier penetration. *Int. J. Pharm.* 416 (1): 331–338.
51 Zhao, J., Liu, Z., Xu, J. et al. (2009). Preparation and in vitro evaluation of brucine-loaded polylacticacid nanoparticles. *Zhongguo Zhong Yao Za Zhi* 34 (18): 2322–2324. Chinese.
52 Hirano, S., Seino, H., Akiyama, Y., and Nonaka, I. (1990). Chitosan: a biocompatible material for oral and intravenous administration. In: *Progress in Biomedical Polymers* (ed. C.H. Gebelein and R.L. Dunn), 283–290. New York: Plenum Press.
53 Chandy, T. and Sharma, C.P. (1990). Chitosan – as a biomaterial. *Biomater. Artif. Cells Artif. Organs* 18: 1–24.
54 Paul, W.S.C. (2000). Chitosan, a drug carrier for the 21st century: a review. *STP Pharm. Sci.* 10: 5–22.
55 Muzzarelli, R. and Muzzarelli, C. (2005). Chitosan chemistry: relevance to the biomedical sciences. *Adv. Polym. Sci.* 186: 151–209.
56 Muzzarelli, R.A.A. (1973). *Natural Chelating Polymers; Alginic Acid, Chitin, and Chitosan.* New York: Pergamon Press/Oxford 254 pp.
57 Braconnot, H. (1811). Sur la nature des champignons. *Ann. Chim. Phys.* 79: 265–304.

58 Jayakumar, R., Nwe, N., Tokura, S., and Tamura, H. (2007). Sulfated chitin and chitosan as novel biomaterials. *Int. J. Biol. Macromol.* 40: 175–181.

59 Rinaudo, M. (2008). Main properties and current applications of some polysaccharides as biomaterials. *Polym. Int.* 57: 397–430.

60 Mourya, V.K. and Inamdar, N.N. (2008). Chitosan-modifications and applications: opportunities galore. *React. Funct. Polym.* 68: 1013–1051.

61 Kurita, K. (2006). Chitin and chitosan: functional biopolymers from marine crustaceans. *Mar. Biotechnol. (NY)* 8: 203–226.

62 Agrawal, P., Strijkers, G.J., and Nicolay, K. (2009). Chitosan-based systems for molecular imaging. *Adv. Drug Deliv. Rev.* 62: 42–58.

63 Suh, J.K. and Matthew, H.W. (2000). Application of chitosan-based polysaccharide biomaterials in cartilage tissue engineering: a review. *Biomaterials* 21: 2589–2598.

64 Chatelet, C., Damour, O., and Domard, A. (2001). Influence of the degree of acetylation on some biological properties of chitosan films. *Biomaterials* 22: 261–268.

65 Felt, O., Buri, P., and Gurny, R. (1998). Chitosan: a unique polysaccharide for drug delivery. *Drug Dev. Ind. Pharm.* 24: 979–993.

66 [110] Muzzarelli, R., Jeuniauk, C., and Gooday, G.W. (1986). *Chitin in Nature and Technology*. New York: Plenum 583 pp.

67 Sjak-Braek, G., Anthonsen, T., and Sandford, P. (ed.) (1989). *Chitin and Chitosan: Sources, Chemistry, Biochemistry, Physical Properties and Applications*. London: Elsevier Applied Sciences.

68 Hou, W.M., Miyazaki, S., Takada, M., and Komai, T. (1985). Sustained release of indomethacin from chitosan granules. *Chem. Pharm. Bull. (Tokyo)* 33: 3986–3992.

69 Miyazaki, S., Ishii, K., and Nadai, T. (1981). The use of chitin and chitosan as drug carriers. *Chem. Pharm. Bull. (Tokyo)* 29: 3067–3069.

70 Genta, I., Pavanetto, F., Conti, B. et al. (1994). Spray drying for the preparation of chitosan microspheres. *Proc. Int. Symp. Control Release Bioact. Mater.* 21: 616–617.

71 Gallo, J.M. and Hassan, E.E. (1988). Receptor-mediated magnetic carriers: basis for targeting. *Pharm. Res.* 5: 300–304.

72 Illum, L., Farraj, N.F., and Davis, S.S. (1994). Chitosan as a novel nasal delivery system for peptide drugs. *Pharm. Res.* 11: 1186–1189.

73 Lueben, H.L., Lehr, C.M., Rentel, C.O. et al. (1994). Bioadhesive polymers for the peroral delivery of peptide drugs. *J. Control. Release* 29: 329–338.

74 Kumbar, S.G., Kulkarni, A.R., and Aminabhavi, T.M. (2002). Crosslinked chitosan microspheres for encapsulation of diclofenac sodium: effect of crosslinking agent. *J. Microencapsul.* 19: 173–180.

75 Higuchi, T. (1963). Mechanism of sustained-action medication. Theoretical analysis of rate of release of solid drugs dispersed in solid matrices. *J. Pharm. Sci.* 52: 1145–1149.

76 Elnaggar, Y.S.R., Etman, S.M., Abdelmonsif, D.A., and Abdallah, O.Y. (2015). Intranasal piperine-loaded chitosan nanoparticles as brain-targeted therapy in Alzheimer's disease: optimization, biological efficacy, and potential toxicity. *J. Pharm. Sci.* 104 (10): 3544–3556.

77 Anter, H.M., Abu Hashim, I.I., Awadin, W., and Meshali, M.M. (2019). Novel chitosan oligosaccharide-based nanoparticles for gastric mucosal administration of the phytochemical "apocynin". *Int. J. Nanomed.* 14: 4911–4929.

78 Kamel, K.M., Khalil, I.A., Rateb, M.E. et al. (2017). Chitosan-coated cinnamon/oregano-loaded solid lipid nanoparticles to augment 5-fluorouracil cytotoxicity for colorectal cancer: extract standardization, nanoparticle optimization, and cytotoxicity evaluation. *J. Agric. Food Chem.* 65 (36): 7966–7981.

79 Smidsrød, O. and Draget, K.I. (1996). Alginates: chemistry and physical properties. *Carbohydr. Eur.* 14: 6–13.

80 Draget, K.I., Skjåk-Bræk, G., and Smidsrød, O. (1994). Alginic acid gels: the effect of alginate chemical composition and molecular weight. *Carbohydr. Polym.* 25: 31–38.

81 Wang, X. and Spencer, H.G. (1998). Calcium alginate gels: formation and stability in the presence of an inert electrolyte. *Polymer* 13: 2759–2764.

82 King, A.H. (1983). *Food Hydrocolloids*, vol. II (ed. M. Glicksman), 115–154. Boca Raton, FL: CRC Press.

83 Takka, S., Ocak, O.H., and Acarturk, F. (1998). Formulation and investigation of nicardipine HCl-alginate gel beads with factorial design based studies. *Eur. J. Pharm. Sci.* 6 (3): 241–246.

84 Fattah, E.A.E., Grant, D.J.W., Gabr, K.E., and Meshali, M.M. (1998). Physical characteristics and release behavior of salbutamol sulfate beads prepared with different ionic polysaccharides. *Drug Dev. Ind. Pharm.* 24 (6): 541–547.

85 Kikuchi, A., Kawabuchi, M., Watanabe, A. et al. (1999). Effect of Ca^{2+} alginate gel dissolution on release of dextran with different molecular weights. *J. Control. Release* 58 (1): 21–28.

86 Choudhary, M., Chhabra, P., Tyagi, A., and Singh, H. (2021). Scar free healing of full thickness diabetic wounds: a unique combination of silver nanoparticles as antimicrobial agent, calcium alginate nanoparticles as hemostatic agent, fresh blood as nutrient/growth factor supplier and chitosan as base matrix. *Int. J. Biol. Macromol.* 178: 41–52.

87 Liu, Q., Jing, Y., Han, C. et al. (2019). Encapsulation of curcumin in zein/caseinate/sodium alginate nanoparticles with improved physicochemical and controlled release properties. *Food Hydrocoll.* 93: 432–442.

88 Ji, M., Sun, X., Guo, X. et al. (2019). Green synthesis, characterization and in vitro release of cinnamaldehyde/sodium alginate/chitosan nanoparticles. *Food Hydrocoll.* 90: 515–522.

89 Sampath Kumar, T.S. (2013). Physical and chemical characterization of biomaterials A2, Chapter 2. In: *Characterization of Biomaterials* (ed. A. Bandyopadhyay and S. Bose). Oxford: Academic Press.

90 Rajendran, M., Parthasarathy, P., Ravi, R. et al. (2020). Synthesis and characterization of copper oxide nano particles from medicinal plant of *Euphorbia cyanthophora*. *Inf. Res.* 9 (1): 103–109.

91 Nandhini, S. and Ilango, K. (2020). Development and characterization of a nano-drug delivery system containing vasaka phospholipid complex to improve bioavailability using quality by design approach. *Res. Pharm. Sci.* 16 (1): 103–117. Published 30 December 2020. https://doi.org/10.4103/1735-5362.305193.

92 Wang, T., Wu, C., Fan, G. et al. (2018). Ginkgo biloba extracts-loaded starch nano-spheres: preparation, characterization, and in vitro release kinetics. *Int. J. Biol. Macromol.* 106: 148–157. https://doi.org/10.1016/j.ijbiomac.2017.08.012. Epub 3 August 2017. PMID: 28780415.

93 Van Holde, K.E., Curtis Johnson, W., and Ho Shing, P. (2006). Thermodynamics and biochemistry. In: *Principles of Physical Biochemistry*, 2e (ed. J. Challice), 72–105. Upper Saddle River, NJ, USA: Pearson Prentice Hall [Google Scholar].

94 Cooper, A., Nutley, M.A., and Walood, A. (2000). Differential scanning microcalorimetry. In: *Protein-Ligand Interactions: Hydrodynamics and Calorimetry* (ed. S.E. Harding and B.Z. Chowdhry), 287–318. Oxford, UK: Oxford University Press [Google Scholar].

95 Cooper, A. (2004). *Biophysical Chemistry*, 103–107. London, UK: Royal Society of Chemistry [Google Scholar].

96 Cooper, A. (2000). Microcalorimetry of protein–DNA interactions. In: *DNA-Protein Interactions* (ed. A. Travers and M. Buckle), 125–139. Oxford, UK: Oxford University Press [Google Scholar].

97 Fultz, B. and Howe, J.M. (2013). *Transmission Electron Microscopy and Diffractometry of Materials*, 4e. Berlin and Heidelberg, Germany: Springer-Verlag.

98 Mazza, G. and Miniati, E. (1993). *Anthocyanins in Fruits, Vegetables, and Grains*, 1e. CRC Press.

99 Pour, P.M., Fakhri, S., Asgary, S. et al. (2019). The signaling pathways, and therapeutic targets of antiviral agents: focusing on the antiviral approaches and clinical perspectives of anthocyanins in the management of viral diseases. *Front. Pharmacol.* 10: 1207.

100 Toufektsian, M.C., De Lorgeril, M., Nagy, N. et al. (2008). Chronic dietary intake of plant-derived anthocyanins protects the rat heart against ischemia–reperfusion injury. *J. Nutr.* 138: 747–752.

101 Cassidy, A., Mukamal, K.J., Liu, L. et al. (2013). High anthocyanin intake is associated with a reduced risk of myocardial infarction in young and middle-aged women. *Circulation* 127: 188–196.

102 Mccullough, M.L., Peterson, J.J., Patel, R. et al. (2012). Flavonoid intake and cardiovascular disease mortality in a prospective cohort of US adults. *Am. J. Clin. Nutr.* 95: 454–464.

103 Mink, P.J., Scrafford, C.G., Barraj, L.M. et al. (2007 Mar). Flavonoid intake and cardiovascular disease mortality: a prospective study in postmenopausal women. *Am. J. Clin. Nutr.* 85 (3): 895–909.

104 Gao, X., Cassidy, A., Schwarzschild, M.A. et al. (2012). Habitual intake of dietary flavonoids and risk of Parkinson disease. *Neurology* 78 (15): 1138–1145.

105 Jomova, K. and Valko, M. (2011). Advances in metal-induced oxidative stress and human disease. *Toxicology* 283 (2-3): 65–87.

106 Speciale, A., Chirafisi, J., Saija, A., and Cimino, F. (2011). Nutritional antioxidants and adaptive cell responses: an update. *Curr. Mol. Med.* 11 (9): 770–789.

107 Speciale, A., Anwar, S., Canali, R. et al. (2013). Cyanidin-3-*O*-glucoside counters the response to TNF-alpha of endothelial cells by activating Nrf2 pathway. *Mol. Nutr. Food Res.* 57 (11): 1979–1987.

108 El-Agamey, A., Lowe, G.M., McGarvey, D.J. et al. (2004). Carotenoid radical chemistry and antioxidant/pro-oxidant properties. *Arch. Biochem. Biophys.* 430 (1): 37–48.

109 Tapiero, H., Townsend, D.M., and Tew, K.D. (2004). The role of carotenoids in the prevention of human pathologies. *Biomed. Pharmacother.* 58 (2): 100–110.

110 Rao, A.V. and Rao, L.G. (2007). Carotenoids and human health. *Pharmacol. Res.* 55 (3): 207–216.

111 Pechinskii, S.V. and Kuregyan, A.G. (2014). Molecular-biological problems of drug design and mechanism of drug action. *Pharm. Chem. J.* 47: 509–513.

112 Stahl, W. and Sies, H. (2005). Bioactivity and protective effects of natural carotenoids. *Biochim. Biophys. Acta* 1740 (2): 101–107.

113 Upadhyaya, K.R., Radha, K.S., and Madhyastha, H.K. (2007). Cell cycle regulation and induction of apoptosis by beta-carotene in U937 and HL-60 leukemia cells. *J. Biochem. Mol. Biol.* 40 (6): 1009–1015.

114 Palozza, P., Calviello, G., Serini, S. et al. (2001). Beta-carotene at high concentrations induces apoptosis by enhancing oxy-radical production in human adenocarcinoma cells. *Free Radic. Biol. Med.* 30 (9): 1000–1007.

115 Palozza, P., Serini, S., Maggiano, N. et al. (2002). Induction of cell cycle arrest and apoptosis in human colon adenocarcinoma cell lines by beta-carotene through down-regulation of cyclin A and Bcl-2 family proteins. *Carcinogenesis* 23 (1): 11–18.

116 Jang, S.H., Lim, J.W., and Kim, H. (2009). Mechanism of beta-carotene-induced apoptosis of gastric cancer cells: involvement of ataxia-telangiectasia-mutated. *Ann. N. Y. Acad. Sci.* 1171: 156–162.

117 Moeller, S.M., Parekh, N., Tinker, L. et al. (2006). Associations between intermediate age-related macular degeneration and lutein and zeaxanthin in the carotenoids in age-related Eye Disease Study (CAREDS): ancillary study of the Women's Health Initiative. *Arch. Ophthalmol.* 124 (8): 1151–1162.

118 Sahin, K., Tuzcu, M., Sahin, N. et al. (2011). Inhibitory effects of combination of lycopene and genistein on 7,12-dimethyl benz(a)anthracene-induced breast cancer in rats. *Nutr. Cancer* 63 (8): 1279–1286.

119 Gupta, S., Jawanda, M.K., Arora, V. et al. (2015). Role of lycopene in preventing oral diseases as a nonsurgical aid of treatment. *Int. J. Prev. Med.* 6: 70. Published 5 August 2015.

120 Bansal, M., Vashisth, S., Gupta, N., and Singh, S. (2012). Antioxidants – its preventive role in oral cancer. *Indian J. Dent. Sci.* 4: 103–105.

121 Perry, E.K., Tomlinson, B.E., Blessed, G. et al. (1978). Correlation of cholinergic abnormalities with senile plaques and mental test scores in senile dementia. *Br. Med. J.* 2 (6150): 1457–1459.

122 Khan, M.T., Orhan, I., Senol, F.S. et al. (2009). Cholinesterase inhibitory activities of some flavonoid derivatives and chosen xanthone and their molecular docking studies. *Chem. Biol. Interact.* 181 (3): 383–389.

123 D'Mello, P., Gadhwal, M., and Joshi, U. (2011). Modeling of COX-2 inhibitory activity of flavonoids. *Int J Pharm Pharm Sci* 3: 33–40.

124 Umamaheswari, M., Madeswaran, A., and Asokkumar, K. (2013). Virtual screening analysis and in-vitro xanthine oxidase inhibitory activity of some commercially available flavonoids. *Iran J. Pharm. Res.* 12 (3): 317–323.

125 Ganugapati, J., Mukkavalli, S., and Sahithi, A. (2011). Docking studies of green tea flavonoids as insulin mimetics. *Int. J. Comput. Appl.* 30: 48–52.

126 Wiseman, H. (2000). The therapeutic potential of phytoestrogens. *Expert Opin. Investig. Drugs* 9 (8): 1829–1840.

127 Metzner, J.E., Frank, T., Kunz, I. et al. (2009). Study on the pharmacokinetics of synthetic genistein after multiple oral intakes in post-menopausal women. *Arzneimittelforschung* 59 (10): 513–520.

128 Wang, H.K., Xia, Y., Yang, Z.Y. et al. (1998). Recent advances in the discovery and development of flavonoids and their analogues as antitumor and anti-HIV agents. *Adv. Exp. Med. Biol.* 439: 191–225.

129 Wu, D., Kong, Y., Han, C. et al. (2008). d-Alanine:d-alanine ligase as a new target for the flavonoids quercetin and apigenin. *Int. J. Antimicrob. Agents* 32 (5): 421–426.

130 Li, M. and Xu, Z. (2008). Quercetin in a lotus leaves extract may be responsible for antibacterial activity. *Arch. Pharm. Res.* 31 (5): 640–644.

131 Moussa, C., Hebron, M., Huang, X. et al. (2017). Resveratrol regulates neuro-inflammation and induces adaptive immunity in Alzheimer's disease. *J. Neuroinflammation* 14 (1): 1.

132 Yang, Q., Wang, B., Zang, W. et al. (2013). Resveratrol inhibits the growth of gastric cancer by inducing G1 phase arrest and senescence in a Sirt1-dependent manner. *PLoS One* 8 (11): e70627.

133 Thordardottir, S., Kinhult Ståhlbom, A., Almkvist, O. et al. (2017). The effects of different familial Alzheimer's disease mutations on APP processing in vivo. *Alzheimers Res. Ther.* 9 (1): 9.

134 Nanjan, M.J. and Betz, J. (2014). Resveratrol for the management of diabetes and its downstream pathologies. *Eur. Endocrinol.* 10 (1): 31–35.

135 Gualdani, R., Cavalluzzi, M.M., Lentini, G., and Habtemariam, S. (2016). The chemistry and pharmacology of citrus limonoids. *Molecules* 21 (11): 1530.

136 Kim, J., Jayaprakasha, G.K., Vikram, A., and Patil, B.S. Cancer chemopreventive properties of Citrus limonoids. *Emerg. Trends Diet. Compon. Prev. Combat. Dis.* 2012 (1093): 37–50. American Chemical Society.

137 Kim, J., Jayaprakasha, G.K., Vikram, A., and B.S. (2012). Patil methyl nomilinate from citrus can modulate cell cycle to induce cytotoxicity in human colon cancer (SW480) regulators cells in vitro. *Toxicol. In Vitro* 26 (7): 1216–1223.

138 Jayaprakasha, G.K., Mandadi, K.K., Poulose, S.M. et al. Novel triterpenoid from *Citrus aurantium* L. possesses chemopreventive properties against human colon cancer cells. *Bioorg. Med. Chem.* 16 (11): 5939–5951.

139 Akram, M., Jabeen, F., Riaz, M. et al. (ed.) (2021). Health benefits of glucosinolate isolated from cruciferous and other vegetables. In: *Preparation of Phytopharmaceuticals for the Management of Disorders*, 361–371. Academic Press.

140 Kestwal, R.M., Lin, J.C., Bagal-Kestwal, D., and Chiang, B.H. Glucosinolates fortification of cruciferous sprouts by sulphur supplementation during cultivation to enhance anti-cancer activity. *Food Chem.* 126: 1164–1171.

141 Ibrahim, M.D., Kntayya, S.B., Mohd Ain, N. et al. (2018). Induction of apoptosis and cytotoxicity by raphasatin in human breast adenocarcinoma MCF-7 cells. *Molecules* 23 (12): 3092.

142 Miao, H.Y., Wang, M.Y., Chang, J.Q. et al. (2017). Effects of glucose and gibberellic acid on glucosinolate content and antioxidant properties of Chinese kale sprouts. *J. Zhejiang Univ. Sci. B* 18 (12): 1093–1100.

143 Tahata, S., Singh, S.V., Lin, Y. et al. (2018). Evaluation of biodistribution of sulforaphane after administration of oral broccoli sprout extract in melanoma patients with multiple atypical nevi. *Cancer Prev. Res. (Phila.)* 11 (7): 429–438.

144 Bahadoran, Z., Tohidi, M., Nazeri, P. et al. (2012). Effect of broccoli sprouts on insulin resistance in type 2 diabetic patients: a randomized double-blind clinical trial. *Int. J. Food Sci. Nutr.* 63 (7): 767–771.

145 Defilippis, A.P., Blaha, M.J., and Jacobson, T.A. (2010). Omega-3 fatty acids for cardiovascular disease prevention. *Curr. Treat Opt. Cardiovasc. Med.* 12 (4): 365–380.

13

Toxicity of Nanostructures and Nanodrugs

Tapash Chakraborty and Anupam Sarma

Department of Pharmaceutics, Girijananda Chowdhury Institute of Pharmaceutical Science (GIPS), Guwahati, Assam, India

13.1 Introduction

Nanostructures are colloidal solid particles having a particle size of less than 1 μm and contain approximately 10–10 000 atoms. When materials used in daily life are formulated into nanostructures, they start to show different magnetic, optical, thermal, mechanical, and electrical characteristics compared to their macromolecular forms. These unique characteristics can be attributed to their nano-size range [1].

Nanostructures have existed in the environment since the beginning of the earth's history. They are very common and are widely distributed throughout the earth's atmosphere, oceans, surface- and groundwater, soil, and even in living organisms. The natural processes that release nanostructures into the atmosphere are forest fires, volcanic activities, weathering, and formation from clay minerals, soil erosion by wind and water, or dust storms from the desert. Atmospheric dust alone is estimated to contain as many as several million tons of natural nanostructures within a year. Naturally occurring nanostructures can be transported over thousands of kilometers and remain suspended in the air for several days [2, 3].

The sources of nanostructures can be broadly divided into three categories.

- *Incidental nanostructures*: These are produced incidentally as a by-product of industrial processes, e.g. vehicle engine exhaust, welding fumes, combustion processes, and even some natural processes such as forest fires, dust storms and cosmic dust, and volcanic eruptions.
- *Engineered nanostructures*: These are fabricated nanostructures to have certain desired properties for different applications, e.g. microprocessors, nano drugs, cigarette smoke, and building demolition.
- *Naturally produced nanostructures*: These can be found in the bodies of organisms, insects, plants, animals, and human bodies, e.g. viruses, nano-bacteria, algae, fungi, yeast, and bacterial spores.

One might think that nanotechnology and nanostructures are new to humankind and are only utilized for the preparation of modern commodities. But interestingly, knowingly, or unknowingly, humankind started using nanostructures long back. According to the evidence found from archaeological excavations, humans started using natural asbestos nanofibers to reinforce ceramic

Phytoantioxidants and Nanotherapeutics, First Edition. Edited by Mithun Rudrapal.

matrices more than 4500 years ago. The ancient Egyptians used nanostructures more than 4000 years ago in the form of synthetic hair dyes [4, 5]. Today, different types of nanostructures find their application in medicine, cosmetics, sunscreens, electronics, energy production, batteries, fuel additives, paints, coatings, pigments, tires, and the cement industry (Table 13.1) [10].

Nanomedicine or nano drugs is the branch of medicine that utilizes the knowledge and tools of nanotechnology for the diagnosis and treatment of diseases. Nanomedicine involves the use of nanoscale materials, such as biocompatible nanostructures and nano-robots, for diagnosis, delivery, sensing, or actuation purposes in a living organism [11]. Nanomedicines preferably have a particle size range of 5–100 nm and possess large surface areas, usually 100–1000 m^2/g. Particles of this size range can easily pass through various biological barriers like the blood–brain barrier [12]. Nanomedicines have been reported to show different advantages over conventional dosage forms like improved efficacy, bioavailability, dose–response, targeting ability, personalization, safety, protection against the degradation of active drugs, decrease in the severity of drug side effects, and inhibiting cellular efflux pumps, countering the drug-resistant mechanisms seen in cancer cells [13, 14].

Though nanostructures have various advantages over conventional materials and medicines, these possess various serious disadvantages too. Due to their small size, nanostructures can translocate active chemical species from biological barriers like skin, lungs, body tissues, and organs. Depending on their composition, nanostructures can lead to irreversible oxidative stress, organelle damage, asthma, and cancer. Acute exposure to nanostructures shows their toxicity by generating reactive oxygen species (ROS), protein denaturation, mitochondrial disconcertion, and perturbation of phagocytic functions. Chronic toxic effects of exposure to nanostructures include uptake of nanostructures by the reticuloendothelial system (RES), nucleus, neuronal tissue, and the generation of neoantigens that cause possible organ enlargement and dysfunction [15].

13.2 Causes of Nanotoxicity and Physicochemical Aspects of Nanostructures Responsible for Nanotoxicity

Reports suggest that the toxicity of nanostructures depends mostly on their origin. Not all nanostructures are toxic to the human body. Many of them are nontoxic and some even show positive health effects. The physicochemical properties of the nanostructures are the main reasons for their toxicity. The physicochemical properties of the nanostructures, that can alter their toxicity, are as follows:

13.2.1 Chemical Composition of Nanostructures

The chemical that is utilized to formulate the nanostructures has the most important role in its toxicity inside the human body. Chemicals that are biodegradable easily in vivo will make the least toxic nanostructures. However, if nanostructures are formulated with chemicals that are slowly biodegradable or are nonbiodegradable, the nanostructures will gradually accumulate inside the organs and tissues on prolonged exposure and lead to toxicity. Moreover, the solubility of the materials of the nanostructures in biological fluids alters its toxicity. Studies have shown that metal nanostructures in their soluble form are more toxic than insoluble ones. This is because soluble nanostructures can reach deeper into cells and tissues as they form molecular dispersions as compared to a colloidal dispersion of the insoluble nano-formulation [16, 17].

Table 13.1 Types of nanomaterials and their production procedure.

Type of nanomaterial	Examples	Production procedure
Carbon-based nanomaterials	• Fullerenes (C60) • Carbon nanotubes (CNTs) • Carbon nanofibers • Carbon black • Graphene (Gr) • Carbon onions	• Laser ablation • Arc discharge • Chemical vapor deposition (CVD)
Inorganic-based nanomaterials	• Metal and metal oxide NPs	• Template method • Electrochemical method • PEGylation • Turkevich method • Brust method • Perrault method • Martin method • Citrate thermoreduction method • Solvent-free photochemical method • Oligonucleotide-functionalized nanoparticles • Seed-mediated method • Non-seed-mediated method • Hot injection technique • Surfactant assisted method • Two-phase methods • Stober process • Sonolysis • γ-Irradiation method • Green biosynthesis method • Sunlight irradiation method • New green chemistry method • Chemical reduction • Microemulsion technology • UV-initiated photoreduction • Photoinduced reduction • Electrochemical method • Electrochemical synthetic method • Microwave-assisted synthesis • Radiolysis • γ-Ray irradiation • From polymers and polysaccharides • Tollens method • Pyrolysis • Physical vapor condensation • Arc-discharge method • Evaporation–condensation • From bacteria, fungi, yeast, algae, and plants

(*Continued*)

Table 13.1 (Continued)

Type of nanomaterial	Examples	Production procedure
Organic-based nanomaterials	• Dendrimers • Micelles • Liposomes • Polymeric NPs	• Precipitation induced by solvent removal • Solvent evaporation • Solvent diffusion • Salting out • Gelation of the emulsion droplets • Polymerization in emulsion • Conventional emulsion polymerization • Surfactant-free emulsion polymerization • Interfacial polymerization • Controlled/living radical polymerization
Composite-based nanomaterials	• Hybrid nanofibers	• Hydrothermal synthesis • Sol–gel synthesis • Polymerized complex method (Pechini process) • Chemical vapor deposition • Microwave synthesis

Source: Based on Refs. [6–9].

13.2.2 Particle Chemistry and Crystalline Structure

The material composition and material chemistry are not the same. Two different nanostructures may have the same chemical composition, but their crystalline structure may be different. The toxicity of nanostructures depends on their crystalline structures as they exert different oxidative stress on the biomolecules. Depending on their chemistry, nanoparticles can show different cellular uptake, subcellular localization, and the ability to catalyze the production of ROS. For example, rutile and anatase are the allotropes of titanium dioxide. So, they have the same chemical composition, but a different crystalline structure, and hence different chemical and physical properties. Studies have shown that rutile nanoparticles induce DNA damage in the absence of light through oxidative stress, but anatase nanoparticles do not damage the DNA. Moreover, nanostructures can change crystal structure on interaction with water or liquids of the biological systems. Studies have shown that zinc sulfide (ZnS) nanoparticles rearrange their crystal structure in the presence of water and become more ordered, becoming like a bulk piece of solid ZnS [17, 18].

13.2.3 Size and Surface Area of Nanostructures

On decreasing the particle size of the nanostructures, the surface area increases exponentially relative to their volume, which makes their surface more reactive to themselves and to their adjacent environment. Moreover, the absorption, distribution, and elimination of the nanostructures have a direct relation to the particle size and surface area of the nanostructures. The biological processes like endocytosis, cellular uptake, and efficiency of particle processing in the endocytic pathway are also dependent on the size of the material [19, 20]. The lung and other epithelial-induced inflammatory responses of nanostructures are directly related to the surface area of the particles [21]. The pharmacological behaviors of nanostructures are also dependent on particle size. Studies have proven that nanostructures administered through an intravenous route having a particle size less than 50 nm cross quickly to nearly all body tissues and show toxicity. On the other hand, nanostructures having a particle size greater than 50 nm are readily engulfed by the RES preventing their entry into other tissues. This engulfment by the RES makes the RES organs like the liver and spleen as main targets of the toxicity of the nanostructures [22].

13.2.4 The Particle Shape of Nanostructures

The nanostructures can be of different shapes, like spherical, triangular, cubic, hexagonal, oval, prism-like, helical, rod-shaped, etc. (Figure 13.1). Shapes can strongly influence the electromagnetic, optical, and catalytic properties of metallic nanostructures [23, 24]. The speed of membrane wrapping processes in vivo during endocytosis or phagocytosis of nanostructures is highly dependent on the shape of the particles. The endocytosis of spherical nanostructures is easier and faster as compared to rod-shaped or helical nanostructures and is relatively less toxic. Nanostructures having other than spherical shapes cannot flow easily through the blood capillaries causing biological consequences [25, 26]. Studies have shown that rod-shaped single-walled carbon nanotubes (SWCNT) can block K^+ ion channels 2–3 times more efficiently than spherical carbon fullerenes. Similarly, TiO_2 nanofibers are more cytotoxic than TiO_2 nanostructures having a spherical shape [27, 28].

13.2.5 Surface Charge of Nanostructures

All biological membranes are electronegative. Hence, the surface charge of nanostructures plays an important role in their toxicity as the type of charge on them defines their interactions with the biological membranes [29]. The surface charge of the nanostructures also regulates the selective adsorption of nanoparticles to different membranes, their dispersion behavior, plasma protein

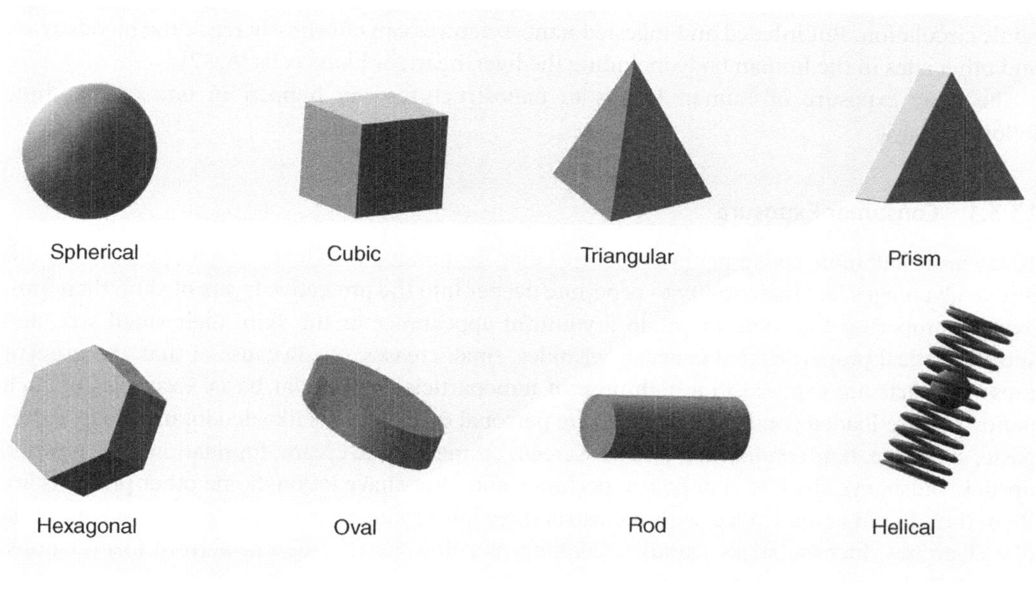

Figure 13.1 Different shapes of nanostructures.

binding, blood–brain barrier permeability, and other transmembrane permeability. Positively charged nanoparticles show significant cellular uptake compared to negatively charged and neutral nanoparticles, as the negatively charged biological membranes attract the positively charged particles leading to high permeability. Positively charged nanostructures form agglomerates more easily compared to the negatively charged particles leading to the alteration of their colloidal behavior, leading to higher cytotoxicity. Moreover, positively charged nanostructures show greater hemolysis and platelet aggregation potential compared to negatively charged particles causing severe toxicity to the system [30–32].

13.2.6 Surface Functionalization

Surface coatings or functionalization of nanostructures can change their surface chemistry, altering their aggregation characteristics that, in turn, change their biomedical potentials. Surface functionalization of nanostructures can change their physicochemical properties like magnetic, electric, optical properties, and chemical reactivity and can change the absorption, distribution, elimination, and toxicity of nanostructures [33]. It is well known that if nanoparticle surfaces contain oxygen, ozone, oxygen radicals, and transition metals, they produce ROS and induce inflammation leading to toxicity [34, 35]. Hence, the surface coating could be employed to reduce the toxicity issues of the nanostructures by covering up those reactive elements. Proper surface coating can stabilize the nanostructures and avoid the release of toxic ions from them [36].

13.3 Different Types of Nanotoxicity

Intentionally or unintentionally, humans encounter nanostructures on daily basis. The exposure can take place during the production, use, disposal, and waste treatment of products containing nanostructures. Healthy human skin can prevent these nanostructures from getting into the systemic circulation. But inhaled and ingested nanostructures can effortlessly reach the bloodstream and other sites in the human body including the liver, heart, or blood cells [6, 37].

The toxic exposure of human beings to nanostructures can happen in one of the three following ways:

13.3.1 Consumer Exposure

Today more and more consumer products are being formulated with loads of nanoparticles for different advantages like their ability to penetrate deeper into the protective layers of skin, their antioxidant properties that help maintain a youthful appearance of the skin, their small size, and specific optical properties that conceals wrinkles, small creases, etc. Because of that, the users of those products are exposed to a high dose of nanoparticles on regular basis. Examples of such nanostructure-loaded consumer products are personal care products like deodorants, soap, toothpaste, shampoo, hair conditioner, and sunscreen, cosmetics like cream, foundation, face powder, lipstick, blush, eye shadow, nail polish, perfume, and after-shave lotion. Some other products are there that do not come loaded with nanostructures but produce them during their consumption like cigarettes, incense sticks, candles, cooking, burning wood, different aerosol formulations, etc. [6]. Prolonged consumption of these products can lead to their accumulation in tissues and cause toxicity.

13.3.2 Occupational Exposure

Among all possible exposure scenarios, occupational exposure is the most likely the earliest and highest exposure than the general population. Exposure to nanostructures may occur through inhalation, ingestion, and skin absorption. While dermal absorption and ingestion are potential entry routes for engineered nanostructures, the inhalation route is the most important one concerning toxicity. Some workplaces generate aerosols at high concentrations and a very high number of the generated particles fall in the inhalable range. For example, during welding, a particulate substance usually referred to as "welding fumes" are formed. Their size falls in the respirable range, having an aerodynamic diameter of less than 1000 nm, mostly even less than 100 nm. In different industrial processes, nanostructures with different particle sizes and morphology are produced. The number of nanostructures produced during those processes depends on the combination of the processes and the materials used, whereas the toxicity of the mixture depends on the material used [38–40].

13.3.3 Environmental Exposure

During the last decade of the twentieth century, the application of nanotechnology and nanoparticles in various fields observed a boom along with the development of new processes for the synthesis of various nanoparticles. As the development and application of nanotechnology increased, nanoparticles have emerged as a new class of environmental pollutants. Increasing application of nanostructures in commercial products and industrial applications has eventually resulted in their release into atmospheric, terrestrial, and aquatic environments significantly affecting the environment and human health. When researchers found the adverse effects of those released nanostructures on the ecosystem, in and around the year 2000, scientists started thinking about safer disposal methods of the nanoparticles [41].

Carbon nanotubes, fullerenes, silver nanoparticles, nano-sized TiO_2, ZnO, Fe_2O_3, CuO, and iron particles are widely used as catalysis, sensors, environmental remediation, in cosmetics, antimicrobial products, and commercial products. These nanoparticles ultimately enter aquatic ecosystems through wastewater discharges and wash-offs during recreational activities such as swimming and water skiing [42, 43]. The presence of these nanoparticles in the ecosystem threatens the health of natural flora and fauna and human beings.

To determine the toxicity of nanoparticles on living creatures, several in vivo and in vitro studies have been carried out with nanostructures prepared from different materials like carbonyl, iron, crystalline silica, amorphous silica, zinc oxide, selenium, silver, carbon nanotubes, fullerenes, quantum dots, TiO_2, etc. Some of these found that the results of in vitro study show a good correlation with in vivo study. But some others reported that results of in vitro study do not implement on in vivo study [44–47].

13.4 Mechanisms of Nanotoxicity

13.4.1 Nanomedicine-Induced Oxidative Stress, Inflammation, and Genotoxicity

ROS are highly reactive chemical molecules formed when an oxygen molecule (O_2) receives an electron during a chemical reaction. In all biological systems, ROS are formed as a natural by-product of the normal aerobic metabolism of oxygen. Singlet oxygen, α-oxygen, peroxides, superoxide, and hydroxyl radical are some of the ROS that are formed in the cells. In normal conditions, when these ROS are present at low and stationary levels in cells, they play an important role in cell signaling and homeostasis. But, when ROS concentration increases in cells above safe levels, they can oxidize and modify some cellular components causing irreversible damage to DNA and preventing them from performing their intended functions [48–50].

Various biochemical reactions that occur during the processes of respiration and photosynthesis in organelles such as mitochondria, peroxisomes, and chloroplasts produce ROS. In normal conditions, the concentration of ROS in cells is controlled by engaging different enzymatic and nonenzymatic antioxidants. A class of enzyme, known as superoxide dismutase, plays a major role in the neutralization process of ROS. Mammalian cells possess three types of superoxide dismutase, viz. SOD1 in the cytoplasm, SOD2 in the mitochondria, and SOD3 in extracellular spaces [51, 52]. Superoxide dismutase controls the cellular level of superoxide ions by catalyzing the disproportionation reaction of superoxide ions into oxygen and hydrogen peroxide [53].

The condition when there is an imbalance between the production of ROS and its elimination in the cell is known as oxidative stress. Oxidative stress can oxidize and modify the cell components and irreversibly damage the DNA, causing chronic inflammation and genotoxicity, which, in turn, can cause cancer, diabetes, and cardiovascular and pulmonary diseases [54]. Nanostructures can cause toxicity by causing oxidative stress in the cells. Nanostructures can lead to oxidative stress in the cells in four different ways. Firstly, the large surface area of nanoparticles can increase the generation of ROS in the cell. Secondly, the nanostructures of transition metal ions can produce more reactive hydroxyl radicals by catalyzing oxygen metabolites. Thirdly, the small-sized nanostructures can trigger mitochondria dysfunction, causing an imbalance in the respiratory chain and disturbed ROS signaling. And, lastly, the nanostructures can activate the macrophages and neutrophils to produce ROS in higher quantities [55, 56]. Though the affected cells try to neutralize the harmful effects of the ROS by engaging the enzymatic antioxidants like glutathione peroxidase, glutathione reductase, catalase, and the nonenzymatic antioxidants like glutathione, vitamins C and D, the overproduction of ROS eventually leads to inflammation and DNA damage. Inflammation increased the production of cytokines and interleukins, which lead to a rush of immune reactions. The consequences of the immune reactions include fibrosis and bronchial granulomas [57, 58].

Nanostructures having particle sizes below 10 nm may show genotoxicity by directly entering the cell nucleus and altering the working of DNA, causing chromosomal fragmentation, breakage of DNA strands, point mutations, oxidative DNA adducts, and alterations in gene expression profiles. On the other hand, nanostructures having larger particle sizes may come into contact with DNA during the mitosis phase of the cell cycle when the nuclear membrane dissolves [59, 60].

13.4.2 Nanomedicine-Induced Dysfunction of Cell Organelles

13.4.2.1 Lysosomal Rupture

To digest and eliminate the nanostructures that have entered a cell, the endocytosis pathway transports the nanostructures and fuses the endosome with lysosomes. Lysosomes are acidic cell organelles essential for degradation, signaling, cell homoeostasis, and programmed cell death. The hydrolytic enzymes and acidic environment of lysosomes degrade the transported nanostructures, along with other damaged organelles and long-lived proteins of the cell. Lysosomotropic agents are some weak bases that penetrate the lysosome in protonated form and increase the intracellular pH. Permeabilization of the lysosomal membrane and release of hydrolytic enzymes to the cytosol generates apoptosis signals and ultimately causes cell death. Nanomaterials with proton-buffering capabilities can serve as lysosomotropic agents to induce nonselective toxicity when uptaken by healthy cells. Studies have shown that several nanomaterials like carbon nanomaterials, metal and metal oxide nanoparticles, and cationic nanoparticles act as lysosomotropic agents and cause cell death. Breakdown of the lysosomes reduces the pH of the cytosol, which, in turn, induces cell death by necrosis [61–65].

13.4.2.2 Mitochondrial Dysfunction

When nanomaterials induce lysosomal membrane permeabilization, the nanomaterials and lysosomal enzymes meet with the cytosol. These released materials can generate inflammation and ROS and damage mitochondria and endoplasmic reticulum. Mitochondria is the powerhouse of cells and play an important role in performing aerobic respiration. The mitochondrion is the only organelle other than the nucleus that has its genome. Mitochondria generate ROS for signaling purposes. It is also responsible for ROS elimination and programmed cell death. Mitochondrial dysfunction caused by nanomaterial releases the enzyme cytochrome-C and activates the enzyme Caspase-9, which is responsible for the apoptosis of the cell [66–69].

13.4.2.3 Disturbance of Endoplasmic Reticulum

Endoplasmic reticulum is the machinery of biosynthesis and is responsible for lipid synthesis and protein secretion and maintains intracellular calcium homeostasis. When nanostructures interrupt the functioning of the endoplasmic reticulum, this causes the accumulation of unfolded protein on the endoplasmic reticulum, leading to activation of a cell-rescue pathway, known as endoplasmic reticulum stress. The discovery of endoplasmic reticulum stress-related biomarkers leads to the realization of the importance of the toxicity of nanostructures' endoplasmic reticulum. Nanotoxicity of several nanomaterials like silver, gold, zinc oxide, polymeric nanoparticles, and anodic alumina nanotubes on endoplasmic reticulum has been determined utilizing this biomarker. Interrupting the functioning of the endoplasmic reticulum can also lead to the disruption of cellular homeostasis due to the upregulation of cytosolic Ca^{2+} and ROS levels [70–73].

13.5 Effect of Nanotoxicity on Different Organs

Due to their size, nanoparticles can reach almost any tissue of the body and cause toxicity. Nanoparticles having particle size less than 10 nm can cross almost all biological barriers. The most significant organ that are affected by nanotoxicity are lungs, liver, kidneys, brain, etc. (Figure 13.2). In this section, we will discuss them one by one.

13.5.1 Neurotoxicity

The brain and spinal cord together are known as the central nervous system. The spinal cord is connected to the brainstem and runs through the spinal canal through the vertebral column. The central nervous system is very delicate and is protected by the skull and the vertebral column from injury and xenobiotics. The brain plays a central role in the control of most bodily functions, including awareness, movements, sensations, thoughts, speech, and memory. Some reflex actions can occur via spinal cord pathways without the participation of the brain. Studies showed that nanoparticles coated with surfactants and polyethylene glycols can cross the blood–brain barrier and accumulate in the brain on administering intravenously [74, 75]. Though there are fewer studies to prove

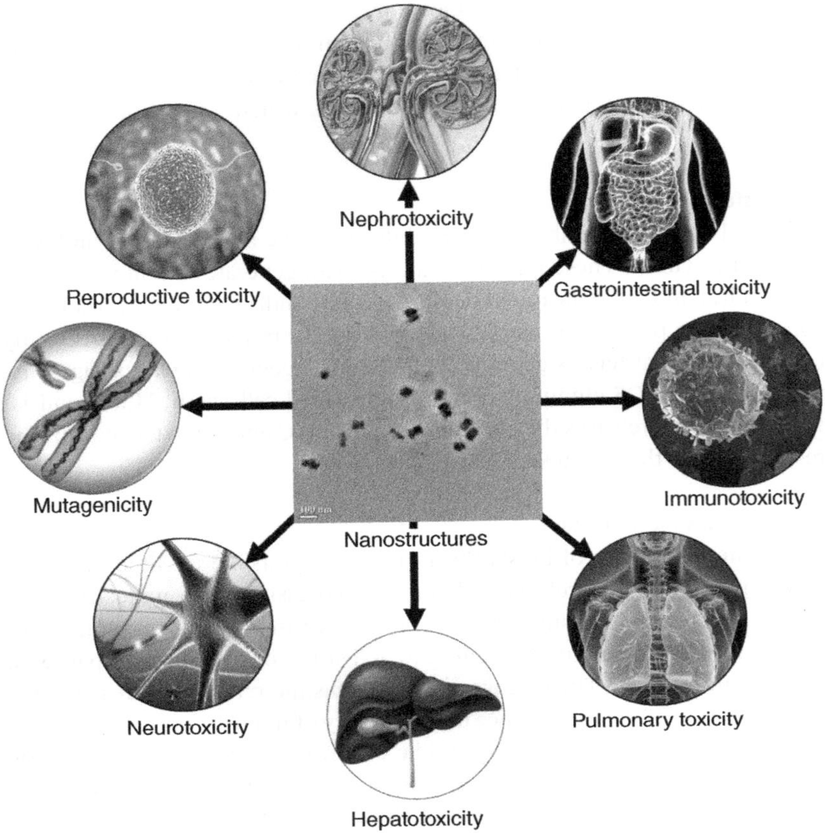

Figure 13.2 Human organs primarily affected by nanotoxicity. *Source:* Crystal light/Adobe Stock, PIC4U/Adobe Stock, dermatzke/Adobe Stock, yodiyim/Adobe Stock, BillionPhotos.com/Adobe Stock, Giovanni Cancemi/Adobe Stock, and abhijith3747/Adobe Stock.

the toxicity, on entering the brain, nanoparticles can show toxicity due to their physicochemical properties, such as large surface area, size, shape, and chemical composition [76, 77].

To use in MRI, FDA has approved colloidal dispersions of superparamagnetic iron oxide nanoparticles (SPIONs). They are widely used for biological, biomedical, and diagnostic purposes for their magnetic properties. Though SPIONs are used on a regular basis, they can induce ROS generation, impair mitochondrial functioning, and cause leakage of the enzyme lactate dehydrogenase (LDH), which can cause neurotoxicity and even worsen pre-existing neuronal damages [78].

13.5.2 Immunotoxicity

The effect of nanoparticles on the immune system has been studied and it has been reported that hydrophilicity, lipophilicity, catalytic activity, composition, electronic structure, capacity to bind or coat surface species, solubility, dimension, surface area are the main factors that contribute to the interactions of nanoparticles with biological tissues and immune system. Dendritic cells take up and process nanoparticles. Hence, they can accumulate in the lymph nodes. Dendritic cells can change the antigenicity of the nanoparticles and altered immune responses leading to autoimmunity. Some nanoparticles can induce contact dermatitis. In vitro studies have shown that nanoparticles can induce cytokine production. Though nanoparticles cannot act as a hapten and induce specific IgE production, they can act as adjuvant and induce the production of a specific type of cytokines causing allergic sensitization. Nanoparticles can also act as pro-inflammatory agents in the lung causing increased expression of IL-1β, MIP-1α, MCP-1, MIP-2, keratinocyte chemoattractant, TARC, GM-CSF, and MIP-1α and activation of the stress-activated MAPKs p38 and JNKs. Data suggest that through the activation of an oxidative stress mechanism, nanoparticles may contribute to pro-inflammatory disease processes in the lungs [79, 80].

13.5.3 Hepatotoxicity

Most of the xenobiotics that enter the systemic circulation are metabolized by the liver [81]. Because of this, the liver is particularly vulnerable to nanoparticle toxicity and can accumulate administered substances, even long after the exposure ends. Hence, determining the toxicity of hepatic cells is very important. To determine the hepatotoxicity on the nanostructures, lipid peroxidation assay, comet assay, and oxidative DNA damage studies are done [82]. The hepatotoxicity of gold nanoparticles, silver nanoparticles, and silica nanoparticles has been intensively studied for their safe clinical applications. The structure and surface modification of nanoparticles are important factors in their interaction with hepatic cells. The liberation of metal ions and oxidative environments are the other reasons for the hepatotoxicity of the metal nanoparticles [83]. The long half-life of the metal nanoparticles has implications for organ toxicity, particularly because of the liver's untoward propensity to heavy metal ion poisoning, which makes exposure to metal nanoparticles potentially very hazardous [84].

13.5.4 Pulmonary Toxicity

Airborne nanoparticles can enter the lungs very easily during respiration. Lungs are by far the most important portal of entry for nanoparticles into the human body as they can reach the alveoli and then to the systemic circulation. During respiration, nanoparticles can also deposit on the olfactory epithelium and directly be translocated to the brain [85]. Lungs-targeted drug delivery systems have been developed using organic nano-polymers including collagen, gelatine, chitosan, alginate, and bovine serum albumin (BSA) to treat diseases like asthma, chronic obstructive pulmonary disease (COPD), tuberculosis (TB), and lung cancer as well as extrapulmonary conditions such as diabetes. However, exposure to nanoparticles prepared with nonbiodegradable materials like silica, metals, etc., can cause severe toxicity to lung tissues. The phagocytic capacity of macrophages is weakened due to the nanotoxicity, which ultimately reduces the lung's clearance ability. Due to this reason, the nanoparticles can reside inside the lung for longer durations [86]. Silica was previously extensively used in the chemical industry, for mechanical polishing, and as an excipient with drug formulations, cosmetics, printer toners, varnishes, and food. In recent times, the use of silica nanoparticles for biomedical and biotechnological applications has increased. They are used as biosensors for simultaneous assay of glucose, lactate, L-glutamate, and hypoxanthine levels in animal models. They are also used as biomarkers for leukaemia cell identification using optical microscopy imaging, cancer therapy, DNA delivery, drug delivery, and enzyme immobilization. When administered at a lower dose, silica nanoparticles show low toxicity. However, they can agglomerate and can cause protein aggregation. Oxidative stress is the main reason for the toxicity of silica nanoparticles [87–89].

13.5.5 Reproductive Toxicity

From the study results on the absorption and distribution of nanoparticles, nanostructures can cross various biological barriers, which are not permeable to other microparticulate materials. Their ability to cross the blood–brain barrier has been proven [90]. Studies have proven the potential danger that nanoparticles have on the foetus, the transfer of nanoparticles across the placenta, including excretion via breast milk by crossing the blood–milk barrier, which could lead to embryotoxicity as a result of exposure to nanoparticles [91]. Though there is the unavailability of robust data addressing the distribution of nanoparticles to the reproductive cells, animal studies show that nanoparticles can show toxic effects on reproductive cells. In addition, no clear data showing the distribution of nanoparticles in the foetus are available. This leads to the recommendation that nephrotoxicity needs to be carefully considered when there is evidence for NP passage of the placenta [92].

13.5.6 Mutagenicity

Once the nanoparticles get inside a cell, it does not appear to be membrane-bound and have direct access to the intracellular proteins, organelles, and DNA of the cell, easily causing toxicity to them. Nanostructures having particle sizes below 10 nm may show genotoxicity by directly entering the cell nucleus and altering the working of DNA, causing chromosomal fragmentation, breakage of DNA strands, point mutations, oxidative DNA adducts, and alterations in gene expression profiles. On the other hand, nanostructures having larger particle sizes may meet DNA during the mitosis phase of the cell cycle when the nuclear membrane dissolves [58, 59]. The interaction of nanoparticles with cell components is needed to be studied well to know the level of toxicity that the nanoparticles have caused to the DNA and other cell components [93].

13.5.7 Nephrotoxicity

Through a process of translocation across biological barriers, nanoparticles can reach different organs other than the target organ and may induce adverse biological reactions. The kidney is one of the highest perfused organs of the human body. A nanoparticle that has reached the systemic circulation is bound to reach the kidney and renal excretion is an expected and possible elimination route of nanoparticles in living organisms. For this reason, kidneys are particularly susceptible to the toxicity of nanoparticles. Studies with carbon, metal, and silica nanoparticles in vitro revealed that different types of nanoparticles can exert significant cytotoxic effects on kidney cells, like decreased cell viability, induction of oxidative stress, mitochondrial or cytoskeleton dysfunction, and cell membrane and DNA damage. In vivo studies have shown that nanoparticles can cause nephrotoxicity at the tubular level, i.e. degeneration of tubular epithelial cells, cellular fragments, and proteinaceous liquid in tubule lumen, renal interstitial fibrosis, and in glomerular level, i.e. swollen glomeruli, changes in Bowman's space and proliferation of mesangial cells. Although the data currently available indicate that nanoparticles may adversely impact the renal system, further studies are needed to clarify all the potential molecular mechanisms of nephrotoxicity induced by these xenobiotics, in particular, at the glomerular level [94–96].

13.5.8 Skin Irritation

The skin is the first-line barrier between the external environment and the internal organs of the human body. It is exposed to various potential toxic materials of the environment as well as cosmetics, sprays, or clothing. A portion of the topically applied nanoparticles can penetrate the skin and reach the systemic circulation and exert adverse effects. Commonly used nanoparticles in cosmetics like TiO_2 can cause toxicity to skin cells affecting cellular functions such as cell proliferation, differentiation, motility, and apoptosis. Similarly, exposure to metal nanoparticles may cause skin irritation, swelling and inflammation in the exposed area [97–99].

13.5.9 Gastrointestinal Toxicity

Nanoparticles can reach the GIT with food, water, cosmetics, drugs, and drug delivery devices, or after mucociliary clearance from the respiratory tract. The complex mixture of enzymes, foods, and bacteria of the GIT can interact with ingested nanoparticles and reduce their toxicity. In vitro studies showed that nanoparticles are less cytotoxic in a medium with high protein content. Nanoparticles have been constantly found in the colon tissue of people affected by cancer, Crohn's disease, and ulcerative colitis, while in healthy humans, nanoparticles were absent. The nanoparticles present in diseased persons had various chemical compositions and are not considered toxic in bulk form. Microscopic and energy-dispersive spectroscopy analysis of colon mucosa of the diseased persons indicated the presence of carbon, ceramic filo-silicates, gypsum, sulfur, calcium, silicon, stainless steel, silver, and zirconium. The size of those nanoparticles varied from 50 to 100 000 nm (0.05–100 μ). The particles were found at the interface between healthy and cancerous tissues. Based on these findings, it was suggested that the gastrointestinal barrier is not efficient for particles smaller than 20 000 nm (200 μ) [100–102].

13.6 Strategies for Making Nontoxic Nano-Carriers

The maximum tolerance doses (MTDs) of nano-carriers are governed by the toxicity of the carrier. Knowledge of the effects of different parameters like surface charge, surface chemistry, etc., on the toxicity of the nano-carriers helps in the successful design of safe nanomaterials. As different types of nanomaterials exhibit toxicity by different mechanisms, strategies to reduce the toxicity of nanomaterials should be discussed on a case-by-case basis.

13.6.1 Nondegradable Inorganic Nanoparticles

Nondegradable inorganic nanoparticles like gold, iron, silica nanoparticles, and quantum dots show toxicity by mainly three mechanisms [103]:

- Organelle dysfunctions are caused by the intracellular accumulation of nanoparticles.
- Oxidative stress due to ROS generation.
- Dissolution of toxic ions.

Studies have proven that nanotoxicity can be minimized by considering different measures while preparing the nanoformulations. For example, to avoid the accumulation of nondegradable nanoparticles inside cells and organs, the sizes and hydrodynamic diameters of nanoparticles should be less than 10 nm to permit complete and rapid elimination from the body. Moreover, an optimized dose of the nanoparticle according to their corresponding MTDs can also prevent the accumulation [104, 105].

Similarly, coating the prepared nanoparticles with biocompatible materials like PEG can prevent ROS generation, nanoparticle aggregation, and dissolution of toxic ions from the nanoparticles. PEGs are FDA-approved polymers and are widely used as a pharmaceutical excipient in different types of formulations. PEGylation can prevent the nanoparticles from coming into direct contact with the biological interfaces. Covalent and non-covalent functionalization of nanoparticle surfaces has also been widely used in the past decades. The surface coating of nanomaterials not only makes them biocompatible but also enables them for attaching other chemical groups for preparing multimodal nanomaterials to utilize in bio-imaging, disease diagnosis, cell targeting, and stimuli-responsive drug release [106].

13.6.2 Biodegradable Polymeric Nanoparticles

Polymeric nanoparticles can undergo biodegradation easily forming different metabolites. For example, upon degradation of PLGA nanoparticles, lactic acid and glycolic acid are produced, which are eliminated by metabolic reaction cycles without inducing significant toxicity in optimum doses. But at higher doses, ROS generation and pro-inflammatory response can be observed in vitro [107]. But the extent of toxicity mainly depends on the raw material used in the production of the nanoparticles as the type of metabolite that is produced upon biodegradation depends on it. For example, the cationic nanoparticles prepared from lipids, like liposomes, solid lipid nanoparticles (SLNs), nanostructured lipid carriers (NLCs), and cationic polymers can trigger toxicities by acidifying the endosomal–lysosomal compartment through the proton sponge effect. The dysfunction of liposome structure has the potential to cause other cellular organelle stresses and trigger inflammation as well as programmed cell death [108].

According to the studies, reducing the cationic charge density on the nanoparticles by replacing the small molecule end-groups of lipid chains can minimize the toxicity of cationic nanoparticles [109]. The cationic charge on the nanoparticles can be reduced by adding hydrophilic monomers and/or polymers, like PEG and prevent unfavorable interactions of the nanoparticles with serum proteins and red blood cells [110].

13.6.3 Precise Manipulation of Protein Corona

The reduction of particle size increases the surface free energy of particles. Nano-sized materials possess very high free energy on their surface. This surface free energy leads to the adsorption of biomolecules like proteins to their surfaces and helps the nanoparticles achieve an equilibrium low-energy state. The formation of this protein layer on the nanoparticle surfaces is known as protein corona formation. Depending on the affinity and binding strength of proteins to the surfaces of nanoparticles, protein corona can be of two types, hard corona and soft corona. Proteins with binding strength bind tightly to the nanoparticle surfaces, forming a relatively stable hard shell, while the proteins with loose binding strength form soft corona on top of the hard one. The proteins of the corona are not static but dynamic, which depends on the environments through which the nanomaterials are being transported. Moreover, the size, surface chemistry, and surface charge of protein corona are also different depending on the physiological properties of nanomaterials [111, 112].

The interactions of nanomaterials with biological barriers like the RES, blood–brain barrier, and blood–placental barrier are dominated primarily by this protein corona [113]. Hence, understanding the impact of materials used in the formulation of nanostructures on the protein corona or the impact of biological conditions on the protein corona holds great significance. But so far, inconsistent reproducibility of nanostructures and bioanalytical results are creating difficulties in the study of protein-corona and bio-nano interactions. Some modern techniques like the nuclear analytical techniques, which have high-throughput and high-sensitivity characterizations need to be used to establish a complete library of protein coronas as a function of different nanomaterials and biological milieus. This will help precisely manipulate the protein corona to better control the biological responses like cell uptake, targeting, and immune response, etc., at the bio-nano interfaces [114].

Though protein corona increases the stability and biocompatibility of nanostructures in biological fluids, it may reduce its biodistribution. When immunoglobulins and complements bind with the nanostructures, it promotes macrophage phagocytosis and activates the immune system. Moreover, protein corona can also inhibit the recognition of ligands that are attached to nanostructure surfaces for targeting purposes. Hence, for in vivo long-circulating targeted nanostructures, corona formation should be reduced.

13.7 Conclusion and Future Perspective

This chapter has summarized the concepts, mechanisms, and recent progress in nanotoxicity research. The studies on the toxicity of nanomaterials both in vitro and in vivo have given us knowledge regarding the mechanisms of nanotoxicity and the pathway by which we can make the next generation of safer nanomedicines. Our increased understanding of nanotoxicity mechanisms can increase the commercial transition of nanomaterial-based drug delivery systems into the market for improving our well-being without being concerned about its toxic effects. This can help produce the next generation of nanomedicines with improved site-targeting, stimuli-responsive drug release, and multimodal capacities with better biocompatibility. Though efforts have been given to produce nanomedicines having lesser or minimum toxicity by proper selection of polymers and surfactants, or by coating them with biocompatible polymers, still so much needs to be done. Researchers have suggested that precise manipulation of protein corona formation by selecting appropriate materials to formulate the nanomedicines can alter nanotoxicity research in a big way [115]. For this, researchers must come forward to work together and make a library correlating the material selection for nanofabrication and characteristics of the corresponding protein corona formation. This may lead to the development of drug delivery techniques for personalized nanomaterial-based therapy which will be able to address the future requirements of individual patients [116].

References

1 Tahir, M.B., Rafique, M., and Rafique, M.S. (ed.) (2020). Nanostructure materials and their classification by dimensionality. In: *Nanotechnology and Photocatalysis for Environmental Applications*, 27–44. Elsevier.

2 Kellogg, C.A. and Griffin, D.W. (2006). Aerobiology and the global transport of desert dust. *Trends in Ecology & Evolution* 21: 638–644.

3 Smita, S., Gupta, S.K., Bartonova, A. et al. (2012). Nanoparticles in the environment: Assessment using the causal diagram approach. *Environmental Health: A Global Access Science Source* 11: S13.

4 Heiligtag, F.J. and Niederberger, M. (2013). The fascinating world of nanoparticle research. *Materials Today* 16: 262–271.

5 Walter, P., Welcomme, E., Hallégot, P. et al. (2006). Early use of PbS nanotechnology for an ancient hair dyeing formula. *Nano Letters* 6 (10): 2215–2219.

6 Jeevanandam, J., Barhoum, A., Chan, Y.S. et al. (2018). Review on nanoparticles and nanostructured materials: History, sources, toxicity and regulations. *Beilstein Journal of Nanotechnology* 9: 1050–1074.

7 Pandey, P. and Dahiya, M. (2016). A brief review on inorganic nanoparticles. *Journal of Critical Reviews* 3 (3): 18–26.

8 Shatrohan Lal, R.K. (2014). Synthesis of organic nanoparticles and their applications in drug delivery and food nanotechnology: a review. *Journal of Nanomaterials & Molecular Nanotechnology* 03 (4): 1–11.

9 Sahay, R., Reddy, V.J., and Ramakrishna, S. (2014). Synthesis and applications of multifunctional composite nanomaterials. *International Journal of Mechanical and Materials Engineering* 9 (1): 1–13.

10 Yah, C.S., Simate, G., and Iyuke, S.E. (2012). Nanoparticles toxicity and their routes of exposures. *Pakistan Journal of Pharmaceutical Sciences* 25: 477–491.

11 Mohapatra, S.S., Ranjan, S., Dasgupta, N. et al. (ed.) (2018). Nano drugs: a critical review of their patents and market. In: *Characterization and Biology of Nanomaterials for Drug Delivery: Nanoscience and Nanotechnology in Drug Delivery*, 523–551. Elsevier.

12 Wagner, W., Sakiyama-Elbert, S., Zhang, G., and Yaszemski, M. (ed.) (2020). Nanoparticles. In: *Biomaterials Science*, 453–483. Elsevier.

13 Ventola, C.L. (2012). The nanomedicine revolution: part 1: emerging concepts. *Pharmacy and Therapeutics* 37 (9): 512.

14 Jabr-Milane, L.S., van Vlerken, L.E., Yadav, S., and Amiji, M.M. (2008). Multi-functional nanocarriers to overcome tumor drug resistance. *Cancer Treatment Reviews* 34: 592–602.

15 Gnach, A., Lipinski, T., Bednarkiewicz, A. et al. (2015). Upconverting nanoparticles: assessing the toxicity. *Chemical Society Reviews* 44 (6): 1561–1584.

16 Griffitt, R.J., Luo, J., Gao, J. et al. (2008). Effects of particle composition and species on toxicity of metallic nanomaterials in aquatic organisms. *Environmental Toxicology and Chemistry* 27 (9): 1972–1978.

17 Gurr, J.R., Wang, A.S.S., Chen, C.H., and Jan, K.Y. (2005). Ultrafine titanium dioxide particles in the absence of photoactivation can induce oxidative damage to human bronchial epithelial cells. *Toxicology* 213 (1–2): 66–73.

18 Xia, T., Kovochich, M., Brant, J. et al. (2006). Comparison of the abilities of ambient and manufactured nanoparticles to induce cellular toxicity according to an oxidative stress paradigm. *Nano Letters* 6: 1794–1807.

19 Powers, K.W., Palazuelos, M., Moudgil, B.M., and Roberts, S.M. (2007). Characterization of the size, shape, and state of dispersion of nanoparticles for toxicological studies. *Nanotoxicology* 1: 42–51.

20 Aillon, K.L., Xie, Y., El-Gendy, N. et al. (2009). Effects of nanomaterial physicochemical properties on in vivo toxicity. *Advanced Drug Delivery Reviews* 61: 457–466.

21 Holgate, S.T. (2010). Exposure, uptake, distribution and toxicity of nanomaterials in humans. *Journal of Biomedical Nanotechnology* 6: 1–19.

22 De Jong, W.H., Hagens, W.I., Krystek, P. et al. (2008). Particle size-dependent organ distribution of gold nanoparticles after intravenous administration. *Biomaterials* 29 (12): 1912–1919.

23 Bansal, V., Li, V., O'Mullane, A.P., and Bhargava, S.K. (2010). Shape dependent electrocatalytic behaviour of silver nanoparticles. *CrystEngComm* 12 (12): 4280–4286.

24 Burda, C., Chen, X., Narayanan, R., and El-Sayed, M.A. (2005). Chemistry and properties of nanocrystals of different shapes. *Chemical Reviews* 105: 1025–1102.

25 Lee, M.K., Lim, S.J., and Kim, C.K. (2007). Preparation, characterization and in vitro cytotoxicity of paclitaxel-loaded sterically stabilized solid lipid nanoparticles. *Biomaterials* 28 (12): 2137–2146.

26 Champion, J.A. and Mitragotri, S. (2006). Role of target geometry in phagocytosis. *Proceedings of the National Academy of Sciences of the United States of America* 103 (13): 4930–4934.

27 Park, K.H., Chhowalla, M., Iqbal, Z., and Sesti, F. (2003). Single-walled carbon nanotubes are a new class of ion channel blockers. *Journal of Biological Chemistry* 278 (50): 50212–50216.

28 Hsiao, I.L. and Huang, Y.J. (2011). Effects of various physicochemical characteristics on the toxicities of ZnO and TiO_2 nanoparticles toward human lung epithelial cells. *Science of the Total Environment* 409 (7): 1219–1228.

29 Sahay, G., Alakhova, D.Y., and Kabanov, A.V. (2010). Endocytosis of nanomedicines. *Journal of Controlled Release* 145: 182–195.

30 Hoshino, A., Fujioka, K., Oku, T. et al. (2004). Physicochemical properties and cellular toxicity of nanocrystal quantum dots depend on their surface modification. *Nano Letters* 4 (11): 2163–2169.

31 Georgieva, J.V., Kalicharan, D., Couraud, P.O. et al. (2011). Surface characteristics of nanoparticles determine their intracellular fate in and processing by human blood–brain barrier endothelial cells in vitro. *Molecular Therapy* 19 (2): 318–325.

32 Goodman, C.M., McCusker, C.D., Yilmaz, T., and Rotello, V.M. (2004). Toxicity of gold nanoparticles functionalized with cationic and anionic side chains. *Bioconjugate Chemistry* 15 (4): 897–900.

33 Yin, H., Too, H.P., and Chow, G.M. (2005). The effects of particle size and surface coating on the cytotoxicity of nickel ferrite. *Biomaterials* 26 (29): 5818–5826.

34 Risom, L., Møller, P., and Loft, S. (2005). Oxidative stress-induced DNA damage by particulate air pollution. *Mutation Research: Fundamental and Molecular Mechanisms of Mutagenesis* 592: 119–137.

35 Brown, D.M., Wilson, M.R., MacNee, W. et al. (2001). Size-dependent proinflammatory effects of ultrafine polystyrene particles: a role for surface area and oxidative stress in the enhanced activity of ultrafines. *Toxicology and Applied Pharmacology* 175 (3): 191–199.

36 Gupta, A.K. and Gupta, M. (2005). Cytotoxicity suppression and cellular uptake enhancement of surface modified magnetic nanoparticles. *Biomaterials* 26 (13): 1565–1573.

37 Buzea, C., Pacheco, I.I., and Robbie, K. (2007). Nanomaterials and nanoparticles: sources and toxicity. *Biointerphases* 2 (4): MR17–MR71.

38 Donaldson, K., Aitken, R., Tran, L. et al. (2006). Carbon nanotubes: a review of their properties in relation to pulmonary toxicology and workplace safety. *Toxicological Sciences* 92 (1): 5–22.

39 Simeonova, P.P. and Erdely, A. (2009). Engineered nanoparticle respiratory exposure and potential risks for cardiovascular toxicity: predictive tests and biomarkers. *Inhalation Toxicology* 21: 68–73.

40 Bergamaschi, E. (2009). Occupational exposure to nanomaterials: present knowledge and future development. *Nanotoxicology* 3: 194–201.

41 Srivastava, V., Gusain, D., and Sharma, Y.C. (2015). Critical review on the toxicity of some widely used engineered nanoparticles. *Industrial and Engineering Chemistry Research* 54 (24): 6209–6233.

42 Mamalis, A.G. (2007). Recent advances in nanotechnology. *Journal of Materials Processing Technology* 181 (1–3 Special Issue): 52–58.

43 Kirchner, C., Liedl, T., Kudera, S. et al. (2005). Cytotoxicity of colloidal CdSe and CdSe/ZnS nanoparticles. *Nano Letters* 5 (2): 331–338.

44 Ghosh, M., Manivannan, J., Sinha, S. et al. (2012). In vitro and in vivo genotoxicity of silver nanoparticles. *Mutation Research, Genetic Toxicology and Environmental Mutagenesis* 749 (1–2): 60–69.

45 Luque-Garcia, J.L., Sanchez-Díaz, R., Lopez-Heras, I. et al. (2013). Bioanalytical strategies for in-vitro and in-vivo evaluation of the toxicity induced by metallic nanoparticles. *TrAC Trends in Analytical Chemistry* 43: 254–268.

46 Beheshti, N., Soflaei, S., Shakibaie, M. et al. (2013). Efficacy of biogenic selenium nanoparticles against Leishmania major: in vitro and in vivo studies. *Journal of Trace Elements in Medicine and Biology* 27 (3): 203–207.

47 Sayes, C.M., Reed, K.L., and Warheit, D.B. (2007). Assessing toxicology of fine and nanoparticles: comparing in vitro measurements to in vivo pulmonary toxicity profiles. *Toxicological Sciences* 97 (1): 163–180.

48 Zorov, D.B., Juhaszova, M., and Sollott, S.J. (2014). Mitochondrial reactive oxygen species (ROS) and ROS-induced ROS release. *Physiological Reviews* 94: 909–950.

49 Sharma, P., Jha, A.B., Dubey, R.S., and Pessarakli, M. (2012). Reactive oxygen species, oxidative damage, and antioxidative defense mechanism in plants under stressful conditions. *Journal of Botany* 2012: 1–26.

50 Murphy, M.P. (2009). How mitochondria produce reactive oxygen species. *Biochemical Journal* 417: 1–13.

51 Tripathy, B.C. and Oelmüller, R. (2012). Reactive oxygen species generation and signaling in plants. *Plant Signaling & Behavior* 7 (12): 1621.

52 Janku, M., Luhová, L., and Petrivalský, M. (2019). On the origin and fate of reactive oxygen species in plant cell compartments. *Antioxidants* 8: 105.

53 Fukai, T. and Ushio-Fukai, M. (2011). Superoxide dismutases: role in redox signaling, vascular function, and diseases. *Antioxidants and Redox Signaling* 15: 1583–1606.

54 Reuter, S., Gupta, S.C., Chaturvedi, M.M., and Aggarwal, B.B. (2010). Oxidative stress, inflammation, and cancer: how are they linked? *Free Radical Biology and Medicine* 49: 1603–1616.

55 Manke, A., Wang, L., and Rojanasakul, Y. (2013). Mechanisms of nanoparticle-induced oxidative stress and toxicity. *BioMed Research International* 2013: 942916.

56 Dayem, A.A., Hossain, M.K., Bin, L.S. et al. (2017). The role of reactive oxygen species (ROS) in the biological activities of metallic nanoparticles. *International Journal of Molecular Sciences* 18 (1): 120.

57 Nita, M. and Grzybowski, A. (2016). The role of the reactive oxygen species and oxidative stress in the pathomechanism of the age-related ocular diseases and other pathologies of the anterior and posterior eye segments in adults. *Oxidative Medicine and Cellular Longevity* 2016: 1–23.

58 Tan, B.L., Norhaizan, M.E., Liew, W.P.P., and Rahman, H.S. (2018). Antioxidant and oxidative stress: A mutual interplay in age-related diseases. *Frontiers in Pharmacology* 9: 1162.

59 Kwon, J.Y., Koedrith, P., and Seo, Y.R. (2014). Current investigations into the genotoxicity of zinc oxide and silica nanoparticles in mammalian models in vitro and in vivo: carcinogenic/genotoxic potential, relevant mechanisms and biomarkers, artifacts, and limitations. *International Journal of Nanomedicine* 9: 271–286.

60 Singh, N., Manshian, B., Jenkins, G.J.S. et al. (2009). Nanogenotoxicology: the DNA damaging potential of engineered nanomaterials. *Biomaterials* 30: 3891–3914.

61 Liu, N. and Tang, M. (2020). Toxic effects and involved molecular pathways of nanoparticles on cells and subcellular organelles. *Journal of Applied Toxicology* 40 (1): 16–36.

62 Villamil Giraldo, A.M., Appelqvist, H., Ederth, T., and Öllinger, K. (2014). Lysosomotropic agents: impact on lysosomal membrane permeabilization and cell death. *Biochemical Society Transactions* 42 (5): 1460–1464.

63 Ashfaq, U.A., Javed, T., Rehman, S. et al. (2011). Lysosomotropic agents as HCV entry inhibitors. *Virology Journal* 8: 163.

64 Ma, X., Wu, Y., Jin, S. et al. (2011). Gold nanoparticles induce autophagosome accumulation through size-dependent nanoparticle uptake and lysosome impairment. *ACS Nano* 5 (11): 8629–8639.

65 Ramachandran, P.V., Savini, M., Folick, A.K. et al. (2019). Lysosomal signaling promotes longevity by adjusting mitochondrial activity. *Developmental Cell* 48 (5): 685–696.e5.

66 Ringwood, A.H., Levi-Polyachenko, N., and Carroll, D.L. (2009). Fullerene exposures with oysters: embryonic, adult, and cellular responses. *Environmental Science and Technology* 43 (18): 7136–7141.

67 Boya, P. and Kroemer, G. (2008). Lysosomal membrane permeabilization in cell death. *Oncogene* 27: 6434–6451.

68 Gallud, A., Klöditz, K., Ytterberg, J. et al. (2019). Cationic gold nanoparticles elicit mitochondrial dysfunction: a multi-omics study. *Scientific Reports* 9 (1): 4366.

69 Chen, Q., Wang, N., Zhu, M. et al. (2018). $TiO_{(2)}$ nanoparticles cause mitochondrial dysfunction, activate inflammatory responses, and attenuate phagocytosis in macrophages: a proteomic and metabolomic insight. *Redox Biology* 15: 266–276.

70 Fiers, W., Beyaert, R., Declercq, W., and Vandenabeele, P. (1999). More than one way to die: apoptosis, necrosis and reactive oxygen damage. *Oncogene* 18: 7719–7730.

71 Onoda, A., Kawasaki, T., Tsukiyama, K. et al. (2020). Carbon nanoparticles induce endoplasmic reticulum stress around blood vessels with accumulation of misfolded proteins in the developing brain of offspring. *Scientific Reports* 10 (1): 10028.

72 Christen, V. and Fent, K. (2016). Silica nanoparticles induce endoplasmic reticulum stress response and activate mitogen activated kinase (MAPK) signaling. *Toxicology Reports* 3: 832–840.

73 Yu, K.N., Chang, S.H., Park, S.J. et al. (2015). Titanium dioxide nanoparticles induce endoplasmic reticulum stress-mediated autophagic cell death via mitochondria-associated endoplasmic reticulum membrane disruption in normal lung cells. *PLoS ONE* 10 (6): e0131208.

74 Clift, M.J.D., Varet, J., Hankin, S.M. et al. (2011). Quantum dot cytotoxicity in vitro: an investigation into the cytotoxic effects of a series of different surface chemistries and their core/shell materials. *Nanotoxicology* 5 (4): 664–674.

75 Olivier, J.C. (2005). Drug transport to brain with targeted nanoparticles. *NeuroRx* 2 (1): 108–119.

76 Sharma, H.S. and Sharma, A. (2007). Nanoparticles aggravate heat stress induced cognitive deficits, blood–brain barrier disruption, edema formation and brain pathology. *Progress in Brain Research* 162: 245–273.

77 Kim, Y.S., Kim, J.S., Cho, H.S. et al. (2008). Twenty-eight-day oral toxicity, genotoxicity, and gender-related tissue distribution of silver nanoparticles in Sprague-Dawley rats. *Inhalation Toxicology* 20 (6): 575–583.

78 Häfeli, U.O., Riffle, J.S., Harris-Shekhawat, L. et al. (2009). Cell uptake and in vitro toxicity of magnetic nanoparticles suitable for drug delivery. *Molecular Pharmaceutics* 6 (5): 1417–1428.

79 Di Gioacchino, M., Petrarca, C., Lazzarin, F. et al. (2011). Immunotoxicity of nanoparticles. *International Journal of Immunopathology and Pharmacology* 24: 75–94.

80 Schulte, P., Leso, V., Niang, M., and Iavicoli, I. (2018). Biological monitoring of workers exposed to engineered nanomaterials. *Toxicology Letters* 298: 112–124.

81 Standaert, F.G. (1988). Absorption and distribution of xenobiotics. *Environmental Health Perspectives* 77: 63.

82 Abdelazeim, S.A., Shehata, N.I., Aly, H.F., and Shams, S.G.E. (2020). Amelioration of oxidative stress-mediated apoptosis in copper oxide nanoparticles-induced liver injury in rats by potent antioxidants. *Scientific Reports* 10 (1): 10812.

83 Li, X., Wang, L., Fan, Y. et al. (2012). Biocompatibility and toxicity of nanoparticles and nanotubes. *Journal of Nanomaterials* 2012: 548389.

84 Yildirimer, L., Thanh, N.T.K., Loizidou, M., and Seifalian, A.M. (2011). Toxicological considerations of clinically applicable nanoparticles. *Nano Today* 6: 585–607.

85 Paur, H.R., Cassee, F.R., Teeguarden, J. et al. (2011). In-vitro cell exposure studies for the assessment of nanoparticle toxicity in the lung – a dialog between aerosol science and biology. *Journal of Aerosol Science* 42 (10): 668–692.

86 Van Rijt, S.H., Bein, T., and Meiners, S. (2014). Medical nanoparticles for next generation drug delivery to the lungs. *European Respiratory Journal* 44 (3): 765–774.

87 Christen, V. and Fent, K. (2012). Silica nanoparticles and silver-doped silica nanoparticles induce endoplasmatic reticulum stress response and alter cytochrome P4501A activity. *Chemosphere* 87 (4): 423–434.

88 Yang, H., Liu, C., Yang, D. et al. (2009). Comparative study of cytotoxicity, oxidative stress and genotoxicity induced by four typical nanomaterials: the role of particle size, shape and composition. *Journal of Applied Toxicology* 29 (1): 69–78.

89 Chen, M. and Von Mikecz, A. (2005). Formation of nucleoplasmic protein aggregates impairs nuclear function in response to SiO_2 nanoparticles. *Experimental Cell Research* 305 (1): 51–62.

90 Sharma, H.S., Hussain, S., Schlager, J. et al. (2009). Influence of nanoparticles on blood–brain barrier permeability and brain edema formation in rats. *Acta Neurochirurgica* 106: 359–364.

91 Melnik, E.A., Buzulukov, Y.P., Demin, V.F. et al. (2013). Transfer of silver nanoparticles through the placenta and breast milk during in vivo experiments on rats. *Acta Naturae* 5 (18): 107–115.

92 Fujimoto, A., Tsukue, N., Watanabe, M. et al. (2005). Diesel exhaust affects immunological action in the placentas of mice. *Environmental Toxicology* 20 (4): 431–440.

93 Bouwmeester, H., Dekkers, S., Noordam, M.Y. et al. (2009). Review of health safety aspects of nanotechnologies in food production. *Regulatory Toxicology and Pharmacology* 53 (1): 52–62.

94 Iavicoli, I., Fontana, L., and Nordberg, G. (2016). The effects of nanoparticles on the renal system. *Critical Reviews in Toxicology* 46: 490–560.

95 Wang, H., Wang, J., Deng, X. et al. (2004). Biodistribution of carbon single-wall carbon nanotubes in mice. *Journal of Nanoscience and Nanotechnology* 4 (8): 1019–1024.

96 Curtis, J., Greenberg, M., Kester, J. et al. (2006). Nanotechnology and nanotoxicology: a primer for clinicians. *Toxicological Reviews* 25: 245–260.

97 Kiss, B., Bíró, T., Czifra, G. et al. (2008). Investigation of micronized titanium dioxide penetration in human skin xenografts and its effect on cellular functions of human skin-derived cells. *Experimental Dermatology* 17 (8): 659–667.

98 Pan, Z., Lee, W., Slutsky, L. et al. (2009). Adverse effects of titanium dioxide nanoparticles on human dermal fibroblasts and how to protect cells. *Small* 5 (4): 511–520.

99 Lademann, J., Weigmann, H.J., Rickmeyer, C. et al. (1999). Penetration of titanium dioxide microparticles in a sunscreen formulation into the horny layer and the follicular orifice. *Skin Pharmacology and Applied Skin Physiology* 12 (5): 247–256.

100 Gatti, A.M. (2004). Biocompatibility of micro- and nano-particles in the colon. Part II. *Biomaterials* 25 (3): 385–392.

101 da Silva, A.B., Miniter, M., Thom, W. et al. (2020). Gastrointestinal absorption and toxicity of nanoparticles and microparticles: Myth, reality and pitfalls explored through titanium dioxide. *Current Opinion in Toxicology* 19: 112–120.

102 Cooper, J.R., Stradling, G.N., Smith, H., and Breadmore, S.E. (1979). The reactions of 10 nanometre diameter plutonium-238 dioxide particles with rat lung fluid. *International Journal of Radiation Biology* 36 (5): 453–466.

103 Soenen, S.J., Parak, W.J., Rejman, J., and Manshian, B. (2015). (Intra)cellular stability of inorganic nanoparticles: effects on cytotoxicity, particle functionality, and biomedical applications. *Chemical Reviews* 115 (5): 2109–2135.

104 Choi, H.S., Ashitate, Y., Lee, J.H. et al. (2010). Rapid translocation of nanoparticles from the lung airspaces to the body. *Nature Biotechnology* 28 (12): 1300–1303.

105 Soo Choi, H., Liu, W., Misra, P. et al. (2007). Renal clearance of quantum dots. *Nature Biotechnology* 25 (10): 1165–1170.

106 Pokorski, J.K. and Steinmetz, N.F. (2011). The art of engineering viral nanoparticles. *Molecular Pharmaceutics* 8: 29–43.

107 Singh, R.P. and Ramarao, P. (2013). Accumulated polymer degradation products as effector molecules in cytotoxicity of polymeric nanoparticles. *Toxicological Sciences* 136 (1): 131–143.

108 Lv, H., Zhang, S., Wang, B. et al. (2006). Toxicity of cationic lipids and cationic polymers in gene delivery. *Journal of Controlled Release* 114: 100–109.

109 Sunshine, J.C., Peng, D.Y., and Green, J.J. (2012). Uptake and transfection with polymeric nanoparticles are dependent on polymer end-group structure, but largely independent of nanoparticle physical and chemical properties. *Molecular Pharmaceutics* 9 (11): 3375–3383.

110 Samal, S.K., Dash, M., Van Vlierberghe, S. et al. (2012). Cationic polymers and their therapeutic potential. *Chemical Society Reviews* 41 (21): 7147–7194.

111 Lima, T., Bernfur, K., Vilanova, M., and Cedervall, T. (2020). Understanding the lipid and protein corona formation on different sized polymeric nanoparticles. *Scientific Reports* 10 (1): 1–9.

112 Wandelf, K. (ed.) (2018). Cluster-assembled materials: from fabrication to function. In: *Encyclopedia of Interfacial Chemistry: Surface Science and Electrochemistry*, 417–427.

113 Gunawan, C., Lim, M., Marquis, C.P., and Amal, R. (2014). Nanoparticle-protein corona complexes govern the biological fates and functions of nanoparticles. *Journal of Materials Chemistry B* 2 (15): 2060–2083.

114 Neagu, M., Piperigkou, Z., Karamanou, K. et al. (2017). Protein bio-corona: critical issue in immune nanotoxicology. *Archives of Toxicology* 91: 1031–1048.

115 Corbo, C., Molinaro, R., Parodi, A. et al. (2016). The impact of nanoparticle protein corona on cytotoxicity, immunotoxicity and target drug delivery. *Nanomedicine* 11: 81–100.

116 Corbo, C., Molinaro, R., Tabatabaei, M. et al. (2017). Personalized protein corona on nanoparticles and its clinical implications. *Biomaterials Science* 5: 378–387.

14

Nanotherapeutics of Phytoantioxidants for Microbial Infections

Ravi Bhushan[1], *Sonal Upadhyay*[1], *and Mithun Rudrapal*[2]

[1] *Centre for Genetic Disorder, Institute of Science, Banaras Hindu University, Varanasi, Uttar Pradesh, India*
[2] *Department of Pharmaceutical Chemistry, Rasiklal M. Dhariwal Institute of Pharmaceutical Education and Research, Pune, Maharashtra, India*

14.1 Introduction

Microbes or microorganisms (from bacteria to multicellular parasites) are extensively present. They have evolved different complex molecular mechanisms, which not only promote to withstand unfavorable nutritional and environmental conditions but also against their competitors, which aid their survival [1]. These microbes though provide an advantage to life, still some lead to serious health issues. Their multiplication inside the host tissue can be termed an infection. Microbial infection-related diseases and complications impose a large threat to public health-leading mortality and morbidity worldwide. This era is witnessing phenomenal changes regarding the health demands of the populations, globally [2].

An increase in infectious health issues has been observed over the last couple of decades, including the ebolavirus, severe acute respiratory syndrome (SARS) virus, COVID-19 virus, pandemic influenza, and the H1N1 epidemic influenza [3]. These microbial diseases have spread rapidly across the globe, revealing the exceptional connectivity indicating modern advancement. Although few countries have stepped toward significant world's surveillance improvement for emerging infectious diseases (EIDs), most of the low- and middle-income countries (LMICs) including fragile healthcare units and numerous systems-based bottlenecks persist vulnerable to the epidemic and thus experience economic and social burdens when they are declared [4].

Evidence regarding the bacterial infection involved in the growth and expansion of several diseases encompassing TB, tetanus, bowel disease, different types of cancer, and cholera has been demonstrated through several findings. Though many improvements involving antibiotic discovery have reduced the burden of infections caused by bacteria, pathogenic pandemics still pose a threat to the global population frequently. The usual example is Haiti cholera epidemic during 2010–2011, where more than thousands were hospitalized and around 4672 individuals died [5].

All the organs of the human body are prone to bacterial infections. For example, *Staphylococcus aureus* is carried on people's skin and mucous membranes causing skin and soft tissue contagions. Likewise, *Neisseria meningitidis* causes meningitis by infecting the meninges, the sheath of the central nervous system. *Salmonella typhi*, a common example of a bacterial infection that causes

Phytoantioxidants and Nanotherapeutics, First Edition. Edited by Mithun Rudrapal.

typhoid fever where individuals being unaware of the situation can become prolonged carriers of the infection. Salmonella lurk sometimes in quiescent conditions or even in a permanent state in different organs like the gallbladder [6].

Bacterial infections can easily be transmitted to human beings either through direct bacterial inhalation from surroundings or through contact with animals or pets or even through insect (acting as a pathogenic vector) bites transmitting the pathogen into the human (as a host). Salmonella causes diarrhoea, which occurs through ingestion of undercooked infected chicken or during turtle handling where hands get contaminated with their fecal matter. *Francisella tularensis* causes tularemia disease in individuals recently involved in skinning a rabbit. Correspondingly, anthrax-like infectious disease caused by *Bacillus anthracis* spreads through spores entry into the wound of an individual or through spores inhalation that are present on hides or dead animals. Similarly, spirochete *Borrelia* leads to Lyme disease that gets transmitted to humans from white-footed mice via deer ticks [7].

Not only bacteria but fungi and eukaryotic microbes are also found to exist on every mucosal and epithelial surface of humans. Failure in public health status also influences fungal contagions negatively. The epidemic of *E. rostratum* infected around 750 individuals, of which 60 patients died. Also, a perception regarding climatic change directly impacts the fungus's ability to cause harm to different body parts of humans [8].

Recent findings suggested that the pathogen *Candida auris*, a multidrug-resistant organism, has provoked a serious threat to individual health globally, found resistant to antifungal drugs, and developing virulence against the host. The latest estimation suggests the occurrence of around 200 000 cases of cryptococcal meningitis, 500 000 cases of *Pneumocystis jiroveci* pneumonia, 700 000 cases of invasive candidiasis, 10 million cases of fungal asthma, and 1 million cases of fungal keratitis globally [9]. These different examples of microbial infections illustrate that concrete steps need to be implemented with combined research and technological advancement to improve the world's health scenario and to create awareness among individuals regarding the widespread infectious disease to improve public health status globally.

Recently, the accessible therapies are limited for the treatment of microbial infections, intended for treating only the symptoms. Adverse effects or contradictions were found associated with these medications that make it so difficult for them in combating microbial infections. Since the existing infectious disease therapeutics are found incapable to delay or stop the spread of infections, so there is a requirement for better therapeutics strategies in controlling microbial infections [10]. In this context, where multi-target action is favorable in averting microbial infections, mother nature endowed us with innumerable phytoantioxidants. These phytoconstituents exhibit anti-inflammatory, antiapoptotic, antioxidant, and antimicrobial efficacy, which prove them as promising candidates for microbial infections.

14.2 Microbial Diseases – Diagnosis and Progression

Microbial infections continue to appear, persist, and reemerge. Some microorganisms cause novel identified human diseases; others are formerly known microbes that are contagious to larger or new groups of population or outspreading into new geographic locations. The occurrence and spread of treats driven by microbes are carried out by a different set of factors, the combination of which can result into the outspread of microbial diseases much powerful than any unique factor might indicates. Biological and genetic factors provide microorganisms to change and adapt, and tend to make humans nearly susceptible to contagions. Alteration in the environment physically can adversely affect the ecology of animal reservoirs and vectors, the transmissibility of microorganisms, and the human activities that unveil them to particular threats. In the occurrence of infectious diseases, human (either individually or collective) behavior is possibly the most composite factor.

14.2.1 Microbial Infections

Various microbes lead to different infectious diseases in humans that pose serious health issues, some of which are discussed in this context.

14.2.1.1 *Candida albicans*

Candida albicans – a fungal disease causal microorganism in humans. It is asymptomatically found colonizing the oral cavity, skin, reproductive tract, gastrointestinal (GI) tract areas of most humans because of being a part of the healthy microbiota [11–14]. In healthy individuals with strong immune systems, *C. albicans* maintains balance with local microbiota members and are sometimes harmless. However, modifications in the host because of antibiotics, or local environmental changes (such as changes in pH levels or variation in nutritional contents), modification in immunosuppressant therapy, infections by other microorganisms or stress can lead *C. albicans* to increase in its number and cause fungal infections. The widespread of these fungal infections varies from the dermal site (due to diaper rash, thrush, and vaginal yeast infection) to disseminated contagions hematogenously with a sizable morbidity rate (approximately 40% in some scenarios) [15–17]. *C. albicans* pose a serious threat especially in immunocompromised patients (like patients undergoing either anticancer or AIDS or immunosuppressive therapies) and in healthy individuals with implanted biomedical devices [18, 19].

14.2.1.2 *Listeria monocytogenes*

The intracellular, ubiquitous Gram-positive pathogen *Listeria monocytogenes*, is a causative microorganism in different epidemics of foodborne diseases and involved for the past 10 years. Listeriosis mainly occurs in pregnant ladies with an approximately 24% mortality rate, along with their fetuses, and also in immunosuppressed individuals, with symptoms of meningitis, abortion, and septicaemia. The major cases of listeriosis in humans are seen in individuals with an underlying problem, which causes T-cell immunity suppression in them. Intracellular facultative microbes such as *L. monocytogenes* should possess routes of overriding the nonspecific immune responses caused due to activated phagocytes. Survival of these microorganisms inside the inactivated mononuclear cells of the phagocytic system, however, is killed during macrophage activation [20]. The formation of superoxide, a toxic-free radical, is a vital part of the series in bacterial phagocytic killing. The superoxide dismutase in bacteria leads to a defense mechanism against the toxic-free radical molecule and thus proves to be a feasible virulence agent of the organism [21].

14.2.1.3 Hepatitis B Virus

It is a prototype of a small DNA virus belonging to the Hepadnaviridae family similar to retroviruses with unusual features. The contagious Hepatitis B Virus (HBV) virion (Dane particle) possesses a 42 nm-diameter spherical-shaped, double-membraned structure, covered with a lipid consisting of HBsAg envelope that covers an inner nucleocapsid formed of hepatitis B core antigen (HBcAg) complexed with DNA viral genome and virally encoded polymerase [22, 23]. Around one-third of people with acute HBV contagions present with clinical signs and symptoms of hepatitis that vary from mild symptoms of nausea and fatigue constitutionally, to more chronic signs such as jaundice, acute liver failure. The incubation time period of acute hepatitis B ranges from a mean of two to three months and after exposure can differ from one to six months clinically [24].

14.2.1.4 *Clostridium botulinum*

Clostridium botulinum, a spore-forming bacterium produces exotoxin that causes botulism. Out of the seven various toxin types, A, B, and E types are implicated commonly in foodborne botulism outbreaks. The most potent toxins released by *Clostridium botulinum* are neurotoxins that are extremely toxic and occur in implicated edible items generally at low concentrations.

The occurrence of botulism in humans and animals causes neuroparalysis that results due to the actions of neurotoxins released by *Clostridium botulinum* and also by *Clostridium butyricum*- and *Clostridium baratii*-like rare bacterial strains. The organisms producing botulinum are widely scattered all over the world from the soils and sediments to the animal's intestine. The most poisonous botulinum neurotoxins (BoNTs) are the toxins familiar by the intravenous and oral routes. Ingestion of BoNTs leads to foodborne botulism as produced in foods. However, colonization with toxin secretion and absorption by the intestine then into the circulation leads to botulism in infants [25, 26].

14.2.1.5 SARS-COV-2

Coronavirus is an acute respiratory ailment that leads to COVID-19 infection. COVID-19 led to a worldwide outbreak in 2020, infecting millions of individuals and the mortality rate crossed globally over 1 million and 210 000 as an early drop in the US. These spherically shaped virus particles are approximately 80–160 mm in diameter. The surface coverge of its envelope is with spike (S) protein, envelope (E) proteins, and the membrane (M) proteins, which are positioned among the S proteins. Within the envelope, it consists of spiral nucleocapsid, which is formed of phosphorylated nucleocapsid (N) protein and the genomic RNA. The first case of this virus was reported in December 2019 from Wuhan, China. High fever, cough, pneumonia, and shortness of breath are some of the symptoms developed in infected individuals [27, 28].

14.2.1.6 Tuberculosis

The term "tuberculosis" was coined by J. L. Schönlein in 1839. Robert Koch discovered *Mycobacterium tuberculosis,* the bacteria which caused TB. The *Mycobacterium tuberculosis* complex is a group of closely related bacterial species which causes TB. TB is an age-old disease, which evolved over time. This bacillus lives and multiplies in macrophages (a type of WBC) and destroys the defense system of the patient. Infection with TB is either asymptomatic latent TB infection (LTBI) or TB disease. There is a 50% chance of the person dying due to TB if it is not treated properly [29]. According to the survey, India is found with an overburdening of TB cases. Generated data from the health ministry department in India revealed that about 3 million new TB cases occur every year. The average death rate in India from TB is 480 000. Moreover, according to WHO, the estimated rate of MDR-TB occurrence in India is about 99 000. It has been found that approximately 3% of new TB cases and 17% of worsened cases in India are MDR. However, still, in many states of India, proper drug-sensitive testing and diagnostic facilities fail to be available for the treatment [30].

14.2.1.7 Gonorrhea

This is an infection occurring by a sexually transmitted (ST) bacterium *Neisseria gonorrhoeae* that infects both females and males. Gonorrhea mostly affects the throat, rectum, and urethra. In females, the cervix can also get infected due to gonorrhea. The infection Gonorrhea most commonly occurs during oral, anal, or vaginal sexual intercourse. Also, the infants of infected mothers during childbirth can get infected. In infants, most often, this infection affects the eyes [31].

14.2.1.8 Ringworm

A common skin fungal infection known as ringworm is named because of its appearance as it can form like a circular ring-shaped rash, which is commonly red and itchy. The occurrence of Tinea corporis is mainly the most common dermatophytosis [32]. Though Tinea corporis is found worldwide, it most frequently occurs in tropical regions. Some dermatophytes like *T. rubrum* consist of Mannans in their cell walls, which possess properties of inhibition to the immune system [33]. This feature allows the stay of fungus on the dermal regions without being sloughed off prior to the skin invasion. The causal fungal microbe can release proteases (keratin-digesting enzyme), keratinases (enzymes that introduce into keratinized tissues), and serine-subtilisins (protein-digesting enzymes with serine residue present in active regions of protein via the nucleophilic attack) causing invasion of the fungus into horny layer of the dermis and proliferate outward [34, 35]. Contagions are commonly cutaneous and restricted toward outer skin and cornified parts of the skin. Because of host defense actions in immunocompetent healthy individuals, like polymorphonuclear leukocytes, inhibitory factors of serum, and the complement system, the penetration of fungus into the deeper tissues becomes difficult or sometimes impossible for it to attack [36].

14.2.2 Recently Developed Strategies Against Microbial Contagions and Their Limitations

Multidrug resistance (MDR), as detected, is one of the main reasons behind human suffering which disturbs the trust and bond among doctors and their patients, accompanied by large socioeconomic deprivation. In the environment of microbe–man coexistence, survival is threatened in the human species with health-destroying microbes, and thus there will be no route for survival from the appearance of MDR superbugs. Therefore, from a health point of view, antibiotics have proved to be our best approach. But the result of antimicrobial resistance (AMR) is a serious concern, which impacts the serious threat to individuals and greatest deprivation to socioeconomic conditions [37]. By 2050, it is estimated that the mortality rate because of AMR will rise to 10 million lives per year at the cost of 100 trillion dollars [38]. Currently, the rapid progress of MDR in microbes is elevating the health issues globally and provides a challenge for the infectious disease therapeutics that scientists declare could return to the position of the pre-antibiotic period [29]. It has also been observed that exposure to antimicrobial agents leads to the production of oxidative stress and ROS (i.e. OH), which are usually connected with the different bactericidal agents' lethal exposure and that stands resulting in the protective role of oxidative stress in bacterial cells, thus promoting resistance to antimicrobial drugs against the microbes.

Some studies have revealed that the removal of ROS defense genes of *E. coli* is associated with elevated susceptibility to antimicrobial agents. Certainly, the protective role of oxidative stress in *E. coli* includes a molecule, indole involved in extracellular signaling produced by antimicrobial drug exposure leading to oxidative stress – ROS producing tnaA, a tryptophanase gene product catalyzing indole synthesis in a microbial cell that results in resistance to antimicrobial drugs. Surprisingly, DMSO, an OH scavenger, reduces antimicrobial induced production of indole, thus revealing the explanation for antimicrobial induced production of indole. However, endogenous production of indole and addition exogenously of indole promotes resistance to antimicrobial agents [39, 40].

Plants have shown a special holistic character as the provider of food, shelter, clothing, drugs, and many more. Natural components have been explored for the determination of novel drugs extensively [41]. Indeed, traditional plants for more than five decades have been in use for various medicinal purposes, as a source of antibiotics, analgesics, cardioprotective, antineoplastic, and others [42]. Recently, plants and their parts are used by humans as natural compounds to kick off microbial infections [43]. About 70–90% of the cohort in developing nations still uses ancient traditional medicines depending upon the plant extracts [44]. The most promising and powerful components of plants are phytochemicals – the secondary metabolites, on which individuals depend for the treatment [45]. Significantly, indigenous plants and their derivatives provide more than half of the drugs being approved by the Food and Drug Administration (FDA). Last two thousand years ago, many efforts and attempts were made to discover new conventional therapeutics against MDR but failed. Later, plant-based therapeutics played a promising role to combat these failures and decrease the threat caused by human infections [46].

14.3 Phytoantioxidants

14.3.1 Roles of Different Traditional Plants Against Microbial Infections

Recently, it was found that for treating different microbial infections, deviation toward the plant-extracted phytochemicals was considered the most potent therapeutics. These phytochemicals, because of synergistic, multidrug targets, and antioxidant properties, were found as effective candidates for treating microbial infections. Moreover, these herbal medicines proved to be safer to use in contrast to conventional medicines. Besides, different research concluded that the phytoconstituent exhibits modulate the numerous cell signaling pathways due to its intense affinity toward different receptors of the brain.

With increasing demands for a human's sustainable living, an uncontrolled widespread microbial infection needs to be taken under consideration. This uncontrolled extensive growth of microbes leads to serious health issues in the human body. Antimicrobial drugs such as antibiotics are used to control microbial infections but they lead to several side effects in the human body like elevating reactive oxygen species (ROS) level. ROS causes serious harm to the well-being of humans. ROS was found to play a pivotal role in causing cancer, putting human health at risk [47].

From the Vedic era, Ayurveda provides much of the information about medicinal plants and their extracts that are even today in use successfully in treating different ailments and diseases. The herbal plants include several medicinal roles such as antiviral, antibacterial, antifungal, antioxidant, anticancer, and many more activities for treating diseases. Medicinal herbs are essential medicines used in developing countries for treating infections. These herbal extracts are even today in use on the verge of finding novel compounds with antioxidant and antimicrobial activities [48]. Several medicinal plants such as clove, turmeric, ginger, eryngium, tribulus, thyme, mint, cinnamon, fennel, portulaca, lemon, garlic, and others are found with different antibacterial, antiviral, and ROS-scavenging properties to treat the contagions.

Cloves with antioxidant and antibacterial properties are traditionally being used as extracts and oils against numerous Gram-positive and negative bacteria like *Listeria monocytogenes*, *Klebsiella pneumoniae*, *S. aureus*, *S. typhimurium*, and *E. coli*. Some findings include clove extracts (1–3 mg/ml) found effective against *E. coli* and *S. typhi*. Their extracts have antioxidant- and antimicrobial-like properties that neutralize free radicals like ROS in lipids, thus preventing further oxidation of lipids. Reported studies revealed its antiviral activity against different viruses like coxsackievirus, poliovirus, adenovirus, and herpesvirus [49].

Cinnamon is a traditional herbal plant used in Asia and Australia. Based on their antimicrobial and antibacterial properties, they are widely in use as medicinal plants. Some findings revealed that the antibacterial activity of cinnamon oil was more effective against *Campylobacter jejuni* than

E. coli. Its essential oil was also found to be effective against *Bacillus*, *S. typhimurium*, *Listeria monocytogenes*, *Pseudomonas aeruginosa*, *E. coli*, *Enterococcus faecalis*, *Staphylococcus aureus*, and *Yersinia enterocolitica* [50].

Turmeric shows antioxidant and antimicrobial activities and is traditionally in use as a herbal medicine. Some literature has revealed antimicrobial (antifungal and antibacterial activities) of turmeric. It also provides strong free radical scavenging activity. Turmeric extracts were found to increase glutathione peroxidases and catalases and reduce plasma malondialdehyde [51]. Some studies also revealed its antiviral activity against the human immunodeficiency virus [52].

Ginger rhizomes are in use for two decades in different native parts across the world. These perennial plants contain antiviral, antibacterial, antioxidant, antiulcer, analgesic, anti-inflammatory, and antihypertensive activities. It also affects insulin secretion. Some studies were conducted in vivo and in vitro using ginger extracts that revealed antimicrobial activities against *E. coli*, *Salmonella typhi*, and *S. aureus*. Moreover, an effective role was seen against the influenza virus [53, 54].

Thyme is a medicinal plant active against different bacterial strains like *Brochothrix thermosphacta*, *Brevibacterium linens*, and *Lactobacillus plantarum*. Its antiviral activity was observed against herpes virus type 1 [55]. Numerous studies revealed its antioxidant activities. Some researchers used thyme extracts, which indicate antimicrobial activity against *E. coli*, *Bacillus*, *Pseudomonas aeruginosa*, *S. aureus*, and *Klebsiella pneumonia* [56].

Mint is an aromatic plant used for several cosmetics and pharmaceutical purposes. It has been found useful in preventing oxidative stress in cellular organisms. Several studies revealed its inhibitory action against Gram-positive bacteria like *S. aureus*. It also showed anti-inflammatory action, thus effective against incurable bowel syndrome. Some studies suggested an antiviral action against HIV and HSV-1 [57]. Some experiments demonstrated revealed antimicrobial activity of mint oil extract against different strains such as *S. aureus*, *Bacillus cereus*, *S. epidermidis*, *E. coli*, *Enterococcus faecalis*, *C. sakazakii*, and *Bacillus cereus* [58].

Garlic is found native to the central Asian region with several biological roles such as antimicrobial, anti-inflammatory, anticancer, cardioprotective, antidiabetic, and antioxidant. Due to being a natural anti-oxidative agent, it prevents the cell from oxidative stress. It neutralizes the charge present on free radicals converting them to harmless products. Garlic extracts being used against different bacterial strains, such as *Klebsiella*, *Pseudomonas*, *Enterococcus faecalis*, *Salmonella typhi*, *S. aureus*, *Proteus*, *Staphylococcus aureus*, and *E. coli* showed antibacterial mechanism [59]. This herbal plant had showed effectiveness against H1N1 influenza virus. Further, many in vivo experiments revealed its powerful antioxidant activities scavenging free radicals, thus protecting damage to the cells [60].

Mallows native to Europe and Asia are medicinal plants that possess a high antimicrobial role. Some studies have revealed its antibacterial activity against *E. coli*, *S. aureus*, *Bacillus cereus*, *Klebsiella pneumoniae*, *Listeria monocytogenes*, *Salmonella typhimurium*, *Proteus vulgaris*, *Micrococcus luteus*, *Streptococcus pyogenes*, *Micrococcus luteus*, *Mycobacterium smegmatis*, and *Pseudomonas aeruginosa*. The antioxidative mechanism of this plant was found to be very effective in treating different microbial infections. Its aqueous and alcoholic extracts possess different antiviral, anticancer, and antioxidant properties [61].

Eucalyptus is also known as fever tree as it strongly holds the antimicrobial activity. This medicinal tree is traditionally used in the treatment of several diseases like pneumonia, influenza, TB, and fungal and bacterial infections. Its oil extracts are being used as antioxidant and antibacterial agents [62]. Different studies have reported free radical scavenging properties of the eucalyptus oil. Its oil extract also revealed the antibacterial activity against different bacterial strains like

E. coli, *S. aureus*, *Bacillus cereus*, *Klebsiella pneumoniae*, *Listeria monocytogenes*, and *Salmonella typhimurium* [63].

Primrose is traditionally used against different skin diseases. This plant extract is used as an antioxidant and antimicrobial agent. Its role as anti-inflammatory was found effective in the treatment of MS diseases. Different studies revealed primrose inhibitory action against different Gram-negative and positive bacteria. It was found effective against Varicella virus, Herpes simplex virus and Epstein Barr virus [64].

The decisive action of these herbal antimicrobial agents is becoming highly acceptable in pharmaceutical industries. It not only increases the demand for developing novel antimicrobial drugs but also is currently used as nanotherapeutics for treating different microbial infections.

14.3.2 Phytoantioxidants as ALTERNATIVES to Microbial Infections

Recently, it was found that for treating different microbial infections, deviation toward the plant-extracted phytochemicals was considered the most potent therapeutics. These phytochemicals, because of synergistic, multidrug targets, and antioxidant properties, were found as effective candidates for treating microbial infections. Moreover, these herbal medicines proved to be safer to use in contrast to conventional medicines. Besides, different research concluded that the phytoconstituent exhibits modulate the numerous cell-signaling pathways due to their intense affinity toward different receptors of the brain [65].

Based on the plants, metabolic role of phytochemicals are classified into two major – primary and secondary metabolites. Primary metabolites are responsible for growth and development of plant parts and secondary metabolites act as defense systems against various environmental stresses. Various secondary metabolites found in different medicinal plant parts and their extracts are responsible for the treatment of diseases [66]. Several studies revealed that these herbal plants consisting of phytochemicals play various biological roles as antimicrobial, antiviral, anticancer, antidiabetic, and antioxidant. Plants have a limitless capacity to synthesize various secondary metabolites that act as antioxidants against different microbial infections. Many are known for their aroma, while, on the other side, many are used in food flavors, which are useful as medicines [67].

Over 80% of the population relies upon herbal medicines for treating different ailments. Reported studies suggested that oxidative stress is responsible for inducing immunoinflammatory responses accompanying different microbial diseases [68]. Herein, the major categories of phytoantioxidants efficient in microbial infections are described.

14.3.2.1 Alkaloids

These are heterocyclic nitrogenous compounds. Diterpenoid alkaloids showed antimicrobial action as reported. Solamargine found in berries of *Solanum khasianum* is highly effective against HIV. Berberine, one of the vital parts of the alkaloid group, is efficient against *L. tropica* due to its antioxidant activity [69]. According to reports regarding the use of pandrug for treating contagions, the requirement of novel antibacterial therapeutics is more crucial than ever. Alkaloids are various large structural groups of compounds that act as scaffolds for vital antibacterial medicines like quinolones and metronidazole. In this context, we focused on various alkaloids with their potent growth.

Different classes of alkaloids are categorized into natural, synthetic, and semisynthetic, which focus on antibacterial and its enhancing activities. An important example involves CJ-13 136, a newly formed actinomycete-mediated quinolone alkaloid against *H. pylori* with MIC of 0.1 ng/ml, and squalamine from dogfish shark, a polyamine alkaloid that is effective against Gram-negative pathogens with 16–>32-fold vulnerable to ciprofloxacin. Some studies suggested regarding virstatin as an isoquinoline alkaloid, which inhibits *Vibrio cholerae* activities by inhibiting its transcriptional regulator ToxT, inhibiting the *Vibrio cholerae* and Fimbriae expression and thus confers defense against intestinal colonization [70].

14.3.2.2 Polyphenols

Plant polyphenols are structurally diverse compounds that have been used for centuries for medicinal purposes, including infection treatment and possess, not only antimicrobial activity, but also antioxidant, anti-inflammatory, and anticancer activities among others. Based on the existing evidence on the polyphenols' antibacterial capacity, polyphenols may be postulated as an alternative or complementary therapy for infectious diseases.

Reported studies suggested that polyphenols obstruct oxidative processes in different ways: Scavenging the reactive species, as an inhibitor of lipoxygenase, and reducing factor for metmyoglobin. Therefore, polyphenols are potent antioxidants found as phytoconstituents in various fruits and vegetables. The capability of polyphenols is to interact with cell wall constituents of bacteria and the cell membrane of bacteria and can control and inhibit the formation of biofilm, along with preventing the development of enzymes of microbes, interfering in the regulation and functioning of the protein, and depriving the bacterial cell metal ions and enzymes for substrates.

The presence of polyphenols can lower the primary and secondary levels of lipid peroxidation, prevent the activity of lipoxygenase and, thus, result in bacterial growth retardation [71]. Some classes and subclasses of polyphenols are described further accompanied by their antimicrobial actions. Phenols are the simplest phytoantioxidants with a single ring substituted on one or more hydroxyl groups [72]. Caffeic acid and cinnamic acids are common examples. Some studies suggested that the pyrogallol and catechol showed toxic activities against microorganisms. These phytoantioxidants show increased hydroxylation and thus elevate the level of toxicity in microbial cells [73].

Reported works revealed that tarragon herbs with caffeic acid components are found to be effective against virus, bacteria, and fungi. These phenolic phytoantioxidants contain oxidized components reacting with sulfhydryl groups or binding nonspecifically with the proteins that inhibit enzyme reactions in microbes and cause toxicity. Some phenolic compounds with C3 side chain at lower levels and containing no oxygen are classified as essential oils. A clove oil extract Eugenol is a phenolic compound that acts as a phytoantioxidant against fungi and bacteria [74, 75].

Flavones and flavonoids are hydroxylated phenolic components with C6–C3 units linked to aromatic rings. These phytoantioxidants synthesized in plants are found to be very effective antimicrobial agents [76]. Lipophilic flavonoids were responsible for microbial cell wall disruption as reported [77]. Catechin, a flavonoid type with reduced C3 unit found in oolong green tea, exerts antioxidant properties against microbes such as *Vibrio cholerae*, *Shigella*, *Streptococcus mutans*, and many other microbial strains as demonstrated [78].

These flavonoids showed inhibition against numerous viruses. Different studies revealed that glycyrrhizin (from licorice), swertifrancheside, and chrysin flavonoids are effective against HIV [79]. Isoflavone phloretin acts as an antioxidant against numerous microbial infections. Galangin is obtained from *Helichrysum aureonitens*, a perennial herb found to be a useful antioxidant against numerous Gram-positive bacteria, fungi, and viruses like coxsackie B virus and HSV-1 [80]. Flavonoids lacking OH group on their beta ring were more effective against microbes [81]. Thus, some findings revealed flavonoids and isoflavonoids as antioxidants against microbes.

Tannin is a phenolic compound found in leather tanning. They are found in different plant parts such as bark, leaves, fruits, woods, and roots. Their presence in beverages is used in curing illness [82]. They are found as anti-infective agents. Scalbert suggested it as a phytoantioxidant against bacteria, yeast, and fungi. They are found to inhibit reverse transcription in viruses [83]. Some studies revealed that tannin in condensed form binds to the bacterial cell walls and inhibits its growth in ruminant animals, thus acting as a phytoantioxidant [84].

Resveratrol belongs to the group of stilbenoid polyphenolic non-flavonoid compound found in cranberries, grapes, pines, peanuts, legumes, and other plant species. They are found active against pathogen invasion and oxidative stress. Different in vitro and in vivo studies revealed its anticancer, antioxidant, cardioprotective, antidiabetic, and anti-inflammatory properties [85]. Resveratrol inhibits CYP1A1, a cytochrome P450 isoenzymes and suppresses the cyclooxygenase activity, cyclins, numerous phosphodiesterase, and inhibition of hypoxia-inducible transcription factor 1 expression (HIF-1α). It was revealed that resveratrol counteracts *Proteus mirabilis* virulence activity due to its antioxidant property [86]. It holds a good impact on human health and is also proven as a potent phytoantioxidant. Piceatannol, a resveratrol analogue found in white tea and grapes, exhibits antioxidant, antimicrobial, and anticancer activities. It is an attractive compound for pharmaceutical industries. Though it is less studied but found to be a promising phytoantioxidant biomolecule [87].

14.3.2.3 Terpenoids

They are also known as Quinta essentia (essential oils), which are phytochemicals based upon isoprene structure. When they hold additional oxygen molecules, they are termed terpenoids. These essential oils are found to be very effective against bacteria and fungi due to their antioxidative properties [88]. Some studies suggested that it is very effective against *Listeria monocytogenes*. Aframodial is a terpenoid present in Cameroonian spice found as antifungal in a broad array [89]. Clover oil yields a terpenoid petalostemum reported to be a phytoantioxidant against *Staphylococcus aureus* and *Bacillus subtilis* [90]. Kadota et al. suggested a diterpene trichorabdal A, a Japanese herb extract, could inhibit directly *H. pylori* [91].

14.3.2.4 Other Phytoconstituents

Organosulfur phytochemicals were found very effective against various microbial infections. For example, garlic (*Allium sativum*) was found to be useful against microbial infections. It also has been demonstrated to be a potent antioxidant since it removes ROS and reduces lipid peroxidation [92]. Coumarins are formed of alpha pyrone rings and fused benzene. These phenolic compounds are anti-inflammatory with vasodilating action. Coumarin is found with an antifungal agent that inhibits *Candida albicans* and helps in the treatment of vaginal infection [93]. Hydroxycinnamic, a type of Coumarin, was reported to show antioxidant properties against Gram-positive bacteria [94]. Woodruff extract (*Galium odoratum*) was found with antibacterial action [95]. Although less evidence was found related to the phytoantioxidant property of Coumarin, still further research continues.

Plants have a limitless capacity to synthesize various secondary metabolites that act as antioxidants against different microbial infections. Many are known for their aroma, while on the other side, many are used in food flavors, which are useful as medicines. Table 14.1 describes different plant parts consisting of different phytochemicals as categorized into different groups and subgroups along with mechanisms of antimicrobial action [96].

Table 14.1 Antimicrobial phytoantioxidants.

Phytochemical name	Plants containing phytochemicals	Phytochemical group	Phytochemical subgroup	Antimicrobial activities
Phenols	Tea, coffee, mango, berries, pigeon pea leaves	1) Phenolic acids and aldehydes	1.1 Gallic acid and *p*-hydroxybenzoic acid.	1.1 *Camplylobacter jejuni*
			1.2 Gallic acid, caffeic acid, chlorogenic acid, and quinic acid	1.2 Herpes simplex virus type 1 (HSV-1)
		2) Coumarins	2.1 Cajanuslactone	2.1 *Staphylococcus aureus*
			2.2 5-Methoxyseselin	2.2 Human immunodeficiency virus (HIV)
		3) Flavanoids	3.1 Genistein	3.1 Parainfluenza type-1 (PI-1)
			3.2 Epigallocatechin gallate	3.2 Gram-positive bacteria
			3.3 Myricetin	3.3 *Lactobacillus*, *E. coli.*
			3.4 Baicalin	3.4 *S. aureus*
		4) Polyphenols	4.1 Ellagitannin extracts	4.1 *Staphylococcus*, *Salmonella*, *Listeria* and *Lactobacillus* strains
			4.2 Tannins	4.2 *S. aureus*, *E. coli*, *Shiegella* spp. and *P. aeruginosa*
			4.3 Prodelphinidin β-2 3-*O*-gallate	4.3 HSV-2
			4.4 Gallotannins	4.4 *Pediococcus acidilactici*

(*Continued*)

Table 14.1 (Continued)

Phytochemical name	Plants containing phytochemicals	Phytochemical group	Phytochemical subgroup	Antimicrobial activities
		5) Naphthoquinones	5.1 Diospyrin and isodiospyrin 5.2 Lapachol and β-lapachone	5.1 *S. pyogenes* and *S. pneumoniae* 5.2 *Candida albicans*, *Candida tropicalis*, and *Cryptococcus neoformans*
Alkaloids	*Strychnos potatorum* leaf	1) Indole 2) Tropane 3) Tropolone	1) Yohimbine and vincamine 2) Scopolamine and atropine 3) Colchicine and allantoin	Herpes simplex (HSV-1) type 1 and RNA virus parainfluenza type-3 (pi-3)
Organosulphur compounds	*Brassica juncea*, *Wasabia japonica* (wasabi), *Armoracia rusticana* (horseradish) and *Brassica oleracea* (cauliflower)	1) Thiosulfinate 2) Glucosinolates	1) Diallyl disulfi de (dds) and diallyl trisulfi de (dts) 2) Indole-3-carbinol	1) *C. albicans*, *C. tropicalis* and *B. capitatus*, Gram-positive bacteria (*S. aureus* and *B. subtilis*) and Gram-negative bacteria (*P. aeruginosa* and *E. coli*). 2) Gram-positive bacteria
Terpenoid	Curcumin, ginger, oil	1) Gingerol		*H. pylori*
Anthranoid	Rhizome	1) Anthraquinones 2) Hydroxyanthraquinones		1) *S. aureus* 2) *B. adolescentis*

Source: Based on Ref. [96].

14.3.3 Challenges of Using Phytoantioxidants Against Microbial Infections

Though phytoantioxidants hold several beneficial activities as antimicrobial agents, still the phytoantioxidant-related treatment's success rate is often poor because of its numerous limitations like the lack of phytoconstituents' stability in the human body or its storage. Moreover, the major challenge is sensitivity of phytoantioxidants to element-induced oxidative stress. Their stability is maintained till their delivery is successful to the microbial target. Then, next is the poor solubility in aqueous solution and less drug bioavailability due to poor oral absorption. Food matrix interaction, gastrointestinal enzyme transformation, the acidic nature of the gut, and microbial enzymes in the gut, all together obstruct phytoantioxidants' absorption. Moreover, rapid metabolic rate, absorption, and elimination cause introduction of drugs in less concentration into the affected site, thus leading to failure in achieving the desired concentration for the treatment of microbial infections [97, 98].

14.4 Nanodelivery Systems/Nanoformulations for Phytomedicine

Recently, the use of herbal medicines has gained insights all over the globe. Because of the natural alternatives with minimal side effects than synthetic drugs, these traditional plants and their phytoconstituents are believed to have potent applications in a large range of health conditions' management. According to the World Health Organization (WHO) norms, in developing and underdeveloped countries, at present, nearly 80% of the global population for initial health care uses herbal medicines. Currently, the focus of the scientific community is on the research study connected with the bioactive components, its chemical formulation, and pharmacological applications of various plant species, to construct pioneering potent constituents that exist moderately with minor adverse effects than present compounds. There are still definite limitations as it reaches toward herbal therapeutics. There are problems of the poor or unknown bioavailability of potent phytoconstituents, though in vitro-revealed possible herbal drugs never always translate directly to in vivo effective curative applications. There is, accordingly, an area for enhancement when it comes to the effective and efficient delivery of herbal drugs to specific body systems and tissues in order to provide therapeutic outcomes [99].

Nanotechnology is a specific solution for the novel drug delivery process under consideration, with the thought of nanoformulations to have a large diversification of benefits as compared to conventional medicines of phytoconstituents, which involves increased solubility, stability, permeability, therapeutic activity, improved distribution within tissues, bioavailability, and sustained drug delivery. Amidst the novel systems for drug delivery, the most important one was considered to be nanoparticles. Nanoparticles can be applied to use herbal drugs to particular organs which enhances the selectivity, effectiveness, and safety of drug delivery, thus reducing administration of dose and increasing compliance of patients [100]. Different nanoparticle formulations harboring herbal medicines have been discussed in detail:

14.4.1 Liposomes

Liposomes consist of one or other concentric lipid bilayers belonging to a nanoparticulate structure that encapsulates a little amount of the solvent. Various properties of liposomes like biocompatibility, amphiphilicity, and biodegradability are vital for herbal drug delivery. Additionally, the advantages involve therapeutic safety and efficacy improvement, increased sustained release, and bioavailability and specific drug delivery to desired locations. They can modify stability and biodistribution [101] and increase therapeutic activity and safety, especially by selected delivery along with perpetuating curative drug levels for extended time periods. There are a vast variety of examples related to the incorporation of the traditional plant constituents into liposomes. Sou et al. demonstrated an altered nanolipid agent filled with curcumin to transport it by injecting it into macrophage tissues intravenously. Encapsulation of curcumin into a phospholipid carrier incorporating 1,2-dimyristoyl-sn-glycero-3-phosphocholine; 1,5-dihexadecyl ester; 1,2-distearoyl-sn-glycero-3-phosphoethanolamine-*N*-[monomethoxy poly(ethylene glycol) (5000)] in a mole ratio of 10 : 1 : 0.06 [102]. Moreover, Samaligy et al. used a buccal liposomes system for drug delivery to enhance the bioavailability of silymarin [103].

14.4.2 Lipid Nanoparticles

Nanostructured lipid carriers (NLCs), solid lipid nanoparticles (SLNs), and lipid drug conjugates all are classified under lipid nanoparticles. SLNs remain in the solid form at room and body temperatures and are formed of lipid molecules. SLN varies from 50 to 1000 nm. Different advantages were observed of SLNs, which involve drug targeting and sustained drug delivery, prevention of drug degradation from chemical exposure, reducing the toxicity of a drug, bioavailability enhancement, biodegradation, superior tolerability, potential to integrate both hydrophilic and lipophilic drugs, and no complications related to mass-scale sterilization and production [104].

The formulations integrating herbal medicines in SLNs involve inhalations (e.g. eucalyptus oil), gargles (e.g. thymol), and mouthwashes (e.g. peppermint oil). Mei et al. suggested SLN incorporated with triptolide and demonstrated the activity for drug delivery transdermally and anti-inflammatory activity [105]. Chen et al. used podophyllotoxin-loaded SLNs and introduced them to the genital epidermal surface to cure genital wart infections [106]. Tiyaboonchai et al. found the effectiveness of curcuminoid (curcumin, bisdemethoxycurcumin, and demethoxycurcumin)-loaded SLNs in the form of cream, since curcuminoids are degraded by oxidation, photodegradation, and acidic and alkaline hydrolysis [107]. Hu et al. developed cryptotanshinone and introduced SLNs for increasing the cryptotanshinone bioavailability [108]. Another work by Shi et al. demonstrated SLNs incorporated with myrrh oil and frankincense [109].

14.4.3 Polymeric Nanoparticles

Polymeric-nanocarrier agents are composed of nanosized tiny particles. Polymeric nanosized micelles are formed by polymer aggregation. The drug provides increased metabolic stability, solubility improvement, and longer circulation time [110]. Nanoparticles have many advantages such as efficiency increment, solubility enhancement, bioavailability increment, improvement in absorption of herbal drug dosage as compared to traditional herbal medicinal dosage, and dose reduction [111]. Liu et al. introduced poly(DL-lactic acid) nanoparticles loaded with triptolide. To overcome the issues of toxicity and poor solubility of triptolide, nanoparticles were constructed with biodegradable and biocompatible poly (DL-lactic acid) polymers. They were spherically shaped with uniform-sized and smooth surfaces [112]. Sahu et al. produced a novel self-assembling and biodegradable polymer, methoxy poly(ethylene glycol)-palmitate loaded with curcumin as therapeutics to cell carcinoma. The carrier consisted of hydrophilic methoxy poly (ethylene glycol), hydrophobic parts such as palmitic acid, and curcumin as incorporated in the polymer micelle's core. The synthesized micellar nanocarriers were found spherical in structure [113].

Hypericin is a particularly lipophilic carrier and thus found insolvable in acceptable media physiologically, which provides its extensive problematic administration and regulates its applications diagnostically. To overcome these issues, a formulation of polymeric nanoparticles loaded with hypericin as an injectable device was fabricated using biocompatible and biodegradable synthetic polymers like polylactic acid (PLA) or polylactic-*co*-glycolic acid (PLGA), which has better photodynamic and photodetection therapy for the initial cancer diagnosis [114].

14.4.4 Microemulsion

The microemulsion is a mixture of amphiphile, oil, and water, which is isotropic optically and stable liquid formulation thermodynamically. Amidst the different drug delivery agents, the microemulsion is found to be an absolute alternative for the oral administration of compounds poorly soluble in water. They have many advantages such as low viscosity, ease of preparation, thermodynamically stable, increased lipophilic drug dissolution, and bioavailability enhancement. Moreover, these can be administered through various routes like pulmonary, transdermal, parenteral, and ocular [115].

Suresh et al. fabricated a novel gel formulation microemulsion-based babchi oil (*Psoralea coryfolia*) [116]. Likewise, Wang et al. also tried to increase the curcumin's anti-inflammatory activity by developing it into a nanoemulsion formulation [117]. In order to decrease the triptolide toxicity, Chen et al. revealed microemulsion agents for transdermal delivery [118].

14.4.5 Phytosomes

Phytosome was fabricated to introduce polar phytochemicals or extracts of plants in a patented technology into phospholipids. It was because of molecular complexes consistent with lipid, thus called phytosomes, which could be developed to improve their absorption and bioavailability. As compared to liposomes, the phytosomes consist of plant components and phosphatidylcholine that actually build molecular complexes comprising chemical bonds [119]. Table 14.2 describes different green synthesized nanotherapeutics as antimicrobial agents.

14.5 Nano Phytoantioxidants and Their Roles in Microbial Diseases

Antioxidants are molecules that can interact with free radicals in a safe manner, terminate the chain reaction, and convert them to a harmless molecule by donating an electron. In the current decade, antioxidants have drawn attention due to their potential to combat microbial diseases.

However, the effectiveness of natural and/or synthetic antioxidants is limited due to their poor absorption, difficulties to cross the cell membranes, and degradation during delivery, hence contributing to their limited bioavailability. To address these issues, antioxidants covalently linked with nanoparticles, entrapped in nanogel, hollow particles, or encapsulated into nanoparticles of diverse origin have been used to provide better stability, gradual and sustained release, biocompatibility, and targeted delivery of the antioxidants with superior antioxidant profiles [121]. Different nano phytoantioxidant-based strategies are currently in use as discussed further in this review to curb the microbial threats.

Table 14.2 Different green synthesized nanotherapeutics as antimicrobial agents.

Plant source	Plant parts used	Metallic nanoparticles	Shape of nanoparticles	Size (nm) of nanoparticles	Affected microbial strains
Glycosmis pentaphylla	Plant extract	Zinc oxide	Spherical	32–36	*S. dysenteriae*; *S. paratyphi*; *S. aureus*; *B. cereus*
Alternanthera bettzickiana	Plant extract	Silver	Spherical	80–120	*S. typhi*; *P. aeruginosa*; *E. aerogenes*; *S. aureus*; *B. subtilis*; *M. luteus*
A. catechu	Plant extract	Gold	Spherical	13.7	*E. coli*, *K. pneumonia*, *P. aeruginosa*, *Enterobacter* and *S. aureus*
M. officinalis	Leaf extract	Silver	Spherical	12	*S. aureus* and *E. coli*
T. procumbens	Plant extract	Iron	Spherical	80–100	*P. aeruginosa*
Punica granatum	Plant extract	Zinc oxide	Spherical and hexagonal shapes	32.98	*E. coli* and *E. faecalis*
H. cordata	Plant extract	Gold	Spherical	30–90	*B. subtillis*, *S. typhimurium*, *B. cereus* and *E. coli* *B. subtilis* IMS/GN13 and *S. aureus*
C. cujete L. (Calabash tree)	Plant extract	Gold	Spherical	32.89	*E. coli* (MTCC 1687), *P. aeruginosa* (MTCC 1688), *V. cholerae*
Suaeda aegyptiaca	Plant extract	Zinc	Spherical	~60	*P. aeruginosa*; *E. coli*; *S. aureus*; *B. subtilis*

Source: Based on Ref. [120].

14.5.1 Nanocurcumin

Curcumin as a polyphenolic component found with the most promising role of free radical scavenging. The anti-inflammatory and antioxidant role of curcumin finds its approach in the prevention of several microbial infections. L. C. Ring and coworkers revealed through their findings that curcumin, a polyphenolic component, is an antimicrobial agent due to its anti-quorum-sensing mechanism but due to its limitation of being poorly soluble in aqueous solution and quick degradation attempts toward the formation of quantum dots of curcumin, which will help increase the solubility and stability and found as antibiofilm and antimicrobial agents.

Since 80% and more of the bacterial contagions were found associated with bacterial biofilm, this study also reported that curcumin quantum dots possess different physicochemical properties such as reduced size, high-energy amorphous state formation, and their intermolecular molecular hydrogen binding, which eventually increases the water solubility and stability and enhances the antimicrobial and antibiofilm-like properties [122].

It has also been detected with improved physicochemical features over the metallic NPs. Moreover, curcumin-based nanoparticles showed antimicrobial properties against *E. coli* strain as reported by Shlar et al. [123]. Thus, this green synthesis approach toward curcumin quantum dots proved as one of the stable nanotherapeutics against antimicrobial infections.

14.5.2 Nanoquercitin

Quercetin is a potent flavonoid occurring in vegetables and fruits with infinite health benefits. According to several studies, it was reported about the use of *M. oleifera* peel consisting of quercetin extract involved in the synthesis of Pd NPs, which were found as potent antibacterial agents against *S. aureus* and *E. coli*. Moreover, these extracts were also used in the synthesis of CeO_2 NPs, which were also found effective against bacteria [124].

P. Roy et al. investigated through their findings the involvement of neem leaf (*A. indica*) extracts in the synthesis of silver nanoparticles. Neem leaves contain quercetin flavonoids as some major phytochemicals, which play an important role in the NPs' synthesis by acting as reducing and capping agents. The aqueous extract of neem leaf reduces silver salt to silver ion, i.e. silver nitrate. The reduction in metal ions was observed through color change. And this caps the NPs along with neem phytoconstituents, which possess antibacterial properties against various Gram-negative and Gram-positive bacterial strains [125].

Vasantharaj et al. found a green facile approach for iron oxide NPs' synthesis using *Ruellia tuberosa* leaf extract found effective antimicrobial activity against *Klebsiella pneumoniae* and *E. coli*. The phytoconstituent quercetin present in leaf extracts was found to be a reducing agent during the synthesis of metallic NPs [126].

Shakeel Ahmed and his coworkers synthesized Ag NPS of spherical structure from *A. indica* leaf extracts. The demonstrated FTIR process revealed that flavonoids as phytoconstituent compounds of these leaf extracts act as stabilizing agents and reducing agents while the NPS synthesis takes place [127].

P. Banerjee and his coworkers also reviewed that these green NPs interfere with bacterial metabolic activities by breaking the disulfide bonds. This provides an eco-friendly and cost-effective antimicrobial agent. In recent years, nonhazardous silver NPs' synthesis based upon the neem leaf extract gained more interest for researchers and was found most effective. Nanotherapeutics against different microbial contagions [128].

Another example, as suggested by Manikandal and coworkers, is the synthesis of TiO_2 NPs using *Prunus yedoensis* leaf extract containing flavonoid component quercetin, which was found as an antioxidant agent against bacteria such as *E. coli* and *S. aureus*. The TiO_2-based NPs lead to the peroxidation of lipids present on the bacterium membrane and thus causes ROS production affecting the integrity and fluidity of the membrane [129].

14.5.3 Nano Ginsenosides

Saponins or ginsenosides, as one of the phytoconstituents, were found effective against microbes. Synthesis of gold NPs using *Euphorbia hirta* leaf extract possesses potent antimicrobial properties. Au NPs using leaf extracts of *Memecylon edule*, a medicinal herb, was synthesized with size range of (10–45 nm). These Au NPs yield more electrons, which leads to ROS generation (O_2 and OH free radicals), which destroys the DNA and other microbial cell components. It also hampers the efflux of transmembrane H^+ ions. Due to its smaller size, it easily adheres to the bacterial cell wall and hampers its growth. Another study by Ahmed and his coworkers is that AuNPs restricts *Candida albicans* transmembrane H^+ efflux [130].

Santosh et al. demonstrated the bactericidal activities of green synthesized TiO_2 NPs using aqueous extracts of *Psidium guajava* containing saponin as a potent phytoantioxidant, which impressively inhibited the growth of *S. aureus* and *E. coli* [131]. Additionally, another work supported the TiO_2 NPs' synthesis using leaf extracts of *T. foenum-graecum* plant and studied its antibacterial and antifungal properties against *P. vulgaris*, *Y. enterocolitica*, *E. faecalis*, *S. faecalis*, *P. aeruginosa*, *B. subtilis*, *E. coli*, *S. aureus*, and *C. albicans*, respectively [132].

14.5.4 Nano EGCG

Anna Regiel-Futyra and her coworkers demonstrated the use of the grape extract which consists of polyphenols such as EGCG (epigallocatechin gallate) that acts as a stabilizing and capping agent and was applied during the formation of chitosan-based Ag nanocomposites. These nanocomposites are found with inhibitory action against *E. coli* [133]. Similarly, Anirudh et al. reported about SeNPs derived from *B. licheniformis* extracts, which were found to be effective against *S. aureus*. An amnestic study was reported regarding CeO_2 NPs bio-fabrication through *G. superba* leaf extracts and found as potent against various Gram-positive and Gram-negative bacterial strains such as *K. pneumoniae*, *S. pneumoniae*, *P. vulgaris*, and *P. aeruginosa*. The synthesis of gold NPs through *M. citrifolia* root extracts were also detected with antimicrobial nature [134].

14.5.5 Nanothymoquinone

The phytoconstituent thymoquinone as a subgroup of terpenoids, also called isoprenoids, belongs to a vast and mixed group of organic compounds occurring naturally extracted from isoprene a 5-carbon component, which were found as effective against microbial infections in the form of nanothymoquinone as suggested. Shehensha et al. reported through their studies the vital role of thymoquinone in the synthesis of silver nanoparticles and also found effective against different bacterial strains.

They used *Nigella sativa* seed extracts, which contain thymoquinone that stabilizes by reducing metal ions via oxidation of aldehyde groups into carboxylic acid groups during the metallic nanoparticles' synthesis and thus provides a large surface area for antibacterial activity [135]. Moreover, Ali et al. suggested that thymoquinone nanoformulation was found effective against various microbial diseases. Moreover, in one of the nanothymoquinone trials against *Candida albicans* yeasts and *Candida biofilm*, it was found to be effective due to its antioxidant nature [136]. Some studies also suggested that thymoquinone nanoformulation provides a critical role as an antibiotic and might be able to minimize the bacterial resistance to specific antibacterial medicines [137].

14.5.6 Nano NRG

Naringenin (NRG) is found to be rich in citrus fruits and belongs to a heterocyclic flavonoid known to possess different pharmacological roles like antibacterial, anticancer, and antioxidant properties. Shanmuganathan et al. explored the nanocomposite formation of NRG@Ag/RGO. These hybrid nanocomposites, where NRGs were used as reducing agents in its fabrication and were found to play a crucial role against bacterial infections since it contains silver ions, which act as free radical scavengers against different bacterial strains [138]. Some studies suggested NRG-mediated nanoliposomes were found effective against bactericidal activities [139].

14.5.7 Nano Apigenin

Apigenin belongs to the flavone group found in various plants. Yan He et al. demonstrated the synthesis of silver nanoparticles using *C. morifolium* extract that consists of apigenin phytoconstituent that is found effective against *S. aureus* and *E. coli*. Apigenin reduces the metallic solution of silver to silver ions, which were found as effective antioxidants against Gram-negative bacterial strains. Moreover, several studies also suggested the antimicrobial activities of nano apigenin because of its free radical scavenging activities and ROS generation, which leads to the control of the microbial growth [140].

14.5.8 Nano Genistein

Genistein belongs to the isoflavone group, which is a naturally occurring group of compounds successfully used as reducing and capping agents in nanomaterial synthesis. L. Wang et al. suggested the use of Genistein present in various plant extracts as a capping agent in silver nanoparticle synthesis. These nano genistein-containing metal ions then cross the membrane of bacteria and accumulate along the metabolic pathway, modifying the structure and function of the bacterial membrane. Then, this nano genistein interacts with the basic components of bacterial cells like lysosomes, DNA, ribosomes, and enzymes causing oxidative stress, alteration in permeability of cell membrane, disorderliness in electrolytic balance, enzymatic inhibition, protein misfolding, and alteration in gene expression [141].

14.5.9 Nano Kaempferol

Silk et al. mentioned the anti-quorum sensing of kaempferol, a phytoconstituent opened with a new antimicrobial strategy. They suggested the NPs loaded with kaempferol as antimicrobial agents against different microbial strains. These nano kaempferols reduce violacein production in microbes and thus act as stable longtime quorum quenchers [142]. Di Ming et al. also found nano kaempferol's role as anti-biofilm formation in *S. aureus* and thus minimizes the risk of antibacterial resistant drugs and thus prevents the effect of infection by *S. aureus* [143].

14.5.10 Nano Tricetin

Various acute and chronic respiratory mediated ailments were found by various microorganisms such as bacteria, viruses, and fungi. MDR was found to be a great challenge in the medical sector for respiratory infection treatment. Thus, nanomedicine was found with great therapeutic efficacy to overcome this issue. K. Alsamhary et al. demonstrated the tricetin-based gold nanoparticle synthesis, which was found very effective against various respiratory infections. Tricetin belongs to the flavonoid group, which exhibits effective nanomaterial against a broad spectrum of bacterial activities [144]. Another group, T. Kamal and his coworkers, also suggested the antibacterial activities of green-synthesized gold nanoparticles using plant extracts that contain triacetin as major flavonoids that act as reducing agents in their formation [145].

14.5.11 Nano Resveratrol

Numerous studies have revealed that resveratrol has different ways of biological functions, involving cardiovascular protective effects, anti-inflammatory, antitumor, treatment of mutations, and antiviral effects. In the current study, resveratrol-loaded nanoparticles were applied to prevent the rhabdomyosarcoma (RD) cells from infection caused due to enterovirus 71 (EV71). Nano resveratrol exhibited the capacity to prevent the increase of oxidative stress intracellularly. The prospective role for the functioning of nano resveratrol revealed that it may prevent the oxidative stress-mediated PERK/eIF2α/ATF4 signaling mechanism, downregulating the autophagy mechanism, and also inhibiting the EV71-induced RD cells injury. Moreover, nano resveratrol therapeutics markedly prevented the inflammatory factor secretion, including tumor necrosis factor-α and interleukin (IL)-6, IL-8 obtained by EV71 infection.

14.6 Conclusion

There is a great reduction of currently conventional drug efficacy because of the drug resistance development and the ability for the formation of biofilm in various pathogenic microbes leading toward serious health issues in the public health domain. Therefore, the nanoparticles serve as an important strategy and a potent alternative toward antibiotics in handling the MDR strains' development posing a serious medical background.

This chapter describes the synthesis of various nano therapeutics using different plants and their extracts, the synthesis processes, and their effective applications of being antioxidants and antimicrobial agents to minimize the problem of multiple drug resistance in various pathogens. Over the past few decades, tremendous growth has been found in the areas of plant-mediated synthesis of nanotherapeutics and its various applications. Owing to rich biodiversity, plants as a biofactory provide a sustainable process and environmentally safe method for the green synthesis of nanoparticles.

Neem, tulsi, curcumin, mint, aloe vera, and others have been successfully introduced for the synthesis of stable metal and metal oxide nanoparticles. Easy cultivation and modification of plant extracts lead to the advantage of manipulating the rate of synthesis and its associated criteria. During the synthesis process, the phytochemicals are used as phytoantioxidants, which aid in nanotherapeutic formation. The biosynthesized nanotherapeutics naturally capped with plant's phytoconstituents impart solubility and stability to these nanosized matter, providing as therapeutics against microbial infections.

Metals and metal oxide nanotherapeutics hold remarkable antiviral, bactericidal, and fungicidal properties, evaluating them as a promising agent with potent antimicrobial therapy. The nanotherapeutics potency nanoparticles are governed particularly by their physical features including size, shape, and stability. Various metallic NPs, such as gold, silver, copper, magnesium, and bimetallics are found with their effective antimicrobial properties. Many nanotherapeutics exhibit antibacterial properties by ROS generation and some penetrate and release metal ions inside the microbe.

Because of their nano-controllable size and features, functionalized nanotherapeutics have emerged as potent drug carrier agents for drug delivery. The recent advancement and the ongoing studies in functionalization provide a way toward novel avenues for antimicrobial applications of nanotherapeutics in the medical and healthcare sectors.

Phytoantioxidant nanotherapeutics provide a significant increment in activity and biocompatibility when conjugated with different phytoconstituents. Nanotherapeutics fabricated with plant extracts exhibit increased antimicrobial action. The green synthesis of nanotherapeutics with antimicrobial activities is being explored with antioxidant properties in controlling various microbial infections.

Further, the application of phyto-based nanoscale antimicrobials found in the controlling life-threatening ailments, facilitated by their effective drug delivery and microbicidal activities, is gaining the interest from researchers and pharmaceutical industries. Further, studies on green synthesized nanotherapeutics possess special physicochemical, optical, and electronic properties that are of great importance for medical applications. Hence, nanotherapeutics in the area of antimicrobial therapy have achieved more attention as an environment-friendly tool to curb microbial threats and inhibit the occurrence of diseases.

References

1 Scheld, W.M. (2012). 286 – introduction to microbial disease: host–pathogen interactions. In: *Goldman's Cecil Medicine*, 24e (ed. L. Goldman and A.I. Schafer), 1761–1762. Philadelphia: W.B. Saunders. https://doi.org/10.1016/B978-1-4377-1604-7.00286-4.

2 Gibson, D.G., Glass, J.I., Lartigue, C. et al. (2010). Creation of a bacterial cell controlled by a chemically synthesized genome. *Science* 329: 52–56.

3 Monsalvo, A.C., Batalle, J.P., Lopez, M.F. et al. (2011). Severe pandemic 2009 H1N1 influenza disease due to pathogenic immune complexes. *Nat. Med.* 17: 195–199.

4 Chattu, V.K. and Yaya, S. (2020). Emerging infectious diseases and outbreaks: implications for women's reproductive health and rights in resource-poor settings. *Reprod. Health* 17 (43). https://doi.org/10.1186/s12978-020-0899-y.

5 Eisenberg, M.C., Kujbida, G., Tuite, A.R. et al. (2013). Examining rainfall and cholera dynamics in Haiti using statistical and dynamic modeling approaches. *Epidemics* 5: 197–207.

6 Dando, S.J., Mackay-Sim, A., Norton, R. et al. (2014). Pathogens penetrating the central nervous system: infection pathways and the cellular and molecular mechanisms of invasion. *Clin. Microbiol. Rev.* 27: 691–726.

7 Ruan, Z., Yu, Y., and Feng, Y. (2020). The global dissemination of bacterial infections necessitates the study of reverse genomic epidemiology. *Brief. Bioinform.* 21: 741–750.

8 Smith, R.M., Schaefer, M.K., Kainer, M.A. et al. (2013). Fungal infections associated with contaminated methylprednisolone injections. *N. Engl. J. Med.* 369: 1598–1609.

9 Bongomin, F., Gago, S., Oladele, R., and Denning, D. (2017). Global and multi-national prevalence of fungal diseases – estimate precision. *J. Fungi (Basel)* 3: 57.

10 Garriss, G., Waldor, M.K., and Burrus, V. (2009). Mobile antibiotic resistance encoding elements promote their own diversity. *PLoS Genet.* 5: e1000775.

11 Weig, M. (1998). Clinical aspects and pathogenesis of Candida infection. *Trends Microbiol.* 6: 468–470.

12 Achkar, J.M. and Fries, B.C. (2010). Candida infections of the genitourinary tract. *Clin. Microbiol. Rev.* 23: 253–273.

13 Kennedy, M.J. and Volz, P.A. (1985). Ecology of *Candida albicans* gut colonization: inhibition of Candida adhesion, colonization, and dissemination from the gastrointestinal tract by bacterial antagonism. *Infect. Immun.* 49: 654–663.

14 Kumamoto, C.A. (2011). Inflammation and gastrointestinal Candida colonization. *Curr. Opin. Microbiol.* 14: 386–391.

15 Nobile, C.J. and Johnson, A.D. (2015). *Candida albicans* biofilms and human disease. *Annu. Rev. Microbiol.* 69: 71–92.

16 Ganguly, S. and Mitchell, A.P. (2011). Mucosal biofilms of *Candida albicans*. *Curr. Opin. Microbiol.* 14: 380–385.

17 Pappas, P.G., Rex, J.H., Sobel, J.D. et al. (2004). Guidelines for treatment of candidiasis. *Clin. Infect. Dis.* 38: 161–189.

18 Calderone, R.A. and Fonzi, W.A. (2001). Virulence factors of *Candida albicans*. *Trends Microbiol.* 9: 327–335.

19 Rajendran, R., May, A., Sherry, L. et al. (2016). Integrating *Candida albicans* metabolism with biofilm heterogeneity by transcriptome mapping. *Sci. Rep.* 6: 35436.

20 Gasanov, U., Hughes, D., and Hansbro, P.M. (2005). Methods for the isolation and identification of *Listeria* spp. and *Listeria monocytogenes*: a review. *FEMS Microbiol. Rev.* 29: 851–875.

21 Drevets, D.A. and Bronze, M.S. (2008). *Listeria monocytogenes*: epidemiology, human disease, and mechanisms of brain invasion. *FEMS Immunol. Med. Microbiol.* 53: 151–165.

22 Liang, T.J. (2009). Hepatitis B: the virus and disease. *Hepatology* 49: S13–S21.

23 Gerlich, W. (1980). Hepatitis B virus contains protein attached to the 5′ terminus of its complete DNA strand. *Cell* 21: 801–809.

24 Gavilanes, F., Gonzalez-Ros, J.M., and Peterson, D.L. (1982). Structure of hepatitis B surface antigen. Characterization of the lipid components and their association with the viral proteins. *J. Biol. Chem.* 257: 7770–7777.

25 Johnson, E.A. (2014). *Clostridium botulinum*. In: *Encyclopedia of Food Microbiology*, 458–462. Elsevier. https://doi.org/10.1016/B978-0-12-384730-0.00072-0.

26 Harris, A. (2016). *Clostridium botulinum*. In: *Encyclopedia of Food and Health*, 141–145. Elsevier. https://doi.org/10.1016/B978-0-12-384947-2.00172-0.

27 Yuen, K.S., Ye, Z.W., Fung, S.Y. et al. (2020). SARS-CoV-2 and COVID-19: the most important research questions. *Cell Biosci.* 10: 40.

28 Yang, Y., Xiao, Z., Ye, K. et al. (2020). SARS-CoV-2: characteristics and current advances in research. *Virol. J.* 17: 117.

29 Sharma, A., De Rosa, M., Singla, N. et al. (2021). Tuberculosis: an overview of the immunogenic response, disease progression, and medicinal chemistry efforts in the last decade toward the development of potential drugs for extensively drug-resistant tuberculosis strains. *J. Med. Chem.* 64: 4359–4395.

30 Sandhu, G. (2011). Tuberculosis: current situation, challenges and overview of its control programs in India. *J. Global Infect. Dis.* 3: 143.

31 Piszczek, J., St. Jean, R., and Khaliq, Y. (2015). Gonorrhea: treatment update for an increasingly resistant organism. *Can. Pharm. J.* 148: 82–89.

32 Teklebirhan, G. and Bitew, A. (2015). Prevalence of dermatophytic infection and the spectrum of dermatophytes in patients attending a tertiary hospital in Addis Ababa, Ethiopia. *Int. J. Microbiol.* 2015: 1–5.

33 Sahoo, A. and Mahajan, R. (2016). Management of tinea corporis, tinea cruris, and tinea pedis: a comprehensive review. *Indian Dermatol. Online J.* 7: 77.

34 Alter, S.J., McDonald, M.B., Schloemer, J. et al. (2018). Common child and adolescent cutaneous infestations and fungal infections. *Curr. Probl. Pediatr. Adolesc. Health Care* 48: 3–25.

35 Shy, R. (2007). Tinea corporis and tinea capitis. *Pediatr. Rev.* 28: 164–174.

36 Leung, A.K., Lam, J.M., Leong, K.F., and Hon, K.L. (2020). Tinea corporis: an updated review. *Drugs Context.* 9: 1–12.

37 Seung, K.J., Keshavjee, S., and Rich, M.L. (2015). Multidrug-resistant tuberculosis and extensively drug-resistant tuberculosis. *Cold Spring Harb. Perspect. Med.* 5: a017863.

38 Prestinaci, F., Pezzotti, P., and Pantosti, A. (2015). Antimicrobial resistance: a global multifaceted phenomenon. *Pathog. Glob. Health.* 109: 309–318.

39 Poole, K. (2012). Bacterial stress responses as determinants of antimicrobial resistance. *J. Antimicrob. Chemother.* 67: 2069–2089.

40 Kuczyńska-Wiśnik, D., Matuszewska, E., and Laskowska, E. (2010). *Escherichia coli* heat-shock proteins IbpA and IbpB affect biofilm formation by influencing the level of extracellular indole. *Microbiology* 156: 148–157.

41 Karunamoorthi, K., Jegajeevanram, K., Vijayalakshmi, J., and Mengistie, E. (2013). Traditional medicinal plants: a source of phytotherapeutic modality in resource-constrained health care settings. *J. Evid. Based Complement. Altern. Med.* 18: 67–74.

42 Yuan, H., Ma, Q., Ye, L., and Piao, G. (2016). The traditional medicine and modern medicine from natural products. *Molecules* 21: 559.

43 War, A.R., Paulraj, M.G., Ahmad, T. et al. (2012). Mechanisms of plant defense against insect herbivores. *Plant Signal. Behav.* 7: 1306–1320.

44 Sofowora, A., Ogunbodede, E., and Onayade, A. (2013). The role and place of medicinal plants in the strategies for disease prevention. *Afr. J. Trad. Compl. Alt. Med.* 10: 210–229.

45 Isah, T. (2019). Stress and defense responses in plant secondary metabolites production. *Biol. Res.* 52: 39.

46 Górniak, I., Bartoszewski, R., and Króliczewski, J. (2019). Comprehensive review of antimicrobial activities of plant flavonoids. *Phytochem. Rev.* 18: 241–272.

47 Vo, T.T.T., Chu, P.-M., Tuan, V.P. et al. (2020). The promising role of antioxidant phytochemicals in the prevention and treatment of periodontal disease via the inhibition of oxidative stress pathways: updated insights. *Antioxidants* 9: 1211.

48 Pandey, M.M., Rastogi, S., and Rawat, A.K.S. (2013). Indian traditional ayurvedic system of medicine and nutritional supplementation. *Evid. Based Complement. Alternat. Med.* 2013: 1–12.

49 Nuñez, L. and D' Aquino, M. (2012). Microbicide activity of clove essential oil (*Eugenia caryophyllata*). Braz. *J. Microbiol.* 43: 1255–1260.

50 Vasconcelos, N.G., Croda, J., and Simionatto, S. (2018). Antibacterial mechanisms of cinnamon and its constituents: a review. *Microb. Pathog.* 120: 198–203.

51 Alizadeh, M. and Kheirouri, S. (2019). Curcumin reduces malondialdehyde and improves antioxidants in humans with diseased conditions: a comprehensive meta-analysis of randomized controlled trials. *Biomedicine* 9: 23.

52 Hewlings, S. and Kalman, D. (2017). Curcumin: a review of its effects on human health. *Foods* 6: 92.

53 Karuppiah, P. and Rajaram, S. (2012). Antibacterial effect of *Allium sativum* cloves and *Zingiber officinale* rhizomes against multiple-drug resistant clinical pathogens. *Asian Pac. J. Trop. Biomed.* 2: 597–601.

54 Dorra, N., El-Berrawy, M., Sallam, S., and Mahmoud, R. (2019). Evaluation of antiviral and antioxidant activity of selected herbal extracts. *J. High Inst. Publ. Health* 49: 36–40.

55 Tzima, K., Makris, D., Nikiforidis, C.V., and Mourtzinos, I. (2015). Potential use of rosemary, propolis and thyme as natural food preservatives. *J. Nutr. Health* 1. https://doi.org/10.13188/2469-4185.1000002.

56 Piccaglia, R., Marotti, M., Giovanelli, E. et al. (1993). Antibacterial and antioxidant properties of Mediterranean aromatic plants. *Ind. Crop. Prod.* 2: 47–50.

57 Ali, H.M., Elgat, W.A.A.A., El-Hefny, M. et al. (2021). New approach for using of *Mentha longifolia* L. and *Citrus reticulata* L. essential oils as wood-biofungicides: GC-MS, SEM, and MNDO quantum chemical studies. *Materials* 14: 1361.

58 Talei, G.-R., Mohammadi, M., Bahmani, M., and Kopaei, M. (2017). Synergistic effect of *Carum copticum* and *Mentha piperita* essential oils with ciprofloxacin, vancomycin, and gentamicin on Gram-negative and Gram-positive bacteria. *Int. J. Pharm. Investig.* 7: 82.

59 Benkeblia, N. (2005). Free-radical scavenging capacity and antioxidant properties of some selected onions (*Allium cepa* L.) and garlic (*Allium sativum* L.) extracts. *Braz. Arch. Biol. Technol.* 48: 753–759.

60 Abkhoo, J. and Jahani, S. (2016). Antibacterial effects of aqueous and ethanolic extracts of medicinal plants against pathogenic strains. *Int. J. Infect.* 4. https://doi.org/10.5812/iji.42624.

61 Paul, D. (2016). A review on biological activities of common mallow (*Malva sylvestris* L.). *Innov. J. Life Sci.* 4: 534626.

62 Luís, Â., Neiva, D., Pereira, H. et al. (2014). Stumps of *Eucalyptus globulus* as a source of antioxidant and antimicrobial polyphenols. *Molecules* 19: 16428–16446.

63 Amakura, Y., Yoshimura, M., Sugimoto, N. et al. (2009). Marker constituents of the natural antioxidant eucalyptus leaf extract for the evaluation of food additives. *Biosci. Biotechnol. Biochem.* 73: 1060–1065.

64 Peschel, W., Dieckmann, W., Sonnenschein, M., and Plescher, A. (2007). High antioxidant potential of pressing residues from evening primrose in comparison to other oilseed cakes and plant antioxidants. *Ind. Crop. Prod.* 25: 44–54.

65 Ayaz, M., Ullah, F., Sadiq, A. et al. (2019). Synergistic interactions of phytochemicals with antimicrobial agents: potential strategy to counteract drug resistance. *Chem. Biol. Interact.* 308: 294–303.

66 Hussein, R.A. and El-Anssary, A.A. (2019). Plants secondary metabolites: the key drivers of the pharmacological actions of medicinal plants. In: *Herbal Medicine* (ed. F.P. Builders). IntechOpen. https://doi.org/10.5772/intechopen.76139.

67 Zhao, Y., Wu, Y., and Wang, M. (2015). Bioactive substances of plant origin. In: *Handbook of Food Chemistry* (ed. C. PCK and B.M. Mehta), 967–1008. Berlin, Heidelberg: Springer. https://doi.org/10.1007/978-3-642-36605-5_13.

68 Ekor, M. (2014). The growing use of herbal medicines: issues relating to adverse reactions and challenges in monitoring safety. *Front. Pharmacol.* 4. https://doi.org/10.3389/fphar.2013.00177.

69 Thawabteh, A., Juma, S., Bader, M. et al. (2019). The biological activity of natural alkaloids against herbivores. *Cancer. Cells Pathog. Toxins* 11: 656.

70 Cushnie, T.P.T., Cushnie, B., and Lamb, A.J. (2014). Alkaloids: an overview of their antibacterial, antibiotic-enhancing and antivirulence activities. *Int. J. Antimicrob. Agents* 44: 377–386.

71 Papuc, C., Goran, G.V., Predescu, C.N. et al. (2017). Plant polyphenols as antioxidant and antibacterial agents for shelf-life extension of meat and meat products: classification, structures, sources, and action mechanisms: polyphenols extending meat shelf-life. *Compr. Rev. Food Sci. Food Saf.* 16: 1243–1268.

72 Kumar, N. and Goel, N. (2019). Phenolic acids: natural versatile molecules with promising therapeutic applications. *Biotechnol. Rep.* 24: e00370.

73 Kocaçalışkan, I., Talan, I., and Terzi, I. (2006). Antimicrobial activity of catechol and pyrogallol as allelochemicals. *Z. Naturforsch. C* 61: 639–642.

74 Hussain, W., Badshah, L., Ullah, M. et al. (2018). Quantitative study of medicinal plants used by the communities residing in Koh-e-Safaid Range, northern Pakistani-Afghan borders. *J. Ethnobiol. Ethnomed.* 14: 30.

75 Goud, M., Komraiah, A., Rao, K. et al. (2008). Antibacterial activity of some folklore medicinal plants from South India. *Afr. J. Trad. Compl. Alt. Med.* 5: 421.

76 Kumar, S. and Pandey, A.K. (2013). Chemistry and biological activities of flavonoids: an overview. *Sci. World J.* 2013: 1–16.

77 Khameneh, B., Iranshahy, M., Soheili, V., and Fazly Bazzaz, B.S. (2019). Review on plant antimicrobials: a mechanistic viewpoint. *Antimicrob. Resist. Infect. Control* 8: 118.

78 Kajiya, K., Hojo, H., Suzuki, M. et al. (2004). Relationship between antibacterial activity of (+)-catechin derivatives and their interaction with a model membrane. *J. Agric. Food Chem.* 52: 1514–1519.

79 Watanbe, H., Miyaji, C., Makino, M., and Abo, T. (1996). Therapeutic effects of glycyrrhizin in mice infected with LP-BM5 murine retrovirus and mechanisms involved in the prevention of disease progression. *Biotherapy* 9: 209–220.

80 Bacanlı, M., Başaran, A.A., and Başaran, N. (2018). Galangin as a plant phenolic and usage in health and disease. In: *Polyphenols: Prevention and Treatment of Human Disease* (ed. R. Watson, V. Preedy and S. Zibadi), 433–438. Elsevier. https://doi.org/10.1016/B978-0-12-813008-7.00034-5.

81 Chabot, S., Bel-Rhlid, R., Chênevert, R., and Piché, Y. (1992). Hyphal growth promotion *in vitro* of the VA mycorrhizal fungus, *Gigaspora margarita* Becker & Hall, by the activity of structurally specific flavonoid compounds under CO_2-enriched conditions. *New Phytol.* 122: 461–467.

82 Widsten, P., Cruz, C.D., Fletcher, G.C. et al. (2014). Tannins and extracts of fruit byproducts: antibacterial activity against foodborne bacteria and antioxidant capacity. *J. Agric. Food Chem.* 62: 11146–11156.

83 Kaczmarek, B. (2020). Tannic acid with antiviral and antibacterial activity as a promising component of biomaterials – a mini review. *Materials* 13: 3224.

84 Lillehoj, H., Liu, Y., Calsamiglia, S. et al. (2018). Phytochemicals as antibiotic alternatives to promote growth and enhance host health. *Vet. Res.* 49: 76.

85 Akinwumi, B., Bordun, K.-A., and Anderson, H. (2018). Biological activities of stilbenoids. *Int. J. Mol. Sci.* 19: 792.

86 Repossi, G., Das, U.N., and Eynard, A.R. (2020). Molecular basis of the beneficial actions of resveratrol. *Arch. Med. Res.* 51: 105–114.

87 Piotrowska, H., Kucinska, M., and Murias, M. (2012). Biological activity of piceatannol: leaving the shadow of resveratrol. *Rev. Mutat. Res.* 750: 60–82.

88 Chouhan, S., Sharma, K., and Guleria, S. (2017). Antimicrobial activity of some essential oils – present status and future perspectives. *Medicine* 4: 58.

89 Ayafor, J.F., Tchuendem, M.H.K., Nyasse, B. et al. (1994). Aframodial and other bioactive diterpenoids from *Aframomum* species. *Pure Appl. Chem.* 66: 2327–2330.

90 Othman, L., Sleiman, A., and Abdel-Massih, R.M. (2019). Antimicrobial activity of polyphenols and alkaloids in Middle Eastern plants. *Front. Microbiol.* 10: 911.

91 Wang, J., Li, P., Li, B. et al. (2014). Bioactivities of compounds from *Elephantopus scaber*, an ethnomedicinal plant from Southwest China. *Evid. Based Complement. Alternat. Med.* 2014: 1–7.

92 Badr, G.M. and AL-Mulhim, J.A. (2014). The protective effect of aged garlic extract on nonsteroidal anti-inflammatory drug-induced gastric inflammations in male albino rats. *Evid. Based Complement. Alternat. Med.* 2014: 1–9.

93 Prusty, J.S. and Kumar, A. (2020). Coumarins: antifungal effectiveness and future therapeutic scope. *Mol. Divers.* 24: 1367–1383.

94 Venugopala, K.N., Rashmi, V., and Odhav, B. (2013). Review on natural coumarin lead compounds for their pharmacological activity. *Biomed. Res. Int.* 2013: 1–14.

95 Wojnicz, D., Kucharska, A.Z., Sokół-Łętowska, A. et al. (2012). Medicinal plants extracts affect virulence factors expression and biofilm formation by the uropathogenic *Escherichia coli*. *Urol. Res.* 40: 683–697.

96 Patra, A.K. (2012). An overview of antimicrobial properties of different classes of phytochemicals. In: *Dietary Phytochemicals and Microbes* (ed. A.K. Patra), 1–32. Dordrecht: Springer. https://doi.org/10.1007/978-94-007-3926-0_1.

97 Forni, C., Facchiano, F., Bartoli, M. et al. (2019). Beneficial role of phytochemicals on oxidative stress and age-related diseases. *Biomed. Res. Int.* 2019: 1–16.

98 Swallah, M.S., Fu, H., Sun, H. et al. (2020). The impact of polyphenol on general nutrient metabolism in the monogastric gastrointestinal tract. *J. Food Qual.* 2020: 1–12.

99 Atanasov, A.G., Waltenberger, B., Pferschy-Wenzig, E.-M. et al. (2015). Discovery and resupply of pharmacologically active plant-derived natural products: a review. *Biotechnol. Adv.* 33: 1582–1614.

100 Patra, J.K., Das, G., Fraceto, L.F. et al. (2018). Nano based drug delivery systems: recent developments and future prospects. *J. Nanobiotechnol.* 16: 71.

101 Ajazuddin, S.S. (2010). Applications of novel drug delivery system for herbal formulations. *Fitoterapia* 81: 680–689.

102 Sivakumar, P.A. and Panduranga Rao, K. (2003). The use of cholesteryl pullulan for the preparation of stable vincristine liposomes. *Carbohydr. Polym.* 51: 327–332.

103 El-Samaligy, M.S., Afifi, N.N., and Mahmoud, E.A. (2006). Increasing bioavailability of silymarin using a buccal liposomal delivery system: preparation and experimental design investigation. *Int. J. Pharm.* 308: 140–148.

104 Ghasemiyeh, P. and Mohammadi-Samani, S. (2018). Solid lipid nanoparticles and nanostructured lipid carriers as novel drug delivery systems: applications, advantages and disadvantages. *Res. Pharma Sci.* 13: 288.

105 Mei, Z., Wu, Q., Hu, S. et al. (2005). Triptolide loaded solid lipid nanoparticle hydrogel for topical application. *Drug Dev. Ind. Pharm.* 31: 161–168.

106 Chen, H., Chang, X., Du, D. et al. (2006). Podophyllotoxin-loaded solid lipid nanoparticles for epidermal targeting. *J. Control. Release* 110: 296–306.

107 Tiyaboonchai, W., Tungpradit, W., and Plianbangchang, P. (2007). Formulation and characterization of curcuminoids loaded solid lipid nanoparticles. *Int. J. Pharm.* 337: 299–306.

108 Hu, L., Xing, Q., Meng, J., and Shang, C. (2010). Preparation and enhanced oral bioavailability of cryptotanshinone-loaded solid lipid nanoparticles. *AAPS PharmSciTech* 11: 582–587.

109 Feng, N. (2012). Preparation and characterization of solid lipid nanoparticles loaded with frankincense and myrrh oil. *Int. J. Nanomed.* 7: 2033–2043.

110 Kim, T., Baek, J., Zhe Bai, C., and Park, J. (2007). Arginine-conjugated polypropylenimine dendrimer as a non-toxic and efficient gene delivery carrier. *Biomaterials* 28: 2061–2067.

111 Aziz, Z.A.A. and Setapar, S.H.M. (2020). Nanotechnology: an effective approach for enhancing therapeutics and bioavailability of phytomedicines. In: *Functional Bionanomaterials* (ed. D. Thangadurai, J. Sangeetha and R. Prasad), 47–71. Cham: Springer International Publishing. https://doi.org/10.1007/978-3-030-41464-1_3.

112 Liu, M., Dong, J., Yang, Y. et al. (2005). Anti-inflammatory effects of triptolide loaded poly(d,l-lactic acid) nanoparticles on adjuvant-induced arthritis in rats. *J. Ethnopharmacol.* 97: 219–225.

113 Sahu, A., Bora, U., Kasoju, N., and Goswami, P. (2008). Synthesis of novel biodegradable and self-assembling methoxy poly(ethylene glycol)-palmitate nanocarrier for curcumin delivery to cancer cells. *Acta Biomater.* 4: 1752–1761.

114 Zeisser-Labouèbe, M., Lange, N., Gurny, R., and Delie, F. (2006). Hypericin-loaded nanoparticles for the photodynamic treatment of ovarian cancer. *Int. J. Pharm.* 326: 174–181.

115 Juškaitė, V., Ramanauskienė, K., and Briedis, V. (2015). Design and formulation of optimized microemulsions for dermal delivery of resveratrol. *Evid. Based Complement. Alternat. Med.* 2015: 1–10.

116 Suresh, K. (2013). Novel topical drug carriers as a tool for treatment of psoriasis: progress and advances. *Afr. J. Pharm. Pharmacol* 7: 138–147.

117 Wang, S., Tan, M., Zhong, Z. et al. (2011). Nanotechnologies for curcumin: an ancient puzzler meets modern solutions. *J. Nanomater.* 2011: 1–8.

118 Chen, H., Chang, X., Weng, T. et al. (2004). A study of microemulsion systems for transdermal delivery of triptolide. *J. Control. Release* 98: 427–436.

119 Amit, G., Ashawat, M.S., Shailendra, S., and Swarnlata, S. (2007). Phytosome: a novel approach toward functional cosmetics. *J. Plant Sci.* 2: 644–649.

120 Sánchez-López, E., Gomes, D., Esteruelas, G. et al. (2020). Metal-based nanoparticles as antimicrobial agents: an overview. *Nanomaterials* 10: 292.

121 Khalil, I., Yehye, W.A., Etxeberria, A.E. et al. (2019). Nanoantioxidants: recent trends in antioxidant delivery applications. *Antioxidants* 9: 24.

122 Leong, C.R., Tong, W.Y., Tan, W.N. et al. (2020). Synthesis of curcumin quantum dots and their antimicrobial activity on necrotizing fasciitis causing bacteria. *Mater. Today Proc.* 31: 31–35.

123 Shlar, I., Droby, S., Choudhary, R., and Rodov, V. (2017). The mode of antimicrobial action of curcumin depends on the delivery system: monolithic nanoparticles vs. supramolecular inclusion complex. *RSC Adv.* 7: 42559–42569.

124 Annu, A.A. and Green, A.S. (2019). Synthesis of metal, metal oxide nanoparticles, and their various applications. In: *Handbook of Ecomaterials* (ed. M. LMT, O.V. Kharissova and B.I. Kharisov), 2281–2325. Cham: Springer International Publishing. https://doi.org/10.1007/978-3-319-68255-6_115.

125 Roy, P., Das, B., Mohanty, A., and Mohapatra, S. (2017). Green synthesis of silver nanoparticles using *Azadirachta indica* leaf extract and its antimicrobial study. *Appl. Nanosci.* 7: 843–850.

126 Vasantharaj, S., Sathiyavimal, S., Senthilkumar, P. et al. (2019). Biosynthesis of iron oxide nanoparticles using leaf extract of *Ruellia tuberosa*: antimicrobial properties and their applications in photocatalytic degradation. *J. Photochem. Photobiol. B Biol.* 192: 74–82.

127 Polash, S.A., Hossain, M.M., Saha, T., and Sarker, S.R. (2021). Biogenic silver nanoparticles: a potent therapeutic agent. In: *Emerging Trends in Nanomedicine* (ed. S. Singh), 81–127. Singapore: Springer. https://doi.org/10.1007/978-981-15-9920-0_4.

128 Banerjee, P., Satapathy, M., Mukhopahayay, A., and Das, P. (2014). Leaf extract mediated green synthesis of silver nanoparticles from widely available Indian plants: synthesis, characterization, antimicrobial property and toxicity analysis. *Bioresour. Bioprocess.* 1: 3.

129 Manikandan, V., Velmurugan, P., Jayanthi, P. et al. (2018). Biogenic synthesis from *Prunus × yedoensis* leaf extract, characterization, and photocatalytic and antibacterial activity of TiO_2 nanoparticles. *Res. Chem. Intermed.* 44: 2489–2502.

130 Nadeem, M., Abbasi, B.H., Younas, M. et al. (2017). A review of the green syntheses and antimicrobial applications of gold nanoparticles. *Green Chem. Lett. Rev.* 10: 216–227.

131 Santhoshkumar, T., Rahuman, A.A., Jayaseelan, C. et al. (2014). Green synthesis of titanium dioxide nanoparticles using *Psidium guajava* extract and its antibacterial and antioxidant properties. *Asian Pac. J. Trop. Med.* 7: 968–976.

132 Subhapriya, S. and Gomathipriya, P. (2018). Green synthesis of titanium dioxide (TiO_2) nanoparticles by *Trigonella foenum-graecum* extract and its antimicrobial properties. *Microb. Pathog.* 116: 215–220.

133 Regiel, A., Irusta, S., Kyzioł, A. et al. (2013). Preparation and characterization of chitosan–silver nanocomposite films and their antibacterial activity against *Staphylococcus aureus*. *Nanotechnology* 24: 015101.

134 Singh, A., Gautam, P.K., Verma, A. et al. (2020). Green synthesis of metallic nanoparticles as effective alternatives to treat antibiotics resistant bacterial infections: a review. *Biotechnol. Rep.* 25: e00427.

135 Shehensha, S. and Jyothi, M.V. (2020). Synthesis of silver nanoparticles using *Nigella sativa* seed extracts and assessment of their antibacterial activity. *Int. J. Res. Pharm. Sci.* 11: 2525–2532.

136 El-Far, A., Al Jaouni, S., Li, W., and Mousa, S. (2018). Protective roles of thymoquinone nanoformulations: potential nanonutraceuticals in human diseases. *Nutrients* 10: 1369.

137 Khan, M.A. (2018). Antimicrobial action of thymoquinone. In: *Molecular and Therapeutic actions of Thymoquinone* (ed. H. Younus), 57–64. Singapore: Springer. https://doi.org/10.1007/978-981-10-8800-1_5.

138 Shanmuganathan, R., Sathishkumar, G., Brindhadevi, K., and Pugazhendhi, A. (2020). Fabrication of naringenin functionalized-Ag/RGO nanocomposites for potential bactericidal effects. *J. Mater. Res. Technol.* 9: 7013–7019.

139 Bhia, M., Motallebi, M., Abadi, B. et al. (2021). Naringenin nano-delivery systems and their therapeutic applications. *Pharmaceutics* 13: 291.

140 He, Y., Du, T., Zheng, Z. et al. (2013). Green synthesis of silver nanoparticles by *Chrysanthemum morifolium* Ramat. extract and their application in clinical ultrasound gel. *Int. J. Nanomed.* 8: 1809–1815.

141 Wang, L., Hu, C., and Shao, L. (2017). The antimicrobial activity of nanoparticles: present situation and prospects for the future. *Int. J. Nanomed.* 12: 1227–1249.

142 Ilk, S., Sağlam, N., Özgen, M., and Korkusuz, F. (2017). Chitosan nanoparticles enhances the anti-quorum sensing activity of kaempferol. *Int. J. Biol. Macromol.* 94: 653–662.

143 Ming, D., Wang, D., Cao, F. et al. (2017). Kaempferol inhibits the primary attachment phase of biofilm formation in *Staphylococcus aureus*. *Front. Microbiol.* 8: 2263.

144 Alsamhary, K., Al-Enazi, N., Alshehri, W.A., and Ameen, F. (2020). Gold nanoparticles synthesized by flavonoid tricetin as a potential antibacterial nanomedicine to treat respiratory infections causing opportunistic bacterial pathogens. *Microb. Pathog.* 139: 103928.

145 Kamal, T., Khan, S.B., and Asiri, A.M. (2020). Antimicrobial activities of metal containing compounds and hybrids. *Curr. Pharm. Des.* 26: 5881–5891.

15

Nanotherapeutics of Phytoantioxidants for Viral Infections

Rakesh Sagar, Sharad P. Pandey, and Abhishek K. Sah

Department of Pharmacy, Shri Govindram Seksariya Institute of Technology and Science (SGSITS), Indore, Madhya Pradesh, India

15.1 Introduction

Viral infection causes serious health issues and keeps financial and socioeconomic burdens on patients and their family members. Globally, it is the major health problems affecting millions of people worldwide [1]. And due to bacterial, viral, fungi, and parasitic infections, about 15 million deaths have been reported globally while human immunodeficiency virus (HIV) along with respiratory infections are among the primary causes of morbidity [2]. However, various synthetic antiviral therapeutic agents have been already available in the market but the major challenges with these agents are systemic toxic effects, and drug resistance especially with HIV [3, 4] and influenza medication [5]. In this context, plant extract and bioactive are better alternatives to synthetic drugs as herbs and plants have been always the prime factors of medicinal sources since ancient ages globally. These herbs and plants not only offerred traditional and ethnic medicines but also showed better, efficient, and novel bioactive molecules [6]. Despite being convenient, it is associated with some other limitations. It includes low solubility and less drug bioavailability. Therefore, there is an urgent need for some novel nanocarrier-based drug delivery system for the improvement of solubility, bioavailability, pharmacokinetics, and therapeutic profile along with safer and better patient compliance. Various plants and their active compounds showed antioxidants and antiviral potential are listed in Table 15.1. For the effective delivery of biomolecules, a novel delivery system (NDS) has been explored, which defeats the limitations of conventional herbal formulations and improved the therapeutic potential along with no side effects [8]. In these NDSs, various phytochemical-loaded nanosystems have been investigated. These include liposomes, polymeric NPs, nanoemulsions, nanospheres, solid lipid NPs (SLNs), etc. [9]. In the present chapter, we have highlighted and compiled the various investigations of novel nano phytoantioxidants and also reported the clinical finding for the nanotherapeutics in the treatment of viral infections.

15.1.1 Challenges in Antiviral Treatment

Treatment of viral infection and, subsequently, its eradication from the host cell is very crucial and it is mainly due to the formation of reservoirs within the cellular tissues and physiological site. It includes the blood–brain barrier (BBB) and the blood–testis barrier [10]. Therefore, they are

Phytoantioxidants and Nanotherapeutics, First Edition. Edited by Mithun Rudrapal.

Table 15.1 Some of the phytochemical for antiviral potential.

Plant name	Phytochemical	Plant part used	Chemical structures
Found in many plants	Apigenin	—	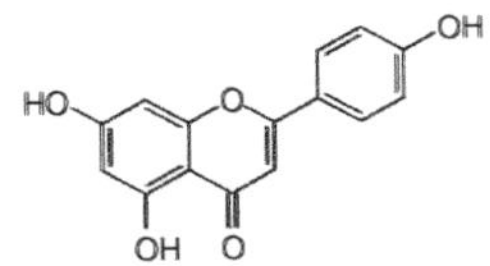
Aloe vera	Aloe emodine	 Leaf	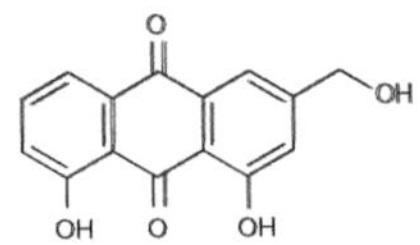 Aloe emodin | 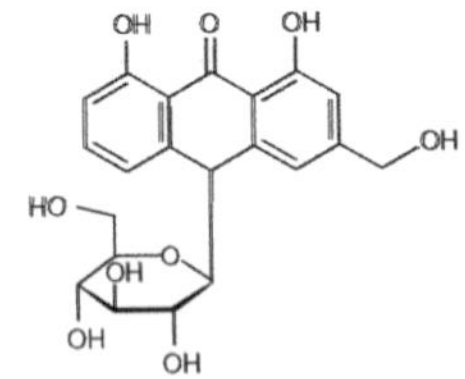 Aloin/Barbaloin | 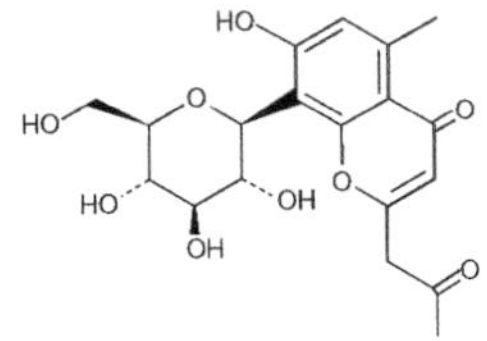 Aloesin | 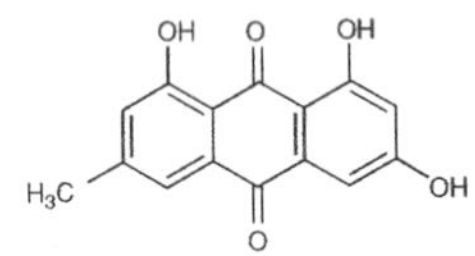 Emodin 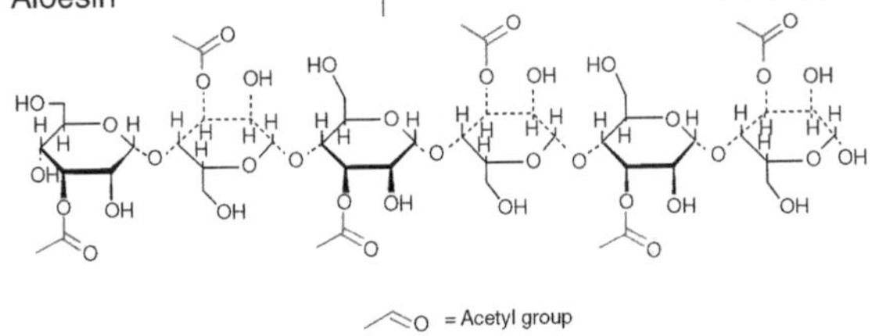 Acemannan

	Baicalein	 Flower	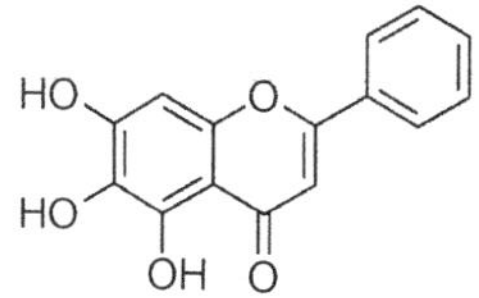
	Biochanin A	 Flower	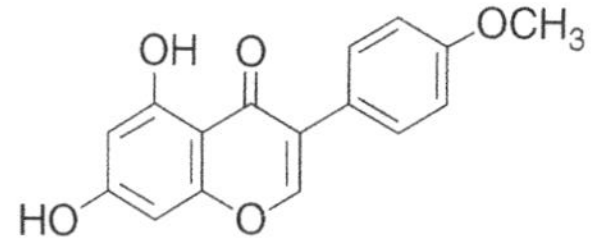
Camellia sinensis	Epigallocathechin-3-gallate	 Leaf	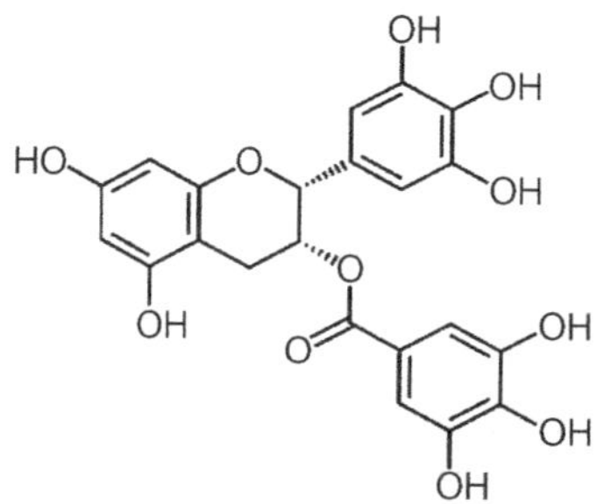

(*Continued*)

Table 15.1 (Continued)

Plant name	Phytochemical	Plant part used	Chemical structures
Capparis spinosa		 Leaf	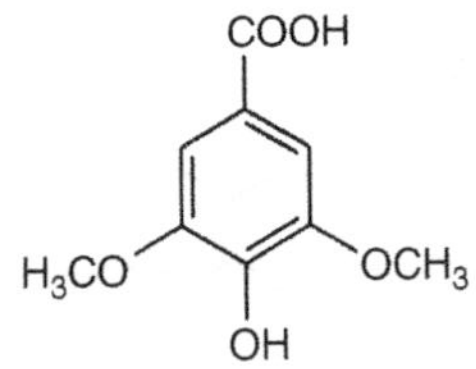
Cassine xylocarpa	Pentacyclic lupane-type triterpenoids	 Leaf and fruits	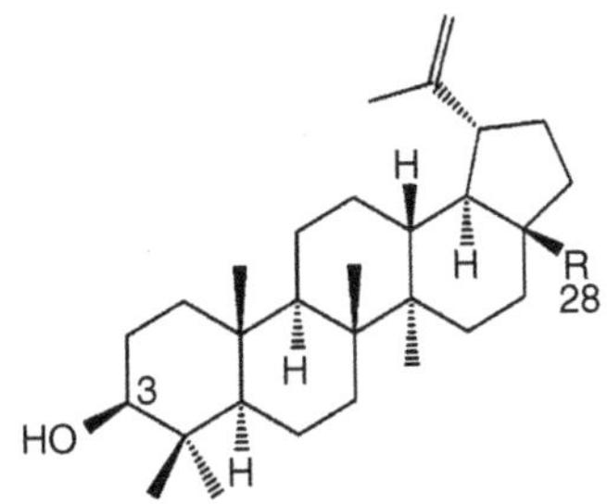 Lupeol (**1**) R = CH_3 Betulin (**2**) R = CH_2OH Betulinic acid (**3**) R = COOH

Curcuma longa	Curcumin	 Root	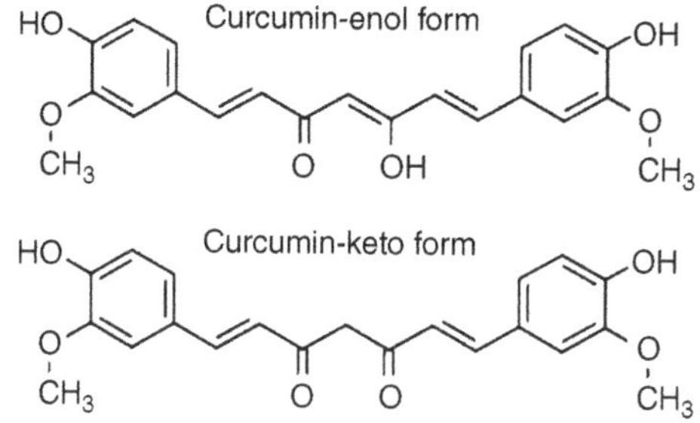
Cyperus rotundus	Cyperene-3, 8-dione, 14-hydroxy cyperotundone, 14-acetoxy cyperotundone, 3β-hydroxycyperenoic acid and sugetriol-3, 9-diacetate	 Whole plant	—
Daphne gnidium	Daphnetoxin, gnidicin, gniditrin and excoecariatoxin	 Leaf and flower	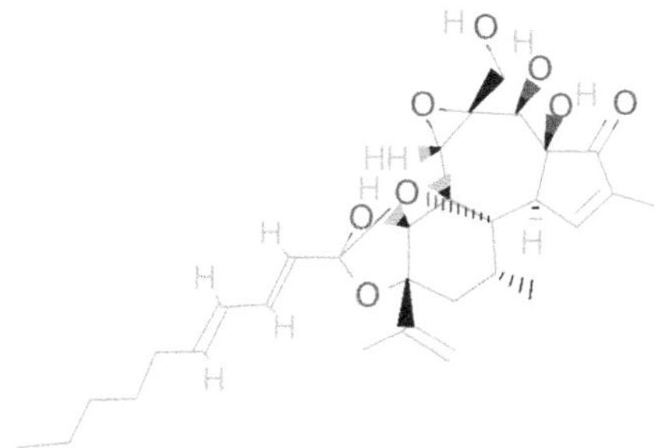

(*Continued*)

Table 15.1 (Continued)

Plant name	Phytochemical	Plant part used	Chemical structures
Diospyros kaki	Licocoumarone, licoflavonol, glyasperin D, 18 β-glycyrrhetinic acid, luteolin, vitexin, apigenin-7-*O*-glucoside	 Leaf and fruits	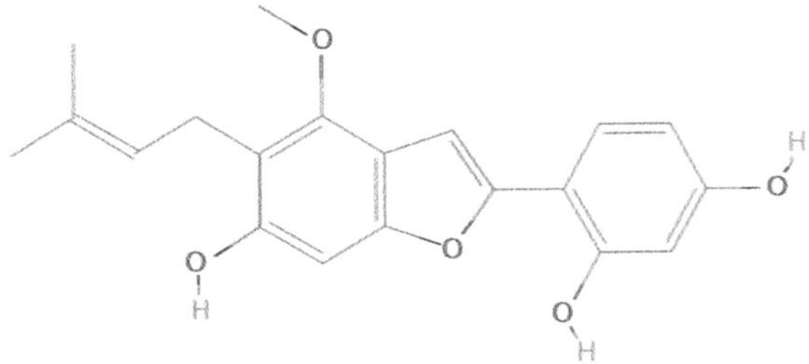
Ficus benjamina	Rutin, kaempferol 3-*O*-rutinoside and kaempferol 3-*O*-robinobioside	 Leaf	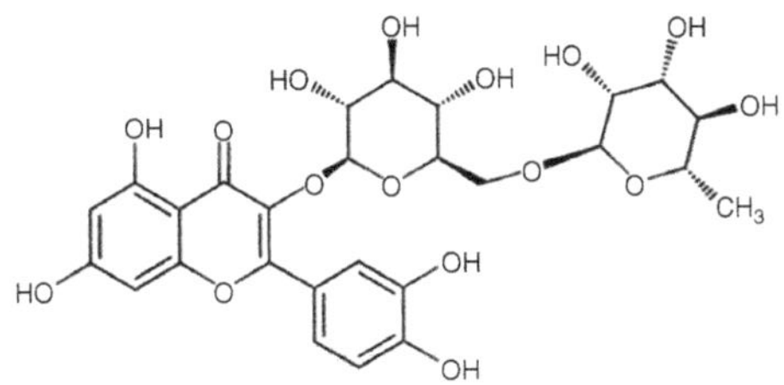

Glycyrrhiza glabra	Liquirtin, isoliquertin liquiritigenin and rhamnoliquirilin	 Root	(a) (b) 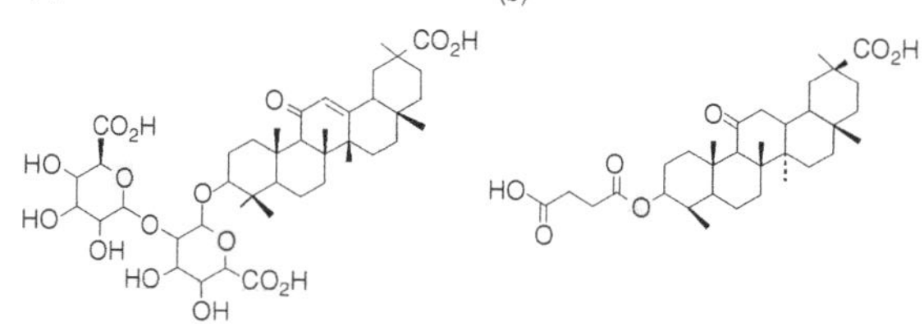
Magnolia officinalis	Honokiol	 Flower	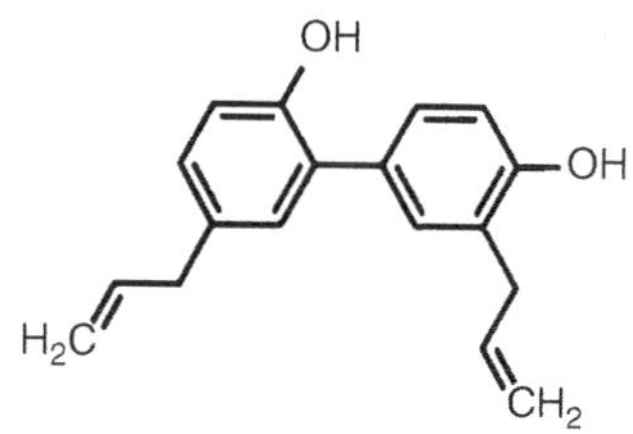
Maytenus cuzcoina	Pentacyclic lupane-type triterpenoids	 Leaf and flower	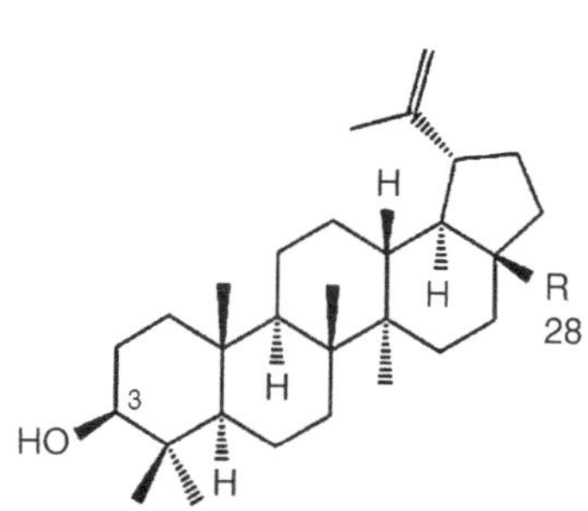 Lupeol (**1**) R = CH_3 Betulin (**2**) R = CH_2OH Betulinic acid (**3**) R = COOH

(*Continued*)

Table 15.1 (Continued)

Plant name	Phytochemical	Plant part used	Chemical structures
Panax ginseng	Epigallocatechin gallate, theaflavin digallate, genistein, hesperidin, neohesperidin, diosmin, pectic polysaccharides	 Root	—
Phyllanthus emblica	Highly oxygenated norbisabolane sesquiterpenoids	 Fruits	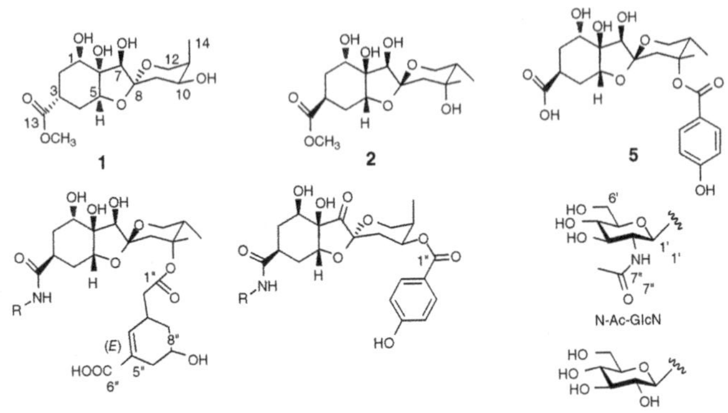 1, 2, 5; 3 R= N-Ac-GlcN; 4 R= N-Ac-GlcN
Vitis labrusca	Resveratrol, piceatannol, *trans*-arachidin-1 and *trans*-arachidin-3	 Fruits	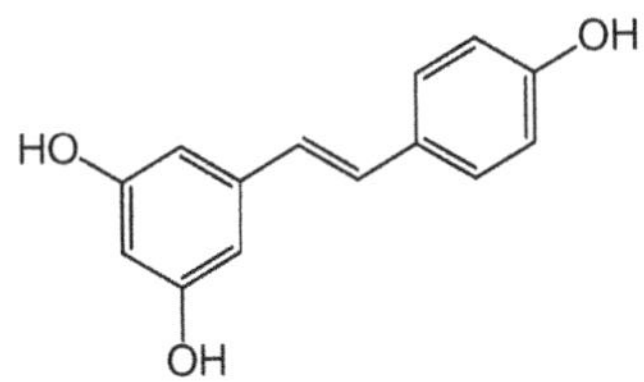

Source: Based on Ref. [7].

unreachable to conventional medications. Nanotherapies have the capability to cross the tight biological membranes and combat these medical complications and, consequently, emerge as a novel tool for effective therapy of viral infection [11]. Additionally, the other challenges for viral therapy in which RNA interference (RNAi) has been used and ultimately incapable to permeate the cellular biomembrane is due to their high molecular weight, negative charges [12, 13], low elimination half-life, engulfed by phagocytes along with lethal action of excited immune responses [14]. All of these challenges restrict the effectiveness clinically. However, loading of the suitable siRNA into the nano-based carrier system leads to circumventing the limitations and possesses success in the intervention of viral infections [15].

15.2 Types of Viral Infections, Drugs, Plant-based Drugs, and Phytoantioxidants Available for the Treatments

15.2.1 Types of Viral Infections

Viruses compel the intracellular parasites, which can survive as an active entity but exhibit static outside of the host cell and all viruses show similar properties for infections (Figure 15.1). Most of the viruses get entered into the host cell by means of respiratory, gastrointestinal routes, venereal contact, or through the bloodstream.

Other pathogenic viruses are in the air. These include common cold (rhinoviruses), measles, and influenza viruses, which get entered into the host by fine droplets when coming in contact with infected persons. Additionally, this pathogen penetrates the biological membrane of the respiratory tract and replicates within the host cells and starts inflammatory reactions, which appear as a symptomatic infection. However, these infections can be spread to other healthy people by means of sneezing, coughing, and coming in contact with susceptible people. Polioviruses can affect the host by the consumption of unhygienic and contaminated food and water. These viruses get infected in the intestinal region followed by local inflammation, diarrheas, and enteric fevers. These infections replicate into the systemic circulation, which affect other vital organs of the body. Some of the virus infections of dengue, encephalitis, and yellow fever are transmitted via blood feed vectors like mosquitoes. On the other hand, hepatitis B and human immunodeficiency virus (HIV) can be spread by the medical procedure of blood transfusion and injections. Some of the

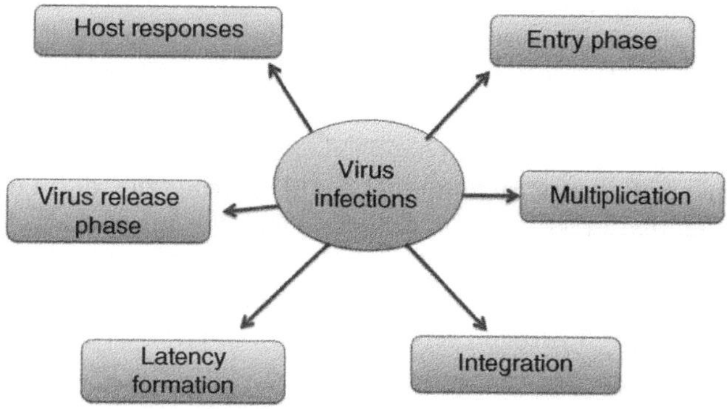

Figure 15.1 Various phases of virus infections.

Table 15.2 Routes of infection and incubation time of some common human viruses.

Types of virus	Mode of infection	Incubation time
Epstein–Barr	Saliva, respiratory	30–50 days
HPT-A	Orofecal, body fluids	15–45 days
HPT-C	Body fluids	Years
HPT-B	Body fluids	1–6 months
HSV	Skin contact	1–2 weeks
HPV	Skin contact	Months to years
HIV	Blood, body fluids	0–10 years
Influenza virus	Respiratory	2–3 days
Measles virus	Respiratory	8–12 days
Mumps virus	Respiratory, direct contact	12–24 days
Norwalk virus	Orofecal	4–48 hours
Parvovirus B19 (fifth disease)	Respiratory	4–20 days
Poliovirus	Orofecal	5–35 days
Rabies virus	Animal bite	3–7 weeks
Rhinovirus (common cold)	Respiratory	3–7 days
SARS coronavirus	Respiratory, direct contact	2–10 days
WNV	Mosquito bite, blood	Variable
YFV	Mosquito bite	3–6 days

HIV, human immunodeficiency virus; HPT, hepatitis; HPV, human papilloma virus; HSV, Herpes simplex virus; WNV, West Nile virus; YFV, yellow fever virus.
Source: Summers [16]/with permission of Elsevier.

viruses are transmitted by venereal routes. These include Herpes simplex virus type 2, certain human papillomavirus (HPV) strains, and HIV [16]. The other viruses and their routes of transmission are depicted in Table 15.2.

15.2.2 Plant-Based Drugs and Phytoantioxdants for the Treatments

In recent years, antioxidants have dragged considerable attention for the effective therapy in the treatment of various medical conditions and it is due to their scavenging activity over free radicals. The fact that whether these radicals' effects are responsible for tissue injury or an accompaniment to or outcome of injury is yet to be established. However, it is obvious that tissue injury goes through a radical effect more readily as compared to healthy tissues, thus worsening the primary injury [17, 18]. During the normal physiological biochemical process, oxidative stress has been generated and it is due to the imbalance between oxidant's production and its scavenging by endogenous antioxidants. Therefore, in this episode, there are increments of various types of reactive oxygen species (ROS). It includes superoxide anions, hydroxyl radicals (HO^-), hydrogen peroxide (H_2O_2), singlet oxygen ($^1O_2$), and reactive nitrogen species (RNS)

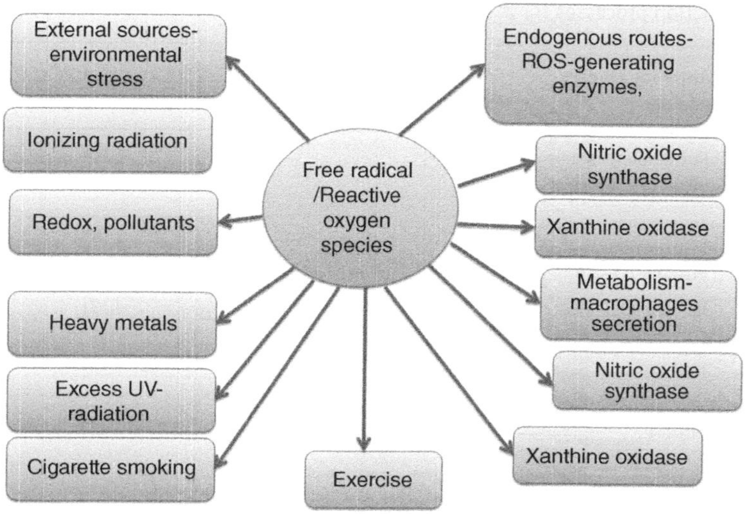

Figure 15.2 Various sources of free radicals/ROS generation. *Source:* Based on Refs. [19, 20].

(peroxynitrite, peroxyl radicals) [19]. There are several factors, which cause the generation of ROS and RNS as presented in Figure 15.2.

And these ROS and RNS lead to the damaging effect at cellular, subcellular, organelles, proteins, and lipids along with genetic levels. Additionally, impairing the normal physiological process of the cell, ultimately, causes mutations, apoptosis, and failure of these functions [21–23]. In this regard, antioxidant supplements are highly advocated and mainly possess two approaches, either by restricting ROS/RNS generation or neutralizing/scavenging. However, various vegetables and fruits have been investigated for their antioxidant potentials such as vitamins (ascorbic acid), carotenoids (lutein), polyphenols (3,6-dihydroxyflavone), and metabolic sensitizers (methyl selenocysteine) [24–26]. Likewise, antioxidant of gallic acid derived from dried rose flower extract of *Rosa rugosa* including plant active from other sources exhibited high antioxidant capacity in scavenging activity [27]. In spite of natural antioxidants, they are subjected to degradations and solubility problems and, ultimately, low bioavailability and patient's incompliance [28]. Some of the synthetic molecules also have been used for antioxidants' purposes. It includes butylated hydroxyanisole (BHA), butylated hydroxytoluene (BHT), its analogs, and GA esters [29], which cause anegative impact on human health. Therefore, naturally occurring dietary antioxidants have been significantly used [30].

15.2.3 Antiviral Medicinal Plants and Phytochemicals

Since ancient ages, various parts of plants have been utilized for the treatment of different medical conditions and most of the plants are active against viral infections. The method of obtaining plant bioactive comprises several steps. It includes collection, drying, extraction, isolation, and purifications for the collection of alkaloids, terpenes, flavonoids, glycosides, and proteins. Some of the plants against viruses and their method of extraction are listed in Table 15.3.

Table 15.3 Some plants and extraction processes for their antiviral efficacy.

Plant	Extraction technique	Antiviral efficacy against	Phytochemical	References
Achillea fragrantissima	Hydro-alcoholic extract	Poliomyelitis-1 virus (POLIO)	—	[31]
Aegle marmelos	Aqueous extract	Human coxsackie viruses B1–B6	—	[32]
Aloe vera	Glycerine extract	HSV-2	—	[33]
Curcuma longa	Aqueous extract	HSV-1	Curcumin	[33]
Boswellia serrata	Oleo-gum-resin	HSV-1	β-Boswellic acid	[34]
Cassine xylocarpa	Aqueous extract	HIV	Pentacyclic lupane-type triterpenoids	[35]
Maytenus cuzcoina	Aqueous extract	HIV	Pentacyclic lupane-type triterpenoids	[35]
Daphne gnidium	Hydro-alcoholic extract	HIV	Daphnetoxin, gnidicin, gniditrin, excoecariatoxin	[36]
Magnolia officinalis	Methanol extract	Dengue virus type 2	Honokiol	[37]
Phyllanthus emblica	Aqueous extract	Influenza A virus strain H3N2	Highly oxygenated norbisabolane sesquiterpenoids	[38, 39]
		HBV	Sesquiterpenoid glycoside dimers	
Tuberaria lignose	An aqueous extract	HIV	Ellagic acid derivative	[40]
Viola diffusa	Ethanol extract	Hepatitis B virus 2	β-Hydroxy-3, 4-seco-friedelolactone-27-oic acid, 2β, 28β-dihydroxy-3,4-seco-friedelolactone27-oic acid, 2β, 30β-dihydroxy-3,4-seco-friedelolactone-27-lactone and stigmastane, stigmast-25-ene-3β,5α,6β-triol	[41]
Zataria multiflora	Methanolic extract	HSV-1	Rosmarinic acid	[42]

HIV, human immunodeficiency virus; HSV, Herpes simplex virus.

15.3 Polymeric-Based Nanophytomedicine Containing Herbal Antioxidants/Phytoantioxidants and Their Antiviral Effects

15.3.1 Polymeric Nanoparticles

Polymeric NPs mainly comprise polymer-based nanoformulation within the size range of 10–100 nm and based on their development technique, they form as nanospheres or nanocapsules [43]. Nanospheres consist of a polymer matrix system in which drugs are adsorbed onto the surface of polymer while in the nanocapsule, it is dissolved within the polymer matrix [44]. Polycaprolactone (PCL) is the major synthetic polymer already been approved by the USFDA for the drug delivery application exclusively for the plant bioactive. In this context, *Syzygium cumini* (ASc) seeds were used for the extraction through solvent evaporation/emulsification technique. The antioxidant potential of aqueous extract of (ASc) seed along with ASc-loaded PCL NP were used for investigation by DPPH method and ferric-reducing antioxidant power assay (FRAP). The results showed that both the plant active had equal and enhanced DPPH radicals scavenging features and low potential in FRAP. These observations suggested that ASc immobilized in PCL NP and were negligible antioxidant activity [45]. Recently Tsai and his team investigated the silymarin and zein (SMN-Z) flavonoid-loaded bacterial cellulose (BC) NPs or nanofiber composite for the improvement of prolonged antioxidant potential. The prepared nanoformulation showed improved wettability, and swelling profile of BC along with improved solubility profile of poorly soluble silymarin drugs with sustained drug release profile. Additionally, SMN-Z-BC exhibited better DPPH, ABTS % with superoxide anion-scavenging efficacy as compared to free SMN [46]. The other nanocarrier-loaded plants active are listed in Table 15.4.

15.3.2 Nanocapsules

These are the colloidal carriers comprising 10–1000 nm in particle diameter with central core in which therapeutic agents are embedded within a polymeric matrix system [43, 44] (Figure 15.3). Owing to various benefits like improved drug-loading efficiency, controlled drug release profile, and targeted drug delivery, it has been used in the delivery of a wide variety of therapeutic agents. Also, it can be prepared by various available techniques. It includes polymer coating, layer-by-layer method, nanoprecipitation, emulsion–diffusion, emulsion coacervation, emulsion evaporation, and double emulsification [62]. In this context, curcumin-loaded nanocapsules (NCs) were prepared with different polymer materials and were compared to the physicochemical properties on coating materials. Different polymers like P80, PEG, Chitosan, Eudragit RS100® and different investigating techniques like AFM, FTIR, DSC, and XRD were used for the study. The results of the investigation displayed the improved stability of the drug molecule within the lipid core of the nanocapsule, which retained its normal physicochemical and pharmacological features after the drug loading along with a controlled drug release profile. Additionally, no cytotoxicity efficacy was observed. These findings considered the drug-loaded nanocapsule as a promising carrier for the delivery of biologically active drugs [63].

15.4 Metallic Nanoparticle

Metallic NPs also have been investigated for potential antiviral therapy, especially gold NPs, which are prominent for interacting with the viruses and restrict the entry into the host cell and it is due to their better specific surface area and which are functionalized with wide groups [64]. The plant extracts and bioactives loaded with herbal gold NP prepared by green synthesis drags considerable attention from the researcher. Owing to its exclusive features, it includes environment-friendly, nonimmuniogenic, nontoxic, cost-effective and ease in scaling up. It has been widely explored. In this context, gold NP was found to be effective against HIV-1, H1N1, H3N2, H5N1, bovine viral

Table 15.4 Some herbal drug-loaded nanoformulations for the treatment of viral infections.

Polymer/lipid	Drug	Novel formulations	Diseases	Major outcomes	References
Tween 40	Apigenin	Microemulsion in gellan gum hydrogels	Enterovirus 71, FMDV, HCV, ASFV, influenza A	Apigenin-loaded microemulsion-laden gellan gum-based hydrogel formulation showed prolonged controlled release profile at pH 1.2 and 7.4 along with the improved bioavailability. These results indicated that apigenin-loaded microemulsion-based hydrogel formulation could be a potential carrier for poorly soluble drug	[47]
Tween 80	Apigenin	Multiple emulsions	Enterovirus 71, FMDV, HCV, ASFV, influenza A	Better stability profile of the formulation, improved oral bioavailability as compared to suspension	[48]
Tween 20, Span 20, isopropyl myristat	Olive leaves extract	Microemulsions	HSV	Improved encapsulation efficiency of drug molecule along with better antiviral potential	[49]
—	Quercetin	SLN, NLC, LNE	—	Particle size within the nanorange from 34–127 nm, high EE > 90% with high stability at simulated gastric fluid. Controlled drug release profile in enzyme-free simulated intestinal fluids.	[50]
—	Resveratrol	NE, SLN, NLC	—	Particle size in the range of 100–150 nm along with better physical stability, controlled the drug release profile	[51]
—	Myricetin	Nanosuspension	HIV, RLV, influenza	β-Cylcodextrin-encapsulated myricetin enhanced the antioxidant potential along with improved bioavailability	[52]
Phospholipon® 90H	Baicalin (BA)	Liposome	Influenza, NDV, enterovirus 71, DENV, RSV, HIV, HBV	Novel BA-LP improved the oral bioavailability as compared to BA-CMC suspension. BA-LP also showed nano size range of 373 ± 15.5 nm, zeta potential of -20.1 ± 0.22 mV, and higher E.E. $82.7 \pm 0.59\%$ along with sustained drug release pattern	[7, 53, 54]
Soya lecithin, TPGS, HP-β-CD	Myricetin	Nanosuspension	HIV, RLV, influenza	Nanodimension of particle 300–500 nm along with better physically stability profile. Improved drug bioavailability as compared with coarse myricetin	[55]

β-Sitosterol-β-D-glucoside (sito-G), Brij 35	Antisense oligonucleotides (asODN)	Cationic liposomes	Hepatitis B virus (HBV) infection	For the treatment of HBV infection, cationic liposomes were developed and bind with ligand of the asialoglycoprotein receptor (ASGPR), β-sitosterol-β-D-glucoside (sito-G). The prepared formulation was investigated for FCM, cellular uptake by CLSM and in vitro cytotoxicity. The FMC and ELISA investigation showed better transfection potential with high antigenic inhibitory efficacy in rat hepatocytes and HepG2.2.15 cells. Improved cellular uptake was found with CLSM along with no cytotoxicity profile. These results strongly recommended cationic liposomes as a promising nanocarrier for the specific transfer of asODN into hepatocytes HBV infection.	[56]
Compritol 888 ATO as lipid, Poloxamer 188 and Miranol Ultra C32 as surfactants	*Artemisia arborescens* essential oil	SLN	HSV-1	Improved drug loading efficacy, nanosize distribution of below 250 nm with better physical stability profile of the formulation and better antiviral potential	[57]
Soy phosphatidylcholine	Essential oil of *Artemisia arborescens* L.	Liposomal	HSV-1 virus	Liposomal incorporated with essential oil of *Artemisia arborescens* L. and were developed by film technology followed by sonication. The prepared formulations exhibited high drug entrapment efficacy along with improved antiviral potential as compared to free oil.	[58]
—	Quercetin	Nanoemulsion	JEV, influenza A, EBV, MAYV, RV, HCV	Quercetin-loaded nanoemulsion formulation showed better anti-inflammatory, antioxidant efficacy as compared to synthetic drug diclofenac along with no hepatic, hematological, renal, and genetic damage	[59]
	PolyICLC (LE-PolyICLC)	Liposome	Dengue viruses (DENVs).	Liposome-loaded PolyICLC exhibited better antiviral potential by improving cytokine expression linked innate immunity (IFN-γ) as compared to plain polyICLC. These results indicated that liposomes-loaded drugs are a potential candidate for antiviral therapeutics as well as prophylactic in DENVs	[60]
Hydrogenated soya phosphatydilcholine and cholesterol	*Santolina insularis* essential oil	liposome	HSV-1	Enhanced drug loading efficacy along with stability profile up to one year, better antiviral potential	[61]

BA, baicalin; CLSM, confocal laser scanning microscopy; CMC, carboxymethyl cellulose; EE, encapsulation efficiency; FCM, flow cytometry; HBV, hepatitis B virus; HSV, Herpes simplex virus; HSV-1, Herpes simplex virus-1; LNE, lipid nanoemulsions; LP, liposome; NLC, nanostructured lipid carriers; PolyICLC, an immunomodulating dsRNA; SLN, solid lipid NPs.

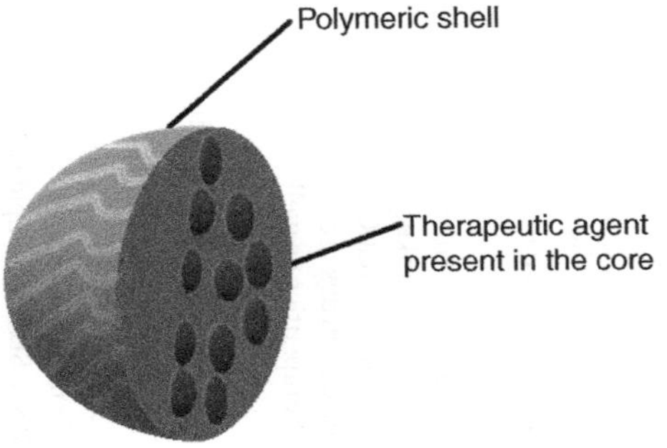

Figure 15.3 Basic internal structure of nanocapsule.

diarrhea virus, dengue virus, and foot-and-mouth disease virus (FMDV) [65]. Additionally, cationic gold NP showed antiviral against H1N1 and H3N2 confirmed by peroxidase enzymatic reaction assay [66]. Some of the other engineered metallic NPs have been investigated for the therapeutic potential in the treatment of viral infection. It includes tannic acid-loaded silver NP, curcumin-loaded silver NP, silver NP/chitosan composite, gallic acid-grafted gold NP, and multivalent gold NPs [53, 54, 67, 68].

15.5 Magnetic Nanoparticles

Quercetin is the most abundant dietary flavonoid available in various herbs and has been clinically reported for the potential anticancer, antiproliferative, proapoptotic, and cell signalling effects, and also growth factor suppression, anti-inflammatory, antioxidant, and antiviral efficacies [69–71]. In this account Verma et al. [72] reported the quercetin-loaded magnetic NP for the aerosolized delivery. The prepared formulation was prepared by the synthesis and surface coating of Fe_3O_4 magnetic NPs (MNPs) along with polymer poly(lactic-*co*-glycolic acid) (PLGA). The particle size was observed within the range of 9.6 to 53.2 nm confirmed by electron microscopy along with no cytotoxicity potential. Additionally, PLGA-MNPs were well tolerated in an experimental mice when administered intranasally by using IL-6 assays along with sustained drug release. These results consider the magnetic NP as a promising nanocarrier for the effective delivery of quercetin via inhalation.

15.6 Lipid-Based Nanophytomedicine Containing Phytoantioxidants and Their Antiviral Effects

15.6.1 Phospholipid Complex

Phospholipid complexes are regarded as pharmacosomes or phytosomes depending upon the nature of active pharmaceutical constituents. It is well known that phospholipids have great significance in drug delivery technology and work as carriers especially for sustained and controlled delivery of medicaments. In phospholipid complexes, the active constituent interacts with the polar head of the phospholipid and the hydrophobic tail group encapsulates the entire system making the surface lipophilic. In this system, the active medicament is an integral part of the system, while in the case of other vesicular systems like liposomes, either drug is encapsulated or is within the membrane. Any of the phosholipids like phosphatidylcholine (PC), phosphatidylethanolamine (PE), phosphatidylserine (PS), etc., can be used for preparing the phospholipid complexes, but PC is the most preferred one due to its good biocompatibility and low toxicity characteristics [73]. There are a number of phytoconstitutents like catechin, epigallocatechin 3-*O*-gallate, silybin, 18-β glycyrrhetinic acid, procyanidins, different polyphenols, terpenes, flavonoids, etc., which have been converted to their phospholipid complexes with improved activity [73–75]. In this context, recently Kalita and Patwary have reported the increased solubility and partition coefficient of hesperidin phospholipid complex in comparison to the crude drug. They suggested that hesperidin phopholipid complex can be used for improving the dissolution behavior and oral bioavailability too [76].

15.6.2 Solid Lipid Nanoparticles (SLNs)

Solid lipid NPs (SLNs) are the submicron-sized newer colloidal particles having the size range of 50–1000 nm. It may contain bioactives embedded in the physiological lipids (triglyceride and glyceride mixtures), which remain solid at room and body temperatures both [77]. SLNs have proven to have several benefits including more stability over conventional and already existing other colloidal systems like emulsion, liposome, noisome, and polymeric NP [78] and it makes the SLNs more suitable for the delivery of bioactives including different phytoconstitutents. Owing to its prominent features, it includes prolonged drug release, improved bioavailability, permeability, lymphatic uptake, and lowering P-gp efflux along with the first-pass metabolism with better tissue distribution [79]. It is most preferable for the encapsulating plant bioactive for the effective drug delivery like antioxidants, antiviral, and anticancer therapeutic agents [80]. In this account, Ban and coworkers investigated the pegylated curcumin-encapsulated SLN for improving the permeability and bioavailability by transforming the surface behavior of prepared SLN [81]. Likewise, recently different essential oils can also be encapsulated in the SLN with better stability and improved activity [82]. Additionally, antiviral essential oil of *Artemisia arborescens*-encapsulated SLNs was investigated as a topical drug application for better physical stability and accumulation in the skin [57]. Furthermore, Mei et al. investigated triptolide obtained from traditional Chinese herb *Tripterygium wilfordii*-loaded SLN for the better therapeutic potential [83].

15.6.3 Nanostructured Lipid Carriers (NLCs)

NLCs are the colloidal nanocarriers prepared by incorporation of two dissimilar lipids like solid lipids along with liquid lipids/oils. Owing to its important benefits such as better biocompatibility, sterility, scale-up, and good bioavailability, it has been widely explored for the drug synthetic molecule as well as plant extracts and active phytochemicals for the efficient therapeutics in the treatment of oral cancers and viral infections. Moreover, various constituents like genistein, curcumin, quercetin, naringenin, and eugenol are used for the above purpose [84]. In this context, Karimi et al.[85] investigated turmeric extract-loaded NLC (T-NLC) for the assessment of its antioxidant, antimicrobial, and physicochemical properties. These novel formulations were prepared by using a high shear homogenizer. The particle size of the NLCs was in the range of 112.4 nm with a better physical stability profile along with a sustained drug release pattern. Additionally, improved antioxidant and antimicrobial potential of T-NLC were found as compared to free turmeric extract (Figure 15.4). These results recommended T-NLC as a promising agent for effective drug delivery. The other drug-loaded NLCs are presented in Table 15.4.

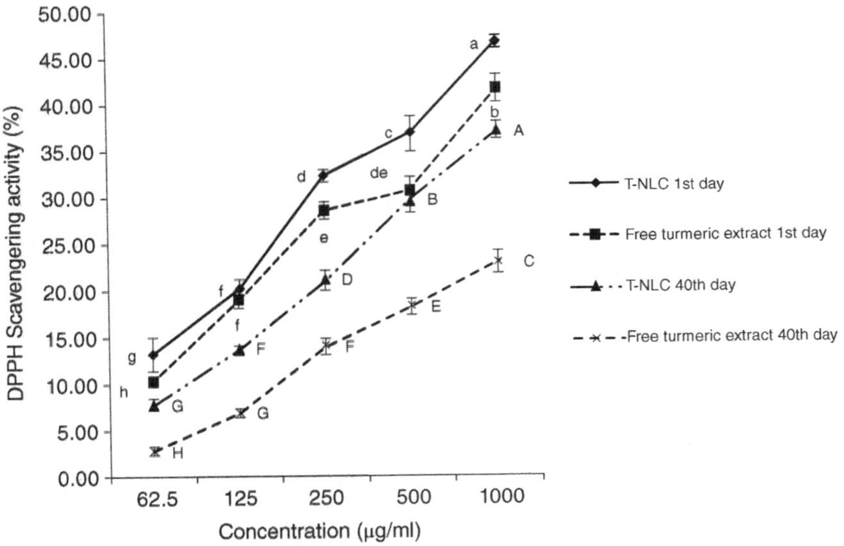

Figure 15.4 Free radical scavenging effects of T-NLC and free turmeric extract in the 1st and 40th days of storage. Bars with different letters in the 1st (a–h) and 40th days (A–H) after production indicate significant differences at the 0.05 level according to the Duncan test ($P < 0.05$). *Source:* Karimi et al. [85]/with permission of Elsevier.

15.7 Vesicular System-Based Nanophytomedicine

15.7.1 Liposomes

These are the lipidic vesicular carriers comprising a self-enclosed phospholipid bilayer with an aqueous core in a spherical shape [86, 87] (Figure 15.5). One of the exclusive features of the liposome is the capacity to encapsulate both hydrophilic as well hydrophobic molecules inside the core. Additionally, it has better features of self-degradability, sustained drug-release profile, protecting loaded therapeutic agents, and biocompatibility. It is also nonantigenic along with improving the drug penetration profile [86, 88]. Aruoma et al. [89] investigated the rosemary extract and oily extract of a mixture of Provencal herbs for antiviral and antioxidant potential in an in vitro model. The extracts were loaded in the liposome and studied the inhibition of peroxidation with 50% inhibition concentration values of 0.0009 and 0.0035% (v/v). The rosemary extract and Provencal herbs (0.2%) were reacted with trichloromethyl peroxyl radical with calculated rates of

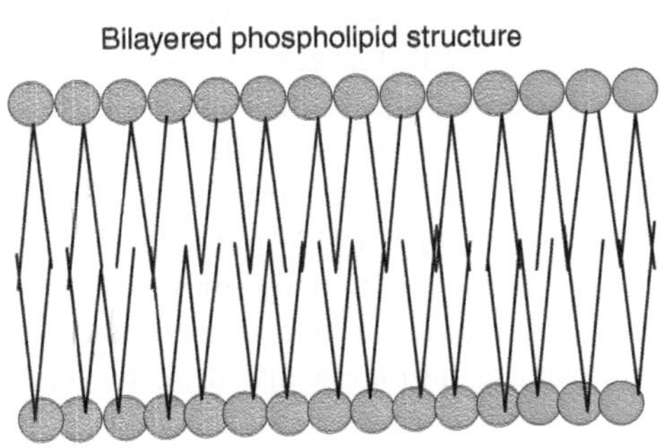

Figure 15.5 Phospolipd bilayer membrane of liposomes.

2.7×10^4 and $1.5\times10^3\ s^{-1}$, respectively. The active constituents in this sample, carnosol and carnosic acid at 0.05% (v/v), react with rate constants of $(1–3)\times10^6$ and $2.7\times10\,M^{-1}\,s^{-1}$, respectively. Both these herbal extracts were found to have good antioxidant potential in the Rancimat test. The rosemary extract and Provencal herbs extract inhibited HIV infection at less concentration, while purified carnosol showed definite anti-HIV potential at 8 μM along with no cell toxic effect. These observations advocated DNA-damaging effect of the copper-phenanthroline and bleomycin-iron system and were recommended for antioxidant efficacy and can be used in food products [89]. The other lipid formulations are listed in Table 15.1.

15.7.2 PEGylated Liposomes

PEGylated liposomes have also been investigated for the improvement of bioavailability of poorly water-soluble herbal drugs and their derivatives. One of the herbal drugs Artemisinin (Qinghaosu) has sesquiterpene lactone attached with an endoperoxide moiety isolated from *Artemisia annua* L. (Qinghao, Asteraceae) and its derivative is reported for their significant antiprotozoals, antitrematodes, and antiviral activities against (human cytomegalovirus and hepatitis B and C viruses). However, its poor solubility causes less oral bioavailability, thus limiting the use and was ineffective in the treatments [90, 91]. In this context, Isacchi et al. investigated artemisinin-loaded plain liposomes and PEGylated liposomes for the improvement of bioavailability in an experimental animal. The bioavailabilities of both the formulations were assessed in mice and were found to be prolonged drug bioavailability up to 3 and 24 hours as compared with the free drug solution, respectively. Additionally, $AUC_{0-24\,hours}$ showed six times high in drug bioavailability in comparison to plain drug solution of artemisinin. The half-life of the drug loaded in liposome and PEGylated liposomes were significantly improved fivefold. These results strongly recommended nanocarriers for the improvement of poorly soluble drugs and in the effective therapy of various diseases [90].

15.7.3 Niosomes

Wilson and his team investigated the leaf extracts of *Psidium guajava* (PG) and *Justicia adhatoda* (JA)-loaded phytoniosomes (PNs) as traditional Indian herbs for the effective treatment of dengue viral infection. These novel formulations (AGN) were developed by high-pressure ultrasonication technique employing tween 80, poloxamer 407, PEG 6000 along with cholesterol. The aqueous extract of PG-, JA-loaded PN formulations were characterized for particle size analysis, in vitro cytotoxicity studies, and antiviral efficacy against dengue virus-2 strain (DEN-2). Nano-sized dimensions of 105.1 ± 3.6 to 279.4 ± 5.2 nm were found for PEGylated niosomes and non-PEGylated niosomes' formulations. Additionally, AGN possesses good antiviral potential against Vero cells infected with DEN-2 virus along with improved drug stability up to three months at room temperature. The formulation code of AGN4 exhibited the highest 40% plaque formation, which significantly combated the viral infection in vitro as compared with AGN 5, blank control, and dummy PN [92].

15.7.4 Cubosomes

Resveratrol is polyphenol stilbene presented naturally in the fermented fruits of grapes, mulberry, red wine, and peanuts having unique therapeutic effects of improving longevity, controlling lipid level, and also potential activity as prophylactic against cancers and associated viral infections [93, 94]. In this context, Ahirrao and Shrotriya investigated the resveratrol-loaded cubosomes for improving the delivery to the brain via transnasal. The novel formulations were prepared by probe sonication technique employing glycerol monooleate and lutrol F127. The formulation was further optimized and characterized for various parameters of particle size, zeta potential, entrapment efficiency (EE), and in vitro release profile. The optimized formulation was found to be nano-dimensional in the range of 161.5 ± 0.12 nm with a zeta potential of −20.9 mV along with 83.08% EE and a controlled drug release profile of 67% up to 24 hours. The formulation was further subjected to incorporation into in situ gelling polymer (poloxamer 407) for nasal administration. The in vivo investigation showed a better transnasal permeation profile with improved drug distribution into the brain as compared with plain drug solution. These results suggested that cubosome could be a potential carrier for drug delivery for brain [95].

15.7.5 Micelles

Micelles are the dispersed system composed of amphiphilic copolymer colloidal particles within the nano range of 5–100 nm [96]. The central part is composed of particles developed by hydrophobic building blocks having the capacity to encapsulate vide variety of therapeutic agents with hydrophobicity in nature and the external surface developed by hydrophilic building blocks along with the potential of functionalizing with other chemical moieties [97] (Figure 15.6). In recent years, the micelles' formulation gained significant interest as a novel drug delivery system for the effective delivery of the therapeutic agents [98]. In this context, quercetin, a potential herbal polyphenolic flavonoid compound, has been investigated for the improvement of therapeutic efficacy by increasing bioavailability. Quercetin contained variety of therapeutic beneficial features. It includes antioxidant [99, 100], anti-inflammatory [101, 102], antiviral [103, 104], and anticancer potential [105]. However, due to poor aqueous solubility, low oral bioavailability, and instability of quercetin, the therapeutic application is very much limited [106, 107]. Zhen and his team investigated the quercetin-loaded sodium taurocholate, pluronic P123 (synthetic polymer)-mixed micelles for the improvement of aqueous solubility. The mixed micelles were developed by thin-film hydration method and were optimized by uniform design. The prepared formulation was assessed for the effect of the formulation on enhancing the aqueous solubility profile. The prepared optimized formulation exhibited improved drug loading and entrapment capacity (12.6%), (95.9%) in

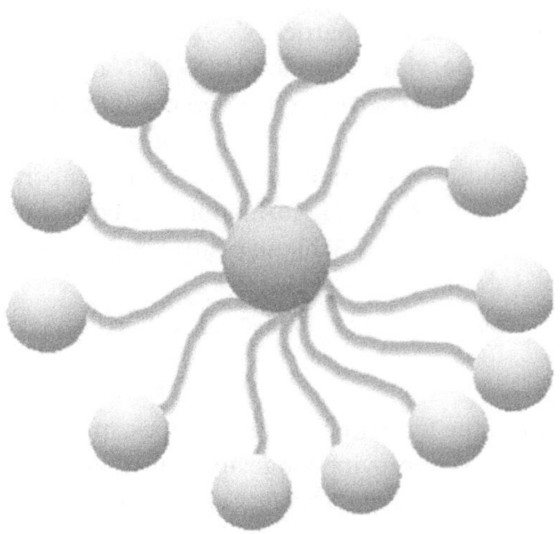

Figure 15.6 Schematic representation of micelles' colloidal dispersion.

nano-sized dimension spherical of 16.20 nm. Sustained drug release profiles along with better stability profiles were found. The bioavailabilities of 1.8-fold and 1.6-fold higher than plain drug solution were found. These results clearly indicated that quercetin-loaded micelles' formulation may be a potential carrier for oral administration of poorly soluble drugs [107].

15.7.6 Microemulsion

Zhao and Wang [47] reported pH-sensitive apigenin-loaded microemulsion-laden gellan hydrogel gum for oral delivery for the controlled drug delivery in the treatment of foot and mouth disease virus (FMDV), hepatitis C virus (HCV), African swine fever virus (ASFV) and influenza A. In this formulation, drugs were incorporated in the O/W microemulsion thereafter incorporate in the hydrogel. The microemulsion released the drug in a controlled manner at 1.2 pH and in the 7.4 pH due to improvement in the cross-linking of the drug release profile exhibiting Fickian diffusion-mechanism behavior. However, in vitro release profile showed erosion-controlled systems. These results suggested that microemulsion-based formulation may be the potential carrier for the effective delivery of hydrophobic drugs. The other emulsion carriers are listed in Table 15.1.

15.8 Hydrogels

Almeida et al. investigated the *Cymbopogon citratus* volatile oil-loaded polymeric NPs with hydrogel for the effective treatment of viral infection. The hydrogel loaded with polymeric NP was prepared by solvent emulsification diffusion technique using poly (D,L-lactide-*co*-glycolide) (PLGA) as a biodegradable polymer and was immersed in the carbomer hydrogel. The prepared formulation was characterized by various parameters. It includes particle size analysis, zeta potential, in vitro drug release pattern, and antiviral potential for Herpes simplex in Vero cells. The particle size was found to be 217.1 nm along with −20.5 mV zeta potential, which showed nano-sized dimension with negative charges on the surface of the nanocarrier. The biphasics with an initial burst release profile were shown by prepared nanocarriers followed by sustained release (Figure 15.7). These nanoparticulate carriers with hydrogel exhibited a high viral inhibitory effect in noncytotoxic concentration, 42.16 times lower than free oil, 8.76 and 2.23 times than loaded NPs and hydrogel-containing free oil, respectively. These results strongly advocated for the antiviral potential of nanogel to protect the oil against volatilization, controlled drug release, and high anti-herpetic activity [108].

15.8.1 Nanogels

Yao and his team investigated the myricetin-loaded chitosan nanogel for the improvement of bioavailability in the treatment of HIV, rhesus lymphocryptovirus (RLV) and influenza [7, 109]. The prepared formulation was characterized for particle-size analysis, rheological studies, sol-gel conversion studies, morphological 3D structures, in vitro cytotoxicity studies, release, and bioavailability profile. The prepared formulation of MYR-CS nanogel showed nano-sized dimension of 100–300 nm, phase conversion was found during exposure to body temperature and scanning electron microscopy (SEM) confirmed the 3D matrix structures (Figure 15.8). The drug release followed by the Fickian's release pattern in acidic buffer media go along with the swelling and erosion mechanism. Additionally, the nanogel was found to be safe as confirmed by MTT analysis along with better drug oral bioavailability observed in experimental rats. These results clearly advocated the nanogel as an excellent carrier for the improvement of oral bioavailability in the treatment of viral infections.

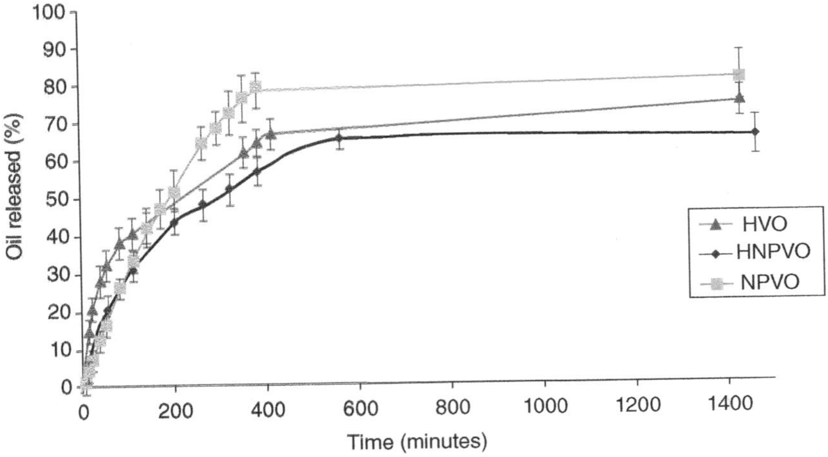

Figure 15.7 *Cymbopogon citratus* volatile oil in vitro release kinetics encapsulated in PLGA NPs (NPVO), incorporated in its free form in hydrogel (HVO), and incorporated into the same formulation with previous nanoencapsulation (HNPVO) (saline–phosphate buffer with sodium lauryl sulfate (1%), pH 6.8, 37 °C, 100 rpm, $n = 3$). *Source:* Almeidam et al. [108]/with permission of Elsevier.

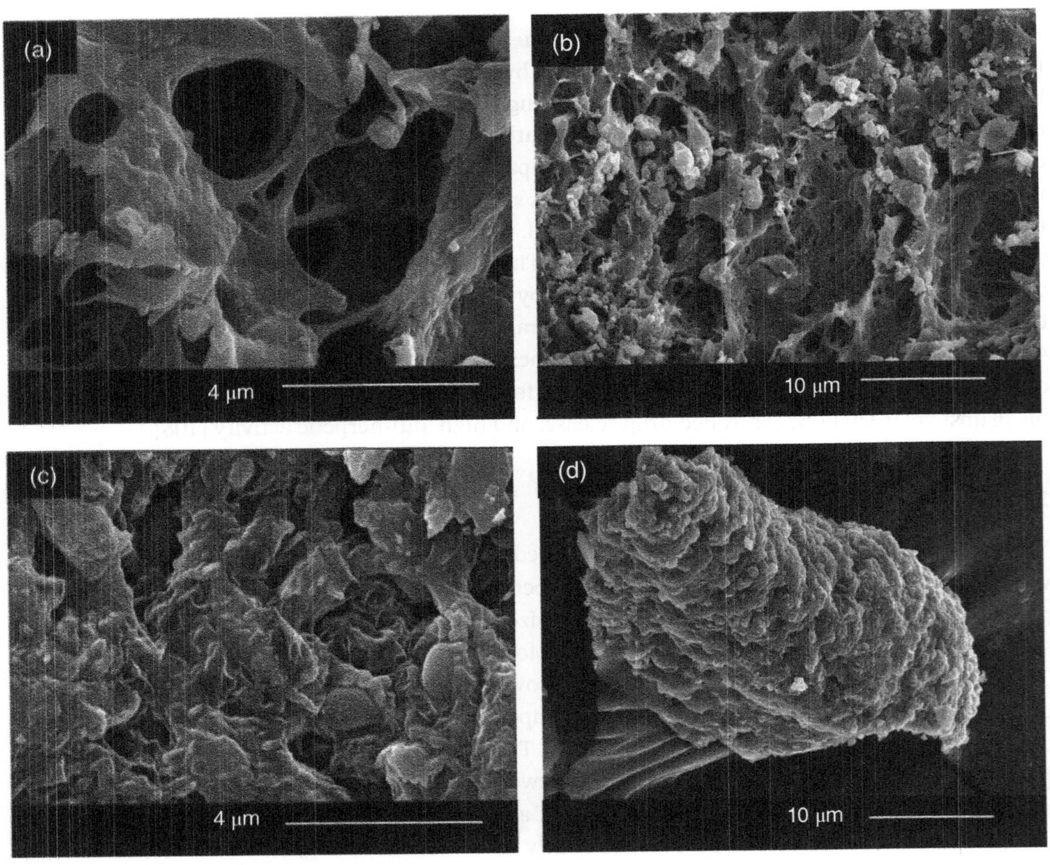

Figure 15.8 SEM morphology of gels showed 3D matrix structures (a) blank CS/β-GP gels, 40 000×; (b) blank CS/β-GP gels, 10 000×; (c) Myr-loaded gels, 40 000×; (d) Myr-loaded gels, 10 000×. *Source:* Yao et al. [109]/adapted with permission of Elsevier.

15.9 Dendrimers

Dendrimers are composed of synthetic polymers and represented as 3D star-like shape with branch macromolecules [110, 111]. It has two layers, internal and external, which participate in the encapsulation of various therapeutic agents as well as in drug conjugation and help in targeting biomolecule along with high drug therapeutic potential, lowering side effect, and sustaining the drug release profile [112, 113]. Additionally, these nanocarriers have other benefits. These include aqueous solubility, biodegradablility, biocompatibility, along with ideal features of less polydispersity index (PDI) [111, 114]. These features led the extensive investigation of dendrimer-loaded phytochemicals for antiviral therapy. However, coumarin, a herbal product obtained from different parts of the plants including seeds, leaves, and roots, belongs to the family *Apiaceae*. *Rutaceae* have been reported as antibacterial and antiviral and also possibilities for chemical modification into novel semisynthetic derivatives [115, 116]. In the same context, Mallik and his team investigated the succinimidyl ester of coumarin dendrimer for conjugating with antibodies and were synthesized for the development of a fluorescent label system for the diagnosis in biological as well as in clinical research. The conjugation was carried out by benzothiazole moiety of coumarin along with propargyl density and the gallic acid-derived linker with three azido groups by using click chemistry. The succinimidyl ester of coumarin dendrimer was synthesized with the acidic part of the gallic acid linker. Additionally, succinimidyl ester of coumarin dendrimer was conjugated with malaria anti-LDH monoclonal antibody D2H specifically. And these fluorescent conjugates were investigated for antigen detection [117].

15.10 Graphene Oxide Composite

Graphene oxide (GO) composite consists of a 2-dimensional crystalline arrangement developed by carbon atoms available in a pattern of the hexagonal lattice. GO has been reported for the various antibacterial, antifungal, and antiviral activities [118–120]. In this context, Sametband et al. investigated the reduction of sulfonated GO, which offered antiviral potential against a Herpes simplex virus type-1 (HSV-1) via cell attachment inhibition [121] along with pseudorabies DNA virus (PRV) and porcine epidemic diarrhea RNA virus (PEDV) by neutralizing virus directly [122]. Curcumin, a natural herbal polyphenolic compound, has been reported for various potential anticancer, antifungal, antibacterial, anti-inflammatory, and antiviral activities [123, 124] along with antiviral activity against the respiratory syncytial virus (RSV) [125]. Yang et al. investigated the synergistic antiviral potential of curcumin-functionalized GO against RSV infection [126]. The main site of action of the RSV is the lower respiratory tract of infants, which leads to serious medical complication as severe lung infection. To assess the antiviral activities of multifunctional GO, they formulated β-cyclodextrin (CD)-functionalized GO composite with a better curcumin-loading capacity. In this investigation, single-strand negative sense RSV RNA virus was used as a model virus and observations were carried out on tissue culture infectious dose assay followed by immunofluorescence assay (Figure 15.9a, b). The formulation of GO composite exhibited improved antiviral activity along with better cell viability in in vitro cytotoxic studies. These results clearly advocated that this curcumin-loaded GO composite could be a better antiviral agent against RSV and can be able to inhibit the virus directly and can be used as prophylactic steps and in therapy against the virus.

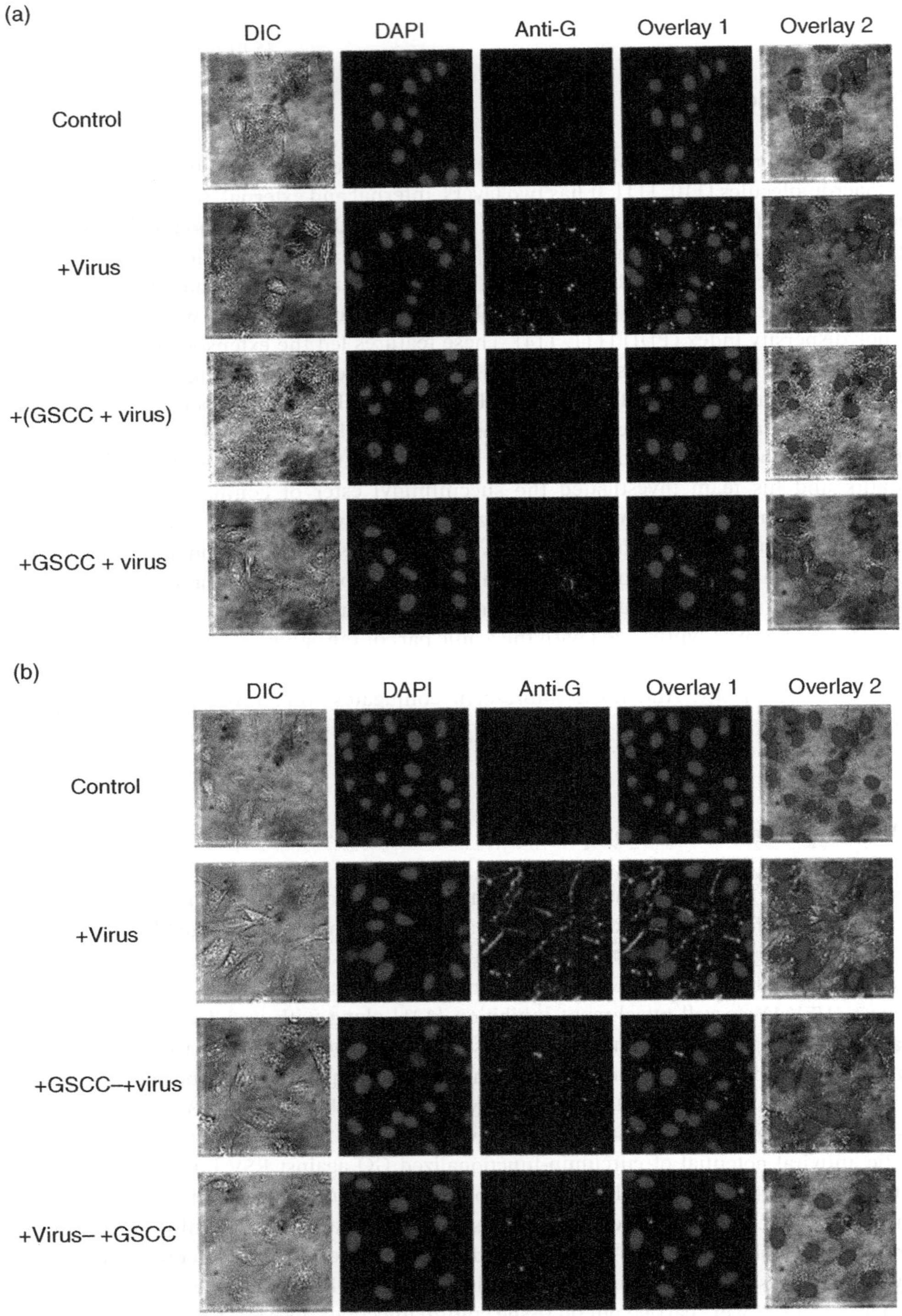

Figure 15.9 (a) The inhibition effects of GSC composite (GSCC) against RSV infection in virus pretreatment and co-addition treatment by immunofluorescence staining (60× oil immersion objective). The first row: Human laryngeal epithelial cells type 2 (HEp-2) only; the second row: HEp-2 cells infected by RSV; the third row: RSV interacted with GSCC for one hour prior to infecting HEp-2 cells; the fourth row: HEp-2 cells co-incubated with RSV and GSCC. *Source:* Yang et al. [126]/adapted with permission from Elsevier. (b) The inhibition effects of GSCC against RSV infection by immunofluorescence staining (60× oil immersion objective). The first row: HEp-2 cells only; the second row: HEp-2 cells infected by RSV; the third row: HEp-2 cells preincubated with GSCC for 1 hour before RSV infection; the fourth row: GSCC was added after RSV infection. The cells were incubated for 24 hours after RSV infection. *Source:* Yang et al. [126]/adapted with permission from Elsevier.

15.11 Self-Nanoemulsifying Drug Delivery Systems (SNEDDS)

SNEDDS have been investigated for the delivery of herbal antiviral agents. These novel drug delivery technologies explored for their special exclusive features of better solubility profile, better oral drug bioavailability, effective therapeutic potential, controlled drug release, and better antiviral activity over the conventional phytomedicine-loaded drug formulations. In the same context, for the improvement of oral bioavailability of myricetin (MYR)-loaded SNEDDS were developed by checking the solubility profile of MYR in the different oil and excipients. Thereafter pseudo-ternary phase graphs were constructed and based on the droplet size and emulsification potential, the formulation was optimized. The different optimized MYR-SNEDDS formulations were prepared like F04 (Capryol 90/Cremophor RH 40/PEG 400 4 : 3 : 3), F08 (Capryol 90/Cremophor RH 40/1,2-propanediol 4 : 3 : 3), F13 (Capryol 90/Cremophor EL/Transcutol HP 4 : 3 : 3), and F15 (Capryol 90/Cremephor RH 40/Transcutol HP 2 : 7 : 1), and were found to be nano-size distribution of less than 200 nm along with rapid release profile (90% in one minute), nontoxic and better solubility with permeability features as compared with free drug. As a result, the oral bioavailability of the MYR for the various formulations was found to be 5.13-, 6.33-, 4.69-, and 2.53-fold for F04, F08, F13, and F15, respectively as compared with plain MYR. These results clearly indicated that SNEDDS is a promising carrier system for the effective delivery of poorly soluble drugs in the treatment of antiviral infection [127].

15.12 Some of the Patents for Nanotherapeutics of Phytoantioxidants

Various patents related to the nano-formulations of phytoconstituents are listed in Table 15.5.

15.13 Conclusion and Future Perspective

Nanotherapeutics-loaded phytoantioxidants have been transformed in the field of nanotechnology. Various synthetic drug-loaded nanocarriers were already prepared and have been reported for their antiviral potential but the main limitation is the serious toxic effect. However, herbal drugs are the safest and the extract has also been reported for antiviral efficacy, but the main challenge is the poor aqueous solubility, which leads to low drug bioavailability. Therefore, it is an urgent need for novel nanocarrier, which improved the therapeutic profile along with better drug bioavailability. In recent years, various nanocarriers have been investigated, which include NPs, SLNs, NLCs, vesicular carriers, liposomes, niosomes, dendrimers, micelles, SNEDDS, GO composite, nanogels, hydrogels, microemulsion, cubosomes, etc. These engineered nanomedicines offer several benefits over conventional formulations, specifically in the prevention, diagnosis, and medication of viral infections. Owing to unique features like nano-dimension, better effective surface area and flexibility for surface modification, specificity, and biocompatibility, these nanocarriers show wide potential in antiviral pharmacotherapeutics and serve to beat the associated problems of drug resistance, poor solubility, low bioavailability, and small therapeutic profile. Nanocarriers have more potential to deliver the vaccine also. Recently, Mönkäre et al. investigated the hyaluronan (HA)-based dissolving micro-needles (MNs) encapsulated with PLGA NP with co-encapsulated ovalbumin (OVA) and poly (I:C) for intradermal application for the immunization. The optimized NP : HA ratio was employed for the dissolving MNs and the formulation was subjected to asymmetrical field flow

Table 15.5 Patents and recent advances related to nano-formulations of phytoconstituents.

Patent no.	Plant constituents	Formulation	References
US20160000714A1	Curcumin	SLN	[128]
US20160228362A1	Silibinin, isosilibinin, silichristin, silidanin, glycyrrhizic acid, glycyrrhetinic acid and combinations	Targeted nano-liposome	[129]
US20170157005A1	Extracts pomegranate oil, fennel oil, rosemary oil, chamomile oil, jojoba oil, rosehip oil	Multilayered nanocapsules	[130]
US5043323A	Quercetin, apigenin	Phytosomes	[131, 132]
US20120148569A1	Terpenoids and fatty acids from *Picrorhiza kurrooa* Royle, *Picrorhiza scrophularflora* Pennell, and *Neopicrorhiza scrophulańiflora*	Nanoparticle	[133]
US20120052126A1	Vitamin A, vitamin E, vitamin K, β carotene, curcumin, Coenzyme Q10, ethoxyquin, ginkgo biloba, glutathione, huperzine, lipoic acid, lycopene, melatonin, resveratrol, quercetin, catechin, epicatechin, hydroxytyrosol, 5-hydroxytryptophan, squalene, apolipoprotein E, bilirubin, creatine, uric acid, carotenoids, flavonoids, lignans	Nanoemulsions	[134]
WO2012000757A1	Curcuminoids, theaflavins, isoflavones, and polymethoxylated flavones	Particles	[135]
US20120128777A1	Hesperitin	Combinations of lipid particles with nano-crystals	[136]
US20150050357A1	Genistein	Nanoparticle	[137]
WO2004080396 US20080253961 WO2010013224	Curcumin	Liposome PLGA nanoparticles Nanoparticles	[138–140]
US20120052126A1	Combination of different herbs	Nanoparticle	[141]
WO2006068759A2	Ellagic acid	Liposomes	[142]
US20050152978A1	Resveratrol	Liposome	[143]
US6756065B1	*Ginkgo biloba* biflavones, catechin and/or epicatechin, cumarin and derivatives	Phospholipids complex	[144]
US4963527A	Extracts of *Vitis winifera*	Phospholipids complex	[145]
EP0283713B1	Saponins from *Aesculus ippocastanum*, *Centella asiatica*, *Ruscus aculeatus*, *Hedera helix*, *Terminalia* sp., *Calendula officinalis*, Soya, Ginseng, *Glycyrriza* sp., *Quillaia saponaria*, *Gypsophylla*	Phospholipids complex	[131]
EP0032556A1	Rosmarinic acid	Phospholipid complex	[146]

fractionation and dynamic light scattering (DLS). The dose of antigen delivered from soluble MNs was assessed in mice by near IR fluorescence imaging. Also, this study was compared with hollow MN-NPs in mice. The result showed better humoral and cellular immune response by NP-loaded MNs as compared with hollow MNs. These results strongly recommend the potential of NP in the development of nano-vaccine and need to be scaled up in upcoming years [147].

Acknowledgments

The authors are thankful to the Director of the Central Library, Shri Govindram Seksaria Institute of Technology and Science (SGSITS), Indore for providing necessary infrastructure facilities and Inflibnet e-resources.

References

1 Mehendale, R., Joshi, M., and Patravale, V.B. (2013). Nanomedicines for treatment of viral diseases. *Crit. Rev. Ther. Drug Carr. Syst.* 30: 1–49.
2 Qasim, M., Lim, D.J., Park, H. et al. (2014). Nanotechnology for diagnosis and treatment of infectious diseases. *J. Nanosci. Nanotechnol.* 14: 7374–7387.
3 Lockhat, H.A., Silva, J.R., Alves, C.N. et al. (2016). Binding free energy calculations of nine FDA-approved protease inhibitors against HIV-1 subtype C I36T↑ T containing 100 amino acids per monomer. *Chem. Biol. Drug Des.* 87: 487–498.
4 Maseko, S.B., Natarajan, S., Sharma, V. et al. (2016). Purification and characterization of naturally occurring HIV-1 (South African subtype C) protease mutants from inclusion bodies. *Protein Expr. Purif.* 122: 90–96.
5 Hayden, F. (2009). Developing new antiviral agents for influenza treatment: what does the future hold? *Clin. Infect. Dis.* 48: S3–S13.
6 Sah, A.K. and Verma, V.K. (2012). Phytochemicals and pharmacological potential of *Nyctanthes arbortristis:* a comprehensive review. *Int. J. Res. Pharmaceut. Biomed. Sci.* 3 (1): 420–426.
7 Shabat, S.B., Yarmolinsky, L., Porat, D. et al. (2020). Antiviral effect of phytochemicals from medicinal plants: applications and drug delivery strategies. *Drug Deliv. Transl. Res.* 10: 354–367. https://doi.org/10.1007/s13346-019-00691-6.
8 Ansari, S.H., Islam, F., and Sameem, M. (2012). Influence of nanotechnology on herbal drugs: a review. *J. Adv. Pharm. Technol. Res.* 3: 142–146.
9 Ambwani, S., Tandon, R., Ambwani, T.K. et al. (2018). Current knowledge on nanodelivery systems and their beneficial applications in enhancing the efficacy of herbal drugs. *J. Exp. Biol. Agric. Sci.* 6 (1): 87–107.
10 Oliveira, D.M.P., Garcion, E., Venisse, N. et al. (2005). Tissue distribution of indinavir administered as solid lipid nanocapsule formulation in mdr1a (+/+) and mdr1a (−/−) CF-1 mice. *Pharm. Res.* 22: 1898–1905.

11 Mahajan, S.D., Aalinkeel, R., Law, W.C. et al. (2012). Anti-HIV-1 nanotherapeutics: promises and challenges for the future. *Int. J. Nanomed.* 7: 5301–5314.

12 Adesina, S.K. and Akala, E.O. (2015). Nanotechnology approaches for the delivery of exogenous siRNA for HIV therapy. *Mol. Pharm.* 12: 4175–4187.

13 Duan, L., Yan, Y., Liu, J. et al. (2016). Target delivery of small interfering RNAs with vitamin E-coupled nanoparticles for treating hepatitis C. *Sci. Rep.* 6: 24867.

14 Chen, B.M., Su, Y.C., Chang, C.J. et al. (2016). Measurement of pre-existing IgG and IgM antibodies against polyethylene glycol in healthy individuals. *Anal. Chem.* 88: 10661–10666.

15 Singh, L., Kruger, H.G., Maguire, G.E.M. et al. (2017). The role of nanotechnology in the treatment of viral infections. *Ther. Adv. Infect. Dis.* 4 (4): 105–131. https://doi.org/10.1177/2049936117713593.

16 Summers, W.C. (2009). *Virus Infection*, 546–557. Elsevier Inc.

17 Halliwell, B. and Gutteridge, J.M.C. (1985). The importance of free radicals and catalytic metal ions in human diseases. *Mol. Asp. Med.* 8: 89–193.

18 Suresh, P.K. and Sah, A.K. (2013). Free radicals: generation, defenses and implications in various diseases. *Res. J. Pharm., Biol. Chem. Sci.* 4 (2): 247–269.

19 Gupta, P., Authimoolam, S.P., Hilt, J.Z. et al. (2015). Quercetin conjugated poly(amino esters) nanogels for the treatment of cellular oxidative stress. *Acta Biomater.* 27: 194–204.

20 Suresh, P.K., Sah, A.K., and Daharwal, S.J. (2014). Role of free radicals in ocular diseases: an overview. *Res. J. Pharm. Tech.* 7 (11): 1330–1344.

21 Halliwell, B. (1996). Antioxidants in human health and disease. *Annu. Rev. Nutr.* 16: 33–50.

22 Sharpe, E., Andreescu, D., and Andreescu, S. (2011). Artificial nanoparticle antioxidants. In: *Oxidative Stress: Diagnostics, Prevention, and Therapy* (ed. S. Andreescu and M. Hepel), 235–253. Washington, DC: ACS Publications.

23 Wattamwar, P.P., Mo, Y., Wan, R. et al. (2010). Antioxidant activity of degradable polymer poly(trolox ester) to suppress oxidative stress injury in the cells. *Adv. Funct. Mater.* 20: 147–154.

24 Anagnostopoulou, M.A., Kefalas, P., Papageorgiou, V.P. et al. (2006). Radical scavenging activity of various extracts and fractions of sweet orange peel (*Citrus sinensis*). *Food Chem.* 94: 19–25.

25 Lim, Y., Lim, T., and Tee, J. (2007). Antioxidant properties of several tropical fruits: a comparative study. *Food Chem.* 103: 1003–1008.

26 Medhe, S., Bansal, P., and Srivastava, M.M. (2014). Enhanced antioxidant activity of gold nanoparticle embedded 3,6-dihydroxyflavone: a combinational study. *Appl. Nanosci.* 4: 153–161.

27 Tułodziecka, A. and Szydłowska-Czerniak, A. (2016). Determination of total antioxidant capacity of rapeseed and its by-products by a novel cerium oxide nanoparticle-based spectrophotometric method. *Food Anal. Methods* 9: 3053–3062.

28 Verma, A.K. (2014). Anti-oxidant activities of biopolymeric nanoparticles: boon or bane! *J. Pharm. Res.* 8: 871–876.

29 Yehye, W.A., Rahman, A.N., Alhadi, A. et al. (2012). Butylated hydroxytoluene analogs: synthesis and evaluation of their multipotent antioxidant activities. *Molecules* 17: 7645–7665.

30 Ibrahim, K., Wageeh, A., and Yehye, A.E. (2020). Nanoantioxidants: recent trends in antioxidant delivery applications. *Antioxidants* 9 (24): 2–30. https://doi.org/10.3390/antiox9010024.

31 Soltan, M.M. and Zaki, A.K. (2009). Antiviral screening of forty-two Egyptian medicinal plants. *J. Ethnopharmacol.* 126 (1): 102–107. https://doi.org/10.1016/j.jep.2009.08.001.

32 Brijesh, S., Daswani, P., Tetali, P. et al. (2009). Studies on the antidiarrhoeal activity of *Aegle marmelos* unripe fruit: validating its traditional usage. *BMC Complement. Altern. Med.* 9: 47. https://doi.org/10.1186/1472-6882-9-47.

33 Moradi, M.T., Rafieian-Kopaei, M., and Karimi, A. (2016). A review study on the effect of Iranian herbal medicines against in vitro replication of herpes simplex virus. *Avicenna J. Phytomed.* 6 (5): 506–515.

34 Goswami, D., Mahapatra, A.D., Banerjee, S. et al. (2018). *Boswellia serrata* oleo-gum-resin and β-boswellic acid inhibits HSV-1 infection in vitro through modulation of NF-κB and p38 MAP kinase signaling. *Phytomedicine* 1 (51): 94–103. https://doi.org/10.1016/j.phymed.2018.10.016.

35 Callies, O., Bedoya, L.M., Beltran, M. et al. (2015). Isolation, structural modification, and HIV inhibition of pentacyclic lupane-type triterpenoids from *Cassine xylocarpa* and *Maytenus cuzcoina*. *J. Nat. Prod.* 78 (5): 1045–1055. https://doi.org/10.1021/np501025r.

36 Vidal, V., Potterat, O., Louvel, S. et al. (2012). Library-based discovery and characterization of daphnane diterpenes as potent and selective HIV inhibitors in *Daphne gnidium*. *J. Nat. Prod.* 75 (3): 414–419. https://doi.org/10.1021/np200855d.

37 Fang, C.Y., Chen, S.J., Wu, H.N. et al. (2015). Honokiol, a lignan biphenol derived from the Magnolia Tree, inhibits dengue virus type 2 infection. *Viruses* 7 (9): 4894–4910. https://doi.org/10.3390/v7092852.

38 Lv, J.J., Yu, S., Wang, Y.F. et al. (2014). Antihepatitis B virus norbisabolane sesquiterpenoids from *Phyllanthus acidus* and the establishment of their absolute configurations using theoretical calculations. *J. Organomet. Chem.* 79 (12): 5432–5447. https://doi.org/10.1021/jo5004604.

39 Lv, J.J., Yu, S., Xin, Y. et al. (2015). Anti-viral and cytotoxic norbisabolane sesquiterpenoid glycosides from *Phyllanthus emblica* and their absolute configurations. *Phytochemistry* 117: 123–134. https://doi.org/10.1016/j.phytochem.2015.06.001.

40 Bedoya, L.M., Abad, M.J., Sanchez-Palomino, S. et al. (2010). Ellagitannins from *Tuberaria lignosa* as entry inhibitors of HIV. *Phytomedicine* 17 (1): 69–74. https://doi.org/10.1016/j.phymed.2009.08.008.

41 Dai, J.J., Tao, H.M., Min, Q.X. et al. (2015). Anti-hepatitis B virus activities of friedelolactones from *Viola diffusa* Ging. *Phytomedicine* 22 (7–8): 724–729. https://doi.org/10.1016/j.phymed.2015.05.001.

42 Arabzadeh, A.M., Ansari-Dogaheh, M., Sharififar, F. et al. (2013). Anti herpes simplex-1 activity of a standard extract of *Zataria multiflora* Boiss. *Pak. J. Biol. Sci.* 16 (4): 180–184.

43 Suresh, P.K. and Sah, A.K. (2014). Nanocarriers for ocular delivery for possible benefits in the treatment of anterior uveitis: focus on current paradigms and future directions. *Expert Opin. Drug Deliv.* 11 (11): 1747–1768.

44 Diebold, Y. and Calonge, M. (2010). Applications of nanoparticles in ophthalmology. *Prog. Retin. Eye Res.* 29: 596–609.

45 Bitencourt, P.E., Ferreira, L.M., Cargnelutti, L.O. et al. (2016). A new biodegradable polymeric nanoparticle formulation containing *Syzygium cumini*: phytochemical profile, antioxidant and antifungal activity and in vivo toxicity. *Ind. Crop. Prod.* 83: 400–407.

46 Tsai, Y.H., Yang, Y.N., Ho, Y.C. et al. (2018). Drug release and antioxidant/antibacterial activities of silymarin-zein nanoparticle/bacterial cellulose nanofiber composite films. *Carbohydr. Polym.* 180: 286–296.

47 Zhao, X. and Wang, Z. (2019). A pH-sensitive microemulsion-filled gellan gum hydrogel encapsulated apigenin: characterization and in vitro release kinetics. *Colloids Surf. B: Biointerfaces* 178: 245–252. https://doi.org/10.1016/j.colsurfb.2019.03.015.

48 Kim, B.K., Cho, A.R., and Park, D.J. (2016). Enhancing oral bioavailability using preparations of apigenin-loaded W/O/W emulsions: in vitro and in vivo evaluations. *Food Chem.* 206: 85–91. https://doi.org/10.1016/j.foodchem.2016.03.052.

49 Khattab, R.A., Hosny, A.E.-D.S., Abdelkawy, M.A. et al. (2016). Anti-HSV type-1 activity of olive leaves extract crude form acting as a microemulsion dosage form. *Afr. J. Microbiol. Res.* 10 (22): 820–828.

50 Aditya, N.P., Macedo, A.S., Doktorovova, S., et al., (2014). Development and evaluation of lipid nanocarriers for quercetin delivery: a comparative study of solid lipid nanoparticles (SLN), nanostructured lipid carriers (NLC), and lipid nanoemulsions (LNE). *LWT Food Sci. Technol.* https://doi.org/10.1016/j.lwt.2014.04.058

51 Sun, R., Zhao, G., Ni, S. et al. (2014). Lipid based nanocarriers with different lipid compositions for topical delivery of resveratrol: comparative analysis of characteristics and performance. *J. Drug Del. Sci. Tech.* 24 (6): 591–600.

52 Chakraborty, S., Basu, S., and Basak, S. (2014). Effect of beta-cyclodextrin on the molecular properties of myricetin upon nano-encapsulation: insight from optical spectroscopy and quantum chemical studies. *Carbohydr. Polym.* 99: 116–125. https://doi.org/10.1016/j.carbpol.2013.08.008.

53 Wei, Y., Guo, J., Zheng, X. et al. (2014). Preparation, pharmacokinetics and biodistribution of baicalin-loaded liposomes. *Int. J. Nanomed.* 9: 3623–3630. https://doi.org/10.2147/IJN.S66312.

54 Wei, J.H., Zheng, L.T., Lv, X. et al. (2014). Analysis of influenza virus receptor specificity using glycan functionalized gold nanoparticles. *ACS Nano* 8: 4600–4607.

55 Hong, C., Dang, Y., Lin, G. et al. (2014). Effects of stabilizing agents on the development of myricetin nanosuspension and its characterization: an in vitro and in vivo evaluation. *Int. J. Pharm.* 477 (1–2): 251–260. https://doi.org/10.1016/j.ijpharm.2014.10.044.

56 Zhang, Y., Qi, X.R., Gao, Y. et al. (2007). Mechanisms of co-modified liver-targeting liposomes as gene delivery carriers based on cellular uptake and antigens inhibition effect. *J. Control. Release* 117 (2): 281–290. https://doi.org/10.1016/j.jconrel.2006.11.006. Epub 2006 Nov 16.

57 Lai, F., Sinico, C., Logu, D. et al. (2007). SLN as a topical delivery system for *Artemisia arborescens* essential oil: in vitro antiviral activity and skin permeation study. *Int. J. Nanomed.* 2 (3): 419–425.

58 Sinico, C., De, L.A., Lai, F. et al. (2005). Liposomal incorporation of *Artemisia arborescens* L. essential oil and in vitro antiviral activity. *Eur. J. Pharm. Biopharm.* 59 (1): 161–168.

59 Hadrich, G., Vaz, G.R., Maidana, M. et al. (2016). Anti-inflammatory effect and toxicology analysis of oral delivery quercetin nanosized emulsion in rats. *Pharm. Res.* 33 (4): 983–993. https://doi.org/10.1007/s11095-015-1844-6.

60 Hu, Y., Hu, Y., Sun, L. et al. (2016). Antiviral effects of liposome-encapsulated PolyICLC against dengue virus in a mouse model. *Biochem. Biophys. Res. Commun.* 478: 913–918.

61 Valenti, D., Logu, D.A., Loy, G. et al. (2001). Liposome-incorporated santolina insularis essential oil: preparation, characterization and in vitro antiviral activity. *J. Liposome Res.* 11 (1): 73–90. https://doi.org/10.1081/LPR-100103171.

62 Mayer, C. (2005). Nanocapsules as drug delivery systems. *Int. J. Artif. Organs* 28: 1163–1171.

63 Santos, R.B.D., Nakama, K.A., Pacheco, C.O. et al. (2021). Curcumin-loaded nanocapsules: influence of surface characteristics on technological parameters and potential antimalarial activity. *Mater. Sci. Eng. C* 118: 111356. https://doi.org/10.1016/j.msec.2020.111356.

64 Nandi, D.K. and Mitra, M. (2020). Herbal gold nanoparticles for attenuating pandemic infection of COVID-19 virus. *J. Nanomed. Nanotech.* 11: 547. https://doi.org/10.35248/2157-7439.20.11.547.

65 Yang, X.X., Li, C.M., and Huang, C.Z. (2016). Curcumin modified silver nanoparticles for highly efficient inhibition of respiratory syncytial virus infection. *Nanoscale* 8: 3040–3048.

66 Mori, Y., Ono, T., Miyahira, Y. et al. (2013). Antiviral activity of silver nanoparticle/chitosan composites against H1N1 influenza A virus. *Nanoscale Res. Lett.* 13 (8): 93.

67 Haider, A., Das, S., Ojha, D. et al. (2018). Highly monodispersed gold nanoparticles synthesis and inhibition of herpes simplex virus infections. *Mater. Sci. Eng. C* 89: 413–421.

68 Papp, I., Sieben, C., Ludwig, K. et al. (2010). Inhibition of influenza virus infection by multivalent sialic-acid-functionalized gold nanoparticles. *Small* 6: 2900–2906.

69 Di Carlo, G., Mascolo, N., Izzo, A. et al. (1999). Flavonoids: old and new aspects of a class of natural therapeutic drugs. *Life Sci.* 65: 337–353. https://doi.org/10.1016/S0024-3205(99)00120-4.

70 Kelly, G.S. (2011). Quercetin. Monograph. *Altern. Med. Rev.* 16: 172–194.

71 Zheng, S.Y., Li, Y., Jiang, D. et al. (2012). Anticancer effect and apoptosis induction by quercetin in the human lung cancer cell line A-549. *Mol. Med. Rep.* 5: 822–826.

72 Verma, N.K., Crosbie, S.K., Satti, A. et al. (2013). Magnetic core-shell nanoparticles for drug delivery by nebulization. *J Nanobiotechnol.* 13: 1–12. https://doi.org/10.1186/1477-3155-11-1.

73 Lu, M., Qiu, Q., Luo, X. et al. (2019). Phyto-phospholipid complexes (phytosomes): a novel strategy to improve the bioavailability of active constituents. *Asian J. Pharm. Sci.* 14: 265–274.

74 Semalty, A., Semalty, M., Singh, D. et al. (2012). Phyto-phospholipid complex of catechin in value added herbal drug delivery. *J. Incl. Phenom. Macrocycl. Chem.* 73: 377–386.

75 Udapurkar, P., Bhusnure, O., Kamble, S. et al. (2016). Phyto-phospholipid complex vesicles for phytoconstituents and herbal extracts: a promising drug delivery system. *Int. J. Herbal Med.* 4: 14–20.

76 Kalita, B. and Patwary, B.N. (2020). Formulation and in vitro evaluation of hesperidin-phospholipid complex and its antioxidant potential. *Curr. Drug Ther.* 15: 28–36.

77 García, P.B., Porras, A.C., Ortega, R.A. et al. (2019). Lipid-based nanoparticles: application and recent advances in cancer treatment. *Nanomaterials* 9: 638–645.

78 Badawi, N., El, S.K., Attia, D. et al. (2020). Development of pomegranate extract-loaded solid lipid nanoparticles: quality by design approach to screen the variables affecting the quality attributes and characterization. *ACS Omega* 5: 21712–21721.

79 Makwana, V., Jain, R., Patel, K. et al. (2015). Solid lipid nanoparticles (SLN) of Efavirenz as lymph targeting drug delivery system: elucidation of mechanism of uptake using chylomicron flow blocking approach. *Int. J. Pharm.* 495: 439–446.

80 Vijayanand, P., Jyothi, V., Aditya, N. et al. (2018). Development and characterization of solid lipid nanoparticles containing herbal extract: in vivo antidepressant activity. *J. Drug Deliv.* 2018: 776–793.

81 Ban, C., Jo, M., Park, Y.H. et al. (2020). Enhancing the oral bioavailability of curcumin using solid lipid nanoparticles. *Food Chem.* 302: 125328.

82 Delshadi, R., Bahrami, A., Mcclements, D.J. et al. (2021). Development of nanoparticle-delivery systems for antiviral agents: a review. *J. Control. Release* 331: 30–44.

83 Mei, Z., Chen, H., Weng, T. et al. (2003). Solid lipid nanoparticle and microemulsion for topical delivery of triptolide. *Eur. J. Pharm. Biopharm.* 56: 189–196.

84 Sah, A.K., Vyas, A., Suresh, P.K. et al. (2018). Application of nanocarrier-based drug delivery system in treatment of oral cancer. *Artif. Cells Nanomed. Biotechnol.* 46 (4): 650–657. https://doi.org/10.1080/21691401.2017.1373284.

85 Karimi, N., Ghanbarzadeh, B., Hamishehkar, H. et al. (2018). Antioxidant, antimicrobial and physicochemical properties of turmeric extract-loaded nanostructured lipid carrier (NLC). *Colloid Interface Sci. Commun.* 22: 18–24. https://doi.org/10.1016/j.colcom.2017.11.006.

86 Ramkar, S., Sah, A.K., Bhuwane, N. et al. (2020). Nano-lipidic carriers as a tool for drug targeting to the pilosebaceous units. *Curr. Pharm. Des.* 26: 3251–3268.

87 Sah, A.K., Dewangan, M., and Suresh, P.K. (2019). Potential of chitosan-based carrier for periodontal drug delivery. *Colloids Surf. B: Biointerfaces* 178: 185–198.

88 Jain, B., Singh, B., Katare, O.P. et al. (2010). Development and characterization of minoxidil-loaded liposomal system for delivery to pilosebaceous units. *J. Liposome Res.* 20: 105–114. https://doi.org/10.3109/08982100903161449.

89 Aruoma, O.I., Spencer, J.P., Rossi, R. et al. (1996). An evaluation of the antioxidant and antiviral action of extracts of rosemary and provençal herbs. *Food Chem. Toxicol.* 34 (5): 449–456.

90 Bilia, A.R., Piazzini, V., Guccione, C. et al. (2017). Improving on nature: the role of nanomedicine in the development of clinical natural drugs. *Planta Med.* 83: 366–381. https://doi.org/10.1055/s-0043-102949.

91 Isacchi, B., Arrigucci, S., Marca, I.G. et al. (2011). Conventional and long-circulating liposomes of artemisinin: preparation, characterization, and pharmacokinetic profile in mice. *J. Liposome Res.* 21: 237–244.

92 Wilson, D.K., Shyamala, G., Paulpandi, M. et al. (2021). Development and characterization of phytoniosome nano vesicle loaded with aqueous leaf extracts of *Justicia adhatoda* and *Psidium guajoava* against dengue virus (DEN-2). *J. Clust. Sci.* 32: 297–304. https://doi.org/10.1007/s10876-020-01788-6.

93 Abba, Y., Hassim, H., Hamzah, H. et al. (2005). Antiviral activity of resveratrol against human and animal viruses. *Adv. Virol.* 184241. https://doi.org/10.1155/2015/184241.

94 Shankar, S., Singh, G., and Srivastava, R.K. (2007). Chemoprevention by resveratrol: molecular mechanisms and therapeutic potential. *Front. Biosci.* 12 (13): 4839–4854.

95 Ahirrao, M. and Shrotriya, S. (2017). In vitro and in vivo evaluation of cubosomal in situ nasal gel containing resveratrol for brain targeting. *Drug Dev. Ind. Pharm.* 43 (10): 1686–1693. https://doi.org/10.1080/03639045.2017.1338721. Epub 2017 Jun 28.

96 Milovanovic, M., Arsenijevic, A., Milovanovic, J. et al. (2017). Nanoparticles in antiviral therapy. In: *Antimicrobial Nanoarchitectonics* (ed. A.M. Grumezescu), 383–410. Elsevier https://doi.org/10.1016/B978-0-323-52733-0.00014-8.

97 Lembo, D., Donalisio, M., Civra, A. et al. (2018). Nanomedicine formulations for the delivery of antiviral drugs: a promising solution for the treatment of viral infections. *Expert Opin. Drug Deliv.* 15: 93–114.

98 Liang, L., Ahamed, A., and Ge, L. (2020). Advances in antiviral material development. *Chem. Plus Chem.* 85: 2105–2128.

99 Bagad, M. and Khan, Z.A. (2015). Poly(*n*-butylcyanoacrylate) nanoparticles for oral delivery of quercetin: preparation, characterization, and pharmacokinetics and biodistribution studies studies in Wistar rats. *Int. J. Nanomed.* 10: 3921–3935.

100 Gomes, I.B., Porto, M.L., Santos, M.C. et al. (2014). Renoprotective, anti-oxidative and anti-apoptotic effects of oral lowdose quercetin in the C57BL/6J model of diabetic nephropathy. *Lipids Health Dis.* 13 (1): 184.

101 Chanput, W., Krueyos, N., and Ritthiruangdej, P. (2016). Anti-oxidative assays as markers for anti-inflammatory activity of flavonoids. *Int. Immunopharmacol.* 40: 170–175.

102 Caddeo, C., Diez-Sales, O., Pons, R. et al. (2014). Topical anti-inflammatory potential of quercetin in lipid-based nanosystems: in vivo and in vitro evaluation. *Pharm. Res.* 31 (4): 959–968.

103 Chiow, K.H., Phoon, M.C., Putti, T. et al. (2016). Evaluation of antiviral activities of *Houttuynia cordata* Thunb. extract, quercetin, quercetrin and cinanserin on murine coronavirus and dengue virus infection. *Asian Pac J Trop Med* 9 (1): 1–7.

104 Johari, J., Kianmehr, A., Mustafa, M.R. et al. (2012). Antiviral activity of baicalein and quercetin against the Japanese encephalitis virus. *Int. J. Mol. Sci.* 13 (12): 16785–16795.

105 Lockhart, J.N., Stevens, D.M., Beezer, D.B. et al. (2015). Dual drug delivery of tamoxifen and quercetin: regulated metabolism for anticancer treatment with nanosponges. *J. Control. Release* 220: 751–757.

106 Chitkara, D., Nikalaje, S.K., Mittal, A. et al. (2012). Development of quercetin nanoformulation and in vivo evaluation using streptozotocin induced diabetic rat model. *Drug Deliv. Transl. Res.* 2 (2): 112–123.

107 Zhen, L., Cuiping, B., Weicheng, H. et al. (2018). Preparation and *in vitro* and *in vivo* evaluation of quercetin-loaded mixed micelles for oral delivery. *Biosci. Biotechnol. Biochem.* 82 (2): 238–246. https://doi.org/10.1080/09168451.2017.1419852.

108 Almeidam, K.B., Araujo, J.L., Cavalcanti, J.F. et al. (2018). In vitro release and anti-herpetic activity of *Cymbopogon citrates* volatile oil-loaded nanogel. *Rev. Bras.* 28: 495–502. https://doi.org/10.1016/j.bjp.2018.05.007.

109 Yao, Y., Xia, M., Wang, H. et al. (2016). Preparation and evaluation of chitosan-based nanogels/gels for oral delivery of myricetin. *Eur. J. Pharm. Sci.* 91: 144–153. https://doi.org/10.1016/j.ejps.2016.06.014.

110 Kim, B.Y., Rutkz, J.T., and Chan, W.C. (2010). Nanomedicine. *N. Engl. J. Med.* 363: 2434–2443.

111 Mhlwatika, Z. and Aderibigbe, B.A. (2018). Application of dendrimers for the treatment of infectious diseases. *Molecules* 23: 2205. https://doi.org/10.3390/molecules23092205.

112 Han, S., Kanamoto, T., Nakashima, H. et al. (2012). Synthesis of a new amphiphilic glycodendrimer with antiviral functionality. *Carbohydr. Polym.* 90: 1061–1068.

113 Nanjwade, B.K., Bechra, H.M., Derkar, G.K. et al. (2009). Dendrimers: emerging polymers for drug-delivery systems. *Eur. J. Pharm. Sci.* 38: 185–196.

114 Gangadharan, D., Dhandhala, N., Dixit, D. et al. (2012). Investigation of solid supported dendrimers for water disinfection. *J. Appl. Polym. Sci.* 124: 138491.

115 Hassan, M.Z., Osman, H., Ali, M.A. et al. (2016). Therapeutic potential of coumarins as antiviral agents. *Eur. J. Med. Chem.* 123: 236–255. https://doi.org/10.1016/j.ejmech.2016.07.056.

116 Mishra, S., Pandey, A., and Manvati, S. (2020). Coumarin: an emerging antiviral agent. *Heliyon* 6 (1): e03217. https://doi.org/10.1016/j.heliyon.2020.e03217.

117 Kumar, M.S., Prasad, B.R., and Minghua, C.U.I. (2011). Synthesis of bio-conjugate using succinimidyl ester of coumarin dendrimer as the potential fluorescent compounds for the detection of specific antigen. *Proceedings of Fall International Convention of the Pharmaceutical Society of Korea*, 247.

118 Perreault, F., de- Faria, A.F., Nejati, S. et al. (2015). Antimicrobial properties of graphene oxide nanosheets: why size matters. *ACS Nano* 9: 7226–7236.

119 Song, Z., Wang, X., Zhu, G. et al. (2015). Virus capture and destruction by label-free graphene oxide for detection and disinfection applications. *Small* 11: 1171–1176.

120 Zou, X., Zhang, L., Wang, Z. et al. (2016). Mechanisms of the antimicrobial activities of graphene materials. *J. Am. Chem. Soc.* 138: 2064–2077.

121 Sametband, M., Kalt, I., Gedanken, A. et al. (2013). Herpes simplex virus type-1 attachment inhibition by functionalized graphene oxide. *ACS Appl. Mater. Interfaces* 6: 1228–1235.

122 Ye, S., Shao, K., Li, S. et al. (2015). Antiviral activity of graphene oxide: how sharp edged structure and charge matter. *ACS Appl. Mater. Interfaces* 7: 21571–21579.

123 Heo, D.N., Ko, W.K., Moon, H.J. et al. (2014). Inhibition of osteoclast differentiation by gold nanoparticles functionalized with cyclodextrin curcumin complexes. *ACS Nano* 4 (8): 12049–12062.

124 Zhang, J., Liang, Y.C., Lin, X. et al. (2015). Self-monitoring and self-delivery of photosensitizer-doped nanoparticles for highly effective combination cancer therapy in vitro and in vivo. *ACS Nano* 9: 9741–9756.

125 Obata, K., Kojima, T., Masaki, T. et al. (2013). Curcumin prevents replication of respiratory syncytial virus and the epithelial responses to it in human nasal epithelial cells. *PLoS One* 8: 70225–70239.

126 Yang, X.X., Li, C.M., Li, Y.F. et al. (2017). Synergistic antiviral effect of curcumin functionalized graphene oxide against respiratory syncytial virus infection. *Nanoscale* 9: 16086–16092.

127 Qian, J., Meng, H., Xin, L. et al. (2017). Self-nanoemulsifying drug delivery systems of myricetin: formulation development, characterization, and in vitro and in vivo evaluation. *Colloids Surf. B: Biointerfaces* 160: 101–109. https://doi.org/10.1016/j.colsurfb.2017.09.020.

128 Diorio, C. and Lokhnauth, J. (2019). Curcumin solid lipid particles and methods for their preparation and use. US Patent 20160000714A1.

129 Amoabediny, G., Ochi, M.M., Rezayat, S.M. et al. (2018). Targeted nano-liposome co-entrapping anti-cancer drugs. US Patent 20160228362A1.

130 Orza, A.I. (2017). Novel anti-wrinkle and anti-aging nano formulations and method of preparation using novel nano co-delivery system. US Patents 20170157005A1.

131 Bombardelli, E., Patri, G.F., and Pozzi, R. (1988). Complexes of saponins with phospholipids and pharmaceutical and cosmetic compositions containing them. Indena Spa, Milano, Italy. US Patent 5043323A.

132 Marakhova, A.I., Stanishevskij, K.P.Y.M., and Shvitko, B.S. et al. (2016). Method of producing a nanoscale phytosome system. US Patent 2016150141.

133 Medasani, M.D.S. (2012). *Picrorhiza kurroa* extract for prevention, elimination and treatment of RNA viruses and infection diseases. US Patent 20120148569A1.

134 Ht, P.Y.T. (2012). Nanoemulsions containing antioxidants and other health-promoting compounds. US Patent 20120052126A1.

135 Patel, A., Tiwari, J. and Velikov, K. (2012). Particles comprising hydrophobic polymer and hydrophobic phenolic compound. WO Patent 2012000757A1.

136 Keck, C., Schwabe, K., and Rimpler, C. (2012). Compositions containing lipid micro-or nanoparticles for the enhancement of the dermal action of solid particles. US Patent 20120128777A1.

137 Edmund, J.E. and Lder, J.M.J.S.J.T.L.Z. (2010). Nanoparticle isoflavone compositions & methods of making and using the same. US Patent 20150050357A1.

138 Kurzrock, R.L., Mehta, K., and Aggarawal, B.B. (2004). Liposomal curcumin for treatment of cancer. World Patent 2004080396.

139 Braden, A.R.C. and Kafka, V.J.E. (2008). Formulation of active agent loaded activated PLGA nanoparticles for targeted cancer nano-therapeutics. US Patent 20080253961.

140 Kumer, K.S., Gopesh, F.A., and Kumar R.P.A. (2010). Curcumin nanoparticles and methods of producing the same. World Patent 2010013224.

141 Kim, J.H. (2010). Nanoparticle composition for prevention of hair loss and promotion of hair growth. US Patent 20120052126A1.

142 John, C.G.S.J. (2005). Liposomes containing phytochemical agents and methods for making and using same. World Patent 2006068759A2.

143 Yatvin, M. (2004). Liposome drug delivery of polycyclic, aromatic, antioxidant or anti-inflammatory compounds. US Patent 20050152978A1.

144 Merizzi, G. (2004). Anti-oxidant preparation based on plant extracts for the treatment of circulation and adiposity problems. US Patent 6756065B1.

145 Bombardelli, E. and Sabadie, M. (1990). Phospholipid complexes of extracts of *Vitis vinifera*, their preparation process and pharmaceutical and cosmetic compositions containing them. US Patents 4963527A.

146 Winkelmann, F.D.W.P.D.P.D. (1980). Rosmarinic acid-phospholipid complexes, process for their preparation and medicines containing them. European Patent 0032556A1.

147 Mönkäre, J., Pontier, M., Eveline, E.M. et al. (2018). Development of PLGA nanoparticle loaded dissolving microneedles and comparison with hollow microneedles in intradermal vaccine delivery. *Eur. J. Pharm. Biopharm.* 129: 111–121. https://doi.org/10.1016/j.ejpb.2018.05.031.

16

Nanotherapeutics of Phytoantioxidants for Parasitic Diseases and Neglected Tropical Diseases

James H. Zothantluanga[1], Zonunmawii[1], Punamjyoti Das[1], Himangshu Sarma[2], and Abd. Kakhar Umar[3]

[1] *Department of Pharmaceutical Sciences, Faculty of Science and Engineering, Dibrugarh University, Dibrugarh, Assam, India*
[2] *Life Sciences Division, Institute of Advanced Study in Science and Technology, Guwahati, Assam, India*
[3] *Department of Pharmaceutics and Pharmaceutical Technology, Faculty of Pharmacy, Universitas Padjadjaran, Jatinangor, Indonesia*

16.1 Introduction

The term "phytoantioxidants" refers to chemicals of plant origin with well-reported antioxidant activities such as flavonoids and polyphenols [1]. Flavonoids are considered phenolic substances whose core structure consisted of a flavan nucleus wherein a total of 15 carbon atoms are arranged to form three rings (A, B, and C) to produce the popular "C_6-C_3-C_6 scaffold" [2]. Flavonoids are further classified into different subgroups based on their structures such as flavones (luteolin, apigenin, chrysin), flavonones (hesperetin, naringenin), flavonols (quercetin, kaempferol, galangin, fisetin, myricetin), flavononol (taxifolin), isoflavones (genistein, genistin, daidzenin, daidzin, biochanin A, formononetin), flavan-3-ols (catechin, epicatechin, epigallocatechin), and flavylium salts (cyanidin, cyanin, pelargonidin) [2].

On the other hand, polyphenols are made up of multiple phenolic rings that are bonded together through different structural elements. The umbrella of polyphenols constitutes different classes of plant secondary metabolites such as flavonoids, phenolic acids (cinnamic acid, *p*-coumaric acid, caffeic acid, ferulic acid, sinapic acid, gentisic acid, vanillic acid, gallic acid, syringic acid, protocatechuic acid), stilbenes (resveratrol, piceatannol, rhapontigenin, isorhapontigenin, pinosylvin, pterostilbene), lignans (niranthin, sesamin, silymarin, rubrifloralignan A, bicyclol, phillygenin, clemastanin B, isatindolignanoside A, diphyllin, hinokinin, yatein, secoisolariciresinol), and tannins (procyanidins, catechin, afzelechin, gallocatechin, ellagic acid, gallic acid, gallotannin, ellagitannin, hexahydroxydiphenic acid) [3–8]. The chemical structures of some phytoantioxidants are given in Figure 16.1.

Phytoantioxidants present in regularly consumed foods is known as dietary polyphenols [5, 8]. Tea, red wine, buckwheat, red pepper, tomato, onion, olive oil, grapefruit, citrus fruits, lemons, oranges, soybean, cherry, raspberry, and strawberry are some examples of regularly consumed foods that are rich in phytoantioxidants [9]. Quercetin, curcumin, rutin, genistein, curcumin, daidzein, resveratrol, proanthocyanidins, and ellagitannins are some examples of the most common dietary phytoantioxidants [5, 10]. The Indian Systems of Medicine (ISM), which mainly comprises Ayurveda, Unani, and Siddha, existed since ancient times and thousands of medicinal plants are

Phytoantioxidants and Nanotherapeutics, First Edition. Edited by Mithun Rudrapal.

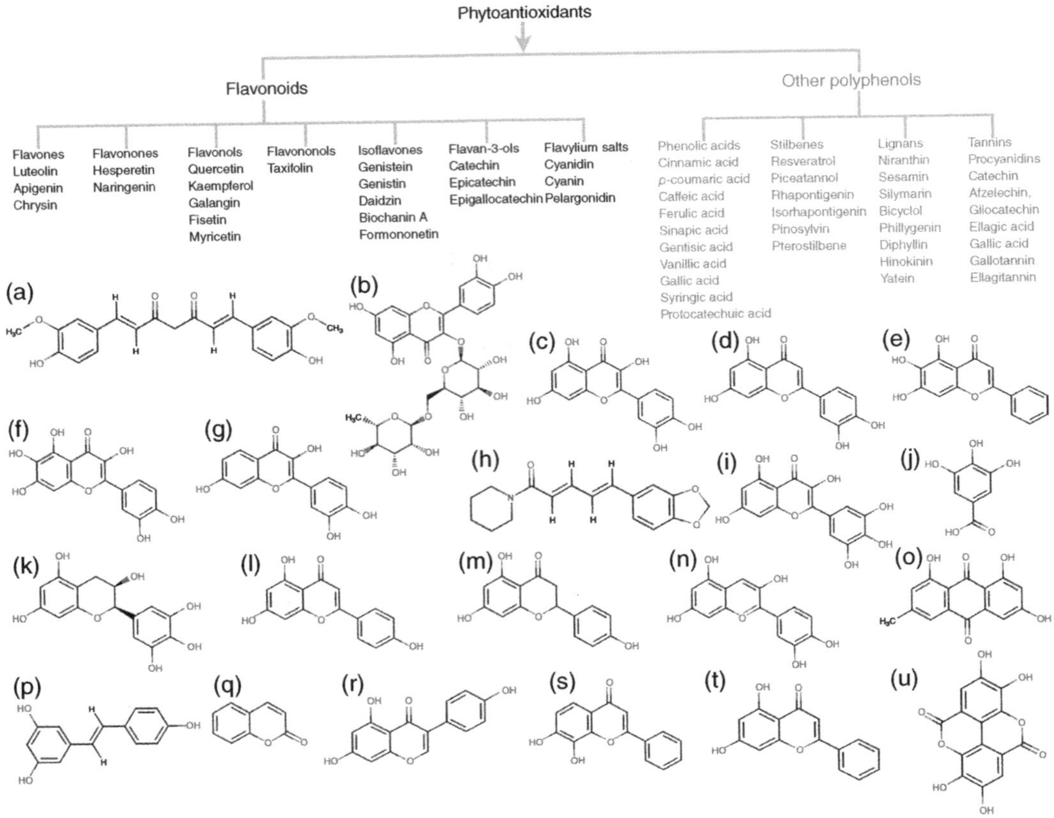

Figure 16.1 Classification of phytoantioxidants and chemical structures of some phytoantioxidants with activity against parasitic diseases and NTDs: (a) curcumin, (b) rutin, (c) quercetin, (d) luteolin, (e) baicalein, (f) quercetagetin, (g) fisetin, (h) piperine, (i) myricetin, (j) gallic acid, (k) epigallocatechin, (l) apigenin, (m) naringenin, (n) cyanidin, (o) emodin, (p) resveratrol, (q) coumarin, (r) genistein, (s) 7,8-dihydroxyflavone, (t) chrysin, (u) ellagic acid.

utilized in ISM. Several studies have also reported the presence of many phytoantioxidants in the medicinal plants utilized in ISM [11]. Similarly, medicinal plants utilized in Traditional Chinese Medicine are also another repository for several phytoantioxidants [12].

Flavonoids are reported to elicit anti-inflammatory, antilipidemic, antihyperglycemic, antiviral, hepatoprotective, gastric antiulcer, cardioprotective, neuroprotective, antioxidant, and anticancer activities [13]. Phenolic acids were reported to have beneficial use in the management of obesity, oxidative stress, type 2 diabetes mellitus, cardiovascular diseases, and neuroprotective and other metabolic syndromes [14]. Stilbenes elicit pharmacological activities such as antioxidant, anti-aging, anti-inflammatory, antidiabetic, and anticancer and are also used to manage obesity [7, 15]. Lignans were reported to show antitumor, hepatoprotective, platelet-activating factor antagonistic, insecticidal, estrogenic, antifungal, antihypertensive, sedative, and antioxidant activities [16]. Activities such as antioxidant, antimicrobial, anticancer, anti-nutritional, and cardioprotective were reported to be elicited by tannins [17].

Parasitic diseases constitute different abnormal health conditions (malaria, dengue, chikungunya, leishmaniasis, Zika virus, and soil-transmitted helminths) that are caused by parasites such as protozoans, helminths, and ectoparasites. On the other hand, bacteria, viruses, and parasites are notorious for causing neglected tropical diseases (NTDs) such as Buruli ulcer, Chagas disease, cysticercosis, dracunculiasis, fascioliasis, trypanosomiasis, leprosy, lymphatic filariasis, mycetoma, onchocerciasis, schistosomiasis, and trachoma. Millions of people are being infected and killed every year due to parasitic diseases and NTDs [18]. There are many drugs used for the treatment of parasitic diseases and NTDs. However, problems of serious concern such as toxicity are associated with these drugs [19–21]. In particular, pathogens (parasites, bacteria, and viruses) have developed resistance against many of the existing drugs [22–25].

Phytoantioxidants are among the most widely utilized natural products for their broad pharmacological action and negligible toxicity [26–28]. Several phytoantioxidants have been reported to inhibit/kill several drug-resistant pathogens that are responsible for causing parasitic diseases and NTDs [29–35]. Even with the emergence of drug-resistant pathogens, phytoantioxidants give a ray of hope as they hold the potential to be developed further into potent therapeutics. However, bioavailability issues associated with phytoantioxidants have limited their use in clinical settings [36].

Phytoantioxidants are poorly absorbed in the gut, intestines, and colon. Bioavailability issues are most prominent with proanthocyanidins, galloyted catechins, and anthocyanins. Isoflavones have the best bioavailability, followed by phenolic acids, flavonols, flavonones, and flavonol [37]. Factors affecting the bioavailability of phytoantioxidants include their physicochemical properties, and chemical structures, along with other factors such as food matrix interactions, the extent of liver metabolisms, and intestinal microflora [36, 37]. Among the different approaches available to enhance the bioavailability of phytoantioxidants, the nanotechnology approach gives the most promising results [38].

The bioavailability of several phytoantioxidants has been enhanced with nanoformulations such as gelatin–cyclodextrin polymeric nanoformulation, graphene oxide-based nanoformulation, liposomes, eudragit-nutriosmes, nanoemulsion, etc. [39–43]. Bioavailability enhancement of phytoantioxidants with nanoformulation increases their therapeutic efficacy, solubility, stability, and distribution. Nanoformulation also allows site-specific delivery of drugs and prevents toxicity issues [44]. The major problems of phytoantioxidants such as low bioavailability and rapid metabolism are solved with nanotechnology/nanoformulations [45].

In this chapter, phytoantioxidants with inhibitory activity against different pathogens responsible for parasitic diseases and NTDs are highlighted. The nanoformulations of phytoantioxidants that are directly used for the treatment of parasitic diseases and NTDs are also highlighted. Those phytoantioxidants with activity against parasitic diseases and NTDs, but whose nanoformulations have not been specifically investigated for parasitic diseases or NTDs, but whose bioavailability has been enhanced through nanotechnology are also highlighted.

16.2 Phytoantioxidants for Parasitic Diseases

Phytoantioxidants that have been reported to inhibit pathogens of parasitic origins are given in Table 16.1. The in vitro and in vivo activities of the phytoantioxidants are also provided according to feasible data. Curcumin, rutin, quercetin, luteolin, baicalein, quercetagetin, fisetin, epigallocatechin gallate, apigenin, piperine, myricetin, naringenin, procyanidin A2, and gallic acid are examples of some polyphenols that have been reported to inhibit the causative agents of at least one of the parasitic diseases.

16.2.1 Malaria

Malaria is caused by *Plasmodium falciparum*, *P. vivax*, *P. knowlesi*, *P. malariae*, and *P. ovale*. These parasites are transmitted by female *Anopheles* mosquitoes. Artesunate, chloroquine, hydroxychloroquine, amodiaquine, quinine, atovaquone, proguanil, mefloquine, primaquine, tafenoquine, doxycycline, and clindamycin are examples of some pharmaceutical drugs that are used for the treatment of malaria [46–49]. Curcumin, rutin, and quercetin are examples of some phytoantioxidants that have been reported to inhibit malarial parasites in in vitro and in vivo experiments (Table 16.1).

Table 16.1 Examples of some phytoantioxidants used against parasitic diseases.

Parasitic diseases	Phytoantioxidants	In vitro	In vivo	References
Malaria	Curcumin	EC_{50} = 1.1–10 μg/ml	—	[46–49]
	Rutin	IC_{50} = 3.53 ± 13.34 μg/ml	—	
	Quercetin	IC_{50} = 4.11 ± 2.05 μg/ml	Quercetin reverts chloroquine-induced toxicity and oxidative stress	
Dengue	Luteolin	DENV1, EC_{50} = 4.36 μM; DENV3, EC_{50} = 5.69 μM; DENV4, EC_{50} = 8.38 μM	Oral administration of luteolin reduces the viral load of DENV1 in infected mice up to 10 times	[29, 32]
	Quercetin	IC_{50} = 35.7 μg/ml	-	
	Baicalein	IC_{50} = 13.5 μg/ml	-	
Chikungunya	Baicalein	IC_{50} = 1.891 μg/ml	-	[33, 35]
	Quercetagetin	IC_{50} = 13.85 μg/ml	-	
	Fisetin	IC_{50} = 8.444 μg/ml	-	
Leishmaniasis	Epigallocatechin gallate	IC_{50} (24 hours' incubation) = 3.7 μM IC_{50} (72 hours' incubation) = 3.4 μM	Oral administration of EGCG at 100 mg/ml for 30 days significantly reduced the lesion size	[29, 50, 51]
	Apigenin	IC_{50} = 4.3 μM	—	
	Piperine	IC_{50} = 4.31 ± 0.44 μg/ml	—	
Zika virus	Epigallocatechin gallate	EC_{50} = 21.4 μM	—	[52, 53]
	Myricetin	IC_{50} = 0.58 ± 0.17 μM	—	
	Naringenin	IC_{50} = 58.79 μM	—	
Soil-transmitted helminths	Luteolin	IC_{50} = 9.7 μg/ml	A single dose of 100 mg/kg induced a 27.6% reduction of worm burden in a mouse model	[54, 55]
	Procyanidin A2	IC_{50} = 12.6 μg/ml	—	
	Gallic acid	At 1000 μg/ml, gallic acid and an unidentified compound displayed an ovicidal activity of 100%	—	

16.2.2 Dengue

Dengue is caused by dengue virus (DENV)-1, DENV-2, DENV-3, and DENV-4 which belong to the *Flaviviridae* family. These viruses are transmitted by *Aedes* species of mosquito (*A. aegypti* and *A. albopictus*). There is no specific pharmaceutical treatment available for dengue. Painkillers, fluid therapy, and bed rest are advised for patients infected with DENV [29, 32]. Luteolin, quercetin, and baicalein are examples of some phytoantioxidants that have been reported to inhibit at least one of the DENV serotypes in in vitro experiments. Studies have also reported that oral administration of luteolin reduces the viral load of DENV1 in mice up to 10 times (Table 16.1).

16.2.3 Chikungunya

Chikungunya is caused by chikungunya virus (CHIKV). The virus is transmitted by day-biting mosquitoes belonging to the *Aedes* species. There are no specific antiviral drugs or vaccines available for treatment. Painkillers, fluid therapy, and bed rest are advised for patients infected with CHIKV [33, 35]. Baicalein, quercetagetin, and fisetin are examples of some phytoantioxidants that have been reported to inhibit CHIKV in in vitro experiments (Table 16.1).

16.2.4 Leishmaniasis

Leishmaniasis is caused by protozoans of the genus *Leishmania*. The parasite is transmitted by the bite of sandflies of the genus *Phlebotomus* and *Lutzomyia*. Sodium stibogluconate, meglumine antimonite, amphotericin B, and miltefosine are examples of pharmaceutical drugs that are used for the treatment of leishmaniasis [29, 50, 51]. Epigallocatechin gallate, apigenin, and piperine are examples of polyphenols that have been reported to inhibit *Leishmania* protozoans in in vitro experiments. Epigallocatechin gallate was also reported to possess the ability to reduce the skin lesion size (Table 16.1).

16.2.5 Zika Virus

Zika virus is caused by a flavivirus. The virus is transmitted primarily by *Aedes* species of mosquito (*A. aegypti* and *A. albopictus*). There is no specific pharmaceutical drug for the Zika virus. Supportive therapy, fluid therapy, analgesics, and antipyretics are given to infected patients [52, 53]. Epigallocatechin gallate, myricetin, and naringenin are examples of some phytoantioxidants that have been reported to inhibit flavivirus in in vitro experiments (Table 16.1).

16.2.6 Soil-Transmitted Helminths

Soil-transmitted helminths are caused by helminthic parasites. These parasites are transmitted through contaminated soil infected with *Ascariasis lumbricoides*, *Trichuris trichiura*, and *Strongyloides stercoralis*. Albendazole, mebendazole, and ivermectin are examples of some pharmaceutical drugs that are used against helminthic parasites [54, 55]. Luteolin, procyanidin A2, and gallic acid are examples of some polyphenols that have been reported to kill the helminthic parasites in in vitro experiments (Table 16.1).

16.3 Phytoantioxidants for NTDs

Phytoantioxidants that have been reported to inhibit pathogens causing NTDs are given in Table 16.2. The in vitro and in vivo activities of the phytoantioxidants are also provided according to feasible data. Emodin, gallic acid, epigallocatechin gallate, resveratrol, coumarin, curcumin, genistein, quercetin, 7,8-dihydroxyflavone, ellagic acid, neobavaisoflavone, sakuranetin, 4,5-diethyl-3′-ethoxy-pyro-flavone, rutin, naringenin, hesperetin, chrysin, naringin, (+)-catechin, (−)-epicatechin, (−)-epigallocatechin, (−)-epicatechin gallate, (−)-epigallocatechin gallate, baicalin, chalcone derivatives, and myricetin are examples of some polyphenols that have been reported to inhibit the causative agents of at least one of the NTDs.

Table 16.2 Examples of some phytoantioxidants used against NTDs.

NTDs	Phytoantioxidants	In vitro	In vivo	References
Buruli ulcer	*Aloe vera* (emodin)	MIC value of 40 mg/ml	–	[34, 56]
	Allium sativum extract (gallic acid)	MIC value of 0.78–6.25 μg/ml	–	
Chagas disease	Epigallocatechin gallate	$IC_{50} = 0.53\,\mu M$	Intraperitoneal administration of Epigallocatechin gallate (0.8 mg/kg) in *T. cruzi*-infected mice decreased the parasitemia level (50%) on day 19 post-infection	[30, 31, 57]
	Resveratrol	–	100 mg/kg is capable to minimize hepatic injury through antioxidant properties in mice infected with *T. cruzi*.	
	Coumarin	At 4 mg/ml $IC_{50} = 64\,\mu M$	–	
Fascioliasis	Curcumin	60 μM concentration inhibit worm motility and suppressed the free radical scavenging ability	–	[58–60]
	Genistein (soybean extract)	Soybean extract caused damage of adult *F. gigantica* DNA after 24 h of incubation	100 mg/kg in rabbit-fecal egg count reduction is 41.6% on the third-day post-treatment and 95.8% on the seventh day post-treatment	
	Quercetin (*Momordica charantia*)	12.5–50 mg/ml shows ovicidal effects cytotoxic effects during the embryonic development of *F. hepatica* eggs	–	
Human African trypanosomiasis	Quercetin	$IC_{50} = <2.0\,\mu M$	–	[61–63]
	7,8-dihydroxyflavone	$IC_{50} = 0.16\,\mu M$	–	
	Gallic acid	$IC_{50} = 15.6\,\mu M$	–	

Table 16.2 (Continued)

NTDs	Phytoantioxidants	In vitro	In vivo	References
Onchocerciasis	(−)-Epigallocatechin 3-*O*-gallate	$LC_{50} = 1.3\,\mu g/ml$	–	[64, 65]
	Ellagic acid	$LC_{50} = 0.03\,\mu M$	In vivo testing demonstrated that there was no alteration in the mean body weight of animals before and after the test period in both the control and acid-treated rats	
Leprosy	Neobavaisoflavone	–	–	[66, 67]
	Sakuranetin	–	–	
Lymphatic filariasis	4,5-Diethyl-3′-ethoxy-pyro-flavone	$IC_{50} = 0.0032\,mg/ml$	–	[68, 69]
	Rutin	Microfilariae: $IC_{50} = 74.3 \pm 7.2\,\mu g/ml$	–	
	Naringenin	Female adult worm: $IC_{50} = 2.5 \pm 0.2\,\mu g/ml$; microfilariae: $IC_{50} = 297.3 \pm 5.2\,\mu g/ml$	Total number of live worms, Naringenin: 10 ± 2.2 at 100 mg/kg, 14 ± 2.1 at 50 mg/kg, and 17 ± 4.2 at 25 mg/kg; standard drug (diethylcarbamazine-citrate): 20 ± 3.5 at 50 mg/kg	
	Hesperetin	Female adult worm: $IC_{50} = 18.2 \pm 2.6$; microfilariae: $IC_{50} = 88.4 \pm 17.0\,\mu g/ml$	–	
	Chrysin	Microfilariae: $IC_{50} = 105.1 \pm 21.1\,\mu g/ml$	–	
	Naringin	Female adult worm: $IC_{50} = 78.8 \pm 11.5\,\mu g/ml$	–	
Trachoma	Polyphenon 70S ((+)-catechin, (−)-epicatechin, (−)-epigallocatechin, (−)-epicatechin gallate, and (−)-epigallocatechin gallate)	Complete inhibition at: 1.6 mg/ml against *Chlamydia trachomatis* D/UW-3/Cx strain; 0.4 mg/ml against *C. trachomatis* L(2)/434/Bu strain	–	[70, 71]
	Baicalin	$IC_{50} = 0.50\,mM$	–	
Schistosomiasis	Myricetin	$LC_{50} = 600\,\mu M$	–	[72, 73]
	Chalcone derivative 1	100, 100, and 50% dead worms were observed at 100, 50, 25 μM, respectively	Significantly reduced worm burden in infected mice	
	Chalcone derivative 3	100, 100, 100, 50% dead worms were observed at 100, 50, 25, 10 μM, respectively	Significantly reduced worm burden in infected mice	

16.3.1 Buruli Ulcer

Buruli ulcer is caused by *Mycobacterium ulcerans*. The mode of transmission is not known although it is believed that the bacteria are transmitted by vectors such as mosquitoes or insects that are found in water. Rifampicin and streptomycin are examples of some pharmaceutical drugs that are used for the treatment of Buruli ulcers [34, 56]. The exact bioactive compound acting against *M. ulcerans* could not be properly identified. However, herbal extracts of *A. vera* and *A. sativum* were reported to act against *M. ulcerans*. Therefore, these two extracts with their major polyphenols (emodin and gallic acid) are included (Table 16.2).

16.3.2 Chagas Disease

Chagas disease is caused by *Trypanosoma cruzi*. The parasite is transmitted through the feces of infected triatomine bugs (kissing bugs), reduviid insects, and congenitally from mother to fetus. Nifurtimox and benznidazole are examples of pharmaceutical drugs that are used for the treatment of Chagas disease [30, 31, 57]. Epigallocatechin gallate, resveratrol, and coumarin are examples of some phytoantioxidants that have been reported to act against *T. cruzi* (Table 16.2).

16.3.3 Fascioliasis

Fascioliasis is caused by *Fasciola hepatica* and *F. gigantica*. They are transmitted through sheep and cattle. Triclabendazole is an example of a pharmaceutical agent that is used for the treatment of fascioliasis [58–60]. Curcumin is an example of a polyphenol that has been reported to act against *F. hepatica* and *F. gigantica*. Soybean extract and *M. charantia* were also reported to act against *F. hepatica* and *F. gigantica*. Although the exact polyphenol of soybean extract and *M. charantia* that act against *F. hepatica* and *F. gigantica* cannot be identified, genistein and quercetin have been included as they are among the major polyphenols of soybean extract and *M. charantia* (Table 16.2).

16.3.4 Human African Trypanosomiasis

Human African trypanosomiasis is caused by *Trypanosoma brucei*. The parasite is transmitted by Tsetse flies (*Glossina* spp.). Pentamidine, suramin, melarsoprol, and eflornithine are examples of some pharmaceutical drugs that are used for the treatment of *T. brucei* infection [61–63]. Quercetin, 7,8-dihydroxyflavone, and gallic acid are examples of phytoantioxidants that have been reported to act against *T. brucei* in in vitro experiments (Table 16.2).

16.3.5 Onchocerciasis

Onchocerciasis is caused by a filarial nematode, i.e. *Onchocerca volvulus*. The nematode is transmitted by repeated bites of a female black fly of the genus, *Simulium damnosum*. Ivermectin and doxycycline are examples of some pharmaceutical drugs that are used for the treatment of onchocerciasis [64, 65]. Epigallocatechin 3-*O*-gallate and ellagic acid are examples of some phytoantioxidants that have been reported to act against *O. volvulus* (Table 16.2).

16.3.6 Leprosy

Leprosy is caused by *Mycobacterium leprae*. Leprosy spreads from person to person if the nasal secretions, droplets, coughs, and sneezes carry the bacteria. *M. leprae* rarely spreads from chimpanzees, monkeys, and armadillos. Studies have suggested that the susceptibility to getting leprosy may be due to specific human genes. Clofazimine, dapsone, and rifampicin are examples of pharmaceutical drugs that are used for the treatment of leprosy [74–76]. Neobavaisoflavone is a polyphenol used in the treatment of leprosy and is predicted to inhibit *M. leprae* by binding to the dihydropteroate synthase enzyme (Table 16.2) [67]. Sakuranetin is a flavone that is reported to be used in the treatment of leprosy (Table 16.2) [66]. *Allium sativum*, *Azadirachta indica*, *Cassia fistula*, *Andrographis paniculata*, *Withania somnifera*, *Asparagus racemosus*, and *Ricinus communis* are examples of herbal medicines containing polyphenols that are used in the treatment of leprosy [77–80].

16.3.7 Lymphatic Filariasis

Lymphatic filariasis is caused by *Wuchereria bancrofti*, *Brugia malayi*, and *Brugia timori*. These thread-like nematode worms are transmitted by *Culex*, *Aedes*, and *Anopheles* mosquitoes. Diethylcarbamazine is the drug of choice for the treatment of lymphatic filariasis [81]. 4,5-Diethy l-3′-ethoxy-pyro-flavone, rutin, naringenin, hesperetin, chrysin, and naringin are examples of polyphenols that have been reported to act against the causative agents of lymphatic filariasis in in vitro experiments. Naringenin was also reported to reduce the number of live worms in in vivo studies (Table 16.2).

16.3.8 Trachoma

Trachoma is caused by *Chlamydia trachomatis*. The bacterium is transmitted through infected eye discharges and nasal secretions. Flies are important to trachoma transmission as they lay their eggs on human feces that lies exposed on the soil. This suggests that access to improved sanitation might help curb the transmission of trachoma. Surgery (for trichiasis), antibiotics, facial cleanliness, and environmental improvement (SAFE strategy) are recommended by the World Health Organization to prevent blinding trachoma [82]. (+)-catechin, (−)-epicatechin, (−)-epigallocatechin, (−)-epicatechin gallate, (−)-epigallocatechin gallate, and baicalin are examples of some polyphenols that have been reported to act against *C. trachomatis* (Table 16.2).

16.3.9 Schistosomiasis

Schistosomiasis is caused by trematode flatworms, mainly by *Schistosoma mansoni* and *S. japonicum*. Schistosomiasis is spread through contaminated freshwater. Snails carrying schistosomes are responsible for contaminating freshwater. It is also spread when infected persons urinate or defecate in the water. Praziquantel, oxamniquine, and artemether are examples of some pharmaceutical drugs that are used for the treatment of schistosomiasis [83]. Polyphenols such as derivatives of chalcone and myricetin have been reported to act against the causative agents of schistosomiasis in in vitro and in vivo experiments (Table 16.2).

16.4 Nanotherapeutics of Phytoantioxidants for Parasitic Diseases and NTDs

In general, nanoformulations increased the solubility, bioavailability, biological activity, cell permeability, and stability of phytoantioxidants. Nanoformulations also allowed sustained release, site-specific drug delivery, target-specific drug activity, and reduced toxicity. Figure 16.2 represents how resistance and bioavailability can be solved with nanophytoantioxidants. The nanoformulation of various phytoantioxidants that have been reported to act against the causative agents of parasitic diseases and NTDs is given in Table 16.3. The nanoformulations of phytoantioxidants such as apigenin, curcumin, luteolin, and quercetin have been studied for leishmaniasis and malaria. The nanoformulations of other phytoantioxidants mentioned in Table 16.3 have been studied for other diseases except for parasitic diseases and NTDs.

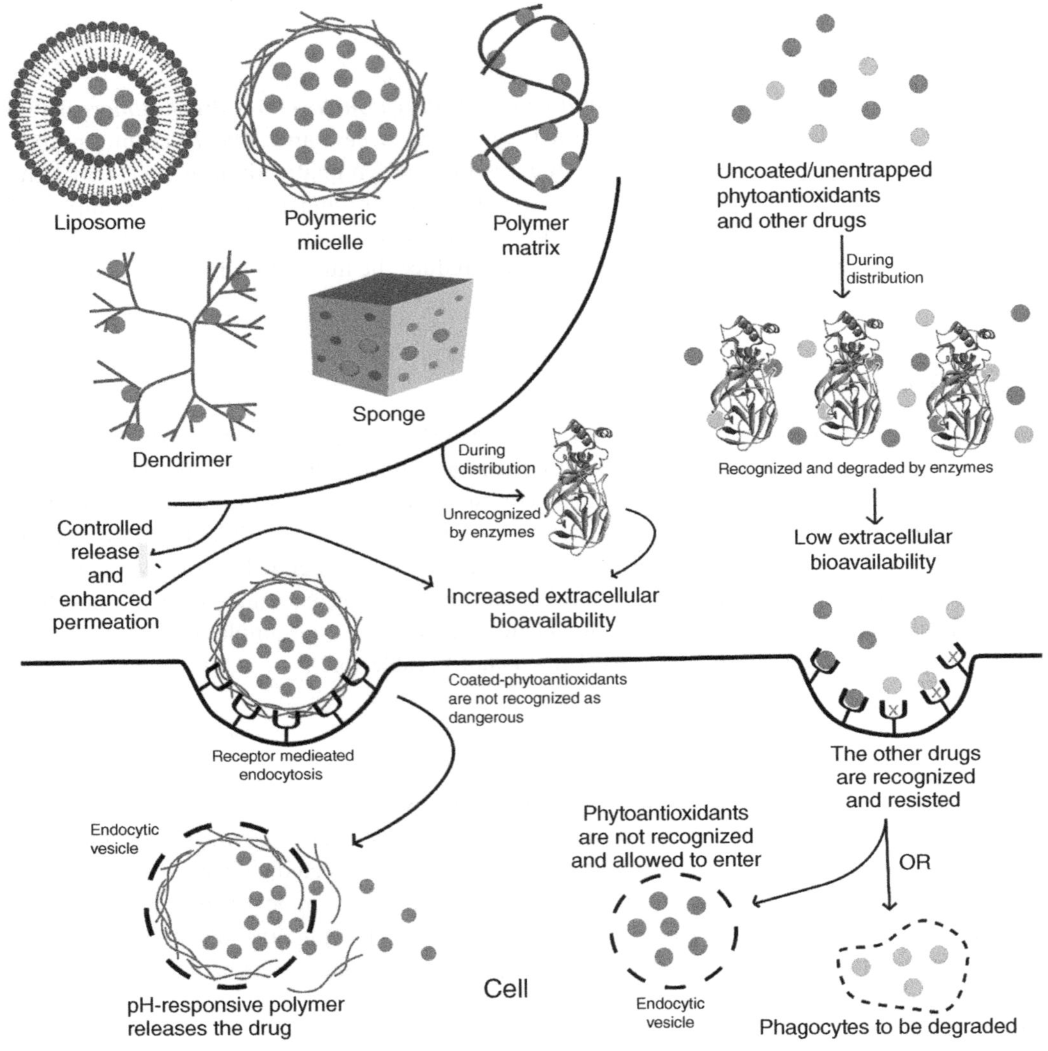

Figure 16.2 Overcoming resistance and bioavailability with nanophytoantioxidants.

Table 16.3 Nanoformulations of phytoantioxidants used against parasitic diseases and NTDs.

Bioactive compounds	Nanoformulations	Improvement after nanoformulation	References
Apigenin	Miltefosin–Apigenin-co-loaded-second-generation nano-transferosomes	Effective novel drug delivery system for the treatment of cutaneous leishmaniasis.	[84]
Baicalin	Baicalin-loaded nano-emulsion	Improved oral bioavailability.	[43]
Baicalein	Baicalein loaded in tocol-nanostructured lipid carriers	Improved stability and blood–brain barrier permeation.	[85]
Catechin	Polylactide nanoparticles	Improved cell permeability and biological activity.	[86]
Chrysin	Polyurea dendrimer nanoparticle	Reduced toxicity, target-specific drug activity.	[87]
Coumarin	Solid lipid nanoparticle	Increased bioavailability.	[88]
Curcumin	Polylactide nanoparticles	Increased bioavailability.	[89]
Curcumin	Eudragit-nutriosomes	Significantly increased the survival of malaria-infected mice.	[41]
Ellagic acid	Cyclodextrin-based nanosponge	Improved solubility, bioavailability, and controlled release.	[90]
Emodin	Polymeric nanoformulation	Sustained release, increased cell permeability, improved biological activity.	[40]
Epigallocatechin gallate	Polymeric nanoformulation	Topically applied nano-epigallocatechin gallate displayed a >20-fold dose advantage.	[91]
Epigallocatechin gallate	Nanoparticles	Increase drug permanence in systemic circulation and brain tissue.	[92]
Fisetin	Polylactide nanoparticles	Potential and efficacious strategy for delivery of fisetin.	[93]
Gallic acid	Graphene oxide-based nanoformulation	Increased biological activity even with low concentration.	[42]
Gallic acid	Polymeric nanoformulation	Improved biological activity, reduced toxicity.	[94]
Genistein	Polymeric micelles	Improved solubility, bioavailability, and bioactivity.	[95]
Hesperetin	Nanocrystals	Increased biological activity and enhanced bioavailability.	[96]
Luteolin	Solid lipid nanoparticle	Highly potential to be used as innovative adjuvant therapy for either artemisinin-resistant or artemisinin-sensitive *P. falciparum* parasite.	[97]
Luteolin	Water-soluble copolymer-encapsulated nanoparticle	Significantly increased biological activity.	[98]
Myricetin	Solid lipid nanoparticles	Increased cellular uptake and increased biological activity.	[99]

(*Continued*)

Table 16.3 (Continued)

Bioactive compounds	Nanoformulations	Improvement after nanoformulation	References
Myricetin	Mesoporous silica nanoparticles	Increased biological activity, site-specific drug delivery.	[100]
Naringenin	Liposomal nanoformulation	Improved solubility and bioavailability.	[101]
Naringenin	Liquid crystalline nanoparticles	Significantly increased biological activity.	[102]
Piperine	Nanoparticles	Controlled drug release, improved bioavailability, site-specific delivery to target organs.	[103]
Quercetagetin	Composite nanoparticle based on Zein and Hyaluronic acid	Prevents quercetin degradation against thermal and UV lights. Provides long-term stability. Offers novel insights into delivering bioactive compounds.	[104]
Quercetin	Polymeric nanoformulation	Higher permeation ability, improved biological activity.	[105, 106]
Quercetin	Nanoparticles	Cellular uptake and antioxidant activity are improved.	[107]
Rutin	Liquid crystalline nanoparticles	Increased biological activity.	[108]
Resveratrol	Lipid nanoparticle	Improved cell permeability and bioavailability.	[109]
7,8-Dihydroxyflavone	Liposome nanoformulation	Increased bioavailability, stability, bioactivity and enables sustained release.	[39]

The major problems faced by antiparasitic drugs are low solubility, low bioavailability, poor cell permeability, nonspecific distribution, and rapid metabolism followed by immediate excretion from the body [110]. Phytoantioxidants that have been reported as promising therapies for parasitic diseases and NTDs are also facing the same challenge [36, 37]. It may also be noted that simply because polyphenols are abundantly present in a particular plant, the polyphenols present in the plant are not necessarily going to exert the required pharmacological activity. Only those that have the best bioavailability will have the best bioactivity profile [111]. Therefore, to unleash the true potential of phytoantioxidants for the treatment of parasitic diseases and NTDs, it is rational to encapsulate phytoantioxidants or to deliver them using nanoformulations. Figure 16.3 represents the possible mechanisms of nanophytoantioxidants in parasitic diseases and NTDs.

16.4.1 Apigenin-Loaded Nano-transferosomes

Nano-transferosomes are lipid bilayered vesicular systems. They are composed of phospholipids and with a single chain surfactant (edge activator). The lipid bilayers can encapsulate hydrophobic compounds (polyphenols) such as apigenin [84]. This nanoformulation of apigenin has been reported to be an effective drug delivery system for the treatment of cutaneous leishmaniasis (Table 16.3).

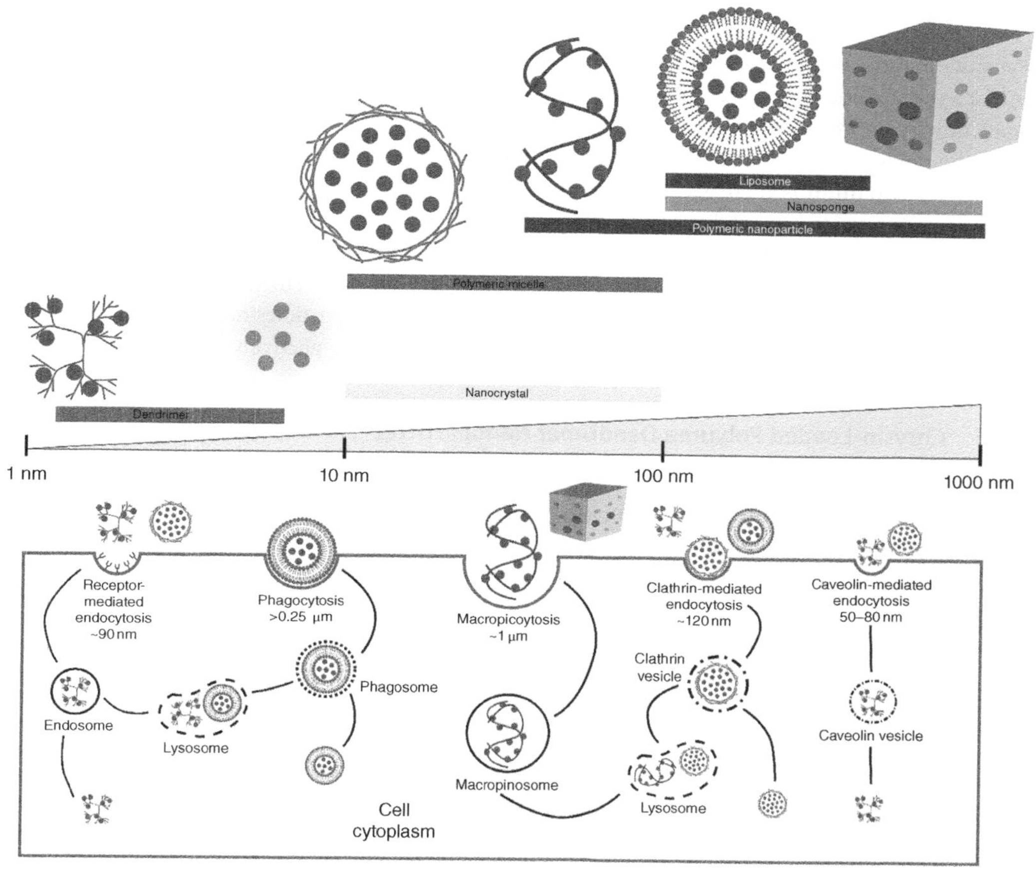

Figure 16.3 Nanophytoantioxidants in parasitic diseases and NTDs with possible mechanisms.

16.4.2 Baicalin-Loaded Nanoemulsions

Nanoemulsions are mixtures of two immiscible liquids that are stabilized with a suitable surfactant and they have a mean droplet diameter of <500 nm [112]. The overall oral bioavailability of baicalin is improved by nanoemulsions (Table 16.3). Baicalin showed an IC_{50} value of 0.50 mM against *C. trachomatis* in in vitro studies (Table 16.2). Therefore, baicalin-loaded nanoemulsion might offer an effective treatment strategy for *C. trachomatis*-induced trachoma.

16.4.3 Baicalein-Loaded Nanostructured Lipid Carriers

Nanostructured lipid carriers are a blend of liquid and solid lipids resulting in a lipid system that is partially crystallized [113]. The stability and membrane permeability of baicalein are increased by nanostructured lipid carriers (Table 16.3). Baicalein showed inhibitory activity against DENV (IC_{50} = 13.5 μg/ml) and CHIKV (IC_{50} = 1.891 μg/ml) (Table 16.2). Therefore, baicalein-loaded nanostructured lipid carriers might offer an effective strategy for the treatment of dengue and chikungunya.

16.4.4 Catechin, Curcumin, and Fisetin-Loaded Polylactide Nanoparticles

Polylactide nanoparticles are a type of polymeric nanoparticles with the ability to maintain drug levels at a therapeutic dose for a prolonged period [114]. The cell permeability and biological activity of catechin are improved when it is loaded with polylactide nanoparticles (Table 16.3). Catechin was among the phytoconstituents of Polyphenon 70S that inhibits *C. trachomatis* (Table 16.2). The bioavailability of curcumin is increased when it is loaded with polylactide nanoparticles (Table 16.3). Curcumin was reported to inhibit *Plasmodium* parasites (EC_{50} = 1.1–10 μg/ml) in in vitro studies (Table 16.1). Polylactide nanoparticles are also considered an effective drug delivery option for fisetin (Table 16.3). Fisetin was reported to inhibit CHIKV (IC_{50} = 8.444 μg/ml) in in vitro studies (Table 16.2). Therefore, catechin-, curcumin-, and fisetin-loaded polylactide nanoparticles might offer an effective strategy for the treatment of trachoma, malaria, and chikungunya, respectively.

16.4.5 Chrysin-Loaded Polyurea Dendrimer Nanoparticles

Polyurea dendrimer nanoparticles are 3-dimensional polymers wherein urea moieties are present in the backbone and amine groups are present peripherally with size <100 nm [87]. Polyurea dendrimer nanoparticles reduce the toxicity of chrysin while it allows target-specific drug activity (Table 16.3). Chrysin has been reported to inhibit the microfilariae (IC_{50} = 105.1 ± 21.1 μg/ml) of lymphatic filariasis (Table 16.2). Therefore, chrysin-loaded polyurea dendrimer nanoparticles might offer an effective strategy for the treatment of lymphatic filariasis.

16.4.6 Coumarin-, Luteolin-, and Myricetin-Loaded Solid Lipid Nanoparticles

Solid lipid nanoparticles are colloidal carriers having sub-micron-sized solid lipids. Their size ranged between 10 and 1000 nm [115]. Solid lipid nanoparticles were reported to increase the bioavailability of coumarin (Table 16.3). At 4 mg/ml IC_{50} = 64 μM, coumarin inhibited *Fasciola* trematodes (Table 16.2). Luteolin-loaded solid lipid nanoparticles are also reported to be an effective alternative treatment regimen for malaria (Table 16.3). Luteolin was reported to inhibit DENVs (Table 16.1) and *Plasmodium* parasites [116]. The cellular uptake and biological activity of myricetin were also reported to be increased by solid lipid nanoparticles (Table 16.3). Myricetin was reported to inhibit *Schistosoma* flatworms in in vitro (LC_{50} = 600 μM) studies (Table 16.2). Therefore, coumarin-, luteolin-, myricetin-loaded solid lipid nanoparticles might offer an effective strategy for the treatment of fascioliasis, dengue, malaria, and schistosomiasis, respectively.

16.4.7 Curcumin-Loaded Eudragit-Nutriosomes

Eudragit-nutriosomes are made up of phospholipid vesicles, water-soluble fibers, and pH-sensitive polymers [41]. Eudragit-nutriosomes were reported to significantly increase the survival time of malaria-infected mice [117]. Curcumin was reported to inhibit *Plasmodium* parasites in in vitro (EC_{50} = 1.1–10 μg/ml) studies (Table 16.1). Therefore, curcumin-loaded eudragit-nutriosomes might offer an effective treatment strategy for the treatment of malaria.

16.4.8 Ellagic Acid-Loaded Cyclodextrin-Based Nanosponge

Cyclodextrin-based nanosponge is a nanosized drug delivery system that has a porous structure and is hyper-crosslinked with cyclodextrin polymers in a 3D network [118, 119]. Cyclodextrin-based nanosponge was reported to increase the solubility, bioavailability as well as facilitate the controlled release of ellagic acid (Table 16.3). Ellagic acid was reported to inhibit *O. volvulus* in in vitro (LC_{50} =0.03 μM) studies (Table 16.2). Therefore, ellagic acid-loaded cyclodextrin-based nanosponge might offer an effective treatment for onchocerciasis.

16.4.9 Emodin, Epigallocatechin Gallate, Gallic Acid, and Quercetin-Loaded Polymeric Nanoparticles

Polymeric nanoparticles (10–1000 nm) are made up of hydrophilic polymers and copolymers on the outer part, while the core-polymeric matrix is designed to encapsulate hydrophobic drugs [120]. Polymeric nanoparticles increased cell permeability and biological activity and facilitated the sustained release of emodin (Table 16.3). Emodin was reported to inhibit *M. ulcerans* (MIC value of 40 mg/ml) in in vitro studies (Table 16.2). Polymeric nanoparticles increased the bioavailability of epigallocatechin gallate to 20 times (Table 16.3). Epigallocatechin gallate was reported to inhibit (IC_{50} 24 hours' incubation = 3.7 μM, IC_{50} 72 hours' incubation = 3.4 μM) *Leishmania* protozoans as well as flavivirus (EC_{50} = 21.4 μM) (Table 16.1). Polymeric nanoparticles improve the biological activity while it reduces the toxicity of gallic acid (Table 16.3). Gallic acid was reported to inhibit helminthic parasites (1000 μg/ml), *M. ulcerans* (MIC value of 0.78–6.25 μg/ml), and *T. brucei* (IC_{50} = 15.6 μM) (Tables 16.1 and 16.2). Polymeric nanoparticles increased the permeation ability and biological activity of quercetin (Table 16.3). Quercetin was reported to inhibit *Plasmodium* parasites (IC_{50} = 4.11 ± 2.05 μg/ml), DENV (IC_{50} = 35.7 μg/ml), *F. hepatica* eggs (12.5–50 mg/ml), and *T. brucei* (IC_{50} = <2.0 μM) (Tables 16.1 and 16.2). Therefore, emodin-loaded polymeric nanoparticles might offer an effective strategy for the treatment of Buruli ulcers. Epigallocatechin gallate-loaded polymeric nanoparticles might offer an effective strategy for the treatment of Leishmaniasis and the Zika virus. Gallic acid-loaded polymeric nanoparticles might offer an effective strategy for the treatment of helminth parasites, Buruli ulcer, and human African trypanosomiasis. Quercetin-loaded polymeric nanoparticles might offer an effective strategy for the treatment of malaria, dengue, fascioliasis, and human African trypanosomiasis.

16.4.10 Epigallocatechin Gallate and Piperine-Loaded Nanoparticles

Nanoparticles are materials within the size range of 1–100 nm [121]. Nanoparticles increase the availability of epigallocatechin gallate for a longer period in the systemic circulation and brain tissue (Table 16.3). Epigallocatechin gallate was reported to inhibit *T. cruzi* (IC_{50} =0.53 μM), *Leishmania* protozoans (IC_{50} 24 hours' incubation = 3.7 μM IC_{50} 72 hours' incubation = 3.4 μM), flavivirus (EC_{50} = 21.4 μM), and *C. trachomatis* (1.6 and 0.4 mg/ml) (Tables 16.1 and 16.2). Nanoparticles facilitate controlled drug release and site-specific drug delivery, and also increases the bioavailability of piperine (Table 16.3). Piperine was reported to inhibit *Leishmania* protozoans (IC_{50} =4.31 ± 0.44 μg/ml) (Table 16.1). Therefore, epigallocatechin gallate-loaded nanoparticles offer an effective strategy for the treatment of leishmaniasis, Zika virus, trachoma, and Chagas disease. Piperine-loaded nanoparticle offers an effective treatment strategy for leishmaniasis.

16.4.11 Gallic Acid-Loaded Graphene Oxide-Based Nanoformulation

Graphene oxide is a carbon-based nanomaterial that is an oxidized form of graphene [122]. The biological activity of gallic acid, even at low concentrations, was reported to be increased by graphene oxide-based nanoformulation (Table 16.3). Gallic acid was reported to inhibit helminthic parasites (1000 μg/ml), *M. ulcerans* (MIC value of 0.78–6.25 μg/ml), and *T. brucei* (IC_{50} = 15.6 μM) (Tables 16.1 and 16.2). Gallic acid-loaded graphene oxide-based nanoformulation might offer an effective strategy for the treatment of helminth parasites, Buruli ulcer, and human African trypanosomiasis.

16.4.12 Genistein-Loaded Polymeric Micelles

Polymeric micelles have amphiphilic block copolymers which make up the nanosized core/shell structures that can encapsulate molecules [123]. The solubility, bioavailability, and bioactivity of genistein were reported to be increased by polymeric micelles (Table 16.3). Genistein was reported to induce DNA damage in *F. gigantica* (Table 16.2). Therefore, genistein-loaded polymeric micelles might offer an effective strategy for the treatment of fascioliasis.

16.4.13 Hesperetin-Loaded Nanocrystals

Nanocrystals (10–800 nm) are colloidal drug delivery systems that are carrier-free and consist of surface-active agents for stability [124]. Nanocrystals increase the biological activity and bioavailability of hesperetin (Table 16.3). Hesperetin was reported to inhibit nematode worms (Table 16.2). Therefore, hesperetin-loaded nanocrystals might offer an effective treatment strategy for the treatment of lymphatic filariasis.

16.4.14 Luteolin-Loaded Water-Soluble Copolymer-Encapsulated Nanoparticle

A water-soluble nanoparticle made up of hyperbranched copolymer wherein luteolin was encapsulated was prepared [98]. This nanoparticle significantly increased the biological activity of luteolin (Table 16.3). Luteolin was reported to inhibit DENV (DENV1, $EC_{50} = 4.36\,\mu M$; DENV3, $EC_{50} = 5.69\,\mu M$; DENV4, $EC_{50} = 8.38\,\mu M$) and helminth parasites ($IC_{50} = 9.7\,\mu g/ml$) (Table 16.1). Therefore, luteolin-loaded water-soluble copolymer-encapsulated nanoparticles might offer an effective treatment strategy against dengue and helminth parasites.

16.4.15 Myricetin-Loaded Mesoporous Silica Nanoparticles

Mesoporous silica nanoparticles are silica-assisted drug delivery systems that have a large uniform pore size with a large surface area [125, 126]. Mesoporous silica nanoparticles increased the biological activity and facilitate site-specific drug delivery of myricetin (Table 16.3). Myricetin inhibited flavivirus ($IC_{50} = 0.58 \pm 0.17\,\mu M$) and trematode flatworms ($LC_{50} = 600\,\mu M$) (Tables 16.1 and 16.2). Therefore, myricetin-loaded mesoporous silica nanoparticles might offer an effective strategy for the treatment of the Zika virus and schistosomiasis.

16.4.16 Naringenin and 7,8-Dihydroxyflavone-Loaded Liposomes

Liposomes are nanosized spherical-shaped vesicles that are made up of only one or more than one phospholipid bilayer [127]. Liposomes improved the solubility and bioavailability of naringenin (Table 16.3). Naringenin was reported to inhibit flavivirus ($IC_{50} = 58.79\,\mu M$) and nematode worms (female adult worm: $IC_{50} = 2.5 \pm 0.2\,\mu g/ml$; microfilariae: $IC_{50} = 297.3 \pm 5.2\,\mu g/ml$) (Tables 16.1 and 16.2). Liposomes increase the bioavailability, stability, and bioactivity and also facilitate the sustained release of 7,8-dihydroxyflavone (Table 16.3). 7,8-dihydroxyflavone was reported to inhibit *T. brucei* ($IC_{50} = 0.16\,\mu M$) (Table 16.2). Therefore, naringenin-loaded liposomes might offer an effective strategy for the treatment of Zika virus and lymphatic filariasis. 7,8-dihydroxyflavone-loaded liposomes might offer an effective treatment strategy for the treatment of human African trypanosomiasis.

16.4.17 Naringenin- and Rutin-Loaded Liquid Crystalline Nanoparticles

Liquid crystal nanoparticles such as cubosomes and hexosomes offer a versatile platform for drug delivery. These types of nanoformulations are an intermediate between liquid and solid phases [128]. Liquid crystalline nanoparticles significantly increased the biological activity of naringenin and rutin (Table 16.3). Naringenin was reported to inhibit flavivirus ($IC_{50} = 58.79\,\mu M$) and nematode worms (female adult worm: $IC_{50} = 2.5 \pm 0.2\,\mu g/ml$; microfilariae: $IC_{50} = 297.3 \pm 5.2\,\mu g/ml$) (Tables 16.1 and 16.2). Rutin was reported to inhibit *Plasmodium* parasites ($IC_{50} = 3.53 \pm 13.34\,\mu g/ml$) and nematode worms (microfilariae: $IC_{50} = 74.3 \pm 7.2\,\mu g/ml$) (Tables 16.1 and 16.2). Therefore, naringenin and rutin-loaded liquid crystal nanoparticles might offer an effective treatment strategy for lymphatic filariasis. Rutin-loaded liquid crystal nanoparticles might offer an effective treatment strategy for malaria.

16.4.18 Quercetagetin-Loaded Zein and Hyaluronic Acid-Based Nanoparticles

Nanoparticles fabricated with zein and hyaluronic acid have increased stability [129]. Composite nanoparticle based on zein and hyaluronic acid protects quercetagetin against heat and UV light. The stability of quercetagetin is also increased (Table 16.3). Quercetagetin inhibits CHIKV ($IC_{50} = 13.85\,\mu g/ml$) (Table 16.1). Therefore, quercetagetin-loaded zein and hyaluronic acid-based composite nanoparticles might offer an effective treatment strategy for chikungunya.

16.5 Future Scope

From the examples that we had highlighted in Tables 16.1 and 16.2, it can be observed that the majority of the study were done up to in vitro level only. This could imply two possibilities, either the animal models for the diseases are difficult to set up, or, the animal studies do not yield fruitful results and, hence, the research findings are not reported. From the examples that we had highlighted in Table 16.3, only a few nanoformulations of phytoantioxidants (apigenin, curcumin, quercetin, and luteolin) have been studied for at least one of the parasitic diseases and NTDs. With this, the research gap existing in this particular line of work is revealed. Thus, there is a huge scope for research to develop nanoformulated phytoantioxidants that will be clinically feasible for the treatment of drug-resistant pathogens responsible for causing parasitic diseases and NTDs.

16.6 Conclusion

Several phytoantioxidants have been reported to act against various pathogens responsible for causing parasitic diseases and NTDs. Some phytoantioxidants were reported to show better activity than the standard drugs. From the available evidence, there are hopes that phytoantioxidants will overcome drug resistance and toxicity issues associated with pharmaceutical drugs. Although phytoantioxidants have bioavailability issues, nanotechnology can solve this particular problem. Nano-phytoantioxidants have better bioavailability, stability, solubility, and bioactivity as compared to conventional phytoantioxidants. With nano-phytoantioxidants, controlled release, site-specific drug delivery, and target-specific drug activity are possible. With a bright future research scope, nano-phytoantioxidants will prove to be a game-changer in the treatment of parasitic diseases and NTDs.

List of Abbreviations

EC_{50}	Effective concentration
IC_{50}	Inhibitory concentration
LC_{50}	Lethal dose
MIC	Minimum inhibitory concentration
NTDs	Neglected tropical diseases
UV	Ultraviolet

References

1 Hsieh, C.-L., Peng, C., Cheng, Y.-M. et al. (2010). Quercetin and ferulic acid aggravate renal arcinoma in long-term diabetic victims. *Journal of Agricultural and Food Chemistry 58* (16): 9273–9280. https://doi.org/10.1021/jf101580j.

2 Pietta, P.-G. (2000). Flavonoids as antioxidants. *Journal of Natural Products 63* (7): 1035–1042. https://doi.org/10.1021/np9904509.

3 Abotaleb, M., Liskova, A., Kubatka, P., & Büsselberg, D. (2020). Therapeutic potential of plant phenolic acids in the treatment of cancer. *Biomolecules*, *10*(2), 221. https://doi.org/10.3390/biom10020221.

4 Cui, Q., Du, R., Liu, M., and Rong, L. (2020). Lignans and their derivatives from plants as antivirals. *Molecules 25* (1): 183. https://doi.org/10.3390/molecules25010183.

5 Luca, S.V., Macovei, I., Bujor, A. et al. (2020). Bioactivity of dietary polyphenols: the role of metabolites. *Critical Reviews in Food Science and Nutrition 60* (4): 626–659. https://doi.org/10.1080/10408398.2018.1546669.

6 Pandey, K.B. and Rizvi, S.I. (2009). Plant polyphenols as dietary antioxidants in human health and disease. *Oxidative Medicine and Cellular Longevity 2* (5): 270–278. https://doi.org/10.4161/oxim.2.5.9498.

7 Reinisalo, M., Kårlund, A., Koskela, A. et al. (2015). Polyphenol stilbenes: molecular mechanisms of defence against oxidative stress and aging-related diseases. *Oxidative Medicine and Cellular Longevity 2015*: 1–24. https://doi.org/10.1155/2015/340520.

8 Serrano, J., Puupponen-Pimiä, R., Dauer, A. et al. (2009). Tannins: current knowledge of food sources, intake, bioavailability and biological effects. *Molecular Nutrition & Food Research 53* (S2): S310–S329. https://doi.org/10.1002/mnfr.200900039.

9 Kumar, S. and Pandey, A.K. (2013). Chemistry and biological activities of flavonoids: an overview. *The Scientific World Journal 2013*: 1–16. https://doi.org/10.1155/2013/162750.

10 Girisa, S., Kumar, A., Rana, V. et al. (2021). From simple mouth cavities to complex oral mucosal disorders – curcuminoids as a promising therapeutic approach. *ACS Pharmacology & Translational Science 4* (2): 647–665. https://doi.org/10.1021/acsptsci.1c00017.

11 Ahmad, S., Zahiruddin, S., Parveen, B., Basist, P., Parveen, A., Gaurav, Parveen, R., & Ahmad, M. (2021). Indian medicinal plants and formulations and their potential against COVID-19 – preclinical and clinical research. *Frontiers in Pharmacology*, 11. https://doi.org/10.3389/fphar.2020.578970.

12 Wang, J., Wong, Y.-K., and Liao, F. (2018). What has traditional Chinese medicine delivered for modern medicine? *Expert Reviews in Molecular Medicine 20*: e4. https://doi.org/10.1017/erm.2018.3.

13 Romano, B., Pagano, E., Montanaro, V. et al. (2013). Novel insights into the pharmacology of flavonoids. *Phytotherapy Research 27* (11): 1588–1596. https://doi.org/10.1002/ptr.5023.

14 Bento-Silva, A., Koistinen, V. M., Mena, P., Bronze, M. R., Hanhineva, K., Sahlstrøm, S., Kitrytė, V., Moco, S., & Aura, A.-M. (2020). Factors affecting intake, metabolism and health benefits of phenolic acids: do we understand individual variability? *European Journal of Nutrition, 59*(4), 1275–1293. https://doi.org/10.1007/s00394-019-01987-6.

15 Rimando, A. and Suh, N. (2008). Biological/chemopreventive activity of stilbenes and their effect on colon cancer. *Planta Medica 74* (13): 1635–1643. https://doi.org/10.1055/s-0028-1088301.

16 Zhang, J., Chen, J., Liang, Z., and Zhao, C. (2014). New lignans and their biological activities. *Chemistry & Biodiversity 11* (1): 1–54. https://doi.org/10.1002/cbdv.201100433.

17 Smeriglio, A., Barreca, D., Bellocco, E., and Trombetta, D. (2017). Proanthocyanidins and hydrolysable tannins: occurrence, dietary intake and pharmacological effects. *British Journal of Pharmacology 174* (11): 1244–1262. https://doi.org/10.1111/bph.13630.

18 CDC (2021). Parasites. https://www.cdc.gov/parasites/index.html (accessed 20 August 2021).

19 Cortez-Maya, S., Moreno-Herrera, A., Palos, I., and Rivera, G. (2020). Old antiprotozoal drugs: are they still viable options for parasitic infections or new options for other diseases? *Current Medicinal Chemistry 27* (32): 5403–5428. https://doi.org/10.2174/0929867326666190628163633.

20 Molyneux, D.H., Savioli, L., and Engels, D. (2017). Neglected tropical diseases: progress towards addressing the chronic pandemic. *The Lancet 389* (10066): 312–325. https://doi.org/10.1016/S0140-6736(16)30171-4.

21 Pink, R., Hudson, A., Mouriès, M.-A., and Bendig, M. (2005). Opportunities and challenges in antiparasitic drug discovery. *Nature Reviews Drug Discovery 4* (9): 727–740. https://doi.org/10.1038/nrd1824.

22 Capela R, Moreira R, & Lopes F. (2019). An overview of drug resistance in protozoal diseases. *International Journal of Molecular Sciences, 20* (22), 5748. https://doi.org/10.3390/ijms20225748.

23 Christaki, E., Marcou, M., and Tofarides, A. (2020). Antimicrobial resistance in bacteria: mechanisms, evolution, and persistence. *Journal of Molecular Evolution 88* (1): 26–40. https://doi.org/10.1007/s00239-019-09914-3.

24 Lampejo, T. (2020). Influenza and antiviral resistance: an overview. *European Journal of Clinical Microbiology & Infectious Diseases 39* (7): 1201–1208. https://doi.org/10.1007/s10096-020-03840-9.

25 Pramanik, P.K., Alam, M.N., Roy Chowdhury, D., and Chakraborti, T. (2019). Drug resistance in protozoan parasites: an incessant wrestle for survival. *Journal of Global Antimicrobial Resistance 18*: 1–11. https://doi.org/10.1016/j.jgar.2019.01.023.

26 Davatgaran-Taghipour, Y., Masoomzadeh, S., Farzaei, M.H. et al. (2017). Polyphenol nanoformulations for cancer therapy: experimental evidence and clinical perspective. *International Journal of Nanomedicine 12*: 2689–2702. https://doi.org/10.2147/IJN.S131973.

27 Sharma, P., Montes de Oca, M.K., Alkeswani, A.R. et al. (2018). Tea polyphenols for the prevention of UVB-induced skin cancer. *Photodermatology, Photoimmunology & Photomedicine 34* (1): 50–59. https://doi.org/10.1111/phpp.12356.

28 Zhao, Y., Zhao, K., Jiang, K. et al. (2016). A review of flavonoids from cassia species and their biological activity. *Current Pharmaceutical Biotechnology 17* (13): 1134–1146. https://doi.org/10.2174/1389201017666160819151153.

29 Boniface, P. K., & Ferreira, E. I. (2019). Flavonoids as efficient scaffolds: recent trends for malaria, leishmaniasis, Chagas disease, and dengue. *Phytotherapy Research*, *33* (10), 2473–2517. https://doi.org/10.1002/ptr.6383.

30 Fracasso, M., Dutra da Silva, A., Bottari, N.B. et al. (2021). Resveratrol impacts in oxidative stress in liver during *Trypanosoma cruzi* infection. *Microbial Pathogenesis 153*: 104800. https://doi.org/10.1016/j.micpath.2021.104800.

31 Güida, M.C., Esteva, M.I., Camino, A. et al. (2007). *Trypanosoma cruzi*: in vitro and in vivo antiproliferative effects of epigallocatechin gallate (EGCg). *Experimental Parasitology 117* (2): 188–194. https://doi.org/10.1016/j.exppara.2007.04.015.

32 Harapan, H., Michie, A., Sasmono, R.T., and Imrie, A. (2020). Dengue: a minireview. *Viruses 12* (8): 829. https://doi.org/10.3390/v12080829.

33 Lani, R., Hassandarvish, P., Shu, M.-H. et al. (2016). Antiviral activity of selected flavonoids against chikungunya virus. *Antiviral Research 133*: 50–61. https://doi.org/10.1016/j.antiviral.2016.07.009.

34 Tsouh Fokou, P., Kissi-Twum, A., Yeboah-Manu, D. et al. (2016). In vitro activity of selected west african medicinal plants against mycobacterium ulcerans disease. *Molecules 21* (4): 445. https://doi.org/10.3390/molecules21040445.

35 Vairo, F., Haider, N., Kock, R. et al. (2019). Chikungunya. *Infectious Disease Clinics of North America 33* (4): 1003–1025. https://doi.org/10.1016/j.idc.2019.08.006.

36 Di Lorenzo, C., Colombo, F., Biella, S. et al. (2021). Polyphenols and human health: the role of bioavailability. *Nutrients 13* (1): 273. https://doi.org/10.3390/nu13010273.

37 Shivashankara, K.S. and Acharya, S.N. (2010). Bioavailability of dietary polyphenols and the cardiovascular diseases. *The Open Nutraceuticals Journal 3* (1): 227–241. https://doi.org/10.2174/1876396001003010227.

38 Patra, J.K., Das, G., Fraceto, L.F. et al. (2018). Nano based drug delivery systems: recent developments and future prospects. *Journal of Nanobiotechnology 16* (1): 71. https://doi.org/10.1186/s12951-018-0392-8.

39 Chen, Y., Xia, G., Zhao, Z. et al. (2020). 7,8-Dihydroxyflavone nano-liposomes decorated by crosslinked and glycosylated lactoferrin: storage stability, antioxidant activity, in vitro release, gastrointestinal digestion and transport in Caco-2 cell monolayers. *Journal of Functional Foods 65*: 103742. https://doi.org/10.1016/j.jff.2019.103742.

40 Ding, W., Sun, J., Lian, H. et al. (2018). The influence of shuttle-shape emodin nanoparticles on the *Streptococcus suis* biofilm. *Frontiers in Pharmacology 9*. https://doi.org/10.3389/fphar.2018.00227.

41 Manconi, M., Manca, M.L., Escribano-Ferrer, E. et al. (2019). Nanoformulation of curcumin-loaded eudragit-nutriosomes to counteract malaria infection by a dual strategy: improving antioxidant intestinal activity and systemic efficacy. *International Journal of Pharmaceutics 556*: 82–88. https://doi.org/10.1016/j.ijpharm.2018.11.073.

42 Shamsi, S., Elias, N., Narti Edayu Sarchio, S., and Md Yasin, F. (2018). Gallic acid loaded graphene oxide based nanoformulation (GAGO) as potential anti-bacterial agent against *Staphylococcus aureus*. *Materials Today: Proceedings 5*: S160–S165. https://doi.org/10.1016/j.matpr.2018.08.059.

43 Zhao, L., Wei, Y., Fu, J. et al. (2013). Nanoemulsion improves the oral bioavailability of baicalin in rats: in vitro and in vivo evaluation. *International Journal of Nanomedicine* 3769. https://doi.org/10.2147/IJN.S51578.

44 Rahman, H.S., Othman, H.H., Hammadi, N.I. et al. (2020). Novel drug delivery systems for loading of natural plant extracts and their biomedical applications. *International Journal of Nanomedicine 15*: 2439–2483. https://doi.org/10.2147/IJN.S227805.

45 Brglez Mojzer, E., Knez Hrnčič, M., Škerget, M., Knez, Ž., & Bren, U. (2016). Polyphenols: extraction methods, antioxidative action, bioavailability and anticarcinogenic effects. *Molecules*, *21*(7), 901. https://doi.org/10.3390/molecules21070901.

46 Basu, S., & Sahi, P. K. (2017). Malaria: an update. *The Indian Journal of Pediatrics, 84*(7), 521–528. https://doi.org/10.1007/s12098-017-2332-2.

47 Ganesh, D., Fuehrer, H.-P., Starzengrüber, P. et al. (2012). Antiplasmodial activity of flavonol quercetin and its analogues in *Plasmodium falciparum*: evidence from clinical isolates in Bangladesh and standardized parasite clones. *Parasitology Research 110* (6): 2289–2295. https://doi.org/10.1007/s00436-011-2763-z.

48 Kumar Mishra, S., Singh, P., and Rath, S.K. (2013). Protective effect of quercetin on chloroquine-induced oxidative stress and hepatotoxicity in mice. *Malaria Research and Treatment 2013*: 1–10. https://doi.org/10.1155/2013/141734.

49 Mohd Abd Razak, M.R., Afzan, A., Ali, R. et al. (2014). Effect of selected local medicinal plants on the asexual blood stage of chloroquine resistant *Plasmodium falciparum*. *BMC Complementary and Alternative Medicine 14* (1): 492. https://doi.org/10.1186/1472-6882-14-492.

50 Torres-Guerrero, E., Quintanilla-Cedillo, M.R., Ruiz-Esmenjaud, J., and Arenas, R. (2017). Leishmaniasis: a review. *F1000 Research* 6: 750. https://doi.org/10.12688/f1000research.11120.1.

51 Vieira-Araújo, F.M., Macedo Rondon, F.C., Pinto Vieira, Í.G. et al. (2018). Sinergism between alkaloids piperine and capsaicin with meglumine antimoniate against *Leishmania infantum*. *Experimental Parasitology 188*: 79–82. https://doi.org/10.1016/j.exppara.2018.04.001.

52 Badshah, S. L., Faisal, S., Muhammad, A., Poulson, B. G., Emwas, A. H., & Jaremko, M. (2021). Antiviral activities of flavonoids. *Biomedicine & Pharmacotherapy*, 140, 111596. https://doi.org/10.1016/j.biopha.2021.111596.

53 Plourde, A.R. and Bloch, E.M. (2016). A literature review of Zika virus. *Emerging Infectious Diseases 22* (7): 1185–1192. https://doi.org/10.3201/eid2207.151990.

54 Jourdan, P.M., Lamberton, P.H.L., Fenwick, A., and Addiss, D.G. (2018). Soil-transmitted helminth infections. *The Lancet 391* (10117): 252–265. https://doi.org/10.1016/S0140-6736(17)31930-X.

55 Liu, M., Panda, S.K., and Luyten, W. (2020). Plant-based natural products for the discovery and development of novel anthelmintics against nematodes. *Biomolecules 10* (3): 426. https://doi.org/10.3390/biom10030426.

56 Tsouh Fokou, P.V., Nyarko, A.K., Appiah-Opong, R. et al. (2015). Ethnopharmacological reports on anti-Buruli ulcer medicinal plants in three West African countries. *Journal of Ethnopharmacology 172*: 297–311. https://doi.org/10.1016/j.jep.2015.06.024.

57 Vieira, P.C., Mafezoli, J., Pupo, M.T. et al. (2001). Strategies for the isolation and identification of trypanocidal compounds from the Rutales. *Pure and Applied Chemistry 73* (3): 617–622. https://doi.org/10.1351/pac200173030617.

58 El Refai Khalil, S., Nassef, N., El-Kersh, W. et al. (2014). In-vitro and in-vivo assessment of the effect of soybean extract on *Fasciola gigantica* infection in comparison with triclabendazole. *Menoufia Medical Journal 27* (1): 93. https://doi.org/10.4103/1110-2098.132768.

59 Pereira, C.A.J., Oliveira, L.L.S., Coaglio, A.L. et al. (2016). Anti-helminthic activity of *Momordica charantia* L. against *Fasciola hepatica* eggs after twelve days of incubation in vitro. *Veterinary Parasitology 228*: 160–166. https://doi.org/10.1016/j.vetpar.2016.08.025.

60 Ullah, R., Rehman, A., Zafeer, M.F. et al. (2017). Anthelmintic potential of thymoquinone and curcumin on *Fasciola gigantica*. *PLoS ONE 12* (2): e0171267. https://doi.org/10.1371/journal.pone.0171267.

61 Hoet, S., Opperdoes, F., Brun, R., and Quetin-Leclercq, J. (2004). Natural products active against African trypanosomes: a step towards new drugs. *Natural Product Reports 21* (3): 353. https://doi.org/10.1039/b311021b.

62 Koide, T., Nose, M., Inoue, M. et al. (1998). Trypanocidal effects of gallic acid and related compounds. *Planta Medica 64* (1): 27–30. https://doi.org/10.1055/s-2006-957360.

63 Mamani-Matsuda, M., Rambert, J., Malvy, D. et al. (2004). Quercetin induces apoptosis of *Trypanosoma brucei* gambiense and decreases the proinflammatory response of human macrophages. *Antimicrobial Agents and Chemotherapy 48* (3): 924–929. https://doi.org/10.1128/AAC.48.3.924-929.2004.

64 Gyasi, M., Okonkwo, O., and Tripathy, K. (2021). *Onchocerciasis*. StatPearls https://www.ncbi.nlm.nih.gov/books/NBK559027/.

65 Ndjonka, D., Abladam, E.D., Djafsia, B. et al. (2014). Anthelmintic activity of phenolic acids from the axlewood tree *Anogeissus leiocarpus* on the filarial nematode *Onchocerca ochengi* and drug-resistant strains of the free-living nematode *Caenorhabditis elegans*. *Journal of Helminthology 88* (4): 481–488. https://doi.org/10.1017/S0022149X1300045X.

66 ChEBI (2021). CHEBI:28927 – sakuranetin. https://www.ebi.ac.uk/chebi/searchId.do?chebiId=CHEBI:28927 (accessed 22 August 2021).

67 Halder, S., Dhorajiwala, T., and Samant, L. (2019). Multiple docking analysis and In silico absorption, distribution, metabolism, excretion, and toxicity screening of anti-leprosy phytochemicals and dapsone against dihydropteroate synthase of *Mycobacterium leprae*. *International Journal of Mycobacteriology 8* (3): 229. https://doi.org/10.4103/ijmy.ijmy_123_19.

68 Lakshmi, V., Joseph, S.K., Srivastava, S. et al. (2010). Antifilarial activity in vitro and in vivo of some flavonoids tested against *Brugia malayi*. *Acta Tropica 116* (2): 127–133. https://doi.org/10.1016/j.actatropica.2010.06.006.

69 Rana, G. (2018). Inhibition efficiency of a newly isolated flavonoid compound from *Vitex negundo* L. leaves against cattle-endosymbiont *Setaria cervi*: phytomedicine for lymphatic filariasis. *Parasite Epidemiology and Control 3* (2): 88–95. https://doi.org/10.1016/j.parepi.2018.03.002.

70 Hao, H., Aixia, Y., Lei, F. et al. (2010). Effects of baicalin on *Chlamydia trachomatis* infection in vitro. *Planta Medica 76* (1): 76–78. https://doi.org/10.1055/s-0029-1185943.

71 Yamazaki, T., Inoue, M., Sasaki, N. et al. (2003). In vitro inhibitory effects of tea polyphenols on the proliferation of *Chlamydia trachomatis* and *Chlamydia pneumoniae*. *Japanese Journal of Infectious Diseases 56* (4): 143–145. http://www.ncbi.nlm.nih.gov/pubmed/14583635.

72 Huang, P., Zhou, M., Cheng, S. et al. (2020). Myricetin possesses anthelmintic activity and attenuates hepatic fibrosis via modulating TGFβ1 and Akt signaling and shifting Th1/Th2 balance in *Schistosoma japonicum*-infected mice. *Frontiers in Immunology* 11. https://doi.org/10.3389/fimmu.2020.00593.

73 Pereira, V.R.D., Junior, I.J.A., da Silveira, L.S. et al. (2018). In vitro and in vivo antischistosomal activities of chalcones. *Chemistry & Biodiversity* 15: e1800398. https://doi.org/10.1002/cbdv.201800398.

74 Chaves, L.L., Patriota, Y., Soares-Sobrinho, J.L. et al. (2020). Drug delivery systems on leprosy therapy: moving towards eradication? *Pharmaceutics 12* (12): 1202. https://doi.org/10.3390/pharmaceutics12121202.

75 Lastória, J.C. and de Abreu, M.A.M.M. (2014). Leprosy: review of the epidemiological, clinical, and etiopathogenic aspects – part 1. *Anais Brasileiros de Dermatologia 89* (2): 205–218. https://doi.org/10.1590/abd1806-4841.20142450.

76 World Health Organization (2011). International classification of functioning, disability, and health – children and youth version. https://books.google.com/books?hl=en&lr=&id=SWFQDXyU-rcC&oi=fnd&pg=PR5&ots=G9KQiBwXAA&sig=hpc9zO_PpSVkZmt36-bdND86E6o (accessed 22 August 2021).

77 Ganie, S.A. and Yadav, S.S. (2014). *Holoptelea integrifolia* (Roxb.) Planch: a review of its ethnobotany, pharmacology, and phytochemistry. *BioMed Research International 2014*: 1–12. https://doi.org/10.1155/2014/401213.

78 Kathirvel, A. and Sujatha, V. (2016). Phytochemical studies, antioxidant activities and identification of active compounds using GC–MS of *Dryopteris cochleata* leaves. *Arabian Journal of Chemistry 9*: S1435–S1442. https://doi.org/10.1016/j.arabjc.2012.03.018.

79 Prasad, N., Yang, B., Kong, K.W. et al. (2013). Phytochemicals and antioxidant capacity from *Nypa fruticans* Wurmb. fruit. *Evidence-based Complementary and Alternative Medicine 2013*: 1–9. https://doi.org/10.1155/2013/154606.

80 Sundar, R.D.V., Settu, S., Shankar, S. et al. (2018). Potential medicinal plants to treat leprosy – a review. *Research Journal of Pharmacy and Technology 11* (2): 813. https://doi.org/10.5958/0974-360X.2018.00153.1.

81 Rajasekaram, S., Anuradha, R., Manokaran, G., and Bethunaickan, R. (2017). An overview of lymphatic filariasis lymphedema. *Lymphology 50* (4): 164–182. http://www.ncbi.nlm.nih.gov/pubmed/30248721.

82 Lansingh, V.C. (2016). Trachoma. *BMJ Clinical Evidence 2016*: 0706. http://www.ncbi.nlm.nih.gov/pubmed/26860629.

83 Chuah, C., Gobert, G.N., Latif, B. et al. (2019). Schistosomiasis in Malaysia: a review. *Acta Tropica 190*: 137–143. https://doi.org/10.1016/j.actatropica.2018.11.012.

84 Dar, M.J., McElroy, C.A., Khan, M.I. et al. (2020). Development and evaluation of novel miltefosine-polyphenol co-loaded second generation nano-transfersomes for the topical treatment of cutaneous leishmaniasis. *Expert Opinion on Drug Delivery 17* (1): 97–110. https://doi.org/10.1080/17425247.2020.1700227.

85 Tsai, M.-J., Wu, P.-C., Huang, Y.-B. et al. (2012). Baicalein loaded in tocol nanostructured lipid carriers (tocol NLCs) for enhanced stability and brain targeting. *International Journal of Pharmaceutics 423* (2): 461–470. https://doi.org/10.1016/j.ijpharm.2011.12.009.

86 Mandal, S., Debnath, K., Jana, N.R., and Jana, N.R. (2020). Trehalose-conjugated, catechin-loaded polylactide nanoparticles for improved neuroprotection against intracellular polyglutamine aggregates. *Biomacromolecules 21* (4): 1578–1586. https://doi.org/10.1021/acs.biomac.0c00143.

87 Santos, I., Ramos, C., Mendes, C. et al. (2019). Targeting glutathione and cystathionine β-synthase in ovarian cancer treatment by selenium–chrysin polyurea dendrimer nanoformulation. *Nutrients 11* (10): 2523. https://doi.org/10.3390/nu11102523.

88 Rivolta, I., Panariti, A., Lettiero, B. et al. (2011). Cellular uptake of coumarin-6 as a model drug loaded in solid lipid nanoparticles. *Journal of Physiology and Pharmacology: An Official Journal of the Polish Physiological Society 62* (1): 45–53. http://www.ncbi.nlm.nih.gov/pubmed/21451209.

89 Oyeyemi, O., Adegbeyeni, O., Oyeyemi, I. et al. (2018). In vitro ovicidal activity of poly lactic acid curcumin-nisin co-entrapped nanoparticle against *Fasciola* spp. eggs and its reproductive toxicity. *Journal of Basic and Clinical Physiology and Pharmacology 29* (1): 73–79. https://doi.org/10.1515/jbcpp-2017-0045.

90 Mady, F.M. and Ibrahim, S.R.-M. (2018). Cyclodextrin-based nanosponge for improvement of solubility and oral bioavailability of ellagic acid. *Pakistan Journal of Pharmaceutical Sciences* 31 (5 Supplementary): 2069–2076. http://www.ncbi.nlm.nih.gov/pubmed/30393214.

91 Chamcheu, J.C., Siddiqui, I.A., Adhami, V.M. et al. (2018). Chitosan-based nanoformulated (−)-epigallocatechin-3-gallate (EGCG) modulates human keratinocyte-induced responses and alleviates imiquimod-induced murine psoriasiform dermatitis. *International Journal of Nanomedicine 13*: 4189–4206. https://doi.org/10.2147/IJN.S165966.

92 Cano, A., Ettcheto, M., Chang, J.-H., Barroso, E., Espina, M., Kühne, B. A., Barenys, M., Auladell, C., Folch, J., Souto, E. B., Camins, A., Turowski, P., & García, M. L. (2019). Dual-drug loaded nanoparticles of epigallocatechin-3-gallate (EGCG)/ascorbic acid enhance therapeutic efficacy of EGCG in a APPswe/PS1dE9 Alzheimer's disease mice model. *Journal of Controlled Release, 301*, 62–75. https://doi.org/10.1016/j.jconrel.2019.03.010.

93 Feng, C., Yuan, X., Chu, K. et al. (2019). Preparation and optimization of poly(lactic acid) nanoparticles loaded with fisetin to improve anti-cancer therapy. *International Journal of Biological Macromolecules 125*: 700–710. https://doi.org/10.1016/j.ijbiomac.2018.12.003.

94 Mahboob, T., Nawaz, M., de Lourdes Pereira, M. et al. (2020). PLGA nanoparticles loaded with gallic acid – a constituent of *Leea indica* against *Acanthamoeba triangularis. Scientific Reports 10* (1): 8954. https://doi.org/10.1038/s41598-020-65728-0.

95 Cheng, Q., Qin, W., Yu, Y. et al. (2020). Preparation and characterization of PEG-PLA genistein micelles using a modified emulsion-evaporation method. *Journal of Nanomaterials 2020*: 1–15. https://doi.org/10.1155/2020/3278098.

96 Paramita, P., Sethu, S.N., Subhapradha, N. et al. (2020). Neuro-protective effects of nano-formulated hesperetin in a traumatic brain injury model of Danio rerio. *Drug and Chemical Toxicology* 1–8. https://doi.org/10.1080/01480545.2020.1722690.

97 Kamarullah, W., Indrajaya, E., and Emmanuella, J. (2018). Potency of luteolin with solid lipid nanoparticle (SLN)-polyethylene glycol (PEG) modification for artemisinin-resistant *Plasmodium falciparum* infection. *Indonesian Journal of Tropical and Infectious Disease 7* (3): 80. https://doi.org/10.20473/ijtid.v7i3.6726.

98 Mahin, M., Ali, A., Elahe, K. et al. (2019). Synthesis of a copolymer carrier for anticancer drug luteolin for targeting human breast cancer cells. *Journal of Traditional Chinese Medicine = Chung i Tsa Chih Ying Wen Pan 39* (4): 474–481. http://www.ncbi.nlm.nih.gov/pubmed/32186094.

99 Nafee, N., Gaber, D.M., Elzoghby, A.O. et al. (2020). Promoted antitumor activity of myricetin against lung carcinoma via nanoencapsulated phospholipid complex in respirable microparticles. *Pharmaceutical Research 37* (4): 82. https://doi.org/10.1007/s11095-020-02794-z.

100 Song, Y., Zhou, B., Du, X. et al. (2020). Folic acid (FA)-conjugated mesoporous silica nanoparticles combined with MRP-1 siRNA improves the suppressive effects of myricetin on non-small cell lung cancer (NSCLC). *Biomedicine & Pharmacotherapy 125*: 109561. https://doi.org/10.1016/j.biopha.2019.109561.

101 Wang, Y., Wang, S., Firempong, C.K. et al. (2017). Enhanced solubility and bioavailability of naringenin via liposomal nanoformulation: preparation and in vitro and in vivo evaluations. *AAPS PharmSciTech 18* (3): 586–594. https://doi.org/10.1208/s12249-016-0537-8.

102 Wadhwa, R., Paudel, K.R., Chin, L.H. et al. (2021). Anti-inflammatory and anticancer activities of naringenin-loaded liquid crystalline nanoparticles in vitro. *Journal of Food Biochemistry 45* (1). https://doi.org/10.1111/jfbc.13572.

103 Ray, L., Karthik, R., Srivastava, V. et al. (2021). Efficient antileishmanial activity of amphotericin B and piperine entrapped in enteric coated guar gum nanoparticles. *Drug Delivery and Translational Research 11* (1): 118–130. https://doi.org/10.1007/s13346-020-00712-9.

104 Chen, S., Sun, C., Wang, Y. et al. (2018). Quercetagetin-loaded composite nanoparticles based on zein and hyaluronic acid: formation, characterization, and physicochemical stability. *Journal of Agricultural and Food Chemistry 66* (28): 7441–7450. https://doi.org/10.1021/acs.jafc.8b01046.

105 Sun, D., Li, N., Zhang, W. et al. (2016). Quercetin-loaded PLGA nanoparticles: a highly effective antibacterial agent in vitro and anti-infection application in vivo. *Journal of Nanoparticle Research 18* (1): 3. https://doi.org/10.1007/s11051-015-3310-0.

106 Zhai, G., Guo, C., Tan, Q., and Liu, W. (2011). Preparation and evaluation of quercetin-loaded lecithin-chitosan nanoparticles for topical delivery. *International Journal of Nanomedicine* 1621. https://doi.org/10.2147/IJN.S22411.

107 Ersoz, M., Erdemir, A., Derman, S. et al. (2020). Quercetin-loaded nanoparticles enhance cytotoxicity and antioxidant activity on C6 glioma cells. *Pharmaceutical Development and Technology 25* (6): 757–766. https://doi.org/10.1080/10837450.2020.1740933.

108 Paudel, K.R., Wadhwa, R., Mehta, M. et al. (2020). Rutin loaded liquid crystalline nanoparticles inhibit lipopolysaccharide induced oxidative stress and apoptosis in bronchial epithelial cells in vitro. *Toxicology in vitro 68*: 104961. https://doi.org/10.1016/j.tiv.2020.104961.

109 Chung, I.-M., Subramanian, U., Thirupathi, P. et al. (2020). Resveratrol nanoparticles: a promising therapeutic advancement over native resveratrol. *Processes 8* (4): 458. https://doi.org/10.3390/pr8040458.

110 Sun, Y., Chen, D., Pan, Y. et al. (2019). Nanoparticles for antiparasitic drug delivery. *Drug Delivery 26* (1): 1206–1221. https://doi.org/10.1080/10717544.2019.1692968.

111 D'Archivio, M., Filesi, C., Varì, R. et al. (2010). Bioavailability of the polyphenols: status and controversies. *International Journal of Molecular Sciences 11* (4): 1321–1342. https://doi.org/10.3390/ijms11041321.

112 Singh, Y., Meher, J.G., Raval, K. et al. (2017). Nanoemulsion: concepts, development and applications in drug delivery. *Journal of Controlled Release 252*: 28–49. https://doi.org/10.1016/j.jconrel.2017.03.008.

113 Khosa, A., Reddi, S., and Saha, R.N. (2018). Nanostructured lipid carriers for site-specific drug delivery. *Biomedicine & Pharmacotherapy 103*: 598–613. https://doi.org/10.1016/j.biopha.2018.04.055.

114 da Luz, C.M., Boyles, M.S.P., Falagan-Lotsch, P. et al. (2017). Poly-lactic acid nanoparticles (PLA-NP) promote physiological modifications in lung epithelial cells and are internalized by clathrin-coated pits and lipid rafts. *Journal of Nanobiotechnology 15* (1): 11. https://doi.org/10.1186/s12951-016-0238-1.

115 Mukherjee, S., Ray, S., and Thakur, R. (2009). Solid lipid nanoparticles: a modern formulation approach in drug delivery system. *Indian Journal of Pharmaceutical Sciences 71* (4): 349. https://doi.org/10.4103/0250-474X.57282.

116 Puttappa, N., Kumar, R.S., and Yamjala, K. (2017). Artesunate-quercetin/luteolin dual drug nanofacilitated synergistic treatment for malaria: a plausible approach to overcome artemisinin combination therapy resistance. *Medical Hypotheses 109*: 176–180. https://doi.org/10.1016/j.mehy.2017.10.016.

117 Martí Coma-Cros, E., Biosca, A., Lantero, E. et al. (2018). Antimalarial activity of orally administered curcumin incorporated in Eudragit®-containing liposomes. *International Journal of Molecular Sciences 19* (5): 1361. https://doi.org/10.3390/ijms19051361.

118 Osmani, R., Hani, U., Bhosale, R. et al. (2016). Nanosponge carriers – an archetype swing in cancer therapy: a comprehensive review. *Current Drug Targets 18* (1): 108–118. https://doi.org/10.2174/1389450116666151001105449.

119 Shrestha, S. and Bhattacharya, S. (2020). Versatile use of nanosponge in the pharmaceutical arena: a mini-review. *Recent Patents on Nanotechnology 14* (4): 351–359. https://doi.org/10.2174/1872210514999200901200558.

120 Ahlawat, J., Henriquez, G., & Narayan, M. (2018). Enhancing the delivery of chemotherapeutics: role of biodegradable polymeric nanoparticles. *Molecules, 23*(9), 2157. https://doi.org/10.3390/molecules23092157.

121 Khan, I., Saeed, K., and Khan, I. (2019). Nanoparticles: properties, applications and toxicities. *Arabian Journal of Chemistry 12* (7): 908–931. https://doi.org/10.1016/j.arabjc.2017.05.011.

122 Rhazouani, A., Gamrani, H., El Achaby, M. et al. (2021). Synthesis and toxicity of graphene oxide nanoparticles: a literature review of in vitro and in vivo studies. *BioMed Research International 2021*: 1–19. https://doi.org/10.1155/2021/5518999.

123 Croy, S. and Kwon, G. (2006). Polymeric micelles for drug delivery. *Current Pharmaceutical Design 12* (36): 4669–4684. https://doi.org/10.2174/138161206779026245.

124 Gigliobianco, M., Casadidio, C., Censi, R., and Di Martino, P. (2018). Nanocrystals of poorly soluble drugs: drug bioavailability and physicochemical stability. *Pharmaceutics 10* (3): 134. https://doi.org/10.3390/pharmaceutics10030134.

125 Tang, F., Li, L., and Chen, D. (2012). Mesoporous silica nanoparticles: synthesis, biocompatibility and drug delivery. *Advanced Materials 24* (12): 1504–1534. https://doi.org/10.1002/adma.201104763.

126 Wang, Y., Zhao, Q., Han, N. et al. (2015). Mesoporous silica nanoparticles in drug delivery and biomedical applications. *Nanomedicine: Nanotechnology, Biology and Medicine 11* (2): 313–327. https://doi.org/10.1016/j.nano.2014.09.014.

127 Torchilin, V.P. (2005). Recent advances with liposomes as pharmaceutical carriers. *Nature Reviews Drug Discovery 4* (2): 145–160. https://doi.org/10.1038/nrd1632.

128 Madheswaran, T., Kandasamy, M., Bose, R.J., and Karuppagounder, V. (2019). Current potential and challenges in the advances of liquid crystalline nanoparticles as drug delivery systems. *Drug Discovery Today 24* (7): 1405–1412. https://doi.org/10.1016/j.drudis.2019.05.004.

129 Chen, S., Han, Y., Huang, J. et al. (2019). Fabrication and characterization of layer-by-layer composite nanoparticles based on zein and hyaluronic acid for codelivery of curcumin and quercetagetin. *ACS Applied Materials & Interfaces 11* (18): 16922–16933. https://doi.org/10.1021/acsami.9b02529.

17

Nanotherapeutics of Phytoantioxidants for Inflammatory Disorders

Santwana Palai

Department of Veterinary Pharmacology & Toxicology, College of Veterinary Science & Animal Husbandry, Orissa University of Agriculture & Technology, Bhubaneswar, Odisha, India

17.1 Introduction

Inflammation is a natural and healthy biological reaction stimulus that the body interprets as having a potentially detrimental consequence, in which the immune system attacks the body's own cells or tissues, producing inappropriate inflammation with obstinate pain, swelling, redness, and tissue damage. Small amounts of inflammation restore homeostasis equilibrium and wound-healing responses, but on a larger scale, inflammation causes pathologic disorders like inflammatory bowel disease, rheumatoid arthritis (RA), gout, atherosclerosis, asthma, and cancer. The methods through which inflammatory mediators regulate tissue damage can be used to discover medicinal targets for medication development to counteract undesired inflammation. Glucocorticoids, histamine blockers, and cyclooxygenase (COX) inhibitors can all help reduce inflammation [1].

In Asia, the prevalence and incidence of inflammatory disorders like RA, acute and chronic lung disease (asthma and COVID-19), chronic obstructive pulmonary disease, cardiovascular disease, inflammatory bowel diseases including Crohn's disease and ulcerative colitis, and atherosclerosis are lesser than in the West. The prevalence and incidence have amplified rapidly over the last 20 years due to greater contact with the West, more antibiotic usage, vaccines, more cleanliness, gut microbiota changes, and genetic variables [2].

Due to the increased reactive oxygen species (ROS) and associated oxidative stress, various clinical illnesses like obesity, cancer, cardiovascular disorders, metabolic syndrome, and neurogenerative disorders arise. ROS are physiologic activators of transcription factors like activator protein-1 (AP-1) or nuclear factor κB (NF-κB). ROS is an intracellular signaling component linked to inflammatory responses and autoimmune disorders. As it produces proinflammatory cytokines like interleukin (IL)-1, IL-8, IL-6, and (TNF-α), it has a critical part in the inflammatory progression resulting in inflammatory disorders [3].

Phytochemicals are potent plants' secondary metabolites like polyphenols, saponins, flavonoids, and anthocyanins with antioxidant and anti-inflammatory properties interfering with various pro-inflammatory mediators. Functional foods with anti-inflammatory properties of plant-derived ingredients or nutraceuticals can be a supplement to anti-inflammatory medicaments reducing dose-minimizing side effects or their substitution [4]. These anti-inflammatory phytochemicals operate successfully as ROS scavengers, but they also have a number of negative side effects,

Phytoantioxidants and Nanotherapeutics, First Edition. Edited by Mithun Rudrapal.

including nonspecific distribution and limited delivery efficiency, low molecular weight, high renal clearance, and renal and hepatic toxicity [5].

Nanotechnological intervention can significantly overcome these constraints inspiring the next phase of technical innovation in phytoantioxidants for inflammatory disorders. Nanotherapeutics have the potential to significantly enhance the pharmacokinetic characteristics of antioxidative drugs while concurrently reducing their negative effects. Furthermore, nanomaterials have shown favorable therapeutic benefits in inflammatory disorders because of their antioxidative and anti-inflammatory properties [6]. This chapter will emphasize the biological benefits of phytochemicals as anti-inflammatory agents against inflammatory disorders and various forms of antioxidant nanoparticles (NPs) effective in the treatment of inflammatory disorders like inflammatory bowel disease (IBD), cardiovascular disorders, neurodegenerative disorders, etc., as well as their key hurdles.

17.2 Phytoantioxidants and Their Antioxidant Role in Inflammatory Disorders

The original supplies of medication for various diseases came from a variety of medicinal plants found all over the planet. Because phytochemicals have superior activities against inflammatory disorders such as cancer, arthritis, diabetes, neurodegenerative disorders, COVID-19, and so on, with negligible side effects, researchers use traditional knowledge, high-throughput drug screening, and computer-aided drug designing, to find potent phytochemicals or combinations of phytochemicals. The phytochemicals, such as polyphenols, terpenoids, glycosides, and so on, are either sparingly soluble or insoluble in water, limiting their bioavailability and also the scope of using purified phytochemicals against specific inflammatory diseases [7, 8]. To boost the bioavailability of phytochemicals, several delivery vehicles, such as nanocarriers, have been developed for the administration of insoluble phytochemicals (Figure 17.1). A rising interest in the repurposing of existing medications or traditional medicines leads to the development of next-generation phytochemicals-based medications for inflammation and oxidative stress-related disorders [9].

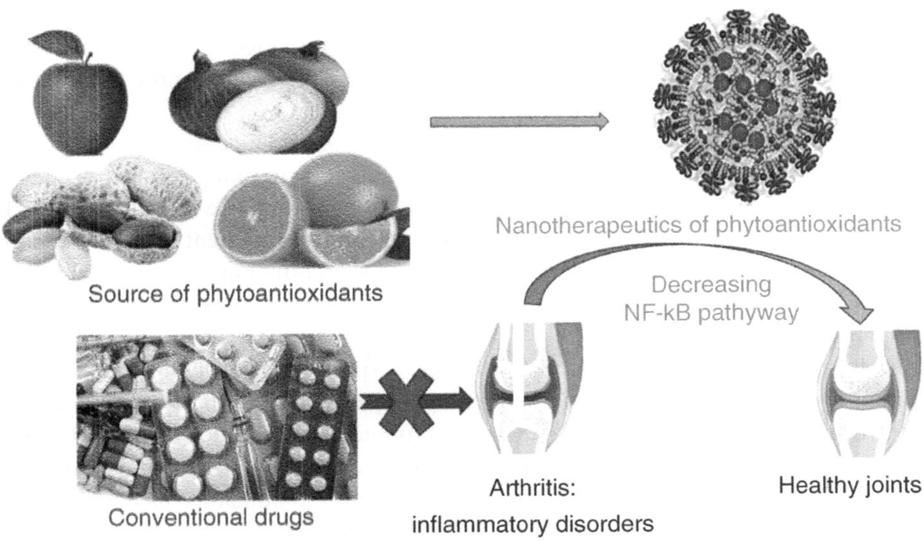

Figure 17.1 Nanotherapeutics of phytoantioxidants for inflammatory disorders are more advantageous than traditional medicines.

Polymeric or lipid-based nanocarriers can be manufactured for a wide range of applications. These extend the drug's circulation period and spatial distribution capabilities by improving penetration and retention (Table 17.1). Thus, nanocarrier-based phytochemicals possess beneficial applications in inflammatory illnesses such as diabetes, cancer, Alzheimer's disease, COPD, and COVID-19 [86].

Table 17.1 describes the antioxidant capacity of phytotherapeutic agent-loaded nanocarriers [10–85].

17.3 Nanodelivery Systems or Nanoformulations, Especially in the Delivery of Herbal Drugs

Because of its easy administration, no pain, patient acceptance, and self-administrable ability, oral administration is the most approved path for drug distribution [87]. Efficacy of oral drug delivery of therapeutic compounds is significantly decreased owing to chemical and enzymatic barriers and its low solubility in the gastrointestinal tract (GIT). So, they are loaded into nanocarriers for oral delivery [88]. Nanodelivery of therapeutic agents has a variety of clinical benefits. These benefits include improved pharmacokinetic clearance, extended circulation time, increased half-life, and enhanced mean residence time inside the body [89].

Various nanodelivery mechanisms for phytotherapeutic oral administration are currently in clinical trials to be used in practice clinically [90]. The main hindrance in the development of novel multifunctional nanomaterials is the ability to move these complex therapeutics through different biological barriers to the targeted cell or organs. Effective nanodelivery systems have ideal features for loading and releasing therapeutic agents, high therapeutic effectiveness, long storage life, and minimal side effects [91].

The nanodelivery systems include liquid (nanoemulsions, nanoliposomes, and nanopolymersomes) and solid (nanocrystals, solid lipid NPs (SLPs), polymeric NPs, and metallic NPs), which are described in detail next.

17.3.1 Nanoemulsions

Nanoemulsions are immiscible liquid mixtures of water and oil. Chemical or mechanical approaches are commonly used to create such nanosystems [92]. Because of the hydrophobic effects of lipophilic molecules with emulsifiers, chemical approaches give rise to the natural creation of emulsion droplets. Mechanical techniques involve high-energy processes that break down large emulsion droplets into smaller ones through a variety of mechanical operations. The primary distinction between nanoemulsions and traditional emulsions is the form and scale of NPs scattered. Droplet volumes in nanoemulsions are usually 20–200 nm [93].

17.3.2 Nanoliposomes

Nanoliposomes are nanosized self-assembled vesicles made up of phospholipid bilayers having more than one aquatic compartment [94]. It is an oil-in-water dispersions droplet of 50–450 nm [95]. Layer-by-layer electrostatic deposition can create such nanodelivery schemes. Here, charged polymers are applied to charged template structure solution. Then, the emulsion droplets are stabilized by ionic emulsifiers. The template structures are hydrogel particles with charged biopolymers or liposomes with charged polar lipids [96]. The gentle hydration method is another method for producing nanoliposomes.

Table 17.1 Phytoantioxidants and their antioxidant role with special reference to inflammatory disorders.

S. no.	Phytoanti-oxidant	Description	Source	Structure	Inflammatory disorder	Mechanism of action	References
1.	Resveratrol	Polyphenol	Red wine	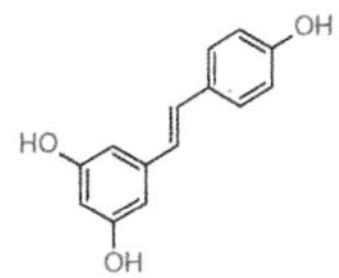	Cardiovascular disorders, ARDS, COPD, neuroinflammation, ischemic stroke, cancer	TNF-α- and IL-6-induced monocyte adhesion increases is inhibited, reduced pro-inflammatory NF-κB levels, COPD-related inflammatory mediatorslike MCP-1, TNF-α, IL-6, IL-8, NF-κB, GM-CSF expression decrease	[10–16]
2.	Curcumin	Polyphenol	Turmeric	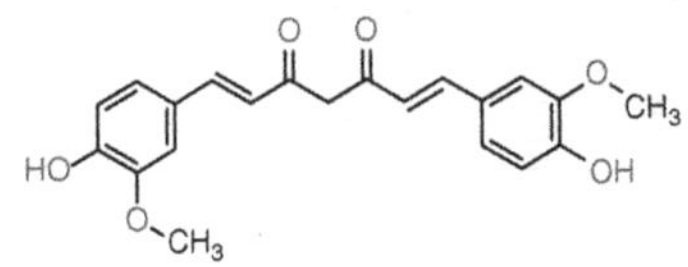	Inflammatory bowel disease, obesity, rheumatoid arthritis, chronic pancreatitis, type 2 diabetes, bronchitis, cancer scleroderma, asthma, allergy, neurodegenerative diseases, chronic kidney disease, cardiovascular diseases, psoriasis	Oxidative stress and inflammation in chronic diseases decreased through the Nrf2-keap1 pathway, pro-inflammatory pathways of chronic diseases, TNF production and cell signaling is blocked.	[17–29]
3.	Butein	Chalcone flavonoid	Dahlia, Butea	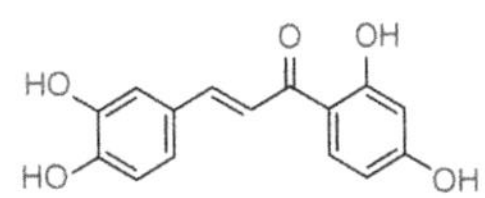	Cancer, hypertension, diabetic, neurodegenerative disorders	Multiple molecular targets like nuclear factor-κB and its downstream molecules are inhibited.	[30]

4.	Luteolin	Flavones	Chamomile tea, celery, green peppers, thyme	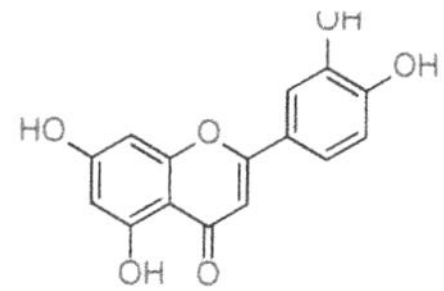	Cancer, hypertension, inflammatory disorders,diabetes	human cytochrome P450 (CYP) 1 family enzymes are inhibited overpowering carcinogen activation	[31]
5.	Eriodictyol	Flavanones	Citrus fruit, peanuts	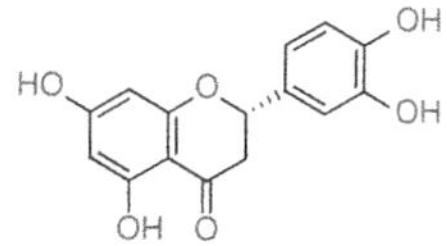	Acute lung injury, rheumatoid arthritis, multiple sclerosis, experimental autoimmune encephalomyelitis	LPS-induced lung injury is alleviated through Nrf2 pathway regulation, expression of inflammatory cytokines in macrophages. cell survival and inflammatory response stopped in RA through AKT/FOXO1 signaling pathway suppressing Th17 cell infiltration	[32–34]
6.	Hesperetin	Flavanones	Oranges, grapes, lemons	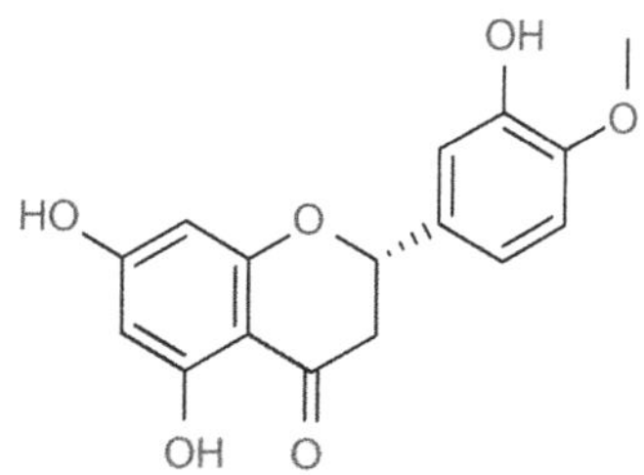	Amyotrophic lateral sclerosis, Alzheimer's disease, neurodegenerative disorders, multiple sclerosis, Parkinson's disease	Oxidative stress reduction through regulating transcription factor Nrf2, NF-kB and mitochondrial apoptosis	[35]
7.	Naringenin	Flavanones	Grapes, citrus fruits	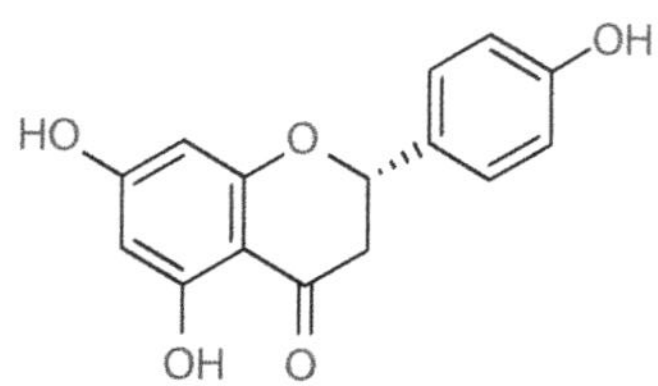	Rheumatoid arthritis, osteoarthritis	Macrophages, dendritic cells, mast cells, T cells activities suppression	[36, 37]

(Continued)

Table 17.1 (Continued)

S. no.	Phytoanti-oxidant	Description	Source	Structure	Inflammatory disorder	Mechanism of action	References
8.	Catechin	Flavanols	Fresh tea leaves, strawberries wines, black grapes	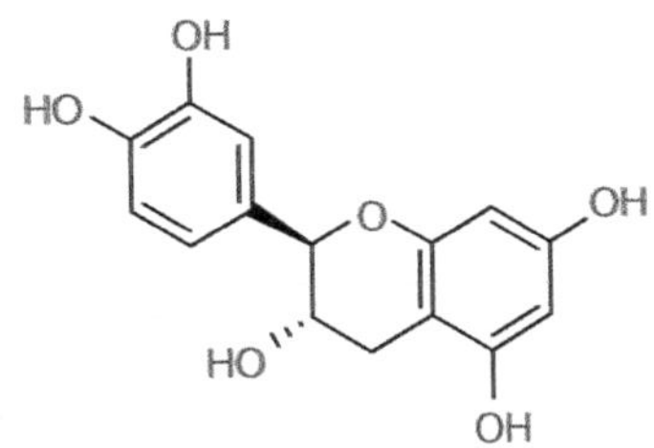	IBD, arthritis, cancer, CVD, diabetes, and obesity	Permeation and production of colonic epithelial cells, macrophages, neutrophils, T lymphocyte are regulated through inflammatory relations reduction protecting neurons	[38, 39]
9.	Epigallo-Catechin gallate	Flavanols	Green tea	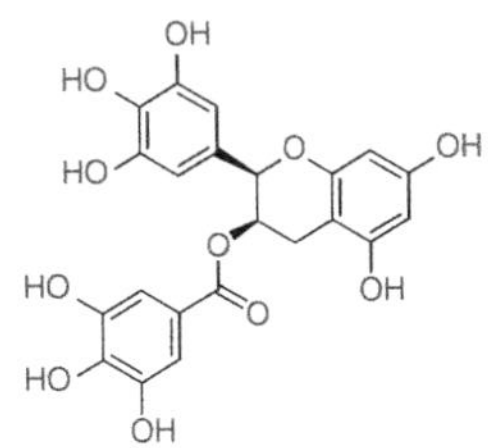	COVID-19, Cancer, Alzheimer's disease, Rheumatoid arthritis, CVD, Parkinson's disease, diabetes, arteriosclerosis	Signal transducer and NF-κB transcription factors is repressed for downstreaming of pro-inflammatory signaling	[40–46]
10.	Keampferol	Flavonols	Citrus fruits, gooseberriesgrape, beans, broccoli, apples, cabbage	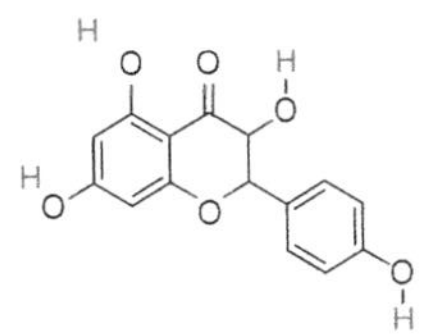	Osteoarthritis, post-menopausal bone loss, colitis, acute inflammatory diseases, cardiovascular diseases, intervertebral disc degeneration, rheumatoid arthritis, acute lung injury, cancer, neurodegenerative diseases	LPS-induced tumor necrosis factor-α and IL-1 expression decrease through activated macrophages; TNF-α suppression translocates NF-κB p65 to the nucleus regulating inflammatory genes expression, inhibition of transcription factors	[47–51]

11.	Morin	Flavanols	Almond, fig	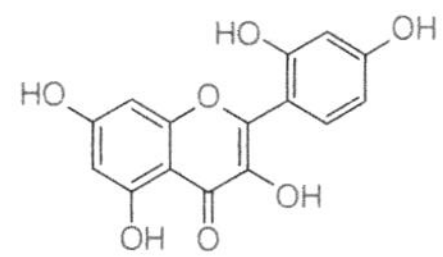 	Osteoarthritis, stroke, heart disease, Parkinson's disease, arthritis, Diabetes, chronic respiratory problems like COPD, cancer	IL-1β-induced inflammatory mediators' inhibition through Nrf2 signaling and IL-1β-induced NF-κB attenuation	[52–54]
12.	Quercetin	Flavanols	Grapes, onions, shallots, tea, and tomatoes	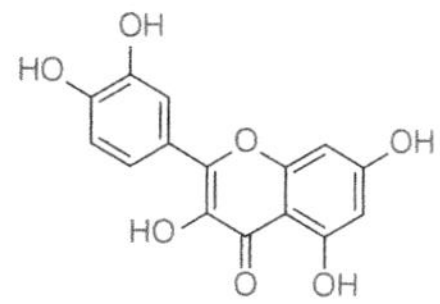 	Heart disease, cancer, gout	TNF is inhibited from activating the powerful inducers of inflammatory gene expression and protein secretion, ERK, JNK, c-Jun, and NF-B. Reduce inflammation by boosting PPAR activity and inhibiting NF-B or AP-1.	[55, 56]
13.	Anthocyanins	Flavonoids	Berries, currants, grapes	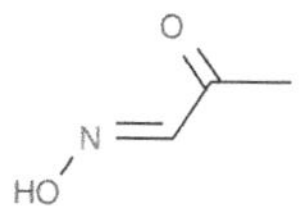 	Ulcerative colitis, obesity, type 2 diabetes mellitus, cancer	Attenuate oxidized LDL-C (oxLDL)-mediated foam cell formation for atherogenesis through inflammatory pathways involving CD36 gene expression regulation	[57–60]
14.	Daidzein	Isoflavonoids	Soybeans, upin, fava beans	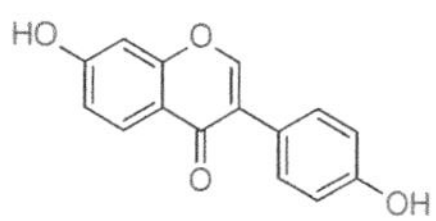 	IBD, ulcerative colitis, indeterminate colitis, Crohn's disease, Obesity-induced inflammation	Signal translator and activator of transcription 1 inhibited and the expression of inducible nitric oxide synthase iNOS reduction. Reduced expression of pro-inflammatory genes via the PPAR/and JNK pathways	[61, 62]

(*Continued*)

Table 17.1 (Continued)

S. no.	Phytoanti-oxidant	Description	Source	Structure	Inflammatory disorder	Mechanism of action	References
15.	Genistein	Isoflavonoids	Lupin, fava beans, soybeans, kudzu	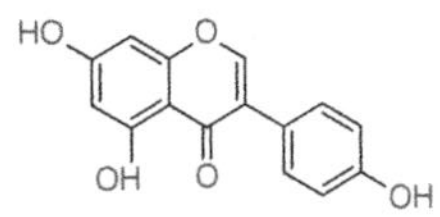	Obesity, metabolic syndrome, cancer, diabetes, atherosclerosis, rheumatoid arthritis	By suppressing the NF-B transcriptional mechanism, the production of these pro-inflammatory components is boosted. Protein levels influence arachidonic acid metabolism and NO generation, and pro-inflammatory enzymes such as PLA2, LOX, COX-2, and iNOS activity is suppressed.	[63, 64]
16.	Caffeic acid	Phenolic acids	Apples, artichoke, berries, olives	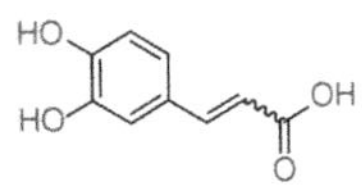	Inflammatory bowel disease, CVD, asthma, rheumatoid arthritis, atherosclerosis cancer	Chemokine, cytokine making is inhibited, lymphokine, T cells creation is proliferated, suppressing inflammation; NF-κB activation is inhibited	[65–67]
17.	Chlorogenic acid	Phenolic acids	Coffee	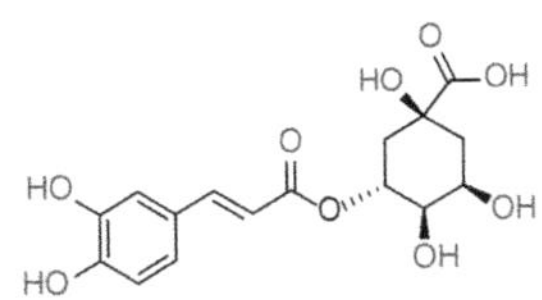	Diabetes mellitus, cancer	manufacture of mediators like IL-1β, IL-6, TNF-α is inhibited	[68, 69]
18.	Ferulic acid	Phenolic acids	Rice bran	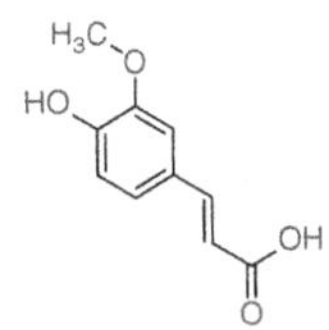	Diabetics, cardiovascular disorder, cancer, Cardiomyopathy, colon, breast, lung, and skin cancer, gouty arthritis	The inflammatory activity through inhibition of xanthine oxidase and COX-2 enzyme providing pro-inflammatory activity is reduced.	[70–72]

19.	Crocin	Carotenoids	Saffron Crocus sativus	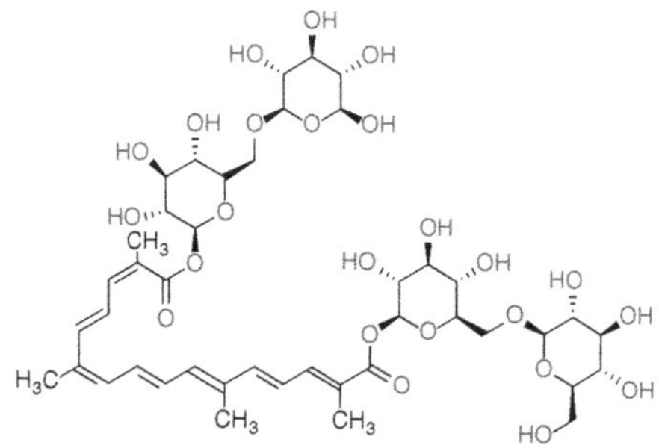	RA, Acute kidney injury	the elevated serum levels of inflammatory mediators, like enzymatic factors like MMP-3, MMP-13, MMP-9 and non-enzymatic factors COX-2, NF-κB, IL-6, IL-1β, TNF-α and ROS are efficiently neutralized methyl methanesulfonate-induced DNA damage, cell necrosis are decreased	[73–76]
20.	Lutein	Carotenoids	Broccoli, collards, spinach, lettuce, sweet potato	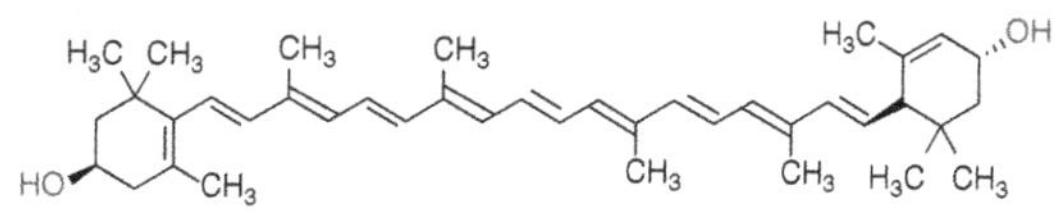	Obesity, atherosclerosis, diabetes, retinal neuronal disorder during inflammation	Pathological pathways of inflammatory cytokines, such as IL-6 and angiotensin II signaling, and neurodegeneration are avoided by suppressing oxidative stress through degradation of functional proteins and DNA damage induced by pro-inflammatory in inflammation or diabetes	[77–81]

(*Continued*)

Table 17.1 (Continued)

S. no.	Phytoanti-oxidant	Description	Source	Structure	Inflammatory disorder	Mechanism of action	References
21.	Lycopene	Carotenoids	Leafy green vegetables	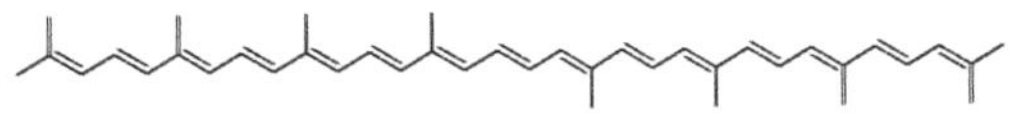	Cancer, atherosclerosis, cataracts, asthma, cardiovascular disease, atherosclerosis	Phosphorylation of p65 and IκB for deactivation of the NF-κB pathway is strongly reduced, expression of IGF-1R and intracellular ROS is decreased, stopping reduces progression of cardiovascular disorders	[82–85]

AP-1, Activator protein-1; c-Jun(JNK), c-Jun NH2-terminal kinase; COX-2, Cuclooxygenase; ERK, Activating extracellular signal-related kinase; IGF-1R, Insulin-like growth factor-1 receptor; iNOS, Inducible nitric oxide synthetase; LOX, Lipoxygenase; LT, Leukotrienes; NF-κB, Nuclear factor-κB; PG, Prostaglandins; PLA2, Phospholipase A2; PPARγ, Peroxisome proliferator-activated receptor.
Source: Based on Refs. [10–85].

17.3.3 Nanopolymersomes

Nanopolymersomes are artificial vesicles of 10 nanometers to 1 micrometer size enclosing aqueous cavities. These are formed through amphiphilic copolymers' self-assembly [97]. Nanopolymersomes have favorable properties allowing them to be personalized for various biomedical applications like artificial organelles or drug delivery vehicles [98]. The method of synthesis is similar to polymeric NPs' synthesis.

The properties of nanopolymersomes like drug encapsulation and functionality during release can be modulated by employing various biodegradable stimuli-responsive block copolymers [97]. Because of their rational properties, nanopolymersome can encapsulate hydrophobic and hydrophilic molecules in an aqueous nucleus or membrane bilayer. Contrary to nanolipid carriers, they have improved stability and mobility, as well as controlled emergence [99]. So, nanopolymersomes are appealing drug carriers in a variety of therapeutic applications.

17.3.4 Nanocrystals

Nanocrystals are 10–800 nm size submicrons' colloidal dispersion structures made up of pure NP drugs [100]. Their synthesis can be done mechanically or chemically.

The primary benefit of these nanosystems is particle size reduction to the nanoscale increasing particle surface area interacting through the dissolution medium [101]. As a result, nanocrystal formulations possess medicinal advantages over traditional pharmaceutical formulations. These advantages include high drug loading, increased dissolution rate, saturation solubility, and alkalinity [102].

17.3.5 Solid Lipid Nanoparticles

SLPs are analogous to nanoemulsions as they have lipids in a solid form. There are submicron colloidal nanocarriers with diameters ranging from 50 to 1000 nm and made up of physiological lipids distributed in water or aqueous surfactant solutions [103]. High-energy processes like microfluidization (high-pressure homogenization) and ultrasonication (high-frequency sound energy technique) are used for breaking particle agglomerates by cavitation, bubble expansion, and implosion during the manufacture of SLP [104]. Small size, high drug-loading efficiency, wide surface area, and phase interaction at the interface are all advantages of SL NPs [105]. With the benefits like selective drug delivery and strong release profile, this nanocarrier addresses the shortcomings of colloidal nanocarriers like liposomes, emulsions, and polymeric NPs. Loading of therapeutic agents into these NPs is done by inserting them into the central matrix or adding to the NP's surface [106].

The SLPs allow for the entrapment of lipophilic molecules in stable particles lacking organic solvents and incorporate both lipophilic and hydrophilic medications. Because of their unusual size-dependent properties, low toxicity, and high bioavailability of loaded drugs, these NPs extend various therapeutic benefits [107].

17.3.6 Polymeric Nanoparticles

Polymeric NPs made of synthetic or natural polymers are stable colloidal NPs from 10 to 1000 nm in size [108]. Because of their high electrochemical activity, poor electrical conductivity, short path lengths, and broad surface area, for ion transport, these nanostructures offer better energy storage than bulk conducting polymers [109]. Nanoprecipitation, salting out, liquid evaporation, supercritical fluid technology, and dialysis produce polymer NPs. They can also be synthesized directly by polymerizing monomers with interfacial polymerization, mini- and microemulsion, and controlled living polymerization technique. The surface of polymeric NPs is modified by self-assembly of block copolymers with segments forming the polymer NP center and surfactant shell outer [110].

Drug-loaded polymer NPs are synthesized with biocompatible and biodegradable polymers or copolymers, encapsulated or entrapped with therapeutic agents within the carriers, chemically bound to or mechanically adsorbed on the NP surfaces. Two types of polymeric NPs are found, i.e. uniformly filled agent nanospheres and only inner oily or aqueous cavities enclosed by polymeric membrane nanocapsules with drugs [111].

Water solubility, compact scale, long shelf life, storage flexibility, biodegradability, and non-toxicity are important properties of these nanostructures. Due to their physicochemical properties, polymeric NPs exhibit increased solubility, good encapsulation efficiency, stability of hydrophobic drugs with less toxicity, and controlled low dose releases at specified locations of the body [112].

17.3.7 Metallic Nanoparticles

Metallic NPs like titanium, gold, silver, magnesium, copper, aluminum, and zinc having diameters from 1 to 100 nm are extensively used for passive or active drug delivery in biomedical privileges. These metallic NPs are synthesized and their chemical functional groups are modified allowing conjugation with various drugs of interest for carrying them to specific targeted cells and tissues [113]. The benefits of therapy are fast biocompatibility, chemical alteration, and changeable biophysical properties [114].

17.3.8 Therapeutic Advantages of Nanodelivery Systems

Nanomedicines can pass chemical and physical blockades in the gut after oral administration, like selectively permeable enterocyte membranes, acidic pH, and intestinal mucosal lining of the stomach. The surface qualities of NPs are linked to their hydrophilicity or hydrophobicity, aid in cellular absorption, plasma protein interactions, and immunological responses [115]. Bioactive substances contained by NPs are released in specific tissues and bodily locations of the gastrointestinal tract or circulatory system where NPs release the medicines [91].

The nanostructures can target either actively or passively [116]. Metallic NPs like gold, silver, and iron oxide NPs aid in their role as drug carriers via active delivery with altered physical surface characteristics [117].

The effectiveness of transport of loaded substances diminishes with increasing molecular weight, resulting in reduced bioavailability. Natural compound-loaded NPs provide advantages such as increased epithelial permeability, increased stability, longer half-life, improved solubility and bioavailability, improved tissue targeting, and fewer side effects [106].

17.4 Nano-phytoantioxidants Extending Antioxidant Effects in Inflammatory Disorders

Because of limited oral bioavailability and poor in vivo biopharmaceutical performance, the application of bioactive phytochemicals is limited. To attain the optimum therapeutic effectiveness, bioactivity, and functionality of phytochemicals with antioxidant properties, they can be encapsulated in nanoformulation for improved targeting with modified absorption while still maintaining their antioxidant benefits [118]. Bioactive phytochemical therapeutic applications include lipid-based nanoformulations like SLNs, NLCs, liposomes, cubosomes, and so on, which increase the therapeutic potential of bioactive phytochemicals [119]. Nanotechnology-enabled biosensors in diagnostics give low-cost, early confirmation of illness kind and severity (Table 17.2). These NPs can deliver targeted drugs with low drug doses and little adverse effects [124]. Nano-phytoantioxidants have the potential to provide therapeutic phytochemicals to decrease inflammation, oxidative stress, and necrosis, hence prolonging antioxidant benefits in inflammatory illnesses [125].

17.5 Nano-phytoantioxidants and Their Therapeutic Importance in Inflammatory Disorders

Inflammation is the innate immune system's defensive reaction to infections or aggravates [126]. Uninhibited inflammation leads to inflammatory disorders like IBD, RA, asthma, pneumonia, and neurological and cardiovascular disorders. Excessive ROS makeup leads to precipitation of acute pathophysiology and chronic inflammation [127]. The uses of antioxidant nanomedicines in the treatment of various inflammatory illnesses are noteworthy.

17.5.1 Inflammatory Bowel Disease

Biocompatible quercetin nanovesicles targeting the colon were created by covering polyethylene glycol-containing vesicles with chitosan and nutriose. Characterization of chitosan/nutriose-coated vesicles was done through cryogenic microscopy, light scattering and X-ray scattering. This chitosan/nutriose protected quercetin during its transfer through the upper GIT and allowed release in colon. This delivery mechanism enabled tiny and numerous dose forms of liposomes to be distributed throughout the mucosa, increasing drug transit to the submucosa, and colonocyte absorption. The prebiotic nutriose was found to be advantageous to the injured colonic mucosa, promoting the repair of physiological states. With the synergistic impact of quercetin as an antioxidant and chitosan/nutriose complex as prebiotic, quercetin-loaded chitosan/nutriose-coated vesicles serve as expected therapy for chronic intestinal inflammatory illnesses [128].

By soaking resveratrol-loaded NPs in 4% CTS suspensions and then gelating by ionic cross-linking, Resveratrol-loaded CTS-NP formulations were effectively produced. The medicine was effectively encapsulated in biocompatible NPs and then incorporated into a chitosan gel matrix. TGA and SEM analysis of the two drug depots revealed porous scaffolds with prominent sameness in their thermal stability patterns. The chitosan network of constructed resveratrol depots significantly decreases the drug release rate, transforming the systems for IBD treatment with the therapeutic advantage of prolonged retention and delivery [129].

Curcumin-loaded solid binary lipid NPs (C-SBLNs) improve curcumin stability, absorption, and therapeutic potential in inflammatory bowel disease. ATR-FTIR physicochemical characterization of C-SBLNs established drug entrapment and thermal and pXRD studies exposed drug crystallinity loss in C-SBLNs. These spherical lyophilized C-SBLNs have high gastrointestinal stability and release time of 24 hours. Curcumin lipid nano-formulation dramatically increased curcumin absorption and in vivo localization in inflamed colon and lowered inflammatory and oxidative stress biomarkers with substantial reductions in oxidative stress, leucocyte infiltration, and release of TNF-α, a pro-inflammatory cytokine. Thus, CSBLNs may be a feasible alternative way for enhancing curcumin as a better option in IBD treatment [130].

Table 17.2 Details of nano-phytoantioxidants extending antioxidant effects in inflammatory disorders.

S. no.	Name of nanoparticle	Group of the polyphenol	Name of the polyphenol and source	Mechanism of antioxidant properties	Drawback of polyphenol	Advantages of nanoformulation	Mechanism of nanoformulation activity	Inflammatory disorders	References
1.	**Nano-Epigallocatechin-3-Gallate**	Catechin	Epigallocatechin-3-Gallate Green tea	Ability to capture free radicals directly	Low bioavailability, poor stability	Better stability and increased potency in oral delivery	Decreases intracellular ROS levels and lipid peroxidation	Cancer, heart diseases, degenerative diseases, diabetes, obesity	[46]
2.	**Nano-Genistein**	Isoflavone	Genistein (4′,5,7-trihydroxyisoflavone) soy	Expression of genes encoding antioxidant Enzymes like SOD and CAT is induced	Low bioavailability, endocrine-Disrupting toxic effects	Overcome side effects, enhanced bioavailability	Higher solubility potentiates its antioxidant effects via of endogenous CAT and SOD modulation	Obesity, type 2 diabetes, osteoporosis, neurodegenerative diseases, cancer	[120]
3.	**Nano-Curcumin**	Polyphenol	Curcumin Turmeric	Can chelate ions Transition metals to metal complexes	Low water solubility Limited bioavailability, gastrointestinal stability, systemic deletion	Better health-span-promoting properties, water solubility, bioavailability	Exert significant cellular antioxidant activity and lower lipid peroxidation	Neurodegenerative disorders, diabetes, COVID-19	[121]
4.	**Nano-Quercetin**	Flavonoid	Quercetin	Metal ion complexes scavenge free radicals	Low energy for chemical activation	Improved bioavailability	Improve antioxidant defence mechanisms	Hypercholesterolemic, antihypertensive, atherosclerotic, diabetic	[122]
5.	**Nano-Resveratrol**	Phytoalexin	Resveratrol grapes skin and seeds	Electron delocalization across its chemical structure	Extensive hepatic and presystemic metabolism, low water solubility causing poor absorption	Pronounced radical scavenging effect, higher antioxidant capability and reduced toxicity	Improved gastrointestinal safety, 6-fold higher bioavailability, endogenous antioxidant enzymes, like GSH, SOD levels were higher	Type 2 diabetes, cardiovascular disorders, dementia, cancers, acute lung injury (ALI), metabolic syndrome, obesity	[123]

Many plant-based bioactive nanodelivery technologies are effective in reducing oxidative stress and accompanying chronic inflammation, which promotes inflammatory diseases. The antioxidant properties of such nanodelivery devices are summarized.
Source: Based on Refs. [46, 120–123].

An amphiphilic curcumin polymer (PCur) was synthesized through a disulfide bond between hydrophobic curcumin (Cur) and hydrophilic poly(ethylene glycol) (PEG). PCur's nano size, solubility, and proximity to the neutral surface potential resulted in better accumulation of the active drug in the inflamed gut as well as restricted drug release and greater resilience under normal physiological gastrointestinal pH. PCur demonstrated minimal cytotoxicity and enhanced transmembrane permeability, leading to higher oral bioavailability. PCur, when taken orally, reduced inflammation of the colon and extended protection against IBD and can be a colon-specific entrant for treatment of IBD [131].

Several inflammatory disorders have therapeutic targets in the hypoxia-inducible factor and nuclear factor-kappa beta pathways. Caffeic acid phenethyl ester (CAPE) lowers p65 and HIF-1, but its therapeutic usage is restricted due to low absorption and a lack of knowledge of the molecular mechanism. Using an inflammation-targeted drug delivery method, CAPE's therapeutic anti-inflammatory capability to affect the hypoxia-inducible factor and nuclear factor-kappa beta pathways can be enhanced. CAPE's anti-inflammatory activity is enhanced by albumin nanoencapsulation, as is its capacity to effectively control inflammation. CAPE's greater therapeutic potential may make it a better alternative for inflammatory bowel diseases [132].

Nigella sativa extract was aerosolized and homogenized in alginate microcapsules for delivering large concentrations of thymoquinone to inflamed areas of the colon directly after oral intake. It was encapsulated by direct loading or diffusion into blank microparticles. Thus, oral alginate microcapsules are very efficient, targeted carriers of *Nigella sativa* extract for treating inflammatory bowel disease [133].

17.5.2 Cardiovascular Diseases

Resveratrol nano- and submicron-sized vesicles were incorporated into emulsions and liposomes using glycerol, soybean lecithin, coconut oil, and nonionic surfactants. The addition of resveratrol to various systems of emulsion-liposome blends including Brij 98 (F5) resulted in the sluggish drug release of resveratrol at zero order. Resveratrol, when used in blended formulations, significantly reduced intimal thickening of vasculature and endothelial damage in many systems, with little or minor liver and kidney effects. Thus, encapsulation through emulsion-liposome mixes is a powerful strategy to boost resveratrol's preventive and therapeutic advantages [134].

The lyophilization process was used to make polymeric superparamagnetic nano-silica (SiN@PLGA) with poly (lactic-co-glycolic acid) layered with quercetin in order to enhance the medication's weak watery solvency and stability in preventing atherosclerosis. The SEM and TEM examinations confirmed the production of the smooth, narrow size dispersion circular nanocomposite. The features of nanobiocarriers increase the practical similarity to the local myocardial state allowing cell attachment, expansion, and cardiac protein articulation. This form of the antioxidant quercetin can be used in the treatment of atherosclerosis and other similar cardiovascular illnesses [135].

Nanotherapeutics can be tried for common causes of cardiovascular mortality like myocardial infarction [139]. The cardioprotective impact of a formulated curcumin and nisin-based polylactic acid NP (CurNisNp) was seen in isoproterenol-induced myocardial infarction in guinea pigs. Atrial fibrillation was produced by myocardial infarction induction and was avoided by pretreatment with metoprolol or CurNisNp. Myocardial infarction induction is linked with elevated expressions of kidney injury molecule-1 (KIM-1) and cardiac troponin I (CTnI) in guinea pigs. The developed curcumin-nisin-based NP is nontoxic and provides considerable cardioprotection in the guinea pig [136].

17.5.3 Obesity

Encapsulation into NPs improves the stability and solubility of phytochemicals (epigallocatechin gallate, trans-resveratrol, curcumin, quercetin), boosts their bioavailability, shields them from an early breakdown in the body, and therefore increases its antiobesity efficacy. Also, there are issues and limits of primary phytochemicals, the primary biocompatible and biodegradable NPs and the efficacy of nanoencapsulated versions of phytochemicals in fighting obesity and its comorbidities [137].

In comparison to free EGCG, EGCG NPs dramatically lowered blood concentrations of triglycerides, low-density lipoprotein (LDL)-cholesterol, total cholesterol as well as the formation of atherosclerotic lesions. The nano-encapsulation of EGCG into polymeric NPs improves anti-atherosclerosis and anti-tumor bioactivity [138].

17.5.4 Neurodegenerative Conditions

17.5.4.1 Curcumin

Curcumin-loaded SLNs demonstrate enhanced bioavailability and dosage reductions of 32–155 times, as well as an increase in curcumin's effectiveness as probable therapy for neurodegenerative disorders (Kakkar et al. 2011b). Curcumin-loaded SLNs improve neuroprotection in aluminum-induced neurodegenerative illness (Kakkar et al. 2013). In a transgenic mouse model of Alzheimer's disease, curcumin NPs at a lower dosage significantly improved working and cue memory (Cheng et al. 2013). Curcumin-loaded solid NPs reduced neurobehavioral and neurochemical impairments in the experimental Huntington's disease model [139].

The anti-transferrin antibody can coat the curcumin-derivative liposomes, which greatly increased their absorption through the blood-brain barier (Mourtas et al. 2014). In amelioration through curcumin-conjugated magnetic NPs made of superparamagnetic iron oxide, MRI can identify the amyloid plaques in Alzheimer's disease [140].

Curcumin-encapsulated PLGA NPs trigger adult neurogenesis and repair cognitive impairments of Alzheimer's disease model through the canonical Wnt/-catenin pathway (Tiwari et al. 2014). N-trimethyl chitosan surface-modified SLNs are more persistent and are unaffected by acid-induced burst release and thus can be used with curcumin for its oral administration [141].

17.5.4.2 Quercetin

The new nanomaterials with phytochemicals extend protection against quercetin degradation, and contribute to improved pharmacokinetic optimization, therapeutic activity, and biodistribution control and decreased cytotoxicity owing to slower but effective release.

In a rat model, nanocapsulated quercetin protects against arsenic-induced ROS-mediated oxidative damage in liver and brain cells [142]. Nanoparticulated quercetin has been shown to be effective in preventing age-related cerebral oxidative damage, with quercetin NPs increasing permeability across the BBB and decreasing oxysterol-induced inflammation in the brain [143]. Encapsulated quercetin in modified silica NPs protects against Cu (II)-induced neurotoxicity with oxidative stress [144].

17.5.4.3 EGCG

Nanolipidic particles improved the bioavailability of EGCG twice and upregulated the α-secretase activity helping in therapeutic measures of neurodegenerative disorders like Alzheimer's disease and HIV-associated dementia [145]. EGCG-stabilized selenium NPs coated with Tet-1 peptide reduced cytotoxicity and amyloid-β aggregation [146].

17.5.4.4 Resveratrol

In rats, a lipid-core nanocapsule formulation containing resveratrol stabilized the medication extending neuroprotective activity against amyloid neurotoxicity, conserving its biotic activity and enhancing bioavailability. Resveratrol-loaded lipid-core nanocapsules reduced Aβ-induced neuroinflammation [147].

Resveratrol-loaded lipid-core nanocapsules inhibited C6 glioma cell viability more than resveratrol alone. In vitro (C6 glioma cell line) and in vivo (brain-implanted C6 cells) models, resveratrol-loaded lipid-core nanocapsules demonstrate anti-glioma action [148]. The nanoencapsulation of resveratrol increases its anti-glioma action, laying the groundwork for clinical trials of resveratrol nanoformulations as a novel chemotherapeutic method for glioma's treatment.

17.5.4.5 Gallic Acid

Nanoencapsulation dramatically enhances the half-life of gallic acid [149]. Gallic acid chitosan NPs coated in tween 80 show cerebroprotective action against scopolamine-induced amnesia owing to its oxidant effects and enhanced cholinergic activities. The increase in this effect after delivery of Tween 80-coated NPs attributed to higher gallic acid absorption. Gallic acid-loaded chitosan NPs also exhibited antidepressive action in mice, mediated by reduced catalase activity, monoamine oxidase-A activity and malondialdehyde levels [150].

17.5.4.6 Sesamol

Sesamol-loaded solid nanoparticles alleviated menopause-related affective and cognitive CNS derangements in a surgical menopause ovariectomy paradigm in rats [151]. Sesamol-loaded solid nanoparticles recover cognitive impairments in rats treated with intracerebroventricular streptozotocin owing to reduced cytokine production and nitrosative stress [152]. Sesamol loading in solid nanoparticles has the potential to be a brain-targeting method for Alzheimer's disease and can be effective in treating neurodegenerative disorders

17.5.5 Gout

Zinc oxide NPs are used to treat gouty arthritis and lower oxidative stress. When hyperuricemia and gouty arthritis mice were administered ZnO-NPs, serum uric acid levels were decreased. The lipid profile tests showed that the NPs therapy was extremely efficient in lowering cholesterol and LDL levels. Additionally, blood counts and LFTs demonstrated the effect of zinc oxide in the decrease of hyperuricemia. ZnO-NPs are effective in depressing cholesterol, blood uric acid, and LDL after oral treatment. ZnO NPs at 10 and 20 ppm concentrations lowered blood uric acid concentrations in the treatment of gouty arthritis [153].

17.5.6 Metabolic Syndrome

Encapsulation in nanoliposomes, nanoemulsions, nanosuspensions, and solid lipid nanoformulations is done for boosting solubility and bioavailability, increasing plasma half-life of phytochemicals like quercetin, curcumin, naringenin, emodin, berberine, scutellarin, gymnemic acid, and resveratrol. They are nanosized and their nanosizing method boosts the efficacy of these natural chemicals in the treatment of metabolic syndrome symptoms like insulin resistance, impaired glucose tolerance, and hyperinsulinemia and alleviates the metabolic syndrome. These can be future candidates for metabolic syndromes with or without pharmaceutical drugs if they pass clinical investigations and are cost-effective [118].

17.5.7 Rheumatoid Arthritis

Nanogold (AuNGs) reduce inflammation and cartilage degradation by preventing the activation of inflammatory intermediaries including NF-kB. AuNGs significantly reduces the disparity in antioxidant and oxidant levels to a normal level nearly. The antiarthritic effect of nanogold in the collagen-induced arthritic (CIA) rat model is demonstrated through the reduction in inflammatory intermediaries like TNF-α, IL-1, COX-2, and NF-kB (nuclear factor-kB) like transcription factor as well as oxidative stress, which was shown to be higher in CIA rats. AuNGs therapy significantly reduced the upregulated mRNA expression of COX-2, NF-kB, iNOS, and TNF-α [154].

This is demonstrated by histology findings of joints, which showed lower inflammation and bone degradation in the nanogold group. This study shows that AuNGs can reduce the generation of inflammatory intermediaries and oxidative stress in CIA rats. AuNGs has shown promising anti-inflammatory and antiarthritic action and has the potential to control various chronic inflammatory disorders including arthritis [155].

The disadvantages of quercetin administration, such as restricted topical application due to low skin permeability, can be resolved by QCT-NE gel, a quercetin (QCT)-loaded nanoemulsion. It demonstrated sufficient rheological behavior, a nice textural profile, and increased drug permeability. QCT-NE shows no toxicity on synoviocytes and a substantial inhibitory impact on TNF generation generated by LPS [156]. A 15:6:6 mixture of oleic acid, Arachis oil, tween 20, and PEG-400 resulted in the spontaneous production of a nanoemulsion and an effective carrier system for quercetin. The modified formulation demonstrated improved physicochemical stability, acceptable mechanical characteristics, and increased skin permeability, resulting in increased skin permeability. The improved therapeutic efficacy demonstrated by its antiarthritic activity makes quercetin-nanoemulsion a potential topical treatment option for rheumatic problems [157].

17.5.8 COPD

Inhaled pharmacologic therapy can be used to treat COPD patients, with benefits such as low drug concentrations for therapeutic effectiveness, bypassing first-pass metabolism, and less adverse effects. However, due to their low inertia, NPs are not the best size for inhalation. As a result, NPs are to be encapsulated within microparticles. The solid matrix can prevent NPs from intermingling, bound their mobility, and promote their long-term stability. This also increases aerosolization qualities and pulmonary administration for a more effective way of medication delivery in COPD patients [158].

Sulforaphane (antioxidant drug)-loaded silk fibroin NPs (SFSNPs) are a multifunctional silk fibroin-based carrier for the transport of antioxidant and imaging mediators that are manufactured in a single process. Self-assembled CeNP-CD@SFSNPs nanocomposites are formed when anionic SFSNPs are combined with CeNPs and PEI-passivated carbon dots (CDs). The nanocomposites CeNP-CD@ SFSNPs destroy toxic ROS like H_2O_2. Sulforaphane activates the Nrf2 pathway, which removes ROS. The mixed valency of cerium oxide NPs (CeNPs) increased antioxidant capacity via mimicking SOD and catalase. This multifunctional CeNP-CD@ SFSNPs can be employed for imaging and treatment of pulmonary illnesses like COPD and other oxidative stress-related illnesses [159].

17.5.9 COVID-19

The nanostructure method against the beta-coronavirus SARS-CoV swiftly utilizes polyphenols such as (-)-gallocatechin gallate and (-)-catechin gallate derived from *Camellia sinensis*, which have a greater inhibitory impact on SARS-CoV N protein. Green tea catechins' antiviral action against other enveloped viruses was credited mostly to their galloyl, pyrogallol, and hydroxyl groups, phenolic OH groups acting at various phases of viral entry, influencing viral antigen production, or preventing genome replication [160].

A lignan from *Cleistanthuscollinus* diphyllin encased in biocompatible block copolymer nanocapsules of PEG-PLGA-inhibited FCoV in Fcwf-4 cells. Diphyllin decreased endosomal acidification, influencing viral cellular vulnerability and inhibiting coronavirus replication downstream [161].

Artemisinin, Artemether, Artesunate, the natural sesquiterpene lactones extracted from *Artemisia annua* coated with antiviral AgNPs, enhanced the permeability and temporal retention of these medications, hence improving their therapeutic activity against COVID-19. Owing to the greater negative charges on the O6 atom of Artesunate, O5 atom of Artemisinin, and O3 atom of Artemether, density functional theory (DFT) studies indicated the maximum affinity of medicines to interact with AgNPs surfaces in the sequence Artesunate > Artemisinin > Artemether [162] (Akbari et al. 2020).

17.6 Future Strategies

Many studies have been conducted for phytoantioxidant nanotherapeutics, still some critical trials should be done for clinical use of these anti-inflammatory nanotherapeutics. The combination processes to produce nanotherapeutics should have upright quality control and batch-to-batch reproducibility. Mass production should be cost-effective benefiting clinical practice. Toxicities (acute and chronic) studies of phytoantioxidant nanotherapeutics for ROS-scavenging activities must be done. The developed phytoantioxidant nanotherapeutics should be biocompatible and easy to excrete [6]. As per varying ROS levels for different inflammatory disorders in the different stages, the dose and dosing frequency of phytoantioxidant nanotherapeutics should be appropriately controlling the intracellular ROS for therapeutic benefits.

17.7 Conclusion

Since oxidative stress is the root of all inflammatory disorders, nanotherapeutics of phytoantioxidant can be an upcoming novel therapeutic strategy. Noteworthy advances are seen in phytoantioxidant nanotherapeutics for different inflammatory diseases using phytochemical-loaded NPs extending the antioxidant activity. The desirable results of ROS-scavenging phytoantioxidant nanotherapeutics in different inflammatory models make them ideal for clinical application. These ideal antioxidative NPs would augment the health of patients and lessen side effects. The thorough hazards and benefits assessment of phytoantioxidant nanotherapeutics will result in its improved, safer, and successful integration into clinical practices for the treatment of inflammatory disorders.

References

1 Vrana, K.E., Karpa, K.D., and Kester, M. (2012). *Elsevier's Integrated Review Pharmacology*. Elsevier Health Sciences.

2 Prideaux, L., Kamm, M.A., De Cruz, P.P. et al. (2012). Inflammatory bowel disease in Asia: a systematic review. *Journal of Gastroenterology and Hepatology* 27 (8): 1266–1280.

3 Phull, A.R., Nasir, B., ul Haq, I., and Kim, S.J. (2018). Oxidative stress, consequences and ROS mediated cellular signaling in rheumatoid arthritis. *Chemico-Biological Interactions* 281: 121–136.
4 Howes, M.J. (2018). Phytochemicals as anti-inflammatory nutraceuticals and phytopharmaceuticals. In: *InImmunity and Inflammation in Health and Disease* (ed. S. Chatterjee, W. Jungraithmayr and D. Bagchi), 363–388. Academic Press.
5 Vysakh, A., Jayesh, K., Helen, L.R. et al. (2020). Acute oral toxicity and anti-inflammatory evaluation of methanolic extract of Rotula aquatica roots in Wistar rats. *Journal of Ayurveda and Integrative Medicine* 11 (1): 45–52.
6 Li, C.W., Li, L.L., Chen, S. et al. (2020). Antioxidant nanotherapies for the treatment of inflammatory diseases. *Frontiers in Bioengineering and Biotechnology* 8: 200. https://doi.org/10.3389/fbioe.2020.00200.
7 Hussain, T., Tan, B., Yin, Y. et al. (2016). Oxidative stress and inflammation: what polyphenols can do for us? *Oxidative Medicine and Cellular Longevity* 2016: 7432797. https://doi.org/10.1155/2016/7432797.
8 Srivastava, A.K., Chaurasia, J.P., Khan, R. et al. (2020). Role of medicinal plants of traditional use in recuperating devastating COVID-19 situation. *Med Aromat Plants (Los Angeles)* 9 (359): 2167–0412.
9 Forni, C., Facchiano, F., Bartoli, M. et al. (2019). Beneficial role of phytochemicals on oxidative stress and age-related diseases. *BioMed Research International* 2019: 1–16. https://doi.org/10.1155/2019/8748253.
10 Li, J., Xie, C., Zhuang, J. et al. (2015). Resveratrol attenuates inflammation in the rat heart subjected to ischemia-reperfusion: role of the TLR4/NF-κBsignaling pathway. *Molecular Medicine Reports* 11 (2): 1120–1126.
11 Liu, X.J., Bao, H.R., Zeng, X.L., and Wei, J.M. (2016). Effects of resveratrol and genistein on nuclear factor-κB, tumor necrosis factor-α and matrix metalloproteinase-9 in patients with chronic obstructive pulmonary disease. *Molecular Medicine Reports* 13 (5): 4266–4272.
12 Chang, G.R., Chen, P.L., Hou, P.H., and Mao, F.C. (2015). Resveratrol protects against diet-induced atherosclerosis by reducing low-density lipoprotein cholesterol and inhibiting inflammation in apolipoprotein E-deficient mice. *Iranian Journal of Basic Medical Sciences* 18 (11): 1063.
13 Bi, X.L., Yang, J.Y., Dong, Y.X. et al. (2005). Resveratrol inhibits nitric oxide and TNF-α production by lipopolysaccharide-activated microglia. *International Immunopharmacology* 5 (1): 185–193.
14 Wang, Q., Xu, J., Rottinghaus, G.E. et al. (2002). Resveratrol protects against global cerebral ischemic injury in gerbils. *Brain Research.* 958 (2): 439–447.
15 Sun, C., Hu, Y., Liu, X. et al. (2006). Resveratrol downregulates the constitutional activation of nuclear factor-κB in multiple myeloma cells, leading to suppression of proliferation and invasion, arrest of cell cycle, and induction of apoptosis. *Cancer Genetics and Cytogenetics* 165 (1): 9–19.
16 de Sá, C.D., Pacheco, M.T., Frozza, R.L., and Bernardi, A. (2018). Anti-inflammatory effects of resveratrol: Mechanistic insights. *International Journal of Molecular Sciences* 19 (6): 1812.
17 Baliga, M.S., Joseph, N., Venkataranganna, M.V. et al. (2012). Curcumin, an active component of turmeric in the prevention and treatment of ulcerative colitis: preclinical and clinical observations. *Food & Function* 3 (11): 1109–1117.
18 Gulcubuk, A., Haktanir, D., Cakiris, A. et al. (2013). Effects of curcumin on proinflammatory cytokines and tissue injury in the early and late phases of experimental acute pancreatitis. *Pancreatology* 13 (4): 347–354.
19 Monroy, A., Lithgow, G.J., and Alavez, S. (2013). Curcumin and neurodegenerative diseases. *Biofactors* 39 (1): 122–132.
20 Wongcharoen, W. and Phrommintikul, A. (2009). The protective role of curcumin in cardiovascular diseases. *International Journal of Cardiology* 133 (2): 145–151.

21 Kurup, V.P. and Barrios, C.S. (2008). Immunomodulatory effects of curcumin in allergy. *Molecular Nutrition & Food Research* 52 (9): 1031–1039.

22 Ramadan, G. and El-Menshawy, O. (2013). Protective effects of ginger-turmeric rhizomes mixture on joint inflammation, atherogenesis, kidney dysfunction and other complications in a rat model of human rheumatoid arthritis. *International Journal of Rheumatic Diseases* 16 (2): 219–229.

23 Soetikno, V., Sari, F.R., Lakshmanan, A.P. et al. (2013). Curcumin alleviates oxidative stress, inflammation, and renal fibrosis in remnant kidney through the N rf2–keap1 pathway. *Molecular Nutrition & Food Research* 57 (9): 1649–1659.

24 Maradana, M.R., Thomas, R., and O'Sullivan, B.J. (2013). Targeted delivery of curcumin for treating type 2 diabetes. *Molecular Nutrition & Food Research* 57 (9): 1550–1556.

25 Aggarwal, B.B. (2010). Targeting inflammation-induced obesity and metabolic diseases by curcumin and other nutraceuticals. *Annual Review of Nutrition* 30: 173–199.

26 Aggarwal, B.B. and Harikumar, K.B. (2009). Potential therapeutic effects of curcumin, the anti-inflammatory agent, against neurodegenerative, cardiovascular, pulmonary, metabolic, autoimmune and neoplastic diseases. *The International Journal of Biochemistry & Cell Biology* 41 (1): 40–59.

27 Sun, J., Zhao, Y., and Hu, J. (2013). Curcumin inhibits imiquimod-induced psoriasis-like inflammation by inhibiting IL-1beta and IL-6 production in mice. *PloS One* 8 (6): e67078.

28 Shehzad, A., Lee, J., and Lee, Y.S. (2013). Curcumin in various cancers. *Biofactors* 39 (1): 56–68.

29 He, Y., Yue, Y., Zheng, X. et al. (2015). Curcumin, inflammation, and chronic diseases: how are they linked? *Molecules* 20 (5): 9183–9213.

30 Sharifi-Rad, J., Rayess, Y.E., Rizk, A.A. et al. (2020). Turmeric and its major compound curcumin on health: bioactive effects and safety profiles for food, pharmaceutical, biotechnological and medicinal applications. *Frontiers in Pharmacology* 11: 1021. https://doi.org/10.3389/fphar.2020.01021.

31 Lin, Y., Shi, R., Wang, X., and Shen, H.M. (2008). Luteolin, a flavonoid with potential for cancer prevention and therapy. *Current Cancer Drug Targets* 8 (7): 634–646.

32 Zhu, G.F., Guo, H.J., Huang, Y. et al. (2015). Eriodictyol, a plant flavonoid, attenuates LPS-induced acute lung injury through its antioxidative and anti-inflammatory activity. *Experimental and Therapeutic Medicine* 10 (6): 2259–2266.

33 Liu, Y. and Yan, X. (2019). Eriodictyol inhibits survival and inflammatory responses and promotes apoptosis in rheumatoid arthritis fibroblast-like synoviocytes through AKT/FOXO1 signaling. *Journal of Cellular Biochemistry* 120 (9): 14628–14635.

34 Yang, T., Li, X., Yu, J. et al. (2020). Eriodictyol suppresses Th17 differentiation and the pathogenesis of experimental autoimmune encephalomyelitis. *Food & Function* 11 (8): 6875–6888.

35 Khan, A., Ikram, M., Hahm, J.R., and Kim, M.O. (2020). Antioxidant and anti-inflammatory effects of citrus flavonoid hesperetin: special focus on neurological disorders. *Antioxidants* 9 (7): 609.

36 Li, Y.R., Chen, D.Y., Chu, C.L. et al. (2015). Naringenin inhibits dendritic cell maturation and has therapeutic effects in a murine model of collagen-induced arthritis. *The Journal of Nutritional Biochemistry* 26 (12): 1467–1478.

37 Chung, T.W., Li, S., Lin, C.C., and Tsai, S.W. (2019). Antinociceptive and anti-inflammatory effects of the citrus flavanone naringenin. *Tzu-Chi Medical Journal* 31 (2): 81.

38 Fei-Yan, F., Li-Xuan, S., and Jiang, M. (2017). Catechins and their therapeutic benefits to inflammatory bowel disease. *Molecules* 22 (3): 484.

39 Reygaert, W.C. (2018). Green tea catechins: Their use in treating and preventing infectious diseases. *BioMed Research International* 2018: 9105261. https://doi.org/10.1155/2018/9105261.

40 Menegazzi, M., Campagnari, R., Bertoldi, M. et al. (2020). Protective effect of epigallocatechin-3-gallate (EGCG) in diseases with uncontrolled immune activation: Could such a scenario Be helpful to counteract COVID-19? *International Journal of Molecular Sciences* 21 (14): 5171.

41 Shankar, S., Ganapathy, S., and Srivastava, R.K. (2007). Green tea polyphenols: biology and therapeutic implications in cancer. *Frontiers in Bioscience* 12: 4881–4899.

42 Lee, S.K., Kim, J.H., Kim, J.S. et al. (2012). Polyphenol (-)-epigallocatechin gallate-induced cardioprotection may attenuate ischemia-reperfusion injury through adenosine receptor activation: a preliminary study. *Korean Journal of Anesthesiology* 63 (4): 340.

43 Weinreb, O., Mandel, S., Amit, T., and Youdim, M.B. (2004). Neurological mechanisms of green tea polyphenols in Alzheimer's and Parkinson's diseases. *The Journal of Nutritional Biochemistry* 15 (9): 506–516.

44 Singh, R., Ahmed, S., Islam, N. et al. (2002). Epigallocatechin-3-gallate inhibits interleukin-1β–induced expression of nitric oxide synthase and production of nitric oxide in human chondrocytes: suppression of nuclear factor κB activation by degradation of the inhibitor of nuclear factor κB. *Arthritis & Rheumatism* 46 (8): 2079–2086.

45 Legeay, S., Rodier, M., Fillon, L. et al. (2015). Epigallocatechin gallate: a review of its beneficial properties to prevent metabolic syndrome. *Nutrients* 7 (7): 5443–5468.

46 Granja, A., Frias, I., Neves, A.R. et al. (2017). Therapeutic potential of epigallocatechin gallate nanodelivery systems. *BioMed Research International* 2017: 5813793. https://doi.org/10.1155/2017/5813793.

47 Zhu, J., Tang, H., Zhang, Z. et al. (2017). Kaempferol slows intervertebral disc degeneration by modifying LPS-induced osteogenesis/adipogenesis imbalance and inflammation response in BMSCs. *International Immunopharmacology* 43: 236–242.

48 Liu-Bryan, R. and Terkeltaub, R. (2015). Emerging regulators of the inflammatory process in osteoarthritis. *Nature Reviews Rheumatology* 11 (1): 35.

49 Suchal, K., Malik, S., Gamad, N. et al. (2016). Kaempferol attenuates myocardial ischemic injury via inhibition of MAPK signaling pathway in experimental model of myocardial ischemia-reperfusion injury. *Oxidative Medicine and Cellular Longevity* 2016: 7580731. https://doi.org/10.1155/2016/7580731.

50 Lei, Y., Chen, J., Zhang, W. et al. (2012). In vivo investigation on the potential of galangin, kaempferol and myricetin for protection of D-galactose-induced cognitive impairment. *Food Chemistry* 135 (4): 2702–2707.

51 Ren, J., Lu, Y., Qian, Y. et al. (2019). Recent progress regarding kaempferol for the treatment of various diseases. *Experimental and Therapeutic Medicine* 18 (4): 2759–2776.

52 Qu, Y., Wang, C., Liu, N. et al. (2018). Morin exhibits anti-inflammatory effects on IL-1β-stimulated human osteoarthritis chondrocytes by activating the Nrf2 signaling pathway. *Cellular Physiology and Biochemistry* 51 (4): 1830–1838.

53 Lee, K.M., Lee, Y., Chun, H.J. et al. (2016). Neuroprotective and anti-inflammatory effects of morin in a murine model of Parkinson's disease. *Journal of Neuroscience Research* 94 (10): 865–878.

54 Sinha, K., Ghosh, J., and Sil, P.C. (2016). Morin and its role in chronic diseases. *Anti-inflammatory Nutraceuticals and Chronic Diseases* 928: 453–471.

55 Bureau, G., Longpré, F., and Martinoli, M.G. (2008). Resveratrol and quercetin, two natural polyphenols, reduce apoptotic neuronal cell death induced by neuroinflammation. *Journal of Neuroscience Research* 86 (2): 403–410.

56 Li, Y., Yao, J., Han, C. et al. (2016). Quercetin inflammation and immunity. *Nutrients* 8 (3): 167.

57 Li, S., Wu, B., Fu, W., and Reddivari, L. (2019). The anti-inflammatory effects of dietary anthocyanins against ulcerative colitis. *International Journal of Molecular Sciences* 20 (10): 2588.

58 Lee, Y.M., Yoon, Y., Yoon, H. et al. (2017). Dietary anthocyanins against obesity and inflammation. *Nutrients* 9 (10): 1089.

59 Kao, E.S., Tseng, T.H., Lee, H.J. et al. (2009). Anthocyanin extracted from Hibiscus attenuate oxidized LDL-mediated foam cell formation involving regulation of CD36 gene. *Chemico-Biological Interactions* 179 (2–3): 212–218.

60 Sancho, R.A. and Pastore, G.M. (2012). Evaluation of the effects of anthocyanins in type 2 diabetes. *Food Research International* 46 (1): 378–386.

61 Wu, Z.Y., Sang, L.X., and Chang, B. (2020). Isoflavones and inflammatory bowel disease. *World Journal of Clinical Cases* 8 (11): 2081.

62 Sakamoto, Y., Kanatsu, J., Toh, M. et al. (2016). The dietary isoflavone daidzein reduces expression of pro-inflammatory genes through PPARα/γ and JNK pathways in adipocyte and macrophage co-cultures. *PLoS One* 11 (2): e0149676.

63 Kao, T.H., Wu, W.M., Hung, C.F. et al. (2007). Anti-inflammatory effects of isoflavone powder produced from soybean cake. *Journal of Agricultural and Food Chemistry* 55 (26): 11068–11079.

64 Nagaraju, G.P., Zafar, S.F., and El-Rayes, B.F. (2013). Pleiotropic effects of genistein in metabolic, inflammatory, and malignant diseases. *Nutrition Reviews* 71 (8): 562–572.

65 Yamamoto, Y. and Gaynor, R.B. (2001). Therapeutic potential of inhibition of the NF-κB pathway in the treatment of inflammation and cancer. *The Journal of Clinical Investigation* 107 (2): 135–142.

66 Armutcu, F., Akyol, S., Ustunsoy, S., and Turan, F.F. (2015). Therapeutic potential of caffeic acid phenethyl ester and its anti-inflammatory and immunomodulatory effects. *Experimental and Therapeutic Medicine* 9 (5): 1582–1588.

67 Chao, P.C., Hsu, C.C., and Yin, M.C. (2009). Anti-inflammatory and anti-coagulatory activities of caffeic acid and ellagic acid in cardiac tissue of diabetic mice. *Nutrition & Metabolism* 6 (1): 1–8.

68 Tajik, N., Tajik, M., Mack, I., and Enck, P. (2017). The potential effects of chlorogenic acid, the main phenolic components in coffee, on health: a comprehensive review of the literature. *European Journal of Nutrition* 56 (7): 2215–2244.

69 Lu, H., Tian, Z., Cui, Y. et al. (2020). Chlorogenic acid: a comprehensive review of the dietary sources, processing effects, bioavailability, beneficial properties, mechanisms of action, and future directions. *Comprehensive Reviews in Food Science and Food Safety* 19 (6): 3130–3158.

70 Nile, S.H., Ko, E.Y., Kim, D.H., and Keum, Y.S. (2016). Screening of ferulic acid related compounds as inhibitors of xanthine oxidase and cyclooxygenase-2 with anti-inflammatory activity. *Revista Brasileira de Farmacognosia* 26 (1): 50–55.

71 Ghosh, S., Basak, P., Dutta, S. et al. (2017). New insights into the ameliorative effects of ferulic acid in pathophysiological conditions. *Food and Chemical Toxicology* 103: 41–55.

72 Srinivasan, M., Sudheer, A.R., and Menon, V.P. (2007). Ferulic acid: therapeutic potential through its antioxidant property. *Journal of Clinical Biochemistry and Nutrition* 40 (2): 92–100.

73 Hemshekhar, M., Santhosh, M.S., Sunitha, K. et al. (2012). A dietary colorant crocin mitigates arthritis and associated secondary complications by modulating cartilage deteriorating enzymes, inflammatory mediators and antioxidant status. *Biochimie* 94 (12): 2723–2733.

74 Hosseinzadeh, H., Abootorabi, A., and Sadeghnia, H.R. (2008). Protective effect of Crocus sativus stigma extract and crocin (trans-crocin 4) on methyl methanesulfonate–induced DNA damage in mice organs. *DNA and Cell Biology* 27 (12): 657–664.

75 Yarijani, Z.M., Pourmotabbed, A., Pourmotabbed, T., and Najafi, H. (2017). Crocin has anti-inflammatory and protective effects in ischemia-reperfusion induced renal injuries. *Iranian Journal of Basic Medical Sciences* 20 (7): 753.

76 Liu, W., Sun, Y., Cheng, Z. et al. (2018). Crocin exerts anti-inflammatory and anti-arthritic effects on type II collagen-induced arthritis in rats. *Pharmaceutical Biology* 56 (1): 209–216.

77 Sluijs, I., Beulens, J.W., Grobbee, D.E., and Van Der Schouw, Y.T. (2009). Dietary carotenoid intake is associated with lower prevalence of metabolic syndrome in middle-aged and elderly men. *The Journal of Nutrition* 139 (5): 987–992.

78 Zou, Z.Y., Xu, X.R., Lin, X.M. et al. (2014). Effects of lutein and lycopene on carotid intima–media thickness in Chinese subjects with subclinical atherosclerosis: a randomised, double-blind, placebo-controlled trial. *British Journal of Nutrition* 111 (3): 474–480.

79 Sam, S., Haffner, S., Davidson, M.H. et al. (2009). Relation of abdominal fat depots to systemic markers of inflammation in type 2 diabetes. *Diabetes Care* 32 (5): 932–937.

80 Zhou, J., Zhao, D., Wang, N. et al. (2020). Effects of lutein supplementation on inflammatory biomarkers and metabolic risk factors in adults with central obesity: study protocol for a randomised controlled study. *Trials* 21 (1): 1–9.

81 Ozawa, Y., Sasaki, M., Takahashi, N. et al. (2012). Neuroprotective effects of lutein in the retina. *Current Pharmaceutical Design* 18 (1): 51–56.

82 Fenni, S., Hammou, H., Astier, J. et al. (2017). Lycopene and tomato powder supplementation similarly inhibit high-fat diet induced obesity, inflammatory response, and associated metabolic disorders. *Molecular Nutrition & Food Research* 61 (9): 1601083.

83 Cheng, H.M., Koutsidis, G., Lodge, J.K. et al. (2017). Tomato and lycopene supplementation and cardiovascular risk factors: a systematic review and meta-analysis. *Atherosclerosis* 257: 100–108.

84 Hong, M.Y., Hartig, N., Kaufman, K. et al. (2015). Watermelon consumption improves inflammation and antioxidant capacity in rats fed an atherogenic diet. *Nutrition Research* 35 (3): 251–258.

85 Mozos, I., Stoian, D., Caraba, A. et al. (2018). Lycopene and vascular health. *Frontiers in Pharmacology* 9: 521.

86 Zhang, M., Gao, S., Yang, D. et al. (2021). Influencing factors and strategies of enhancing nanoparticles into tumors in vivo. *Acta PharmaceuticaSinica B* 11 (8): 2265–2285.

87 Anselmo, A.C. and Mitragotri, S. (2019). Nanoparticles in the clinic: an update. *Bioengineering & Translational Medicine* 4 (3): e10143.

88 Kermanizadeh, A., Powell, L.G., Stone, V., and Møller, P. (2018). Nanodelivery systems and stabilized solid-drug nanoparticles for orally administered medicine: current landscape. *International Journal of Nanomedicine* 13: 7575.

89 Ravindran, S., Suthar, J.K., Rokade, R. et al. (2018). Pharmacokinetics, metabolism, distribution and permeability of nanomedicine. *Current Drug Metabolism* 19 (4): 327–334.

90 Beltrán-GraciaE, L.-C.A., Higuera-Ciapara, I., Velázquez-Fernández, J.B., and Vallejo-Cardona, A.A. (2019). Nanomedicine review: clinical developments in liposomal applications. *Cancer Nanotechnology* 10 (1): 1–40.

91 Patra, J.K., Das, G., Fraceto, L.F. et al. (2018). Nano based drug delivery systems: recent developments and future prospects. *Journal of Nanobiotechnology* 16 (1): 1–33.

92 Che Marzuki, N.H., Wahab, R.A., and Abdul, H.M. (2019). An overview of nanoemulsion: concepts of development and cosmeceutical applications. *Biotechnology & Biotechnological Equipment* 33 (1): 779–797.

93 Gupta, P.K., Bhandari, N., Shah, H. et al. (2019). An update on nanoemulsions using nanosized liquid in liquid colloidal systems. In: *Nanoemulsions-Properties, Fabrications and Applications* (ed. K.S. Koh and V.L. Wong). London: IntechOpen.

94 Vaiserman, A., Koliada, A., Zayachkivska, A., and Lushchak, O. (2020). Nanodelivery of natural antioxidants: an anti-aging perspective. *Frontiers in Bioengineering and Biotechnology* 7: 447.

95 Maurya, V.K., Aggarwal, M., Ranjan, V., and Gothandam, K.M. (2020). Improving bioavailability of vitamin A in food by encapsulation: an update. In: *Nanoscience in Medicine*, vol. 1, 117–145. Cham: Springer.

96 Mohapatra, S., Ranjan, S., Dasgupta, N. et al. (ed.) (2018). *Applications of Targeted Nano Drugs and Delivery Systems: Nanoscience and Nanotechnology in Drug Delivery*. Elsevier.

97 Yang, W., Xia, Y., Zou, Y. et al. (2017). Bioresponsive chimaeric nanopolymersomes enable targeted and efficacious protein therapy for human lung cancers in vivo. *Chemistry of Materials* 29 (20): 8757–8765.

98 Zhang, X.Y. and Zhang, P.Y. (2017). Polymersomes in nanomedicine-a review. *Current Nanoscience* 13 (2): 124–129.

99 Ishak, K.A., Annuar, M.M., and Ahmad, N. (2017). Nano-delivery systems for nutraceutical application. In: *Nanotechnology Applications in Food* (ed. A.E. Oprea and A.M. Grumezescu), 179–202. Academic Press.

100 Gigliobianco, M.R., Casadidio, C., Censi, R., and Di Martino, P. (2018). Nanocrystals of poorly soluble drugs: drug bioavailability and physicochemical stability. *Pharmaceutics* 10 (3): 134.

101 Singh, R. and Lillard, J.W. Jr. (2009). Nanoparticle-based targeted drug delivery. *Experimental and Molecular Pathology* 86 (3): 215–223.

102 Grishkewich, N., Mohammed, N., Tang, J., and Tam, K.C. (2017). Recent advances in the application of cellulose nanocrystals. *Current Opinion in Colloid & Interface Science* 29: 32–45.

103 Borges, A., Freitas, V.D., Mateus, N. et al. (2020). Solid lipid nanoparticles as carriers of natural phenolic compounds. *Antioxidants* 9 (10): 998.

104 da Silva, S.V., Ribeiro, A.P., and Santana, M.H. (2019). Solid lipid nanoparticles as carriers for lipophilic compounds for applications in foods. *Food Research International* 122: 610–626.

105 Mishra, D.K., Shandilya, R., and Mishra, P.K. (2018). Lipid based nanocarriers: a translational perspective. *Nanomedicine: Nanotechnology, Biology and Medicine* 14 (7): 2023–2050.

106 Lin, C.H., Chen, C.H., Lin, Z.C., and Fang, J.Y. (2017). Recent advances in oral delivery of drugs and bioactive natural products using solid lipid nanoparticles as the carriers. *Journal of Food and Drug Analysis* 25 (2): 219–234.

107 Lushchak, O., Strilbytska, O., Koliada, A. et al. (2020). Nanodelivery of phytobioactive compounds for treating aging-associated disorders. *GeroScience* 42 (1): 117–139.

108 Khan, I., Saeed, K., and Khan, I. (2019). Nanoparticles: properties, applications and toxicities. *Arabian Journal of Chemistry* 12 (7): 908–931.

109 Pan, L., Qiu, H., Dou, C. et al. (2010). Conducting polymer nanostructures: template synthesis and applications in energy storage. *International Journal of Molecular Sciences* 11 (7): 2636–2657.

110 Heinz, H., Pramanik, C., Heinz, O. et al. (2017). Nanoparticle decoration with surfactants: molecular interactions, assembly, and applications. *Surface Science Reports* 72 (1): 1–58.

111 Grottkau, B.E., Cai, X., Wang, J. et al. (2013). Polymeric nanoparticles for a drug delivery system. *Current Drug Metabolism* 14 (8): 840–846.

112 Feitosa, R.C., Geraldes, D.C., Beraldo-de-Araújo, V.L. et al. (2019). Pharmacokinetic aspects of nanoparticle-in-matrix drug delivery systems for oral/buccal delivery. *Frontiers in Pharmacology* 10: 1057.

113 Singh, S., Kumar, V., Singh, S. et al. (2020). Biotechnological aspects of nanoparticles driven from natural products for drug delivery system and other applications. In: *Bioactive Natural products in Drug Discovery* (ed. J. Singh, V. Meshram and M. Gupta), 549–583. Singapore: Springer.

114 Iqbal, Z. and Iqbal, S. (2018). Nanobotany and pharmaceuticals. In: *Nanobotany* (ed. S. Javad and A. Butt), 131–159. Cham: Springer.

115 Ensign, L.M., Cone, R., and Hanes, J. (2012). Oral drug delivery with polymeric nanoparticles: the gastrointestinal mucus barriers. *Advanced Drug Delivery Reviews* 64 (6): 557–570.

116 Kim, K.T., Lee, J.Y., Kim, D.D. et al. (2019). Recent progress in the development of poly (lactic-co-glycolic acid)-based nanostructures for cancer imaging and therapy. *Pharmaceutics* 11 (6): 280.

117 Qu, Y.D., Liang, X.L., Kong, X.Q., and Zhang, W.J. (2017). Size-dependent cohesive energy, melting temperature, and Debye temperature of spherical metallic nanoparticles. *Physics of Metals and Metallography* 118 (6): 528–534.

118 Taghipour, Y.D., Hajialyani, M., Naseri, R. et al. (2019). Nanoformulations of natural products for management of metabolic syndrome. *International Journal of Nanomedicine* 14: 5303.

119 Ali, M.Y., Sina, A.A., Khandker, S.S. et al. (2021). Nutritional composition and bioactive compounds in tomatoes and their impact on human health and disease: a review. *Foods* 10 (1): 45.

120 Kwon, S.H., Kim, S.Y., Ha, K.W. et al. (2007). Pharmaceutical evaluation of genistein-loaded pluronic micelles for oral delivery. *Archives of Pharmacal Research* 30 (9): 1138–1143.

121 Valizadeh, H., Abdolmohammadi-Vahid, S., Danshina, S. et al. (2020). Nano-curcumin therapy, a promising method in modulating inflammatory cytokines in COVID-19 patients. *International Immunopharmacology* 89: 107088.

122 Guan, X., Gao, M., Xu, H. et al. (2016). Quercetin-loaded poly (lactic-co-glycolic acid)-d-α-tocopheryl polyethylene glycol 1000 succinate nanoparticles for the targeted treatment of liver cancer. *Drug Delivery* 23 (9): 3307–3318.

123 de Oliveira, M.T.P., de Sá Coutinho, D., de Souza, É.T. et al. (2019). Orally delivered resveratrol-loaded lipid-core nanocapsules ameliorate LPS-induced acute lung injury via the ERK and PI3K/Akt pathways. *International Journal of Nanomedicine 14*: 5215.

124 Barani, M., Rahdar, A., Sargazi, S. et al. (2021). Nanotechnology for inflammatory bowel disease management: detection, imaging and treatment. *Sensing and Bio-Sensing Research* 32: 100417.

125 Bejarano, J., Navarro-Marquez, M., Morales-Zavala, F. et al. (2018). Nanoparticles for diagnosis and therapy of atherosclerosis and myocardial infarction: evolution toward prospective theranostic approaches. *Theranostics* 8 (17): 4710.

126 Karin, M. and Clevers, H. (2016). Reparative inflammation takes charge of tissue regeneration. *Nature* 529 (7586): 307–315.

127 Dandekar, A., Mendez, R., and Zhang, K. (2015). Cross talk between ER stress, oxidative stress, and inflammation in health and disease. *Stress Responses* 1292: 205–214. https://doi.org/10.1007/978-1-4939-2522-3_15.

128 Castangia, I., Nácher, A., Caddeo, C. et al. (2015). Therapeutic efficacy of quercetin enzyme-responsive nanovesicles for the treatment of experimental colitis in rats. *Acta biomaterialia* 13: 216–227.

129 Iglesias, N., Galbis, E., Díaz-Blanco, M.J. et al. (2019). Nanostructured chitosan-based biomaterials for sustained and colon-specific resveratrol release. *International Journal of Molecular Sciences* 20 (2): 398.

130 Sharma, M., Sharma, S., and Wadhwa, J. (2019). Improved uptake and therapeutic intervention of curcumin via designing binary lipid nanoparticulate formulation for oral delivery in inflammatory bowel disorder. *Artificial Cells, Nanomedicine, and Biotechnology* 47 (1): 45–55.

131 Qiao, H., Fang, D., Chen, J. et al. (2017). Orally delivered polycurcumin responsive to bacterial reduction for targeted therapy of inflammatory bowel disease. *Drug Delivery* 24 (1): 233–242.

132 Tambuwala, M.M., Khan, M.N., Thompson, P., and McCarron, P.A. (2019). Albumin nano-encapsulation of caffeic acid phenethyl ester and piceatannol potentiated its ability to modulate HIF and NF-kB pathways and improves therapeutic outcome in experimental colitis. *Drug Delivery and Translational Research* 9 (1): 14–24.

133 Samak, Y.O., Santhanes, D., El-Massik, M.A., and Coombes, A.G. (2019). Formulation strategies for achieving high delivery efficiency of thymoquinone-containing Nigella sativa extract to the colon based on oral alginate microcapsules for treatment of inflammatory bowel disease. *Journal of Microencapsulation* 36 (2): 204–214.

134 Hung, C.F., Chen, J.K., Liao, M.H. et al. (2006). Development and evaluation of emulsion-liposome blends for resveratrol delivery. *Journal of Nanoscience and Nanotechnology* 6 (9–10): 2950–2958.

135 Wang, L., Feng, M., Li, Y. et al. (2019). Fabrication of superparamagnetic nano-silica@ quercetin-encapsulated PLGA nanocomposite: potential application for cardiovascular diseases. *Journal of Photochemistry and Photobiology B: Biology* 196: 111508.

136 Nabofa, W.E., Alashe, O.O., Oyeyemi, O.T. et al. (2018). Cardioprotective effects of curcumin-nisin based poly lactic acid nanoparticle on myocardial infarction in guinea pigs. *Scientific Reports* 8 (1): 1–11.

137 Goktas, Z., Zu, Y., Abbasi, M. et al. (2020). Recent advances in nanoencapsulation of phytochemicals to combat obesity and its comorbidities. *Journal of Agricultural and Food Chemistry* 68 (31): 8119–8131.

138 Hong, Z., Xu, Y., Yin, J.F. et al. (2014). Improving the effectiveness of (−)-epigallocatechin gallate (EGCG) against rabbit atherosclerosis by EGCG-loaded nanoparticles prepared from chitosan and polyaspartic acid. *Journal of Agricultural and Food Chemistry* 62 (52): 12603–12609.

139 Sandhir, R., Yadav, A., Sunkaria, A., and Singhal, N. (2015). Nano-antioxidants: an emerging strategy for intervention against neurodegenerative conditions. *Neurochemistry International* 89: 209–226.

140 Cheng, K.K., Chan, P.S., Fan, S. et al. (2015). Curcumin-conjugated magnetic nanoparticles for detecting amyloid plaques in Alzheimer's disease mice using magnetic resonance imaging (MRI). *Biomaterials* 44: 155–172.

141 Ramalingam, P. and Ko, Y.T. (2015). Enhanced oral delivery of curcumin from N-trimethyl chitosan surface-modified solid lipid nanoparticles: pharmacokinetic and brain distribution evaluations. *Pharmaceutical Research* 32 (2): 389–402.

142 Ghosh, A., Mandal, A.K., Sarkar, S. et al. (2009). Nanoencapsulation of quercetin enhances its dietary efficacy in combating arsenic-induced oxidative damage in liver and brain of rats. *Life Sciences* 84 (3–4): 75–80.

143 Das, S., Mandal, A.K., Ghosh, A. et al. (2008). Nanoparticulated quercetin in combating age related cerebral oxidative injury. *Current Aging Science* 1 (3): 169–174.

144 Nday, C.M., Halevas, E., Jackson, G.E., and Salifoglou, A. (2015). Quercetin encapsulation in modified silica nanoparticles: potential use against Cu (II)-induced oxidative stress in neurodegeneration. *Journal of Inorganic Biochemistry* 145: 51–64.

145 Smith, A., Giunta, B., Bickford, P.C. et al. (2010). Nanolipidic particles improve the bioavailability and α-secretase inducing ability of epigallocatechin-3-gallate (EGCG) for the treatment of Alzheimer's disease. *International Journal of Pharmaceutics* 389 (1–2): 207–212.

146 Zhang, J., Zhou, X., Yu, Q. et al. (2014). Epigallocatechin-3-gallate (EGCG)-stabilized selenium nanoparticles coated with Tet-1 peptide to reduce amyloid-β aggregation and cytotoxicity. *ACS Applied Materials & Interfaces* 6 (11): 8475–8487.

147 Frozza, R.L., Bernardi, A., Hoppe, J.B. et al. (2013). Lipid-core nanocapsules improve the effects of resveratrol against Aβ-induced neuroinflammation. *Journal of Biomedical Nanotechnology* 9 (12): 2086–2104.

148 Figueiró, F., Bernardi, A., Frozza, R.L. et al. (2013). Resveratrol-loaded lipid-core nanocapsules treatment reduces in vitro and in vivo glioma growth. *Journal of Biomedical Nanotechnology* 9 (3): 516–526.

149 Hassani, A., Azarian, M.M., Ibrahim, W.N., and Hussain, S.A. (2020). Preparation, characterization and therapeutic properties of gum arabic-stabilized gallic acid nanoparticles. *Scientific Reports* 10 (1): 1–8.

150 Nagpal, K., Singh, S.K., and Mishra, D.N. (2012). Nanoparticle mediated brain targeted delivery of gallic acid: in vivo behavioral and biochemical studies for improved antioxidant and antidepressant-like activity. *Drug Delivery* 19 (8): 378–391.

151 Kakkar, V., Mishra, A.K., Chuttani, K. et al. (2011). Delivery of sesamol-loaded solid lipid nanoparticles to the brain for menopause-related emotional and cognitive central nervous system derangements. *Rejuvenation Research* 14 (6): 597–604.

152 Sachdeva, A.K., Misra, S., Kaur, I.P., and Chopra, K. (2015). Neuroprotective potential of sesamol and its loaded solid lipid nanoparticles in ICV-STZ-induced cognitive deficits: behavioral and biochemical evidence. *European Journal of Pharmacology* 747: 132–140.

153 Kiyani, M.M., Butt, M.A., Rehman, H. et al. (2019). Antioxidant and anti-gout effects of orally administered zinc oxide nanoparticles in gouty mice. *Journal of Trace Elements in Medicine and Biology* 56: 169–177.

154 Montoya García, T., Sánchez Hidalgo, M., Castejón Martínez, M.L. et al. (2021). Alarcón de la Lastra Romero C. Dietary oleocanthal supplementation prevents inflammation and oxidative stress in collagen-induced arthritis in mice. *Antioxidants* 10 (5): 650.

155 Khan, M.A. and Khan, M.J. (2018). Nano-gold displayed anti-inflammatory property via NF-kB pathways by suppressing COX-2 activity. *Artificial Cells, Nanomedicine, and Biotechnology* 46 (sup1): 1149–1158.

156 Chen-yu, G., Chun-fen, Y., Qi-lu, L. et al. (2012). Development of a quercetin-loaded nanostructured lipid carrier formulation for topical delivery. *International Journal of Pharmaceutics.* 430 (1–2): 292–298.

157 Guan, F., Wang, Q., Bao, Y., and Chao, Y. (2021). Anti-rheumatic effect of quercetin and recent developments in nano formulation. *RSC Advances* 11 (13): 7280–7293.

158 Xu, Y., Liu, H., and Song, L. (2020). Novel drug delivery systems targeting oxidative stress in chronic obstructive pulmonary disease: a review. *Journal of Nanobiotechnology* 18 (1): 1–25.

159 Passi, M., Kumar, V., and Packirisamy, G. (2020). Theranostic nanozyme: Silk fibroin based multifunctional nanocomposites to combat oxidative stress. *Materials Science and Engineering: C* 107: 110255.

160 Roh, C. (2012). A facile inhibitor screening of SARS coronavirus N protein using nanoparticle-based RNA oligonucleotide. *International Journal of Nanomedicine* 7: 2173.

161 Carvalho, A.P. and Conte-Junior, C.A. (2021). Recent advances on nanomaterials to COVID-19 management: a systematic review on antiviral/virucidal agents and mechanisms of SARS-CoV-2 inhibition/inactivation. *Global Challenges* **5** (5): 2000115. https://doi.org/10.1002/gch2.202000115.

162 Akbari, M., Morad, R., and Maaza, M. (2020). First principle study of silver nanoparticle interactions with antimalarial drugs extracted from Artemisia annua plant. *Journal of Nanoparticle Research* 22 (11): 1–9.

163 Pala, R., Anju, V.T., Dyavaiah, M. et al. (2020). Nanoparticle-mediated drug delivery for the treatment of cardiovascular diseases. *International Journal of Nanomedicine.* 15: 3741.

18

Nanotherapeutics of Phytoantioxidants for Cardiovascular Diseases

Himangshu Sarma[1], Pallabi Kashyap[2], James H. Zothantluanga[3], and Rajlakshmi Devi[1]

[1] *Life Sciences Division, Institute of Advanced Study in Science and Technology, Guwahati, Assam, India*
[2] *Department of Pharmaceutical Chemistry, Girijananda Chowdhury Institute of Pharmaceutical Sciences, Guwahati, Assam, India*
[3] *Department of Pharmaceutical Sciences, Faculty of Science and Engineering, Dibrugarh University, Dibrugarh, Assam, India*

18.1 Introduction

Cardiovascular diseases (CVDs) are the most prominent killer in the twenty-first century in the universe. According to the World Health Organization (WHO), approximately 17.9 million people die from CVDs worldwide every year compared to any other cause [1, 2]. Approximately, 80% of universal deaths caused by hypertension, coronary artery disease, heart failure, and strokes occur in developing countries. It is increasing rapidly in Asian countries [3]. These primarily affect low- and middle-income countries [4]. By 2030, the total worldwide cost of CVD is set to rise from approximately US $863 billion in 2010 to the staggering US $ 1044 billion [5]. Diagnosis and treatment prices of CVDs were rising at a higher rate and were expected to escalate more in the next 10 years. The high economic burden related to CVDs is owing to the enhancement in the risk factors of CVDs such as obesity, diabetes, and expansion of the geriatric population [6]. The present statistical index indicates that CVDs have expected to be the single foremost cause of death in the world. The number of deaths owing to CVDs, especially by heart disease and stroke, will exceed 23.3 million by 2030 [7]. A review on "psychosocial factors and CVDs" investigates the effect of psychosocial factors on the morbidity and mortality rates of CVDs. Negative emotional states like depression, anger, anxiety, hostility, social ties, social support, social conflict, and chronic and acute psychosocial stressors have been associated with the raised risk of cardiovascular mortality and morbidity. These psychosocial factors have a direct relationship with pathophysiologic mechanisms that promote atherosclerosis and its clinical manifestations [8]

18.1.1 Hypertension and CVDs

The term "cardiovascular diseases" (CVDs) refers to any disease of the heart, vascular disease of the brain, or disease of the blood vessel. One of the most common incidents in non-communicable CVDs is hypertension or high blood pressure. Hypertension is responsible for the high ratio of mortality contributed by CVDs worldwide. It has been observed in all groups of people irrespective of their age and sex. Generally, hypertension damages most of the vital organs significantly in the

Phytoantioxidants and Nanotherapeutics, First Edition. Edited by Mithun Rudrapal.

body, for instance, the heart, blood vessels, brain, eyes and kidneys. According to a study report published in 2015, approximately 1.13 billion people are affected by hypertension globally [9].

Improving the quality of life by changing the sedentary lifestyle and eating healthy food can only minimize the incidences and mortality rate of CVDs. The response of patients to the current therapeutic methods is limited and is not practical. It is seen that CVDs exert a close connection with diabetes mellitus which increases the morbidity and mortality rate in both man and woman populations [10]. It was observed that CVDs affected populations have raised their mortality rate in adult populations with diabetes compared to non-diabetes. These higher mortality rates of diabetic patients might be owing to the increased risk of myocardial infarction, coronary artery disease and stroke [11].

18.1.2 Types of CVDs and Associated Risk Factors

Atherosclerosis is a major cause of stroke and other CVDs. Lipoprotein retention on the subendothelial extracellular matrix promotes the development of atherosclerotic plaques. Plaques progress as a result of the interplay between lipid buildup and the inflammatory response. Plaques are generally eliminated by oxidative stress, proteases, and inflammation. Proteases immediately degrade the collagen network and cause the plaque to burst. Plaque rupture can result in myocardial infarction and stroke [12]. One of the most severe kinds of atherosclerotic CVD is acute myocardial infarction (AMI) and ischemia death of cardiomyocytes. Coronary artery disease is another significant cause of death worldwide, and it can be fatal if it progresses to AMI [13]. Coronary artery illnesses are caused by coronary atherosclerosis, which is caused by stenosis or blockage of the coronary arteries. It has been shown that mitochondria are important in treatment efforts to minimize the extent of myocardial infarctions and hence avoid heart failure in individuals with ischemic heart disease. Mitochondrial dysfunction can contribute to acute myocardial perfusion injury followed by AMI [14].

The third most common CVD in the world is heart failure. According to the American Heart Association on Heart Disease and Stroke report published in 2017, the prevalence of the heart failure rate is estimated to raise by 46%, i.e. approximately 8 million people by 2030 [15]. It is a complicated pathophysiological condition that results from the heart's inability to fill or expel blood. Heart failure has been linked to a number of clinical symptoms, including hereditary factors, hypertension, hypertrophy, and coronary artery disease [16]. The prevalence of symptomatic heart failure grows with age. The increase in cardiomyocyte size causes hypertrophied heart or myocardial hypertrophy. Adult mammalian cardiomyocytes have a lower proliferative potential, and the size of the heart expands in response to stress or increased workload, such as vascular heart disease, hypertension, aortic stenosis and myocardial infarction. It leads to the development of myocardial hypertrophy or hypertrophied heart. Myocardial hypertrophy is followed by the augmented synthesis of proteins, improved expression of embryonic genes, and assembly of perivascular, sarcomeres, fibrosis, which eventually causes heart failure. Another complication associated with the heart is restenosis, which has an incidence of 30–40% among the population [17].

Vascular calcification is one of the processes that predict the development of cardiovascular illnesses. Vascular calcification is prevalent in people over the age of 60 when calcium deposits in the arteries grow. Vascular calcification reduces arterial and aortic elastance with time, interfering with cardiovascular hemodynamics. Hypertension, congestive heart failure, cardiac hypertrophy, aortic stenosis, and ischemia are all possible outcomes. Mineralization increases, which leads to the development of atherosclerotic plaques and increases the risk of cardiovascular morbidity and death [18].

There are currently no effective methods for reducing vascular calcification or arterial stiffness. Stopping the pathophysiology of vascular calcification is quite challenging and thus necessitates early intervention in targeting vascular calcification [19, 20]. Vascular calcification has been related to CVDs and a number of other illnesses such as end-stage renal failure [21]. One of the major causes of death in individuals with chronic renal disease is vascular calcification [22]. There are a number of classic causes including diabetes, hypertension, and hyperlipidemia and nontraditional risk for vascular calcification. Nontraditional risk factors such as inflammation, oxidative stress, advanced glycation end products, and abnormal mineral metabolisms like hypercalcemia and hyperphosphatemia are also more prevalent and more severe and offer mechanistic insight into the pathogenesis of vascular calcification. Early intervention of vascular calcification can prevent the progression of CVDs [23, 24].

In brief, CVDs can define as abnormalities in morphogenesis, muscle repair, cardiac function, and cardiac rhythm. In order to treat these diseases, an alternative and effective remedy needs to be employed that involves the direct delivery of cardioprotective medicament to the cardiovascular system. The necessity to deliver these medicaments to specific targets site has led to the development of different methods for specific site targeted drug delivery [25]. Treatment approaches for preventing CVDs have been recommended along with a healthy lifestyle, which includes smoking cessation, regular physical activity, a healthy diet, weight reduction, aerobic exercise training, maintaining blood pressure, and lowering blood lipids' dietary interventions [26, 27]. However, several therapeutic approaches and drugs available are based on the utilization of synthetic compounds. This process may lead to different episodes of adverse effects on the body. This topic directs researchers to focus on safer and more promising therapeutic drug delivery methods. In this context, nanotechnology has extended its height to the treatment of CVDs.

Nanoparticles proved their role as an effective and reliable platform for the controlled and site-specific targeted delivery of drugs to cure thrombosis, lipid disorders, inflammation, and angiogenesis in atherosclerotic plaques. This chapter summarizes different phyto-oxidant conjugate nanotherapeutic-based drug carriers available for the management of CVDs.

18.1.3 Mechanism or Pathway Involved in the Progress of CVDs

There are enormous biochemical, molecular, and cellular mechanisms involved in the progress of CVDs. These are mainly involved in prolonged increases in intracellular excessive ROS production. Physiologic levels of ROS (H_2O_2) are signaling molecules vital for normal cardiac muscle cell homeostasis. However, ROS production at too high a level, for too long, or at an inappropriate subcellular location, leads to impaired cellular function and cardiovascular pathology [28]. This stimulates overexpression of the nuclear receptor PPARα and nuclear translocation of forkhead box O1 transcription factor (FOXO1), which inhibits GAPDH, shunting early glycolytic intermediates into pathogenic signaling pathways. ROS and PARP also reduce sirtuin, PGC1α, and AMPK activity. These changes cause decreased mitochondrial biogenesis and disturbed circadian clock synchronization of glucose and lipid metabolism. Finally, it causes cardiomyopathy. Excessive ROS generation also facilitates nuclear transport of pro-atherogenic transcription factors, increases transcription of the neutrophil enzyme initiating NETosis, PAD4, and activates the NRLP3 inflammasome. ROS also shift the balance between mitochondrial fusion and fission in favor of enhanced fission, reducing the metabolic capacity and efficiency of the mitochondrial electron transport chain and ATP synthesis.

Mitochondrial oxidative stress also plays a central role in angiotensin II-induced gap junction remodeling and arrhythmogenesis. ROS contribute to sudden death in diabetics after MI by increasing post-translational protein modifications, which cause increased ryanodine receptor phosphorylation and downregulation of SERCA2a transcription [29]. Moreover, it is also found that a high level of ROS production also depresses autonomic ganglion synaptic transmission by oxidizing the nAch receptor α3 subunit, potentially contributing to the increased risk of fatal cardiac arrhythmias associated with diabetic cardiac autonomic neuropathy [30].

18.2 Different Treatment Strategies for CVDs

Despite recent advances in the diagnosis and treatment of many CVDs, the mortality rate associated with CVDs is still much higher than that of cancer around the world. Controlling and treating CVDs can be done in a variety of ways. Controlling blood pressure, cholesterol, diabetes, weight, and physical activity, as well as depression and selecting the best drugs, are all treatment options for CVDs. A population-wide and individual-level approach has been advocated by the WHO to reduce the burden of CVDs. Among the population, interventions are tobacco control policies, taxes to reduce the consumption of high fat, sugar and salt foods, alcohol consumption reduction tactics and the construction of cycle roads, as well as walking areas, to encourage physical activity (Figure 18.1) as well as heart surgery including valve repair, valve replacement, and artificial heart surgery. For people with high CVD risk, such as hypertension and hypercholesterolemia, medical devices are used to prevent heart attacks and strokes before they occur. Secondary prevention of CVDs associated with other diseases, such as diabetes, can be achieved with a variety of medications. Beta-blockers, statins, aspirin, and angiotensin-converting enzyme inhibitors are some of the medications for it [31]. Additionally, surgical procedures and medical devices, such as balloon angioplasty and heart transplantation, are preferred to treat CVD. These conditions require prosthetic valves and pacemakers, as well as heart patches. As of now, the open aortic repair is the most

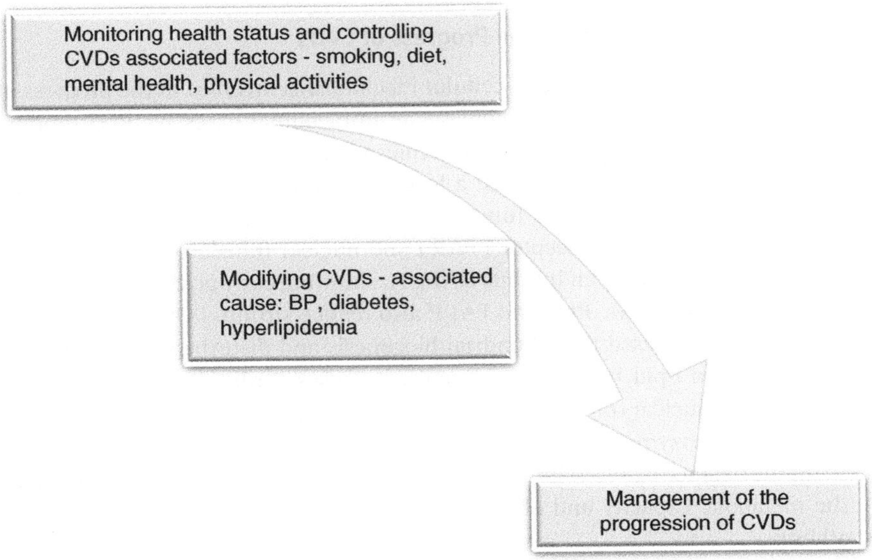

Figure 18.1 Current strategies for the management of CVDs.

common surgical approach for treating thoracic and thoracoabdominal aortic diseases. There has not been a single treatment method developed for the treatment [32].

Vascular gene therapy is a promising treatment for CVD. Cardiovascular gene therapy can target ischemia, restenosis, thrombogenesis, graft failures, in-stent restenosis, atherogenesis, and arterial cytoprotection. Collateral development and blood flow in the ischemic limb and myocardium were improved by the transfer of vascular endothelial (VE) and fibroblast growth factor (GF) genes. Vascular gene therapy of genes coding proteins, for instance, VEGF, thymidine kinase, retinoblastoma and nitric oxide synthase, cyclin or cyclin-dependent kinase inhibitors, growth arrest homoeobox, hirudin, fas ligand, and antisense oligonucleotides against transcription factors exhibited beneficial effects in animal models of vein gene transfer for VEGFs have been studied. However, further improvements in delivery methods and efficient vehicles and valid targets are needed [33].

The market is flooded with new drugs for the prevention and treatment of hypertension and CVDs. However, many drugs such as beta-blockers, calcium channel blockers, nitrates, and ranolazine have been linked to adverse side effects, such as vomiting and nausea. For example, nitrates, calcium channel blockers, and beta-blockers used to treat angina pectoris cause rebound myocardial ischemia, rash, dizziness, fatigue, flushing, abdominal pain, nausea and peripheral oedema, constipation, and headache [34]. Long-term use of available medications and therapeutic procedures for CVDs has severe side effects, including rhabdomyolysis, renal failure, and haemorrhage, regardless of their therapeutic potential [35].

Natural products are gaining popularity these days because they both reduce the advancement of CVDs and are cost-effective. Natural materials used in CVD treatment have several targets and fewer adverse effects than synthetic medications [36]. To prevent and treat CVD, many bioproducts derived from plants or phytotherapy have been approved by the FDA and are available in the market. Already *Ginkobiloba*, garlic, and Crataegus extracts have been suggested for people with CVDs [37, 38].

Polyphenols, the most prevalent antioxidant phytochemicals, are also being investigated for their capacity to protect against degenerative diseases, particularly CVD. Polyphenols have been evaluated for their protective role against CVDs using human trial data [39]. It was established that including polyphenol-rich foods in daily meals reduces the risk of CVDs. Lipemia, inflammation, and oxidative stress have all been linked to CVD biomarkers [40]. The highly polyphenol-rich foods like pomegranate juice, tea extract, grape extract, green tea, black tea, wine polyphenols, red wine, and dealcoholized wine were found to reduce the development of atherosclerotic lesions in apolipoprotein E-/- mice and hamsters [41]. Other than the polyphenols or antioxidants, phytochemicals can be considered potent molecules for the management of CVDs. The complex etiology of CVDs is characterized by the overproduction of ORS and other free radicals that may disrupt endothelial function and other deleterious vasodilator effects [42]. In the experiment, it was found that some flavonoids derived from the pulp of *Euterpeoleracea* demonstrated atheroprotective effects [43]. A flavonoid isolated from *Floschrysanthemi* enhances the vasodilator reactivity by reducing oxidative stress [44]. Quercetin demonstrated protective properties in atherosclerosis in vivo model by interference with the formation of foam cell and pro-inflammatory response [45].

Complementary and alternative medicine (CAM) is approximately 95% recommended new standard treatment modality among patients with CVDs worldwide. CAM includes integral parts of traditional Chinese medicine (TCM). Herbal medicine and acupuncture are essential components of TCM. Clinical trials showed that acupuncture could be successfully utilized in the management of hypertension and in improving the circadian rhythm of patients with hypertension. Chinese herbal therapy has grown in popularity across all forms of CAM. From theory to tests and clinical trials, there has been a lot of progress in the treatment of hypertension with Chinese herbs and formulations. The Niuhuang Jiangya pill, Yangxue Qingnao granule, and Qing Gan JiangYa capsule are a few Chinese patented medicines used in clinics to treat hypertension. Despite the fact that herbal therapy can lower hypertension, improve endothelial function, regulate the renin-angiotensin-aldosterone system, protect target organs, and reverse CVD risk factors, the role of TCM in modern hypertension clinical practice and health care is limited [46]. According to the findings of various randomized controlled trials (RCTs), some TCM medicines may be helpful in controlling CVD risk factors such as hypertension, diabetes, and dyslipidemia. On the other hand, some TCM medications were suggested as complementary and alternative approaches to primary and secondary prevention of CVDs [2]. As a result, CAM and phytotherapy have successfully managed CVDs. CAM continues to prevent CVDs through the roots in traditional folk remedies and ancestral knowledge. Ethnopharmacological medicine will be crucial in the future battle against CVDs [47].

WHO has identified that only 122 compounds from 94 plant species have a potential application in the ethnomedical therapy of CVDs worldwide [48]. For example, papaverine isolated from *Papaver somniferum* is the base for synthesizing antihypertensive medicament verapamil [49]. Commercially, a drug synthesized from natural sources is a viable commodity. Only 39% of the 520 new medicaments were approved from natural molecules or derived from natural molecules from 1983 to 1994. Among them, 60–80% of approved medicaments were antibacterial and anticancer agents [50].

Traditionally, plant-based diets have been associated with the prevention and treatment of CVDs. A fruit-rich diet has a significantly vital role in the management of CVDs. As a result of epidemiological studies, fruits have been shown to reduce the risk of CVDs [51–53]. Grape, pomegranate, blueberry, hawthorn, avocado, and apple are some of the fruits that possess cardiovascular protecting properties [54]. Minimum lipid metabolism and elevated blood pressure, inhibition of thrombosis, oxidative stress and inflammatory response, modulation of molecular events and signaling pathways associated with correcting endothelial dysfunction, suppression of platelet function, and alleviation of I/R injury are some of the mechanisms of action of these fruits for preventing CVDs. Therefore, it is necessary to identify a more significant number of fruits in the potential cardiovascular protection in the future [54].

Curcumin, a phytoantioxidant isolated from *Curcuma longa* (turmeric), has a wide range of therapeutic activities. In addition, curcumin has a pleiotropic effect and is widely applied to manage CVDs [55]. Nuclear factor (erythroid-derived 2)-like 2 (Nrf2), histone acetyltransferases (p300), transcription factor (NF- κB), and angiotensin II type 1 receptor (AT1R) are some of the essential molecular target pathways of curcumin in CVDs. The application of curcumin in the management of CVDs has been restricted owing to its practical insolubility in the aqueous medium at acidic and neutral pH, which causes poor bioavailability. In this context, scientists have worked hard to develop synthetic curcumin derivatives or drug delivery methods that enhance bioavailability and solubility [55]. Hence, curcumin has been utilized as a routine food supplement considering its safety level to prevent or treat CVDs.

CVDs are linked to mitochondrial malfunction, which causes excess production of free radicals in the body. Mitochondrial dysfunction leads to the continuous process of programmed cell death and, ultimately, developed CVDs [56]. As a result, controlling mitochondrial dysfunction is an essential criterion in designing and developing medicament for the management of CVDs. The goal of current research and clinical studies is to figure out how antioxidant phytochemicals affect mitochondrial dysfunction. Coenzyme Q10 has been shown to regulate mitochondrial dysfunction in both animals and humans in several investigations [57]. The coenzymes found in the inner mitochondrial membrane are antioxidant and anti-thrombolytic molecules that can help prevent hyperglycemia and hypertension. This coenzyme can be used alone or in combination with other medications to treat hypertension and heart failure in humans [56]. Berries are currently gaining popularity in the diet chart and functional foods for their pleasing aroma, appearance, and therapeutic characteristics. Natural berries including mulberry, raspberry, blackberry, cranberry, bilberry, currants, and blueberry have commercial and nutritional importance due to their protective and preventive characteristics for some chronic illnesses [58]. The high amounts of anthocyanins and flavanols in berries have been ascribed to their antioxidant potential. These phytochemicals have the ability to scavenge ROS, which is important in the prevention of CVDs. They also play a vital function in blood pressure control, oxidative stress, endothelial function, whole-body metabolism, platelet aggregation, atherosclerosis, and protecting the body against CVDs [59].

Resveratrol is a non-flavonoid polyphenolic molecule, which reduces the risk of hypertension and stroke, heart failure as well as atherosclerosis. It also reduces the risk of chemotherapy-induced cardiotoxicity, ischemic heart disease, and diabetic cardiomyopathy [60]. Resveratrol activates a few vital pathways like cyclic adenosine monophosphate-activated protein kinase, silent information regulator 1 (SIRT1), an endogenous antioxidant enzyme responsible for its positive cardiovascular effects. Furthermore, it has antiplatelet, insulin-sensitizing, anti-inflammatory, and lipid-lowering characteristics that are beneficial to CVDs. There is still a paucity of clinical evidence for these polyphenolic compounds, which has limited their usage in clinics [61].

CVDs may also be treated through targeting by reducing endoplasmic reticulum (ER) stress. Vascular cells and cardiomyocytes would become dysfunctional if ER stress was prolonged. In addition, it may lead to apoptosis [62]. The development of novel natural compounds like alkaloids, terpenoids, sterols, saponins, phenolics, flavonoids, amino acids, C-phycocyanins, and glycosides has increased exponentially in the last five years to improve ER stress and their multiple pathogenic pathways, such as apoptosis, ROS production, and inflammation to treat CVDs has increased exponentially [63]. Baicalein, resveratrol, berberine, quercetin, ginsenoside-standardized extract, anisodamine, notoginsenoside R1, and paeonol are a few examples of natural phytoantioxidants that reduce ER stress-mediated apoptosis, oxidative stress, and inflammation [64, 65]. In addition to its role in regulating many somatic cellular and molecular activities, adenosine offers tremendous therapeutic promise for treating a wide variety of illnesses, from cancer to CVD. Regulation of blood pressure, heart rate, and heart rate variability depend on the A2 adenosine receptor's interaction with adenosine. It has been shown that the malfunctioning of these receptors promotes CVDs, although the specific pathogenic processes behind these A2 receptors remain unknown [66].

Although, the conventional dosage forms and carriers are generally utilized to deliver cardioprotective medicines effectively in patients. Tablets or their novel liquid spray for sublingual administration are standard medication carriers for CVDs. While certain cardioprotective medicines maintain distinct tablet shapes for effective pharmacodynamic action, these oral tablets or transdermal patches can demonstrate long-term absorption through sluggish absorption via the digestive system and skin. So, novel drug delivery through a stent or balloon catheter helps deliver drugs to a small local target area in coronary interventional therapy.

The monolithic nature of conventional drug delivery systems means that only one medicament can be loaded. Therefore, combining multiple medicaments into one drug carrier is impossible, which reduces the overall therapeutic cure rates. Treatment modalities such as interventional therapy, stem cell transplantation, therapeutic drugs, and other treatment modalities are not the most effective ways to manage CVDs. CVD patients must have access to more effective drug delivery and therapeutic strategies. So, nanotherapeutics have been introduced for cardioprotective drug delivery owing to there as excellent and innovative carrier function in order to enhance efficacy with minimal side effects. Nanotherapeutics can eliminate the drawbacks of conventional cardioprotective dosage forms or medicaments, for instance, nonspecific effects, poor solubility and blood-brain barrier impermeability etc.

18.3 The Emergence of Nanomedicine in CVDs

Nanoscience and nanotechnology are new disciplines of study in the pharmaceutical industry, information technology, communication technology, biology, medical technology, and biotechnology. Nanoscience studies the design, manipulation, production, and application of materials with a size less than 100 nm [67]. For example, nanotechnology can be used in pharmaceuticals and medicine for a wide range of purposes, including imaging, diagnosis, gene delivery, targeted drug delivery, and tissue engineering scaffolds [68]. Nanotechnology has received a lot of interest in novel-targeted drug delivery medicine owing to its physicochemical characteristics that promote pharmacokinetics and pharmacodynamics functions in biological systems. It allows for enhancing health and quality of life. Nanotherapeutics have several benefits over conventional and contemporary medicinal techniques (Figure 18.2). These benefits include increased medication half-life, site-specific drug delivery, reduced toxicity, improved bioavailability, and biocompatibility of nanoparticles, as well as reduced therapeutic adverse effects via changing nanoparticle characteristics [69]. Site-specific medication with nanocarriers by either passive or active targeting looks to be a viable therapeutic approach. Active drug targeting needs the conjugation of medicament to the cell-specific ligand attached to the nanocarrier. A high molecular weight polymer has been used in passive targeting to administer medicines that can penetrate and remain in the tissue for improved results [70].

Nanoparticles with sizes ranging from 1 to 100 nm have unique features, including their nanoscale, quantum size, high surface-to-volume ratio, and form, among others, compared to their bigger counterparts [71], as a result of their ability to pass through biological membranes, cells, and tissues. Surface modification of nanoparticles can be used to enhance their physical characteristics. The monodispersed nanoparticles can be prepared by carefully controlling the shape, size, and aggregation of particles that enable cell internalization. As a result of their unique

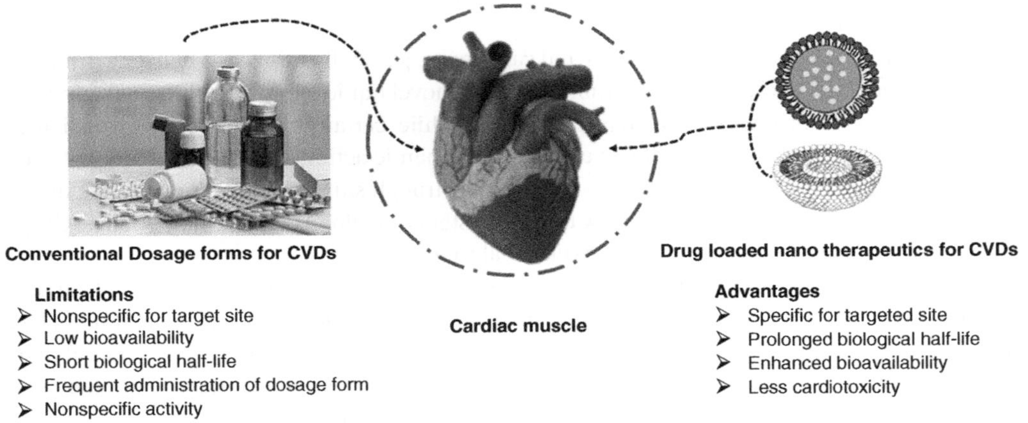

Figure 18.2 Schematic representation of advantages of nanotherapeutics over conventional dosage form on CVD treatment. *Source:* Africa Studio/Adobe Stock.

characteristics, nanomaterials have been used in various biomedical applications as a theranostic, diagnostic, drug carrier, and therapeutic agents.

Nanoparticles with a high surface-to-volume ratio can allow molecules to be conjugated, absorbed, or encapsulated and delivered to the specific target location [72]. Drug delivery carriers are used owing to they can deliver poorly soluble or extremely toxic medicament or drug molecules to the target regions. Rapid advancements in the area of nanotechnology have solved all of the issues related to CVD treatment strategy. Nanotechnology aids in the early diagnosis and treatment of CVDs by lowering the load on the healthcare system. Thus, nanoparticles play a key role in CVDs because of their better medical imaging, specific site-targeted drug delivery and targeted delivery of nanoparticles to destroy sick cells. Nanotherapeutics can be easily excreted from blood and tissue [73].

Nanoparticles used for theranostic purposes in CVDs are able to bridge the gap between experimental data and large-scale clinical evidence. Clinical theranostics in cardiology enables novel site-specific targeted drug delivery. In the pharmaceutical industry, various nanoparticles are used to deliver drugs and further assess their efficacy. To reduce local and systemic side effects, nanoparticle-based medication delivery and its activity in the target location are regulated by either light or an external magnetic field or ultrasound [74].

Nanomedicine in cardiology typically focuses on four key areas: molecular imaging, site-specific targeted drug delivery, tissue engineering, and diagnostics that govern to control or help manage CVDs. Figure 18.3 is showing the frequently utilized nanotherapeutics for phytoantioxidant-mediated drug delivery in CVDs. Nanocarriers have numerous advantages, including controlled release of medicament to site-specific areas, increased bioavailability of therapeutic molecules and minimizes the frequency of the dosage. Liposomes, polymeric nanoparticles, micelles, dendrimers, etc., are the few possible and widely utilized drug delivery carriers for CVDs. For an ideal carrier system, biodegradability, biocompatibility, non-immunogenicity, and nontoxicity have been

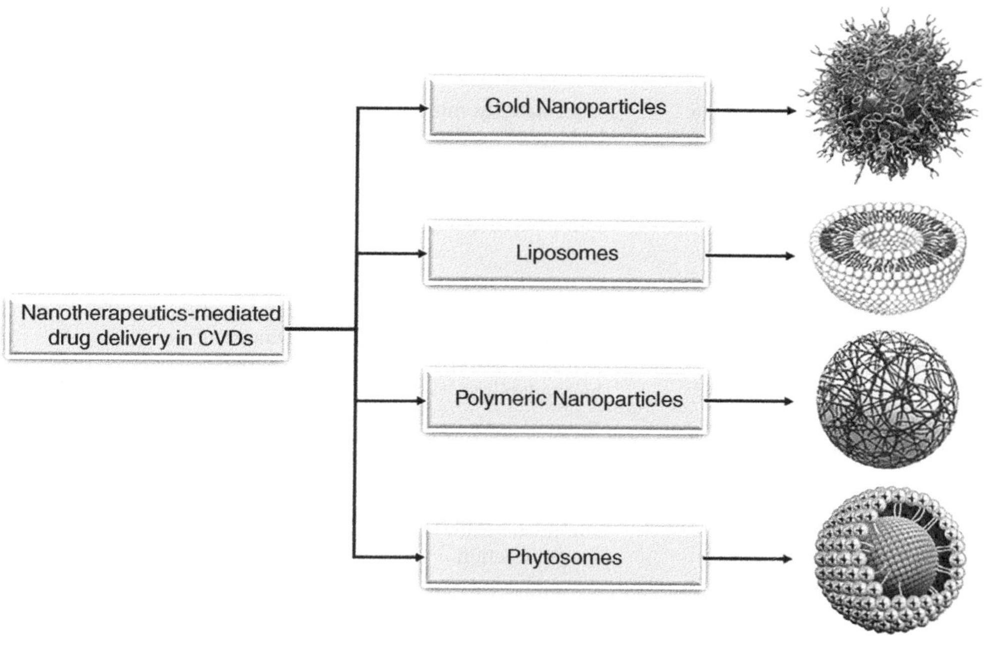

Figure 18.3 Frequently utilized nanotherapeutics for phytoantioxidant-mediated drug delivery in CVDs.

characterized as the essential characteristics of an optimal drug delivery carrier, as well as the ability to escape from the host immune system [75]. For example, polymeric nanoparticles and liposomes can be loaded with a large number of medicaments for a diagnostic or therapeutic approach. Other nanocarriers necessitate additional functional ligands in order to conjugate the medicaments. Due to their high surface ratio, nanoparticles have been utilized as multifunctional carrier agents for detecting and treating CVDs [76].

Nanocarriers allow the conjugation or encapsulation of drug molecules or medicaments, which raise the solubility of drugs, minimize systemic toxicity, and alter the pharmacokinetics and pharmacodynamics parameters of the medicaments [77]. The rationale involved in the design of nanocarriers for cardiovascular medicine is multi-criteria design. Multi-criteria for nanocarrier design and development of nanocarriers for cardiovascular therapy include product physicochemistry, quality of ingredients such as safety, pyrogen content, serializability, manufacturability, including process costs and costs of goods. Other factors to consider include stability, biocompatibility, toxicity, and efficacy, as well as pharmacokinetics and biodistribution [78].

There is a brief description of a few of the nanocarriers designed and prepared to treat CVDs before delving into further depth. Polymeric nanoparticles are the best nanocarriers for targeted drug delivery [79]. These nano range particles have a more excellent absorption in the artery walls, resulting in prolonged drug release to the target location. Cross-linking, molecular weight, polymer use, and their ratio determined the prolonged continuous release of medicaments. Different types of nanocarriers used for the targeted delivery of cardioprotective agents with their features have been described in Tables 18.1 and 18.2.

Table 18.1 Various classes of nanotherapeutics and their features for phytoantioxidant delivery to CVDs.

S. No.	Structure	Nanocarriers	Features	Limitations	References
1.	Nanoparticles, Nano capsules, nano cuboid	Metallic nanotherapeutics	Biocompatible, more stable, site-specific drug delivery	Toxic, instability of particles, impurity	[80, 81]
2.	Core or shell structure formed by itself	Polymeric nanoparticles	High stability, minimum leakage of the drug molecule	High toxicity of intravenous injections	[82]
3.	Lipid bilayer	Liposomes	High biocompatibility, non-immunogenic nature	Less stability, quick leakage of the hydrophilic molecule	[82]
4.	Active molecules surrounded by phospholipids	Phytosomes	Lipid-insoluble phytochemical easily administered, enhanced bioavailability	Leaching of the phytoconstitute from lipid core	[83]
5.	Soft vesicle phospholipids	Ethosomes	Noninvasive nanocarriers, enhanced skin permeation of drug molecules, drug molecules can be easily administered for local or systemic action	Leaching of the phytoconstitute from ethosome core	[84]
6.	Smaller size with adjustable pore size	CNTs	Extremely smaller and lightweight that makes them excellent replacements for metallic wires, resistant to temperature changes	Extremely smaller in size that causes difficulty in handling, toxic nature	[82]

Table 18.2 Various nano encapsulated phytoantioxidants containing herbal formulations.

Phytoantioxidants	Types of nanocarriers	Therapeutic activities	Limitations of active molecules	Model	References
Metallic nanoparticles					
Quercetin	Polymeric superparamagnetic nanosilica	Antioxidant agent, quercetin is utilized to control atherosclerosis and other relative cardiovascular illnesses	Poor aqueous solubility	Mice	[85]
Amaranthus caudatus Linn. Pod extract	Ag-nanoparticles	Significant improvements obtained for phenolic and flavonoid contents of the plant when grown with AgNPs and enhanced antioxidant activity		In vitro DPPH assay	[86]
Polymeric nanoparticles					
Curcumin	Nanoparticles	Improve the efficacy of curcumin to treat atherosclerosis	Fast systemic elimination, alkaline pH breakdown, low gastrointestinal absorption, quick metabolism, and poor water solubility	Rat model	[87]
Lipid vesicles					
Naringenin	Liposomes	Enhanced stability, solubility, and sustained release in vitro and in vivo. Improved bioavailability and tissue distribution after oral administration	Poor aqueous solubility	Mice	[88]
Polydatin	Liposomes	Improved solubility with the sustained release in vitro. Prolonged the drug circulation time and increased the oral bioavailability of the drug by reducing the effect of adriamycin-injured myocardial ultrastructure and cardiomyocytes that showed an evident protective action	Poor solubility and less bioavailability	In the male Sprague-Dawley rat	[89]

(*Continued*)

Table 18.2 (Continued)

Phytoantioxidants	Types of nanocarriers	Therapeutic activities	Limitations of active molecules	Model	References
Green tea extract	Phytosomes	Cholesterol-lowing activity	Poor aqueous solubility	Human model	[90]
Rutin	Phytosomes	Enhanced solubility, stability, releasing dynamics, and bioavailability in vitro. Also fortified to be a good candidate as an antioxidant agent.		In vitro assay	[91]
Butea monosperma flower extract	Phytosomes	It improves solubility, stability, bioavailability, medicament release pattern, and free radical scavenging activity in *in vitro* model.		In vitro DPPH assay	[92]
Resveratrol, quercetin, catechin, procyanidins, epicatechin contain extract from ***Vitis vinifera*** (Grapes)	Phytosomes	Protects endothelial cells from peroxynitrite-induced damage and increased the endothelium-dependent NO release. It increases the antioxidant capacity and enhances the resistance of LDLs; hence it acts as a cardioprotective and nutraceutical.		In vitro assay	[90]

Exosomes are nanosized vesicles prepared from stem cells that are frequently used for site-specific drug delivery. Exosomes are necessary for the delivery to transport mRNA, miRNA, and proteins to destination cells. Heart disease exosomes have both cardioprotective and drug-delivery capabilities. As a result of their resemblance to liposomal drug delivery systems in terms of tissue targeting and drug delivery, they are referred to as natural liposomes. The exosomes are very useful in diagnosing CVDs [93].

Curcumin nanoparticles have been synthesized to improve the efficacy of curcumin, which has been used for many years to treat atherosclerosis. Several experimental reports have shown that curcumin nanoparticles can be used to cure CVDs. Fast systemic elimination, alkaline pH breakdown, low gastrointestinal absorption, quick metabolism, and poor water solubility restrict its clinical use. However, curcumin is an excellent phytoantioxidant for reducing oxidative stress of cardiac tissue. Nevertheless, in nanoformulation, curcumin enhances superior cardioprotection compared to the conventional dosage form. Curcumin nanoparticles reduce myocardium damage in patients with AMI. Curcumin nanoparticles were found to be more cardioprotective than curcumin in a rat model of isoproterenol-induced MI by changing electrocardiogram and myocardial tissue oxidative stress [87]. It also showed improved myocardial function and reduced heart injury after MI due to their increased antioxidant activity and lower levels of pro-inflammatory cytokines and matrix metalloproteinases in blood [94]. Diabetic patients have a more significant effect and incidence of MI than healthy people. When compared to curcumin, curcumin nanoparticles demonstrated improved cardioprotective benefits in a rat model of isoproterenol-induced AMI with diabetes mellitus [95].

18.3.1 Nanoparticles for the Treatment of Coronary Artery Disease

The clinical applications of nanotherapeutics in the area of CVDs are minimum. The buildup of atherosclerotic plaque on the inner wall of the coronary artery causes partially interrupted blood flow and finally developed coronary artery disease [96]. Atherosclerosis is a chronic condition characterized by the thickening of the artery wall and plaque inflammation [97]. When atherosclerosis obstructs coronary arteries, a heart attack ensues.

In this state, numerous cells, extracellular matrix, and cytokines are activated by hypoxia of cardiomyocytes, resulting in complex and interconnected physiological process changes in cellular and extracellular levels. It then led to the loss of cardiac function, which was followed by the fibrous scarring of the myocardial [98]. Nanoparticles can treat atherosclerosis by improving drug solubility, minimizing the medication dosage frequency, lowering drug cytotoxicity, improving site-specific targeted drug delivery, and combining diagnostic and therapeutic approaches to create superior theranostics [99]. Resolving inflammation, preventing ROS formations and defective efferocytosis, preventing plaque formation, targeting macrophages, altering lipid metabolism, and preventing neointimal growth have focused on nanoparticle-mediated drug delivery and treatment for atherosclerosis and associated complications of CVDs (Figure 18.4) [100]. Liposomal preparation of bisphosphonate

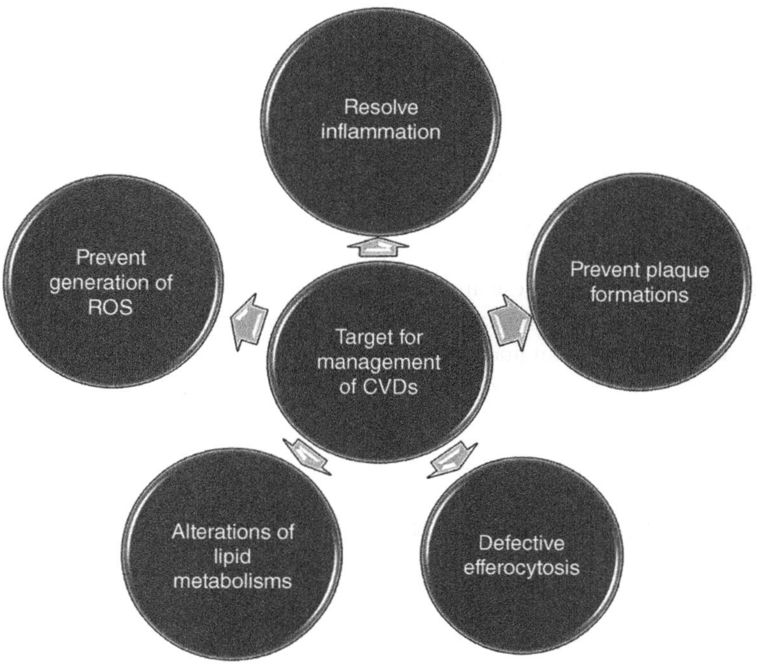

Figure 18.4 Potential targets for the management of CVDs.

alendronate decreased the neointimal formation and then suppressed the circulating monocytes in rabbits with iliac artery stenting. Liposomal alendronate has been found to be safer in early clinical trials for infusion at the time of percutaneous coronary intervention [101]. Pre-resolving nanomedicine is having promising applications in the treatment of atherosclerosis.

18.3.2 Nanoparticles for the Treatment of Ischemic Heart Diseases

Myocardial infarction is an ischemic severe heart condition that leads to cardiac failure and, eventually, death. Cardioprotective medicines have been used to enhance angiogenesis, promoting cardiomyocyte survival, improving heart function, reducing inflammation and myocyte activation, and ultimately preventing fibrosis. In a nutshell, all therapy and therapeutic agents are engaged in heart function protection and preservation. For improved absorption by target cells, specific tags can be covalently bonded to nanoparticles. Another advantage of nanoparticles over viruses is their faster synthesis methods and their ability to cluster receptors to activate signaling pathways [102, 103].

Placental GF (PIGF) is beneficial for stimulating angiogenesis and improving heart function at the site of AMI damage. Chitosan-alginate nanoparticles with biocompatibility, non-toxicity, and biodegradability have been developed for the efficient delivery of PIGF. Chitosan-alginate nanoparticles have been discovered to be an effective vehicle for PIGF administration by intramyocardial injection, enhancing the advantages of PIGF at the damage site. Prior to distribution, these nanoparticles can prevent the decomposition of the GF by enzymes [104].

Mitochondrial malfunction can cause CVDs. As a result, mitochondria are regarded as a significant therapeutic target for preventing heart failure and limiting the extent of myocardial infarctions in individuals with ischemic heart disease. Various mitoprotective medicament (cyclosporine-A (CsA) or TRO40303) have been developed to target cardiac mitochondria. The lack of permeability through the plasma and mitochondrial membranes is a disadvantage of these mitoprotective medicaments. Reduced permeability has been overcome by using nanoparticles and mitoprotective medicaments.

18.3.3 Nanoparticles for the Treatment of Hypertension

Today, a variety of medications are used to treat hypertension, for instance, vascular angiotensin antagonists, angiotensin-converting enzyme inhibitors, adrenergic receptor blockers, diuretics as well as vasodilators [105]. On the other hand, these antihypertensive classes of therapeutic molecules have apparent flaws, such as a short plasma half-life, limited bioavailability and side effects such as upper respiratory tract abstraction, angioedema, reflex tachycardia, severe hypotensive impact, and many more. Therefore, nanocarriers and phytoantioxidants can eliminate the limitation described earlier.

18.4 Nanoparticles as Drug Delivery Agents in the Treatment of CVDs

18.4.1 Gold Nanoparticles

One of the most commonly utilized nanocarriers for the delivery of cardioprotective medicines is gold nanoparticles (AuNPs). The use of AuNPs in conjunction with existing therapeutic medicines has been proposed as a novel method with increased potential for treating cardiac disorders. Gold nanocarriers have been widely employed because of their ease of production, low toxicity, immunogenicity, and stability [106]. Clinical medicines have been shown to be more efficient and precise when conjugated. Simdax is the FDA-approved medicament for the treatment of heart disease. Simdax conjugated on AuNPs exhibits more excellent cardioprotective functions in rats with doxorubicin-induced heart failure owing to the gold nanoconjugates formulation targeting affected tissue more effectively [107].

AuNPs have been shown to be effective drug delivery nanomaterials that do not cause toxicity in healthy tissues. Although AuNPs have been utilized in the laboratory to treat AMI, clinical studies and future research are being investigated. Other consequences of AMI include cardiomyocyte necrosis and apoptosis, fibroblast proliferation, cardiomyocyte conversion to myofibroblasts, collagen deposition in myocytes, and heart hypertrophy. When cardiomyocytes become too large, they might cause heart failure. However, AuNP conjugates or AuNPs alone have favorable benefits in AMI, preventing all other complications associated with other health issues. Protected 10 nm PEGylated AuNPs decrease the size of an infarcted heart by preventing necrosis and apoptosis [108]. Furthermore, they can reduce collagen deposition and, as a result, inflammation. As a result, the study recommends using AuNPs alone or in combination with nanoconjugates to cure CVDs [109].

Ischemia impairs blood flow and damages downstream tissues due to a lack of oxygen and nutrients in the circulation. Therapeutic angiogenesis occurs as a result of the angiogenesis process, in which new blood vessels are formed by developing and branching from existing blood vessels. Ischemia can be treated with therapeutic angiogenesis using GFs such as VEGFs. Clinical trials have shown that the delivery of these GFs is less effective due to the short half-life of GFs in the blood circulation, the lack of specific targeting, and the inability of GFs to remain in the circulation for days to weeks. That is essential to prevent the deterioration of new vessels in a murine hindlimb ischemia model, AuNPs serve as an effective payload for imaging and targeted administration of VEGFs to ischemic muscle tissues. The murine model of ischemia is a typical model for peripheral artery disease in the laboratory. Ischemic muscle can accumulate AuNPs smaller than 200 nm. As a result, therapeutic targeting of exogenous GFs to tissues results in an increase in blood perfusion (1.7-fold). Because of their increased permeability and retention impact, AuNPs target ischemia muscle tissues. The introduction of exogenous GFs stimulates the recovery of a large volume of ischemic tissue as well as increased angiogenesis [110]. AuNPs have gradually established themselves as one of the most promising medication carriers for precise delivery to damaged cardiac tissue [111, 112].

18.4.2 Polymeric and Magnetic Nanoparticles

Since polymeric nanoparticles may be modified and are likely to be reabsorbed in the body, they have attracted considerable interest in cardiovascular nanomedicine. Depending on the site of the action, these nanoparticles can be delivered by oral, dermal, mucosal, or transdermal routes [113]. Polylactic-co-glycolic acid (PLGA), polyglycolic acid (PGA), and polylactic acid are the most widely utilized polymers for the production of biodegradable polymeric nanocarriers (PLA). They are called biodegradable due to their ease in eliminating carbon dioxide and water from the human body. Among these polymers, PLGA has been widely tested as a drug carrier for the treatment of CVDs [13]. Dextran-coated nanoparticles were employed in an experiment to administer poorly water-soluble pharmaceuticals SB431542 and CHIR99021 medicaments to fibroblasts to convert them into cardiomyocytes. They were encapsulated on the surface of AcDX nanoparticles functionalized with spermine, resulting in pH-induced transport into fibroblasts. For the replacement of cardiac cells, cellular reprogramming of damaged myofibroblasts and fibroblasts into healthy myocytes is always a viable therapeutic option [85]. The electrohydrodynamic atomization technique synthesized polymeric nanoparticles containing quercetin to enhance the quercetin water solubility and stability as part of a second study. As a consequence, atherosclerosis plaques were prevented. Because of its low solubility, encapsulating this antioxidant agent on PLGA nanoparticles improved the therapeutic impact of medicines while also reducing the generation of reactive oxygen species [114]. Polymeric nanocarriers are utilized to deliver anti-proliferative medicines to damaged vasculature and are thought to be a safe way to treat coronary artery disease [115]. It is obvious that reactive oxygen species (for example, hydrogen peroxide) produced by ischemia-reperfusion injury cause arterial thrombosis and cardiac cell damage. Directly administered anticoagulants have a modest drug release rate and a short blood circulation time. It also has adverse side effects by hastening the breakdown of clots in thrombo-occluded arteries. As a result, PLGA nanoparticles encapsulated with antioxidant (glutathione) and anticoagulant (heparin) compounds were used to treat ischemia/reperfusion damage. Anticoagulant agents supplied controlled and targeted delivery of drugs occurred within two hours, indicating a potential platform to treat vascular diseases [116].

Recent advances in the design and synthesis of multifunctional nanostructures with magnetic or plasmonic characteristics have the potential to be a formidable weapon in the treatment of CVDs through the use of magnetically assisted targeted medication delivery. Magnetic nanoparticles are being investigated as potential MRI contrast agents and medication delivery agents for the treatment of CVDs. Drug-laden magnetic nanoparticles can be readily monitored and used for targeted administration depending on the contrasting agent utilized. Iron oxide superparamagnetic nanoparticles are known as excellent nanocarriers for treating AMI in this competition.

In an embolic rat model, magnetic nanoparticles loaded with tissue plasminogen activator improved local thrombolysis by magnetically orienting the plasminogen activator. It was discovered that the dose of drug bound to magnetic nanoparticles necessary to break the clot was 100 times lower than that required when the medication was administered free. It indicates that magnetic resonance tracking and medication administration might be used to treat a wide range of occlusive vascular disorders [117]. Magnetic nanoparticles and liposomes are combined to make magnetic liposomes, which are utilized for magnetic resonance imaging and targeted magnetic-based drug delivery in CVDs. Magnetic liposomes might be used to distribute drugs in a variety of ways. They are utilized to treat infarcted myocardial as well as to provide GFs, biomolecules, and cytokines to dying cardiac cells [118]. These nanoparticles effectively targeted the overexpressed angiotensin receptor 1 in the infarcted myocardium [119]. The fibrotic area of the afflicted heart was reduced by nanoparticles encapsulated with p38 inhibitors [120]. Despite the fact that nanoparticles have great promise in cardiology, health hazards and safety should be considered. The impact of nanoparticles on human health and the environment will be evaluated in order to develop safety measures. The effects of nanoparticles on health have been evaluated through the deposition and removal of solid nanoparticles, systemic translocation, biocompatibility, and consequences for the digestive tract, central nervous system, and body distribution. According to research, changes in the structural and physicochemical characteristics of nanoparticles with decreasing material size may result in hazardous consequences. Hence, it is essential to bridge the gap between the toxic impact of nanoparticles and the health benefits that have been employed in CVDs [121].

18.4.3 Liposomal Platforms

Liposomal delivery systems are solid platforms for delivering therapeutic medicines to particular locations. Liposomes are spherical nanoparticles with a bilayer structure comprised of natural or synthetic lipids. The hydrophilic core surrounds the hydrophobic membrane in these spherical forms [122, 123]. At this time, no liposomal pharmacological platforms for the direct treatment of CVDs in humans are available. Several investigations and clinical trials have been conducted to see if liposomal platforms can be used to deliver drugs. PEGylated liposomes encapsulated with therapeutic payloads of the AT1 receptor enhanced the recovery of cardiac cells after the incidence of a heart attack. In vivo experiments indicated that this nanoformulation is capable of precise targeting, long-term drug release, and enhanced accumulation in the injured location while causing no harm to healthy cardiac cells. Myocardial injury was treated more effectively with a nano-delivery method, resulting in more excellent therapeutic responses and more rapid regeneration of heart cells [124]. In early clinical trials, liposomal alendronate platforms were proven to be safer for infusion during the percutaneous coronary intervention [101].

Platelet targeting with liposomal platforms has also been shown to be an effective treatment approach in CVDs. Because of their better and increased medication distribution to the damaged cardiovascular tissues and surroundings, modified liposomes are always appealing in treating CVDs [125]. Surfactant attachment mimics the functions of cell glycocalyx and inhibits cell absorption. Because the peptide binds to the integrin found on platelets, they are more likely to be targeted. As a result, this study showed liposomes' ability to disguise themselves as platelets in order to reduce opsonization and endothelial absorption [126]. Results suggested that liposomal carriers containing peptides may be utilized significantly to target platelets for the therapy of CVDs [127].

18.4.4 Phytosomes

Phytosomes are developed by encapsulating the plant material or plant extract (like flavonoids, terpenoids, tannins, and xanthones) in a nonpolar solvent within the spherical cell-like structure, which is an advanced nanosphere or cell form of herbal formulations that are better absorbed when applied topically or orally and, as a result, develop better results than conventional herbal extracts [83]. Phytosomes produce a better pharmacokinetic and pharmacodynamic profile of drugs than conventional herbal formulations. It is a novel emerging technique applied to phytopharmaceuticals to enhance the bioavailability of natural plant extract for medicinal applications. Phytomedicines, a complex chemical mixture prepared from plants, have been used in India and worldwide since the beginning of human civilization and go forward to have widespread popular use [128].

To examine the various advantages of phytosomes, especially their ability to enhance the bioavailability of polar phytoconstituents, various therapeutic applications of phytosomes have been explored. Maiti and co-workers developed phytosomes of curcumin and naringenin in two different studies. They found that the antioxidant activity of phytosomes of naringenin and curcumin significantly exhibited better antioxidant activity than the free compound with a prolonged duration of action, which may be helpful in reducing the rapid elimination of the molecule from the body[129, 130]. Pierro and coworkers administered phytosomes of green tea extract in 100 numbers of obese individuals and found that the formulation had significantly reduced the body cholesterol level, i.e. LDL, total cholesterol, triglycerides, and body mass index of subjects. Phytosomes of resveratrol, quercetin, catechin, procyanidins, and epicatechin were successfully developed, minimized the free radical level, increased the antioxidant level, and enhanced the resistance of LDLs, which, as a result, causes oxidative modification. Finally, it showed cardiovascular protective effects [90].

18.4.5 Ethosomes

Ethosomes are additional novel soft, malleable vesicle lipid carriers composed mainly of phospholipids, ethanol (relatively high concentration) and water [131]. These are novel vesicular carriers for enhanced delivery through the skin that enables drugs to reach the deeper skin layers or the systemic circulation [84] and improve the skin delivery of various drugs [132]. Ethosomes are noninvasive delivery carriers that enable drugs to reach the deep skin layers or the systemic circulation. This fluidization of intercellular lipid bilayers is considered to occur with the help of ethanol. Soft, pliable vesicles then penetrate these unorganized lipid bilayers. The lipid vesicles serve to solubilize poorly soluble drug molecules and create a regulated transdermal delivery mechanism. Deeper penetration and distribution in the skin lipid bilayers have been attributed to a combination of phospholipids and a high concentration of ethanol in vesicular formulations. Drugs that are both hydrophilic and lipophilic can be incorporated into vesicles. However, while they are theoretically complicated, ethosomal systems are easy to prepare, safe and effective – a combination that has the potential to greatly extend their use When bioactive compounds are delivered by an ethosomal carrier to the skin and cells, there are several difficulties and potential for future study and development of innovative better treatments for CVD [132].

18.4.6 Carbon Nanotubes

Carbon nanotubes (CNTs) are made up of graphite sheets that have been coiled up into tubular shapes and chemically functionalized on the surface to provide selective targeting of therapeutic molecules. The therapeutic active molecules are conjugated in the inner space [13]. CNTs are primarily utilized as drug transporters and can solve poor aqueous solubility, quicker degradation and poor bioavailability issues. However, one of the primary drawbacks of CNTs is the possibility of dissociation in biological fluids [133]. Nonetheless, CNTs are an excellent drug carrier for increased cell penetration as well as privileged pharmacological effects (Zhang et al. 2011). Because of their distinct optical, electrical, and mechanical characteristics, they are a promising contender for future therapeutic uses [134]. Furthermore, a number of studies have verified the promising potentials of CNT in cardiac tissue engineering, for instance, support of cardiomyocyte function and development and acceleration of gap junction creation [135, 136]. There have been other experiments that show the CNT scaffold might be a potential injectable biomaterial to transport therapeutic molecules for cardiac tissue regeneration in infarcted myocardial tissues [134].

18.5 Future Scope of Nano-Based Herbal Formulation in CVDs

It has long been recognized that cardiovascular illnesses and their related problems are among the leading causes of morbidity and death worldwide. Despite the fact that some of the problems are treatable, biomedical experts continue to be concerned about coronary artery diseases due to fast changes in human lifestyle. There are a variety of cardioprotective medicines that can be used to treat and prevent heart disease. Natural chemicals, herbal medicines, and TCM ranked first among all therapeutic methods. Because the adverse effects of these methods and substances are so low, they are favored over manufactured medicines. However, due to a lack of scientific basis, these medicines frequently fail to pass the regulatory affairs. Nonetheless, researchers are continuously looking for new technologies and breakthroughs to help them diagnose and treat CVDs more effectively and efficiently than previous approaches. Conventional treatment modalities that target different pathways, such as lipid metabolism, blood volume, coagulation cascade, and arterial constriction, have drawbacks and ineffective treatment of CVDs with side effects such as thrombosis in stents and unusual immune system activation and systemic toxicity. As a result, nano-drug delivery systems have been used as an alternate vehicle

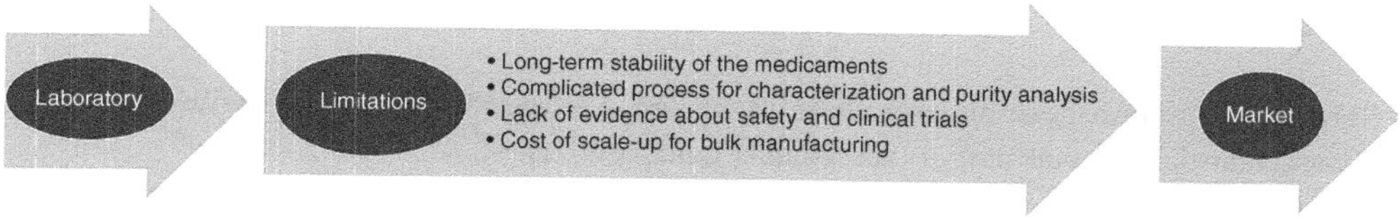

Figure 18.5 Limitations of nanotherapeutics-mediated drug delivery in CVDs from laboratory to the market.

for delivering medicines into damaged cardiac tissue in an efficient and regulated manner. Similarly, nanoparticles have enormous implications in detecting and imaging CVDs since they enable easy CVD diagnosis and real-time tracking during therapy. Nanoimaging in CVD used a variety of multifunctional and multimodal imaging vehicles, which is hard to do with traditional imaging techniques.

Despite the fact that numerous cardiovascular nanoformulations have been evaluated in vitro and in vivo in clinical trials, clinical translation is still in the works. Figure 18.5 shows a few limitations of nanotherapeutics-mediated drug delivery in CVDs from laboratory to market. It addresses issues such as pharmaceutical scale-up, release efficiency, and stringent regulatory criteria for demonstrating safety and efficacy (GMP). In addition, to fully understand the effects of nanoparticles in the body and on cardiac tissues, in vivo and clinical trials are necessary. Future designs and advances in CVD nanocarriers will necessitate further research. Several natural compounds generated from pharmaceuticals and nanoformulations with cardioprotective action are awaiting clinical trials and studies.

Nanomedicine translational research in CVDs has several difficulties that hinder its practical implementation in clinics. Institutional and industrial interests, a lack of sufficient infrastructure and expertise to prepare patients for treatment, proper monitoring of patient outcomes, patient registration in clinical trials, and country-specific remuneration and cost are among the obstacles. Bioengineers, pharmacists, chemists, biologists, and physicians should utilize nanoparticles to collaborate on translational medicine in CVDs. Future studies will focus on using peptides and antibodies to identify CVD indicators and build selective nano-delivery methods. The goal of future nano-cardio medicine is to assess the clinical impacts of new nanosystems in order to enhance the quality of life.

18.6 Conclusions

As theranostic agents, nanotechnology and nanomedicine provide personalized attention to the diagnostic and therapeutic domains of CVDs. Nanoparticles function as nanodevices or nanocarriers to successfully transport medicines to the damaged location. Nanocarriers have the potential to provide effective, targeted drug and gene delivery while avoiding issues with solubility, bioavailability, and other pharmacokinetic characteristics. The current study provides an overview of the many cardiovascular illnesses and therapeutic approaches accessible, as well as recent developments in the field of therapy. It outlines and summarizes the essential efforts and advancements achieved toward targeted drug delivery to cardiovascular tissues and associated biomarkers for therapy. Furthermore, there has been a gap in the realm of nanomedicine for CVDs.

References

1 WHO (2020). Cardiovascular diseases, World Heal. Organ. https://www.who.int/health-topics/cardiovascular-diseases#tab=tab_1 (accessed 23 January 2021).

2 Sarma, H., Upadhyaya, M., Gogoi, B. et al. (2021). Cardiovascular drugs: an insight of *In Silico* drug design tools. *J. Pharm. Innov.* https://doi.org/10.1007/s12247-021-09587-w.

3 Walker, I.F., Garbe, F., Wright, J. et al. (2018). The economic costs of cardiovascular sisease, diabetes mellitus, and associated complications in South Asia: a systematic review. *Value Heal. Reg. Issues.* 15: 12–26. https://doi.org/10.1016/j.vhri.2017.05.003.

4 World Heart Federation (2020). Anonymous, Cardiovascular disease continues to be the leading cause of death and disability in the world today, World Hear. Fed. https://www.world-heart-federation.org/world-heart-day/world-heart-day-2019/cvds/ (accessed 23 January 2021).

5 World Heart Federation (2020). Anonymous, the cost of CVD | champion advocates programme, World Hear. Fed. http://www.championadvocates.org/en/champion-advocates-programme/the-costs-of-cvd (accessed 26 January 2021).

6 Iafisco, M., Alogna, A., Miragoli, M., and Catalucci, D. (2019). Cardiovascular nanomedicine: the route ahead. *Nanomedicine* 14: 2391–2394. https://doi.org/10.2217/nnm-2019-0228.

7 Mathers, C.D. and Loncar, D. (2006). Projections of global mortality and burden of disease from 2002 to 2030. *PLoS Med.* 3: 2011–2030. https://doi.org/10.1371/journal.pmed.0030442.

8 Ross, L., Jennings, P., and Williams, B. (2017). Psychosocial support issues affecting older patients: A cross-sectional paramedic perspective. *Inquiry* 54: 1–6. https://doi.org/10.1177/0046958017731963.

9 Kumar Yadav, B., Lal Khatik, G., Haneef, J., and Subedi, B. (2019). Recent advancement over traditional drug in the treatment of hypertension. *Int. J. Adv. Res. Dev.* 4: 58–61. https://doi.org/xx.xxx/ijariit-v4i1-1163.

10 Martín-Timón, I. (2014). Type 2 diabetes and cardiovascular disease: have all risk factors the same strength? *World J. Diabetes* 5: 470. https://doi.org/10.4239/wjd.v5.i4.444.

11 Einarson, T.R., Acs, A., Ludwig, C., and Panton, U.H. (2018). Prevalence of cardiovascular disease in type 2 diabetes: a systematic literature review of scientific evidence from across the world in 2007-2017. *Cardiovasc. Diabetol.* 17: 83. https://doi.org/10.1186/s12933-018-0728-6.

12 Kim, K.S., Khang, G., and Lee, D. (2011). Application of nanomedicine in cardiovascular diseases and stroke. *Curr. Pharm. Des.* 17: 1825–1833. https://doi.org/10.2174/138161211796390967.

13 Katsuki, S., Matoba, T., Koga, J.I. et al. (2017). Anti-inflammatory nanomedicine for cardiovascular disease. *Front. Cardiovasc. Med.* 4: 87. https://doi.org/10.3389/fcvm.2017.00087.

14 Ong, S.B., Lu, S., Katwadi, K. et al. (2017). Nanoparticle delivery of mitoprotective agents to target ischemic heart disease. *Futur. Cardiol.* 13: 195–198. https://doi.org/10.2217/fca-2017-0012.

15 American Heart Association (2017) Heart failure projected to increase dramatically, according to new statistics. https://www.heart.org/en/news/2018/05/01/heart-failure-projected-to-increase-dramatically-according-to-new-statistics (accessed 23 May 2021).

16 Chahine, J. and Alvey, H. (2020). *Left Ventricular Failure.* StatPearls Publishing LLC https://www.ncbi.nlm.nih.gov/books/NBK537098/ (accessed 23 May 2021).

17 Velagaleti, R.S. and Vasan, R.S. (2007). Heart failure in the twenty-first century: is it a coronary artery disease or hypertension problem? *Cardiol. Clin.* 25: 487–495. https://doi.org/10.1016/j.ccl.2007.08.010.

18 Demer, L.L. and Tintut, Y. (2008). Vascular calcification: pathobiology of a multifaceted disease. *Circulation* 117: 2938–2948. https://doi.org/10.1161/CIRCULATIONAHA.107.743161.

19 Liu, W., Zhang, Y., Yu, C.M. et al. (2015). Current understanding of coronary artery calcification. *J. Geriatr. Cardiol.* 12: 668–675. https://doi.org/10.11909/j.issn.1671-5411.2015.06.012.

20 Cheng, H.-M., Wang, J.-J., and Chen, C.-H. (2017). The role of vascular calcification in heart failure and cognitive decline. *Pulse* 5: 144–153. https://doi.org/10.1159/000484941.

21 Palit, S. and Kendrick, J. (2014). Vascular calcification in chronic kidney disease: role of disordered mineral metabolism. *Curr. Pharm. Des.* 20: 5829–5833. https://doi.org/10.2174/1381612820666140212194926.

22 Disthabanchong, S. (2012). Vascular calcification in chronic kidney disease: pathogenesis and clinical implication. *World J. Nephrol.* 1: 53. https://doi.org/10.5527/wjn.v1.i2.43.

23 Dube, P., DeRiso, A., Patel, M. et al. (2021). Vascular calcification in chronic kidney disease: diversaity in the vessel wall. *Biomedicine* 9: 404. https://doi.org/10.3390/biomedicines9040404.

24 Chen, N.X. and Moe, S.M. (2012). Vascular calcification: pathophysiology and risk factors. *Curr. Hypertens. Rep.* 14: 228–237. https://doi.org/10.1007/s11906-012-0265-8.

25 Liu, M., Li, M., Wang, G. et al. (2014). Heart-targeted nanoscale drug delivery systems. *J. Biomed. Nanotechnol.* 10: 2038–2062. https://doi.org/10.1166/jbn.2014.1894.

26 Buttar, H.S., Li, T., and Ravi, N. (2005). Prevention of cardiovascular diseases: Role of exercise, dietary interventions, obesity and smoking cessation. *Exp. Clin. Cardiol.* 10: 229–249. /pmc/articles/PMC2716237/ (accessed 23 May 2021).

27 Rippe, J.M. (2019). Lifestyle strategies for risk factor reduction, prevention, and treatment of cardiovascular disease. *Am. J. Lifestyle Med.* 13: 204–212. https://doi.org/10.1177/1559827618812395.

28 Nathan, C. and Cunningham-Bussel, A. (2013). Beyond oxidative stress: an immunologist's guide to reactive oxygen species. *Nat. Rev. Immunol.* 13: 349–361. https://doi.org/10.1038/nri3423.

29 Shah, M.S. and Brownlee, M. (2016). Molecular and cellular mechanisms of cardiovascular disorders in diabetes. *Circ. Res.* 118: 1808–1829. https://doi.org/10.1161/CIRCRESAHA.116.306923.

30 Campanucci, V., Krishnaswamy, A., and Cooper, E. (2010). Diabetes depresses synaptic transmission in sympathetic ganglia by inactivating nAChRs through a conserved intracellular cysteine residue. *Neuron* 66: 827–834. https://doi.org/10.1016/j.neuron.2010.06.010.

31 WHO (2017). Cardiovascular diseases (CVDs), World Heal. Orgenaization. https://www.who.int/news-room/fact-sheets/detail/cardiovascular-diseases-(cvds) (accessed 28 May 2021).

32 Shimizu, H., Hirahara, N., Motomura, N. et al. (2019). Current status of cardiovascular surgery in Japan 2015 and 2016: analysis of data from Japan Cardiovascular Surgery Database. 4—Thoracic aortic surgery. *Gen. Thorac. Cardiovasc. Surg.* 67: 751–757. https://doi.org/10.1007/s11748-019-01163-x.

33 Ylä-Herttuala, S. and Martin, J.F. (2000). Cardiovascular gene therapy. *Lancet* 355: 213–222. https://doi.org/10.1016/S0140-6736(99)04180-X.

34 Tarkin, J.M. and Kaski, J.C. (2013). Pharmacological treatment of chronic stable angina pectoris. *Clin. Med. J. R. Coll. Physicians London* 13: 63–70. https://doi.org/10.7861/clinmedicine.13-1-63.

35 Singh, B., Garg, T., Goyal, A.K., and Rath, G. (2016). Recent advancements in the cardiovascular drug carriers. *Artif. Cells, Nanomedicine, Biotechnol.* 44: 216–225. https://doi.org/10.3109/21691401.2014.937868.

36 Shaito, A., Thuan, D.T.B., Phu, H.T. et al. (2020). Herbal medicine for cardiovascular diseases: efficacy, mechanisms, and safety. *Front. Pharmacol.* 11: 422. https://doi.org/10.3389/fphar.2020.00422.

37 Sun, M., Chai, L., Lu, F. et al. (2018). Efficacy and safety of Ginkgo biloba pills for coronary heart disease with impaired glucose regulation: study protocol for a series of N-of-1 randomized,

double-blind, placebo-controlled trials. *Evidence-Based Complement. Altern. Med.* 2018. https://doi.org/10.1155/2018/7571629.

38 Saeed, A., Larik Ali, F., Channar Ali, P., and Muqadar, U. (2015). The heart and herbs: back to the nature. *J. Heal. Med. Informatics.* 6: 1–11. https://www.hilarispublisher.com/open-access/the-heart-and-herbs-back-to-the-nature-2157-7420-1000212.pdf (accessed 28 May 2021).

39 Giglio, R.V., Patti, A.M., Cicero, A.F.G. et al. (2018). Polyphenols: potential use in the prevention and treatment of cardiovascular diseases. *Curr. Pharm. Des.* 24: 239–258. https://doi.org/10.2174/1381612824666180130112652.

40 Behl, T., Bungau, S., Kumar, K. et al. (2020). Pleotropic effects of polyphenols in cardiovascular system. *Biomed. Pharmacother.* 130: 110714. https://doi.org/10.1016/j.biopha.2020.110714.

41 Manach, C., Mazur, A., and Scalbert, A. (2005). Polyphenols and prevention of cardiovascular diseases. *Curr. Opin. Lipidol.* 16: 77–84. https://doi.org/10.1097/00041433-200502000-00013.

42 Zhang, Y.J., Gan, R.Y., Li, S. et al. (2015). Antioxidant phytochemicals for the prevention and treatment of chronic diseases. *Molecules* 20: 21138–21156. https://doi.org/10.3390/molecules201219753.

43 Kang, J., Xie, C., Li, Z. et al. (2011). Flavonoids from acai (Euterpe oleracea Mart.) pulp and their antioxidant and anti-inflammatory activities. *Food Chem.* 128: 152–157. https://doi.org/10.1016/j.foodchem.2011.03.011.

44 He, D., Ru, X., Wen, L. et al. (2012). Total flavonoids of Flos Chrysanthemi protect arterial endothelial cells against oxidative stress. *J. Ethnopharmacol.* 139: 68–73. https://doi.org/10.1016/j.jep.2011.10.043.

45 Cao, H., Jia, Q., Shen, D. et al. (2019). Quercetin has a protective effect on atherosclerosis via enhancement of autophagy in ApoE-/- mice. *Exp. Ther. Med.* 18: 58. https://doi.org/10.3892/etm.2019.7851.

46 Xiong, X., Yang, X., Liu, W. et al. (2013). Trends in the treatment of hypertension from the perspective of traditional chinese medicine, evidence-based complement. *Altern. Med.* 2013: 275279. https://doi.org/10.1155/2013/275279.

47 Burta, O., Tirlea, F., Burta, O.L., and Qadri, S.M. (2008). Phytotherapy in cardiovascular diseases: From ethnomedicine to evidence based medicine. *J. Biol. Sci.* 8: 242–247. https://doi.org/10.3923/jbs.2008.242.247.

48 Farnsworth, N.R., Akerele, O., Bingel, A.S. et al. (1985). Medicinal plants in therapy. *Bull. World Health Organ.* 63: 965–981. https://pubmed.ncbi.nlm.nih.gov/3879679/ (accessed 29 May 2021).

49 Fabricant, D.S. and Farnsworth, N.R. (2001). The value of plants used in traditional medicine for drug discovery. *Environ. Health Perspect.* 109: 69–75. https://doi.org/10.1289/ehp.01109s169.

50 Harvey, A. (2000). Strategies for discovering drugs from previously unexplored natural products. *Drug Discov. Today* 5: 294–300. https://doi.org/10.1016/S1359-6446(00)01511-7.

51 Hodgson, J.M., Prince, R.L., Woodman, R.J. et al. (2016). Apple intake is inversely associated with all-cause and disease-specific mortality in elderly women. *Br. J. Nutr.* 115: 860–867. https://doi.org/10.1017/S0007114515005231.

52 Gunathilake, C. and Considine, M. (2014). Flavonoids rich apple for healthy life. *MOJ Food Process Technol.* 89–91. https://doi.org/10.15406/mojfpt.2018.06.00149.

53 Yamada, T., Hayasaka, S., Shibata, Y. et al. (2011). Frequency of citrus fruit intake is associated with the incidence of cardiovascular disease: the jichi medical school cohort study. *J. Epidemiol.* 21: 169–175. https://doi.org/10.2188/jea.JE20100084.

54 Zhao, C.N., Meng, X., Li, Y. et al. (2017). Bin Li, fruits for prevention and treatment of cardiovascular diseases. *Nutrients* 9. https://doi.org/10.3390/nu9060598.

55 Li, H., Sureda, A., Devkota, H.P. et al. (2020). Curcumin, the golden spice in treating cardiovascular diseases. *Biotechnol. Adv.* 38: 107343. https://doi.org/10.1016/j.biotechadv.2019.01.010.

56 Siasos, G., Tsigkou, V., Kosmopoulos, M. et al. (2018). Mitochondria and cardiovascular diseases—from pathophysiology to treatment. *Ann. Transl. Med.* 6: 256–256. https://doi.org/10.21037/atm.2018.06.21.

57 Hernández-Camacho, J.D., Bernier, M., López-Lluch, G., and Navas, P. (2018). Coenzyme Q10 supplementation in aging and disease. *Front. Physiol.* 9: 44. https://doi.org/10.3389/fphys.2018.00044.

58 Diaconeasa, Z., Iuhas, C.I., Ayvaz, H. et al. (2019). Phytochemical characterization of commercial processed blueberry, blackberry, blackcurrant, cranberry, and raspberry and their antioxidant activity. *Antioxidants* 8: 540. https://doi.org/10.3390/antiox8110540.

59 Yang, H., Tian, T., Wu, D. et al. (2019). Prevention and treatment effects of edible berries for three deadly diseases: cardiovascular disease, cancer and diabetes. *Crit. Rev. Food Sci. Nutr.* 59: 1903–1912. https://doi.org/10.1080/10408398.2018.1432562.

60 Cho, S., Namkoong, K., Shin, M. et al. (2017). Cardiovascular protective effects and clinical applications of resveratrol. *J. Med. Food* 20: 323–334. https://doi.org/10.1089/jmf.2016.3856.

61 Zordoky, B.N.M., Robertson, I.M., and Dyck, J.R.B. (2014). Preclinical and clinical evidence for the role of resveratrol in the treatment of cardiovascular diseases. *Biochim. Biophys. Acta Mol. basis Dis.* 1852: 1155–1177. https://doi.org/10.1016/j.bbadis.2014.10.016.

62 Liu, M., Chen, Z., and Chen, L. (2016). Endoplasmic reticulum stress: a novel mechanism and therapeutic target for cardiovascular diseases. *Acta Pharmacol. Sin.* 37: 425–443. https://doi.org/10.1038/APS.2015.145.

63 Alamgir, A.N.M. (2018). *Biotechnology, In Vitro Production of Natural Bioactive Compounds, Herbal Preparation, and Disease Management (Treatment and Prevention).* Nature Publishing Group. https://doi.org/10.1007/978-3-319-92387-1_7.

64 Liu, H., Yang, J., Li, L. et al. (2016). The natural occurring compounds targeting endoplasmic reticulum stress. *Evid. Based Complement. Alternat. Med.* 2016: 7831282. https://doi.org/10.1155/2016/7831282.

65 Choy, K.W., Murugan, D., and Mustafa, M.R. (2018). Natural products targeting ER stress pathway for the treatment of cardiovascular diseases. *Pharmacol. Res.* 132: 119–129. https://doi.org/10.1016/J.PHRS.2018.04.013.

66 Bahreyni, A., Avan, A., Shabani, M. et al. (2019). Therapeutic potential of A2 adenosine receptor pharmacological regulators in the treatment of cardiovascular diseases, recent progress, and prospective. *J. Cell. Physiol.* 234: 1295–1299. https://doi.org/10.1002/JCP.27161.

67 Ciambelli, P., La Guardia, G., and Vitale, L. (2019). Nanotechnology for green materials and processes. In: *Studies in Surface Science and Catalysis*, 97–116. Elsevier Inc. https://doi.org/10.1016/B978-0-444-64337-7.00007-0.

68 Cancino-Bernardi, J., Marangoni, V.S., and Zucolotto, V. (2017). Nanomedicine. In: *Nanoscience and its Applications* (ed. A.L. Da Róz, M. Ferreira and O.N. Oliveira), 71–92. Elsevier Inc. https://doi.org/10.1016/B978-0-323-49780-0.00003-X.

69 De Jong, W.H. and Borm, P.J.A. (2008). Drug delivery and nanoparticles:applications and hazards. *Int. J. Nanomedicine* 3: 133–149. http://www.ncbi.nlm.nih.gov/pubmed/18686775 (accessed September 24, 2018).

70 Din, F.U., Aman, W., Ullah, I. et al. (2017). Effective use of nanocarriers as drug delivery systems for the treatment of selected tumors. *Int. J. Nanomedicine* 12: 7291–7309. https://doi.org/10.2147/IJN.S146315.

71 Foroozandeh, P. and Aziz, A.A. (2018). Insight into cellular uptake and intracellular trafficking of nanoparticles. *Nanoscale Res. Lett.* 13: 1–12. https://doi.org/10.1186/s11671-018-2728-6.

72 Mauricio, M.D., Guerra-Ojeda, S., Marchio, P. et al. (2018). Nanoparticles in medicine: a focus on vascular oxidative stress. *Oxidative Med. Cell. Longev.* 2018: 6231482. https://doi.org/10.1155/2018/6231482.

73 Wong, I.Y., Bhatia, S.N., and Toner, M. (2013). Nanotechnology: emerging tools for biology and medicine. *Genes Dev.* 27: 2397–2408. https://doi.org/10.1101/gad.226837.113.

74 Patra, J.K., Das, G., Fraceto, L.F. et al. (2018). Nano based drug delivery systems: recent developments and future prospects. *J. Nanobiotechnology* 16: 1–33. https://doi.org/10.1186/S12951-018-0392-8.

75 Su, S. and Kang, P.M. (2020). Systemic review of biodegradable nanomaterials in nanomedicine. *Nanomaterials* 10: 1–21. https://doi.org/10.3390/nano10040656.

76 Pala, R., Anju, V.T., Dyavaiah, M. et al. (2020). Nanoparticle-mediated drug delivery for the treatment of cardiovascular diseases. *Int. J. Nanomedicine* 15: 3741–3769. https://doi.org/10.2147/IJN.S250872.

77 Kyung Kim, D. and Dobson, J. (2009). Nanomedicine for targeted drug delivery. *J. Mater. Chem.* 19: 6294–6307. https://doi.org/10.1039/B902711B.

78 Cicha, I., Chauvierre, C., Texier, I. et al. (2018). From design to the clinic: practical guidelines for translating cardiovascular nanomedicine. *Cardiovasc. Res.* 114: 1714. https://doi.org/10.1093/CVR/CVY219.

79 Westedt, U., Barbu-Tudoran, L., Schaper, A.K. et al. (2002). Deposition of nanoparticles in the arterial vessel by porous balloon catheters: localization by confocal laser scanning microscopy and transmission electron microscopy. *AAPS PharmSci.* 4: E41. https://doi.org/10.1208/PS040441.

80 Sau, T.K., Biswas, A., and Ray, P. (2017). Metal nanoparticles in nanomedicine: advantages and scope. In: *Met. Nanoparticles* (ed. S. Thota and D.C. Crans), 121–168. John Wiley & Sons, Ltd https://doi.org/10.1002/9783527807093.CH6.

81 Kumar, H., Venkatesh, N., Bhowmik, H., and Kuila, A. (2018). Metallic nanoparticle: a review. *Biomed. J. Sci. Tech. Res.* 4: 3765–3774. https://doi.org/10.26717/BJSTR.2018.04.001011.

82 Fan, C., Joshi, J., Li, F. et al. (2020). Nanoparticle-mediated drug delivery for treatment of ischemic heart disease. *Front. Bioeng. Biotechnol.* 8: 687. https://doi.org/10.3389/FBIOE.2020.00687.

83 Bhattacharya, S. (2009). Phytosomes: emerging strategy in delivery of herbal drugs and nutraceuticals. *Pharma Times* 41: 9–12.

84 Bhalaria, M.K., Naik, S., and Misra, A.N. (2009). Ethosomes: a novel delivery system for antifungal drugs in the treatment of topical fungal diseases - PubMed. *Indian J. Exp. Biol.* 47: 368–375. https://pubmed.ncbi.nlm.nih.gov/19579803/ (accessed 30 August 2021).

85 Ferreira, M., Ranjan, S., Kinnunen, S. et al. (2017). Drug-loaded multifunctional nanoparticles targeted to the endocardial layer of the injured heart modulate hypertrophic signaling. *Small* 13: 1–14. https://doi.org/10.1002/SMLL.201701276.

86 Azeez, L., Lateef, A., and Adebisi, S.A. (2017). Silver nanoparticles (AgNPs) biosynthesized using pod extract of Cola nitida enhances antioxidant activity and phytochemical composition of Amaranthus caudatus Linn. *Appl. Nanosci.* 7: 59–66. https://doi.org/10.1007/S13204-017-0546-2.

87 Boarescu, P., Boarescu, I., Bocşan, I. et al. (2019). Curcumin nanoparticles protect against isoproterenol induced myocardial infarction by alleviating myocardial tissue oxidative stress. *Electrocardiogram, and Biological Changes, Mol.* 24: 2802. https://doi.org/10.3390/MOLECULES24152802.

88 Wang, Y., Wang, S., Firempong, C. et al. (2017). Enhanced solubility and bioavailability of naringenin via liposomal nanoformulation: preparation and in vitro and in vivo evaluations. *AAPS PharmSciTech* 18: 586–594. https://doi.org/10.1208/S12249-016-0537-8.

89 Wang, X., Guan, Q., Chen, W. et al. (2015). Novel nanoliposomal delivery system for polydatin: preparation, characterization, and in vivo evaluation. *Drug Des. Devel. Ther.* 9: 1805–1813. https://doi.org/10.2147/DDDT.S77615.

90 Singh, A., Saharan, V.A., Singh, M., and Bhandari, A. (2011). Phytosome: drug delivery system for polyphenolic phytoconstituents. *Iran. J. Pharm. Sci.* 7: 209–219. http://www.ijps.ir/article_1834.html (accessed 3 August 2021).

91 Hooresfand, Z., Ghanbarzadeh, S., and Hamishehkar, H. (2015). Preparation and characterization of rutin-loaded nanophytosomes. *Pharm. Sci.* 21: 145–151. https://doi.org/10.15171/PS.2015.29.

92 Gahandule, M., Jadav, S., Gadhave, M., and Gaikawad, D. (2016). Formulation and development of hepato-protective butea monosperma-Phytosome. *Int. J. Res. Pharm. Pharm. Sci.* 1: 21–27. www.pharmacyjournal.in (accessed 31 August 2021).

93 Zamani, P., Fereydouni, N., Butler, A. et al. (2019). The therapeutic and diagnostic role of exosomes in cardiovascular diseases. *Trends Cardiovasc. Med.* 29: 313–323. https://doi.org/10.1016/J.TCM.2018.10.010.

94 Boarescu, P., Chirilă, I., Bulboacă, A. et al. (2019). Effects of curcumin nanoparticles in isoproterenol-induced myocardial infarction. *Oxidative Med. Cell. Longev.* 2019: 7847142. https://doi.org/10.1155/2019/7847142.

95 Boarescu, P., Boarescu, I., Bocşan, I. et al. (2019). Antioxidant and anti-inflammatory effects of curcumin nanoparticles on drug-induced acute myocardial infarction in diabetic rats. *Antioxidants* 8: 504. https://doi.org/10.3390/ANTIOX8100504.

96 Brito, L. and Amiji, M. (2007). Nanoparticulate carriers for the treatment of coronary restenosis. *Int. J. Nanomedicine* 2: 143–161. https://pubmed.ncbi.nlm.nih.gov/17722543/ (accessed 11 August 2021).

97 Libby, P., Ridker, P., and Hansson, G. (2011). Progress and challenges in translating the biology of atherosclerosis. *Nature* 473: 317–325. https://doi.org/10.1038/NATURE10146.

98 Nugent, H. and Edelman, E. (2003). Tissue engineering therapy for cardiovascular disease. *Circ. Res.* 92: 1068–1078. https://doi.org/10.1161/01.RES.0000073844.41372.38.

99 Lewis, D., Kamisoglu, K., York, A., and Moghe, P. (2011). Polymer-based therapeutics: nanoassemblies and nanoparticles for management of atherosclerosis. *Wiley Interdiscip. Rev. Nanomed. Nanobiotechnol.* 3: 400–420. https://doi.org/10.1002/WNAN.145.

100 Flores, A., Ye, J., Jarr, K. et al. (2019). Nanoparticle therapy for vascular diseases. *Arterioscler. Thromb. Vasc. Biol.* 39: 635–646. https://doi.org/10.1161/ATVBAHA.118.311569.

101 Danenberg, H., Golomb, G., Groothuis, A. et al. (2003). Liposomal alendronate inhibits systemic innate immunity and reduces in-stent neointimal hyperplasia in rabbits. *Circulation* 108: 2798–2804. https://doi.org/10.1161/01.CIR.0000097002.69209.CD.

102 Sung, K., Mosley, D., Peelle, B. et al. (2004). Synthesis of monofunctionalized gold nanoparticles by fmoc solid-phase reactions. *J. Am. Chem. Soc.* 126: 5064–5065. https://doi.org/10.1021/JA049578P.

103 Fu, A., Micheel, C., Cha, J. et al. (2004). Discrete nanostructures of quantum dots/Au with DNA. *J. Am. Chem. Soc.* 126: 10832–10833. https://doi.org/10.1021/JA046747X.

104 Binsalamah, Z., Paul, A., Khan, A. et al. (2011). Intramyocardial sustained delivery of placental growth factor using nanoparticles as a vehicle for delivery in the rat infarct model. *Int. J. Nanomedicine* 6: 2667–2678. https://doi.org/10.2147/IJN.S25175.

105 Deng, Y., Zhang, X., Shen, H. et al. (2020). Application of the nano-drug delivery system in treatment of cardiovascular diseases. *Front. Bioeng. Biotechnol.* 0: 489. https://doi.org/10.3389/FBIOE.2019.00489.

106 Zhang, J., Ma, A., and Shang, L. (2018). Conjugating existing clinical drugs with gold nanoparticles for better treatment of heart diseases. *Front. Physiol.* 9: 642. https://doi.org/10.3389/FPHYS.2018.00642.

107 Spivak, M., Bubnov, R., Yemets, I. et al. (2013). Development and testing of gold nanoparticles for drug delivery and treatment of heart failure: a theranostic potential for PPP cardiology. *EPMA J.* 4: 20. https://doi.org/10.1186/1878-5085-4-20.

108 Yang, C., Tian, A., and Li, Z. (2016). Reversible cardiac hypertrophy induced by PEG-coated gold nanoparticles in mice. *Sci. Rep.* 6: 20203. https://doi.org/10.1038/SREP20203.

109 Tian, A., Yang, C., Zhu, B. et al. (2018). Polyethylene-glycol-coated gold nanoparticles improve cardiac function after myocardial infarction in mice. *Can. J. Physiol. Pharmacol.* 96: 1318–1327. https://doi.org/10.1139/CJPP-2018-0227.

110 Kim, J., Cao, L., Shvartsman, D. et al. (2011). Targeted delivery of nanoparticles to ischemic muscle for imaging and therapeutic angiogenesis. *Nano Lett.* 11: 694–700. https://doi.org/10.1021/NL103812A.

111 Ghosh, P., Han, G., De, M. et al. (2008). Gold nanoparticles in delivery applications. *Adv. Drug Deliv. Rev.* 60: 1307–1315. https://doi.org/10.1016/J.ADDR.2008.03.016.

112 Nguyen, M., Gianneschi, N., and Christman, K. (2015). Developing injectable nanomaterials to repair the heart. *Curr. Opin. Biotechnol.* 34: 225–231. https://doi.org/10.1016/J.COPBIO.2015.03.016.

113 Chandarana, M., Curtis, A., and Hoskins, C. (2018). The use of nanotechnology in cardiovascular disease. *Appl. Nanosci.* 87 (8): 1607–1619. https://doi.org/10.1007/S13204-018-0856-Z.

114 Giannouli, M., Karagkiozaki, V., Pappa, F. et al. (2018). Fabrication of quercetin-loaded PLGA nanoparticles via electrohydrodynamic atomization for cardiovascular disease. *Mater. Today Proc.* 5: 15998–16005. https://doi.org/10.1016/J.MATPR.2018.05.044.

115 Banik, B., Fattahi, P., and Brown, J. (2016). Polymeric nanoparticles: the future of nanomedicine. *Wiley Interdiscip. Rev. Nanomed. Nanobiotechnol.* 8: 271–299. https://doi.org/10.1002/WNAN.1364.

116 Lee, P., Zan, B., Chen, L., and Chung, T. (2019). Multifunctional PLGA-based nanoparticles as a controlled release drug delivery system for antioxidant and anticoagulant therapy. *Int. J. Nanomedicine* 14: 1533–1549. https://doi.org/10.2147/IJN.S174962.

117 Martín Giménez, V., Kassuha, D., and Manucha, W. (2017). Nanomedicine applied to cardiovascular diseases: latest developments. *Ther. Adv. Cardiovasc. Dis.* 11: 133–142. https://doi.org/10.1177/1753944717692293.

118 Namdari, M., Cheraghi, M., Negahdari, B. et al. (2017). Recent advances in magnetoliposome for heart drug delivery. *Artif. Cells, Nanomedicine Biotechnol.* 45: 1051–1057. https://doi.org/10.1080/21691401.2017.1299159.

119 Marcos-Campos, I., Asín, L., Torres, T. et al. (2011). Cell death induced by the application of alternating magnetic fields to nanoparticle-loaded dendritic cells. *Nanotechnology* 22: 205101. https://doi.org/10.1088/0957-4484/22/20/205101.

120 Gamarra, L.F., Brito, G.E.S., Pontuschka, W.M. et al. (2005). Biocompatible superparamagnetic iron oxide nanoparticles used for contrast agents: a structural and magnetic study. *J. Magn. Magn. Mater.* 289: 439–441. https://doi.org/10.1016/J.JMMM.2004.11.123.

121 Jin, S. and Ye, K. (2007). Nanoparticle-mediated drug delivery and gene therapy. *Biotechnol. Prog.* 23: 32–41. https://doi.org/10.1021/BP060348J.

122 Maurer, N., Fenske, D., and Cullis, P. (2001). Developments in liposomal drug delivery systems. *Expert. Opin. Biol. Ther.* 1: 923–947. https://doi.org/10.1517/14712598.1.6.923.

123 Mufamadi, M., Pillay, V., Choonara, Y. et al. (2011). A review on composite liposomal technologies for specialized drug delivery. *J. Drug Deliv.* 2011: 1–19. https://doi.org/10.1155/2011/939851.

124 Dvir, T., Bauer, M., Schroeder, A. et al. (2011). Nanoparticles targeting the infarcted heart. *Nano Lett.* 11: 4411–4414. https://doi.org/10.1021/NL2025882.

125 Lestini, B., Sagnella, S., Xu, Z. et al. (2002). Surface modification of liposomes for selective cell targeting in cardiovascular drug delivery. *J. Control. Release* 78: 235–247. https://doi.org/10.1016/S0168-3659(01)00505-3.

126 Holland, N., Qiu, Q., Ruegsegger, M., and Marchant, R. (1998). Biomimetic engineering of non-adhesive glycocalyx-like surfaces using oligosaccharide surfactant polymers. *Nature* 392: 799–801. https://doi.org/10.1038/33894.

127 Srinivasan, R., Marchant, R., and Gupta, A. (2010). In vitro and in vivo platelet targeting by cyclic RGD-modified liposomes. *J. Biomed. Mater. Res. A* 93: 1004–1015. https://doi.org/10.1002/JBM.A.32549.

128 Rana, M., Kumar, A., and Rana, A.J. (2020). Drug Delivery through Targeted Approach with Special References to Phytosomes. In: *Role of Novel Drug Delivery Vehicles in Nanobiomedicine* (ed. R. Tyagi, N. Garg and S. Rahul), 488. IntechOpen https://doi.org/10.5772/INTECHOPEN.86644.

129 Maiti, K., Mukherjee, K., Gantait, A. et al. (2007). Curcumin-phospholipid complex: Preparation, therapeutic evaluation and pharmacokinetic study in rats. *Int. J. Pharm.* 330: 155–163. https://doi.org/10.1016/J.IJPHARM.2006.09.025.

130 Maiti, K., Mukherjee, K., Gantait, A. et al. (2006). Enhanced therapeutic potential of naringenin-phospholipid complex in rats. *J. Pharm. Pharmacol.* 58: 1227–1233. https://doi.org/10.1211/JPP.58.9.0009.

131 Bodade, S.S., Shaikh, K.S., Kamble, M.S., and Chaudhari, P.D. (2013). A study on ethosomes as mode for transdermal delivery of an antidiabetic drug. 20: 40–46. https://doi.org/10.3109/10717544.2012.752420.

132 Bendas, E.R. and Tadros, M.I. (2007). Enhanced transdermal delivery of salbutamol sulfate via ethosomes. *AAPS PharmSciTech* 8: E107. https://doi.org/10.1208/PT0804107.

133 Raphey, V.R., Henna, T.K., Nivitha, K.P. et al. (2019). Advanced biomedical applications of carbon nanotube. *Mater. Sci. Eng. C* 100: 616–630. https://doi.org/10.1016/J.MSEC.2019.03.043.

134 Gorain, B., Choudhury, H., Pandey, M. et al. (2018). Carbon nanotube scaffolds as emerging nanoplatform for myocardial tissue regeneration: a review of recent developments and therapeutic implications. *Biomed. Pharmacother.* 104: 496–508. https://doi.org/10.1016/J.BIOPHA.2018.05.066.

135 Ahadian, S., Davenport Huyer, L., Estili, M. et al. (2017). Moldable elastomeric polyester-carbon nanotube scaffolds for cardiac tissue engineering. *Acta Biomater.* 52: 81–91. https://doi.org/10.1016/J.ACTBIO.2016.12.009.

136 Martinelli, V., Cellot, G., Toma, F.M. et al. (2013). Carbon nanotubes instruct physiological growth and functionally mature syncytia: nongenetic engineering of cardiac myocytes. *ACS Nano* 7: 5746–5756. https://doi.org/10.1021/NN4002193.

19

Nanotherapeutics of Phytoantioxidants for Diabetes Mellitus

Bui Thanh Tung and Tong Thi Thu Hoa

VNU University of Medicine and Pharmacy, Vietnam National University Hanoi, Cau Giay, Hanoi, Vietnam

19.1 Introduction

19.1.1 Diabetes Mellitus

Diabetes is one of the major chronic and noncommunicable diseases affecting the public health of the whole world. This disease is a group of metabolic diseases characterized by hyperglycemia resulting from defects in insulin secretion, insulin action, or both. There are three major types of diabetes: Type 1, in which the pancreas does not produce insulin; type 2 in which the body cells are resistant to the action of insulin that is being produced and over time, the production of insulin progressively decreases; and type 3 gestational diabetes which occurs in pregnancy and can cause some complications during the pregnancy and at birth and increases the risk of type 2 diabetes in the mother and obesity in the offspring. Additionally, it is known to be associated with the excess generation of reactive oxygen species (ROS), largely due to hyperglycemia, which causes oxidative stress in a variety of tissues. In type 2 diabetic patients, oxidative stress is closely associated with chronic inflammation. Therefore, antioxidant supplementation is regarded as a promising strategy to combat diabetes and associated pathological conditions.

19.1.2 Antioxidants and Their Properties

Antioxidants are molecules that can interact with free radicals in a safe manner, terminate the chain reaction, and convert them into a harmless molecule by donating an electron [1]. In the current decade, antioxidants have drawn attention due to their potential to minimize oxidative stress, which is defined as the pathophysiological response, created due to the imbalance between the production of oxidants and the endogenous antioxidants to counteract. Many different chemical compounds have been evaluated for their antioxidant properties. They may be either endogenous (e.g. glutathione and uric acid) or exogenous, based on the source of origin, but the majority of antioxidants come from our diet [2]. These may be naturally derived, synthetic, or inorganic material at the nanoscale level, or, the most interesting and recently explored nano-encapsulated antioxidant molecules. Natural antioxidants have the ability to potentially modulate oxidative stress [3]. It has been proven that fruits and plants have been a rich source of natural antioxidant

Phytoantioxidants and Nanotherapeutics, First Edition. Edited by Mithun Rudrapal.

compounds for many centuries, which have been widely applied in preventing diabetes because of their therapeutic properties, multi-targeted efficacy, and low toxicity. However, the main limitation of this molecule is its poor oral bioavailability.

19.1.3 Unique Benefits of Nanotechnology

Conventional drug delivery including powder, capsule, tablet, or liquid has crucial limitations including a high-dose requirement, low bioavailability, rapid first-pass metabolism, and poor pharmacokinetics [4]. Recently, novel nanotechnology-based systems were developed for the therapeutic delivery of natural antioxidants with improved bioavailability and, consequently, efficacy in clinical practice [5]. Nanocarrier systems may increase the solubility of certain drugs and protect them against degradation in circulation, further increasing their local bioavailability [6]. By encapsulating bioactive agents into a nanocarrier, selective drug delivery to the desired sites of action may be achieved through a process known as a targeted approach that avoids high or frequent dosing to attain effective drug concentration locally. Nanocarriers preserve encapsulated molecules from degradation and, in turn, enhance stability. Nanocarriers can safely transport the drug across the acidic biological milieu such as the acidic pH of the stomach with improved permeability, thereby delivering maximum efficacy at a relatively lower dose. In addition, due to their malleable physicochemical properties, even a poorly soluble drug candidate can also be effectively translocated to the targeted site with an improved pharmacokinetic/bioavailability profile [6, 7]. Physicochemical characteristics of nanoparticles are responsible for their effects. For instance, they determine the ability to enter the cell. The smaller size makes it easier to do that. Moreover, charge and surface properties such as their hydrophilicity or hydrophobicity help them selectively across the cell membrane [8, 9]. The use of nanocarriers promises to increase drug specificity and bioavailability while reducing unwanted off-target side effects. And in this chapter, natural antioxidants in nano form are summarized, as well as the factors influencing their antioxidant capacity and their health benefits.

19.2 Nanocarried Natural Bioactive Compounds Against Diabetes Mellitus

19.2.1 Quercetin

Flavonoids are polyphenolic secondary metabolites in plants with a common diphenylpropane (C6–C3–C6) skeleton categorized into chalcones, flavonols, flavones, flavan-3-ols, flavanones, isoflavones, and anthocyanins. Quercetin (3,3′,4′,5,7-pentahydroxyflavone) is the most popular flavonoid ubiquitously contained in vegetables and fruits, such as onion, molokheya, broccoli, kale, black tea, orange, blueberry, and apple. The onion is the most important source of quercetin and contains about 300 mg/kg of fresh onion [10]. Quercetin and other related polyphenols possess antioxidative, anti-inflammatory, antidiabetic, and anticancer activities. The presence of a catechol or gallate group in the B-ring, as well as 2,3 unsaturation conjugated with the 4-oxo group in the C-ring, are important for their antioxidative properties [11]. The properties of polyphenols are considered to benefit human health. Many papers have suggested the health-beneficial activity of quercetin to prevent chronic diseases and aging processes in humans. Especially, it is recognized for its effects on preventing diabetes.

19.2.1.1 Anti-Inflammatory Against Diabetes-Induced

It is generally recognized that quercetin is absorbed from the intestines and metabolized into the phase conjugates, glucuronides, and sulfates. For example, it is metabolized after oral intake into its conjugates, such as quercetin-3-O-glucuronide (Q3GA) and quercetin -3′-O-sulfate. In the last decade, scientists have demonstrated the unique actions of quercetin-3-O-glucuronide at sites of inflammation, including specific accumulation in macrophages and the following deconjugation into active aglycone, catalyzed by the macrophage-derived β-glucuronidase [12]. Quercetin plays an important role in anti-inflammatory against diabetes-induced vascular impairment. It could suppress the inflammatory process in aortic tissues in both diabetic models via amelioration of the elevation in TNF-α and CRP serum levels, inhibition of the activation of the transcription factor NF-κB in the aorta of diabetic rats and prevention of leukocyte infiltration in the adventitia and pyknosis of the endothelium. In addition, quercetin was able to prevent collagen deposition in aortic tissues in both diabetic models [13]. The anti-inflammatory effect of Q3GA was evaluated in lipopolysaccharide-treated RAW 264.7 macrophages, and the result demonstrated that Q3GA could significantly inhibit LPS-induced NO release in vitro in a dose-dependent manner compared with the control group [14].

19.2.1.2 Regulatory Effects of Quercetin on Glycemia

In streptozotocin- (STZ-)induced rat models, alloxan-induced diabetes rats, and STZ-nicotinamide-induced rat models, oral administration of various dosages of quercetin were able to reduce glucose levels, hemoglobin (Hb), and glycosylated hemoglobin (HbA1C) in blood and prevent body weight loss in vivo [15].

19.2.1.2.1 Enhancement of Glucose Uptake in Tissues and Organs via AMPK Quercetin activates AMP-activated protein kinase resulting in a variety of downstream antidiabetic effects [11, 16]. AMP-activated protein kinase (AMPK) is a crucial cellular energy sensor [17]. And AMPK is a highly conserved sensor of cellular energy status that is activated under conditions of low intracellular ATP [18]. Once activated by falling energy status, it promotes ATP production by increasing the activity or expression of proteins involved in catabolism while conserving ATP by switching off biosynthetic pathways. AMPK also regulates metabolic energy balance at the whole-body level [17]. Moreover, the activation of AMPK was related to a transient disruption of mitochondrial energy transduction, a mechanism analogous to that of Metformin [19]. Quercetin transiently inhibits mitochondrial ATP synthase, leading to the activation of AMPK, and proposes that quercetin and quercetin glycosides are responsible for the antidiabetic activity. Quercetin-3-O-glycosides enhanced glucose uptake by 38–59% (50 μM; 18 hours' treatment) in the absence of insulin. Quercetin aglycone, a minor constituent, stimulated uptake by 37% [19].

In other studies, L6 myotubes were pretreated with quercetin (100 μM). This pretreatment caused a twofold increase in both AMP to ATP as well as ADP to ATP ratio. Interestingly, besides the AMP:ATP ratio, AMP, ADP, and ATP concentrations significantly increased in treated myotubes. This may be the result of increased ATP utilization in these cells. The AMP:ATP ratio is an essential factor for cellular AMPK activation. The AMPK signaling pathway could contribute to the correction of insulin resistance through bypassing the insulin-regulated system for GLUT4 translocation [20]. This pathway increases the phosphorylation and activity of MAPK family components of which p38 MAPK participates in the full activation of GLUT4. The p38 MAPK (mitogen-activated protein kinase) signaling pathway allows cells to interpret a wide range of external signals and respond appropriately by generating a plethora of different biological effects [21]. P38 MAPK is said to be the downstream target of AMPK but it has also been suggested to have a role in the insulin-stimulated glucose uptake. Furthermore, overexpression of MKK3 and MKK6, the upstream regulatory kinases stimulated GLUT1 and GLUT4 transporters [20].

19.2.1.2.2 Abrogating Hypertension Progression Induced by Diabetes The treatment with quercetin had also the ability to abrogate hypertension progression induced by diabetes together with amelioration of the exaggerated contractile responses of the aorta. Alindo et al. reported that the intravenous injection progressively reduced the mean blood pressure in spontaneously hypertensive rats and that the hypotensive effect of Q3GA was abolished in rats treated with the β-glucuronidase inhibitor [22]. These observations suggest that the β-glucuronidase-mediated deconjugation of glucuronide metabolites, rather than the sulfates, might represent a major pathway for exhibiting the biological activity in vivo after the quercetin intake [12, 23].

19.2.1.2.3 Promoting Releasing Insulin from β-Cells Youl et al. reported that administering 20 $mmol.L^{-1}$ quercetin pancreatic β-cells potentiated insulin secretion induced by glucose and glibenclamide and protected β-cells against oxidative damages [24]. Babacanoglu and other studies found that long-term hyperglycemia reduced the binding of the insulin receptor and the substrate (IRS-1), decreased the expression of endothelial nitric oxide synthase (eNOS), increased the expression of inducible nitric oxide synthase (iNOS), and then led to insulin resistance in diabetic rats. Oral treatment of QE (100 mg/kg) in STZ-treated rats significantly improved the insulin resistance and decreased incremental plasma glucose levels after 30–180 minutes. QE can restore the interference caused by hyperglycemia by regulating the quantities of eNOS and iNOS [15, 25].

19.2.1.3 Effects on Genes and Enzymes

Additionally, it was able to reduce the transcription of genes involved in inflammation, such as pro-inflammatory interleukins and enzymes involved in oxidative stress responses [26]. In conclusion, quercetin exerts a significant hypoglycemic activity might by enhancing insulin sensitivity, promoting glycogen synthesis, inhibiting α-glucosidase activity, and improving insulin resistance.

19.2.1.4 Preclinical Evidences

It is demonstrated that quercetin could improve its poor solubility and bioavailability by encapsulation in a nanosystem. In a study, scientists have developed novel lecithin-based mixed polymeric micelles (LMPMs) for Que and conducted the experiment. In this study, 6 male Sprague Dawley rats, 8–10 weeks old, were randomly divided into 2 groups. Each group contains 3 rats. The first group was given Que-LMPMs at 10 mg/kg and the other was supplemented with free Que solution (que was dissolved in a solvent of DMSO/PEG400 = 3/7) at 10 mg/kg. PK profile of the optimal Que-LMPMs and free Que solution were observed and recorded. The result shows that the area under the curve of Que-LMPMs was higher, which increased 2.4-fold compared to the free Que solution. Moreover, the half-life of Que-LMPMs was also higher, while the CL of Que-LMPMs was slower than those of free quercetin [27]. In another study, there were 5 groups of male Sprague Dawley rats (6 animals per group) to investigate the potential effects of nano quercetin. Group 1: Six normal rats were used as normoglycemic (control group). Group 2: Normoglycemic as before. quercetin suspension 150 mg/kg daily for 15 days. Group 3: Rats were injected with STZ to establish diabetes (diabetic control group). In groups 4 and 5, diabetic rats were given Quercetin suspension 150 mg/kg daily for 15 days and quercetin-loaded NP 150 mg/kg every 5th day for 15 days. The aim of this study was to test the comparative effectiveness of quercetin-loaded nanoparticles with quercetin in experimental DM induced by STZ in rats. The comparative effects of two forms of quercetin were studied on pharmacokinetics, the levels of blood glucose and natural antioxidants, catalase, and superoxide dismutase (SOD) in the pancreas and kidneys. The pharmacokinetics showed that quercetin-entrapped nanoparticles demonstrate about a 5.23-fold increase in oral bioavailability when compared to quercetin suspension. Moreover, the core-shell complex formation markedly increased the stability and controlled release properties of encapsulated quercetin for six days, which indicated reduced dosing frequency. Additionally, blood samples were collected twice weekly for 15 days and evaluated by measuring blood glucose levels. However, upon supplementation with Qu-NP as well as quercetin suspension, blood glucose levels of diabetic rats showed a significant drop compared to diabetic groups without treatment. Interestingly, in comparison to the quercetin suspension-treated group, significant higher hypoglycemia was obtained in Qu-NP-pretreated groups on days 7, 10, and 15 ($P < 0.05$), although Qu-NP was supplemented only every fifth day, whereas daily dosing of quercetin suspension was used. It demonstrated that the same dose of Qu-NP every fifth day was sufficient to bring effect similar to the daily dose of oral quercetin suspension, which showed a reduced dosing frequency and less dose requirement. Then animals were sacrificed after 15 days. The kidneys and pancreas were excised immediately in order to measure antioxidant enzyme levels (CAT and SOD). Both quercetin in the form of drug suspension as well as Qu-NP had good efficiency regarding the improvement of oxidative stress parameters and for antioxidant capacity since quercetin mitigates the action of ROS on SOD and CAT. The result demonstrated that the same effect was observed for catalase and SOD levels in the pancreas and kidneys when treated with Qu-NP (150 mg/kg every fifth day) or Qu-SUS (150 mg/kg daily) [28].

19.2.2 Curcumin

Curcumin, the bioactive component of the curry spice turmeric, and its related structures possess potent antioxidant and anti-inflammatory properties [29, 30]. There is a variety of evidences suggesting that curcumin plays a beneficial role in antidiabetes, both by lowering blood glucose levels and by ameliorating the long-term complications of diabetes. Curcumin is a low molecular-weight polyphenol, first chemically characterized in 1910, with the molecular formula of $C_{21}H_{20}O_6$ [31]. Most of the antioxidants have either a phenolic functional group or a diketone group. Curcumin is a unique antioxidant, which contains a variety of functional groups, including the Β-diketo group, carbon–carbon double bonds, and phenyl rings containing varying amounts of hydroxyl and methoxy substituents, free radical-mediated peroxidation of membrane lipids and oxidative damage of DNA and proteins are believed to be associated with diabetes [31, 32]. Curcumin is also thought to play a vital role against these pathological conditions [33]. Human studies have found some evidence of anti-inflammatory activity of curcumin. The laboratory studies have identified a number of different molecules involved in inflammation that are inhibited by curcumin including phospholipase, lipooxygenase, cyclooxygenase 2, leukotrienes, thromboxane, prostaglandins, nitric oxide, collagenase, elastase, hyaluronidase, monocyte chemoattractant protein-1 (MCP-1), interferon-inducible protein, tumor necrosis factor (TNF), and interleukin-12 (IL-12). Furthermore, curcumin increased the phosphorylation of AMP-activated protein kinase (AMPK) and its downstream target acetyl-CoA carboxylase (ACC) in cells. These results suggest that AMPK-mediated suppression of gluconeogenesis may be a potential mechanism mediating glucose-lowering effects of curcumin [34, 35].

The effects of nano curcumin to prevent diabetes have been proved by so many studies. In a study, scientists dedicated 42 rats, categorized into six groups: control (healthy rats provided with

1 mL saline solution), diabetic rats (STZ-induced DM – 60 mg/100 g body weight a single dose intraperitoneal administration, and no curcumin pretreatment), along with four treated groups via i.p. administration of 1 mg of curcumin/0.1 kg bw (STZ + CC1), 2 mg of curcumin/0.1 kg bw (STZ + CC2), 1 mg of liposomal curcumin/0.1 kg bw (STZ + LCC1) and 2 mg of liposomal curcumin/0.1 kg bw (STZ + LCC2). All groups that received liposomal curcumin or curcumin showed significantly lower serum levels of C-peptide, TNF-α RANTES. The best effects were obtained for two groups that received liposomal curcumin. Indeed, LCC2 is superior to LCC1. The plasma level of C-peptide showed significant improvements when liposomal curcumin was administrated, with better results for LCC2 when compared to LCC1 ($P < 0.003$). LCC2 pretreatment proved to be more efficient in reducing the level of TNF-α ($P < 0.003$) and RANTES ($P < 0.003$) than CC2 pretreatment. Upon comparing LCC2 with LCC1 formulas, the differences were significant for TNF-α ($P = 0.004$), IL-1β ($P = 0.022$), and RANTES ($P = 0.003$) levels [36]. In another study, six groups of seven Wister Bratislava rats were included in the study. C group is the normal control (no diabetes). STZ-C is the group in which STZ-induced DM – 60 mg/100 g body weight a single dose intraperitoneal administration and no treatment. The groups with DM that were provided 1 mg CC/100 g bw and 2 mg CC/100 g bw are called STZ + CC1 and STZ + CC2. The others are two groups with DM and lCC pretreatment 1 mg/100 g bw –STZ + lCC1, 2 mg/100 g bw – STZ + lCC2. The result demonstrated both liposomal curcumin and plain curcumin significantly reduced serum levels of oxidative stress markers (malondialdehyde, the indirect assessment of nitric oxide synthesis, and total oxidative status), compared with STZ control group (STZ-C) (p-values < 0.002), with better results for liposomal curcumin as compared to plain curcumin, for both doses (p-values < 0.005). Liposomal curcumin was found to be more potent than pure curcumin in regard to the antioxidant and anti-inflammatory activities. Moreover, liposomal curcumin pretreatment proved to be more efficient in improving the level of MMP-2 plasma levels than curcumin pretreatment ($p < 0.003$, excepting STZ + CC2 vs. STZ + lCC1 comparison with $p = 0.0553$) [37]. In conclusion, curcumin plays an important role in preventing diabetes and nano curcumin is superior to curcumin due to its improved bioavailability.

19.2.3 Resveratrol

Resveratrol is a naturally occurring polyphenolic compound found largely in the skins of the grapes and the products prepared from it such as red wine. However, presence of resveratrol has been identified in several other plant species except grapes such as peanuts, berries, and legumes [38–40]. Its stilbene-based structure has two phenolic rings linked by a styrene double bond, which allows *cis* and *trans* orientation to generate 3,4′,5-trihydroxystilbene [41]. The antioxidant activity of phenolic compounds depends on the structure, in particular, the number and positions of the hydroxyl groups and the nature of substitutions on the aromatic rings. Resveratrol has three hydroxyl groups known to play crucial roles in ROS scavenging [42, 43]. Moreover, these hydroxyl groups help chelate metal ions, which is an essential feature in the prevention of ROS production.

It is well established that resveratrol decreases blood glucose levels in animals with experimental diabetes [44]. The blood glucose-lowering effect of resveratrol was found in rats with diabetes induced by STZ alone and STZ with nicotinamide [44, 45]. It is well documented that the antihyperglycemic effect of resveratrol in type 1 diabetic animals results from the increase in blood insulin levels, the suppression of hepatic glucose output and the increase in peripheral glucose utilization. Resveratrol improves antioxidant defense in pancreatic tissue, increases activities of antioxidant enzymes (SOD, catalase, glutathione peroxidase, and glutathione-S-transferase) and protects cells from free radical damage [46]. Resveratrol was also found to reverse degenerative changes in β-cells of STZ-NA-induced diabetic rats and to prevent STZ-induced β-cell apoptosis. The decrease in apoptosis is accompanied by the blocking of the activity of caspase-3, is mainly due to the inhibition of PARP cleavage by resveratrol and is similar to the effect of nicotinamide [47].

In a study, eighteen healthy rats (150–200 g) were divided into three groups. Each group included six rats. The normal rats were considered as control (group I). In group II (named GDM challenged group), the rats were injected with STZ (i.p.; 45 mg/kg bw). That six GDM challenged rats were orally provided with CS–ZnO–RS (100 mg/kg bw) were served as group III. In this study, the effect of resveratrol–zinc oxide complex encapsulated with chitosan (CS–ZnO–RS) nanoparticles on GDM was studied. CS–ZnO–RS showed an improvement in stability and effectiveness in GDM. The synthesized CS–ZnO–RS showed an average size of 38 nm. Moreover, the CS–ZnO–RS exposed 95% of drug-releasing activity within 24 hours. The result showed that in GDM-challenged group (group II), blood glucose level was increased, whereas the level of insulin was decreased compared to the control group. Besides, there was a significantly decreased level of blood glucose level in group III. Additionally, it was noticed the GDM rats treated with CS–ZnO–RS (100 mg/kg bw) showed an elevated level of insulin content than the untreated rats. Moreover, to prevent oxidative stress, the main enzymic antioxidants like CAT, SOD, and GPx are very vital. However, it was demonstrated that the level of CAT, SOD, and GPx in group II (the GDM rats) was lower than in group I. Conversely, group III (the GDM-challenged rats treated with CS–ZnO–RS) exhibited a significantly increased level of CAT, SOD, and GPx activities compared to control group [48]. Besides, there is increasing evidence that in certain pathologic states, especially diabetes, the increased production and/or ineffective scavenging of ROS may play a critical role. It has been suggested that enhanced production of free radicals and oxidative stress is a central event in the development of diabetic complications [49–51]. Resveratrol acts as a free radical scavenger and a potent antioxidant in the inhibition of numerous ROS. The antioxidative effects of resveratrol and resveratrol-loaded nanoparticles are proved in a study in which human type-2 alveolar epithelial cells (A549) were prevented against hydrogen peroxide. A549 cells were pretreated with or without nanoparticles (Nano), resveratrol (Res), and resveratrol-loaded nanoparticle (ResN) for four hours, followed by incubation with 400 μM of H_2O_2 for 8 hours. Negative control was non-treated H_2O_2 to the cells, and H_2O_2 (400 μM)-treated cells were used as a positive control to compare with each group of the level of ROS. The generation of ROS at each treatment was represented as a percentage M value. It was shown that the M values between the non-loaded nanoparticle group and positive control group were not different significantly. It was indicated that the non-loaded nanoparticle group had no effect on the formation of intracellular ROS. The M values treated with Res group and with ResN group were low as below 20%, indicating that resveratrol had antioxidant activity. Moreover, the M value of ResN group was significantly lower than Res group. These results indicated that ResN had a higher antioxidant effect than Res when treated to cells, and nanoparticles might increase the cellular uptake of resveratrol [52].

19.2.4 Genistein

Genistein [4′, 5,7-trihydroxyisoflavone hoặc 5,7-dihydroxy-3-(4-hydroxyphenyl) chromen-4-one] is a natural compound belonging to the chemical class of isoflavone and soy is the major source of genistein in the human diet. The main dietary source of genistein is the biologically active glucoside genistein [53, 54]. Fermentation or digestion of soybeans or soy products results in the release of the sugar molecule from the isoflavone glycoside, genistein leaving the isoflavone aglycone. Genistein has a diphenol structure that resembles stereochemically human endogenous estrogen [55, 56].

It is proved that genistein supplementation induces a decrease in hepatic glucose production through preservation of insulin positive β-cells and altering the hepatic glucose metabolic enzyme profile, and the resultant increase in hepatic glucose handling seemed to prevent diabetes onset in mice. Both alloxan- and STZ-induced diabetic rats have shown a significant increase in plasma FFA concentration together with reduced plasma insulin level. Supplementation with genistein significantly lowered the plasma FFA and triglyceride concentrations, whereas it elevated the plasma HDL-C and HDL-C to total cholesterol ratio. Besides, they found that genistein significantly lowered the lipid peroxidation contents [57]. Additionally, the shortage of Estrogen alone or combined with a high-fat diet increased pancreatic inflammation in a rat. However, providing genistein via subcutaneous administration protected rats from these inflammatory changes in the pancreas with or without diabetes. It indicated the unique advantages of genistein in preventing pancreatic injury [58]. In a study, 32 male Sprague Dawley rats were randomly divided into 4 groups. Each group contains 8 rats. Group 1 – control group; group 2 – STZ group: rats received single i.p. injection of STZ; group 3 – STZ-G: diabetic rats treated with genistein (600 mg/kg diet); group 4 – STZ-ISP: diabetic rats treated with isolated soy protein (ISP) (200 g/kg diet). All the supplements of genistein and ISP started on the same day and continued for three weeks. The result showed that elevated blood glucose levels, insulin levels, and dyslipidemia in diabetic rats were significantly improved by genistein and ISP treatments. Plasma insulin levels were lower in the STZ-induced diabetic rats compared with the control rats. The administration of genistein and ISP increased the plasma insulin level of the STZ-induced diabetic rats. Therefore, genistein and ISP-treated diabetic rats showed significant decreases in HbAIC levels compared with diabetic rats. Hepatic SOD, catalase, and glutathione peroxidase activities of the STZ-induced diabetic rats were also dramatically inhibited by these compounds. Additionally, the activities of SOD, CAT, and GSH-Px, which are major antioxidant enzymes that defend us from the effects of ROS, were lower in the STZ-induced diabetic rats compared to the control rats, while genistein and ISP supplements in the STZ-induced diabetic rats appeared to increase its activity [59]. In another study, scientists used two binary micelle systems consisting of HS15 + F127 and HS15 + L61 to increase the solubility and stability of genistein, enhance its absorption, protect it from premature degradation in the body, and prolong its circulation time. HS-15 is a nonionic surfactant including 70% polyglycol mono and diesters of 12-hydroxystearic acid and 30% free polyethylene glycol. F127 is a relatively hydrophilic pluronic and L61 is relatively hydrophobic with a hydrophile–lipophile balance (HLB) value of only 3. They assumed that the combination of pluronic and HS15 will complement each other to form binary mixed micelles, which overcome the limitations of poor solubility and low oral bioavailability of GEN. In this study, 18 SD rats were divided into three groups. Each group contains 6 rats. GEN-F (HS15 + F127 micelle), GEN-L(HS15 + L61 micelle), and free GEN were orally administered at a dose of $60\,mg{\cdot}kg^{-1}$. The results showed that the average Cmax value of genistein was significantly improved by GEN-F and GEN-L compared to GEN. The average Cmax values of GEN-F, GEN-L, and GEN were 2.401 ± 0.480, 3.563 ± 1.121, and $0.445 \pm 0.140\,\mu g/mL$, respectively. It is also indicated that the average area under the curve from time 0 to 8 hours (AUC_{0-8}) of GEN-F, GEN-L was increased than GEN. (AUC_{0-8}) of GEN-F and GEN-L were 10.821 ± 1.538 and $14.929 \pm 5.575\,mg{\cdot}L^{-1}{\cdot}h^{-1}$, respectively, whereas that of free GEN was $3.351 \pm 1.182\,mg{\cdot}L^{-1}{\cdot}h^{-1}$. In conclusion, pharmacokinetics demonstrated the relative oral bioavailability of GEN-F and GEN-L increased by 2.23- and 3.46-fold. A high bioavailability indicates that the drug could be absorbed into the blood, resulting in a good curative effect, including that of preventing diabetes [60]. In another study, the anti-angiogenic effect of genistein/Flt1 peptide–HA micelle in STZ-induced diabetic rats was investigated. Scientists used five kinds of samples including Avastin, Flt1 peptide–HA conjugate, genistein dispersed in water, genistein dissolved in DMSO, and genistein/Flt1 peptide–HA micelle. They were injected into the vitreous of experimental diabetic rats a week after the STZ injection. The results showed all the compounds could improve the pathological state and it is noteworthy that treatment with genistein/Flt1 peptide–HA micelle showed the most preventive effect on the diabetic rats. The anti-angiogenic effect of water-insoluble genistein and Flt1 peptide in an aqueous solution could be drastically enhanced by the formation of drug-loaded micelle-like nanoparticles. Furthermore, the higher bioavailability and a very fine nanoparticle size that could improve absorption and permeation of micelle may take the responsibility for this result [61].

19.2.5 EGCG

Tea has been widely consumed as a beverage besides water. Epigallocatechin-3-gallate [EGCG, (2R, 3R)-5,7-Dihydroxy-2-(3,4,5-trihydroxyphenyl) chroman-3-yl 3,4,5-trihydroxybenzoate], a natural catechin, is mostly found in tea. This catechin is colorless, water-soluble, and contributes to the taste and aroma of tea [62, 63]. It is best known for its antioxidant properties, which have led to its effects in preventing diabetes [63]. EGCG has two triphenolic groups in its chemical structure. These specific structural characteristics affect antioxidative potency. Numerous studies demonstrated that EGCG scavenges a wide range of free radicals including singlet oxygen, superoxide anions, peroxyl radicals, and also the most active hydroxyl radicals, which initiate lipid peroxidation [64, 65].

Oxidative stress is responsible for the development of many diseases that make up a significant cause of death such as diabetes. Interestingly, treatment with EGCG remarkably decreases oxidative stress parameters like MDA, NOx, and TOS. Moreover, antioxidant markers could be increased when treated with EGCG [66, 67]. The next effect of EGCG is that it could decrease the blood glucose level and control glycemia [62]. Supplementation with EGCG significantly elevated plasma insulin levels [68]. Furthermore, EGCG increases glucose tolerance and decreases HbA1c levels in STZ-induced diabetes in rats, contributing to further prevention of diabetic complications [69]. Another suggested mechanism of EGCG's protective effect is the increased glucose uptake due to promoting the glucose transporter-4 (GLUT4) translocation in skeletal muscle [70]. This compound also demonstrates its ability to reduce the number of inflammatory cells in the wounds of diabetic animals, especially in pancreatic tissue [71]. However, the bioavailability of EGCG is low. Encapsulation in special drug delivery systems can improve the low bioavailability of this polyphenol. In this regard, EGCG could be a promising therapeutic option for the prevention of diabetes mellitus and its complications, conferring protection against oxidative damage by nanostructure encapsulation.

Forty-four rats were randomly divided into 4 groups. Each group contains 11 rats. Group 1 was the LFD group (low fat diet, negative control). The name of group 2 was HFD (high-fat diet, positive control). In groups 3 and 4, HFD rats were treated with EGCG at 25 mg/kg (group 3) and 75 mg/kg (group 4). Rats on HFD without administration of EGCG had significantly higher blood glucose levels than the negative control rats on LFD. However, treatment with both EGCG at 25 and 75 mg/kg could decrease the level of blood glucose. Interestingly, this effect appeared to be much better in group 4 (treated with 75 mg/kg EGCG) compared to that in group 3 (treated with 25 mg/kg EGCG). It was also observed that supplementation with EGCG at 75 mg/kg brought insulin levels to nearly identical levels as group 1 (control group). This study proved that EGCG has a potential effect to prevent diabetes by improving insulin concentration and reducing blood glucose levels [72]. In another study, 28 rats were randomly assigned to one of four different treatment groups ($N = 7$/group). Group 1 (C group) contains 7 normal rats. In group 2 (STZ group), rats were injected with STZ that caused diabetes. That seven diabetic rats were provided with EGCG solution (2.5 mg/100 g b.w.) were served as group III (EGCG). And in group 4 (L-EGCG), diabetic rats had i.p. administration of liposomal EGCG at 2.5 mg/100 g b.w. Diabetic rats without administration of EGCG had significantly higher oxidative stress parameters (MDA, NOx, and TOS) than the negative control group, whereas, serum antioxidant capacity (thiol, CAT, TAC, and SOD) was significantly reduced in diabetic rats compared to control animals. Interestingly, these antioxidant parameters significantly increased after providing both EGCG and liposomal EGCG, with the best results for the liposomal form. L-EGCG has a better efficiency regarding the improvement of oxidative stress parameters (highly statistically significant with p-values < 0.001 for MDA, NOx, and TOS) and for antioxidant capacity of plasma (highly significant $p < 0.001$ for thiols and significant for catalase and TAC with $p < 0.05$). Additionally, the increase of MMP plays a crucial role in DM pathogenesis. Administration of free EGCG showed only a slight difference in MDA, MMP values, glycemia, and liver parameters compared to the untreated STZ group. However, upon supplementation with L-EGCG, both transaminases and MMP levels showed a significant drop and glycemic values were improved [73]. In another study, the potential effects of nano EGCG for wound healing in diabetic mice were demonstrated. Thirty male ICR mice of SPF (8 weeks old, 28–30 g) were randomly divided into 5 groups (6 mice/group). The groups were organized as follows: group 1 – control group (C); group 2 – model group: diabetic mice caused by STZ administration; group 3 – no-load NPS group: STZ administration as before + administration of 1 mg/mL no-load NPS; group 4 – EGCG group: STZ administration as before + administration of EGCG at 1 mg/mL and group 5 – EV NPS group: DM mice + 1 mg/mL EV NPS. Full-thickness excisional wounds were established on the back of mice. All the wounds of mice were observed and recorded in 10 days. The results demonstrated that diabetes can really delay the healing of wounds and treatment with EV NPS significantly reduced the wound area and improved the wound-healing rate compared with model group, no-load group, and EGCG group. Moreover, the effect on decreasing the number of inflammatory cells in the wounds of diabetic mice of the EV NPS group is superior to other groups. Additionally, the wounds were almost completely healed in the EV NPS group [74].

19.3 Conclusions and Future Perspectives

Diabetes mellitus is a complex metabolic disease affecting numerous people all over the world and which is characterized by heightened oxidative stress and related systemic inflammation. Moreover, natural antioxidants are known to have numerous health-promoting effects in both animals and humans. Thus, the supplemention of natural antioxidants in the diet is significantly vital to control diabetes but the limitations of poor solubility, stability, and low oral bioavailability, have restricted their development and use in the clinic. Recently, nanotechnology-based systems were developed to improve all these limitations. Quercetin, curcumin, resveratrol, genistein, and EGCG are among the most commonly applied natural compounds, which possess antioxidant, anti-inflammatory properties used to improve pathological conditions of diabetes. Generally, the control of diabetes contains four key facets: reduction of oxidative stress, decreased blood glucose level, protected pancreas organogenesis, and enhancement of insulin action. The data of the published studies summarized in this chapter demonstrated that the wonderful therapeutics of five nanocarried natural antioxidants in relation to diabetes include all these facets with better results for nano form, compared to non-nano form. However, further studies in appropriate animal systems and in clinical research should be done to investigate and elucidate more efficacy of these nanocarried natural antioxidants to prevent and control diabetes. Moreover, nanotherapeutics must be completely explored to determine if these nanomaterials themselves could indicate adverse health effects. In conclusion, nanocarried natural antioxidants are paving the way for advanced strategies and therapeutic approaches for the treatment of diabetes.

References

1 Lobo, V., Patil, A., Phatak, A., and Chandra, N. (2010). Free radicals, antioxidants and functional foods: impact on human health. *Pharmacognosy Reviews* 4 (8): 118.

2 Khan, M.I. and Giridhar, P. (2011). Dietary antioxidants: the insurer of health. *Everyman's Sci* 46: 214–218.

3 Khalil, I., Yehye, W.A., Etxeberria, A.E. et al. (2020). Nanoantioxidants: recent trends in antioxidant delivery applications. *Antioxidants* 9 (1): 24.

4 Mudshinge, S.R., Deore, A.B., Patil, S., and Bhalgat, C.M. (2011). Nanoparticles: emerging carriers for drug delivery. *Saudi Pharmaceutical Journal* 19 (3): 129–141.

5 Kotyla, T., Kuo, F., Moolchandani, V. et al. (2008). Increased bioavailability of a transdermal application of a nano-sized emulsion preparation. *International Journal of Pharmaceutics* 347 (1–2): 144–148.

6 Baruah, S. and Dutta, J. (2009). Nanotechnology applications in pollution sensing and degradation in agriculture: a review. *Environmental Chemistry Letters* 7 (3): 191–204.

7 Singh, R. and Lillard, J.W. Jr. (2009). Nanoparticle-based targeted drug delivery. *Experimental and Molecular Pathology* 86 (3): 215–223.

8 Zhao, Z., Ukidve, A., Krishnan, V., and Mitragotri, S. (2019). Effect of physicochemical and surface properties on in vivo fate of drug nanocarriers. *Advanced Drug Delivery Reviews* 143: 3–21.

9 Amin, M.L., Joo, J.Y., Yi, D.K., and An, S.S.A. (2015). Surface modification and local orientations of surface molecules in nanotherapeutics. *Journal of Controlled Release* 207: 131–142.

10 Arai, Y., Watanabe, S., Kimira, M. et al. (2000). Dietary intakes of flavonols, flavones and isoflavones by Japanese women and the inverse correlation between quercetin intake and plasma LDL cholesterol concentration. *The Journal of Nutrition* 130 (9): 2243–2250.

11 Kawabata, K., Mukai, R., and Ishisaka, A. (2015). Quercetin and related polyphenols: new insights and implications for their bioactivity and bioavailability. *Food & Function* 6 (5): 1399–1417.

12 Kawai, Y. (2018). Understanding metabolic conversions and molecular actions of flavonoids in vivo: toward new strategies for effective utilization of natural polyphenols in human health. *The Journal of Medical Investigation* 65 (3.4): 162–165.

13 Mahmoud, M.F., Hassan, N.A., El Bassossy, H.M., and Fahmy, A. (2013). Quercetin protects against diabetes-induced exaggerated vasoconstriction in rats: effect on low grade inflammation. *PloS One* 8 (5): e63784.

14 Li, F., Sun, X.-Y., Li, X.-W. et al. (2017). Enrichment and separation of quercetin-3-O-β-d-glucuronide from lotus leaves (nelumbo nucifera gaertn.) and evaluation of its anti-inflammatory effect. *Journal of Chromatography B* 1040: 186–191.

15 Shi, G.-J., Li, Y., Cao, Q.-H. et al. (2019). In vitro and in vivo evidence that quercetin protects against diabetes and its complications: a systematic review of the literature. *Biomedicine & Pharmacotherapy* 109: 1085–1099.

16 Hwang, J.-T., Park, I.-J., Shin, J.-I. et al. (2005). Genistein, EGCG, and capsaicin inhibit adipocyte differentiation process via activating AMP-activated protein kinase. *Biochemical and Biophysical Research Communications* 338 (2): 694–699.

17 Hardie, D.G., Ross, F.A., and Hawley, S.A. (2012). AMPK: a nutrient and energy sensor that maintains energy homeostasis. *Nature Reviews Molecular Cell Biology* 13 (4): 251–262.

18 Gwinn, D.M., Shackelford, D.B., Egan, D.F. et al. (2008). AMPK phosphorylation of raptor mediates a metabolic checkpoint. *Molecular Cell* 30 (2): 214–226.

19 Eid, H.M., Martineau, L.C., Saleem, A. et al. (2010). Stimulation of AMP-activated protein kinase and enhancement of basal glucose uptake in muscle cells by quercetin and quercetin glycosides, active principles of the antidiabetic medicinal plant Vaccinium vitis-idaea. *Molecular Nutrition & Food Research* 54 (7): 991–1003.

20 Dhanya, R., Arya, A., Nisha, P., and Jayamurthy, P. (2017). Quercetin, a lead compound against type 2 diabetes ameliorates glucose uptake via AMPK pathway in skeletal muscle cell line. *Frontiers in Pharmacology* 8: 336.

21 Cuadrado, A. and Nebreda, A.R. (2010). Mechanisms and functions of p38 MAPK signalling. *Biochemical Journal* 429 (3): 403–417.

22 Galindo, P., Rodriguez-Gómez, I., González-Manzano, S. et al. (2012). Glucuronidated quercetin lowers blood pressure in spontaneously hypertensive rats via deconjugation. *PloS One* 7 (3): e32673.

23 Patel, K.R., Andreadi, C., Britton, R.G. et al. (2013). Sulfate metabolites provide an intracellular pool for resveratrol generation and induce autophagy with senescence. *Science Translational Medicine* 5 (205): 205ra133–205ra133.

24 Youl, E., Bardy, G., Magous, R. et al. (2010). Quercetin potentiates insulin secretion and protects INS-1 pancreatic β-cells against oxidative damage via the ERK1/2 pathway. *British Journal of Pharmacology* 161 (4): 799–814.

25 Babacanoglu, C., Yildirim, N., Sadi, G. et al. (2013). Resveratrol prevents high-fructose corn syrup-induced vascular insulin resistance and dysfunction in rats. *Food and Chemical Toxicology* 60: 160–167.

26 Derlindati, E., Dall'Asta, M., Ardigò, D. et al. (2012). Quercetin-3-O-glucuronide affects the gene expression profile of M1 and M2a human macrophages exhibiting anti-inflammatory effects. *Food & Function* 3 (11): 1144–1152.

27 Chen, L.-C., Chen, Y.-C., Su, C.-Y. et al. (2016). Development and characterization of self-assembling lecithin-based mixed polymeric micelles containing quercetin in cancer treatment and an in vivo pharmacokinetic study. *International Journal of Nanomedicine* 11: 1557.

28 Chitkara, D., Nikalaje, S.K., Mittal, A. et al. (2012). Development of quercetin nanoformulation and in vivo evaluation using streptozotocin induced diabetic rat model. *Drug Delivery and Translational Research* 2 (2): 112–123.

29 Jayaprakasha, G.K., Rao, L.J., and Sakariah, K.K. (2006). Antioxidant activities of curcumin, demethoxycurcumin and bisdemethoxycurcumin. *Food Chemistry* 98 (4): 720–724.

30 Satoskar, R., Shah, S., and Shenoy, S. (1986). Evaluation of anti-inflammatory property of curcumin (diferuloyl methane) in patients with postoperative inflammation. *International Journal of Clinical Pharmacology, Therapy, and Toxicology* 24 (12): 651.

31 Lestari, M.L. and Curcumin, I.G. (2014). *Profiles of Drug Substances, Excipients and Related Methodology*, vol. 39, 113–204. Elsevier.

32 Esatbeyoglu, T., Huebbe, P., Ernst, I.M. et al. (2012). Curcumin—from molecule to biological function. *Angewandte Chemie International Edition* 51 (22): 5308–5332.

33 Fu, X.-Y., Yang, M.-F., Cao, M.-Z. et al. (2016). Strategy to suppress oxidative damage-induced neurotoxicity in PC12 cells by curcumin: the role of ROS-mediated DNA damage and the MAPK and AKT pathways. *Molecular Neurobiology* 53 (1): 369–378.

34 Kim, J.H., Park, J.M., Kim, E.K. et al. (2010). Curcumin stimulates glucose uptake through AMPK-p38 MAPK pathways in L6 myotube cells. *Journal of Cellular Physiology* 223 (3): 771–778.

35 Kim, T., Davis, J., Zhang, A.J. et al. (2009). Curcumin activates AMPK and suppresses gluconeogenic gene expression in hepatoma cells. *Biochemical and Biophysical Research Communications* 388 (2): 377–382.

36 Bulboacă, A.E., Boarescu, P.M., Bolboacă, S.D. et al. (2019). Comparative effect of curcumin versus liposomal curcumin on systemic pro-inflammatory cytokines profile, mcp-1 and rantes in experimental diabetes mellitus. *International Journal of Nanomedicine* 14: 8961.

37 Bulboacă, A.E., Porfire, A.S., Tefas, L.R. et al. (2019). Liposomal curcumin is better than curcumin to alleviate complications in experimental diabetic mellitus. *Molecules* 24 (5): 846.

38 Gehm, B.D., McAndrews, J.M., Chien, P.-Y., and Jameson, J.L. (1997). Resveratrol, a polyphenolic compound found in grapes and wine, is an agonist for the estrogen receptor. *Proceedings of the National Academy of Sciences* 94 (25): 14138–14143.

39 Iacopini, P., Baldi, M., Storchi, P., and Sebastiani, L. (2008). Catechin, epicatechin, quercetin, rutin and resveratrol in red grape: content, in vitro antioxidant activity and interactions. *Journal of Food Composition and Analysis* 21 (8): 589–598.

40 Zamora-Ros, R., Andres-Lacueva, C., Lamuela-Raventós, R.M. et al. (2008). Concentrations of resveratrol and derivatives in foods and estimation of dietary intake in a Spanish population: European Prospective Investigation into Cancer and Nutrition (EPIC)-Spain cohort. *British Journal of Nutrition* 100 (1): 188–196.

41 King, R.E., Bomser, J.A., and Min, D.B. (2006). Bioactivity of resveratrol. *Comprehensive Reviews in Food Science and Food Safety* 5 (3): 65–70.

42 Szekeres, T., Fritzer-Szekeres, M., Saiko, P., and Jäger, W. (2010). Resveratrol and resveratrol analogues—structure—activity relationship. *Pharmaceutical Research* 27 (6): 1042–1048.

43 Ovesna, Z. and Horvathova-Kozics, K. (2005). Structure-activity relationship of trans-resveratrol and its analogues. *Neoplasma* 52 (6): 450.

44 Movahed, A., Nabipour, I., Lieben Louis, X. et al. (2013). Antihyperglycemic effects of short term resveratrol supplementation in type 2 diabetic patients. *Evidence-Based Complementary and Alternative Medicine* 2013: 1–11.

45 Gencoglu, H., Tuzcu, M., Hayirli, A., and Sahin, K. (2015). Protective effects of resveratrol against streptozotocin-induced diabetes in rats by modulation of visfatin/sirtuin-1 pathway and glucose transporters. *International Journal of Food Sciences and Nutrition* 66 (3): 314–320.

46 Lee, Y.-E., Kim, J.-W., Lee, E.-M. et al. (2012). Chronic resveratrol treatment protects pancreatic islets against oxidative stress in db/db mice. *PLoS One* 7 (11): e50412.

47 Ku, C.R., Lee, H.J., Kim, S.K. et al. (2012). Resveratrol prevents streptozotocin-induced diabetes by inhibiting the apoptosis of pancreatic β-cell and the cleavage of poly (ADP-ribose) polymerase. *Endocrine Journal* 59 (2): 103–109.

48 Du, S., Lv, Y., Li, N. et al. (2020). Biological investigations on therapeutic effect of chitosan encapsulated nano resveratrol against gestational diabetes mellitus rats induced by streptozotocin. *Drug Delivery* 27 (1): 953–963.

49 Newsholme, P., Cruzat, V.F., Keane, K.N. et al. (2016). Molecular mechanisms of ROS production and oxidative stress in diabetes. *Biochemical Journal* 473 (24): 4527–4550.

50 Rendra, E., Riabov, V., Mossel, D.M. et al. (2019). Reactive oxygen species (ROS) in macrophage activation and function in diabetes. *Immunobiology* 224 (2): 242–253.

51 Bhatt, M.P., Lim, Y.-C., Kim, Y.-M., and Ha, K.-S. (2013). C-peptide activates AMPKα and prevents ROS-mediated mitochondrial fission and endothelial apoptosis in diabetes. *Diabetes* 62 (11): 3851–3862.

52 Kim, J.-H., Park, E.-Y., Ha, H.-K. et al. (2016). Resveratrol-loaded nanoparticles induce antioxidant activity against oxidative stress. *Asian-Australasian Journal of Animal Sciences* 29 (2): 288.

53 Doerge, D.R. and Sheehan, D.M. (2002). Goitrogenic and estrogenic activity of soy isoflavones. *Environmental Health Perspectives* 110 (suppl 3): 349–353.

54 Walter, E. (1941). Genistin (an isoflavone glucoside) and its aglucone, genistein, from soybeans. *Journal of the American Chemical Society* 63 (12): 3273–3276.

55 Mazurek, A.P., Kozerski, L., Sadlej, J. et al. (1998). Genistein complexes with amines: structure and properties. *Journal of the Chemical Society, Perkin Transactions* 2 (5): 1223–1230.

56 Cohen, R., Orlova, Y., Kovalev, M. et al. (2008). Structural and functional properties of amylose complexes with genistein. *Journal of Agricultural and Food Chemistry* 56 (11): 4212–4218.

57 Choi, M., Jung, U., Yeo, J. et al. (2008). Genistein and daidzein prevent diabetes onset by elevating insulin level and altering hepatic gluconeogenic and lipogenic enzyme activities in non-obese diabetic (NOD) mice. *Diabetes/Metabolism Research and Reviews* 24 (1): 74–81.

58 Yousefi, H., Alihemmati, A., Karimi, P. et al. (2017). Effect of genistein on expression of pancreatic SIRT1, inflammatory cytokines and histological changes in ovariectomized diabetic rat. *Iranian Journal of Basic Medical Sciences* 20 (4): 423.

59 Lee, J.-S. (2006). Effects of soy protein and genistein on blood glucose, antioxidant enzyme activities, and lipid profile in streptozotocin-induced diabetic rats. *Life Sciences* 79 (16): 1578–1584.

60 Ding, P., Chen, Y., Cao, G. et al. (2019). Solutol® HS15+ pluronicF127 and Solutol® HS15+ pluronicL61 mixed micelle systems for oral delivery of genistein. *Drug Design, Development and Therapy* 13: 1947.

61 Kim, H., Choi, J.-S., Kim, K.S. et al. (2012). Flt1 peptide–hyaluronate conjugate micelle-like nanoparticles encapsulating genistein for the treatment of ocular neovascularization. *Acta Biomaterialia* 8 (11): 3932–3940.

62 Eng, Q.Y., Thanikachalam, P.V., and Ramamurthy, S. (2018). Molecular understanding of Epigallocatechin gallate (EGCG) in cardiovascular and metabolic diseases. *Journal of Ethnopharmacology* 210: 296–310.

63 Balentine, D.A., Wiseman, S.A., and Bouwens, L.C. (1997). The chemistry of tea flavonoids. *Critical Reviews in Food Science & Nutrition* 37 (8): 693–704.

64 Nakagawa, K., Ninomiya, M., Okubo, T. et al. (1999). Tea catechin supplementation increases antioxidant capacity and prevents phospholipid hydroperoxidation in plasma of humans. *Journal of Agricultural and Food Chemistry* 47 (10): 3967–3973.

65 Grzesik, M., Naparło, K., Bartosz, G., and Sadowska-Bartosz, I. (2018). Antioxidant properties of catechins: comparison with other antioxidants. *Food Chemistry* 241: 480–492.

66 Lee, S.-J. and Lee, K.-W. (2007). Protective effect of (−)-epigallocatechin gallate against advanced glycation endproducts-induced injury in neuronal cells. *Biological and Pharmaceutical Bulletin* 30 (8): 1369–1373.

67 Meng, Q., Velalar, C.N., and Ruan, R. (2008). Effects of epigallocatechin-3-gallate on mitochondrial integrity and antioxidative enzyme activity in the aging process of human fibroblast. *Free Radical Biology and Medicine* 44 (6): 1032–1041.

68 Potenza, M.A., Marasciulo, F.L., Tarquinio, M. et al. (2007). EGCG, a green tea polyphenol, improves endothelial function and insulin sensitivity, reduces blood pressure, and protects against myocardial I/R injury in SHR. *American Journal of Physiology-Endocrinology and Metabolism* 292 (5): E1378–E1387.

69 Pramila, K. and Julius, A. (2019). Antihyperglycemic, antidyslipidemic and antifibrotic effect of EGCG in STZ-high fat diet induced DCM rats. *Research Journal of Pharmacy and Technology* 12 (4): 1839–1842.

70 Ueda-Wakagi, M., Hayashibara, K., Nagano, T. et al. (2018). Epigallocatechin gallate induces GLUT4 translocation in skeletal muscle through both PI3K-and AMPK-dependent pathways. *Food & Function* 9 (8): 4223–4233.

71 Niu, Y., Na, L., Feng, R. et al. (2013). The phytochemical, EGCG, extends lifespan by reducing liver and kidney function damage and improving age-associated inflammation and oxidative stress in healthy rats. *Aging Cell* 12 (6): 1041–1049.

72 Sampath, C., Rashid, M.R., Sang, S., and Ahmedna, M. (2017). Green tea epigallocatechin 3-gallate alleviates hyperglycemia and reduces advanced glycation end products via nrf2 pathway in mice with high fat diet-induced obesity. *Biomedicine & Pharmacotherapy* 87: 73–81.

73 Bulboaca, A.E., Boarescu, P.-M., Porfire, A.S. et al. (2020). The effect of nano-epigallocatechin-gallate on oxidative stress and matrix metalloproteinases in experimental diabetes mellitus. *Antioxidants* 9 (2): 172.

74 Sun, M., Xie, Q., Cai, X. et al. (2020). Preparation and characterization of epigallocatechin gallate, ascorbic acid, gelatin, chitosan nanoparticles and their beneficial effect on wound healing of diabetic mice. *International Journal of Biological Macromolecules* 148: 777–784.

20

Nanotherapeutics of Phytoantioxidants for CNS Disorders

Arpita Paul, Nilayan Guha, and Md Kamaruz Zaman

Department of Pharmaceutical Sciences, Faculty of Science and Engineering, Dibrugarh University, Dibrugarh, Assam, India

20.1 Introduction

The central nervous system (CNS) is one of the most intricate and enigmatic systems of an organism. It controls all the activities essential for the normal functioning of the body. Any alterations in the function and structure of the CNS by the factors such as genetics, lifestyle, nutrition, aging, exposure to environmental toxins, chemicals, and viruses give rise to CNS disorders and diseases [1]. Neurological disorders are one of the leading causes of death and disability in the world. These disorders can manifest with the symptoms such as motor dysfunction, cognitive impairment, and difficulty in speech. CNS disorders comprise a set pathological state that is notably challenging to diagnose, assess, monitor, and treat. Amongst them, Alzheimer's disease (AD), Parkinson's disease (PD), multiple sclerosis (MS), amyotrophic lateral sclerosis (ALS), Huntington's disease (HD), ischemic stroke, and neuropathic pain (NP) are worth mentioning. Neuronal loss is the distinctive pathological feature shared by the diverse range of CNS disorders. AD is one of the most common neurological disorders characterized by memory impairment. PD is the second most prevalent CNS disorder, which is expected to double between 2005 and 2030 in Asian countries [2]. In 2019, the World Health Organization (WHO) recorded AD and other forms of dementia as the seventh leading cause of death globally and is the second leading cause of death in high-income countries [3]. It is expected that in Asian countries, neurological diseases will replace infectious diseases and will be the leading cause of death. With the increased incidence of patients with CNS disorders, the socioeconomic burden is also skyrocketing throughout the globe. Currently, available therapies for the treatment of CNS disorders are limited, intended only to treat the symptoms. These medications are associated with either contradictory or adverse effects, which make them unfit for combating neuronal disorders [4]. The existing CNS therapeutics are unable to stop or delay the degeneration of the neurons. So, there is a need to seek better strategies in the management of CNS disorders [5]. In this context, nature provides us with a myriad of phytoantioxidants, which, due to their multi-target mechanism, sound propitious in averting neurodegenerative disorders. The phytochemicals exhibit antioxidant, anti-inflammatory, antiapoptotic, and neuroprotective efficacy that makes them appropriate candidates for CNS disorders. But the blood-brain barrier (BBB) stands as an obstacle and restricts the passage of the majority of small molecules and completely blocks the entry of all the larger molecules to the CNS [6]. Although different

Phytoantioxidants and Nanotherapeutics, First Edition. Edited by Mithun Rudrapal.

approaches for drug delivery to the CNS are reported, these approaches are expensive and bring up issues like poor bioavailability, poor patient compliance, and increased risk of toxicity [7]. In this context, nanotechnology provides superior and effective delivery of drugs to the CNS than the conventional treatment approaches [8]. Nanotechnology-based delivery of some phytoantioxidants such as curcumin, quercetin, and resveratrol could overcome the factors that impede the delivery of these agents across the BBB and have improved the health of patients diagnosed with CNS disorders [9]. Therefore, the advent of nanotherapeutics of phytoantioxidants is an exciting, cutting-edge, and promising proficiency in the treatment of neuronal disorders and has great potential to overcome the limitations posed by conventional drug delivery mechanisms.

20.2 Etiology and Progression of CNS Disorders

A plethora of factors leads to CNS disorders. Apart from nutrition and lifestyle, the growing age of a person and environmental toxicants result in abnormal gene expression and interruption of cellular functions. These epigenetic changes play a critical role in the etiology of CNS disorders. Various literatures suggest that phytochemicals can avert the abnormal epigenetic changes and so can alter the abnormal gene expression [1]. Although the hallmark pathological features common to all CNS disorders are neuronal death, gliosis, and atrophy, however, these disorders differ in specific protein abnormalities and the type of neuronal population lost. Abnormalities of amyloid precursor protein (APP), tau, and α-synuclein protein help in the identification of neuronal disorders [10].

20.2.1 Alzheimer's Disease

The onset of AD includes several factors such as genetic inheritance, sex, aging, environmental factors, and various pathologies like atherosclerosis, hypertension, hypercholesterolemia, type II diabetes, stroke, transient ischemic attack and brain trauma, and epigenetic changes [11]. The key implications include neuronal loss, extracellular β-amyloid plaques, and intraneuronal tau-containing neurofibrillary tangles, which may be triggered by oxidative stress, protein disposition, or synaptic injury [12, 13]. The β-amyloid aggregates disrupt the neuronal communication at their synapses, while the tau tangles block the entry of nutrients into the neurons and, ultimately, route to neuronal death [9]. This leads to elevation of pro-inflammatory cytokines, lipid peroxidation, atrophy, and degeneration of basal forebrain cholinergic neurons in the brain [14]. As a result, there is a reduction in the synthesis of acetylcholine and acetylcholine esterase activity contributing to memory loss, cognitive impairment, behavioral disturbances, and psychiatric symptoms [15]. The polymerization of β-amyloid peptides is reported to interfere with the glutamate pathway and disrupts neurotransmission [16].

20.2.2 Parkinson's Disease

PD is the second most frequent CNS disease and the main predisposing factors of the disease are environmental changes, ageing, chronic diseases, and social anxiety [17]. It is characterized by the continuous loss of dopaminergic neurons within the substantia nigra associated with α-synuclein protein aggregates and Lewy bodies [18]. Consequently, aberrant motor functions, for instance, tremor, limb rigidity, bradykinesia, and gait imbalance are experienced by PD patients [19].

Additionally, elevation of 8-hydroxy-2-deoxyguanosine (8-OHdG), 3-nitro-tyrosine (3-NT), protein carbonyls, and 4-hydroxy-2-nonenal (4-HNE) levels and reduction of glutathione (GSH) levels have been observed in PD brains. On the other hand, 4-HNE covalently binds to α-synuclein and leads to gamma-aminobutyric acid (GABA) dysfunction [20]. The non-motor features include cognitive impairment, depression, anxiety, sleep disturbances, loss of sense of smell, constipation, gastrointestinal motility issues, urinary symptoms, orthostatic hypotension, and sexual dysfunction. Hence, with the course of time, the PD patients develop cognitive decline and potentially full-blown dementia [21].

20.2.3 Multiple Sclerosis

MS is categorized as a chronic neurological disorder as well as an autoimmune disorder [22]. Several factors such as genetic changes, environmental toxins, and viral and microbial infections contribute to the disease onset [23]. It instigates nerve demyelination followed by an axonal loss in the white and gray matter of the brain and spinal cord. This results in the formation of lesions associated with infiltrates of immune T cells, B cells, monocytes, and dendritic cells across the BBB [24, 25]. The intraneuronal communication is disrupted and the immune system is activated, which start to destroy the myelin components, whereas, demyelination further activates the inflammatory and oxidative stress pathways and, consequently, causes tissue damage [24]. In addition to this, the destruction of oligodendrocytes, the cells that build and nourish the myelin sheath, occurs. Furthermore, tau protein, β-amyloid, and APP are present in the plaques and lesions of MS patients [26]. MS disease is triggered by the relapsing-remitting phase which is marked by partial or fully reversible episodes of neurologic dysfunction. With the progression of time typically after 10–20 years, inevitable progressive neurologic dysfunction occurs that eventually leads to impaired mobility and cognition [27].

20.2.4 Amyotrophic Lateral Sclerosis

ALS is a heterogeneous motor neuron disease that causes rapid deterioration of both upper and lower motor neurons projecting from the brain and spinal cord. The two major forms of ALS are sporadic form and familial form. Although the actual cause of the disease is undefined, nevertheless, it is linked to familial history with gene mutations and other associated mechanisms such as oxidative stress, calcium toxicity, inflammation, chronic viral infections, and glutamate excitotoxicity [28]. The symptoms of ALS are localized muscle weakness that starts to appear in the middle-to-late 50s [29]. In the later stages, paralysis and cognitive damage may take place [30]. Death in ALS patients may occur due to respiratory failure within three to five years after the first sign of the disease [31]. The degeneration of the motor neurons leads to disrupted communication between motor neurons and muscle, resulting in gradual muscle weakness and wasting of muscle in the arms, legs, trunk, and bulbar region [32]. Mutation of the gene that encodes superoxide dismutase (SOD1), TAR DNA-binding protein 43 (TDP-43), and fused in sarcoma (FUS) leads to the accumulation of proteins and thereby induces mitochondrial dysfunction, oxidative stress, neurofilament aggregation and disrupted axonal transport, thus contributing to maladaptive motor nerve activity [33]. Furthermore, the inflammatory and autoimmune processes are activated that stimulate the neighbouring glial cells to release pro-inflammatory cytokines, which increases the toxicity and ultimately promotes motor neuron death [34].

20.2.5 Huntington's Disease

HD is an autosomal neurodegenerative disease that progresses with a combination of psychiatric illness, cognitive impairment, and motor dysfunction [35]. The psychiatric illness manifests with apathy, irritability, psychosis, and depression. The motor dysfunctions encompass uncoordinated movements of limbs, face, tongue, and trunk; difficulty in speech; dysphagia; poor eye movement; etc. [36]. Atrophy of the caudate nucleus and the dorsal striatum of the brain takes place due to decreased levels of brain-derived neurotrophic factor (BDNF) [37]. The disease progresses by a repeated expansion of the CAG sequence in the Huntingtin (HTT5) gene. This results in the accumulation of Huntingtin protein, which is associated with neuronal toxicity and death [38]. Abnormality of Huntingtin protein leads to mitochondrial dysfunction, aberrant protein homeostasis, dysregulated gene expression, aberrant synaptic signaling and debarred axonal transport and cytoskeleton functions [39–42]. On the other hand, increased levels of proapoptotic proteins, 8-OHdG, malondialdehyde, lipofuscin, and 3-NT have been observed in HD brains [43, 44].

20.2.6 Ischemic Stroke

Stroke is the third major cause of death in brain disorders worldwide. It is not a neuronal disease but it can lead to the path to the development of neurodegenerative disease [45]. Lack of blood supply to the brain due to a ruptured blood vessel induced by a blood clot during the episode of stroke deprives the brain cells of oxygen and nutrients, thus loss of neuronal functions and ultimately neuronal death take place [46]. Stroke leads to the death of pyramidal neurons of the cornu ammonis (CA1) region of the hippocampus [47]. Ischemia activates a range of pathological processes such as oxidative stress, apoptosis, inflammation, astrocytosis, glutamate excitotoxicity, and microglial activation [48]. As a consequence, there is a depletion of antioxidants and mitochondrial dysfunction causing behavioral impairments. All such events provoke BBB breakdown leading to edema. Furthermore, the pro-inflammatory cytokines like TNF-α, IL-1β, and IL-6 are activated that further contribute to neuronal damage [49]. Reports suggest that every minute delay in treatment of patients with stroke results in the loss of 1.9 million neurons, 13.8 billion synapses, and 12 km of axonal fibers. Further, a delay of an hour causes the loss of as many neurons as a normal person would lose during aging in 3.6 years [50]. Therefore, the main strategy for stroke management is the quick restoration of blood flow to the brain.

20.2.7 Neuropathic Pain

NP is a debilitating form of pain originating from a lesion or a disease related to the somatosensory system [51]. The origin of the pain in the CNS is categorized as a central neuropathic pain. The etiology of the pain ranges from nerve damage due to metabolic disorders such as diabetes, nutritional deficiency, viral infections, exposure to toxins, cancer chemotherapy, ischemia, and autoimmune disease. The signs and symptoms include allodynia, hyperalgesia, and paresthesia [52]. NP is complex, multifactorial, and evolves with the course of time. The pathological conditions start affecting the life of the patients and their physiological state [53]. A phenomenon called central sensitization where abnormal hyperexcitability of central nociceptor neurons occurs in the spinal cord due to peripheral nerve injury and the release of tachykinins and neurotransmitters. These tachykinins and neurotransmitters lead to a series of biochemical reactions that increase the excitability and sensitivity of CNS neurons that lead to hyperalgesia and allodynia [54].

20.3 Current Strategies Developed in the Management of CNS Disorders and Their Limitations

Currently, there are no treatments available to cure neuronal disorders. Although advancement in knowledge and technology has propelled extensive research, these therapies are limited to treating only the symptoms of the disorders. As of now, gene therapy is an emerging field in the treatment of CNS disorders, where limited treatment options are available [55]. Nevertheless, a few randomized clinical trials of symptomatic therapy were found helpful in improving the lives of patients with neuronal disorders.

In AD, Food and Drug Administration (FDA) approved acetylcholinesterase inhibitors such as tacrine, donepezil, rivastigmine, galantamine, and memantine, a N-methyl-D-aspartate receptor antagonist as the first-line drugs to treat cognitive disorders [56–58]. 2, 2-dimethyl dichlorovinyl phosphate, a prodrug of metrifonate, is a long-acting and strong acetylcholinesterase inhibitor used in behavior and cognitive deficits of AD and is under clinical trials [59]. These cholinesterase inhibitors stabilize cholinergic transmission and prevent acetylcholine degradation in the synaptic cleft (Anderson and Mosley 2017). The most common side effects of cholinesterase inhibitors include gastrointestinal complications like nausea, vomiting, and diarrhea [60]. Despite many clinical studies, the safety and efficacy of these agents are short-term and, hence, there is a need for an evidence-based study to determine their long-term efficacy [61]. As in all CNS disorders, the dying neurons lead to the activation of inflammatory cells surrounding them, the nonsteroidal anti-inflammatory drugs (NSAIDs) can be used to reduce inflammation. These agents act via inhibition of either one or both the cyclooxygenase (COX) enzymes [62]. However, in a randomized double-blind study, celecoxib, diclofenac, and rofecoxib failed to prove their efficacy while ibuprofen, indomethacin, and sulindac sulfide were found to be effective [57, 63–65]. Additionally, some synthetic drugs such as ganstigmine, phenserine, eptastigmine, velnacrine maleate, neramexane, and propentofylline are reported to be effective in AD but none of them is exempted from side effects and thus has failed to reach the clinical trial phase [59]. Recently, immunotherapy is new hope for AD treatment, but clinical trials were interrupted in phase II due to the development of aseptic meningitis in approximately 6% of vaccinated patients [66, 67]. A set of anti-PD drugs is prescribed to the patients when their disability starts to affect their daily life. Levodopa or a dopaminergic agonist is the first choice of drug. However, in absence of satisfactory results, a combination of monoamine oxidase-B (MAO-B) inhibitor, catechol-*O*-methyltransferase (COMT) and a small amount of an anticholinergic drug are prescribed [68, 69]. To treat symptoms like tremor, bradykinesia, and rigidity, beta-blockers and muscle relaxants are also used [70]. Immunity-based therapy such as immunophilin-binding proteins are developed for PD. Immunophilin is intracellular receptor protein abundantly present in the brain [71]. When immunophilin ligands bind with immunophilin proteins, they suppress the immune system. The immunophilin ligands, GPI 1046 (Guilford Pharmaceutical) and AMG474-00 or NIL-A (Amgen) were found to promote neuronal growth. Clinical trials were carried out on the stated ligands but their results are unavailable [72, 73]. In 2017, the FDA has approved a number of therapeutic agents that act through different mechanisms to prevent MS progression. Dimethyl fumarate and interferon-β prevent inflammatory responses while teriflunomide, ocrelizumab, and alemtuzumab interfere with the immune system. Likewise, natalizumab blocks the migration of lymphocytes while fingolimod blocks the recirculation of lymphocytes [74]. In addition, Glatiramer acetate (GA, Copaxone, Cop-1), a random mix of four amino acids, is approved by FDA for MS [75]. There are several investigations undertaken to find better remedies for this neurodegenerative disease.

Regardless of a number of studies on therapeutic agents for ALS, only edaravone and riluzole were approved by FDA [76]. Riluzole suppresses presynaptic glutamate release and inactivates postsynaptic voltage-gated sodium channels and N-methyl-D-aspartate (NMDA) receptors, while edaravone suppresses oxidative stress [77].

Tetrabenazine, a synaptic vesicular amine transporter inhibitor is prescribed in HD. The drug is helpful in suppressing the jerky movements of the facial muscles and further counteracts the mood-altering features [36]. Currently, there is no approved therapy to delay or modify HD progression [78].

The goal of the treatment of ischemic stroke comprises the following – reducing of the progression of ischemia and revascularization and reperfusion of the brain parenchyma cells in addition to the reduction of inflammatory responses [79]. Thrombolytic agents such as alteplase, streptokinase, and urokinase activate fibrin-bound and circulating plasminogen that trigger the release of large amounts of plasmin into the circulation, which eventually leads to fibrinogenolysis and degradation of procoagulant proteins [80]. Alteplase is the only FDA-approved medication for ischemic stroke [81]. In addition, antihypertensives, antipyretics, and insulin can be used as supportive care. Further, to prevent recurrent strokes, anticoagulants and antiplatelets are prescribed. But these medications are also associated with side effects such as nausea, diarrhoea, bleeding, headache, rash, etc. [82].

NP management focuses on treating the symptoms rather than the cause of pain. Medication for first-line therapy for NP includes gabapentinoids, tricyclic antidepressants (TCAs), and selective serotonin–norepinephrine reuptake inhibitors (SNRI). On the other hand, lidocaine, capsaicin, and tramadol are the most effective second-line treatment [53]. But these medications come with their own considerable side effects. In case a patient fails to respond to first- and second-line therapies, combinatorial therapy is trailed. Interventional therapies such as epidural injection, pulsed radiofrequency, sympathetic blockade, and adhesiolysis are proposed as third-line therapy. While neurostimulation is recommended as fourth-line therapy, all the other therapies have failed to achieve adequate pain relief [83]. Recently, human trials have been carried out to check the efficacy of new therapeutic approaches. Memantine, ketamine, sativex, dronabinol, etc., have completed their third phase of clinical trials and awaits further investigation to evaluate their efficacy [84]. In addition, nonpharmacological treatments such as exercise are recommended to gain muscle strength and functionality [85]. Other nonpharmacological treatments include physical therapy, psychotherapy, and occupational therapy. Moreover, such approaches are safe and might decrease the use of medications [86].

Apart from Western medicine, traditional and integrative medicines are also found to be beneficial in treating neuronal diseases and disorders. Combinations of traditional Chinese medicine (TCM) with Western medicines were found to improve sleep disorders and non-motor disorders in PD patients [87, 88]. In addition, Ayurvedic medicine is effective in PD with minimal side effects [89]. Certain Ayurvedic herbs such as *Celastrus paniculatus*, *Centella asiatica*, *Withania somnifera*, *Convolvulus pluricaulis*, *Nardostachys jatamansi*, etc., are reported for their anti-AD potential [90]. While in AD, TCM was found to improve the symptoms of dementia [91].

20.4 Phytoantioxidants as Therapeutic Alternatives to CNS Disorders

Recently, there is an inclination toward plant-derived phytoconstituents as potential therapeutic agents in the treatment of CNS disorders. These phytoconstituents due to their multi-target, synergistic, and antioxidant effects serve as better candidates in the treatment of CNS-related etiologies. Moreover, they are comparatively safer than conventional drugs. Besides, a literature survey revealed that these naturally occurring substances exhibit a high affinity for various receptors present in the brain, thereby modulating various molecular signal transduction pathways [92]. Phytoantioxidants like flavonoid supplementation are reported to improve the cognitive function by enhancing existing neuronal activities and stimulating neuronal regeneration [93]. For instance, tea consumption has reduced the risk of AD [94]. Further, alkaloid extracts have been used for over 3000 years as therapeutic agents against various diseases. Apart from their use in modern medicine, alkaloids also possess psychotropic and stimulant activities and hence are used in rituals and recreational purposes [95].

The beneficial effects of phytochemicals on the neuronal system are mainly attributed to their antioxidant, anti-inflammatory, and neuroprotective activities. Herein, the major classes of phytoantioxidants beneficial in CNS disorders are described.

20.4.1 Alkaloids

Alkaloids comprise a class of organic nitrogen-containing compounds present primarily in plants. Based on their chemical structures, they are classified into pyrrole, pyrrolizidine, pyridine, piperidine, indole, quinolone, isoquinoline, tropane, purine, and imidazole alkaloids. Berberine, an isoquinoline alkaloid, can inhibit the activity of acetylcholinesterase (AChE), butyrylcholinesterase (BChE), BACE1, glycogen synthase kinase-3β (GSK-3), and monoamine oxidase-B (MAO-B) enzymes, ameliorate intracellular oxidative stress, attenuate neuroinflammation, and inhibit apoptosis. Further, it enhances the clearance of toxic misfolded and aggregated proteins, inhibits neuroinflammation, and triggers autophagy. Thus, it can act as a promising anti-neurodegenerative agent [96]. Similarly, other isoquinoline alkaloids such as galantamine [97], avicine, nitidine [98], and morphine [99] showed beneficial effects in neurodegenerative diseases. While piperine [100], stepharanine, cyclanoline, N-methyl stepholidine [101], harmane [102], harmine [103], and deoxyvasicine [104] were reported to show significant neuroprotective effects. On the other hand, the β-carboline alkaloids, namely harmol, harmine, and harmaline displayed neurogenesis due to their inhibitory activity toward MAO and dual-specificity tyrosine phosphorylation-regulated kinase 1A (DYRK1A) [105, 106]. Additionally, a study reported that asimilobine, an aporphine alkaloid, not only has significant neurogenesis potential but also has a high penetrating capacity across the BBB [107]. Furthermore, Huperzine A, a sesquiterpene alkaloid isolated from the Chinese medicinal plant *Huperzia serrate* exhibited significant neuroprotective activity in Phase I clinical trial [108, 109]. The alkaloids mainly act as neurodegenerative disease modulators by inhibiting several mechanisms such as neurotransmission, neuroinflammation, oxidative stress, mitochondrial excitotoxicity, apoptosis, β-amyloid aggregation, and tau phosphorylation [110].

20.4.2 Polyphenols

Polyphenols are the largest class of plant secondary metabolites found in fruits and vegetables that show strong neuroprotective activity [111]. The therapeutic activity of polyphenols is attributed due to the presence of phenol rings, hydroxyl groups, and conjugated double bonds, which promote metal-chelating, fibril-destabilizing, estrogen-like, enzyme-binding, and antioxidative properties [112]. A literature survey revealed that dietary polyphenols can cross BBB and thus target the pathological signs of neurological disorders [113]. The dietary polyphenols exert antioxidant effects primarily by activating Nrf2/HO1 pathway and downregulating NFκB, MMPs, PPAR, HIF-1, and STAT pathways. They further modulate the immune response by inhibiting pro-inflammatory mediators and improve cognitive abilities by inhibiting AChE and BChE. These features of polyphenols help prevent the progression of CNS disorders [114].

The flavonoids such as quercetin and epigallocatechin-3-gallate provide neuroprotection by inhibiting reactive oxygen species (ROS), activating the AMP-activated protein kinase (AMPK) pathway and suppressing pro-inflammatory mediator production in various tissues such as liver, sciatic nerve, and brain of diabetic animals. It is shown that phenolic acids like p-coumaric acid, caffeic acid, ferulic acid, gallic acid, vanillic acid, and protocatechuic acid improve neuronal dysfunctions [115, 116]. Furthermore, the non-flavonoid polyphenols, i.e. resveratrol and curcumin are reported for their beneficial effects on CNS disorders. Recently, in the Phase II trials, these compounds were found to improve the condition of AD patients [117, 118]

20.4.3 Terpenoids

Terpenoids are widespread unsaturated organic compounds composed of isoprene units. Ginkgolides, a major terpenoid present in *G. biloba*, exhibited neuroprotective effects via suppression of NLRP3 and caspase-1 expression, and reduction of pro-inflammatory mediator production [119]. While bilobalide, another terpenoid from *G. biloba* also exerted neuronal protection through upregulation of antioxidant enzyme activities and downregulation of ROS and pro-inflammatory cytokines expression [120]. In addition, in phase III clinical trial, ginsenoside Rd was found to improve the disability scores and stroke scales in patients with acute ischemic stroke compared to the placebo group [121]. Thymoquinone from *Nigella sativa* and thymol from *Thymus vulgaris* also show neuroprotective activity.

20.4.4 Other Phytoantioxidants

Polyunsaturated fatty acids such as α-linolenic acid, docosahexaenoic acid, eicosapentaenoic acid, linoleic acid, and arachidonic acid are important regulators of metabolic pathways and are highly susceptible to free radical attacks in the brain. Reports suggest that supplementation of these fatty acids can counteract the age-related cognitive impairments due to their antioxidant, anti-inflammatory, and neuroprotective properties [122]. Another antioxidant compound N-acetyl cysteine derived from onion was reported to show beneficial effects in HD. Further, in Phase II clinical trial, it improved neurological functions and exerted anti-inflammatory activity in patients with IS when compared with the placebo group [123]. Also, vitamin D3 showed significant neuroprotective effects against MS via modulating the expression of anti-inflammatory and pro-inflammatory cytokines [124]. The chemical structures of the phytoconstituents effective against CNS disorders are represented in Table 20.1. Moreover, various factors involved in the progression of CNS disorders and the role of phytoantioxidants in the alleviation of CNS disorders with neuroprotective effects are shown in Figure 20.1.

20.5 Challenges in Administration of Phytoantioxidants in CNS Disorders

Despite the beneficial effects of phytoantioxidants in CNS disorders, the success rate of phytoantioxidant-based therapies is often held back due to several limitations. First is the lack of stability of the phytoconstituents in the body or during storage. As phytoantioxidants are sensitive to oxidative stress-inducing elements, maintaining their stability till their successful delivery to the

Table 20.1 List of some phytoantioxidants as therapeutic agents against CNS disorders.

Phytoantioxidant	Source	Structure	Disease/disorder	References
Alkaloids				
Berberine	*Berberis vulgaris* L.	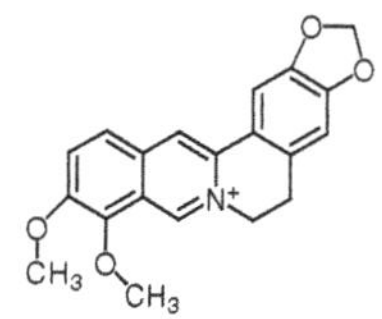	AD	[125]
Arecoline	*Areca catechu* L.	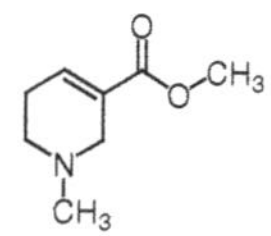	AD	[126]
Nicotine	*Nicotiana tobaccum* L.	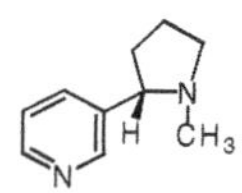	AD	[127]
Lobeline	*Lobelia inflate* L.	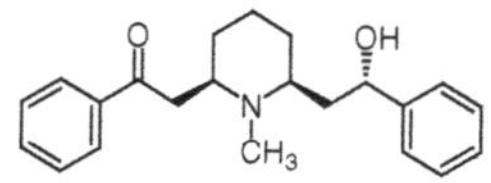	AD, PD	[128, 129]
Morphine	*Papaver somniferum*	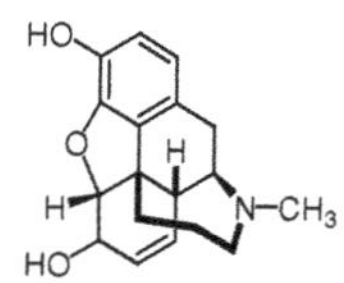	AD	[99]
Piperine	*Piper nigrum*	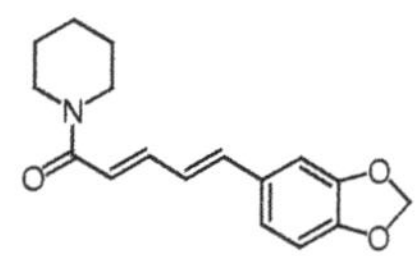	AD, PD, IS	[28, 100]
Caffeine	*Coffee Arabica*	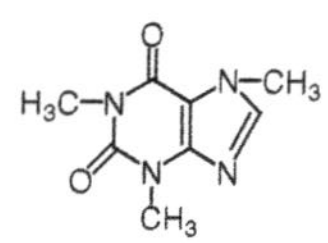	AD, PD	[130, 131]

(*Continued*)

Table 20.1 (Continued)

Phytoantioxidant	Source	Structure	Disease/disorder	References
Capsaicin	*Capsicum annuum*	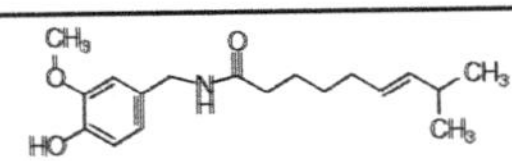	AD,	[95]
Polyphenols				
Quercetin	Many plants	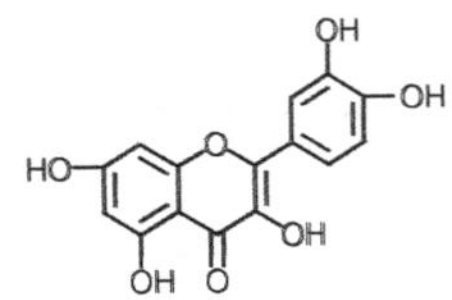	AD, MS, IS, NP	[132–134]
Epigallocatechin-3-gallate	*Camellia sinensis*	S	IS, HD, ALS, NP	[135, 136]
Curcumin	*Curcuma longa* L.	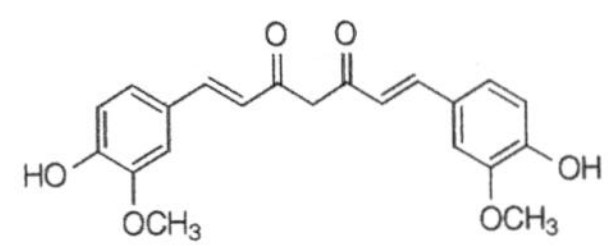	AD, PD, HD, IS, ALS	[28, 137, 138]
Resveratrol	*Vitis vinifera* L.	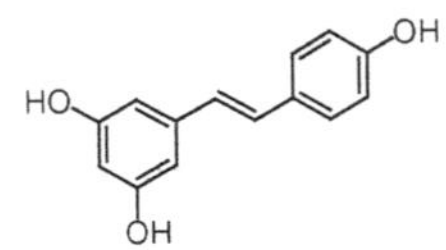	AD, MS, PD, HD, ALS, IS	[28, 136, 139–142]
Naringenin	Citrus fruits	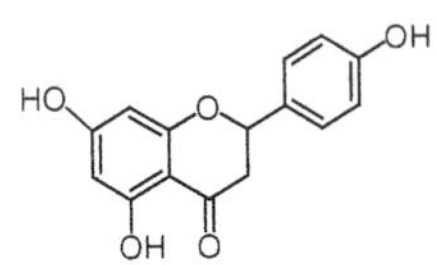	AD, PD, MS, ALS, HD, IS, NP	[143, 144]
Rutin	Citrus fruits	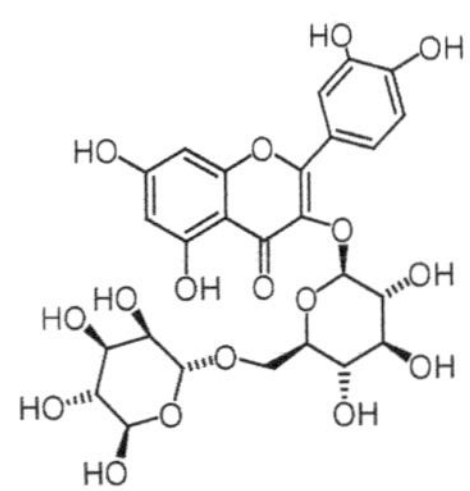	NP, PD	[28]

Silymarin	*Silybum marianum*	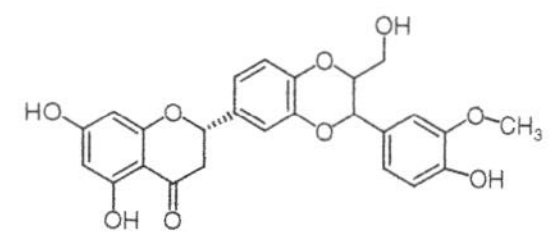	AD, PD	[145]
Genistein	*Glycine max*	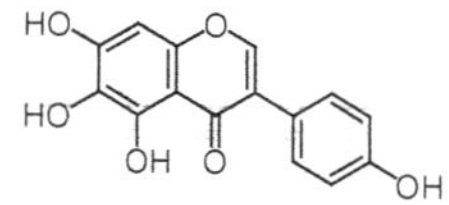	NP, AD, PD, ALS	[28]
Baicalin	*Scutellaria Baicalensis*	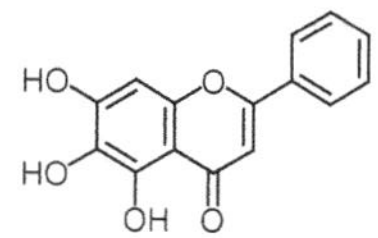	AD, PD, NP, IS	[28, 146]
Luteolin		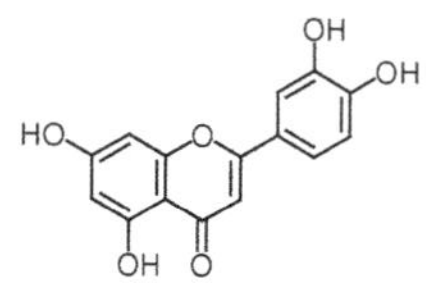	NP, AD, IS	[147, 148]
Terpenoids				
Ginkgolide B	*Gingko biloba*	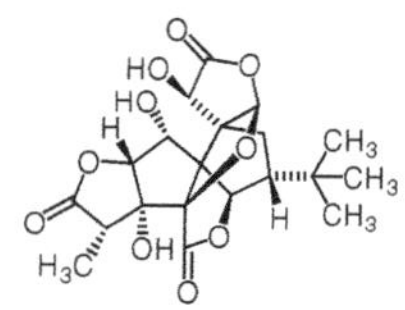	AD, IS	[149, 150]
Thymoquinone	*Nigella sativa*	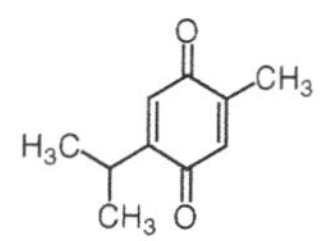	AD	[151]

(*Continued*)

Table 20.1 (Continued)

Phytoantioxidant	Source	Structure	Disease/disorder	References
Andrographolide	*Andrographis Paniculata*	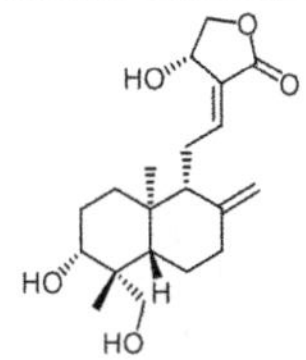	AD, PD, IS	[152, 153]
Asiatic acid	*Centella asiatica*	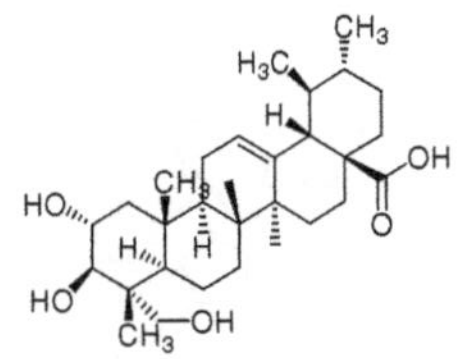	PD	[148]
Linalool	*Rosmarinus ocinalis*	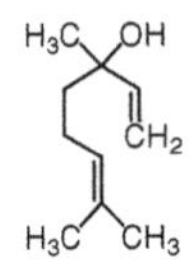	AD	[148]
Other phytoantioxidants				
Vitamin D3		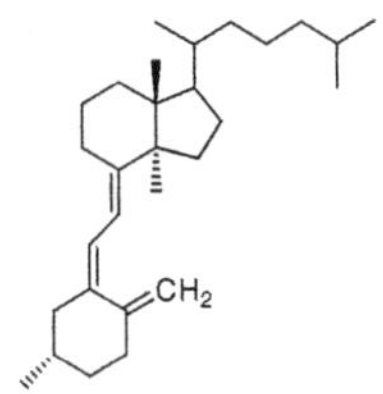	MS, AD, PD, ALS, NP	[149]
N-acetyl cysteine	*Allium sativum*	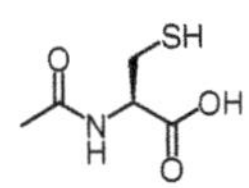	AD, PD, NP, IS	[123, 150]

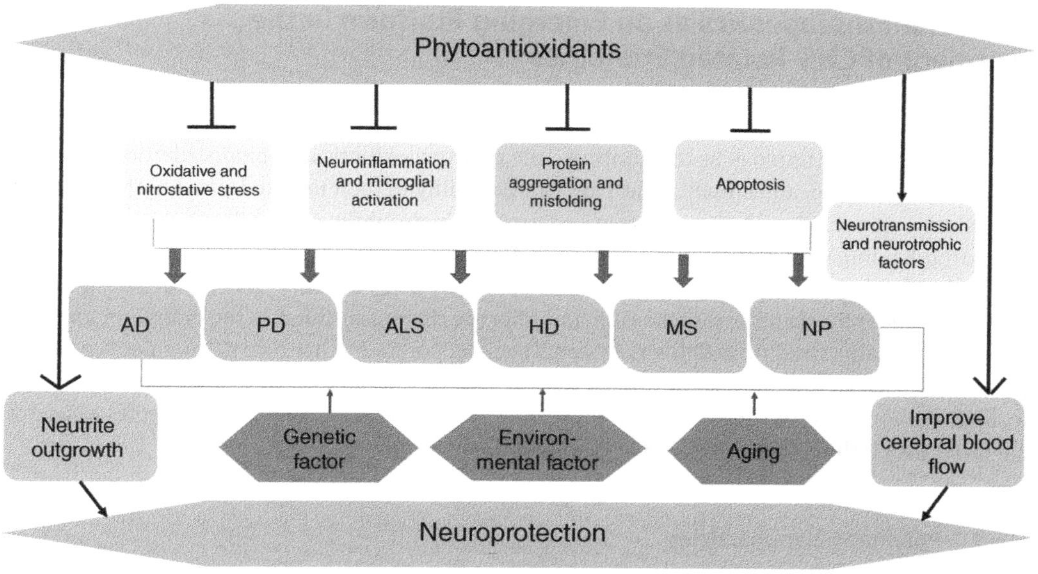

Figure 20.1 Factors involved in the progression of CNS disorders and the role of phytoantioxidants in its alleviation/neuroprotective effects of phytoantioxidants. Aging, genetic, or environmental factors interfere with the neuroprotective pathways and give rise to CNS disorders. The phytoantioxidants provide protection to this pathway by preventing the misfolding and aggregation of proteins; as a consequence, it attenuates neuroinflammation and cellular stresses. By inhibiting the production of inflammatory agents, phytoantioxidants impede neuroinflammation and glial activation. Phytoantioxidants also inhibit the production of apoptotic proteins, provide neurite outgrowth, improve neuronal transmission, and improve cerebral blood flow. AD: Alzheimer's disease; PD: Parkinson's disease; ALS: Amyotrophic lateral sclerosis; HD: Huntington's disease; MS: Multiple sclerosis; NP: Neuropathic pain.

target site is a major challenge. Second is the poor aqueous solubility and poor oral absorption that conflicts with the bioavailability of the drug [121]. Interaction with the food matrix, the acidic condition of the gut, transformation by gastrointestinal enzymes, and gut microbial enzymes, all these factors together hamper the absorption of the phytoantioxidants. Further, poor distribution, rapid metabolism, and elimination lead to less concentration of the drug in the target site. As a result, the desired therapeutic concentration is not achieved [112].

Another most critical hindrance to CNS therapy is the drug delivery to the brain across the BBB. The presence of tight junctions between endothelial cells, efflux transporters such as P-glycoprotein (Pgp), and multidrug resistance-associated protein (MRP), and high levels of metabolizing enzymes restrict the entry of compounds into the brain. Therefore, to circumvent the BBB, certain strategies such as intracranial delivery, transient disruption of the BBB, and chemical modifications of the compounds to increase permeability and intranasal delivery are encouraged. However, each strategy has its own set of drawbacks such as intracranial delivery demands surgical procedures, temporary opening of the BBB might introduce neurotoxic molecules to the CNS, while intranasal delivery might cause rapid mucociliary clearance and undesired mucous toxicity [154].

To overcome the above-stated challenges, nanocarriers have been engineered with better drug-loading capacity to treat various CNS disorders and, thereby, enhance solubility, boost permeability across the BBB, target specificity, improve drug pharmacokinetics, enhance therapeutic efficacy, and reduce toxicity.

20.6 Nanotherapeutics as an Emerging Platform in the Treatment of CNS-Related Etiologies

Nanotechnology is an emerging platform in the development of pharmaceuticals. Nanoparticles are a boon to nanotechnology as they improve the pharmacokinetic and biodistribution of drugs, sustained drug release, and enhance the drug bioavailability in their targeted sites [154]. The nanocarriers are engineered with the ability to encapsulate hydrophilic and hydrophobic drugs, with better drug-loading capacity, better chemical and biological stability, and with the possibility to administer through different routes. Nanoparticle-based drug delivery systems are designed to effectively deliver the drugs across the BBB and improve their absorption in the brain. For effective CNS delivery, nano-sized drug delivery systems such as polymeric nanoparticles, micelles, solid–lipid nanoparticles, liposomes, dendrimers, carbon nanotubes, and exosomes are utilized to bypass the BBB without altering the drug's properties. The nanotechnology-based strategies used for delivering across the CNS are illustrated in Figure 20.2.

20.6.1 Polymeric Nanoparticles

Natural polysaccharides (hyaluronic acid, chondroitin sulfate, and chitosan) and synthetic (poly-lactic acid (PLA), polyglycolic acid (PGA), and poly lactic-co-glycolic acid (PLGA)) polymers are extensively utilized for enhanced delivery of the drug into the brain because of their high drug-loading capacity, stabilization of entrapped molecules, prolonged and sustained drug release, and biodegradability and biocompatibility properties [155].

20.6.2 Micelles

These are amphiphilic nanocarriers with a hydrophobic inner core and hydrophilic outer layer. Lipoidal phytoconstituents can be easily encapsulated into a micelle to enhance their solubility. However, high production cost and high instability are some of the limitations associated with micelles [156].

20.6.3 Solid–Lipid Nanoparticles

These are lipid nanoparticles with surfactants dispersed in an aqueous environment. These are better alternatives to the polymeric nanoparticles due to their high drug-loading capacity, better stability, controlled release, and enhanced bioavailability and biocompatibility. Polyethylene glycol-modified SLNs can penetrate the BBB and permit the delivery of ample amounts of drugs to the CNS. SLNs are promising candidates in the area of CNS delivery of drugs due to their better stability and ease of production [157].

20.6.4 Liposomes

Liposomes are spherical nanocarriers composed of two amphiphilic layers. Liposomes have certain advantages like high drug-loading capacity, ease of surface modification, targeted delivery, prolonged systemic circulation, and biocompatibility. In spite of that, there are certain limitations linked with them such as high production cost, poor solubility, instability, and certain safety issues [158].

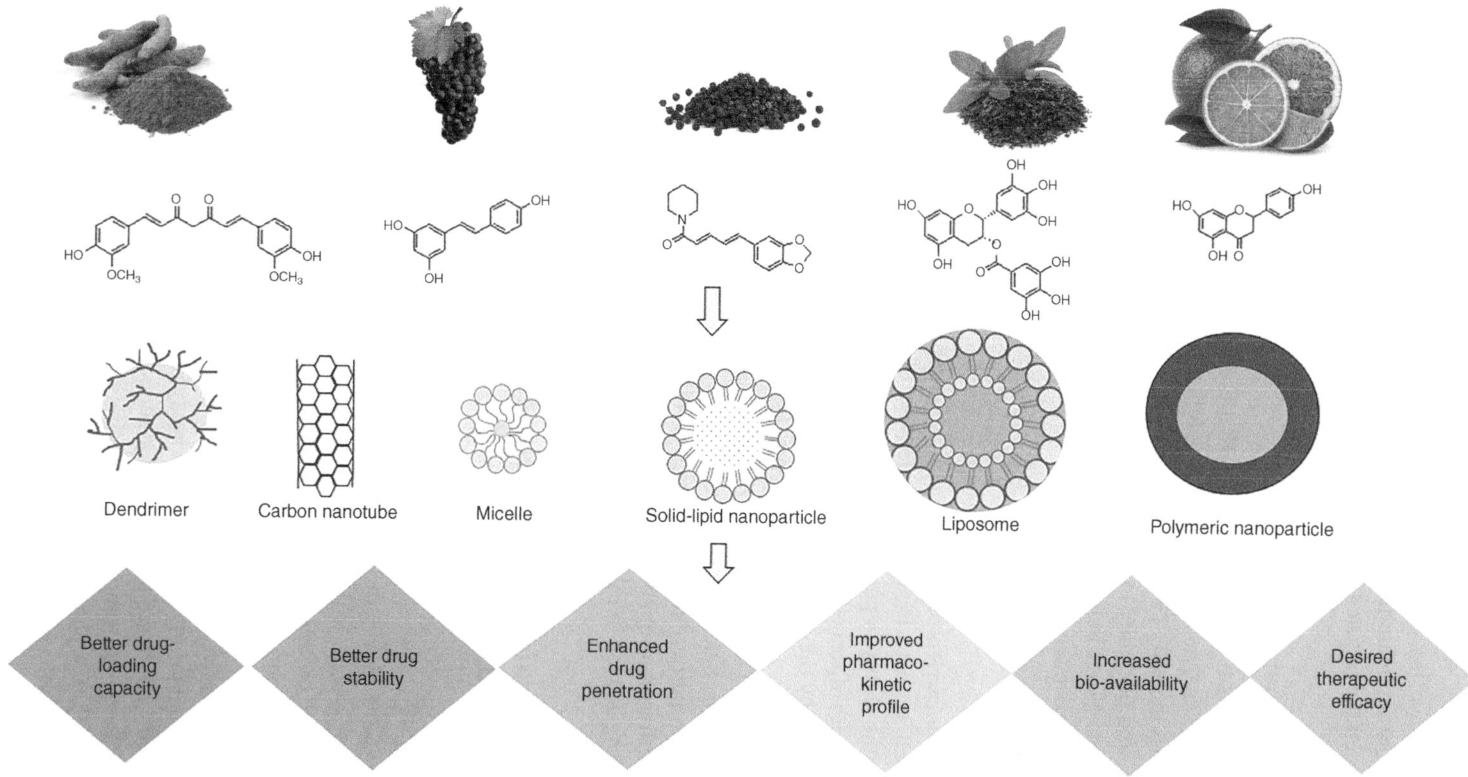

Figure 20.2 Nanotechnology-based strategies used for delivering across the CNS. *Source:* Elzbieta Sekowska/Adobe Stock, Rostislav Sedlacek/Adobe Stock, sommai/Adobe Stock, Dionisvera/Adobe Stock, and Chaotic Design Studio/Adobe Stock.

20.6.5 Dendrimers

These are nanocarriers with multibranched monomers radiating out from the central core. Dendrimers are noteworthy strategies for treating CNS disorders owing to their surface modification properties. Poly amidoamine is one of the widely used dendrimers. However, high production cost, failure to reach the targeted site, and safety issues limit their use [158].

20.6.6 Carbon Nanotubes

These are carbon-based nanocarrier systems that are extensively utilized for delivering drugs in brain-specific therapies. Their ability of surface functionalization with specific moieties enhances their passage across the BBB. The properties such as prolonged and controlled drug release, target-specific delivery, and flexibility to stimuli and stability make them excellent candidates for neurological applications. Nevertheless, high production cost, polydispersity, and toxicity are some of the disadvantages associated with carbon nanotubes [158].

20.6.7 Exosomes

These are novel nanocarriers of 30–150 nm in size, composed of a lipid bilayer, derived from endogenous cells, which are used for enhanced delivery of drugs to the CNS. Exosomes exert certain features such as the ability to penetrate the BBB, controlled drug release, better biocompatibility, ability to interact with other cells in the target area, low immunogenicity, and low toxicity. However, the challenges faced by the exosome-based delivery system are difficulty in isolation and characterization, drug loading, and cost of production [121].

Besides, the above-mentioned nanocarrier systems, inorganic nanoparticles are also utilized for CNS delivery of drugs. Selenium nanoparticles are used to treat neuroinflammation because of their antioxidant property, ability to conjugate with other molecules, ability to pass the BBB, and low toxicity. Furthermore, gold and iron oxide nanoparticles are promising nanocarriers for improving the drug delivery for the treatment of CNS disorders.

20.7 Phytoantioxidants-Based Nanoformulation for Neuroprotection

20.7.1 Resveratrol

Resveratrol (trans-3, 4′, 5-trihydroxystilbene), a phytoalexin, is the most studied natural stilbenoid, present in 70 plant species with remarkable amounts found in red wine [159]. Researchers have reported the potential of RSV in the prevention of CNS disorders. Nevertheless, poor aqueous solubility, lack of stability, rapid metabolism, and low bioavailability limit its therapeutic efficacy [160]. In a study, a group of researchers developed RSV-loaded polymeric nanoparticles that exhibited significant neuroprotective effects against 1-methyl-4-phenyl-1, 2, 3, 6-tetrahydropyridine (MPTP)-induced mouse model for PD [161]. Another study reported that RSV-encapsulated SLNs reduced Aβ fibrillation and were found to be stable for about one month [162]. While Lu et al. formulated micelles that provided protection against Aβ-induced toxicity via reducing oxidative stress and apoptosis in vitro [163]. A pharmacokinetic study of RSV-loaded vitamin E nanoemulsion showed the presence of a significant amount of drug in the brain of PD animals in comparison with the animals treated with free drug [164].

20.7.2 Quercetin

QC is a promising flavonoid found in fruits and vegetables with a myriad of health benefits. The antioxidant and anti-inflammatory properties of QC find their application in the prevention of neurological disorders. Poor solubility and low bioavailability of QC have encouraged researchers to use nanotechnology to improve the pharmacokinetic profile and bioavailability of QC. In a study, QC-loaded nanoparticles prevented neuronal damage at cellular, subcellular, and molecular levels and further restored the levels of tyrosine hydroxylase [165]. In the PD model, QC nanocrystals showed better bioavailability than animals treated with QC alone [166]. QC nanocapsules improved ischemia-induced neuronal damage and enhanced the antioxidant activity [167].

20.7.3 Curcumin

CUR is a polyphenolic compound with potent free radical scavenging properties. Despite its effectiveness in neurological disorders, a set of drawbacks such as poor aqueous solubility, poor bioavailability, and low cellular uptake limits its therapeutic use [168]. In the AD mice model, CUR-selenium PLGA nanospheres were found to be more effective than animals treated with selenium nanospheres without CUR [169]. In another study, CUR-loaded nanocapsules reduced Aβ-aggregation, neuroinflammation, and behavioral impairments in AD animals when compared to the free drug-treated group [170]. Nanoparticles loaded with CUR and piperine suppressed α-synuclein aggregation and alleviated apoptosis in the PD animal model. Further, the bioavailability of CUR was found to be enhanced [171]. In MS patients, CUR-loaded micelle restored the function of Treg cells and diminished the levels of Interleukin-10 (IL-10) [172]. While in ALS patients, nano-CUR micelles, when given as an additional treatment, improved their survival rate [173].

20.7.4 Epigallocatechin-3-Gallate

EGCG, an ester of gallic acid and epigallocatechin, is mostly found in tea. It has better antioxidant potential than vitamins C and E and can modulate various molecular pathways responsible for neuronal disorders. Nanoparticles loaded with EGCG and ascorbic acid showed increased brain accumulation (5-fold higher) when compared with nanoparticles loaded with EGCG alone. Further, the stability of EGCG was enhanced, the pharmacokinetic profile was improved, and brain uptake was enhanced and led to better therapeutic activity [174]. In the parkin Drosophila melanogaster model for PD, EGCG-encapsulated nanoparticles prevented locomotor impairments and improved their life span [175].

Table 20.2 represents phytoantioxidant-based nanoformulation and its beneficial effects on CNS disorders.

20.8 Limitations of Nanotechnology-Based Drug Delivery to the CNS

Despite various advantages associated with nanotechnology-based drug delivery systems, there are certain limitations which restrict their clinical applications. The first and foremost concern is the toxicity associated with nanocarriers. Repeated administration and accumulation of nanocarriers in the brain may induce neurotoxic effects. Secondly, the smaller size of nanoparticles may dislocate them from their target site and lead to blood coagulation, hemolysis, and accumulation of platelets [183]. Thirdly, the inorganic nanoparticles are difficult to metabolize and thus

Table 20.2 Phytoantioxidant-based nanoformulation and their beneficial effects in CNS disorders.

Phyto-antioxidant	Nanocarrier/system	Disease	Activity	References
Curcumin	Nanospheres	AD	Reduce Aβ-aggregation	[169]
	Nanoparticle	PD	Enhanced bioavailability suppression of α-synuclein aggregation	[171]
	Micelle	MS	Restore the function of Treg cells	[172]
	Micelle	ALS	Improve the survival rate	[173]
Resveratrol	Polymeric Nanoparticle	PD	Enhanced neuroprotective properties	[162]
	Micelles	AD	Protection against Aβ-induced damage	[164]
	Nanosuspension	AD	Increased bioavailability, enhanced drug permeability and increased drug residence time in the nasal cavity	[176]
	Nanoemulsion	PD	Increased bioavailability	[165]
	Nanoparticle	PD	Increased bioavailability and pharmacological activity	[177]
Quercetin	Nanoparticle	AD	Reduced neuronal degenerative changes, upregulation of tyrosine hydroxylase	[165]
	Nanoparticle	PD	Enhanced bioavailability	[166, 167]
	Nanocapsules	IS	Enhanced neuronal count and improved antioxidant activity	
EGCG	Nanoparticle	AD	Increased accumulation in the brain, improved memory and learning	[174]
	Nanoparticle	PD	Better drug-loading capacity, improved pharmacokinetic profile	[175]
Naringenin	Nanoemulsion	IS	Enhanced brain bioavailability and nontoxic	[178]
	Nanoparticle	PD	Enhanced antioxidant and neuroprotective activity	[179]
Berberine	Micelle	IS	Increased aqueous solubility and improved anti-inflammatory effects	[180]
Rutin	Nanoparticle	IS	Improved pharmacokinetic profile, improved neurobehavioral activity	[181]
Andrographolide	Nanoparticle	AD	Better tolerance, nontoxic and biocompatible	[182]

accumulate in the brain and lead to maladaptive neuronal functions like redox imbalance, mitochondrial dysfunctions, apoptosis, and activation of neuroinflammatory responses [184]. Fourthly, production cost and use of organic solvents also confine their use [185]. Therefore, these problems should be addressed before clinical implications.

20.9 Conclusions and Future Perspectives

With the advent of time, neurological disorders are increasing due to continuous lifestyle changes. As such, novel strategies should be developed to deal with such diseases and improve the quality of life of the patients. In this context, phytoantioxidants, due to their multi-target and synergistic effects, can serve as potential candidates. For effective delivery of these phytoantioxidants to the CNS, the nanotechnology-based drug delivery system is very promising. The phytoantioxidants, when encapsulated in a nanocarrier, prevent its degradation, enhance drug penetration, improve the pharmacokinetic profile and bioavailability, and achieve the desired therapeutic efficacy. Although nanotechnology is implicated in the formulations of phytoantioxidant-based nanoformulations, still there are limitations in the characterization of nanomaterials and the lack of reliable neurotoxicity reports to define their safety. Therefore, strategies to reduce the toxicity, use eco-friendly nanocarriers, reduce the exposure time, control the shape, size, and morphology of the nanoparticles, and improve surface properties should be adopted.

References

1 Hodjat, M., Rahmani, S., Khan, F. et al. (2017). Environmental toxicants, incidence of degenerative diseases, and therapies from the epigenetic point of view. *Arch. Toxicol.* 91: 2577–2597.

2 Tan, L.C.S. (2013). Epidemiology of Parkinson's disease. *Neurology Asia* 18 (3): 231–238.

3 World Health Organization (2019). *The Top 10 Causes of Death*. Geneva: World Health Organization.

4 Amanzio, M., Palermo, S., Zibetti, M. et al. (2014). Self-unawareness of levodopa induced dyskinesias in patients with Parkinson's disease. *Brain Cogn.* 90: 135–141.

5 Yacubian, T.A. (2017). Neurodegenerative disorders: why do we need new therapies? In: *Drug Discovery Approaches for the Treatment of Neurodegenerative Disorders*, 1–16. https://doi.org/10.1016/B978-0-12-802810-0.00001-5.

6 Zhang, T.T., Li, W., Meng, G. et al. (2016). Strategies for transporting nanoparticles across the blood-brain barrier. *Biomater. Sci.* 4: 219–229.

7 Li, X., Tsibouklis, J., Weng, T. et al. (2017). Nano carriers for drug transport across the blood–brain barrier. *J. Drug Target.* 25: 17–28.

8 Karthivashan, G., Ganesan, P., Park, S.Y. et al. (2018). Therapeutic strategies and nano-drug delivery applications in management of ageing Alzheimer's disease. *Drug Deliv.* 25: 307–320.

9 Moradi, S.Z., Momtaz, S., Bayrami, Z. et al. (2020). Nanoformulations of herbal extracts in treatment of neurodegenerative disorders. *Front Bioeng. Biotechnol.* 8: 238.

10 Rogan, S. and Lippa, C.F. (2002). Alzheimer's disease and other dementias: a review. *Am. J. Alzheimer Dis. Other Dement.* 17: 11–17.

11 Binda, A., Murano, C., and Rivolta, I. (2020). Innovative therapies and nanomedicine applications for the treatment of Alzheimer's disease: a state-of-the-art (2017–2020). *Int. J. Nanomedicine* 15: 6113–6135.

12 Garand, L., Buckwalter, K.C., and Hall, G.R. (2000). The biological basis of behavioral symptoms in dementia. *Issues Ment. Health Nurs.* 21: 91–107.

13 Furcila, D., DeFelipe, J., and Alonso-Nanclares, L. (2018). A study of amyloid-beta and phosphotau in plaques and neurons in the hippocampus of Alzheimer's disease patients. *J. Alzheimers Dis.* 64: 417–435.

14 Ahmad, J., Akhter, S., Rizwanullah, M. et al. (2017). Nanotechnology based theranostic approaches in Alzheimer's disease management: current status and future perspective. *Curr. Alzheimer Res.* 14 (11): 1164–1181.

15 Faustino, C., Rijo, P., and Reis, C.P. (2017). Nanotechnological strategies for nerve growth factor delivery: therapeutic implications in Alzheimer's disease. *Pharmacol. Res.* 120: 68–87.

16 Mi, D.J., Dixit, S., Warner, T.A. et al. (2018). Altered glutamate clearance in ascorbate deficient mice increases seizure susceptibility and contributes to cognitive impairment in APP/PSEN1 mice. *Neurobiol. Aging* 71: 241–254.
17 Schrag, A., Horsfall, L., Walters, K. et al. (2015). Prediagnostic presentations of Parkinson's disease in primary care: a case control study. *Lancet Neurol.* 14: 57–64.
18 Parkinson's Foundation (2018). What is parkinson's?.
19 Bengoa-Vergniory, N., Vingill, S., Attar, M. et al. (2019). Cellular α-synuclein pathology is associated with bioenergetic dysfunction in Parkinson's iPSC-derived dopamine neurons. *Hum. Mol. Genet.* 28: 2001–2013.
20 Md, U.S., Mamun, A.A., Md, K.T. et al. (2020). Neuroprotective role of polyphenols against oxidative stress-mediated neurodegeneration. *Eur. J. Pharmacol.* 886: 173412.
21 Yacoubian, T.A. (2016). *Neurodegenerative Disorders: Why Do We Need New Therapies? Drug Discovery Approaches for the Treatment of Neurodegenerative Disorders*, 1–16. Elsevier.
22 Dobson, R. and Giovannoni, G. (2018). Multiple sclerosis – a review. *Eur. J. Neurol.* https://doi.org/10.1111/ene.13819.
23 Leray, E., Moreau, T., Fromont, A., and Edan, G. (2016). Epidemiology of multiple sclerosis. *Rev. Neurol.* 172 (1): 3–13.
24 Dargahi, N., Katsara, M., Tselios, T. et al. (2017). Multiple sclerosis: immunopathology and treatment update. *Brain Sci.* 7 (7): 78.
25 Polman, C.H., Reingold, S.C., Banwell, B. et al. (2011). Diagnostic criteria for multiple sclerosis: 2010 revisions to the McDonald criteria. *Ann. Neurol.* 69: 292–302.
26 David, M.A. and Tayebi, M. (2014). Detection of protein aggregates in brain and cerebrospinal fluid derived from multiple sclerosis patients. *Front. Neurol.* 5: 251.
27 Baecher-Allan, C., Kaskow, B.J., and Weiner, H.L. (2018). Multiple sclerosis: mechanisms and immunotherapy. *Neuron* 97: 742–768.
28 Khan, A., Jahan, S., Imtiyaz, Z. et al. (2020). Neuroprotection: targeting multiple pathways by naturally occurring phytochemicals. *Biomedicine* 8: 284.
29 Zarei, S., Carr, K., Reiley, L. et al. (2015). A comprehensive review of amyotrophic lateral sclerosis. *Surg. Neurol. Int.* 6: 171.
30 Brown, R.H. and Al-Chalabi, A. (2017). Amyotrophic lateral sclerosis. *N. Engl. J. Med.* 377: 162–172.
31 Boillée, S., Yamanaka, K., Lobsiger, C.S. et al. (2006). Onset and progression in inherited ALS determined by motor neurons and microglia. *Science* 312: 1389–1392.
32 Hardiman, O., Al-Chalabi, A., Chio, A. et al. (2017). Amyotrophic lateral sclerosis. *Nat. Rev. Dis. Primers* 3: 17071.
33 Bonafede, R. and Mariotti, R. (2017). ALS pathogenesis and therapeutic approaches: the role of mesenchymal stem cells and extracellular vesicles. *Front. Cell. Neurosci.* 11: 80.
34 Tripathi, P., Rodriguez-Muela, N., Klim, J.R. et al. (2017). Reactive astrocytes promote ALS-like degeneration and intracellular protein aggregation in human motor neurons by disrupting autophagy through TGF-β1. *Stem Cell Rep.* 9 (2): 667–680.
35 Bates, G.P., Dorsey, R., Gusella, J.F. et al. (2015). Huntington disease. *Nat. Rev. Dis. Primers.* 1: 15005.
36 Snowden, J.S. (2017). The neuropsychology of Huntington's disease. *Arch. Clin. Neuropsychol.* 32: 876–887.
37 Walker, F.O. (2007). Huntington's disease. *Lancet* 369: 218–228.
38 Ross, C.A., Aylward, E.H., Wild, E.J. et al. (2014). Huntington disease: natural history, biomarkers and prospects for therapeutics. *Nat. Rev. Neurol.* 10: 204.

39 Sawa, A., Wiegand, G.W., Cooper, J. et al. (1999). Increased apoptosis of Huntington disease lymphoblasts associated with repeat length-dependent mitochondrial depolarization. *Nat. Med.* 5: 1194–1198.

40 Patassini, S., Begley, P., Xu, J. et al. (2016). Metabolite mapping reveals severe widespread perturbation of multiple metabolic processes in Huntington's disease human brain. *Biochim. Biophys. Acta Mol. Basis Dis.* 1862: 1650–1662.

41 Yu, C., Li, C.H., Chen, S. et al. (2018). Decreased BDNF release in cortical neurons of a knock-in mouse model of Huntington's disease. *Sci. Rep.* 8: 16976.

42 Zhou, X., Li, G., Kaplan, A. et al. (2018). Small molecule modulator of protein disulphide isomerase attenuates mutant huntingtin toxicity and inhibits endoplasmic reticulum stress in a mouse model of Huntington's disease. *Hum. Mol. Genet.* 27: 1545–1555.

43 Almeida, S., Sarmento-Ribeiro, A.B., Januario, C. et al. (2008). Evidence of apoptosis and mitochondrial abnormalities in peripheral blood cells of Huntington's disease patients. *Biochem. Biophys. Res. Commun.* 374: 599–603.

44 Stack, E.C., Matson, W.R., and Ferrante, R.J. (2008). Evidence of oxidant damage in Huntington's disease: translational strategies using antioxidants. *Ann. N. Y. Acad. Sci.* 1147: 79–92.

45 Kumar, A. and Gupta, A.V. (2016). A review on animal models of stroke: an update. *Brain Res. Bull.* 122: 35–44.

46 Kulik, T., Kusano, Y., Aronhime, S. et al. (2008). Regulation of cerebral vasculature in normal and ischemic brain. *Neuropharmacology* 55: 281–288.

47 Kirino, T. (1982). Delayed neuronal death in the gerbil hippocampus following ischemia. *Brain Res.* 239: 57–69.

48 Abas, F., Alkan, T., Goren, B. et al. (2010). Neuroprotective effects of postconditioning on lipid peroxidation and apoptosis after focal cerebral ischemia/reperfusion injury in rats. *Turk. Neurosurg.* 20: 1–8.

49 Jin, W.N., Shi, S.X.Y., Li, Z. et al. (2017). Depletion of microglia exacerbates postischemic inflammation and brain injury. *Br. J. Pharmacol.* 37: 2224–2236.

50 Saver, J.L. (2006). Time is brain – quantified. *Stroke* 37: 263–266.

51 Jensen, T.S., Baron, R., Haanpää, M. et al. (2011). A new definition of neuropathic pain. *Pain* 152 (10): 2204–2205.

52 Haanpää, M., Attal, N., Backonja, M. et al. (2011). NeuPSIG guidelines on neuropathic pain assessment. *Pain* 152 (1): 14–27.

53 Fornasari, D. (2017). Pharmacotherapy for neuropathic pain: a review. *Pain Ther.* 6: 25–33.

54 Pasero, C. (2004). Pathophysiology of neuropathic pain. *Pain Manag. Nurs.* 5 (4): 3–8.

55 Baekelandt, V., De Strooper, B., Nuttin, B., and Debyser, Z. (2000). Gene therapeutic strategies for neurodegenerative diseases. *Curr. Opin. Mol. Ther.* 2 (5): 540–554.

56 Reisberg, B., Doody, R., Stöffler, A. et al. (2003). Memantine in moderate-to-severe Alzheimer's disease. *N. Engl. J. Med.* 348: 1333–1341.

57 Rogers, J., Kirby, L.C., Hempelman, S.R. et al. (1993). Clinical trial of indomethacin in Alzheimer's disease. *Neurology* 43 (8): 1609–1611.

58 Tariot, P.N., Farlow, M.R., Grossberg, G.T. et al. (2004). Memantine treatment in patients with moderate to severe Alzheimer disease already receiving Donepezil: a randomized controlled trial. *JAMA* 291 (3): 317–324.

59 Sahoo, A.T., Dandapat, J., Dash, U.C., and Kanhar, S. (2018). Features and outcomes of drugs for combination therapy as multi-targets strategy to combat Alzheimer's disease. *J. Ethnopharmacol.* 215: 42–73.

60 Ross, C., Taylor, M., Fullwood, N., and Allsop, D. (2018). Liposome delivery systems for the treatment of Alzheimer's disease. *Int. J. Nanomedicine* 13: 8507–8522.

61 Lanctot, K.L., Herrmann, N., Yau, K.K. et al. (2003). Efficacy and safety of cholinesterase inhibitors in Alzheimer's disease: a meta-analysis. *Can. Med. Assoc. J.* 169 (6): 557–564.

62 Drachman, D.B., Frank, K., Dykes-Hoberg, M. et al. (2002). Cyclooxygenase 2 inhibition protects motor neurons and prolongs survival in a transgenic mouse model of ALS. *Ann. Neurol.* 52 (6): 771–778.

63 Aisen, P.S., Schafer, K.A., Grundman, M. et al. (2003). Effects of rofecoxib or naproxen vs placebo on Alzheimer disease progression: a randomized controlled trial. *JAMA* 289 (21): 2819–2826.

64 Scharf, S., Mander, A., Ugoni, A. et al. (1999). A double-blind, placebo-controlled trial of diclofenac/misoprostol in Alzheimer's disease. *Neurology* 53 (1): 197–201.

65 Veld, B.A., Ruitenberg, A., Hofman, A. et al. (2001). Nonsteroidal antiinflammatory drugs and the risk of Alzheimer's disease. *N. Engl. J. Med.* 345 (21): 1515–1521.

66 Orgogozo, J.M., Gilman, S., Dartigues, J.F. et al. (2003). Subacute meningoencephalitis in a subset of patients with AD after Abeta42 immunization. *Neurology* 61 (1): 46–54.

67 Chen, X. and Pan, W. (2014). The treatment strategies for neurodegenerative diseases by integrative medicine. *Integr. Med. Intern.* 1: 223–225.

68 Mizuno, Y. (2014). Recent research progress in and future perspective on treatment of Parkinson's disease. *Integr. Med. Int.* 1: 67–79.

69 Connolly, B.S. and Lang, A.E. (2014). Pharmacological treatment of Parkinson disease: a review. *JAMA* 311 (16): 1670–1683.

70 Fahn, S. (2000). The spectrum of levodopa-induced dyskinesias. *Ann. Neurol.* 47 (4 Suppl 1): S2–S9; discussion S9–11.

71 Guo, X., Dillman, J.F., Dawson, V.L., and Dawson, T.M. (2001). Neuroimmunophilins: novel neuroprotective and neuroregenerative targets. *Ann. Neurol.* 50 (1): 6–16.

72 Gold, B.G., Zeleny-Pooley, M., Chaturvedi, P., and Wang, M.S. (1998). Oral administration of a nonimmunosuppressant FKBP-12 ligand speeds nerve regeneration. *Neuroreport* 9 (3): 553–558.

73 Steiner, J.P., Hamilton, G.S., Ross, D.T. et al. (1997). Neurotrophic immunophilin ligands stimulate structural and functional recovery in neurodegenerative animal models. *Proc. Natl Acad. Sci. USA* 94 (5): 2019–2024.

74 Gironi, M., Arno, C., Comi, G. et al. (2016). Multiple sclerosis and neurodegenerative diseases. In: *Immune Rebalancing*, vol. 4, 63–84. Academic Press.

75 Angelov, D.N., Waibel, S., Guntinas-Lichius, O. et al. (2003). Therapeutic vaccine for acute and chronic motor neuron diseases: implications for amyotrophic lateral sclerosis. *Proc. Natl Acad. Sci. USA* 100 (8): 4790–4795.

76 Wiley, N.J., Madhankumar, A.B., Mitchell, R.M. et al. (2012). Lipopolysaccharide modified liposomes for amyotropic lateral sclerosis therapy: Efficacy in SOD1 mouse mode. *Adv. Nanopart.* 1 (3): 10.

77 Dash, R.P., Babu, R.J., and Srinivas, N.R. (2018). Two decades-long journey from riluzole to edaravone: revisiting the clinical pharmacokinetics of the only two amyotrophic lateral sclerosis therapeutics. *Clin. Pharmacokinet.* 57: 1385–1398.

78 Teixeira, M.I., Lopes, C.M., Amaral, M.H., and Costa, P.C. (2020). Current insights on lipid nanocarrier assisted drug delivery in the treatment of neurodegenerative diseases. *Eur. J. Pharm. Biopharm.* 149: 192–217.

79 Mestre, H., Cohen-Minian, Y., Zajarias-Fainsod, D., and Ibarra, A. (2013). Pharmacological treatment of acute ischaemic stroke. In: *Neurodegenerative Diseases*. IntechOpen https://doi.org/10.5772/53774.

80 Alteplase Monograph Class: Thrombolytic Agents *AHFS Drug Information. 7272 Wisconsin Avenue, Betheda*, 20814. Maryland: American Society of Health-System Pharmacists, Inc.

81 Hacke, W., Kaste, M., Bluhmki, E. et al. (2008). Thrombolysis with Alteplase 3 to 4.5 hours after acute ischemic stroke. *N. Engl. J. Med.* 359: 1317–1329.

82 Bansal, S., Sangha, K.S., and Khatri, P. (2013). Drug treatment of acute ischemic stroke. *Am. J. Cardiovasc. Drugs* 13 (1): 57–69.

83 Bates, D., Schultheis, B.C., Hanes, M.C. et al. (2019). A comprehensive algorithm for management of neuropathic pain. *Pain Med.* 20 (Suppl 1): S2–S12.

84 Cavalli, E., Mammana, S., Nicoletti, F. et al. (2019). The neuropathic pain: an overview of the current treatment and future therapeutic approaches. *Int. J. Immunopathol. Pharmacol.* 33: 2058738419838383.

85 Zilliox, L.A. (2017). Neuropathic pain. *Continuum (Minneap. Minn.)* 23 (2): 512–532.

86 White, C.M., Pritchard, J., and Turner-Stokes, L. (2004). Exercise for people with peripheral neuropathy. *Cochrane Database Syst. Rev.* 4: CD003904. https://doi.org/10.1002/14651858.cd003904.

87 Pan, W., Kwak, S., Liu, Y. et al. (2011). Traditional chinese medicine improves activities of daily living in Parkinson's disease. *Parkinson Dis.* 2011: 789506.

88 Pan, W., Liu, Y., Fang, Z. et al. (2011). A compound belonging to traditional Chinese medicine improves nocturnal activity in Parkinson's disease. *Sleep Med.* 12: 307.

89 Lloret, S.P., Rey, M.V., and Rascol, O. (2013). Ayurveda medicine for the treatment of Parkinson's disease. *Int. Integrat. Med.* 1: 1–6.

90 Dey, A., Bhattacharya, R., Mukherjee, A., and Pandey, D.K. (2017). Natural products against Alzheimer's disease: pharmacotherapeutics and biotechnological interventions. *Biotechnol. Adv.* 35 (2): 178–216.

91 Pan, W., Wang, Q., Kwak, S. et al. (2014). Shen-Zhi-Ling oral liquid improves behavioral and psychological symptoms of dementia in Alzheimer's disease. *Evid. Based Complement. Alternat. Med.* 2014: 913687.

92 Conte, C., Lee, R., Sarkar, M., and Terman, D. (2018). A mathematical model of recurrent spreading depolarizations. *J. Comput. Neurosci.* 44: 203–217.

93 Vauzour, D., Vafeiadou, K., Rodriguez-Mateos, A. et al. (2008). The neuroprotective potential of flavonoids: a multiplicity of effects. *Genes Nutr.* 3: 115–126.

94 Polito, C.A., Cai, Z.Y., Shi, Y.L. et al. (2018). Association of tea consumption with risk of Alzheimer's disease and anti-beta-amyloid effects of tea. *Nutrients* 10: 655.

95 Ng, T.P., Jin, A., Feng, L. et al. (2015). Morality of older persons living alone: Singapore longitudinal ageing studies. *BMC Geriatr.* 15 (1): 1–9.

96 Fan, D., Liu, L., Wu, Z., and Cao, M. (2019). Combating neurodegenerative diseases with the plant alkaloid berberine: molecular mechanisms and therapeutic potential. *Curr. Neuropharmacol.* 17: 563–579.

97 Bhattacharya, S., Haertel, C., Maelicke, A., and Montag, D. (2014). Galantamine slows down plaque formation and behavioral decline in the 5XFAD mouse model of Alzheimer's disease. *PLoS One* 9: e89454.

98 Plazas, E., Hagenow, S., Murillo, M.A. et al. (2020). Isoquinoline alkaloids from the roots of *Zanthoxylum rigidum* as multi-target inhibitors of cholinesterase, monoamine oxidase A and Aβ1-42 aggregation. *Bioorg. Chem.* 98: 103722.

99 Wang, B., Su, C.J., Liu, T.T. et al. (2018). The neuroprotection of low-dose morphine in cellular and animal models of Parkinson's disease through ameliorating endoplasmic reticulum (ER) stress and activating autophagy. *Front. Mol. Neurosci.* 11: 120.

100 Bi, Y., Qu, P.C., Wang, Q.S. et al. (2015). Neuroprotective effects of alkaloids from Piper longum in a MPTP-induced mouse model of Parkinson's disease. *Pharm. Biol.* 10: 1–9.

101 Ingkaninan, K., Changwijit, K., and Suwanborirux, K. (2006). Vobasinyl-iboga bisindole alkaloids, potent acetylcholinesterase inhibitors from Tabernaemontana divaricata root. *J. Pharm. Pharmacol.* 58: 847–852.

102 Park, T.H., Kwon, O., Park, S.Y. et al. (2003). N-methylated β-carbolines protect PC12 cells from cytotoxic effect of MPP+ by attenuation of mitochondrial membrane permeability change. *Neurosci. Res.* 46: 349–358.

103 Yang, Y., Cheng, X., Liu, W. et al. (2015). Potent AChE and BChE inhibitors isolated from seeds of Peganum harmala Linn by a bioassay-guided fractionation. *J. Ethnopharmacol.* 168: 279–286.

104 Deng, G., Wu, C., Rong, X. et al. (2019). Ameliorative effect of deoxyvasicine on scopolamine-induced cognitive dysfunction by restoration of cholinergic function in mice. *Phytomedicine* 63: 153007.

105 Li, Z., Jiang, T., Lu, Q. et al. (2019). Berberine attenuated the cytotoxicity induced by t-BHP via inhibiting oxidative stress and mitochondria dysfunction in PC-12 cells. *Cell. Mol. Neurobiol.* 40: 587–602.

106 Morales-Garcia, J.A., Revenga, M.D.L.F., Alonso-Gil, S. et al. (2017). The alkaloids of Banisteriopsis caapi, the plant source of the amazonian hallucinogen ayahuasca, stimulate adult neurogenesis in vitro. *Sci. Rep.* 7: 1–13.

107 Yano, M., Nakashima, S., Oda, Y. et al. (2020). BBB-permeable aporphine-type alkaloids in Nelumbo nucifera flowers with accelerative effects on neurite outgrowth in PC-12 cells. *J. Nat. Med.* 74: 212–218.

108 Qian, Z.M. and Ke, Y. (2014). Huperzine A: is it an effective disease-modifying drug for Alzheimer's disease? *Front. Aging Neurosci.* 6: 216.

109 Jia, J.Y., Zhao, Q.H., Liu, Y. et al. (2013). Phase I study on the pharmacokinetics and tolerance of ZT-1, a prodrug of huperzine A, for the treatment of Alzheimer's disease. *Acta Pharmacol. Sin.* 34: 976–982.

110 Kong, Y.R., Tay, K.C., Su, Y.X. et al. (2021). Potential of naturally derived alkaloids as multi-targeted therapeutic agents for neurodegenerative diseases. *Molecules* 26 (3): 728.

111 Basli, A., Soulet, S., Chaher, N. et al. (2012). Wine polyphenols: potential agents in neuroprotection. *Oxidative Med. Cell. Longev.* 2012: 805762.

112 Renaud, J. and Martinoli, M. (2019). Considerations for the use of polyphenols as therapies in neurodegenerative diseases. *Int. J. Mol. Sci.* 20: 1883.

113 Faria, A., Pestana, D., Teixeira, D. et al. (2010). Flavonoid transport across RBE4 cells: a blood-brain barrier model. *Cell. Mol. Biol. Lett.* 15 (2): 234–241.

114 Bhullar, K.S. and Rupasinghe, H.P.V. (2013). Polyphenols: multipotent therapeutic agents in neurodegenerative diseases. *Oxidative Med. Cell. Longev.* 2013: 891748.

115 Nabavi, S.F., Devi, K.P., Malar, D.S. et al. (2015). Ferulic acid and Alzheimer's disease: promises and pitfalls. *Mini-Rev. Med. Chem.* 15: 776–788.

116 Szwajgier, D., Borowiec, K., and Pustelniak, K. (2017). The neuroprotective effects of phenolic acids: molecular mechanism of action. *Nutrients* 9: 477.

117 Ullah, F., Liang, A., Rangel, A. et al. (2017). High bioavailability curcumin: an anti-inflammatory and neurosupportive bioactive nutrient for neurodegenerative diseases characterized by chronic neuroinflammation. *Arch. Toxicol.* 91 (4): 1623–1634.

118 Moussa, C., Hebron, M., Huang, X. et al. (2017). Resveratrol regulates neuro-inflammation and induces adaptive immunity in Alzheimer's disease. *J. Neuroinflamm.* 14: 1.

119 Chen, A., Xu, Y., and Yuan, J. (2018). Ginkgolide B ameliorates NLRP3 inflammasome activation after hypoxic-ischemic brain injury in the neonatal male rat. *Int. J. Dev. Neurosci.* 69: 106–111.

120 Jiang, M., Li, J., Peng, Q. et al. (2014). Neuroprotective effects of bilobalide on cerebral ischemia and reperfusion injury are associated with inhibition of pro-inflammatory mediator production and down-regulation of JNK1/2 and p38 MAPK activation. *J. Neuroinflamm.* 11: 167.

121 Khadka, B., Lee, J., Park, D.H. et al. (2020). The role of natural compounds and their nanocarriers in the treatment of CNS inflammation. *Biomolecules* 10 (10): 1401.

122 Wang, J., Song, Y., Gao, M. et al. (2016). Neuroprotective effect of several phytochemicals and its potential application in the prevention of neurodegenerative diseases. *Geriatrics* 1: 29.

123 Sabetghadam, M., Mazdeh, M., Abolfathi, P. et al. (2020). Evidence for a beneficial effect of oral N-acetylcysteine on functional outcomes and in flammatory biomarkers in patients with acute ischemic stroke. *Neuropsychiatr. Dis. Treat.* 16: 1265–1278.

124 Hashemi, H., Behzadi, S., Ghanaati, H. et al. (2014). Evaluation of plaque detection and optimum time of enhancement in acute attack multiple sclerosis after contrast injection. *Acta Radiol.* 55 (2): 218–224.

125 Tan, Z., Chen, Y., Xie, W. et al. (2017). Nimodipine attenuates tau phosphorylation at Ser396 via miR-132/GSK-3β pathway in chronic cerebral hypoperfusion rats. *Eur. J. Pharmacol.* 819: 1–8.

126 Molinengo, L., Orsetti, M., Pastorello, B. et al. (1995). The action of arecoline on retrieval and memory storage evaluated in the staircase maze. *Neurobiol. Learn. Mem.* 63 (2): 167–173.

127 Salomon, A.R., Marcinowski, K.J., Friedland, R.P., and Zagorski, M.G. (1996). Nicotine inhibits amyloid formation by the β-peptide. *Biochemist* 35: 13568–13578.

128 Decker, M.W., Majchrzak, M.J., and Arnerić, S.P. (1993). Effects of lobeline, a nicotinic receptor agonist, on learning and memory. *Pharmacol. Biochem. Behav.* 45 (3): 571–576.

129 Carradori, S., D'Ascenzio, M., Chimenti, P. et al. (2014). Selectiv MAO-B inhibitors: a lesson from natural products. *Mol. Divers.* 18 (1): 219–243.

130 Choudhury, B., Saytode, P., and Shah, V. (2014). Neurodegenrative disorders: past, present and future. *Int. J. Appl. Pharm. Biotechnol.* 5 (2): 14–28.

131 Liu, Y.-J., Peng, W., Hu, M.-B. et al. (2016). The pharmacology, toxicology and potential applications of arecoline: a review. *Pharm. Biol.* 54 (11): 2753–2760.

132 Davis, J.M., Murphy, E.A., Carmichael, M.D., and Davis, B. (2009). Quercetin increases brain and muscle mitochondrial biogenesis and exercise tolerance. *Am. J. Phys.* 296 (4): R1071–R1077.

133 Jimenez-Aliaga, K., Bermejo-Bescos, P., Benedi, J., and Martin-Aragon, S. (2011). Quercetin and rutin exhibit antiamyloidogenic and fibril-disaggregating effects in vitro and potent antioxidant activity in APPswe cells. *Life Sci.* 89 (25–26): 939–945.

134 Lee, J.K., Kwak, H.J., Piao, M.S. et al. (2010). Quercetin reduces the elevated matrix metalloproteinases-9 level and improves functional outcome after cerebral focal ischemia in rats. *Acta Neurochir.* 112 (6): 1477–1487.

135 Park, J.W., Hong, J.S., Lee, K.S. et al. (2010). Green tea polyphenol (-)-epigallocatechin gallate reduces matrix metalloproteinase-9 activity following transient focal cerebral ischemia. *J. Nutr. Biochem.* 21 (11): 1038–1044.

136 Ho, D.J., Calingasan, N.Y., Wille, E. et al. (2010). Resveratrol protects against peripheral deficits in a mouse model of Huntington's disease. *Exp. Neurol.* 225 (1): 74–84.

137 Yang, F., Lim, G.P., Begum, A.N. et al. (2005). Curcumin inhibits formation of amyloid beta oligomers and fibrils, binds plaques, and reduces amyloid in vivo. *J. Biol. Chem.* 280: 5892–5901.

138 Hickey, M.A., Zhu, C., Medvedeva, V. et al. (2003). Improvement of neuropathology and transcriptional deficits in CAG 140 knock-in mice supports a beneficial effect of dietary curcumin in Huntington's disease. *Mol. Neurodegener.* 4 (7): 7–12.

139 Feng, Y., Wang, X., Yang, S. et al. (2009). Resveratrol inhibits beta-amyloid oligomeric cytotoxicity but does not prevent oligomer formation. *Neuro-Toxicology.* 30: 986–995.

140 Yun, Y.C., Jeong, S.-G., Kim, S.H., and Cho, G.-W. (2019). Reduced sirtuin 1/adenosine monophosphate-activated protein kinase in amyotrophic lateral sclerosis patient-derived mesenchymal stem cells can be restored by resveratrol. *J. Tissue Eng. Regen. Med.* 13: 110–115.

141 Khan, M.M., Ahmad, A., Ishrat, T. et al. (2010). Resveratrol attenuates 6-hydroxydopamine-induced oxidative damage and dopamine depletion in rat model of Parkinson's disease. *Brain Res.* 1328: 139–151.

142 Fonseca-Kelly, Z., Nassrallah, M., Uribe, J. et al. (2012). Resveratrol neuroprotection in a chronic mouse model of multiple sclerosis. *Front. Neurol.* 3: 84.

143 Nouri, Z., Fakhri, S., El-Senduny, F.F. et al. (2019). On the Neuroprotective effects of naringenin: pharmacological targets, signaling pathways, molecular mechanisms, and clinical perspective. *Biomolecules* 9: 690.

144 Jung, U.J. and Kim, S.R. (2014). Effects of naringin, a flavanone glycoside in grapefruits and citrus fruits, on the nigrostriatal dopaminergic projection in the adult brain. *Neural Regen. Res.* 9: 1514–1517.

145 Borah, A., Paul, R., Choudhury, S. et al. (2013). Neuroprotective potential of silymarin against CNS disorders: insight into the pathways and molecular mechanisms of action. *CNS Neurosci. Ther.* 19: 847–853.

146 Soares, T.B., Loureiro, L., Carvalho, A. et al. (2018). Lipid nanocarriers loaded with natural compounds: potential new therapies for age related neurodegenerative diseases? *Prog. Neurobiol.* 168: 21–41.

147 Wang, H., Wang, H., Cheng, H., and Che, Z. (2016). Ameliorating effect of luteolin on memory impairment in an Alzheimer's disease model. *Mol. Med. Rep.* 13: 4215–4220.

148 Sharif-Rad, M., Kumar, N.V.A., Zucca, P. et al. (2020). Lifestyle, oxidative stress, and antioxidants: back and forth in the pathophysiology of chronic diseases. *Front. Physiol.* 11: 694.

149 Tchantchou, F., Xu, Y., Wu, Y. et al. (2007). EGb 761 enhances adult hippocampal neurogenesis and phosphorylation of CREB in transgenic mouse model of Alzheimer's disease. *FASEB J.* 21: 2400–2408.

150 Zheng, J.-S., Zheng, P.-D., Mungur, R. et al. (2018). Ginkgolide B promotes the proliferation and dierentiation of neural stem cells following cerebral ischemia/reperfusion injury, both in vivo and in vitro. *Neural Regen. Res.* 13: 1204–1211.

151 Khan, A., Vaibhav, K., Javed, H. et al. (2012). Attenuation of A-induced neurotoxicity by thymoquinone via inhibition of mitochondrial dysfunction and oxidative stress. *Mol. Cell. Biochem.* 369: 55–65.

152 Yang, C., Yen, T., Hsu, C. et al. (2017). Multi-targeting andrographolide, a novel NF-κβ inhibitor, as a potential therapeutic agent for stroke. *Int. J. Mol. Sci.* 18 (8): 1638.

153 Lu, J., Ma, Y., Wu, J. et al. (2019). A review for the neuroprotective effects of andrographolide in the central nervous system. *Biomed. Pharmacother.* 117: 109078.

154 Devalapally, H., Chakilam, A., and Amiji, M.M. (2007). Role of nanotechnology in pharmaceutical product development. *J. Pharm. Sci.* 96 (10): 2547–2565.

155 Kreuter, J., Alyautdin, R.N., Kharkevich, D.A., and Ivanov, A.A. (1995). Passage of peptides through the blood–brain barrier with colloidal polymer particles (nanoparticles). *Brain Res.* 674 (1): 171–174.

156 Ganesan, P., Ko, H.-M., Kim, I.-S., and Choi, D.-K. (2015). Recent trends in the development of nanophytobioactive compounds and delivery systems for their possible role in reducing oxidative stress in Parkinson's disease models. *Intern. J. Nanomed.* 10: 6757.

157 Barchet, T.M. and Amiji, M.M. (2009). Challenges and opportunities in CNS delivery of therapeutics for neurodegenerative diseases. *Expert Opin. Drug Deliv.* 6 (3): 211–225.

158 Ahlawat, J., Barroso, G.G., Asil, S.M. et al. (2020). Nanocarriers as potential drug delivery candidates for overcoming the blood-brain barrier: challenges and possibilities. *ACS omega* 5 (22): 12583–12595.

159 Paul, A., Rajiung, M., Zaman, K. et al. (2020). Quantification of the bioactive marker resveratrol in *Morus alba* Linn. fruits by high–performance thin–layer chromatography. *J. Planar Chromatogr.* https://doi.org/10.1007/s00764-020-00062-9.

160 Richard, T., Pawlus, A.D., Iglesias, M. et al. (2011). Neuroprotective properties of resveratrol and derivatives. *Ann. N. Y. Acad. Sci.* 1215 (1): 103.

161 Da Rocha, L.G., Santos, D.B., Colle, D. et al. Improved neuroprotective effects of resveratrol-loaded polysorbate 80-coated poly (lactide) nanoparticles in MPTP induced Parkinsonism. *Int. J. Pharm. Sci. Res.* 10: 1127–1138. https://doi.org/10.2217/nnm.14.165.

162 Loureiro, J.A., Andrade, S., Duarte, A. et al. (2017). Resveratrol and grape extract-loaded solid lipid nanoparticles for the treatment of Alzheimer's disease. *Molecules* 22 (2): 277.

163 Lu, X., Ji, C., Xu, H. et al. (2009). Resveratrol-loaded polymeric micelles protect cells from Aβ-induced oxidative stress. *Brain Res.* 375: 89–96.

164 Pangeni, R., Sharma, S., Mustafa, G. et al. (2014). Vitamin E loaded resveratrol nanoemulsion for brain targeting for the treatment of Parkinson's disease by reducing oxidative stress. *Nanotechnology* 25: 485102.

165 Rifaai, R.A., Mokhemer, S.A., Saber, E.A. et al. (2020). Neuroprotective effect of quercetin nanoparticles: a possible prophylactic and therapeutic role in alzheimer's disease. *J. Chem. Neuroanat.* 107: 101795.

166 Ghosh, A., Sarkar, S., Mandal, A.K., and Das, N.J.P.O. (2013). Neuroprotective role of nanoencapsulated quercetin in combating ischemia-reperfusion induced neuronal damage in young and aged rats. *PLoS One* 8: e57735.

167 Ghaffari, F., Moghaddam, A.H., Zare, M.J.B., and Neuroscience, C. (2018). Neuroprotective effect of quercetin nanocrystal in a 6-hydroxydopamine model of parkinson disease: biochemical and behavioral evidence. Basic. *Clin. Neurosci.* 9: 317.

168 Tosi, G., Pederzoli, F., Belletti, D. et al. (2019). Nanomedicine in Alzheimer's disease: amyloid beta targeting strategy. *Prog. Brain Res.* 245: 57–88.

169 Huo, X., Zhang, Y., Jin, X. et al. (2019). A novel synthesis of selenium nanoparticles encapsulated PLGA nanospheres with curcumin molecules for the inhibition of amyloid b aggregation in Alzheimer's disease. *J. Photochem. Photobiol. B Biol.* 190: 98–102.

170 Hoppe, J.B., Haag, M., Whalley, B.J. et al. (2013). Curcumin protects organotypic hippocampal slice cultures from Aβ1-42-induced synaptic toxicity. *Toxicol. In Vitro* 27 (8): 2325–2330.

171 Kundu, P., Das, M., Tripathy, K., and Sahoo, S.K. (2016). Delivery of dual drug loaded lipid based nanoparticles across the blood–brain barrier impart enhanced neuroprotection in a rotenone induced mouse model of parkinson's disease. *ACS Chem. Neurosci.* 7 (12): 1658–1670.

172 Dolati, S., Babaloo, Z., Ayromlou, H. et al. (2019). Nanocurcumin improves regulatory T-cell frequency and function in patients with multiple sclerosis. *J. Neuroimmunol.* 327: 15–21.

173 Nasab, N.A., Kumleh, H.H., Beygzadeh, M. et al. (2018). Delivery of curcumin by a pH-responsive chitosan mesoporous silica nanoparticles for cancer treatment. *Artif Cell Nanomed. B.* 46 (1): 75–81.

174 Cano, A., Ettcheto, M., Espina, M. et al. (2021). Epigallocatechin-3-gallate PEGylated poly(lactic-co-glycolic) acid nanoparticles mitigate striatal pathology and motor deficits in 3-nitropropionic acid intoxicated mice. *Nanomedicine (London)* 16 (1): 19–35.

175 Sánchez-Giraldo, V., Monsalve, Y., Palacio, J. et al. (2020). Role of a novel (−)-epigallocatechin-3-gallate delivery system on the prevention against oxidative stress damage *in vitro* and *in vivo* model of Parkinson's disease. *J. Drug Deliv. Sci. Tec.* 55: 101466.

176 Hao, J., Zhao, J., Zhang, S. et al. (2016). Fabrication of an ionic-sensitive *in situ* gel loaded with resveratrol nanosuspensions intended for direct nose-to-brain delivery. *Colloids Surf. B: Biointerfaces* 147 (1): 376–386.

177 Palle, S. and Neerati, P. (2018). Improved neuroprotective effect of resveratrol nanoparticles as evinced by abrogation of rotenone-induced behavioral deficits and oxidative and mitochondrial dysfunctions in rat model of Parkinson's disease. *Naunyn Schmiedeberg's Arch. Pharmacol.* 391 (4): 445–453.

178 Ahmad, N., Ahmad, R., Ahmad, F.J. et al. (2020). Poloxamer-chitosan-based Naringenin nanoformulation used in brain targeting for the treatment of cerebral ischemia. Saudi. *J. Biol. Sci.* 27 (1): 500–517.

179 Md, S., Alhakamy, N.A., Aldawsari, H.M., and Asfour, H.Z. (2019). Neuroprotective and antioxidant effect of naringenin-loaded nanoparticles for nose-to-brain delivery. *Brain Sci.* 9 (10): 275.

180 Azadi, R., Mousavi, S.E., Kazemi, N.M. et al. (2019). Anti-inflammatory efficacy of berberine nanomicelle for improvement of cerebral ischemia: formulation, characterization and evaluation in bilateral common carotid artery occlusion rat model. *Research Square* https://doi.org/10.21203/rs.3.rs-124520/v1.

181 Ahmad, N., Ahmad, R., Naqvi, A.A. et al. (2016). Rutin-encapsulated chitosan nanoparticles targeted to the brain in the treatment of Cerebral Ischemia. *Int. J. Biol. Macromol.* 91: 640–655.

182 Bilia, A.R., Nardiello, P., Piazzini, V. et al. (2019). Successful brain delivery of andrographolide loaded in human albumin nanoparticles to TgCRND8 mice, an Alzheimer's disease mouse model. *Front. Pharmacol.* https://doi.org/10.3389/fphar.2019.00910.

183 Ramanathan, S., Archunan, G., Sivakumar, M. et al. (2018). Theranostic applications of nanoparticles in neurodegenerative disorders. *Intern. J. Nanomed.* 13: 5561–5576. https://doi.org/10.2147/IJN.S149022.

184 Niu, X., Chen, J., and Gao, J. (2019). Nanocarriers as a powerful vehicle to overcome blood-brain barrier in treating neurodegenerative diseases: focus on recent advances. *Asian J. Pharm. Sci.* 14: 480–496.

185 Sharma, A., Goyal, A.K., and Rath, G. (2017). Recent advances in metal nanoparticles in cancer therapy. *J. Drug Target.* 26 (8): 617–632.

21

Nanotherapeutics of Phytoantioxidants for Aging and Neurological Disorders

Ganesh B. Shevalkar[1], *Mahendra K. Prajapati*[2], *Bhushan A. Bhairav*[3], *and Chandrakantsing V. Pardeshi*[1]

[1] *Department of Pharmaceutics, R. C. Patel Institute of Pharmaceutical Education and Research, Shirpur, Maharashtra, India*
[2] *Department of Pharmaceutics, School of Pharmacy and Technology Management, SVKM's NMIMS, Shirpur, Maharashtra, India*
[3] *Department of Pharmaceutical Quality Assurance, R. C. Patel Institute of Pharmaceutical Education and Research, Shirpur, Maharashtra, India*

21.1 Introduction

Aging is a degenerative biological process. It takes a long time to develop and is considerably more challenging to quantify. With growing human life expectancy, aging and neurological diseases are becoming an increasingly important social and economic concern for modern society in most developed countries [1]. Neurological disorders like Alzheimer's disease (AD), Parkinson's disease (PD), amyotrophic lateral sclerosis (ALS), multiple sclerosis (MS), neurological tumors, and ischemic stroke are all characterized by progressive loss of neuron cells, impaired motor or cognitive functions, and abnormal protein aggregation. With growing age, bioenergetics deficiencies and changes in the reduction-oxidation (redox) homeostasis are observed in the brain. The brain comprises highly differentiated cells distributed across several anatomical areas and requires around 20% of body basal oxygen to operate [2]. Most cellular energy is generated by oxidative phosphorylation in mitochondria, which are critical organelles for various cellular activities, including energy metabolism, calcium homeostasis, lipid biosynthesis, and apoptosis. As a result, it is no surprise that changes in brain energy metabolism result in neurodegeneration [3].

Our understanding of the fundamental mechanisms underlying aging has grown at an extraordinary rate during the last three decades [4]. Chronic oxidative stress and systemic inflammation linked with age are unquestionably significant contributors to cellular senescence and aging [5]. Damages to essential biomolecules, including lipids, proteins (enzymes), and nucleic acids, caused by reactive oxygen species (ROS) and other free radicals produced in mitochondria during metabolic activities are one of the leading causes of aging [6]. In most reports, ROS found in primary aging-related processes such as telomere attrition, autophagy, and stem-cell fatigue [4]. Oxidative stress and inflammation, which are associated with it, also show their impact on age-related pathological conditions like atherosclerosis, diabetes, and most cardio-metabolic and neurodegenerative diseases [2]. Considering these theoretical assumptions, many authors believe that dietary supplementation of antioxidants can reduce oxidative stress levels. They can neutralize the free radicals by donating the electron and delay or prevent age-related pathological conditions [6].

Phytoantioxidants and Nanotherapeutics, First Edition. Edited by Mithun Rudrapal.

However, the therapeutic potential of dietary antioxidants is limited in most cases due to the concern of solubility and stability in gastrointestinal fluids, which results in their low bioavailability. With the advancement in research, various innovative nanotechnology-based systems have been developed by researchers with the aim to improve oral bioavailability and, accordingly, the therapeutic effectiveness of phytochemicals and other natural antioxidants. Hence, this chapter provides detailed information about recent advances in applying phytoantioxidant-based nanodelivery systems to combat aging-related oxidative stress and associated pathological conditions.

21.2 Reactive Oxygen Species and Oxidative Stress

ROS are generally produced in living organisms due to normal cellular metabolism and are essential in maintaining cellular homeostasis. In physiological conditions such as immune response, inflammation, synaptic plasticity, learning, and memory, a low to moderate concentration of ROS is involved [7]. However, its excess production can be harmful, producing adverse oxidative modifications to cell components, mainly the mitochondria [8]. The human body has many antioxidants, such as superoxide dismutase (SOD) and glutathione (GSH), that serve to counteract the effects of oxidants [9]. When an imbalance between free radical production and detoxification occurs, ROS production may overwhelm antioxidant defences, leading to a harmful condition called oxidative stress and impairment of cellular functions. This phenomenon is observed in many pathological cases involving mitochondrial dysfunction and aging [10] (Figure 21.1). The brain is particularly vulnerable to oxidative stress and damage because of its

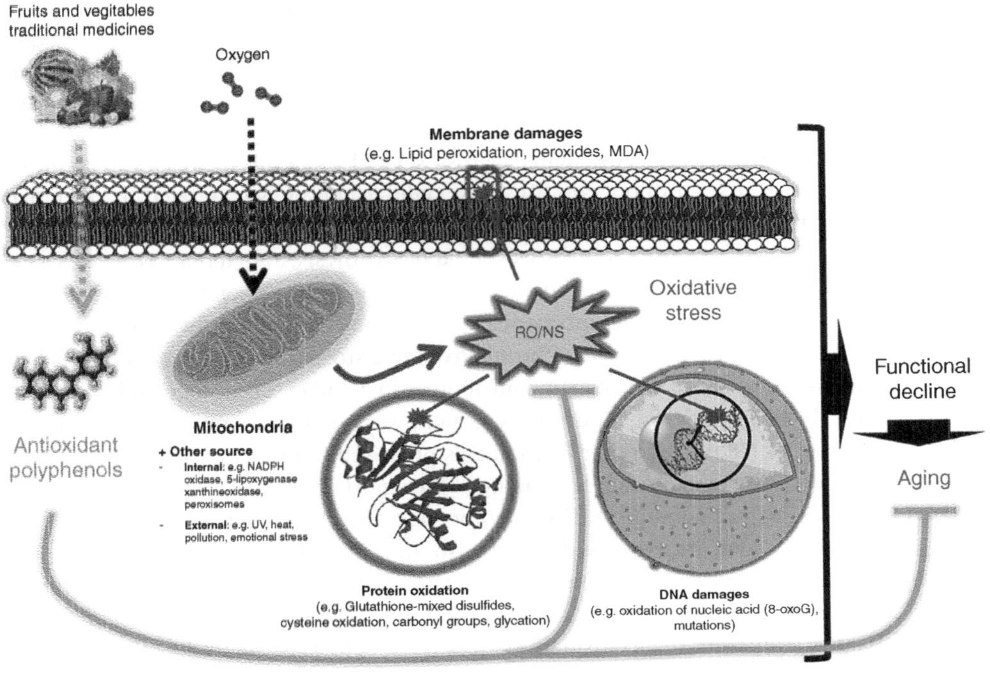

Figure 21.1 Role of reactive oxygen species and oxidative stress in aging and neurological disorders. *Source:* alinamd/Adobe Stock.

high oxygen consumption, low antioxidant defences, and high content of polyunsaturated fats very prone to be oxidized [11].

Biological molecules such as proteins, lipids, nucleic acids, and carbohydrates are generally prone to oxidation, leading to consistent oxidative damage of the biomolecules like the change of their structures and, consequently, their functions.

Lipid peroxidation is another characteristic process of oxidative stress that affects lipids and leads to the formation of relative markers. More specifically, lipid peroxidation derives from the damage of cellular membranes by ROS that generates a heterogeneous group of relatively stable end products such as malondialdehyde (MDA), 4-hydroxy-2-nonenal (HNE), acrolein, and isoprostanes. MDA, HNE, and acrolein are able to bind proteins and DNA, leading to the alteration of conformation and function [12]. Carbohydrates are also affected by ROS. Indeed, reducing sugars plays a pivotal role in modifying proteins by forming advanced glycation end products (AGEs) in a nonenzymatic reaction called glycation [13]. AGEs are involved in some diseases such as diabetes mellitus, cardiac dysfunction, and neurodegenerative diseases [14].

The hydroxyl radical (·OH) is the most hazardous of all the free radicals due to its high reactivity and limited diffusion from the production site. It has been found to damage biological molecules, including nucleic acid [15]. The involvement of nucleic acid oxidation in neurodegenerative diseases might cause the reduction of protein level and translation errors in vivo with alteration of protein structure and function [16].

The antioxidant molecules are divided into two groups: enzymatic and nonenzymatic compounds. The enzymatic group includes SOD, catalase (CAT), glutathione peroxidase (GPx), and glutathione reductase (GR). SOD is one of the first protective mechanisms against ROS and catalyzes the conversion of $\cdot O_2$- to H_2O_2 and O_2 [17]. The generated H_2O_2 is converted to water and O_2 by CAT. The nonenzymatic group involves GSH, the most abundant antioxidant in most brain cells, thioredoxin (Trx), vitamins A, E, and C, and selenium. GSH reacts with ROS generating glutathione disulphide and enters a cycle together with GPx and GR. Vitamin E (also called α-tocopherol) is a lipophilic molecule acting against lipid peroxidation [18]. Vitamin C (also called ascorbic acid) is one of the most important water-soluble antioxidants. Selenium is an essential trace element and a crucial cofactor for GPx and thioredoxin reductase (TrxR). Altogether, they act and balance ROS and reactive nitrogen species (RNS) levels to avoid the onset and propagation of harmful effects in the nearby tissues.

21.3 Antioxidants

ROS has a role in carrying out vital physiological processes. Although elevated ROS levels lead to stress and pathology, reduced ROS levels are associated with normal physiological conditions [19]. The presence of free radicals near tissues and cells induces a series of reactions and activates multiple internal defence cellular processes to eliminate ROS [20]. The cells and the organism have a specific anti-oxidative protection system to defend themselves. The oxidation of substrate is prevented by antioxidants present in smaller amounts in the body and, hence, have a principal role in the defence mechanism [21]. Antioxidants quench or inhibit the reactions caused by free radicals and ultimately prevent or delay cellular damage. The existence of the antioxidant defence mechanism is universal, although there are variations in antioxidant defences between different species [22]. In humans, three lines of defence have been evolved by nature. They function through various mechanisms as inhibiting, catching, and restoring.

The first line of defence that scavenges ROS/RNS involves enzymatic and nonenzymatic antioxidants. They can delay or prevent the initiation of different oxidative stresses directly, including

GSH, uric acid, vitamin E, and vitamin C. The nonenzymatic substances in blood plasma such as transferrin, ferritin, ceruloplasmin, and albumin are preventive antioxidants. They inhibit new reactive species formation by binding transition metal ions (e.g. copper and iron). Another defence line consists of nonenzymatic antioxidants that serve as an intermediate defence to inactivate oxidants and radicals. The third and last line of defence involves repairing and damage-removing, causing regeneration of damaged bio-molecules from oxidative injury [23]. Antioxidants terminate oxidative chain reactions by binding with free radicals and donating their electrons, making free radicals unavailable for further attack. After donating their electrons, the free radical state is attained by antioxidants. Hence, they are not harmful because they have the ability to accommodate the change in electrons without themselves being reactive [24].

21.4 Phytoantioxidants for Aging and Neurological Disorders

Many researchers and medical practitioners believe that supplementation of dietary antioxidants has the ability to delay or prevent aging and related pathological conditions. However, the research outcomes on this subject are divisive. On the one hand, dietary antioxidant supplementation has positively impacted aging and longevity in several animal models [25]. On the other hand, many randomized controlled studies of antioxidant supplements in people have failed to enhance health or other age-related pathological disorders [26]. Furthermore, synthetic antioxidants like vitamins have been found potentially harmful. Some epidemiological studies suggest that frequent intake of β-carotene, vitamin A, and vitamin E may even increase mortality, particularly in well-nourished people [27]. The disparity between outcomes achieved in animal models and clinical studies, commonly known as the antioxidant paradox because of the dual function of antioxidants in ROS production [28]. Indeed, dietary antioxidants can function not just as ROS scavengers, but they can also be quickly oxidized and behave as prooxidants, causing harm to essential biomolecules when consumed in high doses [29]. Furthermore, mounting evidence shows that ROS plays critical roles as key second messengers in various necessary activities, such as intracellular signaling, cell survival, proliferation, differentiation, and death, as well as immunological and stress responses [30, 31]. Based on the above data, targeting ROS-related disorders employing pharmacologically synthesized antioxidants is a highly complex and challenging task.

Clinical experiments using synthetic antioxidants have had mixed results, raising questions about their suitability for combating ROS-mediated degenerative illnesses [32]. As a result, dietary supplementation with natural antioxidants, such as polyphenols and carotenoids, obtained chiefly from plant sources, has been recommended as a viable alternative to synthetic antioxidant consumption [23, 33]. Plants generate phytochemicals as secondary metabolites to defend themselves from stress conditions such as microbial infections, pollution exposure, temperature fluctuations, and drought. Because of these characteristics, they are thought to be attractive candidates for creating health- and lifespan-promoting treatments [34]. Phytochemicals such as resveratrol, curcumin, catechins, genistein, and quercetin have been shown to be effective in combating a variety of pathological diseases caused by aging-related oxidative stress and associated chronic inflammation [35]. The most commonly used phytotherapeutic compounds with potent antioxidant properties are presented in Figure 21.2, and their applications in aging and neurological diseases are mentioned in Table 21.1.

However, bioavailability is one of the critical issues in the therapeutic potential of orally given phytochemicals due to insufficient water solubility and intestinal permeability. The popular

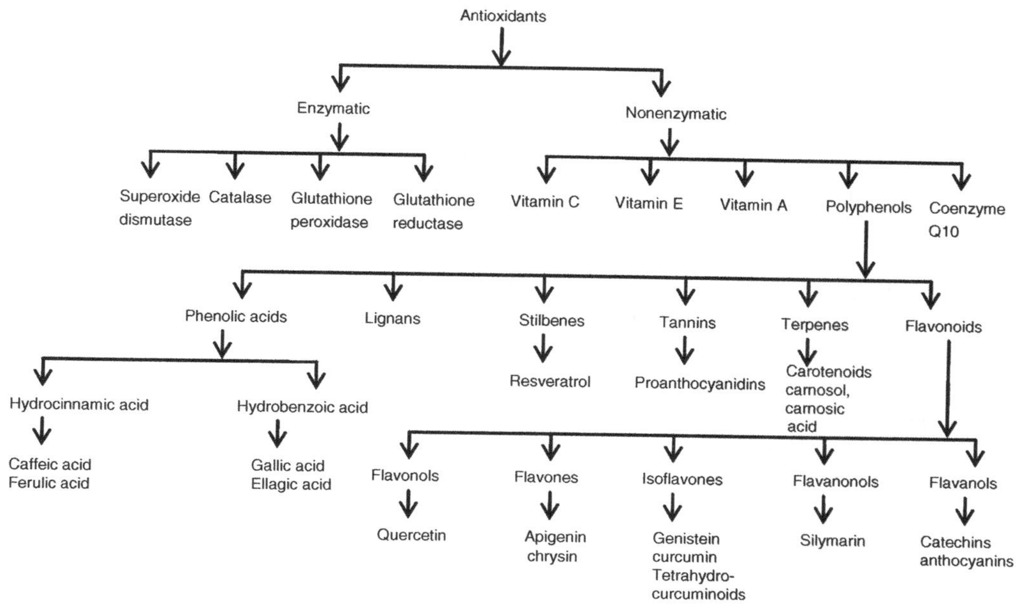

Figure 21.2 Commonly used phytotherapeutic compounds with potent antioxidant properties.

Table 21.1 List of phytoantioxidants used for the treatment of aging and neurological diseases [36].

Classification	Antioxidants	Chronic disease
Carotenoids	Lutein	Macular degeneration
	Zeaxanthin	
Flavonols	Quercetin	AD
	Catechin	
	Epigallocatechin gallate	Aging
Flavanones	Eriodictyol	AD
	Hesperetin	AD
	Naringenin	AD
Organosulfur compounds	Allicin	Aging
Polyphenols	Curcumin	Aging; AD
	Resveratrol	Aging AD
Stilbenes	Tetrahydroxystilbene glucoside	Aging

AD, Alzheimer's disease.
Source: Based on Ref. [36].

phytoantioxidants like quercetin, resveratrol, and curcumin have shown only 1–2% bioavailability. Hence, innovative nanotechnological solutions have recently been developed to address this issue by increasing the bioavailability of phytochemicals after oral administration.

21.5 Nanonization of Phytoantioxidants

The most crucial feature of nanonization for orally administered medicines is their ability to overcome the physical (tight intestinal mucosal lining and selectively permeable membranes of enterocytes) as well as chemical (acidic pH) barriers in the gastrointestinal tract (GIT) [37]. Particle size, surface charge, and hydrophobicity are the critical parameters of nanoparticles (NPs) that impact their physicochemical characteristics like absorption, distribution, metabolism, and excretion. The NP size influences its pharmacokinetics, cell entrance, and immune system interactions [38]. Whereas, the charge present on the surface of NPs significantly affects their cellular absorption and cytotoxicity [39]. Similarly, the hydrophilicity or hydrophobicity of NPs is responsible for various biological reactions like cellular absorption, plasma protein interactions, particle elimination, and immunological responses [40]. These properties of NPs preserve the loaded active molecule from degradation during gastrointestinal digestion and cellular metabolism. The active molecule transported by NPs is eventually released into the gastrointestinal tract, the circulatory system, or various cells [41]. The fate of these biologically active substances depends on their physicochemical properties and the site where they get released, and this release of active molecules at a particular location can be controlled by altering the surface chemistry of NPs [42].

The NPs target the site either by active or passive mechanism [43]. In passive targeting, the loaded drug is released either by diffusion or erosion of the nanocarrier. Whereas in active targeting, the release of loaded drug is controlled at target sites via different mechanisms. This mechanism is commonly used to target RNAs, proteins, lipids, carbohydrates, and small molecules at specific sites [44]. Selective tissue targeting can be accomplished by integrating stimulus-responsive components into the nanocarrier, which are activated by certain stimuli such as electric or magnetic fields, light, pH, heat, ultrasound, and contact with concentrated ionic solutions or specific enzymes [45]. Besides, it could be achieved by employing metallic NPs, such as silver, gold, and iron oxide, as carriers for delivering the drug. Modifying the surface properties of these metallic NPs helps in targeting the drug at a specific site [46]. However, the application of organic nanocarriers is favored because their physicochemical properties could be perfectly altered by varying their chemical composition, shape, size, structural morphology, and surface properties [44]. The effectiveness of the nanoparticulate system also depends on the molecular weight of the loaded drug. High molecular weight compound shows less entrapment in the nanocarrier system, resulting in insufficient loading [47]. Nanoparticulate drug delivery systems provide different health-related benefits based on their properties, loaded drugs, and desired therapeutic application [48]. Notably, the therapeutic potential of natural compounds loaded in NPs provides many benefits over its conventional formulations like better epithelium permeability, higher stability, longer half-life, increased solubility and bioavailability, improved tissue targeting, and reduced side effects [49]. Hence, nanotechnology-based systems are rapidly being used to prevent and treat aging-related pathological illnesses, such as AD and PD [50], cardiovascular diseases [51], obesity, type 2 diabetes [52, 53], and malignancies [54].

21.6 Nanotechnologies Developed for Phytoantioxidants

Nowadays, nanotherapeutics have shown their potential in overcoming critical challenges associated with effective and targeted medication delivery into the brain to diagnose and treat neurological diseases [55, 56]. Several nanotherapeutics are produced utilizing nanotechnology to deliver desirable therapeutic molecules into brain tissues and other tissues near the site of action [57, 58].

The use of biodegradable materials as a carrier also showed effective drug delivery near the site of action. As a result, these formulations and therapies are expected to prevent, heal, and control CNS tissue damage [59]. Furthermore, several nanotherapeutics and polymers like polylactic acid, polyglycolic acid, polylactic-co-glycolic acid, polycaprolactone, and chitosan are widely utilized in drug delivery systems by coating with the surfactant polysorbate 80, allowing them to penetrate the BBB via receptor-mediated endocytosis readily [19, 25]. These nanotherapeutics have additional properties since their surfaces may be modified and engineered with a hydrophilic polyethylene glycol layer to shield the drug from enzyme degradation and immune system recognition [60]. Thus, these salient features enable the consideration of nanotherapeutics to diagnose and treat aging and other neurological diseases [57].

One major problem that restricts the therapeutic application of most NP formulations is their propensity to produce oxidative stress at the cellular level [61]. However, the amount of ROS generated by NPs in the cell depends on the concentration of NPs in the cell. NPs affect the cellular redox environment by either promoting or suppressing ROS generation. This biphasic dose-response relationship is characterized by low-dose stimulation and high-dose inhibition [62]. Excess ROS production and overloading of endogenous antioxidant systems are expected after exposure to high NP concentrations, resulting in cytotoxicity and inflammation [63]. As a result, determining the maximum permissible dosages is seen as essential to avoid adverse health effects. However, research shows that low-level NP exposure may surprisingly boost antioxidant defenses and reduce oxidative stress [64]. Some nanomaterials have been shown to have enzyme-like antioxidant characteristics, such as the ability to scavenge ROS and other free radicals, therefore decreasing oxidative damage [65]. These nanomaterials are commonly referred to as nano-antioxidants. Nonorganic NPs having inherent antioxidant capabilities, such as metallic NPs and NPs functionalized with antioxidant enzymes or natural antioxidants, are examples of nano-antioxidants. The antioxidant potential of nanocarriers loaded with phytotherapeutic compounds is discussed in more depth in the following subsections.

21.6.1 Liposomes

Liposomes are bilayer vesicles made up of phospholipids and cholesterol. They are not toxic and biocompatible, hence widely used for the delivery of therapeutic molecules. Liposomes exhibit hydrophilic core and hydrophobic bilayer structure, which makes them capable of delivering hydrophilic and hydrophobic molecules. Though the biological system recognizes them as foreign particles, they do not show any negative response after entering the system. Besides that, they are biodegradable, non-immunogenic, noncarcinogenic, and non-thrombogenic [66]. They also protect the encapsulated therapeutics from enzymatic breakdown and elimination by the reticuloendothelial system. The essential characteristics of liposomes are fusing with biological membranes, moving across the cell membrane, and penetrating the BBB. The liposome's half-life can be easily increased by treating its surface with PEG [67].

Mourtas et al. synthesized multifunctional liposomes containing a curcumin derivative and anti-transferrin antibody showed a high affinity for amyloid deposits in post-mortem brain samples from AD patients. Thioflavin analysis was performed to determine the liposome's capacity to delay Aβ1-42 peptide aggregation. However, coating the curcumin-derivative liposomes with the anti-transferrin antibody greatly enhanced their absorption by the BBB [68]. In animal models, the Aβ oligomers having a high affinity for liposomes can be utilized to deliver therapeutic molecules [69]. In an in vitro research, the cell membrane fluidity was enhanced using phosphatidylcholine liposomes containing omega-3 fatty acid and docosahexenoic acid inamyloid precursor protein (APP)-overexpressing cells [70].

21.6.2 Solid Lipid Nanoparticles

SLNs are also being used as efficient and alternative carriers for drug delivery as they have better advantages with improved characteristics. SLNs are known as an attractive colloidal drug carrier system for brain targeting. The accumulation of SLNs in the reticuloendothelial system limits their use for targeted drug delivery in the brain. The lipid matrix is solid at room temperature with a unique size and better advantages to be used as nanocarriers, allowing better release and stability of drugs without causing cytotoxic effects in the tissue [71]. The SLNs have better advantages of reproducibility by using multiple strategies and larger scale-up feasibility. It is also a good option for other formulations that lack organic solvents. This also reduces the chance of residual contaminations. Based on these characteristics, SLN provides one of the most promising systems for drug delivery against many neurodegenerative diseases and cancer treatment [72].

The polysorbate-coated SLN has shown promising outcomes for drug delivery across the BBB. The use of SLN can improve drug stability in the blood and absorption across the BBB because the polysorbate activates serum proteins by acting as an anchor for apolipoproteins. The interaction of lipoproteins with brain capillary endothelial cell receptors and apolipoproteins enhances BBB crossing. Phagocytosis can also be avoided by covering the surface of SLN with hydrophilic polymers or surfactants [57]. Additionally, ligands to the SLN surface boost drug concentration, stability, and availability across the BBB for neurological therapies [73].

Kakkar et al., in their work, created a curcumin-loaded SLN for neuroprotection. The developed SLN demonstrated increased bioavailability with the reduction in dose size, indicating a role for SLN in neurological disorders [74, 75]. However, the primary barrier is drug burst release in acidic conditions following oral administration of curcumin SLNs. To stabilize the SLNs, Ramalingam et al. recommended an acid-resistant coating. Curcumin distribution in the brain was more significant in N-trimethyl chitosan surface-modified SLNs than in free curcumin, chitosan, and non-coated SLN [76]. In another work, Kakkar et al. developed sesamol-loaded SLN as a therapeutic agent for CNS disorders. The superiority of orally given sesamol-loaded SLNs was proven using biochemical and behavioral paradigms. Scintigraphy and confocal microscopy of rabbit brain sections treated with radiolabeled and fluorescent SLN, respectively, support the findings [77].

21.6.3 Polymeric Nanoparticles

Polymeric NPs are solid colloidal particles containing macromolecular materials to attach, adsorb, dissolve, and encapsulate the drugs or therapeutic compounds. Degradable polymeric NPs of 10–100 nm are a standard drug delivery system for neurological disease treatments. These particles exist in two variable units, nanocapsules and nanospheres [78, 79]. Nanocapsules are made of core-shell NPs, whereas nanospheres contain homogeneous matrices. These particle sizes facilitate fine-tuning to acquire desired properties like active compound protection with easy delivery and permeability of drugs into the target cells with higher efficacy and efficiency at low-cost preparation [80].

21.6.3.1 Curcumin-loaded Polymeric NPs

Nanoprecipitation of curcumin with polyethylene glycol-polylactic acid co-block polymer and polyvinylpyrrolidone polymer followed by freeze-drying with β-cyclodextrin was performed by Cheng et al. to produce dry nano curcumin. Dry nano curcumin showed significant improvements in memory enhancement at a reduced dose in the transgenic mice model of AD [81]. Curcumin was also encapsulated in PLGA NPs by Tiwari et al. The developed curcumin-PLGA NPs were found to induce adult neurogenesis and reverse cognitive deficits in the AD model [82]. Similarly, Joseph et al. developed PLGA-PEG-Curcumin NPs, ensuring neuroprotection in neonatal rats with hypoxic-ischemic encephalopathy [83]. Curcumin-encapsulated PLGA NPs were modified with Tet-1 peptide by Mathew et al. loaded to improve the interaction of NP with motor neurons and undergo fast retrograde transportation. The developed NP was beneficial in AD as it destroys amyloid aggregates [84].

21.6.3.2 Quercetin-loaded Polymeric Nanoparticle

Quercetin-loaded PLGA nanocapsule was prepared by Ghosh et al. and showed improved efficacy in combating arsenic-induced oxidative damage in the liver and brain of the rat model [85]. Further, nano-encapsulated quercetin prevents the loss of pyramidal neurons from the hippocampal CA1 and CA3 subfields in ischemia reperfusion-induced young and aged rats by downregulating inducible nitric oxide synthase and caspase-3 activities [86]. Das et al. also encapsulated quercetin into the NP and was found to be beneficial in combating age-related cerebral oxidative injury [87]. In another study, Testa et al. loaded quercetin into b-CD-dodecyl carbonate NPs to improve quercetin efficacy. The developed quercetin NPs were found to have increased permeability across the BBB and reduced oxysterol-induced inflammation in the brain [88].

21.6.3.3 Resveratrol-loaded Polymeric NPs

Resveratrol-loaded poly(lactide) NPs modified with polysorbate 80 were reported to be neuroprotective in MPTP-induced PD in mice by mitigating oxidative stress [89]. In another study, resveratrol loaded into PEG-PLA NPs and modified with transferrin has been shown to have a therapeutic effect on gliomas [90]. Resveratrol NPs made up of poly(N-vinylpyrrolidone)-b-poly(ε-caprolactone) polymer have been reported protective against H_2O_2-induced oxidative stress in rat cortical cells [91].

21.6.3.4 Gallic Acid-loaded Polymeric NPs

Nagpal et al. fabricated GA-loaded chitosan NPs showed anti-depressant-like activity and considerable improvement in the drug's tolerance limits compared with the free drug [92]. Further, chitosan NPs of gallic acid were modified with Tween 80, showing Cerebro protective activity against scopolamine-induced amnesia [93]. This increased efficacy may be due to the increased uptake of GA. The maximum tolerated dose of Tween 80-coated chitosan NPs (cGANP) in Wistar rats was observed to be 825 mg/kg body weight [94].

21.6.4 Mesoporous Silica Nanoparticle

Ozlem Yesil-Celiktas et al. synthesized silica-poly(amidoamine) dendrimer hybrid NPs loaded with black carrot anthocyanins using the sol-gel technique. The author investigated the nanocarrier of the lowest particle size (134.8 nm) for inhibiting the proliferative effects of neuroblastoma. It was observed that NPs were not toxic to the cells, and 87.9% inhibition was achieved, which was associated with anthocyanins. This dendrimer-mediated formation of biomimetic silica NPs might provide options to be applied in the industry [95]. Quercetin encapsulation in modified silica NPs can be used against Cu(II)-induced oxidative stress in neurodegeneration [96]. Kalashnikova et al. explored the anticancer properties of curcumin-loaded nanoceria (CNP-Cur) and dextran-nanoceria (Dex-CNP-Cur) in MYCN-amplified and non-amplified cell lines of neuroblastoma models. Increased tumor cell death with no or minor toxicity in healthy cells was observed. Dextran coating of CNPs, therefore, not only helps reduce the cellular viability in cancer cells but also assisted in preventing the opsonization and clearance of the nanoformulations from the circulation by the phagocytes [97].

21.6.5 Magnetic Nanoparticles

Cheng et al. developed curcumin-conjugated magnetic NPs for detecting amyloid plaques in AD mice using magnetic resonance imaging (MRI) [98]. Curcumin-conjugated magnetic NPs (made of superparamagnetic iron oxide) have been used for detecting amyloid plaques in AD mice using MRI [81]. Nanoencapsulation of GA has been reported to increase its half-life significantly [99].

21.7 Conclusion and Future Perspectives

It is a well-known fact that high mortality and morbidity rates are attributed to aging. The time-dependent and natural aging process cannot be stopped, but it can be converted into a healthy aging process by taking specific preventive measures. As environmental parameters also contribute to aging, we should regulate the ROS-producing environmental factors. Similarly, our diet should be rich in antioxidants such as fresh fruits and vegetables that may help preserve the equilibrium between oxidants and antioxidants and the ultimate healthy aging process. A lot of work is needed to produce effective exogenous antioxidants from natural sources, though they cannot stop it but may lessen the process of aging.

Oxidative stress, induced by an imbalance between ROS generation and scavenging, is undoubtedly one of the major contributors to most aging-related clinical diseases. Hence, one of the most significant challenges for the researcher is creating treatment methods to counteract the adverse effects of oxidative stress, such as damage to key biomolecules, increased telomere attrition, and systemic inflammation. For decades, dietary supplementation with chemically produced antioxidants has been considered the most promising to prevent and treat ROS-related diseases. However, recent studies have produced somewhat contentious and disappointing results, such as increased mortality in those who regularly take synthetic antioxidants such as β-carotene, vitamin A, and vitamin E. In aging-related pathogenic disorders, oxidative stress and chronic inflammation coexist. In clinical studies, synthetic antioxidants have shown their inability to address both these conditions simultaneously. Based on this information, dietary supplementation with natural antioxidants (mostly polyphenols) remains a viable alternative to synthetic antioxidants, as natural phytoantioxidants are known to combat oxidative stress and inflammation successfully. However, the main issue with this phytoantioxidant is its low bioavailability. Hence, nowadays, research has been focused on developing nanotherapeutics for these phytoantioxidants, as nanocarriers have the ability to enhance solubility, stability, absorption, and protection from enzymatic degradation. Studies have shown that these nano-encapsulated phytoantioxidants exhibit 5–10 times greater bioavailability than native formulations. Furthermore, this form of oral administration prolongs the circulation time of phytoantioxidants and thereby reduces their potential toxicity and side effects. Nanotherapeutic methods would most likely provide a way to overcome many of the limitations of traditional treatment modalities. Furthermore, the nano delivery method has the advantage of allowing phytochemicals to be delivered to specific organs, such as the brain, following oral ingestion.

To summarize, although nanotherapeutics of phytoantioxidants appear to have health-promoting potential, numerous significant hurdles have to be overcome. Nanocarriers containing phytotherapeutics must be extensively studied for their health effects in patients, mainly if employed for the long term. Indeed, because the constituent may or may not be soluble in biological matrices, the toxicology of applied nanocomposites may differ significantly from that of loaded bioactive chemicals; this might substantially impact the results for various internal organs. A burst drug release in some nanoformulations may raise concerns about cellular toxicity, but a slower drug release, on the other hand, may result in inadequate therapeutic effectiveness. As a result, developing nanoformulations with optimal release profiles based on the physicochemical characteristics of transported therapeutic agents is a significant research challenge. Besides that, some questions need to be addressed, like do applied nanomaterials have the potential to bioaccumulate in the human body? Is it possible that they will be transformed into possibly harmful metabolites? Can they be destroyed and expelled once the therapeutic agent has been delivered? Are they harmful after being excreted (in the urine and feces of treated patients) and accumulating in live human environments? After looking into these obstacles, it is possible to conclude that, even though significant progress has been made toward bringing these nanotherapeutic approaches closer to clinical applications, more research is needed to improve the efficacy and long-term safety of phytoantioxidants-loaded nanotherapeutics in the treatment of aging and neurological disorders.

References

1 Vaiserman, A., Koliada, A., Zayachkivska, A., and Lushchak, O. (2020). Nanodelivery of natural antioxidants: an anti-aging perspective. *Front. Bioeng. Biotechnol.* 7: 1–19. https://doi.org/10.3389/fbioe.2019.00447.

2 Cenini, G., Lloret, A., and Cascella, R. (2019). Oxidative stress in neurodegenerative diseases: from a mitochondrial point of view. *Oxidative Med. Cell. Longev.* 2019: 1–18. https://doi.org/10.1155/2019/2105607.

3 Ambesh, A.D.G. (2015). Nanotechnology in neurology: Genesis, current status, and future prospects. *Ann. Indian Acad. Neurol.* 18: 382. https://doi.org/10.4103/0972-2327.169535.

4 Campisi, J., Kapahi, P., Lithgow, G.J. et al. (2019). From discoveries in ageing research to therapeutics for healthy ageing. *Nature* 571: 183–192. https://doi.org/10.1038/s41586-019-1365-2.

5 Liguori, I., Russo, G., Curcio, F. et al. (2018). Oxidative stress, aging, and diseases. *Clin. Interv. Aging* 13: 757–772. https://doi.org/10.2147/CIA.S158513.

6 Warraich, U.-A., Hussain, F., and Kayani, H.U.R. (2020). Aging - oxidative stress, antioxidants and computational modeling. *Heliyon* 6: e04107. https://doi.org/10.1016/j.heliyon.2020.e04107.

7 Kishida, K.T. and Klann, E. (2007). Sources and targets of reactive oxygen species in synaptic plasticity and memory. *Antioxid. Redox Signal.* 9: 233–244. https://doi.org/10.1089/ars.2007.9.ft-8.

8 Rego, A.C. and Oliveira, C.R. (2003). Mitochondrial dysfunction and reactive oxygen species in excitotoxicity and apoptosis: implications for the pathogenesis of neurodegenerative diseases. *Neurochem. Res.* 28: 1563–1574. https://doi.org/10.1023/a:1025682611389.

9 Indo, H.P., Yen, H.-C., Nakanishi, I. et al. (2015). A mitochondrial superoxide theory for oxidative stress diseases and aging. *J. Clin. Biochem. Nutr.* 56: 1–7. https://doi.org/10.3164/jcbn.14-42.

10 Grimm, A., Friedland, K., and Eckert, A. (2016). Mitochondrial dysfunction: the missing link between aging and sporadic Alzheimer's disease. *Biogerontology* 17: 281–296. https://doi.org/10.1007/s10522-015-9618-4.

11 Cobley, J.N., Fiorello, M.L., and Bailey, D.M. (2018). 13 reasons why the brain is susceptible to oxidative stress. *Redox Biol.* 15: 490–503. https://doi.org/10.1016/j.redox.2018.01.008.

12 Winczura, A., Zdżalik, D., and Tudek, B. (2012). Damage of DNA and proteins by major lipid peroxidation products in genome stability. *Free Radic. Res.* 46: 442–459. https://doi.org/10.3109/10715762.2012.658516.

13 Vistoli, G., De Maddis, D., Cipak, A. et al. (2013). Advanced glycoxidation and lipoxidation end products (AGEs and ALEs): an overview of their mechanisms of formation. *Free Radic. Res.* 47: 3–27. https://doi.org/10.3109/10715762.2013.815348.

14 Ahmed, N. (2005). Advanced glycation endproducts—role in pathology of diabetic complications. *Diabetes Res. Clin. Pract.* 67: 3–21. https://doi.org/10.1016/j.diabres.2004.09.004.

15 Cooke, M.S., Evans, M.D., Dizdaroglu, M., and Lunec, J. (2003). Oxidative DNA damage: mechanisms, mutation, and disease. *FASEB J.* 17: 1195–1214. https://doi.org/10.1096/fj.02-0752rev.

16 Dukan, S., Farewell, A., Ballesteros, M. et al. (2000). Protein oxidation in response to increased transcriptional or translational errors. *Proc. Natl. Acad. Sci.* 97: 5746–5749. https://doi.org/10.1073/pnas.100422497.

17 Rodriguez-Rocha, H., Garcia-Garcia, A., Pickett, C. et al. (2013). Compartmentalized oxidative stress in dopaminergic cell death induced by pesticides and complex I inhibitors: distinct roles of superoxide anion and superoxide dismutases. *Free Radic. Biol. Med.* 61: 370–383. https://doi.org/10.1016/j.freeradbiomed.2013.04.021.

18 Brigelius-Flohé, R. and Traber, M.G. (1999). Vitamin E: function and metabolism. *FASEB J.* 13: 1145–1155.

19 Bodega, G., Alique, M., Puebla, L. et al. (2019). Microvesicles: ROS scavengers and ROS producers. *J. Extracell. Vesicles* 8: 1626654. https://doi.org/10.1080/20013078.2019.1626654.

20 Mirończuk-Chodakowska, I., Witkowska, A.M., and Zujko, M.E. (2018). Endogenous non-enzymatic antioxidants in the human body. *Adv. Med. Sci.* 63: 68–78. https://doi.org/10.1016/j.advms.2017.05.005.

21 Suleman, M. (2018). Antioxidants, its role in preventing free radicals and infectious diseases in human body. *Pure Appl. Biol.* 7: 380–388. https://doi.org/10.19045/bspab.2018.700197.

22 Nimse, S.B. and Pal, D. (2015). Free radicals, natural antioxidants, and their reaction mechanisms. *RSC Adv.* 5: 27986–28006. https://doi.org/10.1039/C4RA13315C.

23 Neha, K., Haider, M.R., Pathak, A., and Yar, M.S. (2019). Medicinal prospects of antioxidants: a review. *Eur. J. Med. Chem.* 178: 687–704. https://doi.org/10.1016/j.ejmech.2019.06.010.

24 Ahmad, R. (2018). Introductory chapter: basics of free radicals and antioxidants. In: *Free Radicals, Antioxidants Diseases*. InTech https://doi.org/10.5772/intechopen.76689.

25 Sadowska-Bartosz, I. and Bartosz, G. (2014). Effect of antioxidants supplementation on aging and longevity. *Biomed. Res. Int.* 2014: 1–17. https://doi.org/10.1155/2014/404680.

26 Gruber, J. and Halliwell, B. (2017). Approaches for extending human healthspan: from antioxidants to healthspan pharmacology. *Essays Biochem.* 61: 389–399. https://doi.org/10.1042/EBC20160091.

27 Bjelakovic, G., Gluud, L.L., Nikolova, D. et al. (2014). Vitamin D supplementation for prevention of mortality in adults. *Cochrane Database Syst. Rev.* 2014. https://doi.org/10.1002/14651858.CD007470.pub3.

28 Halliwell, B. (2000). The antioxidant paradox. *Lancet* 355: 1179–1180. https://doi.org/10.1016/S0140-6736(00)02075-4.

29 Milisav, I., Ribarič, S., and Poljsak, B. (2018). Antioxidant vitamins and ageing. *Subcell. Biochem.* 90: 1–23. https://doi.org/10.1007/978-981-13-2835-0_1.

30 Bardaweel, S.K., Gul, M., Alzweiri, M. et al. (2018). Reactive oxygen species: the dual role in physiological and pathological conditions of the human body. *Eurasian J. Med.* 50: 193–201. https://doi.org/10.5152/eurasianjmed.2018.17397.

31 Milkovic, L., Cipak Gasparovic, A., Cindric, M. et al. (2019). Short overview of ROS as cell function regulators and their implications in therapy concepts. *Cell* 8: 793. https://doi.org/10.3390/cells8080793.

32 Toledo-Ibelles, P. and Mas-Oliva, J. (2018). Antioxidants in the fight against atherosclerosis: is this a dead end? *Curr. Atheroscler. Rep.* 20: 36. https://doi.org/10.1007/s11883-018-0737-7.

33 Forni, C., Facchiano, F., Bartoli, M. et al. (2019). Beneficial role of phytochemicals on oxidative stress and age-related diseases. *Biomed. Res. Int.* 2019: 1–16. https://doi.org/10.1155/2019/8748253.

34 Leonov, A., Arlia-Ciommo, A., Piano, A. et al. (2015). Longevity extension by phytochemicals. *Molecules* 20: 6544–6572. https://doi.org/10.3390/molecules20046544.

35 Martel, J., Ojcius, D.M., Ko, Y.-F. et al. (2019). Antiaging effects of bioactive molecules isolated from plants and fungi. *Med. Res. Rev.* 39: 1515–1552. https://doi.org/10.1002/med.21559.

36 Zhang, Y.-J., Gan, R.-Y., Li, S. et al. (2015). Antioxidant phytochemicals for the prevention and treatment of chronic diseases. *Molecules* 20: 21138–21156. https://doi.org/10.3390/molecules201219753.

37 Moss, D.M., Curley, P., Kinvig, H. et al. (2018). The biological challenges and pharmacological opportunities of orally administered nanomedicine delivery. *Expert Rev. Gastroenterol. Hepatol.* 12: 223–236. https://doi.org/10.1080/17474124.2018.1399794.

38 Hoshyar, N., Gray, S., Han, H., and Bao, G. (2016). The effect of nanoparticle size on in vivo pharmacokinetics and cellular interaction. *Nanomedicine* 11: 673–692. https://doi.org/10.2217/nnm.16.5.

39 Fröhlich, E. (2012). The role of surface charge in cellular uptake and cytotoxicity of medical nanoparticles. *Int. J. Nanomedicine* 7: 5577. https://doi.org/10.2147/IJN.S36111.

40 Ajdary, M., Moosavi, M., Rahmati, M. et al. (2018). Health concerns of various nanoparticles: a review of their in vitro and in vivo toxicity. *Nanomaterials* 8: 634. https://doi.org/10.3390/nano8090634.

41 Martínez-Ballesta, M.C., GIL-Izquierdo, Á., García-Viguera, C., and Domínguez-Perles, R. (2018). Nanoparticles and controlled delivery for bioactive compounds: outlining challenges for new "Smart-Foods" for health. *Foods* 7: 72. https://doi.org/10.3390/foods7050072.

42 Patra, J.K., Das, G., Fraceto, L.F. et al. (2018). Nano based drug delivery systems: recent developments and future prospects. *J. Nanobiotechnol.* 16: 71. https://doi.org/10.1186/s12951-018-0392-8.

43 Kydd, J., Jadia, R., Velpurisiva, P. et al. (2017). Targeting strategies for the combination treatment of cancer using drug delivery systems. *Pharmaceutics* 9: 46. https://doi.org/10.3390/pharmaceutics9040046.

44 Conte, R., Marturano, V., Peluso, G. et al. (2017). Recent advances in nanoparticle-mediated delivery of anti-inflammatory phytocompounds. *Int. J. Mol. Sci.* 18: 709. https://doi.org/10.3390/ijms18040709.

45 Gu, M., Wang, X., Toh, T.B., and Chow, E.K.-H. (2018). Applications of stimuli-responsive nanoscale drug delivery systems in translational research. *Drug Discov. Today* 23: 1043–1052. https://doi.org/10.1016/j.drudis.2017.11.009.

46 Kong, F.Y., Zhang, J.W., Li, R.F. et al. (2017). Unique roles of gold nanoparticles in drug delivery, targeting and imaging applications. *Molecules* 22: 1445. https://doi.org/10.3390/molecules22091445.

47 Ganesan, P., Ramalingam, P., Karthivashan, G. et al. (2018). Recent developments in solid lipid nanoparticle and surface-modified solid lipid nanoparticle delivery systems for oral delivery of phyto-bioactive compounds in various chronic diseases. *Int. J. Nanomedicine* 13: 1569–1583. https://doi.org/10.2147/IJN.S155593.

48 Gupta, R. and Xie, H. (2018). Nanoparticles in daily life: applications, toxicity and regulations. *J. Environ. Pathol. Toxicol. Oncol.* 37: 209–230. https://doi.org/10.1615/JEnvironPatholToxicolOncol.2018026009.

49 Date, A.A., Hanes, J., and Ensign, L.M. (2016). Nanoparticles for oral delivery: design, evaluation and state-of-the-art. *J. Control. Release* 240: 504–526. https://doi.org/10.1016/j.jconrel.2016.06.016.

50 Poovaiah, N., Davoudi, Z., Peng, H. et al. (2018). Treatment of neurodegenerative disorders through the blood–brain barrier using nanocarriers. *Nanoscale* 10: 16962–16983. https://doi.org/10.1039/C8NR04073G.

51 Li, T., Liang, W., Xiao, X., and Qian, Y. (2018). Nanotechnology, an alternative with promising prospects and advantages for the treatment of cardiovascular diseases. *Int. J. Nanomedicine* 13: 7349–7362. https://doi.org/10.2147/IJN.S179678.

52 Bahadori, E., Farjami, Z., Rezayi, M. et al. (2019). Recent advances in nanotechnology for the treatment of metabolic syndrome. *Diabetes Metab. Syndr. Clin. Res. Rev.* 13: 1561–1568. https://doi.org/10.1016/j.dsx.2019.03.002.

53 Tsou, Y., Wang, B., Ho, W. et al. (2019). Nanotechnology-mediated drug delivery for the treatment of obesity and its related comorbidities. *Adv. Healthc. Mater.* 8: 1801184. https://doi.org/10.1002/adhm.201801184.

54 Qiao, Y., Wan, J., Zhou, L. et al. (2019). Stimuli-responsive nanotherapeutics for precision drug delivery and cancer therapy. *Wiley Interdiscip. Rev. Nanomedicine Nanobiotechnology* 11: e1527. https://doi.org/10.1002/wnan.1527.

55 Wilson, B. (2009). Brain targeting PBCA nanoparticles and the blood–brain barrier. *Nanomedicine* 4: 499–502. https://doi.org/10.2217/nnm.09.29.

56 Lauzon, M.-A., Daviau, A., Marcos, B., and Faucheux, N. (2015). Nanoparticle-mediated growth factor delivery systems: a new way to treat Alzheimer's disease. *J. Control. Release* 206: 187–205. https://doi.org/10.1016/j.jconrel.2015.03.024.

57 Hadavi, D. and Poot, A.A. (2016). Biomaterials for the treatment of Alzheimer's disease. *Front. Bioeng. Biotechnol.* 4: 49. https://doi.org/10.3389/fbioe.2016.00049.

58 Béduneau, A., Saulnier, P., and Benoit, J.-P. (2007). Active targeting of brain tumors using nanocarriers. *Biomaterials* 28: 4947–4967. https://doi.org/10.1016/j.biomaterials.2007.06.011.

59 Orive, G., Anitua, E., Pedraz, J.L., and Emerich, D.F. (2009). Biomaterials for promoting brain protection, repair and regeneration. *Nat. Rev. Neurosci.* 10: 682–692. https://doi.org/10.1038/nrn2685.

60 Locatelli, E. and Comes Franchini, M. (2012). Biodegradable PLGA-b-PEG polymeric nanoparticles: synthesis, properties, and nanomedical applications as drug delivery system. *J. Nanopart. Res.* 14: 1316. https://doi.org/10.1007/s11051-012-1316-4.

61 Zuberek, M. and Grzelak, A. (2018). Nanoparticles-caused oxidative imbalance. *Adv. Exp. Med. Biol.* 1048: 85–98. https://doi.org/10.1007/978-3-319-72041-8_6.

62 Iavicoli, I., Leso, V., Fontana, L., and Calabrese, E. (2018). Nanoparticle exposure and hormetic dose–responses: an update. *Int. J. Mol. Sci.* 19: 805. https://doi.org/10.3390/ijms19030805.

63 Nel, A. (2006). Toxic potential of materials at the nanolevel. *Science* (80-) 311: 622–627. https://doi.org/10.1126/science.1114397.

64 Abdal Dayem, A., Hossain, M., Lee, S. et al. (2017). The Role of Reactive Oxygen Species (ROS) in the biological activities of metallic nanoparticles. *Int. J. Mol. Sci.* 18: 120. https://doi.org/10.3390/ijms18010120.

65 Lushchak, O., Zayachkivska, A., and Vaiserman, A. (2018). Metallic nanoantioxidants as potential therapeutics for type 2 diabetes: a hypothetical background and translational perspectives. *Oxidative Med. Cell. Longev.* 2018: 1–9. https://doi.org/10.1155/2018/3407375.

66 Siddiqi, K.S., Husen, A., Sohrab, S.S., and Yassin, M.O. (2018). Recent status of nanomaterial fabrication and their potential applications in neurological disease management. *Nanoscale Res. Lett.* 13: 231. https://doi.org/10.1186/s11671-018-2638-7.

67 Spuch, C. and Navarro, C. (2011). Liposomes for targeted delivery of active agents against neurodegenerative diseases (Alzheimer's disease and Parkinson's disease). *J. Drug Deliv.* 2011: 1–12. https://doi.org/10.1155/2011/469679.

68 Mourtas, S., Lazar, A.N., Markoutsa, E. et al. (2014). Multifunctional nanoliposomes with curcumin-lipid derivative and brain targeting functionality with potential applications for Alzheimer disease. *Eur. J. Med. Chem.* 80: 175–183. https://doi.org/10.1016/j.ejmech.2014.04.050.

69 Gobbi, M., Re, F., Canovi, M. et al. (2010). Lipid-based nanoparticles with high binding affinity for amyloid-β1–42 peptide. *Biomaterials* 31: 6519–6529. https://doi.org/10.1016/j.biomaterials.2010.04.044.

70 Eckert, G.P., Chang, S., Eckmann, J. et al. (2011). Liposome-incorporated DHA increases neuronal survival by enhancing non-amyloidogenic APP processing. *Biochim. Biophys. Acta Biomembr.* 1808: 236–243. https://doi.org/10.1016/j.bbamem.2010.10.014.

71 Shevalkar, G. and Vavia, P. (2019). Solidified nanostructured lipid carrier (S-NLC) for enhancing the oral bioavailability of ezetimibe. *J. Drug Deliv. Sci. Technol.* 53: 101211. https://doi.org/10.1016/j.jddst.2019.101211.

72 Shevalkar, G., Pai, R., and Vavia, P. (2019). Nanostructured lipid carrier of propofol: a promising alternative to marketed soybean oil-based nanoemulsion. *AAPS PharmSciTech* 20: 1–14. https://doi.org/10.1208/s12249-019-1408-x.

73 Cacciatore, I., Ciulla, M., Fornasari, E. et al. (2016). Solid lipid nanoparticles as a drug delivery system for the treatment of neurodegenerative diseases. *Expert Opin. Drug Deliv.* 13: 1121–1131. https://doi.org/10.1080/17425247.2016.1178237.

74 Kakkar, V., Singh, S., Singla, D., and Kaur, I.P. (2011). Exploring solid lipid nanoparticles to enhance the oral bioavailability of curcumin. *Mol. Nutr. Food Res.* 55: 495–503. https://doi.org/10.1002/mnfr.201000310.

75 Kakkar, V. and Kaur, I.P. (2011). Evaluating potential of curcumin loaded solid lipid nanoparticles in aluminium induced behavioural, biochemical and histopathological alterations in mice brain. *Food Chem. Toxicol.* 49: 2906–2913. https://doi.org/10.1016/j.fct.2011.08.006.

76 Ramalingam, P. and Ko, Y.T. (2015). Enhanced oral delivery of curcumin from N-trimethyl chitosan surface-modified solid lipid nanoparticles: pharmacokinetic and brain distribution evaluations. *Pharm. Res.* 32: 389–402. https://doi.org/10.1007/s11095-014-1469-1.

77 Kakkar, V., Mishra, A.K., Chuttani, K. et al. (2011). Delivery of sesamol-loaded solid lipid nanoparticles to the brain for menopause-related emotional and cognitive central nervous system derangements. *Rejuvenation Res.* 14: 597–604. https://doi.org/10.1089/rej.2011.1193.

78 Fornaguera, C. and Solans, C. (2016). Polymeric nanoparticles for drug delivery in neurological diseases. *Curr. Pathobiol. Rep.* 4: 189–197. https://doi.org/10.1007/s40139-016-0118-2.

79 Soppimath, K.S., Aminabhavi, T.M., Kulkarni, A.R., and Rudzinski, W.E. (2001). Biodegradable polymeric nanoparticles as drug delivery devices. *J. Control. Release* 70: 1–20. https://doi.org/10.1016/S0168-3659(00)00339-4.

80 Jain, K.K. (2012). Nanobiotechnology-based strategies for crossing the blood–brain barrier. *Nanomedicine* 7: 1225–1233. https://doi.org/10.2217/nnm.12.86.

81 Cheng, K.K., Yeung, C.F., Ho, S.W. et al. (2013). Highly stabilized curcumin nanoparticles tested in an in vitro blood-brain barrier model and in Alzheimer's disease Tg2576 mice. *AAPS J.* 15: 324–336. https://doi.org/10.1208/s12248-012-9444-4.

82 Tiwari, S.K., Agarwal, S., Seth, B. et al. (2014). Curcumin-loaded nanoparticles potently induce adult neurogenesis and reverse cognitive deficits in Alzheimer's disease model via canonical Wnt/β-catenin pathway. *ACS Nano* 8: 76–103. https://doi.org/10.1021/nn405077y.

83 Joseph, A., Wood, T., Chen, C.C. et al. (2018). Curcumin-loaded polymeric nanoparticles for neuroprotection in neonatal rats with hypoxic-ischemic encephalopathy. *Nano Res.* 11: 5670–5688. https://doi.org/10.1007/s12274-018-2104-y.

84 Mathew, A., Fukuda, T., Nagaoka, Y. et al. (2012). Curcumin loaded-PLGA nanoparticles conjugated with Tet-1 peptide for potential use in Alzheimer's disease. *PLoS One* 7: e32616. https://doi.org/10.1371/journal.pone.0032616.

85 Ghosh, A., Mandal, A.K., Sarkar, S. et al. (2009). Nanoencapsulation of quercetin enhances its dietary efficacy in combating arsenic-induced oxidative damage in liver and brain of rats. *Life Sci.* 84: 75–80. https://doi.org/10.1016/j.lfs.2008.11.001.

86 Ghosh, A., Sarkar, S., Mandal, A.K., and Das, N. (2013). Neuroprotective role of nanoencapsulated quercetin in combating ischemia-reperfusion induced neuronal damage in young and aged rats. *PLoS One* 8: e57735. https://doi.org/10.1371/journal.pone.0057735.

87 Das, S., Mandal, A.K., Ghosh, A. et al. (2008). Nanoparticulated quercetin in combating age related cerebral oxidative injury. *Curr. Aging Sci.* 1: 169–174. https://doi.org/10.2174/1874609810801030169.

88 Testa, G., Gamba, P., Badilli, U. et al. (2014). Loading into nanoparticles improves quercetin's efficacy in preventing neuroinflammation induced by oxysterols. *PLoS One* 9: 1–12. https://doi.org/10.1371/journal.pone.0096795.

89 da Rocha, L.G., Bonfanti Santos, D., Colle, D. et al. (2015). Improved neuroprotective effects of resveratrol-loaded polysorbate 80-coated poly(lactide) nanoparticles in MPTP-induced Parkinsonism. *Nanomedicine (Lond)* 10: 1127–1138. https://doi.org/10.2217/nnm.14.165.

90 Guo, W., Li, A., Jia, Z. et al. (2013). Transferrin modified PEG-PLA-resveratrol conjugates: in vitro and in vivo studies for glioma. *Eur. J. Pharmacol.* 718: 41–47. https://doi.org/10.1016/j.ejphar.2013.09.034.

91 Lu, X., Xu, H., Sun, B. et al. (2013). Enhanced neuroprotective effects of resveratrol delivered by nanoparticles on hydrogen peroxide-induced oxidative stress in rat cortical cell culture. *Mol. Pharm.* 10: 2045–2053. https://doi.org/10.1021/mp400056c.

92 Nagpal, K., Singh, S.K., and Mishra, D. (2013). Influence of the formulation on the maximum tolerated doses of brain targeted nanoparticles of gallic acid by oral administration in Wistar rats. *J. Pharm. Pharmacol.* 65: 1757–1764. https://doi.org/10.1111/jphp.12068.

93 Nagpal, K., Singh, S.K., and Mishra, D.N. (2012). Nanoparticle mediated brain targeted delivery of gallic acid: in vivo behavioral and biochemical studies for improved antioxidant and antidepressant-like activity. *Drug Deliv.* 19: 378–391. https://doi.org/10.3109/10717544.2012.738437.

94 Nagpal, K., Singh, S.K., and Mishra, D.N. (2013). Nanoparticle mediated brain targeted delivery of gallic acid: in vivo behavioral and biochemical studies for protection against scopolamine-induced amnesia. *Drug Deliv.* 20: 112–119. https://doi.org/10.3109/10717544.2013.779330.

95 Yesil-Celiktas, O., Pala, C., Cetin-Uyanikgil, E.O., and Sevimli-Gur, C. (2017). Synthesis of silica-PAMAM dendrimer nanoparticles as promising carriers in neuro blastoma cells. *Anal. Biochem.* 519: 1–7. https://doi.org/10.1016/j.ab.2016.12.004.

96 Nday, C.M., Halevas, E., Jackson, G.E., and Salifoglou, A. (2015). Quercetin encapsulation in modified silica nanoparticles: potential use against Cu(II)-induced oxidative stress in neurodegeneration. *J. Inorg. Biochem.* 145: 51–64. https://doi.org/10.1016/j.jinorgbio.2015.01.001.

97 Kalashnikova, I., Mazar, J., Neal, C.J. et al. (2017). Nanoparticle delivery of curcumin induces cellular hypoxia and ROS-mediated apoptosis via modulation of Bcl-2/Bax in human neuroblastoma. *Nanoscale* 9: 10375–10387. https://doi.org/10.1039/C7NR02770B.

98 Cheng, K.K., Chan, P.S., Fan, S. et al. (2015). Curcumin-conjugated magnetic nanoparticles for detecting amyloid plaques in Alzheimer's disease mice using magnetic resonance imaging (MRI). *Biomaterials* 44: 155–172. https://doi.org/10.1016/j.biomaterials.2014.12.005.

99 Dorniani, D., Kura, A.U., Hussein-Al-Ali, S.H. et al. (2014). In vitro sustained release study of gallic acid coated with magnetite-PEG and magnetite-PVA for drug delivery system. *Sci. World J.* 2014. https://doi.org/10.1155/2014/416354.

22

Nanotherapeutics of Phytoantioxidants in Cancer

Ashish Shah[1], Vaishali Patel[2], and Ghanshyam Parmar[1]

[1] *Department of Pharmacy, Sumandeep Vidyapeeth, Vadodara, Gujarat, India*
[2] *Department of Pharmaceutics, Laxminarayan Dev College of Pharmacy, Bharuch, Gujarat, India*

22.1 Introduction

Cancer is a group of diseases characterized by the uncontrollable growth of abnormal cells that have the ability to spread in various parts of the body. Cancer normally grows in body cells and multiplies as per the need of the body. By the time cells grow old or become damaged or die, new cells replace the old cells. Sometimes this normal response process breaks down and instead of normal new cells, damaged or abnormal cells grow and multiply, which may result in the formation of tumors. Tumors may be benign or metastatic. Tumor cells have an ability to spread and invade nearby tissues. Benign tumors does not spread or invade nearby tissues while malignant tumors can spread. There are more than a hundred types of cancers and the nomenclature of cancer was done based on organs or tissues where the cancer forms. The major classes of cancer include carcinoma, sarcoma, and lymphoma [1].

Cancer is the leading cause of death. According to global cancer statistics, there were about 19.3 million new cancer cases and 10 million cancer deaths reported in the year 2020 globally. The pharmacological treatment of cancer includes hormone therapy, immunotherapy, chemotherapy or a combination of all. Among these, the major obstacle in chemotherapy is multidrug resistance (MDR). Continuous doses of chemotherapeutic agents reduce efflux outside the cell, which hampers permeation across the cell membranes. To understand the permeation of cell across the cell, various nanocarriers are used in the treatment of cancer [2].

Many years ago, phytomedicines have been widely explored for the treatment of different types of cancer, which provide alternative therapy to the conventional one. The major advantage of using phytochemicals as anticancer agents is that they do not show any deleterious effect during the course of treatment. Phytochemicals are found in fruits, vegetables, plants, dietary fibers, and natural products, which can regulate many molecular processes. Plant-based phytochemicals like oils, gums, terpenes, saponins, minerals, glycosides, flavonoids, lignins, podophyllotoxins, polyphenolic compounds, vinca alkaloids, and taxol constitute primary and secondary metabolites, which play a major role in the inhibition. Phytochemicals do not show any toxic effect on normal cells [3].

Phytoantioxidants and Nanotherapeutics, First Edition. Edited by Mithun Rudrapal.

22.1.1 Oxidative Stress in Cancer

Oxidative stress is considered as the situation in which reactive oxygen species (ROS) formation exceeds cellular antioxidant capacity. ROS produce continuously in aerobic cells as by-products of normal oxygen metabolism. ROS are generated as a response to endogenous and exogenous stimuli. ROS such as superoxide radical, hydroxyl radical, singlet oxygen, and hydrogen peroxide. Free radicals are highly reactive chemicals that have the potential to harm cells. Free radicals are generated when atoms or molecules gain or lose an electron. Although free radical generation is a normal response of the human body and plays an important role in various biological processes, at high concentrations, they prove hazardous to the body [4, 5]. The free radicals have the ability to damage various components such as DNA, proteins, and cell membranes. The damage caused by free radicals, especially with respect to DNA, may play a role in the development of cancer. Various types of carcinogens mediate their effect through the formation of ROS during their metabolism. DNA damage via oxidative stress causes mutations and therefore plays an important role in the initiation and progression of carcinogenesis [6].

Therefore, in the diseases where ROS levels are elevated, antioxidants provide an important role in the balance of ROS. Oxidative stress is one of the main components involved in the development and growth of cancer cells. Elevated ROS levels cause down-signaling of various antioxidant enzymes that result in the abnormal activity of different molecular targets like NF-KB and Nrf2. Overactivation of these targets causes cancer cell proliferation, angiogenesis, metastasis, and suppression of the cell death process [7].

22.1.2 What Are Antioxidants?

Antioxidants are chemicals that have an ability to interact with free radicals and have the potential to neutralize their effect. Antioxidants are also known as "free radical scavengers." The antioxidants synthesized directly by the body mechanism are known as endogenous antioxidants and those which are taken from external sources are known as exogenous antioxidants. Exogenous antioxidants are also referred to as dietary antioxidants, which are obtained from various sources like fruits, vegetables, and grains. The antioxidant may obtain either from natural resources or synthetic products. Synthetic antioxidants are useful to improve health but at the same time, the adverse effect of synthetic compounds is a serious issue. Natural products can provide a solution for the said purpose. There are many natural herbs/phytochemicals, which have been reported for the potential antioxidant activity with few or no side effects [8, 9].

In this chapter, we have discussed in detail the role of ROS in cancer and the application of antioxidants with respect to the cancer disease. The different types of nanoformulations using the different antioxidant natural products with their application are also discussed in detail (Figure 22.1).

22.2 Phytoantioxidants

22.2.1 Structure of Phytochemicals and Occurrence in Plant Kingdom

Phytochemicals are plant-derived natural products that play an important role in various biological functions of an animal body. A large number of natural compounds have been reported for antioxidant activity. These phytochemicals belong to diverse classes that include polyphenols,

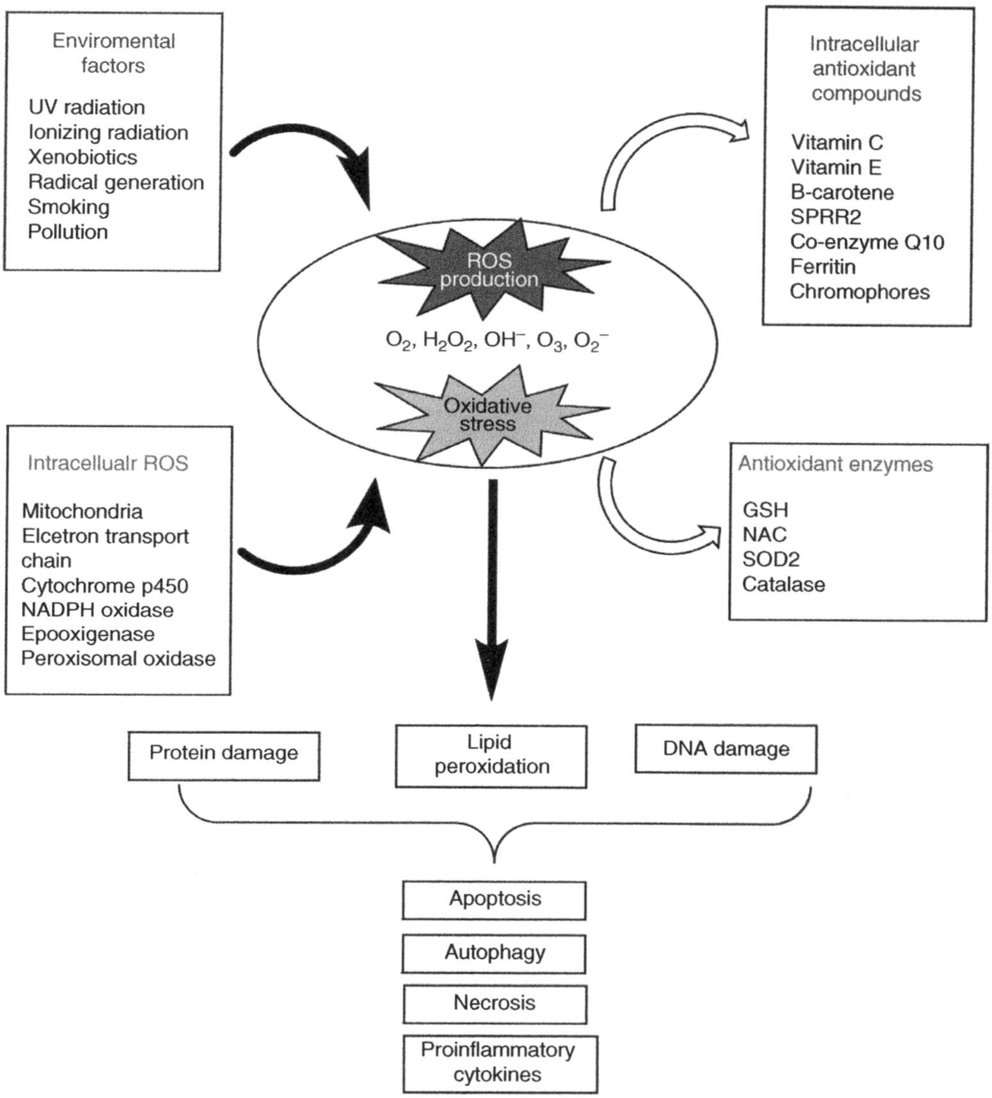

Figure 22.1 Oxidative stress generation and its biological effect. Environmental factors and intracellular ROS increase oxidative stress. Antioxidant enzymes and intracellular antioxidant compounds decrease oxidative stress.

flavonoids, saponins, organosulfur compounds, and vitamins. Antioxidant phytochemicals exist in a variety of sources like berries, grape, guvava, sweetsop, etc [10–12]. Wild fruits like *eucalyptus robusta, eurya nitida, melastoma sanguineum, melalecuca lecucadendron*, etc., have been reported for high antioxidant activity [13]. Peel and seed of fruits also contain high content of phytochemicals like catechin, cyanidin-3-glucoside, epicatechin, gallic acid (GA), kaempferol, and chlorogenic acid [14]. Some vegetables like a penile leaf, cowpea, caraway lotus root, soya bean, ginseng leaf, and broccoli also contain high antioxidant activity [15]. Some grains like pigmented rice, black rice, red rice, and purple rice possess high antioxidant capacities [16].

Table 22.1 Various plant secondary metabolites and their sources.

Secondary metabolites	Common dietary sources
Polyphenols	Fruit, vegetables, tea, coffee, tea, cereals
Carotenoids	Carrots, tomato, pumpkin, pepper
Anthocyanins	Strawberries, black rice, berries
Flavones	Blueberries, orange juice
Flavonols	Cherries, chokeberry, elderberries, wolfberry
Resveratrol	Purple wine, peanut
Theaflavins	Black tea

Polyphenols and carotenoids are two main phytochemicals that contribute to most of the antioxidant properties of foods/plants. Natural polyphenolic compounds found in the human diet have antioxidant activities due to radical scavenging properties. Dietary polyphenols are classified into five different groups: flavonoids, stilbenes, tannins, phenolic acids, and coumarins [17]. Total phenolic content and antioxidant activity have a direct relationship. The literature data suggest that phytochemical extract which has higher phenolic content possesses higher antioxidant activity [18]. Carotenoids are a group of phytochemicals responsible for like red, yellow, and orange color in the food. Various fruits and vegetables are the major sources of carotenoids. The different carotenoids like α-carotene, β-carotene, cryptoxanthin, lycopene, and lutein are found commonly in the diet, which has antioxidant activity [19] (Table 22.1).

22.2.2 Antioxidant Activity of Phytochemicals

In order to fight against ROS, cells develop a different mechanism to maintain redox homeostasis. This mechanism could provide direct and indirect antioxidant activities. Based on the mechanism, classification of antioxidants can be done. The direct antioxidant administered in high doses neutralizes the deleterious effect of ROS [20]. The indirect antioxidant enhances the antioxidant capacity of specific genes of antioxidant proteins through various transcription factors and due to these physiological effects, last longer than direct antioxidants. The key transcription factor, nuclear factor-like 2 (Nrf2) is a basic zipper-containing transcription factor that activates many other enzymes that play an important role in cell defence by improving the removal of ROS. Antioxidant activity of phytochemicals derived from various fruits, vegetables, species, and herbal medicines have been reported in various studies. The antioxidant activity is due to the activation of Nrf2 and other phase-II detoxifying enzymes that help in scavenging different types of free radicals [21].

22.2.3 Some Examples of Antioxidant Phytochemicals

22.2.3.1 Catechin

Catechins are monomer of flavonols with a variety of similar compounds that include catechin, epigallocatechin, epicatechin gallate (EGC), and epigallocatechin gallate (EGCG). EGCG is found in green tea as a major component of polyphenols. EGCG was reported for a variety of biological activities including antioxidant and anticancer [22]. Sahin et al. [23] reported the mode of action of EGCG to fight against oxidative stress. In that study, it was noted that oxidative biomarkers are strongly correlated with hepatic nuclear transcription factors. In another study, Romeo et al. [24] reported that EGCG increases HO-1 activity in rat neurons, which is associated with Nrf2 activation and translocation. The resulting effect provides protection against oxidative stress. In various studies, it was found that catechin, in combination with curcumin, gives synergistic activity against human colon adenocarcinoma and larynx carcinoma.

22.2.3.2 Lycopene

Lycopene is a type of carotenoid, which has powerful antioxidant activity and provides protection against free radical-induced damage to the cell. The singlet oxygen quenching capacity of lycopene is two times higher than β-carotene and tenfold higher than α-tocopherol. Lycopene has the ability to regulate redox-sensitive molecular targets such as mitogen-activated protein kinase and Nrf2 [25]. The supplementation of lycopene in the form of tomato powder increases the expression of Nrf2, HO-1, gluthione, and other antioxidant enzymes. Lycopene has the ability to stimulate various transcription factors that induce antioxidant activity; in addition, lycopene metabolites also increase intracellular GSH levels and decrease ROS levels. A combination of lycopene with β-carotene could inhibit cell proliferation, and arrest the cell cycle [26]. Lycopene along with β-carotene could induce apoptosis in human breast cancer. Additionally, lycopene is also useful against various types of cancer such as colon, tongue, and gastric [27].

22.2.3.3 Curcumin

Curcumin, a polyphenolic found in turmeric, has been reported for anti-inflammatory, antioxidant, and anticancer activity. The antioxidant activity of curcumin is due to the modulation of the antioxidant enzyme by modulating Nrf2 [28]. It has been reported that curcumin increases ARE binding of Nrf2 and induces expression of GST and NQO1. It had also been reported that the liver and lung of mice treated with dietary curcumin reduced oxidative stress [29]. Sahin et al. reported that the Nrf2 level linearly increases as the supplemental level of curcumin increases [30]. The anticancer potential of curcumin has been tested against a variety of cancer cell line that includes: cervical, colorectal, and breast [31–33]. The anticancer potential of curcumin is due to its ability to inhibit cancer cell proliferation and induce cancer cell death.

22.2.3.4 Resveratrol

Resveratrol is a polyphenolic compound found in peanuts, cranberries, and grapes reported for diverse biological actions [34]. Resveratrol provides an antioxidant effect by reducing oxidative stress and prevents the action of free radicals. The antioxidant activity is due to increased Nrf2 expression and its accumulation in the nucleus and due to this, deleterious effect of free radicals is reduced or nullified. The treatment of resveratrol can increase the expression of Nrf2-targeted genes such as CAT, SOD, NQO1, HO-1, and γ-GCS. Furthermore, research data suggest that resveratrol increase Nrf2 expression upregulate antioxidant/phase II enzymes that reduced the expression of heat shock protein (HSP) [35]. Resveratrol has a potent cytotoxic effect on cancer cells. The evidence is supported by transglutaminase type II inhibition in cholangiocarcinoma and gallbladder carcinoma [36].

22.2.3.5 Genistein

Genistein is a type of isoflavone found in soybean that has structural similarity with mammalian estradiol. Genistein reported for antioxidant and anticancer activity. Genistein induces the expression of antioxidant enzymes, manganese superoxide dismutase and catalase, which may associate with AMP-activated protein kinase (AMPK) and phosphate. The anticancer activity of genistein is observed in a variety of cancers such as breast, ovary, prostate, urinary bladder, colon, liver, and

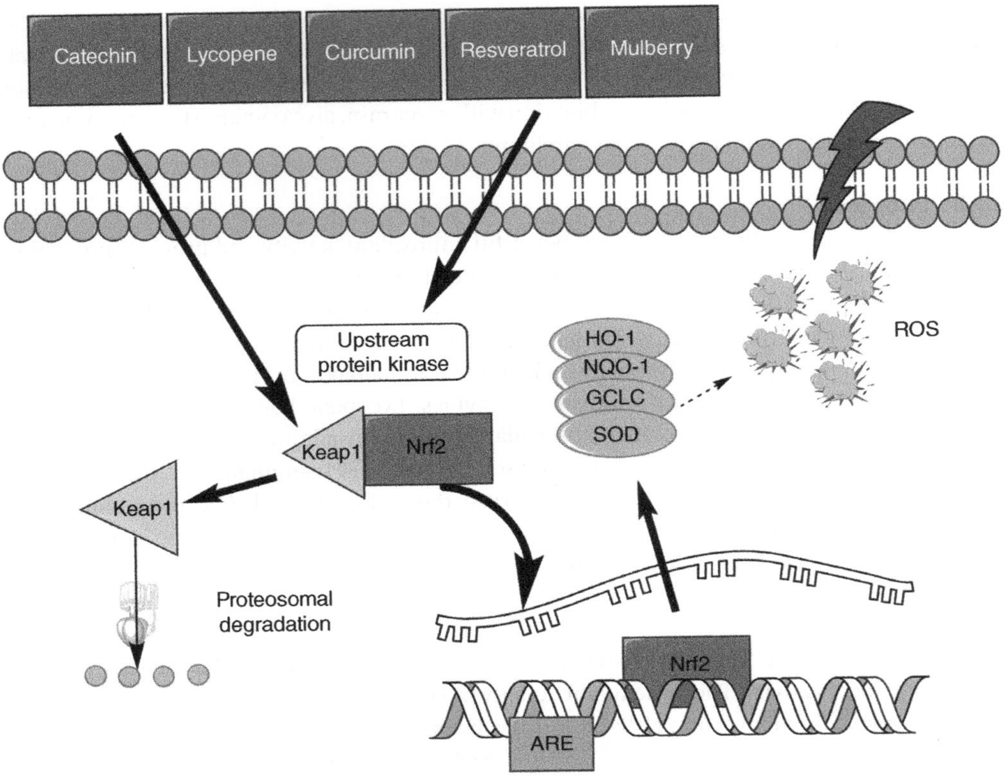

Figure 22.2 The potential molecular mechanism of catechin, lycopene, curcumin, resveratrol, and mulberry leaf modulating Nrf2/ARE pathway and its downstream antioxidant protein. Nrf2/ARE, nuclear factor (erythroid-derived 2)-like 2/antioxidant response element.

stomach. In several studies, it has been demonstrated that dietary soy supplementation reduces the inflammatory process related to prostate carcinogenesis [37, 38]. Apart from this, many of the phytoantioxidants have been tested for anticancer activity and some of those molecules are currently used in clinical practice. Still, rigorous studies and preclinical investigations are needed to clarify their potential chemopreventive and anticancer activities (Figure 22.2).

22.2.4 Antioxidant Activity of Nutraceuticals

Nutrition and genomics both have significant effects on human health. Dietary intake and nutrition have the ability to create genetic polymorphism which leads to the development of various types of diseases. Today, nutrition research is not restricted to the effect of nutrition on health but also focusing on the genomic level to establish the role of nutrition in gene modification. The advancement in technology in association with the different scientific applications has created an ample opportunity to understand how nutrients modulate gene and protein expressions.

Nutrition may have the ability to modify gene expression either by directly binding with nuclear receptors or increasing epigenetic modification. Apart from this, nutrition also has various functional applications such as coenzymes in biological reactions, antioxidants, and energy sources and can be converted into bioactive products. Dietary factors have the ability to interact with multiple biological processes. Nutrition interacts with genes and alters the function through epigenetic changes, with the influence of ligands for nuclear receptors, modifying proteins and other metabolic changes. Alteration in the gene that encodes the protein, which is involved in the above process, affects both the production of protein and its function. Genetic modification leads to modification in nutrition value that may be responsible for the development of various diseases like obesity, diabetes, cardiovascular disease, cancer, etc. (Figure 22.3).

Micronutrients are important for the regulation of various biological processes. The deficiency of micronutrients may cause illness as they have a specific effect on protein metabolism and gene expression. For example, ω-3 polyunsaturated fatty acids have a positive effect on cardiac arrhythmia. The effect of micronutrients depends on various physicochemical factors. Various metal ions like zinc and selenium and various vitamins like D, E, B2, B6, and B12 obtained from the nutrition are involved in various biological responses like oxidative stress, DNA metabolism, and immune response. Transcription is the main factor through which influence on gene expression occurs by nutrition. The organs that are metabolically active (like liver, intestine, and adipose tissue) act as nutrition sensors that identify nutrition changes using transcription factors by changing the level of DNA transcription of specific genes [39, 40].

Vitamin C provides an antioxidant activity by neutralizing toxic effects upon chemical reaction with singlet oxygen, ozone, and enzymatic reaction with hydrogen peroxide. Besides this, vitamin C also helps in the regeneration of carotenoids and vitamin E. Vitamin E also acts as an antioxidant and provides protection against lipid peroxidation [41]. The in vitro studies suggest that β-carotene can regenerate tocopherol from tocopheroxyl radical; resulting in carotenoid radical cation repaired by Vitamin C. The antioxidant activity of carotenoids is based on the scavenging of peroxyl radicals. The number of conjugated double bonds is correlated with quenching efficiency. The molecules like α-carotene, β-carotene, cryptoxanthin, and zeaxanthin are active quenchers of oxygen radicals due to the presence of conjugated double bonds. Lycopene, an intermediate product of the biosynthetic pathway of carotenoids, acts as a scavenger of ROS [19, 42].

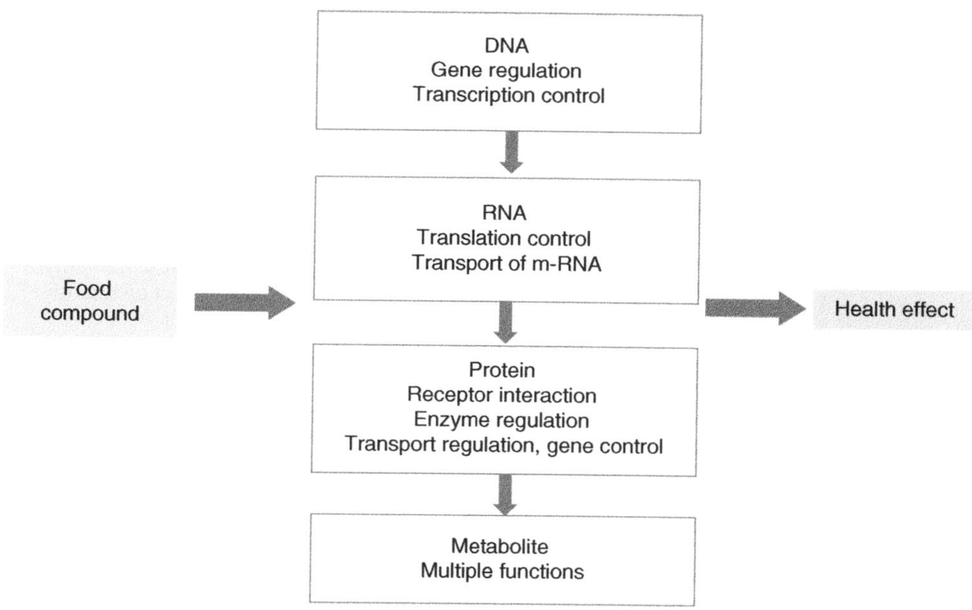

Figure 22.3 Nutritional genomics.

22.2.5 Clinical Evidence of Phytochemicals

Exogenous antioxidants have been shown to prevent the types of free radical damage that have been associated with cancer development [43]. Therefore, researchers have investigated whether taking dietary antioxidant supplements can help lower the risk of developing or dying from cancer in humans. Many observational studies, including case-control studies and cohort studies, have been conducted to investigate whether the use of dietary antioxidant supplements is associated with reduced risks of cancer in humans. Overall, these studies have yielded mixed results. Because observational studies cannot adequately control for biases that might influence study outcomes, the results of any individual observational study must be viewed with caution (Table 22.2).

Table 22.2 Phytoantioxidants in clinical trial.

S. No.	Trial name	Dose	Types of cancer patients	Results
1.	Linxian	β-carotene (15 mg) + α-tocopherol (30 mg) + selenium (50 μg) daily	Oesophageal and gastric	Initial: no effect on the risk of developing either cancer; decreased risk of dying from gastric cancer only. Later: no effect on the risk of dying from gastric cancer Later: no effect on the risk of dying from gastric cancer
2.	α-tocopherol β-carotene cancer prevention study	β-carotene (20 mg) + α-tocopherol (50 mg) daily	Middle-aged male smokers	Initial: increased incidence of lung cancer for those who took beta-carotene supplements Later: no effect of either supplement on the incidence of urothelial, pancreatic, colorectal, renal cell, or upper aerodigestive tract cancers
3.	Carotene and retinol efficacy trial	15 mg beta-carotene and 25 000 international units	People at high risk of lung cancer because of a history of smoking or exposure to asbestos	Initial: increased risk of lung cancer and increased death from all causes – trial ended early Later: higher risks of lung cancer and all-cause mortality persisted; no effect on the risk of prostate cancer
4.	Physicians' Health Study I (PHS I), United States	Beta-carotene supplementation (50 mg every other day for 12 years)	Male physicians	No effect on cancer incidence, cancer mortality, or all-cause mortality in either smokers or nonsmokers
5.	Women's Health Study (WHS), United States	Beta-carotene supplementation (50 mg every other day), vitamin E supplementation (600	Women ages 45 and older	Initial: no benefit or harm associated with two years of beta-carotene supplementation Later: no benefit or harm associated with two years of vitamin E supplementation

Table 22.2 (Continued)

S. No.	Trial name	Dose	Types of cancer patients	Results
6.	Supplémentation en Vitamines et Minéraux Antioxydants (SU.VI.MAX) Study, France	Daily supplementation with vitamin C (120 mg), vitamin E (30 mg), beta-carotene (6 mg), and the minerals selenium (100 μg) and zinc (20 mg) for a median of 7.5 years	Men and women	Initial: lower total cancer and prostate cancer incidence and all-cause mortality among men only; increased incidence of skin cancer among women only Later: no evidence of protective effects in men or harmful effects in women within five years of ending supplementation
7.	Heart Outcomes Prevention Evaluation–The Ongoing Outcomes (HOPE–TOO) Study, International	Daily supplementation with alpha-tocopherol (400 IU) for a median of seven years	People diagnosed with cardiovascular disease or diabetes	No effect on cancer incidence, death from cancer, or the incidence of major cardiovascular events
8.	Selenium and Vitamin E Cancer Prevention Trial (SELECT), United States	Daily supplementation with selenium (200 μg), vitamin E (400 IU), or both	Men aged 50 and older	Initial: no reduction in the incidence of prostate or other cancers – trial stopped early Later: more prostate cancer cases among those who took vitamin E alone
9.	Physicians' Health Study II (PHS II), United States	400 IU vitamin E every other day, 500 mg vitamin C every day, or a combination of the two	Male physicians aged 50 years and older	No reduction in the incidence of prostate cancer or other cancers

22.3 Nanotechnology in Cancer Diagnosis and Target-Specific Treatment as Phyto-Nanoformulations

22.3.1 Redox-Based Combinational Nano Therapy: A Novel Anticancer Strategy

Oxidative stress is termed as the involvement of ROS in physiological and pathophysiological phenomena [44, 45]. The main source of ROS levels in cells is dysregulation in the functioning of mitochondria [46]. However, high levels of ROS are responsible for the alteration of the redox status of cancer cells. This strategy can be employed as a novel approach for the phyto-nanotherapeutics as an anticancer drug. Recently, two opposite therapeutic approaches to redox status modulation have been clinically tested. Firstly, to reduce the levels of ROS in cancer cells, the antioxidants have been used and, consequently, there is a cell cycle arrest and, subsequently, the apoptosis of tumor cells occurs. In contrast, to increase the ROS level in tumor cells, the prooxidative approach is used to increase cell survival, which results in increasing the apoptotic rate. These two strategic approaches imbalance of the antioxidant and prooxidative system along with high ROS levels in tumor cells manifestation into the detrimental oxidative stress. This stress level impaired the defence mechanism of tumor cells, which leads to the cytotoxicity of cells and ends in necrosis [47].

Healthy cells maintain the increased ROS level through their antioxidant system. Chemotherapeutical drug-like anthracyclines can serve as a prooxidant. However, healthy and stromal cells are damaged by these compounds, which results in serious adverse reactions. Hence, to overcome this challenge, Mn-SOD, Cu-phenanthroline, Se-quinone-based triazoles, or iridium conjugate are examples of molecules/catalysts being potential anticancer drugs as nanotherapy [48–51]. The NPs less than 100 nm are the most suitable carrier system for the specific targeted drug delivery to the mitochondria as anticancer therapy [52]. Recently, pH-sensitive doxorubicin (DOX)-loaded PEGylated gold nanoparticles targeted the glioma cancer cells resulting in the rapid release of DOX at low pH and increase the ROS level [53]. Cerium NPs (nanoceria) have redox modulating and enzyme-like promising action, which acts as prooxidants or antioxidants depending on the microenvironment of tumor cells [54–58].

There are plenty of approaches available as anticancer therapy mainly focused on tumor cells, which answer the question of how the anticancer agents manifest the tumor cell and metastasis [59–62]. The tumor stroma interaction, which is known as communication between the tumor cell and tumor microenvironment, is neglected during these answers [63–65]. This stroma and its components play active participation in tumor invasion, angiogenesis, drug resistance, and metastasis [66, 67]. Besides these components, tumor pericytes, the extracellular matrix (ECM), soluble matters, tumor-associated macrophages, and various other cells, i.e. immune, endothelial, and fibroblasts are also associated with the tumor cells [68]. In present days, the conventional options for the treatment of cancer include chemotherapy, radiotherapy, and surgery. These traditional therapies result in very harmful effects on non-tumor cells (healthy), which alter the normal behavior of these cells and consequences in the late secondary cancer.

The ECM challenged the delivery of nano-chemotherapeutics to target tumor stroma. After the eruption, the stiff ECM is represented as a barrier to the diffusion of nano chemotherapeutics in the interstitial space. Particle size is the dependable factor for the diffusional movement of nanoparticles. The particle size is restricted to less than 20–40 nm permissible to diffuse across the solid tumor environment and penetrate easily. Interstitial space and its tortuosity in tumor cells increase the diffusion path length of NPs. The abundant amount of collagen alters the hydraulic conductivity, which reduces the convection flow in the interstitium [69]. Hence, enzymes such as collagenase and hyaluronidase are employed to enhance the dispersion of NPs through the intertumoral matrix via the disruption of this matrix. However, collagenase was found more efficient in the formulation of NPs as compared to hyaluronidase [70]. Cyclooxygenase-2 (COX2) is highly expressed at the tumor site. Hence, celecoxib, a selective COX2 inhibitor, can enhance the localization of paclitaxel (PTX)-loaded polyethene glycol-polylactide (PEG-PLA) micelles in the human lung cancer cells (A549) xenograft mice model through modulation of the tumor microenvironment [71].

22.3.2 ROS-Scavenging Inorganic Nanoparticles

Oxidative stress is referred to as the generation of ROS and reactive nitrogen species (RNS) as by-products of metabolic processes and environmentally induced factors. Both these reactive species cause damage to DNA, proteins, cell membranes, and lipids which ultimately results in signal nucleotide polymorphism, apoptosis, and necrosis. Antioxidants primarily functioned to inhibit and restore the damage caused by ROS and RNS, which might be the main causes of various ailments such as metabolic syndrome, cancer, cardiovascular disease, neurodegenerative disorders, and inflammatory diseases. Antioxidants mainly function through either prevention or scavenging of ROS/RNS [72–74]. There are plenty of natural antioxidants available, which mainly function as scavengers of free radicals while other antioxidants neutralize the ROS/RNS and make them unharmful [75].

A wide variety of inorganic NPs have displayed the scavenging of free radicals and preventing the ROS/RNS in vitro and in vivo systems and ameliorating oxidative stress. This potential application of inorganic NPs has been relevant to its redox and catalytic action, electronic charges, high surface to volume ratio and compatibility with the biological system [76].

22.3.2.1 Transitional Metal NPs

The transition metal oxide NPs demonstrated the antioxidant properties through the aforementioned mechanism and its repetition. Transition metal oxide state TiO_2 to ZnO can possess catalytic activity with minimal typical transitional issues. Hence, these metal oxide NPs are being employed as antibacterial, anticancer, anti-inflammatory, and against neurodegeneration [77–79]. The TiO_2-CeO_x NPs mimic catalytic activity like the peroxidase enzyme by switching redox of Ce^{4+} to Ce^{3+} [80]. Similarly, Fe_3O_4 NPs also displayed the catalytic activity as peroxidase-like enzymes through oxidation of a series of substrates H_2O_2 in cells [81]. Vanadium pentoxide (V_2O_5) nanowires demonstrated potent antioxidant activity via catalyzing the destruction of H_2O_2 in association with reduced glutathione in the biological system [82]. Copper-ZnO and Co_3O_4 NPs also possess peroxidase-like activity against the substrates in presence of H_2O_2 [83, 84]. Iridium-phosphorescent transition metal conjugated with mesoporous silica NPs can sense oxygen concentrations and demonstrate a potential agent for mitochondrial-targeted anticancer along with photodynamic therapy [85, 86]. Polymeric dots with iridium phosphorescent complex are employed in radiometric imaging and time-gated luminescence imaging for cancer. The limitation of the above-mentioned complex is poor water solubility. Alternatively, the ruthenium complex with mesoporous silica NPs has high water solubility and intense phosphorescence emission [87].

22.3.2.2 Gold Nanoparticles (AuNPs)

Gold nanoparticles (AuNPs) possess typical characteristics like electrical, optical, and biochemical sensing properties. Hence, AuNPs are significantly explored for the possible application in photodynamic imaging for the diagnostic utility and tumor therapy. A gold nanoparticle is a widely investigated NP having typical therapeutic activity, inert, and nontoxic. Hence, they attract many scientists and researchers from different fields to explore various applications of AuNPs in various fields [88]. AuNPs are synthesized from various synthetic schemes and a variety of shapes and possess antioxidant activity. Morphologically, AuNPs are synthesized in various shapes like nanosphere, nanocages, nanocluster, nanorod, nanocube, nanostar, and nanoshell ranging from 1 to 100 nm along with multiple diagnostic and therapeutic utility.

AuNPs are ideally adhered to on the surface or conjugated with natural antioxidants to establish antioxidant or antiradical activity in tumor cells [89, 90]. For instance, AuNPs-cellulose fiber shows more significant DPPH scavenging activity than pure fiber in both light and dark conditions because of the catalytic activity of AuNPs [91]. Salvianolic acid A-polyethylene glycol-coated AuNPs showed nine times more significant antiradical activity than salvianolic acid A alone. Phyto antioxidants functionalized with AuNPs own significant kinetic effect on ROS scavenging action via O_2-centered free radicals and ROS in living cells [89]. AuNPs functionalized with thiol ligand in Trolox showed a significant antioxidant as compared to Trolox monomer, the mechanism behind the enhanced antioxidant activity is a well-organized assembly of Trolox on the surface of AuNPs [92]. The multiple phyto antioxidants therapy of selenium methyl selenocysteine, 3,6-dihydroxyflavone, and lutein embedded with AuNPs showed significant free radical scavenging activity against single phytoconstituents [93]. Natural antioxidant GA conjugated with AuNPs enhanced the inhibition of cervical cancer cells proliferation as compared to GA monotherapy through induction of the apoptosis process [94]. Recently, curcumin-capped AuNPs conjugated with reduced graphene oxide nanocomposite showed prominent antioxidant and anticancer action against colon and liver cancer cells [95]. AuNPs-capped *Cassia tora* leaf extract enhanced bioavailability, antioxidant, and anticancer activity against colon cancer cells [96]. Epigallocatechin-3-gallate conjugate with AuNPs possesses prominent anticancer and free radical scavenging activity against mouse bladder tumor cell lines (MBT-2) [97].

22.3.2.3 Silver Nanoparticles (AgNPs)

Since the last decade, AgNPs catch the attention due to their remarkable applications in the field of medical science. Recently, AgNPs are synthesized through the green synthesis approach, which makes them nontoxic and eco-friendly capping agents [98]. For instance, lignin, *Sida cordifolia* extract, *Clerodendrum phlomidis* extract, seabuckthorn extract, and PVA-capped green synthesized AgNPs showed remarkable free radical scavenging antioxidant activity [99–101].

22.3.2.4 Iron Oxide Magnetic Nanoparticles (Fe_2O_3NPs)

Recently, iron oxide magnetic nanoparticles (MNPs) have been popular because of their multifunctional properties such as superparamagnetic property for MRI, noninvasiveness for drug targeting and biocompatibility for drug carriers. The Fe_2O_3NPs sizes ranging from 1 to 100 nm have been found potential for their applicability in various scientific fields due to their high surface volume ratio and magnetic properties [102]. Antioxidant and anticancer properties of Fe_2O_3NPs have already been explored in accordance with free radical neutralization [103]. Tailor-made Fe_2O_3NPs with natural antioxidants and with curcumin was evaluated for antioxidant and cytotoxicity studies for the targeted drug delivery [104]. Magnetite (Fe_3O_4) and maghemite (Fe_2O_3) are termed superparamagnetic iron oxide nanoparticles (SPIONs), which have been potentially employed for targeted drug delivery with induction of magnetic force. Functionalized SPIONs overcome the aggregation and prevent the oxidation of MNPs and provide the surface to conjugate natural antioxidants and targeting molecules. Different anticancer drug molecules, proteins, antibodies, and peptides are easily embedded on the surface of SPIONs and facilitate the specific receptor targets that are expressed on tumor cells.

22.3.2.5 Quantum Dots

Quantum dots (QDs) are synthesized from the group II and group VI molecules like CdS, CdTe, and ZnS. These QDs have fluorescent properties which make them suitable for drug delivery and cellular imaging application. Two approaches such as bottom-up or top-down are implicated in the synthesis of QDs. A majority of QDs are composed of three layers with 2–10 nm in diameters of core from semiconductor material CdSe covered with another semiconductor ZnS. Lastly, these two layers are covered with various materials for encapsulation. Because of their size and quantum effect, QDs typically are used to visualize tumor in a real-time manner during the drug release phenomena.

22.3.2.6 Carbon Nanotubes

Carbon nanotubes with the size range of 50–100 nm are suited ideally for the biological system in context with structure, purity, surface charge, size, and chemistry. These nanotubes are designed as a carrier for targeted cancer therapy drugs through encapsulation or adsorption of anticancer agents on the surface [105].

22.4 Recent Advances in Phyto Antioxidant-Based Nanoformulations in Cancer

Cancer is a major health issue leading to morbidity and mortality worldwide despite growing advances and preventive strategies. The abnormal proliferation of the cells which can be controlled or stopped is the main characteristic of cancer [106]. In addition, the cells of a malignant tumor have the tendency to attack tissues adjacent to it. Common types of cancers are skin cancer, breast cancer, colon cancer, rectal cancer, liver cancer, lung cancer, pancreatic cancer, and prostate cancer. The non-pharmacological treatments of cancer include radiation, removal of tumor with surgery, stem cell therapy, and hyperthermia. The pharmacological treatment of cancer includes hormone therapy, immunotherapy, chemotherapy or a combination of all. Among these, the major obstacle to chemotherapy is MDR. Continuous doses of chemotherapeutic agents reduce efflux outside the cell which hampers permeation across the cell membranes. To understand the permeation of cell across the cell, various nanocarriers are used in the treatment of cancer.

Nanotechnology created the most important carrier system for the delivery of targeted material to the site of action. The active drug molecule is encapsulated or enclosed in the carrier system in such a way that it protects the drug from degradation as well as improves stability [107]. The nanocarrier system stays in the blood circulatory system for a prolonged period of time. The structure penetrates the tissue system, facilitates easy uptake of the active molecule by the cells, permits an efficient drug delivery system, and ensures action at the targeted site. The nanocarriers could be surface functionalized or grafted using biomolecules, which are able to find active targets for delivery. The various cell barriers can be crossed by the surface-engineered nanoformulations. The drugs having poor solubility and low absorption can be targeted to form nanoparticles [108].

Many years ago, phytomedicines have been widely explored for the treatment of different types of cancer, which provides alternative therapy to conventional treatment. The major advantage of using phytochemicals as anticancer agents is that they do not show any deleterious effect during the course of treatment. Phytochemicals are found in fruits, vegetables, plants, dietary fibers, and natural products, which can regulate many molecular processes. Plant-based phytochemicals like oils, gums, terpins, saponins, minerals, glycosides, flavonoids, lignins, podophyllotoxins, polyphenolic compounds, vinca alkaloids, and taxol constitute primary and secondary metabolites, which play a major role in the inhibition. Phytochemicals do not show any toxic effect on normal cells. The phytochemicals act by inhibiting various pathways like CDC2, CDK2, CDK4 kinases, topoisomerase, mTOR, TNK, PI3K, Akt, Bcl-2, Bcl-1, cytokines, and interleukins. Phyto antioxidants, which are readily taken up by the healthy tissue, may be recognized as supplements. Phytochemicals are isolated from plant sources and they have low solubility, which directly affects the bioavailability. Phyto antioxidant has a larger volume of distribution and also leads to accumulation in the cell. Moreover, the problem related to the MDR can be solved by the use of phyto antioxidants [3]. To overcome shortfalls related to chemotherapy, nanotherapeutics are being implemented. Nanotechnology-based phyto antioxidants have numerous advantages like passive diffusion of the molecule across the biological membrane, improved permeation, improved bioavailability, site-specific drug delivery, protection from biological and environmental degradation, controlled drug delivery, etc. Many phyto antioxidants can be entrapped in the biodegradable polymers, which can increase their stability, cell uptake, and absorption [109] (Table 22.3).

Table 22.3 Phytochemicals used in cancer [110].

General plant name or family	Phyto antioxidants	Cancer type
Curcuma longa	Curcumin	Colon cancer
Soybean	Genistein	Skin cancer, breast cancer
Thyme, basil, rosemary	Ursolic acid	Skin cancer
Mulberries, peanuts, grapes	Resveratrol	Skin cancer
Onion	Quercetin	Lung cancer, breast cancer
Sophora flavescens	Matrine	Bladder cancer

Source: Based on Ref. [110].

The phytoconstituents act via various mechanisms. The phytochemicals like chrysin, silymarin, resveratrol, and kaempferol inhibit the cellular function through G1, G2, M-phase arrest, and cell division. Programmed cell death via apoptotic or antiapoptotic activity was proved by many plant-based antioxidants like ursolic acid, quercetin, kaempferol, and fisetin. The polyphenolic compounds act through multiple mechanisms [111]. It can act by reducing oxidative stress, inflammation, cell division, angiogenesis, and metastasis. For instance, Fisetin downregulates or upregulates certain types of apoptotic or antiapoptotic proteins. It can be combined with other drugs for colon cancer and lung cancer. These phytochemicals can be formulated as polymeric nanoparticles, liposomes, micelles, dendrimers, gold nanoparticles, solid lipid nanoparticles, and carbon nanotubes.

22.4.1 Polymeric Nanoparticles

Nanoparticles are ultrafine microscopic particles with a size of less than 100 nm. They are further classified as nanospheres or nanocapsules. Nanospheres constitute the matrix system where the drug is dispersed in the polymer matrix and nanocapsules are the reservoir-based system in which the drug is enclosed in the inner cavity of the polymeric membrane. Nanoparticles are obtained from natural sources and their surface can be modified with polyethylene glycol to improve the cellular uptake. Several polymers such as poly(lactide-co-glycolide), (PLGA), PLA, polyglycolide, polycaprolactone (PCL), chitosan, PLGA-polyethylene glycol have been developed for passive and ligand-targeted delivery of active ingredient [112]. Various phyto antioxidants are used in combination with different polymers. Several findings suggest that quercetin has proven anticancer activity against C6 glioma cells. Polylactic glycolic acid (PLGA) is used as a carrier for loading quercetin. Curcumin-loaded nanoparticles demonstrated a sustained release property and enhanced cellular bioavailability of curcumin, and further decreased cell viability in in vitro studies. The method of preparing nanoparticles includes solvent evaporation method, nanoprecipitation, salting out, supercritical fluid, and dialysis.

22.4.2 Liposomes

Liposomes are lipid vesicles, which consist of a lipid bilayer as an outer membrane and an aqueous vehicle enclosed in the inner membrane. The internal core serves as a carrier for drugs, protein, vaccines, and enzymes delivered to the targeted site. The liposomal delivery of betulinic acid enhances the penetration across the cell membrane and shows anticancer activity. Icariin-loaded liposomes formulated by anti-solvent evaporation technique showed potent cytotoxic activity against ovarian cancer. Liposomes of curcumin were found to be active against breast cancer cells, adenocarcinoma, and other tumor cells. Curcumin-loaded liposomes act by targeting MDF-7 and SKRB3 cells [113]. Further, the in vivo and in vitro effects of curcumin on proliferation, apoptosis, signaling were carried out on the human pancreatic cells. The curcumin was found to suppress its binding and decreased expression of this protein complex regulated gene products including COX2-linked immunoassay, both of which can be implicated in tumor growth/invasiveness. The liposomes containing resveratrol and curcumin in order to increase the bioavailability of curcumin were developed. The study was conducted on the mice by taking serum and prostate tissue. The concentration of the combination was found to be increased as compared to the curcumin alone. Quercetin liposomes were investigated to have an antiproliferative effect in numerous tumor cells like breast, leukemia, colon, ovary, squamous cell, endometrial cells, and non-small lungs [114]. The inhibition of tyrosine kinase was shown by the parenteral preparation of quercetin. The combination of quercetin and vincristine into the liposomal formulation exhibited effectiveness against the tumor growth of breast cells. The resveratrol in the pegylated state with and without glycine was evaluated for its anticancer activity. The results showed increased drug-loading efficiency without glycine.

22.4.3 Solid Lipid Nanoparticles

Solid lipid nanoparticles consist of solid and lipid phases in the formulation. The particle size of solid lipid nanoparticles ranges from 10 to 1000 nm. They are stabilized by salts, sterols, surfactants, phospholipids, etc. The nanoparticles fabricated from biocompatible polymers can incorporate hydrophilic and lipophilic drugs in their structure. The trans-resveratrol coated with chitosan has shown effectiveness for target cells HT-29 in colon cancer when combined with folic acid. The combination of piperine and quercetin shows a cytotoxic effect against caco-2 and C26 cells. The myricetin-loaded nanoparticles showed a cytotoxic effect against adenocarcinoma cancer cells. The encapsulation efficiency directly inhibits the growth of cancer cells. Sesamol is the phenol component of the sesame oil and it is produced due to hydrolysis of sesamolin during thermal oxidation. Sesamol is the phenolic component of sesame oil whose antioxidant activity is attributed to the presence of the benzodioxole group in its ring structure [115]. Solid lipid nanoparticles of sesame oil prepared using lecithin and polysorbate 80 showed a hepatoprotective effect.

22.4.4 Nano Micelles

If the solvent is polar, the hydrophilic portion of the molecule orients toward the outer surface to maximize contact with a polar solvent while the hydrophobic parts are clustered in the core region to minimize the contact with a polar solvent. This kind of molecular arrangement is referred to as normal micelles. Alternatively, the assembly is formed in the nonpolar solvent, which is known as reverse nano micelles. Insoluble drugs are loaded in the normal nano micelles while the soluble drugs are loaded in the reverse nano micelles. Here hydrophobic part of the copolymer provides the core, which allows encapsulation of drugs, whereas the hydrophilic portion of the polymer forms a shell of the micelles. This property prevents its uptake by the reticuloendothelial system and thereby prolongs its circulation into the bloodstream. Polymer derives nano micelles showing greater stability and lower critical micellar concentration value as compared to surfactant-based nanoparticles. Micelles of crosslinked N-isopropylacrylamide with N-vinyl-2-pyrrolidone and PEG monoacrylate encapsulate curcumin. These nanoparticles demonstrated better efficacy comparable to that of pure curcumin against human pancreatic cancer. Marked improvement in the in vivo antitumor efficacy with two times decrease in the tumor growth was shown by folate conjugated micelles as compared to the non-targeted micelles. Folate-anchored pluronic-P105 and L-101 were investigated as micellar carriers for the delivery in the human breast cancer cells [116]. It showed enhanced internalization of the entrapped therapeutic ingredient. Chitosan quercetin nano micelles, which demonstrated a strong release profile in the gastrointestinal environment, have shown rapid cellular uptake as compared to the free drug [117]. The polymeric micelles of luteolin with PEG-PCL by self-assembly method showed increased bioavailability in the C-26 colon carcinoma cells.

22.4.5 Dendrimers

A dendrimer is a nano-sized repetitively branched, tree-like structure comprising a central core, interior branches, and exterior multifunctional surface groups. The role of branches is to accumulate as many as possible molecules like proteins, amino acids, etc. The functional groups like carboxyl, amine, or hydroxyl-terminated can be easily incorporated with the help of surface functional groups [118]. Dendrosome is an amphipathic and biodegradable that has been used in the delivery of genes into cell lines. Poly (glycerol-succinic acid) dendrimers exhibited potential carriers for camptothecin. Increased cytotoxicity was observed in the human breast adenocarcinoma, colorectal carcinoma, and glioblastoma. The ursolic acid and folic acid conjugated with polyamidoamine (PAMAM) dendrimer can target HepG2 cells and increase cellular uptake. Dendrimers formulated by using GA with PAMAM exhibited an enhanced cytotoxic effect on MCF-7 breast cancer cells [119].

22.4.6 Gold Nanoparticles

Gold is one of the noble metals which is known for its unique optical properties and this popular phenomenon is known as localized surface plasmon resonance. Gold can easily make bonds with thiol and amine groups so that it would be preferable for gold to be chosen as a carrier. Gold nanoparticles can be easily tagged with ligands for site targeting. Biocompatibility, surface property, unique size, and shape characteristics with internalization has shown the capability of different gold nanoparticles in retinal pigment epithelium cells [120]. The resveratrol-coated gold nanoparticles showed better effect in diabetic retinopathy with the size of 20 nm. Curcuma wenyujin-loaded gold nanoparticles showed good anticancer activity against breast cancer. Gold nanoparticles from ethanolic extract of plectranthus amboinicus showed their anticancer effect in adenocarcinoma of lung cancer [121].

22.4.7 Polyphenolic Nanoparticles

The nano delivery system of polyphenolic compounds is widely used as an anticancer agent. The polyphenolic compounds perform a key role in the suppression of ROS. The ROS activity in the cancerous cells is higher and polyphenolic compounds inhibit ROS activity along with oxidative stress regulation. The PLGA nanoparticles consist of different amounts of lactic acid and glycolic acid. They are biocompatible, biodegradable, and hydrophobic in nature. Nanoformulations of polyphenols have been prepared using a wide variety of techniques that use carriers such as liposomes, nanocapsules, dendrimers, nanosheets, micelles, etc. Polyphenol nanoformulations perform their anticancer activity via several cellular mechanisms, cell cycle arrest at different phases of cancer, activation of caspase enzymes, reduction of tumor vascularization, reducing tumor cell invasion, and metastasis. Apoptosis is the major indication of anticancer therapy. Many polyphenolic drugs exert their action through apoptosis, which is a programmed cell death for the removal of unwanted cells [122]. For higher bioavailability, better antineoplastic activity and to improve passive targeting to the cells, polyphenolic nanoformulations are widely used.

22.4.8 Carbon Nanotubes

Carbon nanotubes consist of rolled graphite sheets having wider length, narrow diameter, and larger surface area, which can accommodate the active pharmaceutical ingredient. The surface modification is possible due to the presence of oxygen-containing functional groups. The property of carbon nanotubes is that they absorb near-infrared light waves and pass harmlessly through the cells. The polyphenolic compounds can be easily loaded into the walls of nanotubes. Curcumin was loaded successfully in the nanotubes having a modified surface with phosphatidylcholine and polyvinylpyrrolidone. The lipid and polymeric coating increase the stability of curcumin-loaded nanoparticles. Multi-walled carbon nanotubes using folic acid as a carrier show increased bioavailability, improved aqueous solubility, and high ability of cell targeting (Table 22.4).

22.4.9 Magnetic Nanoparticles

MNPs have gained immense attention in the treatment of cancer due to their physicochemical properties, magnetic resonance contrast, easy surface decorations, low toxicity, and good biodegradability. MNPs have been developed using nickel, cobalt, Prussian blue, and gadolinium.

Table 22.4 Various polyphenolic phytoconstituents used in cancer therapy [122].

Phytoconstituent category	Phytoconstituent used	Nanoformulations	Targeted cells	Anticancer activity
Flavonoid	Baicalein	Targeted ligands of folate and hyaluronic acid	Human lung cancer cells	Decreases cell viability
	Chrysin	Nanosuspension	Human hepatocellular carcinoma	Decreases cell growth
	ECEG	Nanoparticles	Human melanoma	Increases apoptosis, Bax, Caspase-9
	Quercetin	Liposomes	Human breast cancer cells	Increased apoptosis
Flavonolignan	Silibinin	Pegylated nanoliposomes	Human hepatocellular carcinoma	Decreases cell viability
Stilbene	Resveratrol	Pegylated nanoliposomes	Human head and neck cancer	Anticancer effect at high concentration
Naphthoquinone	Plumbagin	Silver nanoparticles	Human skin cells	Increased apoptosis by production of free radicals

Source: Based on Ref. [122].

MNPs consist of a magnetic core-shell and a polymer coating. The stability is increased by surface coating and functionalization [123]. Recently, for the synthesis of nanoparticles, microfluidic systems are provided, which have advantages of screening, automation, increased control over reaction time, temperature, and concentration. Tumor ablation therapy, which includes magnetic hyperthermia, photothermal therapy, and photodynamic therapy showed an outstanding effect in the treatment of cancer. Increased accumulation and increased magnetic relaxivity using folic acid on the mice placed in an external magnetic field showed a reduction in the tumor size. Gold and silver metals are extensively used in the regulation of cancer immunity. Nanoparticles formed using abundant 6-gingerol and 6-shogaol from ginger showed ideal tissue stability and cancer therapy potential in the mice [124]. The nanoparticles isolated from the citrus lemon are effective in inhibiting the growth of tumor in the leukaemia mice model.

22.5 Future Perspective and Conclusion

Oxidative stress is responsible for the development of a variety of diseases including cancer. The use of antioxidants is to subside the deleterious effect of free radicals in the living system. Synthetic antioxidants are useful to improve health but at the same time, the adverse effect of synthetic compounds is a serious issue. Natural products can provide a solution for the said purpose. There are many natural herbs/phytochemicals, which have been reported for potential antioxidant activity with few or no side effects. However, the application of phyto antioxidants is restricted due to many reasons, inability to cross the blood–brain barrier, which, in turn, has rendered them less selective in many neurodegenerative diseases. Due to the aforementioned limitations, the use of natural and synthetic antioxidants is less preferable.

Conventional chemotherapy has plenty of limitations and to overcome shortfalls related to chemotherapy, nanotherapeutics are being implemented. Inorganic and organic nanoparticles are widely used in various drug delivery systems. Nanotechnology-based phyto antioxidants have numerous advantages like passive diffusion of a molecule across the biological membrane, improved permeation, improved bioavailability, site-specific drug delivery, protection from biological and environmental degradation, controlled drug delivery, etc. Nanophytoantioxidants therapy has given many positive outcomes that include targeted drug delivery, greater cellular free radical scavenging activity, and higher sensitivity.

Despite various advantages of nanoparticles, the complexity of the biological environment impacts the functionality of nanomaterials, which restricts their effective use for therapeutic treatment. The drawbacks of current nanomedicine are mainly due to limited absorption. The lower absorption of nanomedicine demands higher dosing and due to this chance of adverse events may increase. To overcome this limitation, thorough knowledge and understanding are required. Moreover, detailed toxicological screening is also required for the metal nanoparticles and insoluble nanoparticles.

The use of novel technological formats using machine learning and integrated analytical systems may provide an innovative solution to improve the specificity and selectivity of targeted therapy. Artificial intelligence demonstrates significant potential in "closing the loop" of nanoparticle synthesis, characterization, refinement, and testing or predicting activity in vitro and in vivo. The journey initiated from experimental laboratories is still lined with multiple challenges and obstacles along the way.

References

1 Shewach, D.S. and Kuchta, R.D. (2009). Introduction to cancer chemotherapeutics. *Chemical Reviews* 109 (7): 2859–2861.

2 Sung, H., Ferlay, J., Siegel, R.L. et al. (2021). Global cancer statistics 2020: GLOBOCAN estimates of incidence and mortality worldwide for 36 cancers in 185 countries. 71 (3): 209–249.

3 Lagoa, R., Silva, J., Rodrigues, J.R., and Bishayee, A. (2020). Advances in phytochemical delivery systems for improved anticancer activity. *Biotechnology Advances* 38: 107382.

4 Hayes, J.D., Dinkova-Kostova, A.T., and Tew, K.D. (2020). Oxidative stress in cancer. *Cancer Cell* 38 (2): 167–197.

5 Fiaschi, T. and Chiarugi, P. (2012). Oxidative stress, tumor microenvironment, and metabolic reprogramming: a diabolic liaison. *International Journal of Cell Biology* 2012: 762825.

6 Gonzalez-Hunt, C.P., Wadhwa, M., and Sanders, L.H. (2018). DNA damage by oxidative stress: measurement strategies for two genomes. *Current Opinion in Toxicology* 7: 87–94.

7 Kurutas, E.B. (2016). The importance of antioxidants which play the role in cellular response against oxidative/nitrosative stress: current state. *Nutrition Journal* 15 (1): 71.

8 Bouayed, J. and Bohn, T. (2010). Exogenous antioxidants--double-edged swords in cellular redox state: Health beneficial effects at physiologic doses versus deleterious effects at high doses. *Oxidative Medicine and Cellular Longevity* 3 (4): 228–237.

9 Dias, T.R., Martin-Hidalgo, D., Silva, B.M. et al. (2020). Endogenous and exogenous antioxidants as a tool to ameliorate male infertility induced by reactive oxygen species. *Antioxidants & Redox Signaling* 33 (11): 767–785.

10 Fu, L., Xu, B.T., Xu, X.R. et al. (2011). Antioxidant capacities and total phenolic contents of 62 fruits. *Food Chemistry* 129 (2): 345–350.

11 Manganaris, G.A., Goulas, V., Vicente, A.R., and Terry, L.A. (2014). Berry antioxidants: small fruits providing large benefits. *Journal of the Science of Food and Agriculture* 94 (5): 825–833.

12 Xia, E.Q., Deng, G.F., Guo, Y.J., and Li, H.B. (2010). Biological activities of polyphenols from grapes. *International Journal of Molecular Sciences* 11 (2): 622–646.

13 Fu, L., Xu, B.T., Xu, X.R. et al. (2010). Antioxidant capacities and total phenolic contents of 56 wild fruits from South China. *Molecules (Basel, Switzerland)* 15 (12): 8602–8617.

14 Deng, G.F., Shen, C., Xu, X.R. et al. (2012). Potential of fruit wastes as natural resources of bioactive compounds. *International Journal of Molecular Sciences* 13 (7): 8308–8323.

15 Deng, G.-F., Lin, X., Xu, X.-R. et al. (2013). Antioxidant capacities and total phenolic contents of 56 vegetables. *Journal of Functional Foods* 5 (1): 260–266.

16 Deng, G.F., Xu, X.R., Zhang, Y. et al. (2013). Phenolic compounds and bioactivities of pigmented rice. *Critical Reviews in Food Science and Nutrition* 53 (3): 296–306.

17 Yang, J. (2009). Brazil nuts and associated health benefits: a review. *LWT- Food Science and Technology* 42 (10): 1573–1580.

18 Song, F.L., Gan, R.Y., Zhang, Y. et al. (2010). Total phenolic contents and antioxidant capacities of selected chinese medicinal plants. *International Journal of Molecular Sciences* 11 (6): 2362–2372.

19 Stahl, W. and Sies, H. (2003). Antioxidant activity of carotenoids. *Molecular Aspects of Medicine* 24 (6): 345–351.

20 Tan, B.L., Norhaizan, M.E., Liew, W.-P.-P., and Sulaiman, R.H. (2018). Antioxidant and oxidative stress: a mutual interplay in age-related diseases. *Frontiers in Pharmacology* 9: 1162.

21 Ma, Q. (2013). Role of nrf2 in oxidative stress and toxicity. *Annual Review of Pharmacology and Toxicology* 53: 401–426.

22 Beecher, G.R. (2003). Overview of dietary flavonoids: nomenclature, occurrence and intake. *The Journal of Nutrition* 133 (10): 3248s–3254s.

23 Sahin, K., Orhan, C., Tuzcu, M. et al. (2010). Epigallocatechin-3-gallate prevents lipid peroxidation and enhances antioxidant defense system via modulating hepatic nuclear transcription factors in heat-stressed quails. *Poultry Science* 89 (10): 2251–2258.

24 Romeo, L., Intrieri, M., D'Agata, V. et al. (2009). The major green tea polyphenol, (-)-epigallocatechin-3-gallate, induces heme oxygenase in rat neurons and acts as an effective neuroprotective agent against oxidative stress. *Journal of the American College of Nutrition* 28 (Suppl): 492s–499s.

25 Palozza, P., Catalano, A., Simone, R., and Cittadini, A. (2012). Lycopene as a guardian of redox signalling. *Acta Biochimica Polonica* 59 (1): 21–25.

26 Linnewiel, K., Ernst, H., Caris-Veyrat, C. et al. (2009). Structure activity relationship of carotenoid derivatives in activation of the electrophile/antioxidant response element transcription system. *Free Radical Biology & Medicine* 47 (5): 659–667.

27 Jiang, L.N., Liu, Y.B., and Li, B.H. (2018). Lycopene exerts anti-inflammatory effect to inhibit prostate cancer progression. *Asian Journal of Andrology* 21 (1): 80–85.

28 He, H.J., Wang, G.Y., Gao, Y. et al. (2012). Curcumin attenuates Nrf2 signaling defect, oxidative stress in muscle and glucose intolerance in high fat diet-fed mice. *World Journal of Diabetes* 3 (5): 94–104.

29 Tapia, E., Zatarain-Barrón, Z.L., Hernández-Pando, R. et al. (2013). Curcumin reverses glomerular hemodynamic alterations and oxidant stress in 5/6 nephrectomized rats. *Phytomedicine: International Journal of Phytotherapy and Phytopharmacology* 20 (3–4): 359–366.

30 Sahin, K., Orhan, C., Tuzcu, Z. et al. (2012). Curcumin ameloriates heat stress via inhibition of oxidative stress and modulation of Nrf2/HO-1 pathway in quail. *Food and Chemical Toxicology: An International Journal Published for the British Industrial Biological Research Association* 50 (11): 4035–4041.

31 de Matos, R.P.A., Calmon, M.F., Amantino, C.F. et al. (2018). Effect of curcumin-nanoemulsion associated with photodynamic therapy in cervical carcinoma cell lines. *BioMed Research International* 2018: 4057959.

32 Jalili-Nik, M., Soltani, A., Moussavi, S. et al. (2018). Current status and future prospective of curcumin as a potential therapeutic agent in the treatment of colorectal cancer. *Journal of Cellular Physiology* 233 (9): 6337–6345.

33 Danafar, H., Sharafi, A., Kheiri, S., and Kheiri, M.H. (2018). Co-delivery of sulforaphane and curcumin with PEGylated iron oxide-gold core shell nanoparticles for delivery to breast cancer cell line. *Iranian Journal of Pharmaceutical Research: IJPR* 17 (2): 480–494.

34 Calabrese, V., Cornelius, C., Mancuso, C. et al. (2008). Cellular stress response: a novel target for chemoprevention and nutritional neuroprotection in aging, neurodegenerative disorders and longevity. *Neurochemical Research* 33 (12): 2444–2471.

35 Palsamy, P. and Subramanian, S. (2011). Resveratrol protects diabetic kidney by attenuating hyperglycemia-mediated oxidative stress and renal inflammatory cytokines via Nrf2-Keap1 signaling. *Biochimica et Biophysica Acta* 1812 (7): 719–731.

36 Sundar Dhilip Kumar, S., Houreld, N.N., and Abrahamse, H. (2018). Therapeutic potential and recent advances of curcumin in the treatment of aging-associated diseases. *Molecules (Basel, Switzerland)* 23 (4): 835–848.

37 Mukund, V., Mukund, D., Sharma, V. et al. (2017). Genistein: its role in metabolic diseases and cancer. *Critical Reviews in Oncology/Hematology* 119: 13–22.

38 Kim, S.H., Kim, C.W., Jeon, S.Y. et al. (2014). Chemopreventive and chemotherapeutic effects of genistein, a soy isoflavone, upon cancer development and progression in preclinical animal models. *Laboratory Animal Research* 30 (4): 143–150.

39 Afman, L. and Müller, M. (2006). Nutrigenomics: from molecular nutrition to prevention of disease. *Journal of the American Dietetic Association* 106 (4): 569–576.

40 Chen, Y., Michalak, M., and Agellon, L.B. (2018). Importance of nutrients and nutrient metabolism on human health. *The Yale Journal of Biology and Medicine* 91 (2): 95–103.

41 Kasote, D.M., Katyare, S.S., Hegde, M.V., and Bae, H. (2015). Significance of antioxidant potential of plants and its relevance to therapeutic applications. *International Journal of Biological Sciences* 11 (8): 982–991.

42 Ames, B.N. (2018). Prolonging healthy aging: Longevity vitamins and proteins. *Proceedings of the National Academy of Sciences of the United States of America* 115 (43): 10836–10844.

43 NIH (2017). Antioxidants and cancer prevention. https://www.cancer.gov/about-cancer/causes-prevention/risk/diet/antioxidants-fact-sheet (accessed 26 August 2021).

44 Sies, H. (2015). Oxidative stress: a concept in redox biology and medicine. *Redox Biology* 4: 180–183.

45 Sies, H., Berndt, C., and Jones, D.P. (2017). Oxidative stress. *Annual Review of Biochemistry* 86: 715–748.

46 Dröse, S. and Brandt, U. (2012). Molecular mechanisms of superoxide production by the mitochondrial respiratory chain. *Advances in Experimental Medicine and Biology* 748: 145–169.

47 Gorrini, C., Harris, I.S., and Mak, T.W. (2013). Modulation of oxidative stress as an anticancer strategy. *Nature Reviews Drug Discovery* 12 (12): 931–947.

48 Jardim, G.A.M., Lima, D.J.B., Valença, W.O. et al. (2017). Synthesis of selenium-quinone hybrid compounds with potential antitumor activity via Rh-catalyzed C-H bond activation and click reactions. *Molecules (Basel, Switzerland)* 23 (1): 83–97.

49 Laws, K., Bineva-Todd, G., Eskandari, A. et al. (2018). A copper(II) phenanthroline metallopeptide that targets and disrupts mitochondrial function in breast cancer stem cells. *Angewandte Chemie (International Ed. in English)* 57 (1): 287–291.

50 Liu, Z. and Sadler, P.J. (2014). Organoiridium complexes: anticancer agents and catalysts. *Accounts of Chemical Research* 47 (4): 1174–1185.

51 Nicco, C. and Batteux, F. (2017). ROS modulator molecules with therapeutic potential in cancers treatments. *Molecules (Basel, Switzerland)* 23 (1): 84–99.

52 Pathak, R.K., Kolishetti, N., and Dhar, S. (2015). Targeted nanoparticles in mitochondrial medicine. *Wiley Interdisciplinary Reviews. Nanomedicine and Nanobiotechnology* 7 (3): 315–329.

53 Cheng, Y., Dai, Q., Morshed, R.A. et al. (2014). Blood-brain barrier permeable gold nanoparticles: an efficient delivery platform for enhanced malignant glioma therapy and imaging. *Small* 10 (24): 5137–5150.

54 Das, S., Dowding, J.M., Klump, K.E. et al. (2013). Cerium oxide nanoparticles: applications and prospects in nanomedicine. *Nanomedicine (London, England)* 8 (9): 1483–1508.

55 Karakoti, A., Singh, S., Dowding, J.M. et al. (2010). Redox-active radical scavenging nanomaterials. *Chemical Society Reviews* 39 (11): 4422–4432.

56 Kodiha, M., Wang, Y.M., Hutter, E. et al. (2015). Off to the organelles - killing cancer cells with targeted gold nanoparticles. *Theranostics* 5 (4): 357–370.

57 Pešić, M., Podolski-Renić, A., Stojković, S. et al. (2015). Anti-cancer effects of cerium oxide nanoparticles and its intracellular redox activity. *Chemico-Biological Interactions* 232: 85–93.

58 Singh, S. (2016). Cerium oxide based nanozymes: Redox phenomenon at biointerfaces. *Biointerphases* 11 (4): 04b202.

59 Atkinson, V. (2015). Medical management of malignant melanoma. *Australian Prescriber* 38 (3): 74–78.

60 Flaherty, K.T. (2007). Sorafenib: delivering a targeted drug to the right targets. *Expert Review of Anticancer Therapy* 7 (5): 617–626.

61 Langdon, S.P. and Cameron, D.A. (2013). Pertuzumab for the treatment of metastatic breast cancer. *Expert Review of Anticancer Therapy* 13 (8): 907–918.

62 Rajanna, S., Rastogi, I., Wojdyla, L. et al. (2015). Current molecularly targeting therapies in NSCLC and melanoma. *Anti-Cancer Agents in Medicinal Chemistry* 15 (7): 856–868.

63 Cat, B., Stuhlmann, D., Steinbrenner, H. et al. (2006). Enhancement of tumor invasion depends on transdifferentiation of skin fibroblasts mediated by reactive oxygen species. *Journal of Cell Science* 119 (Pt 13): 2727–2738.

64 De Vlieghere, E., Verset, L., Demetter, P. et al. (2015). Cancer-associated fibroblasts as target and tool in cancer therapeutics and diagnostics. *Virchows Archiv* 467 (4): 367–382.

65 Fang, H. and Declerck, Y.A. (2013). Targeting the tumor microenvironment: from understanding pathways to effective clinical trials. *Cancer Research* 73 (16): 4965–4977.

66 De Wever, O., Demetter, P., Mareel, M., and Bracke, M. (2008). Stromal myofibroblasts are drivers of invasive cancer growth. *International Journal of Cancer* 123 (10): 2229–2238.

67 Mei, L., Du, W., and Ma, W.W. (2016). Targeting stromal microenvironment in pancreatic ductal adenocarcinoma: controversies and promises. *Journal of Gastrointestinal Oncology* 7 (3): 487–494.

68 Desmoulière, A., Guyot, C., and Gabbiani, G. (2004). The stroma reaction myofibroblast: a key player in the control of tumor cell behavior. *The International Journal of Developmental Biology* 48 (5–6): 509–517.

69 Nichols, J.W. and Bae, Y.H. (2014). EPR: Evidence and fallacy. *Journal of Controlled Release* 190: 451–464.

70 Au, J.L., Yeung, B.Z., Wientjes, M.G. et al. (2016). Delivery of cancer therapeutics to extracellular and intracellular targets: determinants, barriers, challenges and opportunities. *Advanced Drug Delivery Reviews* 97: 280–301.

71 Zhang, B., Hu, Y., and Pang, Z. (2017). Modulating the tumor microenvironment to enhance tumor nanomedicine delivery. *Frontiers in Pharmacology* 8: 952.

72 Guo, M., Perez, C., Wei, Y. et al. (2003). Iron-binding properties of plant phenolics and cranberry's bio-effects. *Dalton Transactions (Cambridge, England)* 2007 (43): 4951–4961.

73 Koltover, V.K. (2010). Antioxidant biomedicine: from free radical chemistry to systems biology mechanisms. *Russian Chemical Bulletin* 59 (1): 37–42.

74 Lehninger, A.L., Nelson, D.L., Cox, M.M., and Cox, M.M. (2005). *Lehninger Principles of Biochemistry*, 130–132. Macmillan.

75 Pellegrini, N., Serafini, M., Colombi, B. et al. (2003). Total antioxidant capacity of plant foods, beverages and oils consumed in Italy assessed by three different in vitro assays. *The Journal of Nutrition* 133 (9): 2812–2819.

76 Schubert, D., Dargusch, R., Raitano, J., and Chan, S.-W. (2006). Cerium and yttrium oxide nanoparticles are neuroprotective. *Biochemical and Biophysical Research Communications* 342 (1): 86–91.

77 Ivanov, V.K., Shcherbakov, A.B., and Usatenko, A.V. (2009). Structure-sensitive properties and biomedical applications of nanodispersed cerium dioxide. *Russian Chemical Reviews* 78 (9): 855–871.

78 Patil, S., Sandberg, A., Heckert, E. et al. (2007). Protein adsorption and cellular uptake of cerium oxide nanoparticles as a function of zeta potential. *Biomaterials* 28 (31): 4600–4677.

79 Tsai, Y.-Y., Oca-Cossio, J., Lin, S.-M. et al. (2008). Reactive oxygen species scavenging properties of ZrO2–CeO2 solid solution nanoparticles. *Nanomedicine* 3 (5): 637–645.

80 Artiglia, L., Agnoli, S., Paganini, M.C. et al. (2014). TiO_2@CeO_x core–shell nanoparticles as artificial enzymes with peroxidase-like activity. *ACS Applied Materials & Interfaces* 6 (22): 20130–20136.

81 Gao, L., Zhuang, J., Nie, L. et al. (2007). Intrinsic peroxidase-like activity of ferromagnetic nanoparticles. *Nature Nanotechnology* 2 (9): 577–583.

82 Vernekar, A.A., Sinha, D., Srivastava, S. et al. (2014). An antioxidant nanozyme that uncovers the cytoprotective potential of vanadia nanowires. *Nature Communications* 5 (1): 5301.

83 Mu, J., Wang, Y., Zhao, M., and Zhang, L. (2012). Intrinsic peroxidase-like activity and catalase-like activity of Co_3O_4 nanoparticles. *Chemical Communications* 48 (19): 2540–2542.

84 Nagvenkar, A.P. and Gedanken, A. (2016). Cu0.89Zn0.11O, a new peroxidase-mimicking nanozyme with high sensitivity for glucose and antioxidant detection. *ACS Applied Materials & Interfaces* 8 (34): 22301–22308.

85 Feng, Z., Tao, P., Zou, L. et al. (2017). Hyperbranched phosphorescent conjugated polymer dots with iridium(III) complex as the core for hypoxia imaging and photodynamic therapy. *ACS Applied Materials & Interfaces* 9 (34): 28319–28330.

86 Lv, W., Zhang, Z., Zhang, K.Y. et al. (2016). A mitochondria-targeted photosensitizer showing improved photodynamic therapy effects under hypoxia. *Angewandte Chemie International Edition* 55 (34): 9947–9951.

87 Zhu, C., Liu, L., Yang, Q. et al. (2012). Water-soluble conjugated polymers for imaging, diagnosis, and therapy. *Chemical Reviews* 112 (8): 4687–4735.

88 Khan, M.S., Vishakante, G.D., and Siddaramaiah, H. (2013). Gold nanoparticles: a paradigm shift in biomedical applications. *Advances in Colloid and Interface Science* 199–200: 44–58.

89 Du, L., Suo, S., Wang, G. et al. (2013). Mechanism and cellular kinetic studies of the enhancement of antioxidant activity by using surface-functionalized gold nanoparticles. *Chemistry* 19 (4): 1281–1287.

90 Esumi, K., Takei, N., and Yoshimura, T. (2003). Antioxidant-potentiality of gold–chitosan nanocomposites. *Colloids and Surfaces B: Biointerfaces* 32 (2): 117–123.

91 Bumbudsanpharoke, N., Choi, J., Park, I., and Ko, S. (2015). Facile biosynthesis and antioxidant property of nanogold-cellulose fiber composite. *Journal of Nanomaterials* 2015: 146460.

92 Nie, Z., Liu, K.J., Zhong, C.J. et al. (2007). Enhanced radical scavenging activity by antioxidant-functionalized gold nanoparticles: a novel inspiration for development of new artificial antioxidants. *Free Radical Biology & Medicine* 43 (9): 1243–1254.

93 Medhe, S., Bansal, P., and Srivastava, M.M. (2014). Enhanced antioxidant activity of gold nanoparticle embedded 3,6-dihydroxyflavone: a combinational study. *Applied Nanoscience* 4 (2): 153–161.

94 Daduang, J., Palasap, A., Daduang, S. et al. (2015). Gallic acid enhancement of gold nanoparticle anticancer activity in cervical cancer cells. *Asian Pacific Journal of Cancer Prevention* 16 (1): 169–174.

95 Al-Ani, L.A., Yehye, W.A., Kadir, F.A. et al. (2019). Hybrid nanocomposite curcumin-capped gold nanoparticle-reduced graphene oxide: anti-oxidant potency and selective cancer cytotoxicity. *PLoS One* 14 (5): e0216725.

96 Abel, E.E., John Poonga, P.R., and Panicker, S.G. (2016). Characterization and in vitro studies on anticancer, antioxidant activity against colon cancer cell line of gold nanoparticles capped with Cassia tora SM leaf extract. *Applied Nanoscience* 6 (1): 121–129.

97 Hsieh, D.-S., Lu, H.-C., Chen, C.-C. et al. (2012). The preparation and characterization of gold-conjugated polyphenol nanoparticles as a novel delivery system. *International Journal of Nanomedicine* 7: 1623–1633.
98 Vilas, V., Philip, D., and Mathew, J. (2016). Essential oil mediated synthesis of silver nanocrystals for environmental, anti-microbial and antioxidant applications. *Materials Science & Engineering. C, Materials for Biological Applications* 61: 429–436.
99 Kalaiyarasan, T., Bharti, V.K., and Chaurasia, O.P. (2017). One pot green preparation of Seabuckthorn silver nanoparticles (SBT@AgNPs) featuring high stability and longevity, antibacterial, antioxidant potential: a nano disinfectant future perspective. *RSC Advances* 7 (81): 51130–51141.
100 Sriranjani, R., Srinithya, B., Vellingiri, V. et al. (2016). Silver nanoparticle synthesis using Clerodendrum phlomidis leaf extract and preliminary investigation of its antioxidant and anticancer activities. *Journal of Molecular Liquids* 220: 926–930.
101 Marulasiddeshwara, M.B., Dakshayani, S.S., Sharath Kumar, M.N. et al. (2017). Facile-one pot-green synthesis, antibacterial, antifungal, antioxidant and antiplatelet activities of lignin capped silver nanoparticles: a promising therapeutic agent. *Materials Science & Engineering. C, Materials for Biological Applications* 81: 182–190.
102 Wu, K., Su, D., Liu, J. et al. (2019). Magnetic nanoparticles in nanomedicine: a review of recent advances. *Nanotechnology* 30 (50): 502003.
103 de Cristo Soares Alves, A., Mainardes, R.M., and Khalil, N.M. (2016). Nanoencapsulation of gallic acid and evaluation of its cytotoxicity and antioxidant activity. *Materials Science & Engineering. C, Materials for Biological Applications* 60: 126–134.
104 Shah, S.T., Yehya, W.A., Saad, O. et al. (2017). Surface functionalization of iron oxide nanoparticles with gallic acid as potential antioxidant and antimicrobial agents. *Nanomaterials (Basel)* 7 (10): 306–322.
105 Chen, Z., Zhang, A., Wang, X. et al. (2017). The advances of carbon nanotubes in cancer diagnostics and therapeutics. *Journal of Nanomaterials* 2017: 3418932.
106 Gupta, M., Chauhan, D.N., Sharma, V., and Chauhan, N.S. (2019). *Novel Drug Delivery Systems for Phytoconstituents*. CRC Press.
107 Vaiserman, A., Koliada, A., Zayachkivska, A., and Lushchak, O. (2020). Nanodelivery of natural antioxidants: an anti-aging perspective. *Frontiers in Bioengineering and Biotechnology* 7: 447.
108 Patra, J.K., Das, G., Fraceto, L.F. et al. (2018). Nano based drug delivery systems: recent developments and future prospects. *Journal of Nanobiotechnology* 16 (1): 1–33.
109 Khan, T. and Gurav, P. (2018). PhytoNanotechnology: enhancing delivery of plant based anti-cancer drugs. *Frontiers in Pharmacology* 8: 1002.
110 Chanphai, P. and Tajmir-Riahi, H. (2018). Encapsulation of micronutrients resveratrol, genistein, and curcumin by folic acid-PAMAM nanoparticles. *Molecular and Cellular Biochemistry* 449 (1): 157–166.
111 Rizeq, B., Gupta, I., Ilesanmi, J. et al. (2020). The power of phytochemicals combination in cancer chemoprevention. *Journal of Cancer* 11 (15): 4521.
112 Zhou, H., Zheng, B., and McClements, D.J. (2021). Encapsulation of lipophilic polyphenols in plant-based nanoemulsions: impact of carrier oil on lipid digestion and curcumin, resveratrol and quercetin bioaccessibility. *Food & Function* 12 (8): 3420–3432.
113 Shindikar, A., Singh, A., Nobre, M., and Kirolikar, S. (2016). Curcumin and resveratrol as promising natural remedies with nanomedicine approach for the effective treatment of triple negative breast cancer. *Journal of Oncology* 2016: 1–13.

114 Van Tran, V., Moon, J.-Y., and Lee, Y.-C. (2019). Liposomes for delivery of antioxidants in cosmeceuticals: challenges and development strategies. *Journal of Controlled Release* 300: 114–140.

115 Bhardwaj, R., Sanyal, S., Vaiphei, K. et al. (2017). Sesamol induces apoptosis by altering expression of Bcl-2 and Bax proteins and modifies skin tumor development in Balb/c mice. *Anti-Cancer Agents in Medicinal Chemistry (Formerly Current Medicinal Chemistry-Anti-Cancer Agents)* 17 (5): 726–733.

116 Arora, R., Malhotra, P., Chawla, R. et al. (2010). Herbal drugs for oncology: current status and future directions in cancer chemoprevention. In: *Herbal Medicine: A Cancer Chemopreventive and Therapeutic Perspective* (ed. A. Rajesh), 3–41. Delhi: Jaypee Brothers Medical Publishers (P) Ltd.

117 Hu, B., Ting, Y., Zeng, X., and Huang, Q. (2013). Bioactive peptides/chitosan nanoparticles enhance cellular antioxidant activity of (−)-epigallocatechin-3-gallate. *Journal of Agricultural and Food Chemistry* 61 (4): 875–881.

118 Abderrezak, A., Bourassa, P., Mandeville, J.-S. et al. (2012). Dendrimers bind antioxidant polyphenols and cisplatin drug. *PLoS One* 7 (3): e33102.

119 Shen, Z., Li, B., Liu, Y. et al. (2018). A self-assembly nanodrug delivery system based on amphiphilic low generations of PAMAM dendrimers-ursolic acid conjugate modified by lactobionic acid for HCC targeting therapy. *Nanomedicine: Nanotechnology, Biology and Medicine* 14 (2): 227–236.

120 Roy, A.D., Varghese, K.K., Ghosh, S. et al. (2021). Metallic nanoparticles from natural products as a potential anti-cancer. *International Journal of Pharmaceutical Sciences Review and Research* 67 (1): 171–177.

121 Stakleff, K.S., Sloan, T., Blanco, D. et al. (2012). Resveratrol exerts differential effects in vitro and in vivo against ovarian cancer cells. *Asian Pacific Journal of Cancer Prevention* 13 (4): 1333–1340.

122 Abbaszadeh, H., Keikhaei, B., and Mottaghi, S. (2019). A review of molecular mechanisms involved in anticancer and antiangiogenic effects of natural polyphenolic compounds. *Phytotherapy Research* 33 (8): 2002–2014.

123 Wei, Q.-Y., He, K.-M., Chen, J.-L. et al. (2019). Phytofabrication of nanoparticles as novel drugs for anticancer applications. *Molecules (Basel, Switzerland)* 24 (23): 4246.

124 Wei, Q., Yang, Q., Wang, Q. et al. (2018). Formulation, characterization, and pharmacokinetic studies of 6-gingerol-loaded nanostructured lipid carriers. *AAPS PharmSciTech* 19 (8): 3661–3669.

23

Challenges and Regulatory Issues of Nanophytotherapeutics

Smriti Sharma[1] *and Vinayak Bhatia*[2]

[1] *Department of Chemistry, Miranda House, University of Delhi, Delhi, India*
[2] *Department of Glaucoma, ICARE Eye Hospital and Postgraduate Institute, Noida, Uttar Pradesh, India*

23.1 Introduction

K. Eric Drexler [1] titled "Engines of creation: the coming era of nanotechnology" brought to forefront the term "Nanotechnology" based on the famous lecture of Richard Feynman "There's Plenty of Room at the Bottom" [2–5]. Initially, its applications ranged from chemical synthesis to storing information at the atomic level. But now it has grown exponentially with hardly any area of life it does not touch upon, be it medical field, high-performance batteries, biology, or treatment of wastewater. In medical field, there have been important breakthroughs in the field of diagnostics, cancer therapy, drug delivery, and tissue engineering. Such is the pervasion of nanotechnology in the area of medicine, that a new term "Nanomedicine" has been coined for it (Figure 23.1).

So, when nanotechnology is applied to medical field, it is called nanomedicine. Not only has it included medical use of nanomaterials but also nano-based biological devices, and nano-electronic biosensors. The challenges of toxicity and environmental impact of nanotechnology also come under the purview of nanomedicine. The reason for nanotechnology being so popular for biological and, more specifically, for medical applications, is the similarity of size of nanomaterials with biological molecules making them suitable for integration with biological materials both in in vitro and in vivo applications [7–17]. It is no surprise that nanomedicine has seen such an exponential rise with it contributing in the design of drug delivery systems, in vivo imaging, diagnostic devices, analytical tools, contrast agents, and other myriad uses in pharmaceutical industry. The rise of this field can be gauged from the fact that global funding for emerging nanotechnology increased by 45% per year in recent years and is having an appreciable impact on the economy [18, 19] (Table 23.1).

The advantages of nanodrugs over normal drugs are as follows:

- Nano-sized drugs, i.e. nanodrugs have been known to show better safety and efficacy. In addition to that, they demonstrate better bioavailability and enhanced active concentration.
- One can control the rate of drug release over the desired timescale.

Phytoantioxidants and Nanotherapeutics, First Edition. Edited by Mithun Rudrapal.

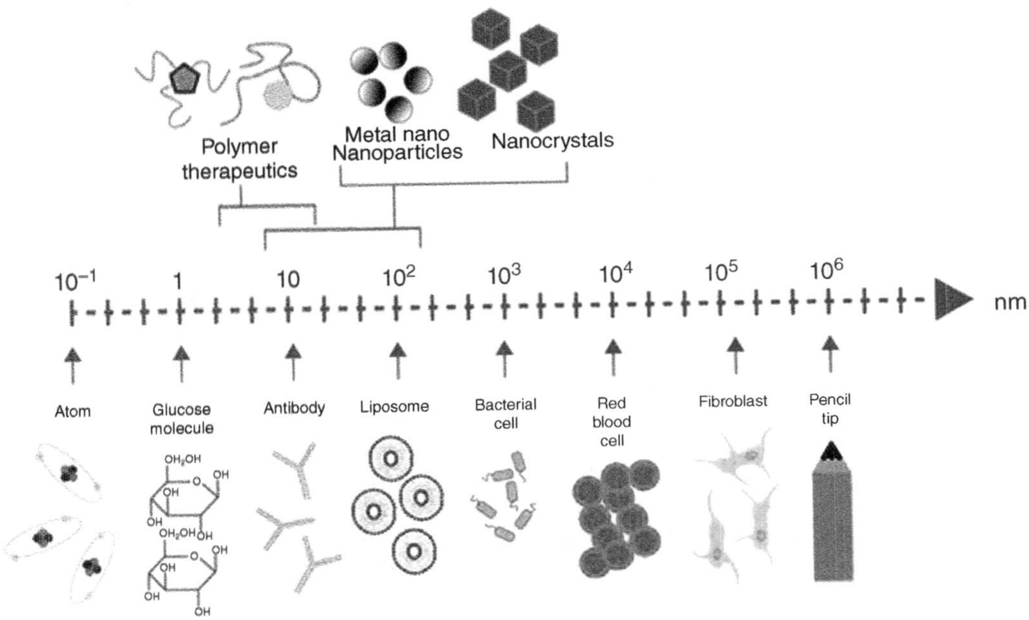

Figure 23.1 Different types of nanoformulations in medicine. *Source:* From Ref. [6]/with permission of Royal Society of Chemistry.

Table 23.1 Examples of nanotechnological interventions approved for clinical use.

Clinical use	Name	Approved for	Class of nanomedicine	Use
Cancer	Doxil	Ovarian cancer/HIV-associated Kaposi's sarcoma	Liposome	Drug delivery
	NBTXR3/ Hensify	Locally advanced sarcoma	Metallic	Radiation enhancer
	Vyxeos	Myeloid leukemia	Bilamellar liposomes	Combination therapy
	Abraxane	Pancreatic cancer, breast cancer, non-small cell lung cancer	Albumin-bound	Drug delivery
	Onivyde	Pancreatic cancer	Liposome	Drug delivery
	DaunoXome	HIV-associated Kaposi's sarcoma	Liposome	Drug delivery
	Myocet	Breast cancer	Liposome	Drug delivery
Antifungal	AmBisome	Cryptococcal meningitis, aspergillus, candida infections, and visceral leishmaniasis	Liposome	Drug delivery
Other	Patisiran/ ONPATTRO	Transthyretin amyloidosis	Lipid	siRNA delivery
	Diafer	Iron-deficient anemia	Metallic	Iron replacement
	Diprivan	Anesthesia	Liposome	Anesthetic

Source: From Ref. [6]/with permission of Royal Society of Chemistry.

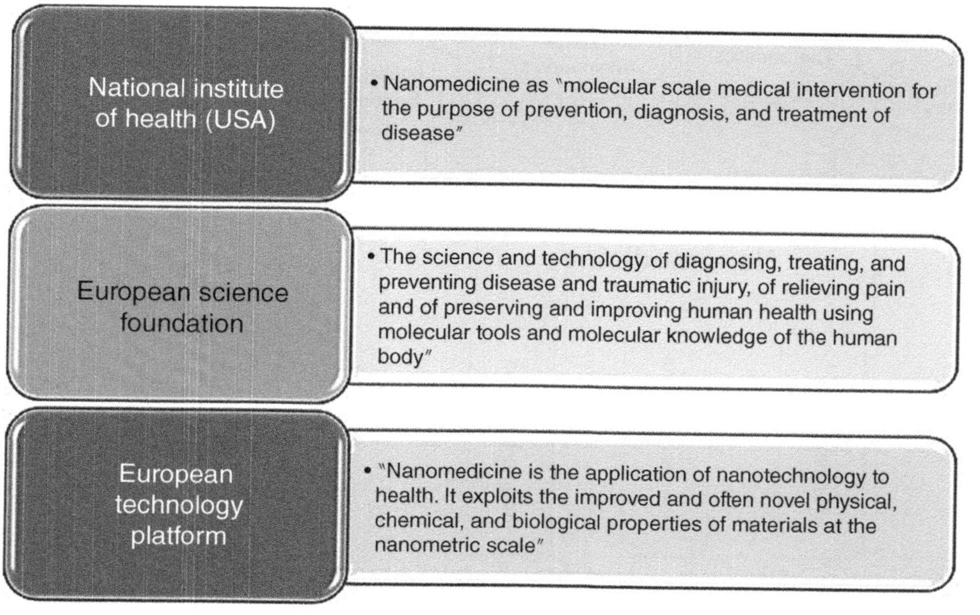

Figure 23.2 Different definitions of nanomedicine. *Source:* From Ref. [20]/with permission of Walter de Gruyter.

- Because of their small size, it is easy to attach these small molecule therapeutic agents to NPs acting as nanocarriers. Thus, nanoformulation of a drug would help in the precise delivery to the target tissue reducing the toxicity.
- The accumulation of nanodrugs at the pathological and malignant sites is possible because of the enhanced permeability and retention (EPR) effect.
- Conventional drugs are much cheaper than nanodrugs (Figure 23.2).

These advantages do not mean that there are no challenges for using nanodrugs with respect to minimization of adverse effects and maximization of clinical advantages. In addition to that, the nanodrugs should be biocompatible and should have a broader application. Also, the synthesis of nanodrugs sometimes entails the usage and generation of potentially toxic reagents, which are a concern for the environment. Also, nanodrug storage is a big challenge due to the nanomaterials being fragile and unstable [21].

23.1.1 Historical Perspective

The roots of nanomedicine go back to the usage of colloidal gold in primeval times. Metchnikov and Ehrlich's seminal research on phagocytes for which they were awarded the Nobel Prize for Medicine in 1908 is considered the foundation of nanomedicine in modern times. Last 30 years of the twentieth century have seen an exponential rise in this field with strides reported in DNA-drug complexes, polymer-drug conjugates, liposomes, antibody-drug conjugates, polymer nanocapsules, and many more [22].

In Figure 23.3, various nanomaterials in terms of their effectiveness in nanomedicine are compared.

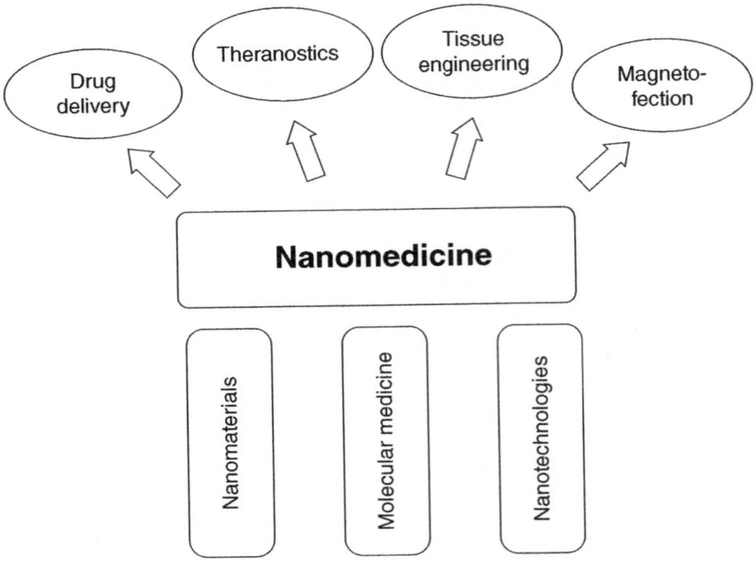

Figure 23.3 Nanomedicine and its future course of action. *Source:* From Ref. [23]/Walsh Medical Media/ CC BY 4.0.

23.1.2 Need for Regulations

There is no doubt about the role of nanophytotherapeutics in providing solutions for complex diseases but there are associated risks also. For instance, the mechanism of interaction of nanodrugs with it is not fully understood and concerns have been raised for their safety in the biosystems. Thus, there is an urgent need for introducing regulatory measures in nanomedicine (Figure 23.4).

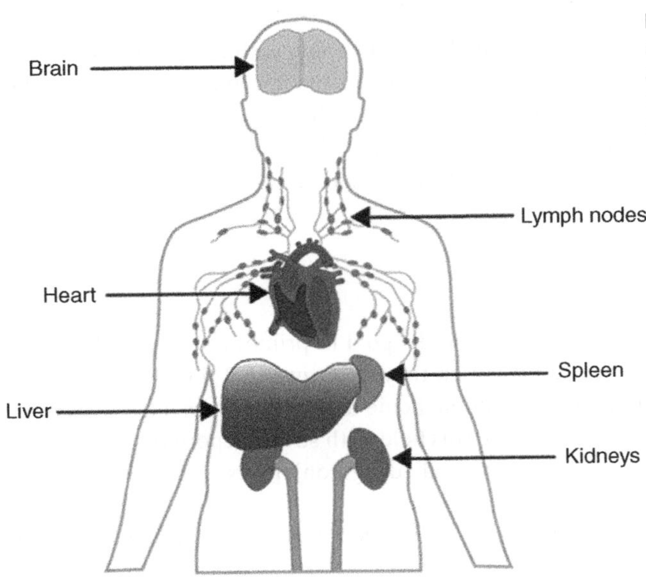

Figure 23.4 Main areas of nanoparticle translocation and accumulation after administration. *Source:* From Ref. [6]/with permission of Royal Society of Chemistry.

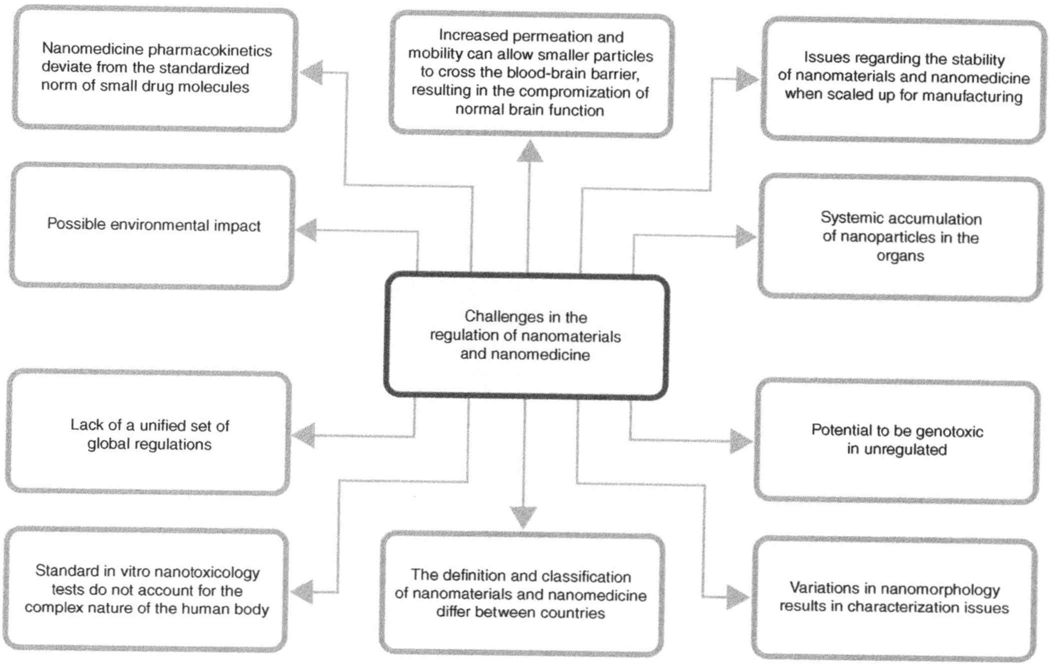

Figure 23.5 Major challenges faced in the regulation of nanomaterials. *Source:* From Ref. [6]/with permission of Royal Society of Chemistry.

Presently, the status of the nano pharmaceutical regulation is not very impressive. Not much has been done to address the complexity of the nano-specific medical field. For instance, combinatorial products like nanodevices and drugs are not covered under one umbrella legislation. Also, there is a lack of uniformity and a lot of uncertainty with respect to challenges arising because of combining electronics and biological products. Thus, it is very important that there should be clear-cut regulations. Otherwise, there is some fear that this may go the way of genetically modified crops where a lot of investments were done but consumers have not accepted the end product due to socio-ethical reservations. Therefore, it is critical to have a clear regulatory pathway. Though these issues seem to be universal, still the legislations have to be framed keeping in mind the challenges specific to a country. It is to be noted that countries like the United States (USA), the United Kingdom (UK), Canada, Russia, Australia, and Europe and Asian countries like India, Thailand, Taiwan, Japan, and China have recognized this need are working to regulate the field of nanopharmaceuticals (Figure 23.5).

23.2 Regulatory Issues of Nanophytotherapeutics

The nanotoxicological classification system (NCS) by Keck and Müller [24] classifies the nanomaterials into four classes and is based on their ability to biodegrade and respective sizes. As the size of the NM decreases, it becomes more soluble which increases its toxicity [25, 26] (Figures 23.6 and 23.7).

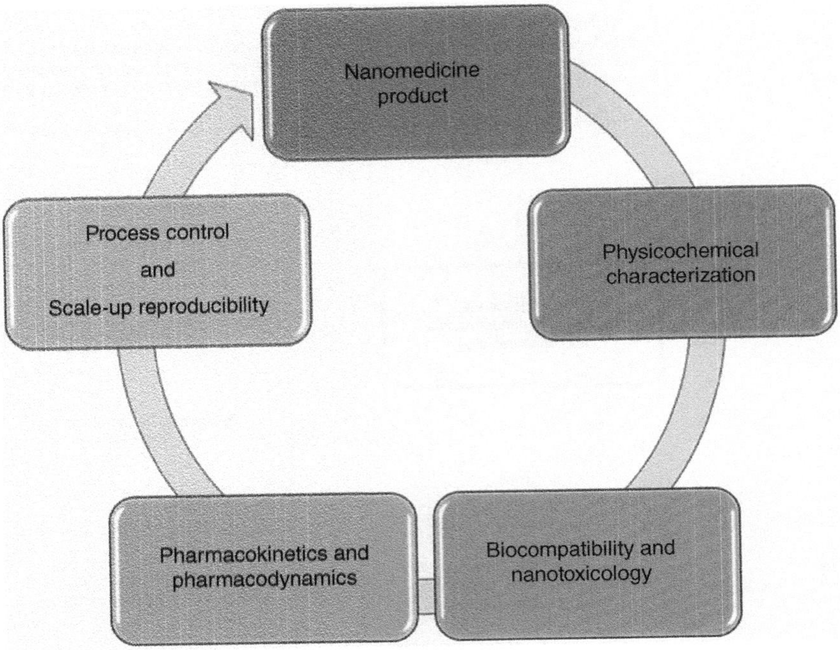

Figure 23.6 Regulatory issues of nanomedicine. *Source:* From Ref. [27]/with permission of Taylor & Francis.

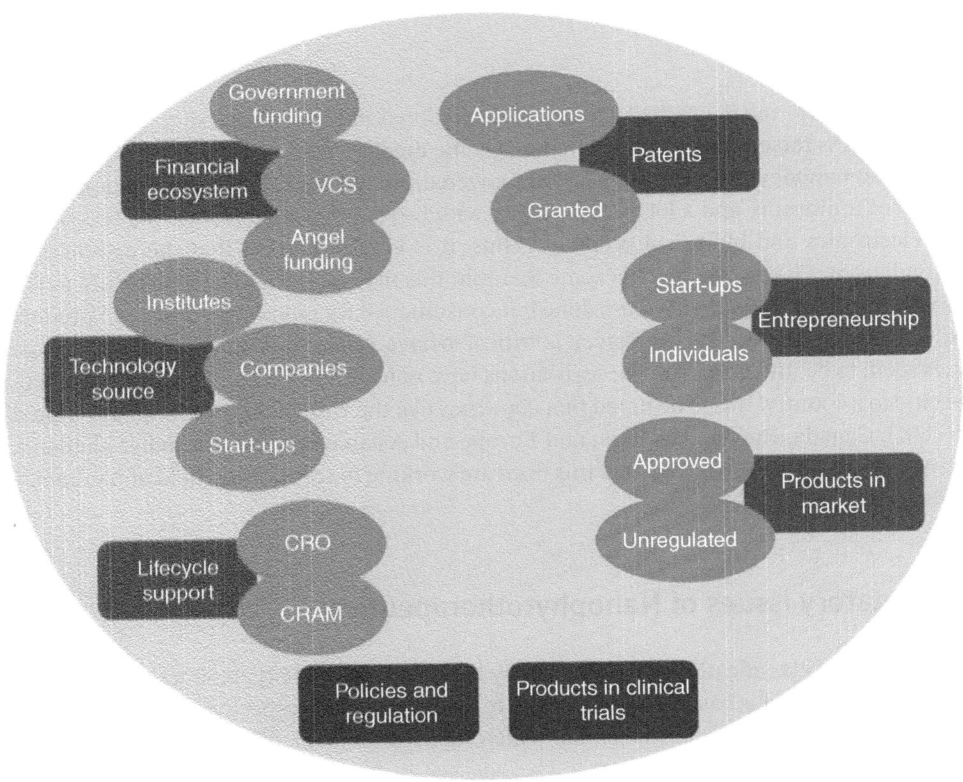

Figure 23.7 Parameters for the nanomedicine innovation landscape. *Source:* From Ref. [28]/with permission of Walter de Gruyter.

23.3 Regulatory Issues of Nanophytotherapeutics in the USA

The regulation of nanotechnology products, more specifically, nanophytotherapeutics, is controlled by FDA. FDA is a public-funded organization and uses scientific data and information to make informed decisions about regulating novel technologies. Nanophytotherapeutics, like any novel technology, carry a lot of promise but there is equal, if not more, amount of indeterminacy and risks involved. At the nanoscale, the nanophytotherapeutics can exhibit changed physicochemical properties that can alter the biosafety profile of the drug.

As quoted on the website of FDA [29],

> However, FDA does not categorically judge all products containing nanomaterials or otherwise involving the application of nanotechnology as intrinsically benign or harmful. FDA will regulate nanotechnology products under existing statutory authorities, in accordance with the specific legal standards applicable to each type of product under its jurisdiction.
>
> Consistent with Executive Order 13563 on improving regulation, as well as with the White House policy statements on regulating emerging technologies and applications of nanotechnology, FDA supports innovation and the safe use of nanotechnology in FDA-regulated products under appropriate and balanced regulatory oversight. By enhancing its scientific expertise and tools necessary to assess the safety and, as applicable, the effectiveness of products (see FDA's nanotechnology regulatory science research plan), FDA can enable the responsible development of products with new and beneficial properties. FDA intends to ensure transparent and predictable regulatory pathways grounded in the best available science.

Main points to be considered by FDA in its regulation strategy:

- ***Product-focused, science-based regulatory policy.*** Any kind of technical evaluation of nano-related products will not follow a general guideline. It will, however, be product-specific and will judge the effect in the mechanical and biological context of each product and its application. And also, the policies will be dictated by the statutory authorities of that area. It is advised that the policies are substantive and follow due procedures. It is also advised that the manufacturers consult the FDA well enough in time so that a thorough and mutual understanding of the regulatory and scientific aspects can be facilitated.
- ***Respecting differences in legal criteria for diverse product classes.***
  There is a difference in terms of the evaluation of drugs and food additives. Drugs are evaluated based on their risk profile balanced with their predicted benefits as opposed to food additives, which are considered safe if there is a certainty of no reasonable harm from them. So, the legal criterion or standard is different in both cases because the context is different. So, it may appear that there are divergent regulations for different nano-products but that is because of the inherent nature of how context works in the case of regulations.
- ***Premarket review.*** Premarket review means the applicant has to submit data regarding the efficacy and regulatory status of the product. If such a premarket data exists, it can be incorporated into the standing procedures for novel drugs, biologics, color additives, certain human devices, novel animal drugs, food additives, and certain novel dietary ingredients in dietary supplements.
- ***Consultation in the case of nonexistence of premarket review.*** If premarket review is not available due to statutory restrictions, it is encouraged to consult with the agency to decrease the

risk of unintentional harm to human or animal health. The agency will rely on information in the public domain or information submitted by volunteers or the reportage of side effects and keeping surveillance on the post-market fate of the drug.

23.4 Regulatory Issues of Nanophytotherapeutics in Europe

The "European Commission's Joint Research Center (JRC)" came out with a white paper titled "Anticipation of regulatory needs for nanotechnology-enabled health products" which acknowledges that right now no precise and detailed regulatory outline is there for nano-based health products and the present regulations need extra guidance to fully appreciate the peculiarities of these nano-based health products [30]. The main issues according to the white paper with regulating these products are as follows:

- It becomes difficult to determine the mode of action of a nano-based pharmaceutical, as in, whether it is a medical device or a medicinal product. This is because such products are very complex and have varying size-based properties.
- The major problem in formulating guidelines is the lack of strong data sets and a deluge of information. Also, the testing methods have to be standardized for the determination of efficacy, quality, and safety of nano-based products.
- Also, after the expiry of the patent of some nanomedicinal plants, generic versions of the same will come into play. But equating both products will not be a good practice as the complex and intricate physicochemical characteristics of a nano-based product may not be achieved by a generic manufacturing unit. This is a very significant issue.
- The European Union (EU) definition of nanomaterials will be applicable for nano-based medical devices and will outline their classification and regulatory necessities.

The white paper [11] states that the regulatory challenges it highlights "should guide the research projects and the involved communities willing to advance the regulatory science in the area of nanotechnology-enabled health products."

23.5 Regulatory Issues of Nanophytotherapeutics in India

Presently, there are no nano-specific regulatory guidelines in place. Considering the unprecedented growth of this field, now, the Government of India has come up with various initiatives to regulate this sector. That includes all types of nano-products and thus includes nanopharmaceuticals also (Table 23.2).

The Department of Science and Technology (DST), Government of India has constituted a working group to set up guidelines for regulating nano-based products. This was done under a program titled "Nanomission" under which it was announced that "National Regulatory Authority Framework Roadmap for Nanotechnology" will be established. In addition to that, nanomission has already framed guidelines for best practices in handling nanomaterials. Also, the Council for Scientific and Industrial Research (CSIR) started a project nano-SHE that is "Nanomaterials: Application and Impact on Safety, Health, and Environment" to evaluate the toxicity of nanomaterials. Also, the Department of Pharmaceuticals (DOP), Government of India, in the year 2006, gave the responsibility of coming out with guidelines for nanomedicines to the "National Institute of Pharmaceutical Education and Research (NIPER) Mohali" and then to NIPER Kolkata in the

Table 23.2 Contract research organizations in India working on nanomedicine.

Ogene Systems (I) Pvt. Ltd., India, Hyderabad
Vanta Bioscience, Chennai
Quintiles Technologies (India) Pvt. Ltd., Bengaluru
Cliantha Research Ltd., Ahmedabad
GVK Bio Pvt. Ltd., Hyderabad
Espin Nanotech Pvt. Ltd., Kanpur
Lambda Therapeutic Research Ltd., Ahmedabad
Consortium Clinical Research Pvt. Ltd., Bengaluru

Source: From Ref. [28]/with permission of Walter de Gruyter.

year 2012. DOP is planning to establish "National Center for Pharmaceutical Nanotechnology" at NIPER Kolkata. This institute will be responsible for nanotoxicology assessment and regulation of nanodrugs and devices (Table 23.3).

Many researchers have tried to identify the issues related to the patentability of nanotechnology, and the patentability of nanoparticles for therapeutic and diagnostic uses. There are many challenges like the dearth of integrated research, low priority to risk research, pitiable policy coordination, the restricted role of the private sector and funding restraints. There is an urgent need to identify stakeholders who will facilitate this process. It is to be noted that all drugs that include traditional medicine, delivery systems, diagnostics, and medical devices come under the purview of the Drugs and Cosmetics Act, 1940. In addition to that, the Drugs and Cosmetics (Amendment) Bill, 2013 includes vaccines, living modified organisms, stem cells, recombinant deoxyribonucleic acid-derived products, gene therapeutic products, etc. But there still are a lot of shortcomings in the regulation of nanophytotherapeutics [20].

The current Drugs and Cosmetics Act, 1940 and Drug Price Control Order, 1995 do not include nanoscale medicines. It is the need of the hour that these acts be amended to include nanomedicine and nano as an ingredient and make sure that the product containing nanomaterial is labelled.

23.6 Conclusions and Future Perspectives

The regulation of nanophytotherapeutics is very challenging as it is an interdisciplinary field and there are many variables that need to be clearly defined before the regulatory framework is set. Most countries have recognized the importance of setting regulations in this field. But nanotechnology is a vast field and has many facets. Therefore, the regulations cannot cover all the areas of nanophytotherapeutics. For the promotion of global trade, it is imperative that there is an internationally accepted definition of nanomedicine and nanophytotherapeutics. There should be interoperability and harmonization of standards of products and procedures. Nanotoxicity assays should be standardized. The International Standard Organization and Organization for Economic Co-operation and Development (OECD) gave standards for handling nanomaterials which are valid for all the OECD member countries. The most important point is that merely drafting regulatory guidelines is not enough. One would need to implement and enforce the said guidelines. Since nanomedicine is an upcoming field, it would be great if a nano-certification system can be adopted. It is advised that integration of all three aspects involving regulatory space, policy, and governance is needed with efficient implementation and enforcement.

Table 23.3 Nano-based healthcare products in India.

Product	Company	Type	Source
Analgesics and pain management			
Volini	Ranbaxy Laboratories Ltd, Gurgaon	Pain relieving	http://www.volini.com/
Trois	Venus Remedies Ltd., Panchkula	Arthritis	http://www.myiris.com/newsCentre/storyShow.php?fileR=20120822134008715&dir=2012/08/22&secID=corporates
Oxalgin Nanogel	Zydus Cadila Ltd., Ahmedabad	Joint and muscular pains	http://www.mims.com/India/drug/info/OXALGIN%20NANOGEL/OXALGIN%20NANOGEL%20gel
Heal-o-Kind	Mankind Pharma, Delhi	Nanocrystalline silver gel	http://www.financialexpress.com/article/pharma/latest-updates/mankind-pharma-launches-heal-o-kind-for-topical-wounds/30274/
Zyflex Nanogel	Zydus Cadila Ltd., Ahmedabad	Pain	http://zyduscadila.com/
Antimicrobials and anti-infectives			
Fungisome	Lifecare Innovations Pvt Ltd, Gurgaon	Drug delivery system with liposomal amphotercin B	http://www.fungisome.com/
Receptol	Biomix Network Ltd., Gujarat	HIV/AIDS	http://biom.cc/?q=category/5/31
Zyclin® Nanogel™	Zydus Cadila Ltd., Ahmedabad	Acne	http://www.ncbi.nlm.nih.gov/pmc/articles/PMC4190779/
Cancer treatment			
Nanoxel	Dabur (now Fresenius-kabi)	For cancer treatment nanoparticle drug delivery system	http://www.ncbi.nlm.nih.gov/pubmed/23975704
Taxedol	Venus Remedies Limited, Panchkula	Formulation for cancer in nanoparticulate form	http://www.indiaprwire.com/pressrelease/medical/20120709124559.htm
Albupax	Natco Pharma Ltd., Hyderabad	Cancer treatment	http://articles.economictimes.indiatimes.com/2010-01-11/news/28477692_1_abraxane-natco-pharma-albupax
PacliALL	Panacea Biotec Ltd., New Delhi	Albumin-bound paclitaxel nano particle formulation	http://www.panaceabiotec.com/press_releases/PR13052011.pdf
DoceAqualip	Intas Pharmaceutical, Ahmedabad	Anticancer	http://www.intaspharma.com/index.php?option=com_content&view=article&id=119

Ayurvedic nanoformulations			
Nanocurcumin	Biotexlife, India	Capsules of curcumin, with help of nanotechnology	http://www.biotexlife.com/nano-curcumin/
N-Curcumin	Konark Herbals and Health Care, Mumbai	Curcumin in nanoparticle size	http://www.konarkherbal.com/curcumin.html
En-Tube	Prakruthik Health Care Pvt. Ltd., Hyderabad	Nano Svarjika (mineral) and Plantago ovata husk (herb)	http://www.indiaprwire.com/pressrelease/health-care/2010040847670.htm
Biomaterials and implants			
SyboGraf	Eucare Pharmaceuticals Pvt. Ltd., Chennai	Synthetic nanocrystalline hydroxyapatite	http://www.eucareindia.com/product

Source: From Ref. [28]/with permission of Walter de Gruyter.

References

1 Drexler, K.E. (1986). *Engines of Creation*. Anchor books.

2 Rajiv Saini, S.S. and Sharma, S. (2010). Nanotechnology: the future medicine. *J. Cutan. Aesthet. Surg.* 3 (1): 32–34. https://doi.org/10.4103/0974-2077.63301.

3 Goyal, G., Garg, T., Rath, G., and Goyal, A.K. (2014). Current nanotechnological strategies for treating glaucoma. *Crit. Rev. Ther. Drug Carrier Syst.* 31 (5): 365–405. https://doi.org/10.1615/critrevtherdrugcarriersyst.2014010123.

4 Zarbin, M.A., Montemagno, C., Leary, J.F., and Ritch, R. (2010). Nanotechnology in ophthalmology. *Can. J. Ophthalmol.* 45 (5): 457–476. https://doi.org/10.3129/i10-090.

5 Feynman, R.P. (1960). There's plenty of room at the bottom. California Institute of Technology, Engineering and Science magazine.

6 Foulkes, R., Man, E., Thind, J. et al. (2020). The regulation of nanomaterials and nanomedicines for clinical application: current and future perspectives. *Biomater. Sci.* 8 (17): 4653–4664. https://doi.org/10.1039/d0bm00558d.

7 Cai, X., Conley, S., and Naash, M. (2008). Nanoparticle applications in ocular gene therapy. *Vis. Res.* 48 (3): 319–324. https://doi.org/10.1016/j.visres.2007.07.012.

8 Kumar, A., Gayakwad, A., and Nagale, B. (2014). a Review: nano membrane and application. *Int. J. Innov. Res. Sci. Eng. Technol.* 3 (1): 8373–8381.

9 Suri, S.S., Fenniri, H., and Singh, B. (2007). *J. Occup. Med. Nanotechnol. -Based Drug Delivery Sys.* 6: 1–6. https://doi.org/10.1186/1745-6673-2-16.

10 Tiwari, A. and Tiwari, A. (ed.) *Nanomaterials in Drug Delivery, Imaging, And Tissue Engineering*. John Wiley & Sons.

11 https://statnano.com/news/67458/A-Regulatory-Framework-for-Nanotechnology-enabled-Health-Products-in-the-EU#ixzz6ualIdfrf.

12 De Jong, W.H. (2008). Drug delivery and nanoparticles: applications and hazards. *International Journal of Nanomedicine* 3 (2): 133–149.

13 Cascone, M.G., Lazzeri, L., and Carmignani, C. (2002). Gelatin nanoparticles produced by a simple W/O emulsion as delivery system for methotrexate. *Journal of Materials Science: Materials in Medicine* 3: 523–526.

14 Kakkar, R., Sharma, S., and Badhani, B. (2014). Density functional study of functionalization of carbon nanotubes with carbenes. *Can. Chem. Trans.* 2 (4): 434–449. https://doi.org/10.13179/canchemtrans.2014.02.04.0132.

15 Kakkar, R. and Sharma, S. (2011). DFT study of interactions of carbenes with boron nitride nanotubes. *Chemistry Journal* 01 (01): 9–20.

16 Sharma, S. and Bhatia, V. (2020). Phytochemicals for drug discovery in Alzheimer's disease: in silico advances. *Current Pharmaceutical Design* 1–17.

17 Sharma, S. (2020). Nanoscale drug delivery systems for glaucoma: in silico and experimental advances. *Current Topics in Medicinal Chemistry* 21 (2): 115–125.

18 Dreaden, E.C., Alkilany, A.M., Huang, X. et al. (2012). The golden age: gold nanoparticles for biomedicine. *Chem. Soc. Rev.* 41 (7): 2740–2779. https://doi.org/10.1039/c1cs15237h.

19 Wood, S., Geldart, A., and Jones, R. (2003). The social and economic challenges of nanotechnology. *TATuP - Zeitschrift für Tech. Theor. und Prax.* 12 (3–4): 72–73. https://doi.org/10.14512/tatup.12.3-4.72.

20 Bhatia, P. and Chugh, A. (2017). A multilevel governance framework for regulation of nanomedicine in India. *Nanotechnol. Rev.* 6 (4): 373–382. https://doi.org/10.1515/ntrev-2016-0083.

21 Iavicoli, I., Leso, V., Ricciardi, W. et al. (2014). Opportunities and challenges of nanotechnology in the green economy. *Environ. Heal. A Glob. Access Sci. Source* 13 (1): 1–11. https://doi.org/10.1186/1476-069X-13-78.

22 Astruc, D. (2015). Introduction to nanomedicine _ Enhanced Reader.pdf. *Molecules* 21 (4): 1–6.

23 Krukemeyer, M.G., Krenn, V., and Huebner, F. (2015). History and possible uses of nanomedicine based on nanoparticles and nanotechnological progress. *J. Nanomed. Nanotechnol.* 06 (06). https://doi.org/10.4172/2157-7439.1000336.

24 Cornelia, R.H.M. and Keck, M. (2013). Nanotoxicological classification system (NCS) – a guide for the risk-benefit assessment of nanoparticulate drug delivery systems. *Eur. J. Pharm. Biopharm.* 84 (3): 445–448.

25 Fu, P.P., Xia, Q., Hwang, H.M. et al. (2014). Mechanisms of nanotoxicity: generation of reactive oxygen species. *J. Food Drug Anal.* 22 (1): 64–75. https://doi.org/10.1016/j.jfda.2014.01.005.

26 Fischer, H.C. and Chan, W.C. (2007). Nanotoxicity: the growing need for in vivo study. *Curr. Opin. Biotechnol.* 18 (6). Elsevier Current Trends: 565–571. https://doi.org/10.1016/j.copbio.2007.11.008.

27 Saroya, A.S. (2017). *Nanophytomedicine*. CRC Press.

28 Vasaikar, S., Wali, A., and Bhatia, P. (2018). A landscape of nanomedicine innovations in India. *Nanotechnol. Rev.* 7 (2): 131–148. https://doi.org/10.1201/b11126-15.

29 "https://www.fda.gov/science-research/nanotechnology-programs-fda/fdas-approach-regulation-nanotechnology-products.".

30 Bremer, S., Halamoda-Kenzaoui, B., and Borgos, S.E. (2016). Identification of regulatory needs for nanomedicines. 1st EU-NCL survey with the "Nanomedicine" working group of the international pharmaceutical regulators.

Index

a

ABTS method 146
Acetylcholinesterase 186
Acetylcholinesterase inhibitors 251, 453
N-Acetyl cysteine 456, 460
Achillea fragrantissima 328
Actinomycetes 21
Acute inflammation 181
Acute myocardial infarction (AMI) 406, 419
Adiponectin 183
Advanced glycation end products (AGEs) 479
Aegle marmelos 328
Aerobic metabolism 77
AFM. *See* Atomic force microscopy (AFM)
Aframodial 298
Aging 7, 18
Aging and neurological disorders, 477. *See also* CNS disorders
  nanotherapeutics of phytoantioxidants for 482–483
    limitations of 483
    liposomes 483
    magnetic nanoparticles 486
    mesoporous silica nanoparticle 485–486
    polymeric NPs 484–485
    SLNs 484
  phytoantioxidants for 480–481
  ROS and oxidative stress in 478–479
AgNPs. *See* Silver nanoparticles (AgNPs)
Ajwain 58
Alginates 244–245
Alkaloids 296–297, 455, 457–458
Allicin 37, 42, 43, 45
Alliin 37
Allopathic drugs 239
Allspice
  antioxidant compounds in 58, 68, 69
  therapeutic effects of 58, 69
Aloe vera 318, 328
Aloe vera gel-loaded nanofibers 132
Aloin 130
Aluminosilicate clay 132
Alvocidib 254
Alzheimer's disease (AD) 91, 106, 212–215, 244, 449, 450, 453
American Diabetes Association (ADA) 8
AMP-activated protein kinase (AMPK) 435
β-Amyloid peptides 450
Amyotrophic lateral sclerosis (ALS) 451, 454
Anacardic acid 38
Andrographolide 460, 466
Angelica 58
Anise 58

Phytoantioxidants and Nanotherapeutics, First Edition. Edited by Mithun Rudrapal.

Annual herbs/spices 57
Anthocyanidin 63, 84, 101
Anthocyanin 41, 92, 250, 383
Antiaging 18
Antibiotics 179
Anticancer drugs 180
Anti-inflammatory drugs 182
Antimicrobial phytoantioxidants 299–300
Antimicrobial resistance (AMR) 293
Antioxidant response elements (AREs) 106
Antioxidants 5, 77, 145, 146, 191, 303, 433–434, 496
  defense mechanisms 13
  dietary intake of 6
  endogenous 1, 11–12, 77, 79, 89, 496
  exogenous 1, 77, 496, 502
  in fruits and vegetables 2, 6
  as medications 5
  role in ameliorating oxidative stress 80, 82
  role of 13
  supplements 5, 18
Antiviral medicinal plants 327–328
Antiviral treatment, challenges in 317, 325
Antoxid HC capsule 44
Apigenin 41, 113, 129, 131, 306–307
Apigenin-loaded nano-transferosomes 362
Apoptotic pathways 80
Apple, antioxidants in 2
ARE-binding proteins (ARE-BP) 214
Arecoline 457
AREs. *See* Antioxidant response elements (AREs)
Argyreia nervosa 185, 235
Arjunolic acid (AA) 16
Aromatic herbs 58, 74
Artemisia annua 335
Artemisinin 335
Arthritis
  nanophytomedicine for 181–183
  phytoantioxidants in 90–91
Artichokes 72
Ascorbic acid 3, 12, 16, 221
  bioactivity of 85
  sources of 84
Asiatic acid 460
Asimilobine 455
Aspergillus terreus 256
Astringent herbs 58
Atherosclerosis 17, 406, 417
Atherosclerotic plaques 9
Atomic force microscopy (AFM) 246
AuNPs. *See* Gold nanoparticles (AuNPs)
Aurone 63
Ayurveda 22, 135, 257, 294, 454
Ayush 257
Azadirachta indica 305

b

Bacillus anthracis 290
Bacterial infections 289–290
Baicalein-loaded nanostructured lipid carriers 363
Baicalin 188, 459
Baicalin-loaded nanoemulsions 363
Banana
  antioxidants in 2
  peel 4
Basil leaves 65–66, 68
Basil seeds 65
Bay leaves
  antioxidant compounds in 59, 68, 69
  therapeutic effects of 59, 69
Beans 73
Beetroot 72
Belotecan hydrochloride 254
Benzoic acid 33
Berberine 187, 296, 455, 457, 466
Berries 72, 250, 411
β-carotene 12, 34, 250, 499
β-cells 7

β-glucans 18
Betula pubescens 255
Betulinic acid 255
Biennial herbs 57
Bioactive flavonoids
  as phytoantioxidants 99–113
  semisynthetic alterations in 112
Biodegradable polymeric nanoparticles 280
Biomolecules, ROS-mediated damage on 79–81
Bitter herbs 58
Black garlic 152
Black pepper
  antioxidant compounds in 59, 68, 69, 147
  antioxidant properties of 152–153
  chemical constituents in 149
  therapeutic effects of 59, 69
Bleekeria vitensis 256
Blood–brain barrier (BBB) 188, 449
Blueberries 72
Body mass index (BMI) 7
Boronic ester groups 207, 209
Boswellia serrata 328
Bottom-up approach 232–233
Botulinum neurotoxins (BoNTs) 292
Brassica oleracea 255
Broccoli sprouts 252
Buruli ulcer 356, 358
Butein, for inflammatory disorders 380
Butyrylcholinesterase 251

C

Caffeic acid (CA) 33, 36, 39, 42, 158, 297, 384
Caffeic acid phenethyl ester (CAPE) 158–160, 391
Caffeine 457
Callicarpa maingayi 129
CAM. *See* Complementary and alternative medicine (CAM)
Camellia sinensis 127, 256, 319, 394
Camptotheca acuminate 127, 254
Camptothecin 128, 129
Cancer 495
  clinical trial of phytopharmaceutical formulation 253–256
  lung 127
  nanocarrier system 507–508
    carbon nanotubes 511
    dendrimers 510
    drawbacks of 510
    gold nanoparticles 510
    liposomes 508–509
    magnetic nanoparticles 511, 512
    nano micelles 509–510
    polymeric NPs 508
    polyphenolic nanoparticles 510–511
    SLNs 509
  nano(phyto)medicine for 180–181, 210, 211
  natural nanotherapeutics in 214–217
  oxidative stress in 10–11, 496
  phytoantioxidants in 91–92
  phytochemicals for 496–500, 507–508
  prevalence of 180
  prevention of
    flavonoids 64
    turmeric 151
  protective effect of natural antioxidants 16
  redox-based combinational nano therapy 503–504
  ROS-scavenging inorganic NPs 504–506
  skin 133
  types of 507
Candida albicans 291, 298, 306
Candida auris 290
Cannabidiol 39, 43
CAPE. *See* Caffeic acid phenethyl ester (CAPE)
Capparis spinosa 320
Capsaicin 131, 458

Capsicum 60
Caraway 59
Carbon-based nanomaterials 269
Carbon nanotubes (CNTs) 130–131, 159, 161, 164, 176, 232, 422, 506, 511
Carcinogenesis 203
Cardamom
  antioxidant compounds in 60, 68, 69
  antioxidant properties of 154–155
  chemical constituents in 150
  pharmacology of 67
  therapeutic effects of 60, 69
Cardioprotective medicines 418
Cardiovascular diseases (CVDs)
  associated risk factors 406–407
  conventional drug delivery systems 411
  dyslipidemia and 9
  hypertension and 405–406
  lifestyle changes for 407
  nanophytomedicines for 186–188
  nanotherapeutics of phytoantioxidants for 391
    advantages of 412–413
    AuNPs 418–419
    classes of 414–416
    CNTs 422
    coronary artery disease 417–418
    ethosomes 421–422
    frequently utilized techniques 413
    hypertension 418
    ischemic heart diseases 418
    limitations of 423
    liposomes 420–421
    magnetic nanoparticles 420
    phytosomes 421
    polymeric NPs 414, 419–420
  phytoantioxidants in 88–89
  prevalence and incidence of 405
  progression mechanism/pathway 407–408
  protective effect of natural antioxidants 14–15
  psychosocial factors 405
  treatment strategies for 408–412
    berries 411
    CAM 409–410
    coenzyme Q10 410–411
    curcumin 410
    natural products 409
    plant-based diets 410
    polyphenols 409
    resveratrol 411
    vascular gene therapy 409
β-Carotene 12, 34, 250, 499
Carotenoids 4, 12, 34, 84, 85, 140, 250, 481, 498
Carrot, antioxidants in 2
Carvedilol-A 131
Cassia fistula 130
Cassine xylocarpa 320, 328
Catalases 11, 13–14
Catechin 2, 3, 40, 43, 85, 108, 297, 498–499
  in green tea 87–88
  for inflammatory disorders 382
  loaded with polylactide nanoparticles 364
Catharanthus roseus 218
Cell organelles, dysfunction of 274–275
Cell signaling pathways, in oxidative stress 80, 81
Cellular dysfunction 80, 81
Cellular redox balance 78–79
Cellular senescence 6
Centella asiatica 126, 127
Central nervous system (CNS) disorders. *See* CNS disorders
Central neuropathic pain 452
Cephalotaxus harringtonia 256
Cerium NPs 504
Cerivastatin 17
CGA. *See* Chlorogenic acid (CGA)
Chagas disease 356, 358
Chalcone 63, 84
Chamaecostus cuspidatus 185

Chervil 59
Chikungunya 354, 355
Chili pepper, antioxidant compounds in 147
Chinese traditional medicine 257
Chitin 244
Chitosan 243–244
Chitosan-alginate nanoparticles 418
Chitosan NPs 156
Chives 59
Chloramphenicol acetyltransferase (CAT) 14
Chlorogenic acid (CGA) 36, 40, 43, 384
Chronic diseases
  phytomedicines for 42–44
  role of oxidative stress in 34–35
Chronic inflammation 10, 11, 181, 183
Chrysin-loaded polyurea dendrimer nanoparticles 364
Cilantro 59
Cinnamic acid 33, 88, 297
Cinnamon
  antimicrobial and antibacterial properties of 294–295
  antioxidant compounds in 60, 70
  therapeutic effects of 60, 70
Citrus aurantium 252
Citrus fruits 32, 42, 252
Clostridium botulinum 292
Clove essential oil (CEO) 156
Cloves
  antioxidant and antibacterial properties of 153, 294
  antioxidant compounds in 59, 68, 69, 147
  chemical constituents in 150
  eugenol in 66
  health benefits of 66
  therapeutic effects of 59, 69
CNS disorders, 188–190, 449–450. *See also* Aging and neurological disorders
  Alzheimer's disease 450
  amyotrophic lateral sclerosis 451
  etiology and progression of 450–452
  existing therapeutics for 449
  Huntington's disease 452
  ischemic stroke 452
  management of 453–454
  multiple sclerosis 451
  nanotherapeutics of phytoantioxidants for 462, 463
    beneficial effects of 466
    carbon nanotubes 464
    curcumin 465
    dendrimers 464
    EGCG 465
    exosomes 464
    limitations of 465, 466
    liposomes 462
    micelles 462
    polymeric NPs 462
    quercetin 465
    resveratrol 464
    SLNs 462
  neuropathic pain 452
  Parkinson's disease 450–451
  phytoantioxidants for 454–455, 457–460
    N-acetyl cysteine 456
    alkaloids 455
    challenges in administration of 456, 461
    polyphenols 455–456
    polyunsaturated fatty acids 456
    terpenoids 456
    vitamin D3 456
CNTs. *See* Carbon nanotubes (CNTs)
Coenzyme Q10 410–411
Coffee, antioxidants in 2
Colonic flavonoidal metabolism 109
Combretum caffrum 255
Complementary and alternative medicine (CAM) 409–410
Composite-based nanomaterials 270
Conjugated diene assay 147

COPD 394
Coriander
  antioxidant properties of 152
  chemical constituents in 149
  pharmacology of 67
Coronary artery disease 9, 406, 417–418
Coronavirus 292
Corticosteroids 182
Costus speciosus 185
p-Coumaric acid 40
Coumarin 298, 299
  dendrimer 339
  loaded with SLNs 364
COVID-19 394–395
Cow urine 22
C-reactive protein 9
Crocin 38, 385
Cruciferous vegetables 252
Cryptophycin 52 256
C-SBLNs. *See* Curcumin-loaded solid binary lipid NPs (C-SBLNs)
Cubosomes 335–336, 367
Cucumber, antioxidants in 2
Culinary herbs and spices 58
  in daily diet 65–67
  flavonoids in 63
Cumin
  antioxidant compounds in 147
  antioxidant properties of 154
  chemical constituents in 150
Curcuma longa 141, 256, 321, 328
Curcuma plus herbal capsule 44
Curcumin 20, 40, 85, 88, 130, 185, 191, 218, 256, 321
  anticancer activity of 499, 509
  antidiabetic effect of 437–438
  anti-inflammatory activity of 303
  antioxidant properties of 148, 151, 499
  in cancer nanomedicines 216
  delivery of 133, 134
  for inflammatory disorders 380, 390
  liposomes of 509
  loaded with nanocapsules 329
  loaded with eudragit-nutriosomes 364
  loaded with polylactide nanoparticles 364
  loaded with polymeric NPs 484–485
  mitochondrial redox potential 221
  nanoformulation 141, 302, 304
  for neurodegenerative conditions 392
  neuroprotective effect of 456, 458, 465, 466, 484–485
  NPs 186, 416–417
  PCur 391
  preventive effect on cancer 92
  protective role against CVDs 410
Curcumin and nisin-based polylactic acid NP (CurNisNp) 391
Curcumin-loaded solid binary lipid NPs (C-SBLNs) 389
Curcuminoids, sources of 84
CuZn-SOD 11
Cyclodextrin-based nanosponge 364–365
Cyclodextrins 111
Cyclooxygenase-2 (COX2) 504
Cymbopogon citratus 328, 337
Cyperus rotundus 321
Cytosolic GST 12

d

Daidzein 216, 383
Daily diet, culinary herbs and spices in 65–67
Daphne gnidium 321, 328
Dark chocolate 72
Dendrimers 131–132, 174, 189, 339, 464, 510
Dendrosome 510
Dengue 354, 355
Department of Science and Technology (DST), India 528
Diabetes mellitus (DM) 7–8, 130, 442
  nanophytomedicine for 183–185
  nanotherapeutics of phytoantioxidants for

benefits of 434
curcumin 437–438
EGCG 441–442
genistein 439–441
quercetin 434–437
resveratrol 438–439
phytoantioxidants in 87–88
types of 433
Diallyl sulfide 37
Dibenzoylmethane 216
Dietary ACNs 250
Dietary constituents, nano-drug delivery vehicles for 155–164
Dietary pattern 10
Dietary phytoantioxidants 57, 82, 84
Diethylcarbamazine 359
Differential scanning calorimetry (DSC) 248
Dihydroartemisinin 216
Dill 60, 68
2, 2-Dimethyl dichlorovinyl phosphate 453
Diospyros kaki 322
Disease and therapy, nanodelivery/ nanoparticulate systems in, 175. *See also* Nano(phyto)medicines
carbon nanotubes 176
liposomes 175–176
nanobiosensors 176–177
nanoparticles 175
nanowires 177
Disease management, natural antioxidants in 4–5
Disease-modifying antirheumatic drug (DMARD) 182
Diseases treatment, phytopharmaceuticals in 249–253
DMARD. *See* Disease-modifying antirheumatic drug (DMARD)
DNA
methylation status of 89
ROS-mediated damages to 80
DNA damage 274
assessment of 20
prevention of 20
ROS-induced 11
Doxil® 193
DPPH assay method 146, 148
DPPH, reaction mechanism with antioxidants 148
Drug delivery systems (DDSs) 205
ROS-responsive 206
Drug development 20–21
Drugs and Cosmetics Act of 1940 253
DSC. *See* Differential scanning calorimetry (DSC)
Dynamic light scattering (DLS) 247
Dyslipidemia
of metabolic syndrome 8
risk factor of cardiovascular diseases 9
Dysoxylum binectariferum 254

e

ECM. *See* Extracellular matrix (ECM)
E. coli 293
EC-SOD 11
EGCG. *See* Epigallocatechin gallate (EGCG)
Electrophile-responsive elements (EpREs) 106
Ellagic acid 130, 216
Ellagic acid-loaded cyclodextrin-based nanosponge 364–365
Elliptinium acetate 256
Emodin-loaded polymeric nanoparticles 365
Endogenous antioxidants 1, 496
defense mechanism 11–12, 77, 79, 89
Endoplasmic reticulum (ER)
disturbance of 275
stress 411
Endothelial dysfunction 9
Engineered nanostructures 267

Enhanced permeability and retention (EPR) effect 181
Entrapment efficiency 247
Enzymatic antioxidants 11–12, 479
Epicatechin 251
Epicatechin gallate (ECG) 85
Epigallocatechin gallate (EGCG) 16, 18, 85, 216, 256
  antidiabetic effect of 441–442
  antioxidant activity of 498–499
  for inflammatory disorders 382, 390
  loaded with nanoparticles 365
  loaded with polymeric nanoparticles 365
  nano EGCG 141–142, 306
  for neurodegenerative conditions 392
  neuroprotective effect of 456, 458, 465, 466
  PEG–PLA NPs of 181
Epigenetic diet 18
EPR effect. *See* Enhanced permeability and retention (EPR) effect
Eriodictyol 381
Erythrocyte membranes 20
Ethosomes 421–422
Eucalyptus 295–296
Eudragit
  application of 242
  grades of 241, 242
  oral drug release mechanisms of 242
  preparation of 241
Eudragit-nutriosomes 364
Eugenol 66, 216, 297
Euphorbia peplus 255
European Commission's Joint Research Center (JRC) 528
Europe, regulatory issues of nanophytotherapeutics in 528
Exercise 10
Exogenous antioxidants 1, 77, 496, 502
Exosomes 416, 464
Extracellular matrix (ECM) 504

f
Fasciola gigantica 358
Fasciola hepatica 358
Fascioliasis 356, 358
FDA-approved nano-medicines 174
Fennel 60, 68
Fenugreek 60, 73
Ferulic acid (FA) 33, 39, 43, 216, 384
Ficus benjamina 322
Ficus religiosa 184
Fisetin, loaded with polylactide nanoparticles 364
Fish, omega-3 fatty acids in 252–253
Flanones 63
Flavandiol 63
Flavanones 32, 42, 84, 101, 481
Flavan-3-ols 101
Flavones 63, 84, 101, 251, 297
Flavonoids 55, 82, 297, 434
  anticancer activity of 511
  anti-inflammatory activities of 17
  antimicrobial activities of 299
  antioxidant property of 99, 112
  antiviral activity of 251
  bioavailability of 42, 107, 109, 111
  cardioprotective activity of 112
  in culinary herbs and spices 63
  effects on human health 2
  inhibition of acetylcholinesterase 251
  mechanisms of antioxidant effect of 102–107
    free radical scavenging action 105
    lipophilic antioxidant protection 107
    metal ion chelation effect 105–106
    modulation of endogenous enzyme activities 106–107
  medical importance of 63, 64
  metabolism of 42, 108
    colonic 109
    enterohepatic circulation in 108, 111
    phase I biotransformation 108
    phase II conjugation 108–109

natural 100
pharmacology of 64
semisynthetic alterations in 112
sensitivity 192
sources of 19
structure of 102–103
subclasses of 63, 100, 101, 351
in therapeutics 112–113
Flavonolignan 511
Flavonols 32, 42, 63, 100, 101, 108, 251, 481
Fluvastatin 17
Folic acid 131
Food and Drug Administration (FDA) 527–528
Fosbretabulin tromethamine 255
Fourier transform infrared spectroscopy (FTIR) 246
Francisella tularensis 290
Fraxinus angustifolia 218
Free fatty acids (FFAs) 8, 9
Free radicals 4, 31, 34, 222, 496
Free radical scavenging action 105
Fruits
antioxidants in 2, 5, 6, 36
residues 3
waste 139
Functional foods 15, 190, 377

g

Galangin 297
Gallic acid (GA) 39, 43, 130, 132, 327
loaded with graphene oxide-based nanoformulation 365
loaded with polymeric nanoparticles 365, 485
for neurodegenerative conditions 393
neuroprotective effect of 485
Gallocatechin (GE) 4
Gambogic acid 216
Garlic
antimicrobial activity of 295, 298
antioxidant compounds in 60, 68, 70
antioxidant properties of 152
black 152
chemical constituents in 149
health benefits of 73
therapeutic effects of 60, 70
Garlic powder, toxicity study of 45
Gas condensation technique 232
Gastrointestinal disease 17
Gastrointestinal (GI) toxicity 279
GE. *See* Gallocatechin (GE)
Gelatin 126
Gene therapy 453
Genetic modification 501
Genistein 216, 251, 255, 307, 459
anticancer activity of 499, 500
antidiabetic effect of 439–441
for inflammatory disorders 384, 390
loaded with polymeric micelles 366
nano-genistein 141
Genotoxicity 274
Ginger 43
antimicrobial activity of 295
antioxidant compounds in 60, 68, 70, 147
antioxidant properties of 148
chemical constituents in 149
health benefits of 66, 73
nano-drug delivery systems 185
therapeutic effects of 60, 70
Gingerol 38
Ginkgo biloba 127, 456
Ginkgolide B 459
Ginsenoside 131, 305–306
Ginsenoside Rg3 (Rg3) 187
Glucocorticoids 182
Glucose homeostasis 7
α-Glucosidase inhibitors 8
Glucosinolates 252
Glucuronidation 109
Glutathione (GSH) 12, 14, 17
Glutathione reductase (GR) 14

Glutathione-*S*-transferases (GST) 12–14
Glycin max 255
Glycyrrhiza glabra 323
Gold nanoparticles (AuNPs) 130, 157, 163, 175, 185, 232, 305, 329, 418–419, 505–506, 510
Gonorrhea 292
Gossypium herbaceum 256
Gout 393
GR. *See* Glutathione reductase (GR)
Grapes
  antioxidants in 2
  phytonutrients in 73
Grape seed proanthocyanidin extract (GSPE) 19
Graphene oxide-based nanoformulation 365
Graphene oxide (GO) composite 339–340
Green nanotechnology 233
Green synthesis of NPs 233–236
Green tea 127, 251
  catechins in 43, 87–88
  polyphenols 89
  synthesis of NPs 234
Growth factors (GFs) 419
GSC composite (GSCC) 340
GSSG 17
Guava, antioxidants in 2

h

Halloysite clay nanotube 132–133
HBA-containing copolyoxalate (HPOX) 209
HD. *See* Huntington's disease (HD)
Heart diseases, nanophytomedicines for 186–188
Heart failure 406
Helichrysum aureonitens 297
Hepatic metabolism, of bioactive flavonoids 108, 111
Hepatitis B Virus (HBV) 291–292
Hepatotoxicity 15–16, 277
Herbal medicines, 239–240. *See also* Phytomedicines
Herbs 145
  antioxidant compounds in 57–62, 147
  antioxidant properties of 55, 62, 148–155
  application of 64
  chemical composition of 62
  chemical structure of 69–72
  classification of 57, 58
  definition of 56
  extraction of active compounds 74
  health benefits of 56, 73
  medicinal 294
  nano-drug delivery vehicles for 155–164
  therapeutic effects of 58–62
  traditional and commercial value of 73–74
Hesperetin 381
Hesperetin-loaded nanocrystals 366
Hexosomes 367
High-density lipoprotein-cholesterol (HDL-c) 8, 9
H_2O_2 11, 12, 14, 204, 211, 212
Honey 15–16, 18, 234
Honokiol 216
Human African trypanosomiasis 356, 358
Human diseases, oxidative stress and 6–14
Human immunodeficiency virus (HIV) 325
Huntington's disease (HD) 452
Huperzine A 455
Hydrogel 164, 337–338
Hydrogen peroxide 78
Hydroxybenzoic acids 33
10-Hydroxy camptothecin 254
Hydroxycinnamic acids 33
Hydroxyl 78, 479
Hydroxyl radical scavenging activity 148
Hydroxytyrosol 192
Hypercholesterolemia 17

Hyperglycemia 7–8
Hypericin 303
Hyperlipidemia 8–9
Hypertension
 and CVDs 405–406
 and MetS 6, 9
 nanoparticles for treatment of 418
 nanophytomedicines for 186–188
 and T2DM 9–10

i

IBD. *See* Inflammatory bowel disease (IBD)
Immune system, effect of nanoparticles on 276–277
Immunotherapy 453
Immunotoxicity 276–277
Incidental nanostructures 267
India
 ASU (Ayurveda, Siddha, Unani) 257
 Ayurveda system 135
 phytopharmaceutical market in 257–258
 regulatory issues of nanophytotherapeutics in 528–531
 traditional medicinal practices 22–23
 turmeric usage in 151
Indian diet
 ginger-garlic paste 66
 health benefits of 65
 herbs and spices
  antioxidants in 67–68
  culinary 65–67
  dietary 67
Indian Systems of Medicine (ISM) 351
Infection 289
Infectious diseases 20, 179–180
Inflammation 273–274, 377
 nanomedicine targets in 214
 phytoantioxidants in 90
Inflammatory bowel disease (IBD) 389, 391
Inflammatory disorders 181–183
 nanotherapeutics of phytoantioxidants for 378–379
  COPD 394
  COVID-19 394–395
  CVDs 391
  extended antioxidant effects 388–390
  gout 393
  IBD 389, 391
  metabolic syndrome 393
  metallic NPs 388
  nanocrystals 387
  nanoemulsions 379
  nanoliposomes 379
  nanopolymersomes 387
  neurodegenerative conditions 392–393
  polymeric NPs 387–388
  rheumatoid arthritis 394
  solid lipid nanoparticles 387
  summary of 380–386
  therapeutic advantages 379, 388
 prevalence and incidence of 377
Inorganic-based nanomaterials 269
Inorganic NPs 504–506
Insulin resistance 7, 9, 183
in vivo drug release studies 249
Irinotecan hydrochloride 254
Iron oxide magnetic nanoparticles (MNPs) 506
Ischemia–reperfusion (IR) injury 212, 213
 natural nanotherapeutics in 218, 220–221
Ischemic heart diseases 418
Ischemic stroke 452, 454
Isoflavones 63, 353
Isoflavonoids 84
Isoquercitrin 37

k

Kaempferol 85, 307
Kale 72
Kanglaite (*Coix* seed oil) 256

KBr pellets approach 246
Keampferol 382
Kidney, nanotoxicity to 278

l

Lactase phlorizine hydrolase (LPH) 107
β-Lapachone 216
Lecithin-based mixed polymeric micelles (LMPMs) 436
Legumes 65
Leishmaniasis 354, 355
Leonotis leonurus 184
Leprosy 357, 359
Leucocyanidin 2–3
Lignans, sources of 84
Limonoids 252
Linalool 460
Linseed oil 192
Linxian 502
Lipid-based nanoparticles 125, 127–129, 134
Lipid microspheres (LMs) 182–183
Lipid nanoparticles (NPs) 189, 302
Lipid peroxidation 479
  prevention of 19–20
Lipids, ROS-mediated damages to 80
α-Lipoic acid 16
Lipophilic antioxidant protection 107
Liposomes 175–176, 301–302, 334–335, 366, 420–421, 462, 483, 508–509
Lipotoxicity 15
Liquid crystals (LC) 127, 129
  nanoparticles 367
Listeria monocytogenes 291, 298
Listeriosis 291
Lobeline 457
Lovastatin 17, 256
Low-density lipoprotein-cholesterol (LDL-c) 8, 9
LPH. *See* Lactase phlorizine hydrolase (LPH)
Lungs, nanotoxicity to 277
Lupeol 15
Lutein 385
Luteolin 459
  for inflammatory disorders 381
  loaded-with SLNs 364
  loaded-with water-soluble copolymer-encapsulated nanoparticles 366
Lycium barbarum 132
Lycopene 34, 251
  antioxidant activity of 499
  for inflammatory disorders 386
Lymphatic filariasis 357, 359
Lysosomal rupture 274–275

m

Mace, antioxidant compounds in 147
Magnetic nanoparticles (MNPs) 129, 175, 329, 420, 511, 512
  iron oxide 506
  neuroprotective effect of 486
Magnolia officinalis 323, 328
Malaria 354
Mallows 295
Mango, antioxidants in 2
Marjoram 68, 70
  antioxidant compounds in 147
Marrubiin 184
Marsilea quadrifolia 185
Maytansine 255
Maytenus cuzcoina 323, 328
Maytenus ovatus 255
MC. *See* Myricetin (MC)
Medical nanorobots 178
Medicinal herbs 22, 57, 58
Medicinal plants 74, 294
Medihoney 18
Melatonin 14, 16
Mesoporous silica nanoparticles 366, 485–486
Metabolic disorders 87–88
Metabolic syndrome (MetS) 393
  dyslipidemia of 8
  nanophytomedicine for 183–185

risk factors 6
dietary habit 10
hypertension 9
obesity 7
sedentary lifestyle 10
Metal ion chelation effect 105–106
Metallic nanoparticles 129–130, 156–158, 231, 235, 329
Micelles 157–160, 189, 336–337, 366, 462, 509–510
Microbes 289, 290
Microbial infections 289–290
Candida albicans 291
Clostridium botulinum 292
gonorrhea 292
HBV 291–292
Listeria monocytogenes 291
phytoantioxidants for 294–301
ringworm 293
roles of traditional plants against 294–296
SARS-COV-2 292
tuberculosis 292
Microemulsion 303, 337
Micronutrients 501
Microwave cooking 107
Mint 295
health benefits of 73
in Indian diet 66
Mitochondrial dysfunction 275, 410, 418
Mitochondrial oxidative stress 497
Mitogen-activated protein kinases (MAPKs) 17
MNPs. *See* Magnetic nanoparticles (MNPs)
Mn-SOD 11
Modulation of endogenous enzyme activities 106–107
Monounsaturated fatty acid (MUFA) 10
Morin 383
Morphine 457
Mucilaginous herbs 58
Multidrug resistance (MDR) 293
Multiple sclerosis (MS) 451
Multiwalled CNTs (MWNTs) 176, 181
Mustard 60
Mutagenicity 278
Mycobacterium leprae 359
Mycobacterium tuberculosis 292
Mycobacterium ulcerans 358
Myocardial hypertrophy 406
Myocardial infarction 418
Myricetin (MC) 40, 42, 43
loaded with mesoporous silica nanoparticles 366
loaded with SLNs 364

n

NADPH oxidase (NOX2) 211
Nanoantibiotics 179
Nano apigenin 306–307
Nanobiosensors 176–177
Nanocapsules (NCs) 329, 332, 484
Nanocrystals 366, 387
Nanocurcumin 141, 304
Nanodelivery systems
in disease and therapy 175
for herbal/phyto-antioxidants 140
challenges 142
future perspectives 142
nano-curcumin 141
nano-genistein 141
nano-quercetin 141
nano-resveratrol 140
therapeutic advantages of 142
of herbs, spices, and dietary constituents 155, 163–164
CNTs 159, 161
metallic NPs 156–158
micelles 157–160
nanoemulsions 161–162
NPs 156
patents 164, 165
SLNs 159, 161
nanocarriers for phytoconstituents 124–126

Nanodelivery systems (*cont'd*)
benefits of 123
carbon nanotubes 130–131
dendrimers 131–132
lipid-based 125, 127–129
metallic 129–130
nanofibers 132
nanoparticular carriers 124
nanovesicular carriers 124
polymeric 125–127
for skin cancer 133
target specificity and toxicity concerns of 133–134
unsolved issues with 135
Nano drugs, 268. *See also* Nano(phyto) medicines
Nano EGCG 306
Nanoemulsions 161–163, 190
for inflammatory disorders 379
for parasitic diseases and NTDs 363
Nanoencapsulation technique 191
Nano-epigallocatechin-3-gallate (EGCG) 141–142
Nanofibers 132, 164
Nanoflavanoids 191–192
Nanogels 190, 337–338
Nanogenerators 178–179
Nano genistein 141, 307
Nano ginsenosides 305–306
Nanogold (AuNGs) 394
Nanoids 178
Nano kaempferol 307
Nanoliposomes 379
Nanomaterials, types of 269–270
Nano(phyto)medicines 175, 222, 268
advantages of 521, 523
in Alzheimer's disease 212–214
for arthritis and inflammatory disorders 181–183
as biomedical devices for disease treatment 177–179
for cancer 180–181, 210, 211
in CVDs and heart diseases, hypertension 186–188
definitions of 523
for diabetes and MetS 183–185
dysfunction of cell organelles 274–275
FDA-approved 174
future course of action 524
against infectious diseases 179–180
in inflammation 214
in ischemia–reperfusion injury 212, 213
in neurological disorders 188–190
as nutraceuticals/functional foods 190–192
opportunities, scopes, and challenges 192–193
oxidative stress, inflammation, and genotoxicity 273–274
in rheumatoid arthritis 214
as therapeutics 179–190
in wound healing 210–212
Nano micelles 509–510
Nanomission 528
Nanoparticles (NPs) 239
AgNPs 129–130, 157, 185, 218, 235, 506
antimicrobial 179
AuNPs 130, 157, 163, 175, 185, 232, 305, 329, 418–419, 505–506, 510
biodegradable polymeric 280
cationic charge on 280
chitosan 156, 244
curcumin 186, 416–417
environmental pollution 273
green synthesis of 233–235
from plants 235–236
using biological entities 236
hydrophilicity/hydrophobicity of 482
inorganic 504–506
lipid 125, 127–129, 189, 302
liquid crystal 367
magnetic 130, 163, 332
metal-based 218

- metallic 156–158, 231, 235–236, 308, 329, 388
- nanodelivery of herbs/spices/dietary constituents 156
- nondegradable inorganic 279–280
- perfluorocarbon 175
- platinum 231, 235–236
- PLGA 181
- polylactide 364
- polymeric 125–127, 189, 206, 302–303, 328–329, 337, 365, 387–388, 414, 419–420, 462, 484–485, 508
- polyphenolic 510–511
- production methods of 232
  - bottom-up approach 232–233
  - top-down approach 232
- silica 277
- SLNs 159, 161, 182, 184, 189, 192, 302, 364, 387
- TiO_2 279, 305
- toxicity of 273
- ZnNPs 157
- ZnS 270

Nanoparticulate carriers 124
Nanophytoantioxidants 191–192, 303–308
Nanophytopharmaceuticals. *See also* Phytopharmaceuticals
- fabrication of, polymers in 240–241
  - alginates 244–245
  - chitosan 243–244
  - Eudragit 241–242
  - PLA and PLGA 242–243

Nanophytotherapeutics
- need for regulations 524–525
- regulatory issues of 525, 526
  - in Europe 528
  - in India 528–531
  - in USA 527–528

Nanopolymersomes 387
Nanoquercitin 141, 305
Nano resveratrol 140, 307–308
Nanorobots 178
Nanospheres 508
Nanosponges 179
Nanostructured lipid carriers (NLCs) 127, 192, 333–334, 363
Nanostructures 267
- chemical composition of 268
- crystalline structure 270
- engineered 267
- exposure to 272–273
- incidental 267
- naturally produced 267
- particle chemistry 270
- particle shape of 271
- size and surface area of 270–271
- sources of 267
- surface charge of 271, 272
- surface functionalization of 272

Nanotechnology 111, 173–174, 231, 434
- delivery of phytoconstituents 122–124
- green 233
- in nutraceuticals 191

Nanotherapeutics of phytoantioxidants, for aging and neurological disorders 482–483
- limitations of 483
- liposomes 483
- magnetic nanoparticles 486
- mesoporous silica nanoparticle 485–486
- polymeric NPs 484–485
- SLNs 484

Nanotherapeutics of phytoantioxidants, for cancer 507–508
- carbon nanotubes 511
- dendrimers 510
- drawbacks of 510
- gold nanoparticles 510
- liposomes 508–509
- magnetic nanoparticles 511, 512
- nano micelles 509–510
- polymeric NPs 508
- polyphenolic nanoparticles 510–511
- SLNs 509

Nanotherapeutics of phytoantioxidants, for CNS disorders, 462, 463. *See also* Aging and neurological disorders
beneficial effects of 466
carbon nanotubes 464
curcumin 465
dendrimers 464
EGCG 465
exosomes 464
limitations of 465, 466
liposomes 462
micelles 462
polymeric NPs 462
quercetin 465
resveratrol 464
SLNs 462
Nanotherapeutics of phytoantioxidants, for CVDs 391
advantages of 412–413
AuNPs 418–419
classes of 414–416
CNTs 422
coronary artery disease 417–418
ethosomes 421–422
frequently utilized techniques 413
hypertension 418
ischemic heart diseases 418
limitations of 423
liposomes 420–421
magnetic nanoparticles 420
phytosomes 421
polymeric NPs 414, 419–420
Nanotherapeutics of phytoantioxidants, for diabetes mellitus
benefits of 434
curcumin 437–438
EGCG 441–442
genistein 439–441
quercetin 434–437
resveratrol 438–439
Nanotherapeutics of phytoantioxidants, for inflammatory disorders 378–379
COPD 394
COVID-19 394–395
CVDs 391
extended antioxidant effects 388–390
gout 393
IBD 389, 391
metabolic syndrome 393
metallic NPs 388
nanocrystals 387
nanoemulsions 379
nanoliposomes 379
nanopolymersomes 387
neurodegenerative conditions 392–393
polymeric NPs 387–388
rheumatoid arthritis 394
solid lipid nanoparticles 387
summary of 380–386
therapeutic advantages of 379, 388
Nanotherapeutics of phytoantioxidants, for parasitic diseases and NTDs 360–362
apigenin-loaded nano-transferosomes 362
baicalein-loaded nanostructured lipid carriers 363
baicalin-loaded nanoemulsions 363
catechin, curcumin, and fisetin-loaded polylactide NPs 364
chrysin-loaded polyurea dendrimer NPs 364
coumarin, luteolin, and myricetin-loaded SLNs 364
curcumin-loaded eudragit-nutriosomes 364
7,8-dihydroxyflavone-loaded liposomes 366
EGCG
loaded with nanoparticles 365
loaded with polymeric nanoparticles 365
ellagic acid-loaded cyclodextrin-based nanosponge 364–365

emodin-loaded polymeric NPs 365
gallic acid
loaded with graphene oxide-based nanoformulation 365
loaded with polymeric nanoparticles 365
genistein-loaded polymeric micelles 366
hesperetin-loaded nanocrystals 366
luteolin-loaded water-soluble copolymer-encapsulated NPs 366
myricetin-loaded mesoporous silica NPs 366
naringenin
loaded with liposomes 366
loaded with liquid crystalline nanoparticles 367
piperine-loaded NPs 365
quercetagetin-loaded zein and hyaluronic acid-based NPs 367
quercetin-loaded polymeric NPs 365
rutin-loaded liquid crystalline NPs 367
Nanotherapeutics of phytoantioxidants, for viral infections 330–331
dendrimers 339
GO composite 339–340
lipid-based
NLCs 333–334
phospholipid complexes 332–333
SLNs 333
magnetic NPs 329
metallic NPs 329
nanocapsules 329, 332
nanogels 337–338
patents for 341, 342
polymeric NPs 328–329
SNEDDS 341
vesicular system-based
cubosomes 335–336
liposomes 334–335
micelles 336–337
microemulsion 337
niosomes 335
PEGylated liposomes 335
Nanothymoquinone 306
Nanotoxicity
consumer exposure 272
effect on different organs 275, 276
GI toxicity 279
hepatotoxicity 277
immunotoxicity 276–277
mutagenicity 278
nephrotoxicity 278
neurotoxicity 275, 276
pulmonary toxicity 277
reproductive toxicity 278
skin irritation 278–279
environmental exposure 273
nanomedicine-induced
dysfunction of cell organelles 274–275
oxidative stress, inflammation, and genotoxicity 273–274
occupational exposure 273
physicochemical aspects of nanostructures responsible for 268–272
Nano-transferosomes 362
Nano tricetin 307
Nanotubes
CNTs 130–131, 159, 161, 164, 176, 232
halloysite clay 132–133
MWNTs 176, 181
SWCNTs 159, 161
Nanowires (NWs) 177
Nanozymes 221
Naphthoquinone 511
Naringenin 41, 85, 130, 216, 306, 359, 458, 466
for inflammatory disorders 381
loaded with liposomes 366
loaded with liquid crystalline nanoparticles 367

Natural antioxidants 65, 139–140, 433, 504–505
chemical structures of 1–4
classification of 1, 3, 33, 100
demand for 146
health benefits of 4–5
molecular level protective mechanism 19–20
as nutraceuticals 5–6
protective effect of
in cancer 16
in CVDs 14–15
in hepatoxicity 15–16
in lipotoxicity 15
nephroprotective effect 19
in obesity 15
in T2DM 14
restorative effect of
in aging and skin health 18
in atherosclerosis 17
in gastrointestinal disease 17
in inflammation 16–17
in NDs 16
in wound healing 18–19
sources of 1, 2, 33
in therapeutic drug development 20–21
Naturally produced nanostructures 267
Natural nanotherapeutics
in cancer 214–217
in ischemia–reperfusion injury 218, 220–221
in wound healing 217–220
Natural product drug discovery 21
Natural products 121–122
NCs. *See* Nanocapsules (NCs)
Neem oil 161, 162
Neglected tropical diseases (NTDs), 353. *See also* Nanotherapeutics of phytoantioxidants, for parasitic diseases and NTDs
phytoantioxidants for 356–357
Buruli ulcers 358
Chagas disease 358
fascioliasis 358
Human African trypanosomiasis 358
leprosy 359
lymphatic filariasis 359
nanoformulation 360–367
onchocerciasis 358
schistosomiasis 359
trachoma 359
Neisseria gonorrhoeae 292
Neisseria meningitidis 289
Neobavaisoflavone 359
Nephrotoxicity 278
Neurodegenerative diseases (NDs) 5, 16, 91, 392–393
Neurological disorders, 188–190. *See also* Aging and neurological disorders; CNS disorders
Neuropathic pain (NP) 452, 454
Neurotoxicity 275, 276
Neutraceuticals 113
Neutrophils 220
NFkB 10, 17, 86, 90
Nicotinamide adenine dinucleotide phosphate (NADPH) 34
Nicotine 457
Nigella sativa 234, 235, 306, 391
Niosomes 335
Nitric oxide 78
NLCs. *See* Nanostructured lipid carriers (NLCs)
Nobiletin 216
Noisome 164
Nondegradable inorganic nanoparticles 279–280
Nonenzymatic antioxidants 12, 479
Nonsteroidal anti-inflammatory drugs (NSAIDs) 151, 182–183, 453
Nontoxic nanocarriers 279–282
Novel drug delivery systems (NDDSs) 239, 240
NP. *See* Neuropathic pain (NP)

NTDs. *See* Neglected tropical diseases (NTDs)
Nuclear analytical techniques 280
Nuclear factor-like 2 (Nrf2) 106, 498, 499
Nutmeg
 antioxidant compounds in 60, 68, 71, 147
 therapeutic effects of 60, 71
Nutraceuticals
 antioxidant activity of 500–501
 benefits of 191
 nanophytomedicine as 190–192
 nanotechnology in 191
 natural antioxidants as 5–6
Nutritional genomics 501
Nutritive herbs 58
Nutrova kera strength capsule 44
NWs. *See* Nanowires (NWs)

o

Obesity 7, 392
 prevalence of 15
 protective effect of natural antioxidants 15
Obesity-induced inflammatory stress 15
Oleuropein 192
Olivenol™ 44
Omacetaxine mepesuccinate 256
Ombrabulin 255
Omega-3 fatty acids in 252–253
Omega woman 44
ω-3 polyunsaturated fatty acids 501
Onchocerca volvulus 358
Onchocerciasis 357, 358
Onion, 434. *See also* Quercetin
 antioxidant compounds in 71
 antioxidant properties of 151
 chemical constituents in 150
 therapeutic effects of 71
Oregano 61, 68
 antioxidant compounds in 147
 antioxidant properties of 153–154
 chemical constituents in 150
 health benefits of 73
Organic-based nanomaterials 270
Organization for Economic Co-operation and Development (OECD) 529
Organosulfur compounds 84, 481
Organosulfur phytochemicals 298, 300
Orientin 41
Ornamental herbs 58
Osteoarthritis 90–91
Oxidants 31, 78
Oxidative stress 5, 55, 77, 204, 274, 441, 486, 512
 acute 79
 antioxidants role in ameliorating 80, 82
 in cancer 496
 cell signaling pathways in 80, 81
 cellular-level consequences of 80
 chronic 80
 development of chronic disease 34–35
 H_2O_2-induced 218
 and human diseases 6–14
 injuries 205
 management 205
 role of oxidants in 78
 ROS as the main inducer of 79–90

p

Paclitaxel 128, 131, 256
Padina gymnospora 234, 236
Panax ginseng 324
Panax notoginsenoside (PNS) 187
Papaverine 410
Parasitic diseases 352–353
 phytoantioxidants for 353, 354
  chikungunya 355
  dengue 355
  leishmaniasis 355
  malaria 354
  soil-transmitted helminths 355
  Zika virus 355

Parkinson's disease (PD) 5, 6, 16, 91, 106, 449–451, 454
Patents 164, 165, 341, 342
PCur 391
Pecans 72
Peepal tree 184
PEGylated liposomes 335
Peppermint
  antioxidant compounds in 61
  pharmacology of 66–67
  therapeutic effects of 61
Percent yield 247
Perennial herbs 57
Perfluorocarbon NPs 175
Peroxidation of lipids 19–20
Peroxynitrite 107
Phenolic acids 33, 84, 85, 352
Phenolic compounds 1–4, 15, 17
Phosphatidylcholine (PC) 333
Phospholipase C (PLC) 89
Phospholipid complexes 332–333
Phyllanthus emblica 324, 328
Physical (in)activity 10
Phytoantioxidants, 82, 139–140, 294–301. *See also* Nano phytoantioxidants
  as activators of endogenous antioxidant pathways 86
  for aging and neurological disorders 480–481
  in arthritis 90–91
  bioactive flavonoids as 99–113
  in cancer 91–92
  in cardiovascular diseases 88–89
  as cellular oxidative stress signaling modulator 86–87
  chemical structures of 352
  classification of 82, 83
  in clinical trial 502–503
  for CNS disorders 454–455, 457–460
    N-acetyl cysteine 456
    alkaloids 455
    challenges in administration of 456, 461
    polyphenols 455–456
    polyunsaturated fatty acids 456
    terpenoids 456
    vitamin D3 456
  in diabetes and metabolic disorders 87–88
  as free radical scavengers 86
  in inflammation 90
  for microbial infections
    challenges of using 301
    for microbial infections 294–300
    nanoformulations 301–303
  nanodelivery systems for 140
    challenges 142
    future perspectives 142
    nano-curcumin 141
    nano-genistein 141
    nano-quercetin 141
    nano-resveratrol 140
    therapeutic advantages of 142
  nanoformulation of 353, 360–367
  in neurodegenerative disorders 91
  for NTDs 356–357
    Buruli ulcers 358
    Chagas disease 358
    fascioliasis 358
    Human African trypanosomiasis 358
    leprosy 359
    lymphatic filariasis 359
    onchocerciasis 358
    schistosomiasis 359
    trachoma 359
  for parasitic diseases 353
    chikungunya 355
    dengue 355
    leishmaniasis 355
    malaria 354
    soil-transmitted helminths 355
    Zika virus 355
  as phytopharmaceuticals 92–93

polyphenols as 85
protective mechanisms of 86–87
sources of 83, 84
Phytochemicals 1, 294, 296, 308, 377, 496–498
as anticancer agents 496–500, 507–508
antioxidant activity of 498
clinical evidence of 502–503
Phytoconstituents 121
delivery of
challenge for 121–122
nanotechnology in 122–124
difficulty of isolation 122
nanocarriers for 124–126
benefits of 123
carbon nanotubes 130–131
dendrimers 131–132
lipid-based nanoparticles 125, 127–129
metallic 129–130
nanofibers 132
nanoparticulate carriers 124
nanovesicular carriers 124
polymeric nanoparticles 125–128
skin cancer 133
target specificity and toxicity concerns of 133–134
unsolved issues with 135
Phytolacca decandra 127
Phytomedicines 31
with antioxidant potentials 36–41
bioavailability and metabolism of 36, 42
marketed preparations of 44
in prevention/mitigation of chronic diseases 42–44
sources of 32–34
toxicity and long-term risk assessment of 44–45
Phytonutrients 72–73
Phytopharmaceuticals, 92–93. *See also* Nanophytopharmaceuticals
characterization of
AFM 246
DSC 248
FTIR 246
particle size 247
polydispersity 247–248
SEM 245–246
solubility studies 248–249
TEM 246
TGA 248
UV–Vis spectroscopy 246
in vivo drug release studies 249
X-ray diffraction 249
zeta potential 248
formulations 253
clinical aspects of 253
clinical trial of 253–256
market in India 257–258
in treatment of diseases 249–253
Phytosomes 303, 421
Phytotherapy 82
Piceatannol 298
Piperine 457
Piperine-loaded nanoparticles 365
Pirarubicin 131
Placental GF (PIGF) 418
Plant-based diets 410
Platinum nanoparticles (PtNPs) 231, 235–236
Pleurotus ostreatus 256
PLGA NPs 181
PNS. *See* Panax notoginsenoside (PNS)
Polioviruses 325
Poly(ε-caprolactone) (PCL) 156
Polyamidoamine 131
Polycaprolactone (PCL) 329
Polycyanoacrylate NPs 189
Poly (glycerol-succinic acid) dendrimers 510
Polydispersity index (PDI) DLS 247
Polyesters 242–243
Poly lactic acid (PLA) 242–243, 303

Polylactic-*co*-glycolic acid (PLGA) 141, 242–243, 303, 419, 420, 508
Polylactide nanoparticles 364
Polymeric micelles 366
Polymeric nanoparticles 125–128, 189, 206, 302–303, 328–329, 337, 365, 387–388, 414, 419–420, 462, 484–485, 508
Polymers
  fabrication of nanophytopharmaceuticals
    alginates 244–245
    chitosan 243–244
    Eudragit 241–242
    PLA and PLGA 242–243
  for fabrication of nanosystems 240–241
Poly(*n*-butyl cyanoacrylate) (PBCA) NPs 189
Polyoxalate 207, 209
Polyphenolic nanoparticles 510–511
Polyphenols 1, 6, 15, 21, 140, 297–299, 351, 481, 498
  in cancer 91–92
  as cellular oxidative stress signaling modulator 86
  classification of 82, 83
  as major phytoantioxidants 85
  in neurodegenerative disorders 91
  protective role against CVDs 409
  radical scavenging activities of 86
Poly-(1,4-phenyleneaceton dimethylene thioketal) (PPADT) 209
Polyunsaturated fatty acids (PUFA) 10, 456
Polyurea dendrimer nanoparticles 364
Pomegranate, antioxidants in 2
Poppy 61
Potato, antioxidants in 2
Primary metabolites 296
Primrose 296
Proanthocyanidins 2, 3, 19, 33, 63
Prostaglandin synthase 2
Protein corona
  hard corona 280
  precise manipulation of 280–281
  soft corona 280
Protein modification 20
Proteins, ROS-mediated damages to 80
Proteus mirabilis 298
Provencal herbs 334–335
Provitamin A 34
Prunus × yedoensis 234
Puerarin 132, 187
Puerarin lobata 132
Pulmonary toxicity 277
Punica granatum 234

q

Quantum dots (QDs) 175, 177, 506
Quantum effects 222
Que-LMPMs 436
Quercetagetin-loaded zein and hyaluronic acid-based composite nanoparticles 367
Quercetin 32, 39, 43, 64, 85, 89, 109, 130, 188, 192, 251
  activation of AMPK 435
  antidiabetic effect of 434–437
  anti-inflammatory property of 435
  antioxidant properties of 151
  bioavailability 111
  in cancer nanomedicines 217
  delivery of 128, 131
  for inflammatory disorders 383, 390
  liposomes 509
  loaded with polymeric NPs 365, 485
  micelles' formulation 336
  nanoquercitin 141, 305
  for neurodegenerative conditions 392
  neuroprotective effect of 456, 458, 465, 466, 485
  regulatory effects on glycemia 435–437
  semisynthetic alterations in 112
Quercetin-3-O-glucuronide (Q3GA) 435

r

RA. *See* Rheumatoid arthritis (RA)
Raphanus sativus 252
Raphasatin 252
Raspberries 72
Reactive nitrogen species (RNS) 55, 56, 204, 327, 504
Reactive oxygen species (ROS), 10–11, 273–274, 326–327, 377. *See also* Oxidative stress
  acute overproduction of 205
  in carcinogenesis 203
  chronic overproduction of 205
  elevated ROS levels 214
  generation of 1, 5, 56, 327
  level 13, 17
  in physiological processes 203, 204
  production
    endogenous 203–206
    exogenous 204
  release of 77
  risk of 294
  role in aging and neurological disorders 477–479
  ROS-responsive DDSs 206
Red cabbage 73
Redox balance 78–79, 82
Redox-based combinational nano therapy 503–504
Red pepper 68, 71
  antioxidant compounds in 147
Reperfusion, 212. *See also* Ischemia-reperfusion (IR) injury
Reproductive toxicity 278
Residues, fruit 3
Resveratrol 85, 88–92, 130, 188, 192, 251–252, 298
  anticancer activity of 499
  antidiabetic effect of 438–439
  antioxidant activity of 499
  in cancer nanomedicines 217
  cubosomes 335–336
  for inflammatory disorders 380, 390, 391
  loaded with polymeric NPs 485
  nano resveratrol 140, 307–308
  for neurodegenerative conditions 393
  neuroprotective effect of 252, 456, 458, 464, 466, 485
  protective role against CVDs 411
Retinoic acids (RAs) 12
Reverse nano micelles 509–510
Rheumatoid arthritis (RA) 90, 182, 214, 232, 394
Ringworm 293
Risk assessment of phytomedicines 44–45
RNA, ROS-mediated damages to 80
RNA interference (RNAi) 325
Rosemary
  antioxidant compounds in 61, 68, 71, 147
  therapeutic effects 61, 71
Rosemary oil 133
Rose oil 16
ROS-induced DNA damages 11
Rosmarinic acid 41
ROS-mediated damages to biomolecules 79–81
Rosmeric acid 154
ROS-prompted drug release 206
ROS-receptive chemical structures 206–207
ROS-scavenging inorganic NPs 504–505
  AgNPs 506
  AuNPs 505–506
  carbon nanotubes 506
  iron oxide MNPs 506
  quantum dots 506
  transition metal NPs 505
ROS-targeting nano-materials
  boronic ester groups 209
  polyoxalate 209
  polypropylene sulfide 207–208

ROS-targeting nano-materials (*cont'd*)
selenium-containing polymers 208
tellurium-containing polymers 208
thioketals 208–209
Rutin 37, 85, 42, 44, 45, 458, 466

S
SA. *See* Sinapic acid (SA)
Saffron 44
antioxidant compounds in 61, 72
therapeutic effects of 61, 72
Safranal 38
Sage
antioxidant compounds in 61, 68, 71, 147
therapeutic effects of 61, 71
Salmonella typhi 289–290
SARS-COV-2 292
Saturated fatty acid (SFA) 10
Scanning electron microscopy (SEM) 245–246
Schistosomiasis 357, 359
Secondary hyperlipidemia 9
Secondary metabolites 296, 498
Sedentary lifestyle 10
Selenide 207
Selenium 479
Selenium-containing polymers 208
Selenium (Se) NPs 163
Self-nanoemulsifying drug delivery systems (SNEDDS) 341
SEM. *See* Scanning electron microscopy (SEM)
Sensors 176–177
Sesame seed, antioxidant compounds in 147
Sesamol 509
loaded with SLNs 159
for neurodegenerative conditions 393
Shake flask approach 248–249
Sheep milk 234
Shogaol 38, 43
Silibinin 133
Silica nanoparticles 277
Silver nanoparticles (AgNPs) 129–130, 157, 185, 218, 235, 506
Silybum marianum 133
Silymarin 459
Simvastatin 17, 256
Sinapic acid (SA) 40, 42, 45
Single-walled CNTs (SWCNTs) 159, 161
Skin cancer 133
Skin health 18
Skin irritation 278–279
SNEDDS, 185. *See* Self-nanoemulsifying drug delivery systems (SNEDDS)
SOD. *See* Superoxide dismutase (SOD)
Soil-transmitted helminths 354, 355
Solanum khasianum 296
Solid lipid nanoparticles (SLNs) 127, 128, 159, 161, 182, 184, 189, 192, 302, 333, 364, 387, 462, 484, 509
Solubility studies 248–249
Soy 439, 499
Soy-conglycinin 191
Spices 145
antioxidant compounds in 57–62, 147
antioxidant properties of 55, 62, 148–155
application of 64
chemical composition of 62
chemical structure of 69–72
classification of 57, 58
definition of 56
extraction of active compounds 74
health benefits of 56, 73
nano-drug delivery vehicles for 155–164
therapeutic effects of 58–62
traditional and commercial value of 73–74
Spinach 72
Staphylococcus aureus 289, 295
Star anise 61
Statins 17
Steroid drugs 182
Sterylamine-based SLNs 189

Stigmasterol 40
Stilbenes 34, 352, 481, 511
Strawberries 72
Streptomyces species 234
Stroke 452
Sulfides 207–208
Sulforaphane 252, 255
Sulforaphane (antioxidant drug)-loaded silk fibroin NPs (SFSNPs) 394
Sulfur-containing polymers 208
Superoxide 78
Superoxide dismutase (SOD) 11, 13, 17, 159, 203, 221, 274
Superparamagnetic iron oxide nanoparticles (SPIONs) 276, 506
Supplementations, antioxidant 5, 18
Surface charge of nanostructures 271, 272
Synthetic antioxidants 65, 146, 480, 496, 512
Syzygium cumini 184, 329

t

Tabebuia avellanedae 256
Tannin 84, 298
Taxifolin 217
Taxus brevifolia 256
Tea, quercetin in 32
Tellurium-containing polymers 208
TEM. *See* Transmission electron microscope (TEM)
Terpenes 6, 34, 84
Terpenoids 298, 300, 456, 459–460
Tetrabenazine 454
Theaflavins 101
Therapeutic drug development 20–21
Thermogravimetric analysis (TGA) 248
Thioketals 207–209
Thyme
  antimicrobial activity of 295
  antioxidant compounds in 61, 68, 72, 147
  therapeutic effects of 61, 72
Thyme oil 133
Thymoquinone 217, 306, 459
Tilianin 187
TiO_2 nanoparticles 279, 305
Tissue damage 212
Tocopherol 12, 37, 84, 107, 479
Tomato, antioxidants in 2
Top-down approach 232
Topotecan 254
Toxicity. *See also* Nanotoxicity
  of nanomedicine produts 193
  of phytomedicines 44–45
Trachoma 357, 359
Traditional Chinese medicine (TCM) 409–410, 454
Transcription 501
Transition metal NPs 505
Transmission electron microscopy (TEM) 246
Triptolide 217
TRRs 11–12
TRXs 11–12
Trypanosoma brucei 358
Trypanosoma cruzi 358
Tuberaria lignose 328
Tuberculosis 292
Tumor ablation therapy 511
Tumor necrosis factor-alpha (TNF-α) 220
Tumor stroma interaction 504
Turmeric 185
  antimicrobial activities of 295
  antioxidant compounds in 61, 72, 147
  antioxidant properties of 148, 151
  chemical constituents in 149
  therapeutic effects of 61, 72
Type 1 diabetes mellitus (T1DM) 7, 433
Type 2 diabetes mellitus (T2DM) 433
  dietary habit 10
  and hypertension 9–10
  obesity and 7
  prevalence of 7
  protective effect of natural antioxidants 14

Type 3 diabetes mellitus 433
Tyrosol 192

u

Ultracentrifugation method 247
Ultra-high-molecular-weight polyethylene (UHMWPE) 159
Ursolic acid 129, 217
USA, regulatory issues of nanophytotherapeutics in 527–528
UV–Vis-spectroscopic analysis 246

v

VA-containing copolyoxalate (PVAX) 209
Vascular calcification 406–407
Vascular gene therapy 409
Vegetables
  antioxidants in 2, 5, 6, 36
  phytonutrients in 73
Very low-density-lipoprotein-cholesterol (VLDL-c) 9
Vesicular system-based nanophytomedicine 334–337
Vibrio cholerae 297
Vicenin 41
Vinca rosea 254–255
Vincristine sulfate 254
Vinflunine ditartrate 255
Vinorelbine tartrate 254
Viola diffusa 328
Viral infections 317
  antiviral medicinal plants 327–328
  antiviral treatment, challenges in 317, 325
  nanotherapeutics of phytoantioxidants 330–331
    dendrimers 339
    GO composite 339–340
    lipid-based 332–334
    magnetic NPs 329
    metallic NPs 329
    nanocapsules 329, 332
    nanogels 337–338
    patents for 341, 342
    polymeric NPs 328–329
    SNEDDS 341
    vesicular system-based 334–337
  phases of 325
  plant-based drugs and phytoantioxdants for 326–327
  routes of infection and incubation time 326
  types of 325–326
Viruses 325, 326
Vitamin C 12, 221, 501
  for DM patients 14
  sources of 84
Vitamin D3 456, 460
Vitamin E 5, 6, 12, 221, 479, 501
  for DM patients 14
  sources of 84
Vitis labrusca 324

w

Watermelon, antioxidants in 2
Wound healing 18–19
  natural nanotherapeutics in 217–220
  nanomedicines in 210–212
Xanthine dehydrogenase 106–107
X-ray diffraction (XRD) analysis 249

z

Zataria multiflora 328
Zeaxanthin 34
Zeta potential 248
Zika virus 354, 355
Zinc oxide nanoparticles (ZnO-NPs) 393
Zinc sulfide (ZnS) nanoparticles 270

Printed and bound by CPI Group (UK) Ltd, Croydon, CR0 4YY
17/09/2026
14964257-0008